D. Moore, P. Bie, H. Oser (Eds.) - Biological and Medical Research in Space

Springer
Berlin
Heidelberg
New York
Barcelona
Budapest
Hong Kong
London
Milan
Paris
Santa Clara
Singapore
Tokyo

D. Moore P. Bie H. Oser (Eds.)

Biological and Medical Research in Space

An Overview of Life Sciences Research in Microgravity

With 33 Figures

Springer

DAVID MOORE
University of Manchester
School of Biological Sciences
1.800, Stopford Building
Manchester M13 9PT
United Kingdom

PETER BIE
University of Copenhagen
Institute of Medical Physiology
Panum Institute
Blegdamsvej 3
2200 Copenhagen
Denmark

HEINZ OSER
European Space Agency
8-10 rue Mario-Nikis
75738 Paris Cédex 15
France

ISBN 3-540-60636-X Springer-Verlag Berlin Heidelberg New York

Library of Congress Cataloging-in-Publication Data
Biological and medical research in space:
an overview of life sciences research in microgravity/ed. by David Moore ... – Berlin; Heidelberg; New York; Barcelona; Budapest; Hong Kong; London; Milan; Paris; Santa Clara; Singapore; Tokyo: Springer, 1996
ISBN 3-540-60636-X
NE: Moore, David [Hrsg.]

Typesetting: K & V Fotosatz, Beerfelden
Cover design: E. Kirchner, Heidelberg
Production: PRO EDIT GmbH, D-69126 Heidelberg

SPIN: 10497160 27/3136 - 5 4 3 2 1 0 – Printed on acid-free paper

Preface

Life Science studies in space were initially driven by the need to explore how man could survive spaceflight conditions; the effects of being launched under high accelerations, exposed to weightlessness and radiation for different periods of time, and returned to Earth in safety. In order to substantiate the detailed knowledge of potentially adverse effects, many model experiments were launched using organisms which ranged from bacteria, plants, invertebrates, rodents and primates through to man. Although no immediate life threatening effects were found, these experiments can be considered today as the precursors to life science research in space. Many unexplained effects on these life forms were attributed to the condition of weightlessness. Most of them were poorly recorded, poorly published, or left simply with anecdotal information.

Only with the advent of Skylab, and later Spacelab, did the idea emerge, and indeed the infrastructure permit, weightlessness to be considered as an extended tool for research into some fundamental mechanisms or processes associated with the effect of gravity on organisms at all levels. The initial hypothesis to extrapolate from hypergravity through $1 \times g$ to near $0 \times g$ effects could no longer be retained, since many of the experiment results were seen to contradict the models or theories in the current textbooks of biology and physiology.

The past decade has been dedicated primarily to exploratory research. Recently however, there has been a move towards controlled experiments, probing into the effect of microgravity on all levels of organisms, and employing in-flight and ground-based $1 \times g$ controls.

It is against this background and the move to rigorous in-depth research that the ESA Microgravity Advisory Committee (MAC) proposed that in the future, ESA should endeavour to focus the scientific research programmes in microgravity ('weightlessness') towards what are perceived as the most promising areas, and also seek to exploit current strengths. At the same time, ESA was required to remain open to new ideas in microgravity research, as befits a recent and developing scientific activity.

This document was prepared originally, essentially as an internal report, to provide ESA with the essential critical review of the whole Space Life Sciences research field, preparatory to establishing the future research directions. It was recognised, however, that the document also provides a very timely overview of the field as this initial exploratory research phase comes to an end to be replaced by the in-depth experimental programmes which will become more accessible in the era of International Space Station Alpha. The review is, therefore, offered to the wider scientific community as a published reference work in the hope that it will provide a foundation for, and stimulate, future research.

The last chapter of this book is different from the rest. It provides a summary overview of the microgravity life science experiments which have been carried out since 1980 by ESA, by ESA member states, and by the ESA international partners. This is accompanied by a summary of the current and future research plans and missions, where known, for each of the ESA member states, and the international partners. This statistical information is presented to allow future plans to be placed in their natural context.

Chapters 1 to 7 are scientific reviews which have been prepared by recognized experts in the areas concerned. These cover the main areas of current space life sciences research, Biology, Physiology, Radiation Biology and Exobiology. An important point is that, from the outset, the intention was to provide a *critical* analysis of what has been accomplished so far in each of these research areas, coupled with *critical* assessment of the manner in which it has been accomplished and the *reliability* of the conclusions drawn (paying attention to the distinction between peer-reviewed and non-peer-reviewed publications).

Critical assessment is an ethos which runs through the entire preparative phase of this book. The original document provided to ESA was critically examined in depth, on a chapter-by-chapter basis, by appropriate experts from ESA's Life Sciences Working Group (LSWG) and the MAC. The authors were asked to take account of the resulting criticisms and suggestions in revising their contributions to the document, and finally, the LSWG met to consider the document as a whole and to make their own recommendations for the future focused life science research programme. Subsequently, the authors were invited to present their views and recommendations at a meeting with MAC which included members of the LSWG. Rarely can authors have faced so much peer review, and it must be even rarer for so much of the peer review process to be face-to-face!

The Editors wish to take this opportunity to thank all those who contributed advice and constructive criticism during the preparation of the text, but most of all we thank the authors for turning the ideas and plans into reality.

David Moore, School of Biological Sciences, University of Manchester
Peter Bie, Institute of Medical Physiology, University of Copenhagen
Heinz Oser, European Space Agency, Paris

Contents

Contributors

Dr M. H. Bestle, Institute of Medical Physiology, Panum Institute, Blegdamsvej 3, University of Copenhagen, DK-2200 Copenhagen, Denmark

Prof. P. Bie, Institute of Medical Physiology, Panum Institute, Blegdamsvej 3, University of Copenhagen, DK-2200 Copenhagen, Denmark

Prof. Elizabeth H. Burger, ACTA-VU, Department of Oral Cell Biology, v.d. Boechorststraat 7, 1058 VJ, Amsterdam, The Netherlands

Dr Gilles Clément, Laboratoire de la Perception et de l'Action, Collège de France, 15 rue de l'Ecole de Medicine, F-75270 Paris Cedex 06, France

Prof. Augusto Cogoli, Space Biology Group, ETH Technopark, Pfingstweidstrasse 30, CH-8005 Zürich, Switzerland

Dr Brian Fitton, Hoofdstraat 157, 2171 BB, Sassenheim, The Netherlands

Prof. Dr med. Helmut G. Hinghofer-Szalkay, Volume Regulation & Space Medicine Research Group, Department of Physiology, University of Graz, Harrachgasse 21, A-8010 Graz, Austria

Dr Gerda Horneck, DLR Institut für Flugmedizin, Porz-Wahnheide, Linder Höhe, D-51147 Köln, Germany

Dr L. B. Johansen, Institute of Medical Physiology, Panum Institute, Blegdamsvej 3, University of Copenhagen, DK-2200 Copenhagen, Denmark

Prof. Dr Jürgen Kiefer, Strahlenzentrum der Justus-Liebig-Universität, Leihgesterner Weg 217, D-35392 Giessen, Germany

Dr M. Kost, Strahlenzentrum der Justus-Liebig-Universität, Leihgesterner Weg 217, D-35392 Giessen, Germany

Dr David Moore, School of Biological Sciences, 1.800 Stopford Building, The University of Manchester, Manchester M13 9PT, United Kingdom

Dr Heinz Oser, European Space Agency, 8-10 Rue Mario Nikis, F-75738 Paris Cedex 15, France

Dr M. F. Reschke, Space Biomedical Research Institute, Code SD5, NASA-Johnson Space Center, Houston, Tx 77058, U.S.A.

Dr K. Schenk-Meuser, Strahlenzentrum der Justus-Liebig-Universität, Leihgesterner Weg 217, D-35392 Giessen, Germany

Dr Jack J. W. A. van Loon, Bradford Engineering B.V., P.O.Box 120, 4645 ZJ, Putte,
The Netherlands

Dr J. Paul Veldhuijzen, ACTA-VU, Department of Oral Cell Biology, v.d. Boechorststraat 7, 1058 VJ, Amsterdam, The Netherlands

Abbreviations, acronyms and conventions

AChR	Acetylcholine receptor
AFP	Alpha fetoprotein
ASGSB	American Society for Gravitational and Space Biology
CAT	Chloramphenicol acetyltransferase
CEF	Chicken embryo fibroblasts
CELSS	Controlled ecological life support system
CNS	Central nervous system
Con A	Concanavalin A
Da	Dalton (unified atomic mass unit)
DCCS	Dynamic cell culture system
DMSO	Dimethyl-sulfoxide
EGF	Epidermal growth factor
FITC	Fluorescein-isothiocyanate
GCR	Galactic cosmic radiation
IAA	Indole acetic acid (Auxin)
IFN	Interferon
IL	Interleukin
IML	International Microgravity Laboratory
IND	Indomethacin
ipMAPs	Immunoprecipitable microtubule-associated proteins
LBNP	Lower body negative pressure
LET	Linear energy transfer
mAB	Monoclonal antibodies
MASER	Material sciences experiment rocket
MAXUS	Combination of MASER & TEXUS
NARP	Nerve associated acethylcholine receptor protein
NIZEMI	Niedergeschwindigkeit Zentrifugenmikroskop
OD	Optical Density
PBL	Peripheral blood lymphocytes
PGE	Prostaglandin E
PKA	Protein kinase A
PKC	Protein kinase C
PMA	Phorbol myristate acetate
RME	Receptor mediated endocytosis
SLJ	Spacelab Japan
SLS	Spacelab Life Sciences
SMS	Space motion sickness

SRE	Serum response element
TEM	Transmission electron microscopy
TEXUS	Technologische Experimente unter Schwerelosigkeit
TLI	Thymidine pulse-labelling index
TPA	Tetradecanoyl-phorbol-acetate
t-PA	Tissue plasminogen activator
TNF	Tumour necrosis factor
ZARM	Zentrum für Angewandte Raumfahrttechnologie und Mikrogravitation

Although there is an internationally accepted value of the normal acceleration due to gravity ($9.80665\ m\ s^{-2}$), there is no agreed symbol for this 'Earth gravitational acceleration' when it is used as a unit. In many printed texts it is symbolised in ways which can be confused with the Metric Unit for gramme (g) or the gravitational constant (*G*). Throughout this text we have adopted the convention of using '$\times$ g', which was suggested by Prof. Dr Donat-P. Häder, following correspondence between editors of space life science texts (Prof. Dr W. Haupt & Prof. Dr A. Sievers) for dealing with multiples (e.g. $5 \times g$) or fractions (e.g. $0.2 \times g$) of normal Earth gravitational acceleration.

Throughout the text and reference lists of the Chapters which follow, **emboldened type face** is used to indicate a reference to a journal which is known to be peer reviewed; citations to references in books, technical manuals, conference proceedings and other sources not known to subject manuscripts to peer reviewing are shown with a normal weight type face.

Chapter 1

Gravitational and space biology at the cellular level

David Moore & Augusto Cogoli

1.1 General Introduction

Space biology can be said to have begun in the late 1940s, with the launch of monkey Albert on a U.S. rocket called 'Blossom 3' in 1948. Several other sub-orbital flights with monkeys and mice followed before the first manned space flight took place in 1961. Scientists in the then Soviet Union began their programme in space biology with the launches of nine dogs in 1951–52 before the historic flight of the dog Laika which survived one week in orbit in 1957.

On the other hand, interest in gravitational biology began well over a century earlier, as reports on the effects of hypergravity (growth in a centrifuge) on plants and oocytes were published in the last century (**Knight 1806**; Pfluger 1883). Indeed, plant scientists have been using clinostats to study plant gravitropism for most of the past century, and are still using them.

On Earth, there is no way to perform an experiment without gravity, i.e. in the true weightless conditions of free fall, so space flight is essential to provide the opportunity to perform microgravity research. Progress in gravitational biology in general, and space biology in particular, was initially slow because the interest of space agencies in the U.S. and the former Soviet Union was focused on means to monitor and protect the health of astronauts. Experiments with bacteria, fungi or mammalian cells, such as lymphocytes, were seen as probes to detect effects of radiation, acceleration, vibration and/or weightlessness on living cells within the 'space medicine' context, and the experiments were devised accordingly. Nevertheless, these pioneering experiments supplied scientists involved in basic research with sufficient results to awaken their interest. The primary motive inducing the first few scientists to study single cells in space was scientific curiosity; the wish to see how relatively simple living systems react to an environment they had never experienced before throughout evolution. More recently, space flight research has been applied in a more focused analytical manner, the weightless environment being used to study specific biological phenomena, and there have been some notable and important discoveries.

Gravitational biology as a new discipline, then, started with the advent of biosatellites in the early 1960s. In 1960, the U.S. satellite Discoverer XVII and the Soviet satellite Biosputnik II, were loaded with bacteria, plant and animal cells. Failure due to technical difficulties was common in these early experiments, and although some missions (particularly Skylab III (1973)) were outstandingly successful, many of the biological results were of limited value because of engineering constraints on the scope of the experiments and the frequent lack of meaningful controls.

Systematic and extensive investigation became possible in 1983 when the first Shuttle–Spacelab mission was realized. Since then, important steps to increase the credibility and scientific gain from biological experimentation in space have been the integration of the Spacelab and, especially, the Biorack concepts in the Shuttle programme. Biorack provides a wide (and widening) range of laboratory facilities which, though commonplace on Earth, are new to orbital vehicles and provide experimenters with the necessary scope for more meaningful experimentation. Biorack has been flown successfully in three Spacelab missions, D–1 in 1985, IML–1 in 1992 and IML–2 in 1994. Other important biological experiments were carried out in Spacelab SLS–1 in 1991, SLJ in 1992 and D–2 in 1993.

A wide range of cell types have been subjected to space flight, including bacteria and protists as well as fungal, plant and animal cells. In many of these experiments the cells have shown alterations in behaviour compared to ground based samples (review in Cogoli & Gmünder 1991). However, ground based samples are not the optimal control as it is not always possible to attribute observed changes solely to alterations in the gravitational field strength. Space flight introduces a number of additional factors that need to be considered other than microgravity. These include high hypergravity levels at launch and re–entry (brief exposure to 3–25 × g), strong vibration and acoustic levels, and high–energy radiations, all of which may influence the function of the system under investigation (see Hamacher *et al.* 1987 and Chapter 4 of this volume). The full spectrum of effects cannot be simulated adequately on earth, so the only way to eliminate them from the investigation is to provide a 1 × g centrifuge facility for control experiments aboard the spacecraft.

This necessary device has only recently been incorporated into orbital vehicles (in equipment such as Biorack). Consequently, many of the earlier space experiments which are reported in the literature could not include an on-board 1 × g control. These experiments must be interpreted with caution, but they still make a useful contribution.

The reported experiments also vary enormously in the biological material involved, as well as in the gravitational conditions encountered. Contrasts are obviously to be expected between experiments involving different organisms, different organs or different cells, but there are also disparities in the nature of the gravitational acceleration to which the experimental material is exposed. **Stavrinidis *et al.* (1991)** discuss both the opportunities for use of hypogravity and, importantly, the quality of the microgravity environments the various installations provide. The terms *microgravity*, μg, 0 g, *low gravity*, *hypogravity*, and *weightlessness* do not have a quantitative meaning but are used in this chapter to refer to the gravitational conditions provided by manned space laboratories (10^{-4} to $10^{-2} \times g$), automatic satellites (10^{-5} to $10^{-3} \times g$), sounding rockets (10^{-5} to $10^{-4} \times g$), parabolic flights (10^{-1} to $10^{-2} \times g$) and drop towers (10^{-6} to $10^{-5} \times g$), respectively (see Hamacher *et al.* 1987). Experiments with single cells in microgravity investigate changes in molecular organization, genetics, growth, proliferation and differentiation, and morphological characteristics. Space experiments represent the most spectacular segment in the field of gravitational cell biology in which the effects of gravity on living systems are studied, but experimentation in space must be interpreted in the context of related research. Gravitational biology also encompasses experiments carried out under hypergravity in centrifuges and ultracentrifuges. Low gravity experiments in the range of $0.02 \times g$ can be simulated on Earth in clinostats, which can be accommodated in any biological laboratory. However, gravity conditions below $0.01 \times g$ need elaborate and costly equipment such as drop towers, aircraft flying through parabolic trajectories, sounding rockets, and orbital vehicles.

The clinostat. Although other machines have been suggested (Mesland 1990, 1992), the key device among earthbound 'hypogravity' equipment is the clinostat. This valuable tool has a long history of use though it seems to us that its users do more than their fair share to maintain an unnatural and spurious division between microgravity research on animals and similar research on plants. For example, **Brown *et al.* (1976)** state that '...the clinostat must be considered an essentially botanical device...' because animals are seen to respond to gravity so much faster than plants. Those who use animal cells for their research are currently most likely to rely on the 'fast-rotating' clinostat although the original reference to which this is usually ascribed (**Muller 1958**) referred specifically to the human *organism* and a speed of rotation at the '...psycho-physiological optimum for disengaging his vestibular apparatus from an effective pull by gravity...' Preference for such a device over the conventional one for animal *cells* would clearly need other justification than the psychological. Fortunately, the basic theory upon which clinostat operation depends is well represented in the literature, very similar detailed mathematical treatments being published at about the same time by botanists (**Dedolph & Dipert 1971**), a zoologist (**Silver 1976**), though publication of this was delayed by the death of the author, and in association with a mycological study (Tobias *et al.* 1973). The main points in these slightly different treatments will be brought together briefly below.
The clinostat provides circular rotation at uniform speed about the horizontal axis. This does not remove the subject from the effects of gravity and care must be exercised to use descriptive terms which do not carry unwarranted implications. A test subject mounted on the clinostat experiences altered vector direction; the normal gravity vector sweeps through 360° in each revolution, so in comparison with a stationary subject which is placed horizontally (and thereby experiences unilateral gravity stimulus) the clinostat subject experiences a continuously shifting omnilateral stimulation. The crucial point, though, is that the effect of the treatment is *relative*. It depends

on the rate of rotation, on the mass of the object considered, on its size and density and on the viscosity and density (specific gravity) of the surrounding fluid.

In complete contrast to the situation on an orbiting space craft, in which all objects experience the same microgravity environment, it is inevitable that different components of an experimental object on a clinostat experience different conditions. In the literature the clinostat has been variously described as providing gravity nullification, gravity compensation, or a time averaged gravitational force of zero. None of these descriptions can be true for all components of the object on the clinostat. It *is* true for all components that the gravitational stimulus is confused relative to the normal. The clinostat does not provide an exact analogue of the orbital (microgravity) environment but results of experiments using the two approaches are often broadly similar (e.g. **Moore, 1991**), though care must be exercised in interpreting clinostat experiments as differences of detail have been noted (**Brown *et al.* 1976**). The use of clinostats in gravitational biological research was discussed in detail at a symposium organized in 1991 by the American Society for Gravitational and Space Biology. The reports presented at the symposium appear in Halstead *et al.* (1992) and the papers (three presenting the physical properties and two reviewing the results) related to the clinostat are briefly discussed here.

Great caution in the use of the clinostat, in particular with single cells, is recommended by Albrecht-Buehler (1992) because '... it (a) generates centrifugal forces, (b) generates particle oscillations with mass-dependent amplitudes of speed and phase shifts relative to the clinorotation, (c) is unable to remove globally the scalar effects of gravity such as hydrostatic pressure, which are independent of the direction of gravity in the first place, and, (d) generates more convective mixing of gaseous or liquid environment of the test object, rather than eliminating it, as would true weightlessness'. Therefore, accurate control of the conditions is required and tests at different rotation speed must be conducted. Also, Kessler (1992a) points out a number of artifacts that may affect the results of clinorotation. Important advice on how to avoid artifacts in the use of the clinostat is given by **Briegleb (1992)**. 'Functional weightlessness' can be achieved (i) with *small* systems, i.e. when the correlation between the quality of the simulated weightlessness and the size of the specimen depend on the threshold of gravitational sensitivity of the process, (ii) when there are only low density differentials between the components involved and (iii) when the viscosity of the medium is sufficiently high.

In a historical overview of the application of the fast rotating clinostat to single cells, M. Cogoli (1992) concludes that there is a good correlation between the results obtained in space and in the clinostat (see following sections). Sievers & Hejnowicz (1992) describe the effect of clinorotation and microgravity on plant cells and organs by distinguishing two types of gravistimulation: static and dynamic. While static as well as dynamic gravistimulation are switched off by microgravity, clinorotation can switch off the first but not the latter if the rotation is not sufficiently fast. In addition, one must take into account that plants are extremely sensitive to the mechanical stresses which may appear during clinorotation and, therefore, influence the results.

Clinostat dynamics. In a stationary cell an intracellular organelle or particle which is more dense than the cytoplasm will sink under the effect of the normal gravity vector; one which is relatively less dense will tend to float. Sedimentation and flotation are physically and mathematically equivalent except for the sign reversal of the vector (see note in Tobias *et al.* 1973). Any organelle which occupies a characteristic position within a cell under normal conditions must do so through the active control of its position by regulation of its density and/or by structural constraint by the membrane or cytoskeletal systems of the cell. The statements above, and those which follow, must, because of ignorance of the forces involved, neglect the possible effects of active cytoplasmic movements (streaming or cyclosis; but see Kessler, 1992b) and any specific restraint imposed by the cytoskeletal or membrane architecture.
Dedolph & Dippert (1971) and **Silver (1976)** provide detailed mathematical treatments of the theoretical behaviour of sedimenting particles subjected to clinostat rotation. The fundamental purpose of the clinostat is to equalise the effects of gravity on a particle through its circular rotation with uniform speed about a horizontal axis. For any particle there is a theoretical optimum rotation rate at which the rotation sweeps the gravity vector around the particle too swiftly for any sedimentation to occur. Both higher and lower speeds result in the particle executing circular motions; in the former case because of the centripetal accelerations due to centrifugal forces and in the latter case because of sedimentation during the slow sweep of the gravity vector. A major problem in interpreting clinostat experiments is that the cell must contain many particles exhibiting a variety of density differentials with their suspending medium with a consequent potential for a variety of responses to any given clinostat treatment. However, the clinostat can be used in an analytical manner. Theoretically, any quantifiable effect produced by clinostat treatment can be optimised at a specific rotation rate and the physical characteristics of the particle(s) deduced from the experimental circumstances which achieve this. Apart from the early work on plant organs (Larsen, 1962; **Audus, 1962**) this experimental approach has been applied rarely, most experimenters being content to work at one speed of rotation.
Dissimilar responses to different speeds of rotation have been noted. Increased fresh weight of cell cultures of *Haplopappus gracilis* occurred when they were cultivated on a 50 rpm clinostat, but not when rotated at 2 rpm (Klimchuk, 1984). A decrease in cell division in rapeseed protoplast cultures was recorded on a 2 rpm clinostat, but not on a 50 rpm clinostat (**Rasmussen, Baggerud & Iversen, 1989**). **Lyon (1970)** used nine clinostat rotation rates to study root and coleoptile development in wheat seedlings (*Triticum aestivum*) and curvatures of leaves and branches of *Coleus blumei*. The former showed the same effect over the whole range of rotation rates tested,

while curvatures in *Coleus* were maximal at 0.3 to 1 rpm. Evidently, different organisms, and perhaps even separate phenotypes in the same organism, depend upon particles with diverse physical characteristics so experiment at different rates of rotation is crucial. As with many other experimental devices, comparisons of responses to clinostat treatment must be made with care, and this is especially true if the device is to be used as an analogue of orbital conditions. Ideally, some direct evidence should be obtained to show that such a comparison is valid.

Fortunately, there is now a sufficient range of facilities available to the experimenter for him/her to match the vehicle or facility to the specific biological functions and structures investigated. While long-lasting processes (lasting hours or days), such as cell division, differentiation and, sometimes, genetic expression can be studied only in orbiting laboratories, other events, like binding of ligands to the cell membrane, signal transduction, cell movements and morphological aspects can be investigated easily in sounding rockets (which provide 7–12 min of weightlessness), parabolic flights (serial episodes of microgravity of 15–30 seconds each) and even in drop towers (3–4 seconds of microgravity). On the ground, there is no limit to the exposure time when centrifuges or clinostats are used. Clearly the latter instrument is a useful tool to verify or prepare experiments in space providing it is always remembered that real microgravity is attainable only under free fall conditions. As stressed above, it is essential to appreciate that the clinostat simulates microgravity only through rotational compensation of the gravity vector (**Silver 1976**; Briegleb *et al.* 1976; Cogoli & Gmünder 1991). As the instrument is still within the Earth's gravitational field, gravity-driven phenomena (like density differentials which are absent in free fall) still occur.

In this overview, experiments conducted in hypergravity in centrifuges, in simulated microgravity in clinostats, and in real microgravity in drop towers, parabolic flights, sounding rockets, and in manned as well as automatic orbital flights, will be discussed. Section 1.2 will focus on bacteria, slime moulds, protozoa and mammalian cells, whereas research on protozoal motility and fungi is discussed in Section 1.3 and the gravitational biology of plants and animals will be dealt with in Sections 1.4 and 1.5.

Data have now accrued to provide ample evidence for gravity dependent responses in a wide variety of cell types from simple prokaryotes to complex animal cells in culture (see earlier reviews in **Gmünder & Cogoli 1988**; Cogoli *et al.* 1989; Cogoli & Gmünder 1991). This is of great interest since such responses imply that gravity may influence biological processes through its action on physical phenomena within individual cells, suggesting that gravisensing at the cellular level could be an ancient and ubiquitous faculty.

1.2 Cell and molecular biology

1.2.1 Introduction

The two main objectives of today's cell biology in microgravity are:

(i) basic science, i.e. the study of significant changes of cellular behaviour. In certain instances (see Section 1.2.3) experiments in microgravity may help to understand better the complicated systems regulated by cytokines which are expressed by genes under differential control. In fact, at $0 \times g$, specific genes may be switched off thus permitting identification of the sequence of events leading to cell differentiation or to signal transduction;

(ii) medical diagnostics, i.e. the use of cellular systems to assess the status of certain physiological conditions of humans in space.

Certain preliminary observations made in early missions encouraged hypotheses and speculations on the profitability of bioprocessing in space. So a third goal, namely bioprocessing, i.e. the use of microgravity to produce substances of pharmaceutical interest, might also prompt some experiments. However, it is doubtful that the, still theoretical, gains would balance the high cost of operations in space.

Several experiments described here have been carried out with peripheral blood lymphocytes (PBL), monocytes or with cell lines derived from PBL, monocytes or hybridoma cells. All these experiments appear here under the title 'Cells of the immune system'. In addition, it is important to notice that mitogenic activation of T cells always takes place in the presence of monocytes as accessory cells. More details on the lymphocyte–mitogen system are presented in Section 1.2.4.8.

1.2.2 Proliferation

1.2.2.1 Background

Cell proliferation or division is the result of a complex series of processes which constitute the cell cycle. In eukaryotes, the cell cycle is subdivided into four discrete stages as shown in Fig. 1: the G1 period, the S phase, the G2 period and the M period. In the synthetic, S phase, DNA replication and synthesis of histone proteins occur. The S phase (DNA replication phase) is preceded by G1 and followed by G2, two 'gap' periods for what concerns DNA synthesis activity. Cell division occurs in the mitotic, M, period. Under certain circumstances, e.g. when the nutrients are consumed, the cell cycle can come to a stop at a so called Go stage. Other cells, like lymphocytes, are in quiescent Go status unless a stimulatory signal (an antigen or a mitogen) triggers the onset of the cell cycle. The regulation and timing of the cell cycle phases varies from cell to cell. While the S, G2 and M periods show little variation for mammalian cells (7–10 h, 2–4 h, and 1–2 h, respectively), G1

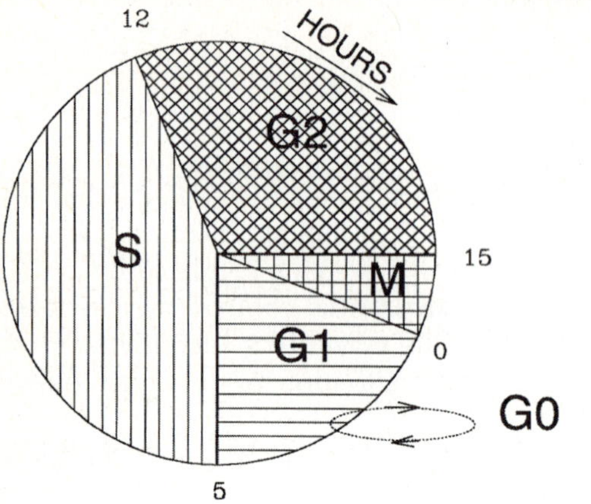

Fig. 1. Mammalian cell cycle. While certain cells, like cancer cells go trough an indefinite number of cycles, other cells can enter a quiescent Go status. Resting lymphocytes are in a Go status until activation is triggered either by the corresponding antigen or by mitogens.

may vary from 1 h to many days or even be absent. Whereas yeast cells can double their number within 2 h, mammalian cells may require a minimum of 10–20 h, for instance, HeLa cells, a cell line derived from a human cervical carcinoma, take approximately 16 h to double their number.

Methods commonly used to measure cell proliferation are:

(i) cells are counted electronically with a Coulter Counter or by eye with a haemacytometer;

(ii) the optical density of a culture suspension is measured spectrophotometrically and related to the number of cells;

(iii) DNA content. Sometimes, especially when cells form aggregates that cannot be counted, the rate of synthesis of DNA can be related to cell proliferation. For instance, mitogenic response of lymphocytes is measured by incorporation of radiolabelled thymidine into DNA, which, with this material and under these conditions indeed reflects the ^{3}H-thymidine pulse-labelling index (TLI). Immunologists thus often use the term 'proliferation' to indicate the rate of DNA synthesis in lymphocytes exposed to mitogens. This might not necessarily apply to all systems or conditions. In particular, data obtained in microgravity which are based on the TLI procedure are subjected to strong criticism and must be interpreted with great caution because microgravity, or other space flight factors, may change the labelling pattern of the precursor pool and thus indirectly affect the incorporation of label into nucleic acids. Independent assays are required to confirm the data.

1.2.2.2 Hypergravity

In parallel to experimentation in microgravity, cell proliferation and other cell functions have been studied under hypergravity conditions. The rationale is that if one wants to test the hypothesis that the transition from 1 × g on Earth to 0 × g in space causes alteration of cellular behaviour, the transition from 1 × g to hypergravity in a centrifuge should cause changes as well. Therefore, inexpensive and easy experiments in centrifuges could allow us to speculate whether changes may or may not occur also in microgravity. There is, however, an obvious and fundamental qualitative difference between the transition from 1 × g to hypergravity and from 1 × g to 0 × g. Experiments with single cells in centrifuges are reviewed here only as a complement to investigations in space, microgravity remaining that interesting environment not reproducible on Earth. Nevertheless, several interesting findings have been made in hypergravity at g forces between 2 and 2000 × g, except for some experiments with *E. coli* which were conducted at 50,000 and 110,000 × g, respectively. Human lymphocytes, HeLa, MC3T3, JTC-12 cells and the ciliate *Paramecium tetraurelia* were among the most extensively studied cells in centrifuges (Cogoli & Gmünder 1991).

Cells of the immune system. Gravitational effects were first discovered in murine spleen lymphocytes exposed to 2 × g and activated with the T cell mitogen concanavalin A (Con A) (Cogoli *et al.* 1979). Studies with human lymphocytes purified from peripheral blood and cultured with Con A at 10 × g showed an increase of the proliferation rate of 20–30% (**Tschopp & Cogoli 1983; Cogoli 1993b**). The effect was not due to an increase of hydrostatic pressure, as 1 × g controls in which paraffin was over-layered on the culture medium failed to show the effect. When the lymphocytes were kept in whole blood cultures, obtained from fresh blood diluted 1:10 with culture medium, the increase in proliferation rate at 10 × g could be as high as 300% of the 1 × g controls (**Lorenzi *et al.* 1986**). Interestingly enough, the effect appeared also when the cultures were first centrifuged for 72 h without mitogen and then cultured at 1 × g with Con A. It could be shown that the increase at high g was attributable mainly to the concomitant activation of B cells by Con A molecules attached to the erythrocytes which are present at 1–5% in purified cultures and in overwhelming amounts in whole blood cultures (Lorenzi *et al.* 1988; Cogoli, A *et al.* 1990). It is known that when Con A is presented to B lymphocytes attached to a solid matrix, B cells may also be activated. Maybe this effect is strongly enhanced at 10 × g, however, there is no explanation yet and why the effect occurs at high-g and not at 1 × g remains to be clarified.

HeLa cells. The proliferation rate was increased by 14% and 30% at 10 × g after 24 h and 48 h of culture, respectively. Cell counts and radioactive thymidine incorporation data correlated fairly well (**Tschopp & Cogoli 1983**). In other studies, conducted by Japanese scientists, it was seen that at 35 × g proliferation was 60–80% higher than at 1 × g after 4 days. At 18 × g and 70 × g the increase was 20% and 30% of the respective controls. It was shown that the change was due to a shorter G1 phase, 10.4 h at 1 × g and 8.0 h at 35 × g. Thus, the cell generation time was 22.4 h at 1 × g and 19.5 h at 35 × g (**Kumei *et al.* 1987, 1989**). In a further study (Sato *et al.* 1992) proliferation rate increased by 34% after 72 h. In contrast to the response of MC3T3-E1 cells (see below), no effect of indomethacin (IND), a suppressor of prostaglandin (PG) synthesis, on the proliferation of HeLa cells was detected.

Friend cells. An experiment conducted at 10 × g with this murine erythroleukemic cell line in the presence of dimethylsulfoxide (DMSO), an inducer of haemoglobin expression, for 6 days showed an increase of 35% in cell number, while glucose consumption remained unchanged (**Tschopp & Cogoli 1983**). Friend cells were chosen for their interesting peculiarity of differentiating along erythroid lines in the presence of DMSO. Indeed, Friend Cells are a widely used *in vitro* model of murine erythropoiesis. When incubated in minimum essential medium, the cell cycle lasts approximately 16 h. If 1.3% DMSO is added to the culture cell proliferation stops and haemoglobin is produced and stored intracellularly. Induction of haemoglobin synthesis is measured spectrophotometrically after lysing the cells with the detergent Triton X-100 and treatment of the supernatant with tetramethylbenzidine hydrochloride. Haemoglobin-positive cells are counted after staining with benzidine hydrochloride. The induction of Friend cells with DMSO can be regarded as a typical *in vitro* differentiating system which is analogous to the mitogenic activation of T lymphocytes.

Human lung adenocarcinoma cells A 549. The cells have the property to grow as nodules of 3.5 mm in diameter and to express the normal differentiation of pneumocytes. Incubation at 2 and 15 × g for up to 27 d did not reveal any change of cell growth, protein and DNA content. Also the parameters indicating cell differentiation such as phospholipid composition, particularly phosphatidyl content, were the same as in the controls (**Croute *et al.* 1990; Gaubin *et al.* 1991**). However, structural changes were detected as discussed in section 1.2.4.2.

Human embryonic lung fibroblasts (HEL). Exposure to 15 × g resulted in a significant decrease (17%, $P < 0.05$) of proliferation rate determined as number of cells on the 5th and 7th day of culture. Exposure to 5 × g gave less evident differences. The data correlate fairly well with the DNA content determined spectrofluorometrically (**Gaubin *et al.* 1991**). Surprisingly, there was no difference in protein content.

Human dermic fibroblasts (F 106). Exposure to 5 and 15 × g for up to 7 days at various stages of the life span of the cells did not show differences in cell growth rate measured in terms either of number of cells or as DNA content (**Gaubin *et al.* 1991**).

Other mammalian cells. There are reports (reviewed by Cogoli & Gmünder 1991) on other mammalian cells exposed to hypergravity conditions. Cells like V-79 (Chinese hamster), sarcoma Galliera (rat), CEF (chicken) showed increases in proliferation rate between 20% (sarcoma) and 70% (V-79) compared to the controls. Another erythroleukemic cell line, K-562 cells (human), showed no change or reduction of the proliferation rate at 10 × g.

Hypergravity (2.6 × g) depressed the development of limb elements in cultured mouse limb rudiments. There was a proximo-distal gradient of sensitivity to excess gravity in the limb with proximal structures being less susceptible than distal ones (**Duke 1983**). In some cases, proximal limb elements present prior to explantation disappeared upon exposure to excess gravity. The effect of hypergravity was assumed to operate via changes in mechanical tension and/or pressure on the cells, accompanied by alterations in cell size and morphology and membrane properties.

Three cell lines (MC3T3-E1 from neonatal mouse calvaria, HeLa (see above), and JTC-12 from monkey renal tubules) were investigated in hypergravity (Sato *et al.* 1992). Cell growth was determined in cultures of all cell lines kept at 5 × g, 10 × g, 20 × g, 40 × g for 72 h, respectively. The effect of conditioned medium on MC3T3-E1 cells was studied as follows: cells were incubated at 40 × g for 24 h, the supernatant was used as conditioned medium for stationary cultures for 48 h. The effect of IND (indomethacin) on all cell lines was studied during 48 h at 40 × g. The effect of PGE2 was studied on MC3T3-E1 cells for 8 h.

Proliferation of MC3T3-E1 cells was depressed (– 8%) after 72 h at 10 × g, whereas at 20 × g and 40 × g it was increased by 11 and 27% respectively. In JTC-12 cells, centrifugation at 40 × g for 72 h caused an 8% increase in proliferation. Conditioned medium favoured slightly but significantly more (+16%) growth of MC3T3-E1 cells at 1 × g suggesting that humoral factors may play a role in gravitational effects. IND significantly suppressed proliferation of MC3T3-E1 cells at 1 × g as well as at 40 × g. At 40 × g the suppression brought proliferation at the level seen at 1 × g without IND (i.e. the hypergravity effect was abolished by IND). IND abolished the 40 × g effect in JTC-12 cells also. IND and PGE effects seem to be related to humoral factors. Other authors reported that low PGE concentrations (10^{-7}M) suppressed DNA synthesis whereas high PGE (10^{-5}M) activated it. However, the concentrations used by Sato *et al.* (1992) were lower than those used by the other authors in MC3T3-E1 cells. In addition, the amounts of PGE measured by Sato *et al.* (1992) at 20 × g and 40 × g are not consistent with those of the other authors. It is concluded that some other unknown factors must be involved. These interesting data need further verification.

Hypergravity (5 × g) stimulates proliferation (possibly through enhanced production of prostaglandin E2) but suppresses differentiation of osteoblast-like cells. The signalling pathway involved was independent of the activation of protein kinase C and the production of cyclic nucleotides, and distinct from the pathway through which IGF-I stimulates proliferation of these cells (**Miwa *et al.* 1991**). Further evidence for hypergravity enhancement of the proliferation of the same osteoblast-like cell line being mediated via a prostaglandin E2-mediated mechanism is provided by Nakajima (1991). In this study cells were exposed to 5 and 18 × g for 1 to 3 days, or cultured at 5, 10, 20 and 40 × g. Migration was increased by 18 × g, and proliferation was enhanced at 20 and 40 × g but reduced at 10 × g.

Paramecium tetraurelia. This protozoan lives in ponds and can swim against gravity. There is a negative linear correlation between gravitational field strength and proliferation time measured at 1 × g, 2.5 × g, 5 g, 7.2 × g, 10 × g, and 20 × g. At 20 × g the number of cells after 3 days of incubation was one fifth of the value at 1 × g (Tixador *et al.* 1984b). When the generation time is determined 24 h after the centrifugation was stopped, the effect disappeared.

E. coli. Incubation in an ultracentrifuge at 50,000 × g for 2 h did not affect DNA, RNA or protein synthesis (Pollard 1971). However, at 110,000 × g, **Montgomery *et al.* (1963)** observed increased lag phases and prolonged generation times (between 1.41 and 1.58 h in the centrifuge and between 0.71 and 0.9 h in the control, respectively) after 24 h of incubation. A control to test whether the effect was due to packing of the bacteria at the bottom of the tube was conducted at 1,000 × g. No change of growth pattern was detected thus ruling out simple mechanical effects.
In conclusion, the data obtained in hypergravity strongly suggest that alterations of the gravitational environment have an important impact on several cell functions and thus justify the investment of resources to conduct similar studies in space.

1.2.2.3 Hypogravity in clinostats

Cells of the immune system. Reduction of proliferation by 50%, measured as the rate of DNA synthesis as TLI, was observed in cultures of purified as well as of whole blood lymphocytes in the fast rotating (90 rpm) clinostat in the presence of Con A (Cogoli A. *et al.* 1980, 1990; **Cogoli 1993b**). The data were confirmed by two independent qualitative analyses. First, blast cells (e.g. activated lymphocytes) appeared in large amount in the 1 × g controls, whereas none or very few were seen in the clinostated cultures. A quantitative analysis was not possible since cells cannot be counted when clustered in the typical aggregates formed by the polyvalent intercellular binding of Con A to the α-o glycosides on membrane proteins. Second, activated lymphocytes were identified in large amount also in transmission electron micrographs of the 1 × g controls whereas very few or none were found in the clinostated samples.

Other mammalian cells. Proliferation did not change in the erythroleukemic human cell line K-562 exposed to haemin (Wiese *et al.* 1988). This was expected since haemin is known to trigger the synthesis of haemoglobin with concomitant halt of proliferation. Conversely, proliferation was increased in Friend cells exposed to DMSO (Lorenzi *et al.* 1988).

Paramecium tetraurelia. Growth rate during 3 days at rotation speeds between 50 and 110 rpm in the fast rotating clinostat was increased by approximately 20% at 90 rpm, lower (or no) differences were seen at the other rotation speeds (Ayed *et al.* 1992). In earlier work (Tixador *et al.* 1984b) no effect was seen in a slow rotating (1 rpm) clinostat. These data support calculations indicating that, for particles of the size of *Paramecium*, low-gravity simulation with the clinostat needs 30–100 rpm to generate the desired tumbling motion (Schatz *et al.* 1992; Björkman 1992.).

Salmonella typhimurium. Growth was increased by 62% in the clinostat (Mattoni *et al.* 1971).

1.2.2.4 Hypogravity – in orbit

The proliferation of representatives of nearly all classes of organism has been investigated in space.

Cells of the immune system. Extensive studies on several space flights were conducted in the last decade with lymphocytes and monocytes from human peripheral blood as well as with derived cell lines (for review see **Cogoli 1993b**). Activation of T lymphocytes by Con A, purified from human peripheral blood, was carried out for the first time in space in an experiment in Spacelab-1 in 1983 (**Cogoli *et al.* 1984; Cogoli & Tschopp 1985**). Surprisingly, it was discovered that activation, measured as the rate of DNA synthesis as TLI, was inhibited by 93%, compared to synchronous 1 × g controls on the ground, despite the fact that the cells formed aggregates in microgravity. Again, the results were confirmed by qualitative light and electron microscopic analyses. Unfortunately, due to the limited resources, a 1 × g control in space was not conducted. However, the results were confirmed in two experiments performed in Biorack in Spacelab D-1 (Bechler & Cogoli 1986; Cogoli *et al.*1988) which included 1 × g controls on board.
A step towards understanding the strong inhibition of activation in microgravity was taken in an experiment conducted on Spacelab SLS-1 in 1991 (**Bechler *et al.* 1992; Cogoli *et al.* 1993**). Using a new technology, lymphocytes and monocytes were attached to Cytodex 1 microcarriers prior to exposure to Con A in microgravity. Activation was more than doubled compared to the flight and ground controls at 1 × g. As described in Section 1.2.3.6, analysis of the cytokines secreted established that the lack of activation in free cells is due to the failure of monocytes to deliver IL-1 as the second signal required for T lymphocyte activation. Not yet explained is the fact that attached T cells double their activation in microgravity when sufficient IL-1 is available. A summary of the changes observed on lymphocyte activation under altered gravitational conditions is shown in Fig. 2.
Proliferation of hybridoma cells (derived from the fusion of myeloma cells and B lymphocytes) has been determined in two Spacelab flights. In both ex-

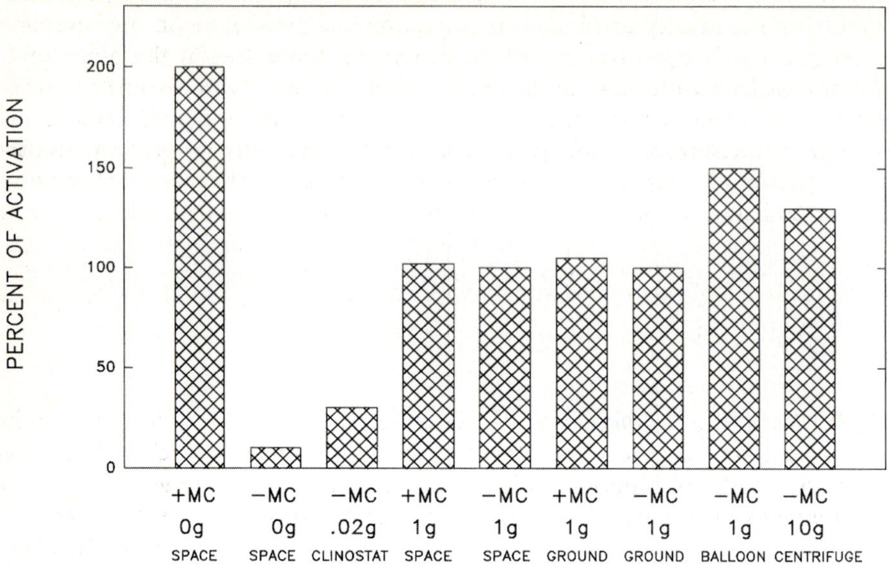

Fig. 2. Gravitational effects on lymphocyte activation. The diagram gives an overview of what we know to date on the effects of space flight (0 × g/space and 1 × g/space respectively), hypergravity (10 × g/centrifuge), simulated low gravity (0.02 × g/clinostat) and cosmic radiation (1 × g/stratospheric balloon) on mitogenic activation by Con A of T lymphocytes. The experiments were conducted either with resuspended (-MC) or microcarrier-attached cells (+MC). The bars give the percent of activation (measured on day 3 after exposure to Con A at 37°C as counts min^{-1} of tritiated thymidine incorporated into DNA) with respect to the relative control (taken as 100%) on the ground in the absence of MC beads (-MC/1 × g/ground). Since the number of independent experiments shown in the diagram varies from one to 20, standard errors are not given here.

periments the results in 0 × g were compared to in flight 1 × g control. AM2 mouse hybridoma cells, cultured in Biorack in D-1, did not show differences in cell counts after 5 days in space (Beaure d'Augéres *et al.* 1986).

The cell line 7E3-N, producing monoclonal antibodies against a lipopolysaccharide binding protein, was cultured in Biorack in IML-1 (**Bechler *et al.* 1993**). Being derived from lymphocytes, it was thought that hybridoma cells may also be gravity sensitive. The number of cells ml^{-1} after 4 days in culture (independent duplicates) was increased from 7.0×10^5 (1 × g in flight) to 9.8×10^5 (0 × g). This difference is significant ($P = 0.05$). No significant difference was detected between the 1 × g controls in flight and on the ground. The increase of the proliferation rate (+ 40%, $P = 0.05$) at 0 × g was confirmed by the DNA synthesis data. Conversely, no important differences were seen after 2 days of incubation. This may indicate that microgravity effects became important only after a longer exposure (48 h) to 0 × g.

Friend cells. Friend leukaemia virus transformed cells were chosen for their interesting peculiarity of differentiating along erythroid lines in the presence of DMSO. The major objective of the experiment was to test the hypothesis that, in analogy with another *in vitro* differentiating system, namely T lymphocytes exposed to mitogens, important cellular functions would change in microgravity (**Bechler *et al.* 1993**). The numbers of cells found in 1 ml volumes of culture after 140 h of incubation in the presence of DMSO were 12 $\times 10^5$ in both the in-flight 0 × g and the 1 × g centrifuge controls, and 10 × 10^5 in the 1 × g ground controls; these differences were not significant. In addition, the cells grew normally, the starting concentration being 0.75 ×5 cells ml^{-1} (contrast these results with the increase in cell number of cells cultured at 10 × g discussed in Section 1.2.2.2).

WI38 human embryonic lung cells. These cells were cultured for 28 days in Skylab, in a sophisticated instrument consisting of incubation chambers with automatic medium supply, fixation and time-lapse cine-camera. No effect was observed on the lag phase or growth rate, although glucose consumption from the medium was 20% higher in the flight cultures than in the ground control (**Montgomery *et al.* 1978**).

HeLa cells. Cultures were exposed successively to five to six orbital flights, though the duration of the flights and the intervals between them are not specified (Zhukov-Verezhnikov *et al.* 1971). No alteration of the mitotic index was detected.

HaK cells. Hamster embryonic kidney cells were grown for 7 days in Spacelab IML-1 in a newly developed cell culture chamber (DCCS - dynamic cell culture system). No change was detected in cell proliferation, measured as cell counts, in comparison with in-flight 1 × g controls as well as ground controls (**Gmünder *et al.* 1988; Lorenzi *et al.* 1993**).

Paramecium aurelia. On Spacelab D-1, microgravity was found to have a stimulatory effect on proliferation (Richoilley *et al.* 1986, 1988a, 1988b). Maximal growth rate of zero gravity cultures was 0.67 per day compared to 0.41 per day of 1 × g controls both in the Biorack centrifuge and on the ground. The *Paramecium* cells grown at 0 × g had a 15–20% larger cell volume during early log-phase than cells grown at 1 × g (both in the gravity simulating centrifuge and on the ground). In contrast, the cells of 0 × g cultures were 10–20% smaller in the late log-phase than the 1 × g controls. These findings are in line with earlier experiments carried out on Salyut 7 (Richoilley *et al.* 1988a & b; **Tixador *et al.* 1981**). Radiation was found to have a weak growth-stimulating effect (Tixador *et al.* 1984b). On a stratospheric balloon flight, the number of cells per culture was significantly higher than in earth-based synchronous cultures (19 vs 17 cells per culture, means of 64 cultures) (Tixador *et al.* 1984b).

Pelomyxa carolinensis. Of particular interest is the behaviour in microgravity of this giant amoeba, 5 mm in length when fully extended and containing 100 to 1,000 nuclei. No significant difference in growth rate was detected in two experiments on Biosatellite II (Ekberg *et al.* 1971; Abel *et al.* 1971). Unfortunately, the flight had to be shortened from 3 to 2 days due to weather conditions at the landing site. Thus the exposure to microgravity was too short to provide conclusive data on the growth rate. It is surprising that this unicellular organism which, due to its size, appears to be an ideal object for gravitational studies has not been flown in space again so far.

E. coli. In an experiment conducted on board the US Biosatellite II and designed mainly to investigate effects of radiation, viable bacterial densities were 19% higher in flight samples than in ground controls when measured by plate counts, but not different when measured by Coulter counter, after a flight lasting two days (Mattoni *et al.* 1971). In samples exposed to 2 Roentgen of chronic γ-radiation, the density of the flight samples was 45% and 41% higher as determined by the same methods, respectively. Samples of *E. coli* were cultured for 24 h at 37°C in an automated instrument during the flight of the Biocosmos 2044 satellite which lasted 15 days (**Bouloc & D'Ari 1991**). Cultures were inoculated by breaking ampoules containing the bacteria. The medium contained glucose or glycerol as limiting carbon source. Growth was stopped by lowering the temperature to 4°C. Analysis revealed that cell growth, calculated from the OD of triplicate cultures at 600 nm, did not change in space. Using a different procedure (cell counts of cultures held at 5° after incubation for various times in flight), **Gasset *et al.* (1994)** also came to the conclusion that microgravity (during the STS-42 mission) has no effect on exponential growth of *E. coli.*

Bacillus subtilis. Increases of yield and growth rate were noted in Spacelab D-1 (Mennigmann & Lange 1986, 1988). The authors were not able to give exact numbers, because the optical density reading of the $0 \times g$ culture went out of range. However, they found that *B. subtilis* yielded far fewer spores under microgravity conditions than at $1 \times g$ (0.5×10^5 ml^{-1} against 8×10^5 ml^{-1}). Thus, under microgravity the spore forming bacillus does not develop spores at the end of the stationary growth phase as easily as it normally does. This suggests that under microgravity conditions cell differentiation processes may not proceed normally.

Salmonella typhimurium. In an investigation in Biosatellite II (Mattoni *et al.* 1971) the increase in viable density of the flight samples was 30% (from plate counts) and 15% (by Coulter counting), respectively, thus confirming the results in the clinostat. In addition, in samples irradiated with 265 R, 654 R, and 1630 R the increase in cell density was 91%, 88%, and 84% (from plate counts), and 88%, 93% and 62 % (from Coulter counts), respectively.

Proteus vulgaris. The number of cells grown for 2 days in Soyuz-12 was 7 times higher than the corresponding controls (Kordyum *et al.* 1974).

Space experiments fall into two categories. Ideally, experiments should be systematically performed in space, should include 1 × g controls on board and be accompanied by ground experiments, preferably involving use of the clinostat. The most valid conclusions on 0 × g effects on cell proliferation are drawn from this group, but the criteria have been met only rarely and many experiments fall into the second category – being conducted only once and with little or no supporting ground-based investigation.

(a) Changes of the proliferation rate of single cells in microgravity are now well documented. However, no general pattern of behaviour can be recognised, sometimes opposite effects are observed (e.g. lymphocytes vs *Paramecium*), and in the majority of the cells investigated there is no effect.

(b) The fast rotating clinostat delivers results which are in fair agreement with those obtained in space as seen in lymphocytes, *Paramecium* and *Salmonella*.

(c) Among mammalian cells, the lymphocyte/monocyte system appears to be one of the most dramatically affected. Interestingly enough, Friend cells, induced to differentiate with DMSO, are not sensitive to 0 × g in space. Among other unicellular organisms, *Paramecium* shows the greatest effects.

(d) The partial failure of the Biosatellite II mission in the early seventies, may have discouraged the continuation of studies on interesting objects like the giant amoeba *Pelomyxa carolinensis*.

1.2.3. Genetic expression and signal transduction

1.2.3.1 Background

After the discovery of gravitational effects on the proliferation of single cells, the next step was to deepen the investigations to genetic and metabolic effects at the molecular level. This section deals (i) with effects on the expression of cell-specific products, like cytokines as well as on the early expression of genes like *c-fos*, *c-jun* and *c-myc*; (ii) with effects on the transfer of genetic material between cells; and (iii) with effects on the consumption of nutrients as well as on the secretion of waste products. All these effects are related through signal transduction within the cell. Normally, biochemical signals are received via recognition and binding of ligands to specific cell membrane receptors, so experiments on the binding of ligands in microgravity are also described in this section. Since the process takes only a few seconds, interesting experiments can be conducted in parabolic flights with episodes of microgravity of 15–30 sec.

The gene products of the c-*fos* and c-*jun* proto-oncogene family are known for their prominent role in cell proliferation and differentiation. Their expression is usually rapidly induced by growth factors and can be induced also by a variety of agents that by-pass the receptor and mimic the partial activation of signal transduction pathways. Lymphocytes and monocytes, A431 epidermoid cells, and HeLa cells are the most extensively studied systems with respect to signal transduction and genetic expression under changed gravitational conditions. The hypothetical mechanism of the T lymphocyte activation by Con A is shown in Fig. 3 as an example of signal transduction in which the G proteins/phospholipase C/phosphoinositol bisphosphate as well as the protein kinase C pathways are involved. A third important mechanism involves the adenylate cyclase/protein kinase A/cAMP pathway. Phorbol esters (e.g. TPA) are known to activate PKC and thus mimic the second signal (IL-1) in T cell activation. Ionophores (e.g. A23187) may bypass the first signal (Con A or antigen) by releasing calcium ions into the cytoplasm from the endoplasmic reticulum. Forskolin is an activator of adenyl cyclase.

1.2.3.2 Hypergravity

HeLa cells. The induction of the proto-oncogene *c-myc* in response to hypergravity was determined in cells grown in monolayer and centrifuged at 18 $\times$ g, 35 $\times$ g and 70 $\times$ g at 37°C for between 15 and 360 min. Induction of c-*myc* mRNA was determined by electrophoresis of the extracted RNA, followed by hybridization with ^{32}P-labelled c-*myc* probes and densitometry of Northern blots. Elevated levels of c-*myc* mRNA were observed at all hypergravity values tested. The effect was most evident at 35 $\times$ g, levels of c-*myc* mRNA were 3.0- and 3.8- fold higher than the control after 30 and 120 min of centrifugation, respectively. The elevated level of c-*myc* mRNA was maintained (3.3-fold greater than the control) even after 360 min exposure to 35 $\times$ g (**Kumei *et al.* 1989**).
In a later study conducted at 35 $\times$ g, **Kumei *et al.* (1991)** observed that hypergravity increased the production of inositol 1,4,5,-triphosphate (IP_3) 1.5-fold after 2 min and 2.1-fold after 5 min and reported that intracellular level of cAMP decreased by 11% after 10 min and by 16% after 20 min at 35 $\times$ g. Phosphorylation of proteins immunoprecipitated by antibodies recognizing microtubule-associated proteins (ipMAPs) was determined. Phosphorylation of a 115-kDa detergent-insoluble protein was enhanced by 100% after 5-min exposure to hypergravity but did not differ from the control after 80 min. Phosphorylation of a 200-kDa detergent-soluble protein was observed after 20 min exposure to 35 $\times$ g. These results suggest that IP_3 and c-AMP may act as second messengers in hypergravity signal transduction. Phosphorylation of ipMAPs in both detergent-soluble and -insoluble fractions suggests that cytoskeletal structures may be influenced by gravity (**Kumei *et al.* 1991**). No similar data are available on microgravity effects (though sounding

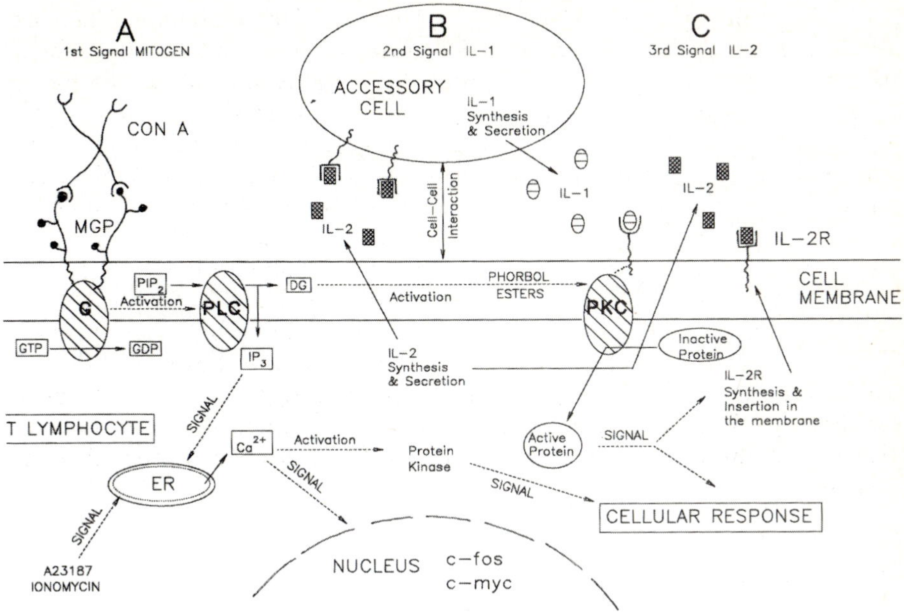

Fig. 3. Mechanism of T lymphocyte activation by Con A. Three phases of T cell activation can be distinguished:

Phase A: Binding of Con A (1st signal) to membrane glycoproteins (MGP), induction of protein G (G) to activate phospholipase C (PLC). Cleavage of phosphoinositol bisphosphate (PIP2) into diacylglycerol (DG) and inositol triphosphate (IP3). Release of calcium ions from the endoplasmic reticulum (ER). Activation of a cytoplasmic kinase. Triggering of synthesis of IL-2 following expression of the oncogenes *c-fos* and *c-myc*. The release of calcium ions from the ER can be induced by calcium ionophores thus bypassing and mimicking the transduction of the 1st signal.

Phase B: Secretion of limited amounts IL-2 and cell-cell interaction between T and accessory cells (AC), usually monocytes, triggering of the production and secretion of IL-1 (2nd signal), activation of protein kinase C (PKC) upon interaction with IL-1. Probably there is a synergistic effect of DG in the activation of PKC. Activation of a cytoplasmic protein, production and insertion of IL-2 receptors (IL-2R) in the cell membrane and amplification of synthesis and secretion of IL-2. The action of IL-1 can be by-passed and mimicked by phorbol esters.

Phase C: IL-2 binds (3rd signal) to the IL-2 receptor. Full activation of T lymphocytes is triggered. Cells start to divide and two populations, the effector and the memory cells, are generated

rocket experiments have implicated involvement of actin microfilaments in gravity detection in plants, see Section 1.4.5.7).

Erythropoietic cell lines. K-562 and Friend cells were exposed to haemin and DMSO, respectively, at 10 × g (Lorenzi *et al.* 1988). K-562 cells showed depression of haemoglobin production, but no effects on glucose consumption and proportion of haemoglobin-producing cells were detected. In

Friend cells, glucose consumption was reduced though haemoglobin production remained unchanged.

A431 cells. At 10 × g a slight increase (+ 18%) of epidermal-growth factor (EGF) induced *fos* expression and no change of constitutive expression was found by **de Groot *et al.* (1991)**, see following section for more details.

Human dermic fibroblasts (F 106). Cell differentiation in terms of synthesis of elastase was studied at 15 × g, and a moderate but significant increase of elastase enzymatic activity (+8% and +13% on the 8th and 9th day of culture, respectively) was recorded. Since elastase degrades elastic fibres the data suggest that remodelling of the basement membrane reconstituted by fibroblasts is enhanced in hypergravity (**Gaubin *et al.* 1991**).

Osteoblastic cells MC3T3-E1. In a recent study conducted with cultures kept at 50, 200, 900 and 2000 × g for 5 min the expression of c-*fos*, c-*jun*, and *egr*-1 genes as well as protein kinase C activity were determined (**Nose & Shibanuma 1994**). A qualitative analysis by electrophoresis and autoradiography of cellular RNA hybridized with molecular probes showed that exposure to 50 × g was sufficient to stimulate transiently osteoblastic cells to induce the early response genes c-*fos* and *egr*-1, whereas c-*jun* was only marginally affected. Induction of c-*fos* in hypergravity was almost completely inhibited by staurosporin (an inhibitor of PKC) whereas that of *egr*-1 was only partially inhibited. Also, TPA inhibited almost completely the induction of c-*fos*, whereas *egr*-1 was not affected. A quantitative analysis showed that PKC activity is increased at 900 × g but not at 5 × g. Altogether these result confirm that hypergravity stimulates multiple signal transduction cascades connected with early genetic expression.

1.2.3.3 Hypogravity in clinostats

K-562 cells. Human K-562 cells were cultured in the fast rotating clinostat in the presence of haemin. A 10% decrease in glucose consumption and a 50% decrease in haemoglobin production (measured in the supernatants of lysed cells) were observed, while the number of haemoglobin-producing cells (stained with benzidine) remained unchanged (Wiese *et al.* 1988).

A431 cells. The induction of so-called 'immediate early genes', like c-*fos*, is the earliest detectable nuclear indication of a normally functioning signal transduction cascade. The EGF induced expression of c-*fos* proto-oncogene in A431 cells was studied in the clinostat (60 rpm) (**de Groot *et al.* 1991**). Under normal laboratory conditions at 1 × g, a 20-fold increase of c-*fos* mRNA level is seen after 10 min exposure to EGF, after 30 min it is 50-fold (maximum). The induction is transient, returning to pre-stimulation levels after 2–2.5 h. With a more sensitive method, induction can be detected after only

3–6 min. With cycloheximide, the EGF effect is longer-lasting as a result of blocking the expression of proteins responsible for the repression of *fos* transcription and degradation of *fos* mRNA. c-*fos* induction is temperature dependent: there being no induction at 4°C, the maximum being at 30–37°C. Various agents like TPA (phorbol esters), calcium ionophores (A23187, ionomycin), and mitogenic neuropeptides (bradykinin, histamine, bombesin) can induce the *fos* gene in A431 cells. The efficiencies are different at excess concentrations. TPA and A23187 mimic aspects of the signalling cascade initiated by EGF or mitogenic neuropeptides.

Constitutive *fos* mRNA levels did not change in the clinostat. EGF-induced *fos* expression was slightly depressed (–20%). Based also on results obtained in the centrifuge, in which the opposite effect was found, it is concluded that the results clearly show the EGF-induced *fos* expression in A431 cells is sensitive to change in gravitational acceleration. 'The induction is a rapid nuclear response following activation of the signal transduction cascade by extracellular factors, and is therefore a good indicator to study the influence of gravity changes on this process' (**de Groot *et al.* 1991**).

In another study by the same team (**de Groot *et al.* 1990**), it was possible to show, by coupling c-*fos* serum response element (SRE) with the CAT (chloramphenicol acetyl transferase) gene and transfecting into A431 cells, that the reduction of expression of c-*fos* gene is caused by a specific decrease in SRE activity in microgravity. To identify which subset(s) of signal transduction is/are affected in microgravity, induction of c-*fos* expression was tested in the clinostat in the presence of EGF, TPA, activator of protein kinase C (PKC), A23187 (by-passing phospholipase C) and forskolin, activator of protein kinase A (PKA) respectively. EGF and TPA induced expression were reduced by 25% and 30% respectively, no significant difference (from the $1 \times g$ control) was found with A23187 or forskolin.

In parallel with experiments in sounding rockets (see below, Section 1.2.3.5), the question whether microgravity just delays c-*fos* expression or really depresses it, was addressed in an experiment conducted in the clinostat (**de Groot *et al.* 1991**). Maximum expression was reached after 30 min, incubations were performed for 10, 20, 30, 60, and 90 min (at this time it decreased by 20%). The $1 \times g/0 \times g$ ratio of expression remained greater than 1 throughout the incubation, showing that microgravity really depresses rather than delays the response. However, the effect remained low.

1.2.3.4 Hypogravity in parabolic flights

Inflammatory cells. In a device specially developed for experiments lasting a few seconds, it was possible to show that production of superoxide anion induced by phorbol ester (PMA) in peritoneal murine neutrophils was enhanced by a factor of four in microgravity (**Fleming *et al.* 1991**).

Rat osteosarcoma. A rat osteosarcoma cell line (ROS 17/2.8 osteogenic cells) was exposed to gold labelled epidermal growth factor at 37°C during 20 sec microgravity in a parabolic flight (Schmitt *et al.* 1991). EGF was added 3 sec before 0 × g, glutaraldehyde was added immediately before the end of the microgravity period. Gold particles were counted on electronmicrographs. No difference in binding of EGF was observed in microgravity. Internalization of EGF via receptor mediated endocytosis (RME) was observed during the 25 min flight (8 parabolas). RME is an active mechanism closely related to the cytoskeleton.

In another study with the same cell line, the effect of gravitational stress on RME was studied with cells exposed continuously from 18 min to 5 parabolas consisting of 5 periods of microgravity and 10 periods of hypergravity (1.8 × g) (Malouvier *et al.* 1992). Three phases of RME were determined: binding, clathrin-mediated internalization of the ligand-receptor complex, and recycling or degradation via endosomes and lysosomes. Membrane binding (1st phase) of EGF was lower in gravity-stimulated cells (–19%). No difference was seen for the coated structures (2nd phase). More cytoplasmic labelling (21%) was seen in the gravity-stimulated cells (3rd phase).

Blood platelets. The PKC signal transduction pathway has been studied on the PKC-dependent phosphorylation of the 40 kDa protein plekstrin. Phosphorylation occurs within a few seconds upon activation of ^{32}P-labelled platelets with thrombin or phorbol esters like PMA (**Schmitt *et al.* 1993**). Under the experimental conditions the 40 kDa band was clearly visible 10 sec after activation on SDS-PAGE autoradiographs. Quantitative analysis of the radiolabelled bands from samples exposed to 1 × g, 1.8 × g and microgravity did not show significant differences. The results indicate that neither Ca^{2+}-calmodulin-mediated activation nor the PKC-dependent pathways are inhibited during these short periods of microgravity.

Gap junction channelling. Gap junctions extracted from cardiac tissue (chicken) were inserted in liposomes containing FITC-microperoxidase and incubated at 1 and 0 × g with Azure C/H_2O_2 as substrate (**Claassen & Spooner 1989**). The oxidation of Azure C was followed spectrophotometrically at 612 nm. No effect was observed during 20 sec microgravity.

Rhizobia. In an experiment on the binding of lectin to the cell membrane, no effect of microgravity was observed on the interaction between lectin and rhizobia (**Henry *et al.* 1990**).

Antigen-antibody binding. The binding of α-ofetoprotein (AFP) to immobilized monoclonal antibodies (mAb) was determined (**Spooner *et al.* 1990**). Microgravity lasted 20 sec, AFP was mixed with monoclonal anti-AFP bound to latex microparticles. Bound AFP was determined with mAb anti-

AFP marked with alkaline phosphatase using nitrophenylphosphate as substrate. No effect of microgravity on binding was observed.

1.2.3.5 Hypogravity in sounding rockets

Cells of the immune system. The hypothesis that changes in membrane function and, consequently, in the availability of membrane glycoproteins to Con A may occur in microgravity, was tested in two experiments conducted on sounding rockets (Cogoli, M. *et al.* 1990). The instruments provided automatic injection of fluorescence-labelled Con A, followed by fixation at given times with paraformaldehyde during seven minutes in microgravity. Both experiments showed that binding of the mitogen to the membrane was not affected, only a slight delay of patching and capping was observed.

A431 epidermoid cells. Important studies were carried out by Dutch investigators with A431 cells exposed to EGF on MASER 3 and MASER 4, respectively. In MASER 3, the parameters measured were expression of c-*fos* and c-*jun*, and, in the clinostat (see above), activity of SRE. Expression of c-*fos* gene was depressed by 50% as well as that of c-*jun* (the product of c-*jun* forms a heterodimeric complex with the product of c-*fos*) (de Groot *et al.* 1990). Expression of the β-2-microglobulin gene, which is not modulated by EGF, did not change indicating that the effect on proto-oncogenes is a specific one. SRE is present in the 5′ regulatory region of the *fos* gene.
In MASER 4, the protocol was essentially the same as that in MASER 3, except that incubation with EGF, TPA, A23187 and forskolin were performed in real microgravity on the sounding rocket and not in the clinostat. EGF- and TPA-induced expression of c-*fos* was decreased by 47% and 26% respectively. No effect with A23187 or forskolin, nor effects on the expression of β-microglobulin were detected. Similar results were obtained with c-*jun*: with EGF and TPA, depression was 56% and 51% respectively, no effect was seen with A23187. Forskolin was not tested because it does not induce the expression of c-*jun* (**de Groot *et al.* 1991**).

1.2.3.6 Hypogravity in orbit

Cells of the immune system. The first to study cultures of lymphocytes in Space were Tálas *et al.* (1983) who discovered that the production of interferon-α (IFN-α), induced by polynucleotides in human lymphocytes cultured on the Soviet space station Salyut 7 was increased by 500% compared to the ground controls. However, the experimental conditions were far from optimal because the incubator was switched off during the sleep time of the crew for safety reasons. In addition, no $1 \times g$ reference was available on board.

Dramatic effects on ^3H-thymidine incorporation were discovered in T lym phocytes, as discussed in Section 1.2.2.4. It was with the experiment conducted in Spacelab SLS-1, that a systematic determination of the production of cytokines by free floating as well as by microcarrier-attached cells contributed to understanding of the nature of the effects (**Bechler *et al.* 1992; Cogoli *et al.* 1993**). In the resuspended cells, the production of interleukin-1 (IL-1) (believed to be the second signal required for T cell activation) by monocytes was almost nil, thus indicating that the function of the accessory cells in delivering the second signal for activation had failed (see also Fig. 3). In correlation with the increased activation of T lymphocytes on microcarriers, the production of IFN-γ and of IL-2 were also remarkably increased, IFN-γ2-fold and IL-2 two-fold compared to the $1 \times g$ flight and ground controls (Cogoli *et al.* 1993). In addition, the pattern of tumor necrosis factor production by monocytes was different from that of IL-1. The most relevant data on the secretion of cytokines by T lymphocytes and monocytes in space are summarised in Fig. 4.

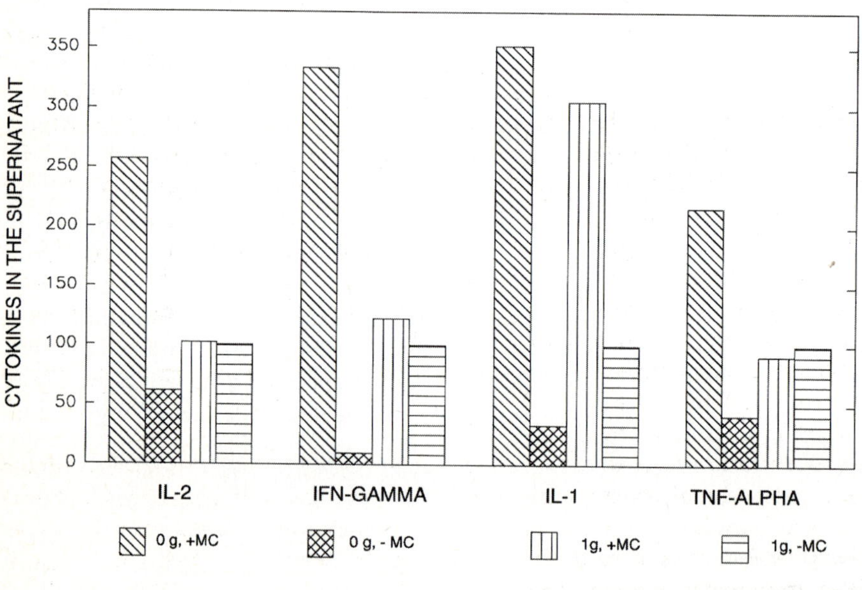

Fig. 4. Production of cytokines by T lymphocytes and monocytes in space. Cultures of lymphocytes with monocytes as accessory cells, either attached to microcarrier beads (+MC) or free floating (−MC), were incubated with Con A in space either at $0 \times g$ or at $1 \times g$ in a reference centrifuge. The data relative to IL-2 are from samples frozen with dimethylsulfoxide 46 h after incubation with Con A at 37°C, those relative to IFN-γ, IL-1 and TNF-α are from samples cultured for 65 h. The results are expressed as a % of the respective control ($1 \times g$, −MC) which is taken as 100%.

In conclusion, the data from the experiment on mitogenic activation of the T lymphocyte/monocyte system support the following hypotheses:
(i) IL-2 is produced independently of IL-1 in Con A-induced activation;
(ii) IL-1 production is low in cultures of free cells at 0 × g, an effect that is reversed by attachment to microcarrier beads;
(iii) the expression of IL-2 receptors depends on IL-1;
(iv) provided sufficient IL-1 is available, activation is enhanced in microgravity;
(v) it appears to be a selective effect of microgravity on the secretion of IL-1 and tumor necrosis factor (TNF) by monocytes and in the production of IL-2, IL-2R and IFN-γ by T lymphocytes.

A number of interesting effects were observed by Chapes *et al.* (**1992**, 1994) in cultures of three types of immune cells in space. However, the cultures were kept in the mid-deck of the Space Shuttle at ambient temperature throughout the incubation time instead of in an incubator at 37°C (an obvious requirement in work with mammalian cells), so the results must be interpreted with caution. The anchorage-dependent bone marrow-derived macrophage cell line B6MP102 secreted, upon activation with lipopolysaccharide, significantly more IL-1 and TNF-α in space than on the ground. Murine spleen cells, stimulated with poly I:C released significantly more IFN-α in space than on earth. Also, human PBL as well as murine lymph node T cells activated with Con A produced significantly more IFN-γ in space than on earth. Experiments on Shuttle flight STS-50 found that cellular cytotoxicity caused by TNF-α was inhibited. This was confirmed in experiments on later flights (STS-54 and STS-57) and it was found that TNF-mediated cytotoxicity was restored to levels observed in the ground controls in the presence of inhibitors of PKC (**Woods & Chapes 1994**). The authors conclude that space flight ameliorates the action of TNF by affecting PKC in target cells, but none of these experiments were accompanied by on-board 1 × g controls (only ground controls were available) so what aspect of space flight is effective has not been established.

Another important investigation (**Limouse *et al.* 1991**) was carried out on the Soviet biosatellite Cosmos 2044 with THP-1 monocytes and Jurkat T cells, phenotypically similar to T cells. Cells were incubated at 37°C at a low starting concentration (2×10^5 cells ml^{-1}) in plastic bags for 5 days before launch (due to operational constraints). Activation occurred 6 h after launch. Jurkat T-derived lymphocytes were activated to produce IL-2 either by monocyte-derived THP-1 cells + mAb anti CD3/T cell receptor, or by A23187 + PMA. THP-1 monocytes were activated to produce IL-1 either by Jurkat cells + anti-CD3 mAb or by PMA. Although in the 1 × g controls IL-2 production under normal culture conditions was 10-fold higher than under the experiment conditions, it was still 20 to 40 times above the detection limit of the test. IL-2 production in the presence of THP-1 cells + mAb anti CD3/T was not different from 1 × g ground controls, but it was inhibited by 100% with A23187 + PMA. Similarly, IL-1 production in the presence of Jurkat cells + anti-CD3 mAb was not different from the 1 × g control, but was inhibited by

85% when PMA alone was present. Glucose consumption was nearly identical (85%) in all cultures, thus showing that cell growth and metabolism were not affected. A control experiment conducted on the ground showed that 85% of the cells were still viable after 5 days in culture. The results show that (1) cell–cell contacts (Jurkat/THP–1) work normally in microgravity; (2) binding of mAb to the T–receptor works normally; (3) cosmic irradiation effects are unlikely; (4) PMA, an activator of PKC, binds to cells in microgravity as shown in parabolic flights. However, anti–CD3 mAb are also potent activators of PKC in T–lymphocytes. Nevertheless, data in the literature do not implicate a PKC–dependence of IL–2 production by anti–CD3 mAb. The data are compared to those of **de Groot *et al.* (1990, 1991)** and direct effects of microgravity are considered to be the likely cause of the effects (**Limouse *et al.* 1991**).
The metabolic data of the experiment with Hybridoma 7E3–N cells in Spacelab IML–1 and described in Section 1.2.2.4 (**Bechler *et al.* 1993**) reveal another interesting behaviour pattern: the production per cell of monoclonal antibodies, the glucose and glutamine consumption per cell as well as the secretion per cell of waste products like lactate and ammonia were lower at 0 × g than at 1 × g. In fact, the lack of significant differences of metabolite concentrations in the supernatants at 0 × g and 1 × g is only apparent since approximately 40% more cells were present in the cultures at 0 × g than in those at 1 × g. Although there is not yet an explanation, the data show that gravitational unloading had significant effect on hybridoma cell metabolism. It appears that the transition from a two–dimensional configuration, as in the case of cells sedimented to the flat bottom of the culture flask at 1 × g, to a three–dimensional configuration, as for free floating cells at 0 × g, increased cell proliferation despite a lower metabolic turnover. It appears also that the biosynthesis of a specific cell product was coupled to the glucose/glutamine consumption and to the lactate/ammonia secretion rather than to the proliferation rate.

Friend cells. In the experiment in Spacelab IML–1, mentioned in Section 1.2.2.4 (**Bechler *et al.* 1993**), the amount of haemoglobin produced upon induction with DMSO was the same in the flight 0 × g and ground 1 × g samples. The counts of haemoglobin–positive cells show that 60 to 70% of the cells were induced to express haemoglobin upon exposure to DMSO. Again, there were no significant differences between cultures at 1 × g and 0 × g. The metabolic analyses on glucose and glutamine consumption as well as on lactate and ammonia production, clearly reflected the fact that Friend cells do not change their behaviour in microgravity.

HaK cells. Production of tissue plasminogen activator (t–PA) by these cells cultured in the DCCS chamber was determined in Spacelab IML–1 (see Section 1.2.2.4). Tissue plasminogen activator is a substance of high pharmaceutical value since it is used to prevent the formation of blood clots where there is risk of thrombosis. There was no difference in metabolic data on t–

PA production, data on the consumption of glucose and glutamine from the medium, nor on the secretion of waste products like ammonium and lactate between the cultures kept at 0 × g and those at 1 × g in the reference centrifuge in flight or in the ground laboratory (**Lorenzi *et al.* 1993**).

Bone cells and tissues. Astronauts lose significant amounts of skeletal calcium during space flight (see Section 2.7). Orbital experiments with rats has shown that even short term space flight results in dramatic inhibition of bone formation but no detectable change in bone resorption (Wakley & Turner 1993). Normal gene expression (in terms of mRNA levels of bone matrix proteins) is not restored in such animals even 48 hours after return to Earth (Westelind *et al.* 1993). Such bone remodelling could be responsible for space flight induced bone loss, and this type of result has prompted considerable interest in the development of cell and tissue cultures to study, *in vitro*, the behaviour of the specialised cells which are responsible for the formation (osteoblasts) and resorption (osteoclasts) of bone.
Experiments on the Biobox facility on Biocosmos 10 (including on–board 1 × g centrifuge controls) showed retarded growth of embryonic bone cultures and altered morphology of osteoblasts in culture. After six days of incubation, the cell bodies displayed a round and compressed morphology, with abnormal extensions protruding from these contracted cell bodies (Demets 1993; Genty *et al.* 1993). Very similar morphological changes have been reported in cultured osteoblasts subjected to clinostat treatment (Al–Ajmi *et al.* 1994).
Something else which might be of immediate interest to astronauts is the question of wound repair during orbital flight. The Animal Enclosure Module on the Space Shuttle has been used for initial experiments on this topic and preliminary results indicate that microgravity reduced the chemotactic effect of two cytokines and diminished collagen formation in a model of deep wound healing, suggesting that the orbital environment retards growth factor responses and reduces wound healing in the rat (Davidson *et al.* 1994).

Sf9 cells. These cells (of the army worm, *Spodoptera frugiperda*) were flown in the BioServe Fluids Processing Apparatus aboard Shuttle flights STS-50, 54, and 57. Cell proliferation was measured with and without exposure to a cell regulatory sialoglycopeptide inhibitor (CeReS). Cell proliferation in flight samples was the same as ground controls (there was no on–board 1 × g control), the CeReS inhibitor bound to its specific receptor and its signal transduction cascade was not gravity sensitive (**Moos *et al.* 1994**).

E. coli. Mechanisms of genetic recombination were investigated in Biorack on Spacelab D–1 with a 1 × g control centrifuge in flight (Ciferri *et al.* 1986, 1988). There are three types of interaction between bacteria: the exchange of chromosomal DNA via sex pili (conjugation), the transfer of short stretches of DNA by a bacteriophage (transduction), and the uptake of extracellular DNA fragments (transformation). No gravity effect was found with respect to transduction and transformation (though the authors point out that at-

tempted transformation of *Bacillus subtilis* with chromosomal DNA fragments, rather than plasmids, would have been a more sensitive test system). On the other hand, conjugation in *E. coli* was enhanced in microgravity, there being a 3 to 4-fold increase in conjugations lasting 40 minutes or more over the 1 × g in-flight centrifuge and ground-based controls (Ciferri *et al.* 1986, **1988**). The same investigators observed an increased growth rate in microgravity, though this does not account for the observed increase in the frequency of recombinants.
In another experiment on Biocosmos 2044 (**Bouloc & D'Ari 1991**, see also Section 1.2.2.4) DNA damage (cosmic radiation) was determined by analysing the expression of SOS response using a gene fusion in which *lac*Z, the structural gene of *β*-galactosidase is transcribed solely from the promoter of the SOS gene *sfi*A; *β*-galactosidase has the advantage of being extremely stable, thus permitting long-term storage of the cells (the flight lasted 15 days, the experiment 270 min). No effects of space flight (microgravity and cosmic radiation) were observed on the three parameters measured.

1.2.4 Morphology and motility

1.2.4.1 Background

Two questions have always intrigued researchers in gravitational biology: first, do shape and morphology of a cell change in microgravity? Second, are free floating cells which are not provided with locomotion organelles like cilia, capable of autonomous movements in microgravity? Biophysical calculations of the forces necessary to alter the shape of a cell show that gravity is not strong enough to influence cell shape. This means that the flattening of a cell sedimented to the bottom a cell culture flask at 1 × g is not due to gravitational forces but rather to the spreading of the cell on the surface. Nevertheless, shape and morphology changes could be the result of small alterations in the pressure that single organelles with higher density than the cytosol may exert on the cytoskeleton. In this case, shape and morphological changes may occur when the gravitational environment is altered. The answer to the second question is: yes, cells may display autonomous movements in microgravity generated either by Brownian motion or by Marangoni convection (**Napolitano 1984; Scriven & Sternling 1960**). While Brownian motion is difficult to observe, due to the large size of the cells compared to the length of each impulse, Marangoni convection is generated by temperature or concentration gradients in a fluid and causes macroscopic displacements of the cell. It is conceivable that the microenvironment of a cell in a culture medium changes continuously due to the cell metabolism with consequent generation of concentration gradients of nutrients as well as of waste products.

Shape changes and motion may alter important cell functions including cell–cell interactions and processes related to cell structures, like the formation of the mitotic spindle.
Of great importance for morphological and dynamic studies has been the advent of NIZEMI, the low-speed centrifuge microscope which was used for the first time in space in Spacelab IML-2 in 1994.

1.2.4.2 Hypergravity

HeLa cells. Movements on a substratum coated with colloidal gold were tracked after 48 h in dark-field illumination at $1 \times g$ and $10 \times g$ (**Tschopp & Cogoli 1983**). While at $1 \times g$ the cells showed normal patterns of migration, at $10 \times g$ the cells did not change their position. In addition, whilst after mitosis the daughter cells go in opposite directions following a symmetrical pattern at $1 \times g$, at $10 \times g$ the cells remained almost motionless forming aggregates by successive division. The focal contacts between cell and substratum did not differ between $1 \times g$ and $10 \times g$., and no shape differences were observed by light microscopy. Absence of shape changes was also reported by **Kumei *et al.* (1987)** after 4 days at $35 \times g$.

V79 and JTC-12 cells. The lack of shape changes in hypergravity was confirmed also in V79 Chinese hamster lung cells (3 days at $18 \times g$) and in JTC-12 monkey kidney cells (4 days at $35 \times g$) respectively (**Kumei *et al.* 1987**).

Human lung adenocarcinoma cells A 549. As reported in section 1.2.2.2, **Croute *et al.* (1990)** carried out experiments at 2 and $15 \times g$ lasting up to 27 d. While no changes in cell proliferation were measured, interesting morphological alterations were detected. Electron micrographs of the A549 were composed at the start of the experiment of small translucent areas (alveolar structures) evenly distributed among a dense cellular mass. After 27 d of centrifugation, $1 \times g$ controls exhibited larger alveoli which protruded at the periphery. Conversely, in the hypergravity cultures large alveoli developed only in the middle part of the nodule, surrounded by a dense cellular mass encircled by a thin layer of cells.

Paramecium. The swimming behaviour has been studied recently in the low-speed centrifuge microscope, NIZEMI at $1–5 \times g$ using a computer-controlled image analysis system (Hemmersbach-Krause *et al.* 1992). Cells retained their swimming capability, did not sediment, and even increased the precision of their negative gravitaxis but reduced their mean swimming velocity.
In an earlier experiment, conducted in culture tubes separated by a septum and fixed in a centrifuge so that the direction of the vector of the centrifugal force coincided with the longitudinal axis of the tube, it was possible to

show that *Paramecium* was able to swim even against an acceleration of 8 × g (Tixador *et al.* 1984b).

Amoeba proteus. The growth, shape and movements were observed without interruption (except for daily feeding) during 36 days in a microscope centrifuge at 40 × g. Despite a partial stratification of cytoplasmic particulate material, no effects on growth rate and motility were observed (Montgomery *et al.* 1965).

1.2.4.3. Hypogravity in clinostats

Cells of the immune system. Human peripheral blood lymphocytes were exposed to Con A in the fast rotating clinostat (**Cogoli *et al.* 1980**). Electron transmission micrographs of cells cultured in the clinostat were compared to those of control cells after fixation with glutaraldehyde. Cells activated for 3 days at 1 × g were characterized by the presence in the cytoplasm of widely spread vacuoles, by the formation of several pseudopodia and by a rather low number of mitochondria. The formation of vacuoles was more marked on day 4 of culture and can be interpreted as an indicator of cell aging. Cells tended to swell up to a diameter of 10–15 µm. Cells grown in the clinostat appeared to be equally distributed in two populations. The first population was very similar to the control cells. The second populations appeared to be remarkably different and, after 3 days, it was characterized by the presence of a high number of tightly packed and well-developed mitochondria. Cinematographic recording of the rotated cells showed amoeboid movements as well as cytoplasmic streaming within the cells.

A431 epidermoid cells. It is known that EGF causes rapid rounding of A431 human epidermoid carcinoma cells. The effect is temperature dependent. At 37°C lower concentrations of EGF are required for 80% rounding than at 20°C. A431 cells were exposed to EGF in the clinostat and in the centrifuge at 10 × g at 20°C and 37°C. At 1 × g 72% of the cells showed rounding, and at 0 × g rounding was significantly increased to 85%. A slight, non-significant depression of rounding was seen at 10 × g (**Rijken *et al.* 1991, 1992**).
Epithelial cell adhesion seems to be unaffected by simulated microgravity (Jessup *et al.* 1994). Human colorectal carcinoma cells were cultured in the Rotating Wall Vessel – a clinostat which provides randomization of the gravity vector to mammalian cell cultures but with low shear stress. After 6–7 days, cells were assayed for binding to various substrates and compared to cells grown in standard tissue culture flasks and static suspension cultures. No differences in epithelial cell adhesion were observed.

Neurones. In a first study (**Gruener & Hoeger 1990**) neurones and myocytes were isolated from *Xenopus laevis* embryos. Myocytes were seeded on collagen-coated glass fibre cover slips. Spinal neurones were added 6 h later.

The formation of nerve-associated acetylcholine receptor patches (NARPs) was observed by phase contrast microscopy and by fluorescence microscopy of myocytes stained with rhodamine α-bungarotoxin. Nerve-associated accumulation of acetylcholine receptor (AChR) patches was characterized by fluorescence bands along the neuritic paths. Samples were kept at $1 \times g$ (static), 1 rpm and 10 rpm in a horizontal clinostat. Controls for mechanical perturbations were kept (i) in a vertically rotating clinostat, (ii) at 100 rpm in a centrifuge, (iii) in an inverted configuration, and (iv) on an oscillating platform. Three types of synapses were considered: (a) mature, cultured for 36 h before rotation, (b) immature, cultured for 16 h before rotation, and (c) *de novo*, cultured for 6 h before rotation. Significant reduction of NARP incidence and NARP area was observed in immature (both -80%) and *de novo* (-45%, -80%, respectively) NARPs. In contrast, the incidence and the area of NARPs did not differ significantly in the mature synapses. The data suggest the existence of a window of sensitivity to microgravity during synaptogenesis.

In another study by the same authors (**Gruener & Hoeger 1991**) co-cultures of myocytes and neurones were observed at rotation rates between 1 and 50 rpm. When spinal neurones were co-cultured with myocytes, nerve cells produced robust neurites which frequently made physical and functional contacts with myocytes. No significant alterations were observed in survival of neurones or in the formation of neuritic extensions, or in establishment of contacts with myocytes. However, neurones in the clinostat frequently showed large, bean-shaped swellings along their neuritic shafts. In control neurones, swelling had a mean major diameter of 8.6 µm, in rotated samples the mean major diameter was 11.9 µm at 1 rpm and 12.3 µm at 10 rpm. Swellings were significantly larger (120%) in rotated cultures.

Myocytes. In the same experiments described above (**Gruener & Hoeger 1991**) rotated myocytes were larger, the nuclei were often more prominent and their nucleoli were larger. Whilst control cells changed their shape as a function of time from a rounded to an elongated shape, rotated myocytes appeared rounded at all times. Surface area increased significantly at 1 and 5 rpm, cell thickness remained unchanged.

Physarum polycephalum. An interesting model to study gravity effects on cells is the slime mould *Physarum polycephalum*, which forms side branches that can extend over a distance of more than one metre. These branches consist of an ectoplasmic tube in which the endoplasm streams back and forth in regular oscillations. This intracellular transport is driven by actin-myosin interactive filaments. Particles in the stream can move as fast as 10 mm h^{-1}. Briegleb and coworkers (Briegleb *et al.* 1986; Block *et al.* **1986**, 1988) devised an experimental system to detect gravity effects on the amplitude and frequency of these oscillations. In the fast rotating clinostat (30 - 100 rpm) which was connected to a microscope, they were able to monitor the intracellular fluctuations of endoplasmic streaming under conditions of simulated low gravitational acceleration. An alternative to the clinostat was to turn the samples by 180° relative to the gravity vector. Distinct disturbances and changes in the kinetics were detected. In the initial phase of $0 \times g$ simulation (20 min) and after the 180° turn, the rhythmic contractions showed an average increase in frequency of 10%. This initial phase was followed by further increase in the frequency of contractions and, later, by oscillations around the mean frequency. In addition, a 40% increase of the velocity of cytoplasmic streaming compared to the $1 \times g$ control was measured. These data show the existence of gravisensitivity in *Physarum*, without suggesting a mechanism.

By applying inhibitors of respiration, it was possible to show in the clinostat that the receptor site of gravisensitivity may be correlated with the mitochondria (Block & Briegleb 1989; **Block *et al.* 1986**).

Dictyostelium discoideum. The height of the fruit body of this slime mould was measured after clinorotation or centrifugation at $2 \times g$ and $3 \times g$ for 4 days, with the acceleration vector oriented both normally and inversely to the normal direction of the gravity vector (Kawasaki *et al.* 1987). The height of the fruiting bodies was least (less than 1 mm) under simulated microgravity conditions and greatest as acceleration was increased, both in normal and inverse directions (heights were up to 1.4 mm in inverted cultures at $3 \times g$ and 1.6 mm in normally-oriented cultures at $2 \times g$). The differences were significant at $P<0$. The authors suggest that the height of the fruiting bodies is controlled by two factors, one is gravity-dependent the other gravity independent.

1.2.4.4 Hypogravity in parabolic flights

Blood platelets. It is known that stimulation with thrombin and ADP causes shape changes within 15 sec. The platelets are transformed to spherocytes with appearance of pseudopodia via a calmodulin-dependent mechanism. The effect of microgravity on cell shape and aggregation has been studied recently (**Schmitt *et al.* 1993**). The activators were injected into the cell suspensions after 5 sec in microgravity, fixation followed 15 sec later. Scanning electron micrographs of platelets activated by phorbol ester PMA, ADP or thrombin showed no difference between $1 \times g$ controls and microgravity samples. Also, transmission electron micrographs of platelets exposed to thrombin did not reveal differences.

1.2.4.5 Hypogravity in drop towers

Paramecium. The swimming behaviour was recorded with a CCD-camera in drop tower experiments conducted by ZARM at the University of Bremen during free fall episodes lasting 4.5 sec each (**Machemer *et al.* 1992**). The data support the following conclusions: (i) at $1 \times g$, the rates of upward swimming are lower than that of downward swimming. The difference is clear although it is only 20% of the gross speed. (ii) During microgravity, differences in swimming rate did not subside instantaneously. Only after 3 sec did the upward and downward rates of swimming adjust toward a common value indicating that relaxation of the gravikinetic response takes several seconds. The results are in agreement with an electrophysiological hypothesis on graviperception in cells. Similar data were obtained with *Loxodes*, but not with *Didinium*. The results, although preliminary, show that drop towers may be useful in certain studies in gravitational biology (see also Section 1.3.1).

1.2.4.6 Hypogravity in sounding rockets

Cells of the immune system. Interesting results were obtained recently in two experiments conducted on MAXUS I which showed that free floating cells are capable of autonomous movements and of aggregation in microgravity, implying that cell–cell contacts occur as well, though the number of aggregates is reduced at $0 \times g$ (**Cogoli-Greuter *et al.* 1994**). In the first experiment, movements of purified human resting lymphocytes kept at 37°C were recorded during 12 min microgravity by use of a microscope telemanipulated from the ground station. Focus regulation and object selection could be operated manually during flight. Images were recorded by an on-board camera. The recorded images clearly show that the free floating cells were able to display autonomous motion in random directions. A more accurate analysis (one image every 13 sec) shows that the movements were much more complex. The cells often changed direction, moved back and forwards and sometimes crossed the same point several times. The average velocity, calculated from the displacement in the 13 sec increment, was 0.14 ± 0.02 μm sec^{-1}, with a range of 0–0.49 μm sec^{-1}. Also of interest is the observation that the cells in microgravity were not all rounded. Very often they exhibited longitudinal forms, rotated around their axis and also showed contraction waves similar to those described in the literature for lymphocytes moving under $1 \times g$ conditions. This result is of primary importance to interpretation of the behaviour of T lymphocytes in microgravity. The origin of the movement can be attributed to two major causes: First, as discussed above, the changes of the cytoskeleton may determine at least in part the random displacements observed. Second, Marangoni convection due to differences in the concentration of components dissolved in the medium may generate the movements of resuspended cells in microgravity (Section 1.2.4.1). Concentration gradients are generated by the metabolism of the cells which are consuming nutrients (mainly glucose and glutamine) and producing waste (like lactate and ammonia).

In the second experiment, tubulin and vimentin were marked in Jurkat cells exposed to microgravity with FITC-labelled monoclonal antibodies at prese-

lected times followed by fixation with formaldehyde. Cytoskeleton analysis of Jurkat cells revealed the formation of large and compact bundles of intermediate filaments of vimentin. Analogous structures appeared in the 1 × g controls, but to a much lesser extent. The changes occurred after 30 sec exposure to microgravity and remained stable throughout the flight. Similar but less evident changes occurred also in tubulin. These data are in favour of direct effects of gravity on the cell. Thus, gravitational unloading does not cause changes of the cell shape. However, changes of the cytoskeleton, probably due to the change in pressure of dense organelles may occur (**Cogoli–Greuter *et al.* 1994**).
The fusion of the antibody–producing hybridoma cell line G8 with the hypoxanthine–aminopterine–thymidine sensitive SP2/0–UZ cell line was attempted in microgravity on a TEXUS flight (**Schnettler *et al.* 1989**). The fusion rate was substantially increased and the yield of viable cell hybrids was 2–fold enhanced compared to the ground control experiment. The higher yield is related to a better alignment of the parent cells resuspended in a weightless electric field compared to a system at 1 × g in which cells rapidly sediment to the bottom of the fusion chamber.

A431 epdermoid cells. The clustering of EGF receptors was used as a marker to investigate whether the effects detected on the EGF–induced c–*fos* and c–*jun* expression in microgravity (see Section 1.2.3.5) are due to the inhibition of processes occurring at the beginning of the EGF–induced signal transduction. The clusters were visualized in the electron microscope by immunogold labelling with mAb of samples in a sounding rocket flight at the beginning of and after 5 min exposure to microgravity (**Rijken *et al.* 1993**). No difference was detected between the flight and ground control samples. This suggests that the influence of microgravity on EGF signal transduction occurs downstream of EGF binding and receptor redistribution, but upstream of early gene expression; inducer binding and clustering of signal receptors are not affected in microgravity.

1.2.4.7 Hypogravity in orbit

Cells of the immune system. Samples of lymphocytes activated for 3 days with Con A and frozen in flight in Spacelab 1 with hydroxyethylstarch showed the formation of cell aggregates in microgravity (**Cogoli *et al.* 1984**). Transmission electron microscopic analysis of samples activated with Con A and fixed in flight (Spacelab D–1) with glutaraldehyde confirmed the formation of aggregates. As shown in Fig. 5, lymphocytes appeared to be highly damaged under microgravity conditions whereas cells cultured at 1 × g in flight undergo mitosis and blastogenesis. Ultrastructural changes observed in lymphocytes may be related to increased apoptosis in microgravity (Cogoli *et al.* 1988; **Cogoli *et al.* 1993**). This is in agreement with the data on DNA synthesis (see Section 1.2.2.4). Monocyte ultrastructure appeared to be intact,

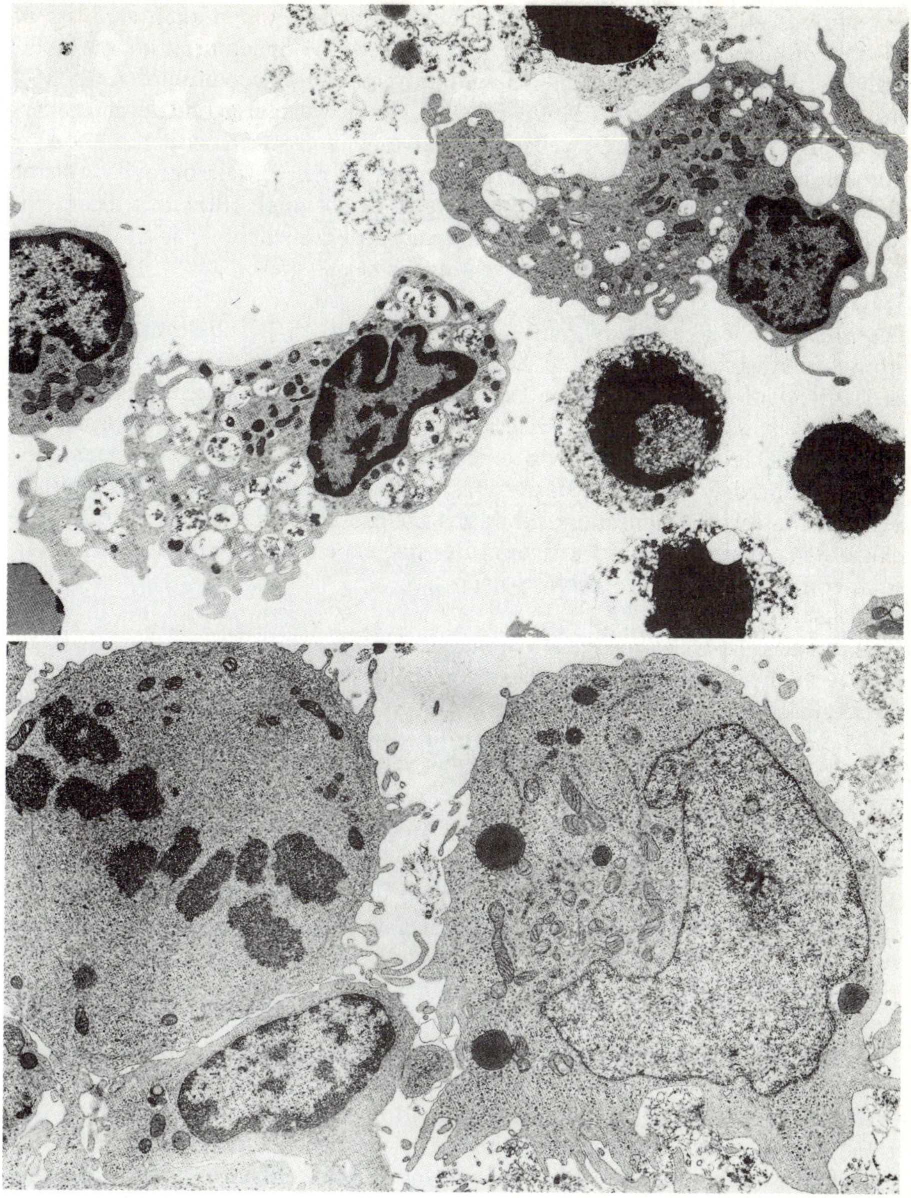

Fig. 5. Electron micrographs of lymphocytes and monocytes exposed to Con A in Spacelab D–1. Top, cells in microgravity fixed after 36 h of culture (× 5100): dead lymphocytes and two monocytes (larger cells) with a large number of vacuoles. The monocyte on the right is displaying intense membrane activity. Bottom, cells in the 1 × g reference centrifuge fixed after 96 h of culture (× 8200): one lymphocyte is in metaphase (left) and another is transformed into a lymphoblast (right) following mitogenic activation. Micrographs by O. Müller & E. Hunzinger from Cogoli *et al.* 1988.

though with a stronger membrane activity, i.e. display of a large number of pseudopodia, in the 0 × g samples (Cogoli *et al.* 1988; Cogoli *et al.* 1993). Finally, as shown in Fig. 6, transmission and scanning electronmicrographs of cultures flown in Spacelab SLS-1, in which cells were attached to microcarrier beads, revealed that (i) cells had a strong interaction with the substratum; and (ii) that attached cells did not deteriorate in microgravity as free floating cells did (**Bechler *et al.* 1992**; **Cogoli *et al.* 1993**). Ultrastructural analysis of an experiment with hybridoma cells conducted in Spacelab D-1 did not give conclusive results (Beaure d'Augéres *et al.* 1986).

Human embryonic kidney cells. The attachments to a substratum of adhesion-dependent cells was tested in microgravity in an experiment carried out in an incubator installed in the flight deck of the Space Shuttle flight STS-8 (**Tschopp *et al.* 1984**). Microcarriers were added in flight to the cells in culture at 37°C. Scanning electron microscopy showed that attachment took place qualitatively and quantitatively as in the ground controls, thus confirming that the related membrane functions are not altered at 0 × g. Similar conclusions have been arrived at from clinostat experiments with human colorectal carcinoma cells (see Section 1.2.4.3).

Friend cells. In an experiment conducted in Biorack in Spacelab IML-1, extensive analysis (scanning, transmission, volume measurements) of the ultrastructure of cells cultured for 6 days in the presence of DMSO did not reveal differences between the cell cultures at 0 × g and in the reference centrifuge at 1 × g (**Bechler *et al.* 1993**).

WI38 human embryonic lung cells. In an experiment carried out in Skylab and described in Section 1.2.2.4, cinematographic recording, phase, electron and scanning microscopy produced no observable differences in ultrastructure and in cell migration between flight and ground controls (**Montgomery *et al.* 1978**).

Erythrocytes. The peripheral blood from several donors (either healthy or with history of a number of diseases) was diluted with autologous plasma. Storage at ambient temperature on two Space Shuttle flights showed a dramatic decrease of red blood cell aggregation compared to the ground controls. Without specifying the nature of the effect, the authors concluded that membrane function is altered in microgravity (Dintenfass *et al.* 1986; **Dintenfass 1990**).

L8 rat myoblasts. Several permanent phenotypic alterations were recorded in cell cultures of rat myoblasts which were recultured on Earth after return from a 10 day Space Shuttle flight (STS-45). The differences included altered morphology and failure to fuse and differentiate into myotubes (**Kulesh *et al.* 1994**). Unfortunately, the space flight cultures were accommodated in an automated cell culture apparatus in a mid-deck locker on the Space Shuttle,

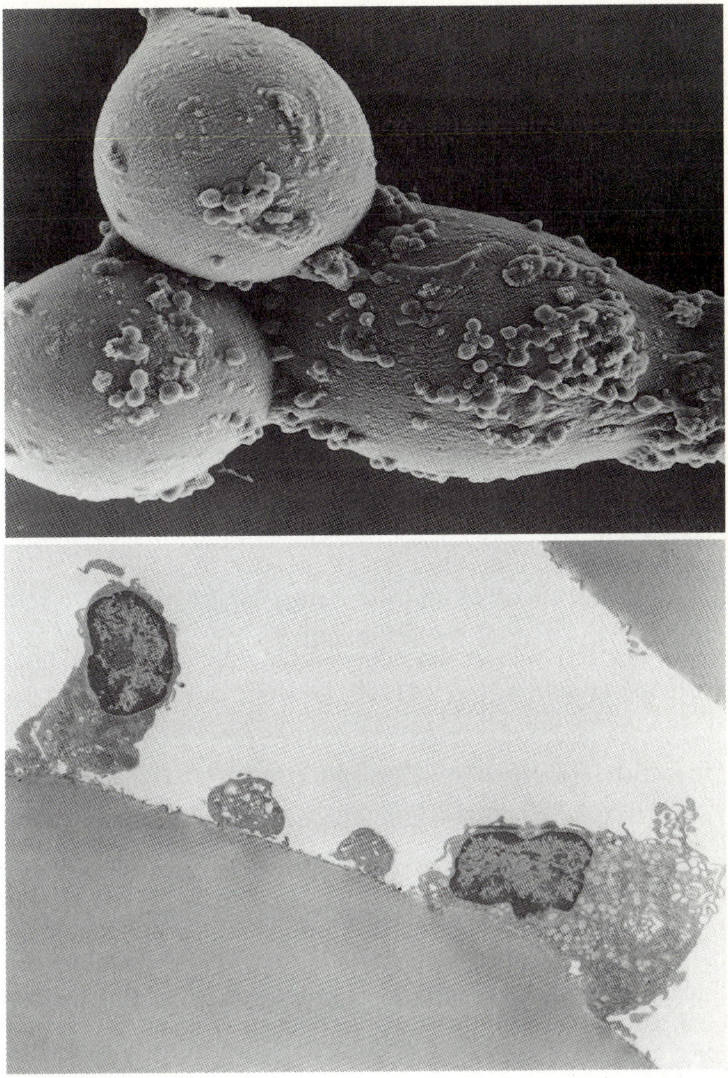

Fig. 6. Electron micrographs of microcarrier-attached lymphocytes and monocytes exposed to Con A in Spacelab SLS-1. Cells were cultured for 72 h in microgravity before freezing at -20°C in the presence of dimethylsulfoxide. After thawing in the ground laboratory, the cells were fixed with glutaraldehyde. Control samples were kept at 1 × g in-flight (not shown here). Top, scanning electron micrograph showing cells on three microcarrier beads (× 730). Bottom, transmission electron micrograph showing, from left to right, one lymphocyte, two blood platelets and one monocyte (× 3400). Micrographs by O. Müller & E. Hunzinger from Cogoli *et al.* 1993.

and there was no on-board 1 × g control. Consequently, the cause of the altered phenotype is unknown.

Paramecium. Interesting observations are reported from experiments carried out in Spacelab D-1 and in Soviet space laboratories (Tixador *et al.* 1984b; Richoilley *et al.* 1986, 1988a & b). The cell membrane of cells exposed to microgravity appeared thicker than that of the controls in transmission electron micrographs. Changes were seen also in exocytosis which under normal conditions is characterized by the extrusion of trichocysts docked beneath the plasma membrane. Trichocysts are spindle-like organelles the function of which is unknown. In a number of cells cultured at 0 × g exocytosis was reversed and decondensed trichocysts were detected inside the cytoplasm.

Pelomyxa carolinensis. Transmission electron micrographs of cell cultures aboard Biosatellite II for two days did not show differences between flight and ground control samples (Ekberg *et al.* 1971). Locomotion on a substratum seems to be independent from gravity. In this context, it is interesting to note that the speed of amoeba *Naegleria gruberi* was the same on the top surface and the bottom surface of a horizontal perfusion chamber. This shows that cell-substratum adhesive forces must be considerably greater than gravitational forces (**Davies *et al.* 1981**).

Physarum polycephalum. Two experiments were conducted in Biorack in Spacelab D-1 and IML-1 respectively, using a photodiode and a microscope with camera (Briegleb *et al.* 1986, 1987; **Block *et al.* 1986, 1992**). In space, the cells were first adapted to 1 × g during 6 h in the Biorack centrifuge and then the cytoplasmic streaming was recorded at 0 × g. The experiment in D-1 was only partially successful due to operational problems in flight. Nevertheless, the data obtained in the clinostat, i.e. increase of frequency of the cytoplasmic oscillations and of the streaming velocity (Section 1.2.4.3), were qualitatively confirmed. In the second experiment, carried out in IML-1, the effect of white light stimulation was studied. White light is known to have an opposite effect to microgravity, i.e. a decrease of cytoplasmic frequency and velocity in *Physarum*. The transition from 1 × g to 0 × g induced - as seen before - an increase of frequency that was maximum 18-26 min after onset of 0 × g. The mean period decreased from 88 sec to 60 sec, after backregulation it rose to 98 sec. This high value persisted for nearly 2 h with mean period values of 98-99 sec. Except for a more pronounced reaction to 0 × g, the data from the clinostat were qualitatively and quantitatively confirmed. When the opposing effects of microgravity and white light were measured simultaneously, a mutual suppression of both responses was observed. Another test, in which *Physarum* was first adapted to 0 × g and then exposed to white light, provided further evidence that the two responses have a strong mutual influence on each other.

1.2.4.8 Experimental models

As described in the previous Sections, few single cell systems have showed significant changes in space. In this section, the characteristics of systems which are good candidates for future research in gravitational biology are described.

Cells of the immune system. Several investigators in different laboratories have used either peripheral blood lymphocytes (PBL) or derived cells lines for experiments in gravitational biology, especially the lymphocyte-monocyte system (Beaure d'Augéres *et al.* 1986; **Chapes *et al.* 1992; Cogoli & Tschopp 1982, 1985; Cogoli, 1993b; Gmünder *et al.* 1990, 1992; Limouse *et al.* 1991**; Tálas *et al.* 1983). Lymphocytes are easily separated as purified resting (i.e. non-activated) cells from peripheral blood of healthy donors and upon exposure in culture to mitogens, can be activated polyclonally to proliferate and, in the case of T cells, to produce a number of lymphokines. The separation from fresh blood is based on centrifugation on Ficoll/metrizoate gradients and leads to a cell population consisting of 90–80% lymphocytes (T and B cells), 10–15% monocytes and 5% granulocytes. Alternatively, activation can be triggered in samples of whole blood obtained by diluting fresh blood with culture medium. In some cases, lymphocytes can be isolated from the spleen of mice or rats. The terms activation, stimulation or proliferation refer to the measurement of the amount of tritiated thymidine incorporated into DNA (precipitated with cold trichloroacetic acid) after a 2–5 h pulse (see Section 1.2.2.1).

The mitogenic activation of lymphocytes *in vitro* constitutes a system of interest, since the activation *in vitro* mimics the events occurring *in vivo* during antigenic challenge, the mitogenic activation assay is also used as a diagnostic tool to assess the efficiency of the specific immune system. The test was introduced to determine the effects of space flight on the immune system of astronauts more than two decades ago and it is currently used in space. A discussion of the effects observed on T lymphocytes drawn from space crew members prior to, during and after flight goes beyond the scope of this Chapter. For reviews see Cogoli (1993a & c), Cogoli & Tschopp (1985), Gmünder & Cogoli (1995, in press). It could be useful to detect alterations of the immune system on future long-duration space flights and to investigate *in vitro* neuroimmunological interactions under the physical and psychological stress of space flight.

Moreover, the transition from the status of resting to that of activated cells can be considered an example of cell differentiation that can be controlled *in vitro* by addition of the mitogen. T lymphocytes are activated by a number of mitogens of different origins. Most of the results presented here were obtained with the lectin Con A from lentil seeds. The mechanism of activation is very complex and not yet completely understood. Studies in microgravity may contribute to the understanding of a basic mechanism underlying lymphocyte differentiation. In fact, certain steps of activation, such as the production of IL–1 (the second signal required for T cell activation as shown in Fig. 3) may be switched off at 0 × g, and thus the process leading to activation might be dissected into separate phases which could be analyzed in detail (**Cogoli *et al.* 1993**).

An alternative to the lymphocyte/monocyte system from PBL are derived cell lines as the Jurkat (T cells) or THP–1 (monocytes) cell lines. Cell lines have the advantage of consisting of homogeneous and well characterized cell populations (**Limouse *et al.* 1991**; **Chapes *et al.* 1992**). PBL, in contrast, comprise several T and B cell populations. Nevertheless, while the possibility of regulating *in vitro* the function of derived cell lines is rather limited, the PBL system is more flexible to mitogenic activation and certainly much closer to the *in vivo* function of the cells of the immune system.

Less interesting for experiments in microgravity and possible bioprocessing applications (in terms of proliferation rate and antibody production) appear to be hybridoma cells. This is not surprising, since hybridoma cells are ultimately committed to divide and to secrete antibodies with no, or little, possibility of regulation by external agents (Beaure d'Augéres *et al.* 1986; **Bechler *et al.* 1993**). Conversely, the electrofusion of immune cells to produce antibodies gave promising results in microgravity (**Schnettler *et al.* 1989**).

A431 epidermoid cells. The use of these carcinoma cells in combination with EGF proved to be very useful to study gravitational effects (**de Groot *et al.* 1990, 1991**; **Rijken *et al.* 1991**). In particular, important data were gathered, first, on the intracellular signal transduction cascade involving either protein kinase A or protein kinase C; second, on the binding of EGF to the cell membrane; and, third, on the early expression of oncogenes. Similar studies are certainly needed also on the T lymphocyte/monocyte system.

HeLa cells. Similar conclusions can be drawn for these cells from the work of **Kumei *et al.* (1987, 1989)** on the expression of oncogenes and on the study of the cell cycle. HeLa cells however, were not subject to regulation by mitogens or growth factors as PBL and A431 cells, respectively.

Neurones. In vitro systems using neuronal cell cultures are particularly important for inclusion in experiments involving change in gravitational acceleration because of the anticipated rapid developments in neurobiology. Studies with the clinostat which led to the identification of windows of gravity sensing are very promising (**Gruener & Hoeger 1990, 1991**; see Sections 1.2.4.3 and 1.5.2).

Physarum. Regulation of rhythms and cytoplasmic streaming are important elements of gravitational biology and *Physarum* may be a good experimental system (Briegleb *et al.* 1986; Block *et al.* **1986**, 1988, **1992**; Block & Briegleb 1989). The discovery of its gravisensitivity is an important step in this analysis.

1.3 Taxes and tropisms in microbes

Much of the early research on the space biology of microorganisms was concerned with microbial assay of lunar samples (Taylor *et al.* 1971; **Taylor & Wooley 1973**); with the microbial biota of spacecraft and astronauts (**Taylor 1974a & b**; Taylor ***et al.*** **1974**; **Carmichael & Taylor 1977**); and with the use of microorganisms as test organisms for the mutagenic effects of exposure to the space environment (**Hotchin *et al.* 1965**; **Orlob & Lorenz 1968**; De Serres 1969; **De Serres *et al.* 1969**; **Jerger & Volz 1977**; **Horneck 1992**; and see Chapters 4 and 5 of this book).

Over the last 20 years, more attention has been given specifically to the effect of gravitational disturbance on the behaviour of microbial cells, in part as contribution to study of their taxes and tropisms (Carlile 1975), but with increasing emphasis on study of the overall gravitational biology of relatively simple multicellular systems and, most importantly, single cells. Most of this work is reviewed in Section 1.2 but there are a few other observations which are worth recording here.

Extremes of hypergravity (10^3 to $10^5 \times g$) increased enzyme activities but reduced endogenous respiration in *Staphylococcus aureus* (Nemec *et al.* 1983); perhaps not surprising at this level of acceleration stress! Sporulation of *Bacillus subtilis* was suppressed in microgravity, but higher growth yields were recorded (Mennigmann & Lange 1988). In contrast, **Bouloc & D'Ari (1991)** and **Gasset *et al.* (1994)** report no increase in growth in space-flown cultures of *Escherichia coli* (see also Section 1.2.2.4). Greater growth has also been observed in space flight cultures of *Nocardia meditterranei* (Gu *et al.* 1989), though antibiotic titres and the stability of recombinant plasmids were unaffected. Minimum inhibitory concentrations of the antibacterial colistin (= polymyxin E, which inserts into the bacterial membrane and collapses the proton gradient) have been claimed to double, implying increased bacterial (*E. coli*) resistance to or tolerance of the antibiotics in microgravity conditions during Salyut and Spacelab D-1 missions (Tixador *et al.* 1984a, **1985**; Lapchine *et al.* 1986, 1988; Moatti & Lapchine 1986). As alterations in membrane structure are a common response to acceleration stress (see Section 1.4.5.2), some such change might be suspected as a cause of this antibiotic tolerance. However, it is disturbing that the only differences observed were between ground based control cultures and space flown cultures. The authors do not discuss the fact that there were no differences between in-flight static and in-flight centrifuged cultures (Lapchine *et al.* 1986).

Cellulose synthesis by *Acetobacter xylinum* is affected by the 20 sec of microgravity during parabolic flights. Controls produced compact ribbons of microbial cellulose, but during the parabola regime cellulose ribbons became splayed. This observation suggests that the crystalline cellulose is normally produced under strain, and the microgravity exposure relieves the stress, perhaps by reduction of intermicrofibrillar hydrogen bonding, to produce splayed ribbons (**Brown *et al.* 1992**). This observation may be relevant to the changes in plant cell wall composition and structure discussed in Section 1.4.5.1.

The majority of research on eukaryotic microbes has been done with protozoa, including the cellular slime mould, *Dictyostelium discoideum* (Kawasaki *et al.* 1987) and, especially, the plasmodial slime mould *Physarum polycephalum*. The latter has been chosen because of the readily-visible cytoplasmic movements which occur in the plasmodia (**Hejnowicz & Wohlfarth-Botterman 1980**; Block ***et al.* 1986**). The effect of the gravitational field on these movements has been studied (Block & Briegleb 1989; and see Sections 1.2.4.3 and 1.2.4.7) in rather similar ways to the investigations which have been made of cytoplasmic streaming in plants (Section 1.4.5.7). The *Physarum* plasmodium is essentially an abnormally large amoeboid cell. An unfortunate consequence of its size and its abnormality is the perpetual question of whether phenomena observed in this organism relate to (widely applicable) general cell biology, or result from the peculiar specialisations of the organism. Ironically, *Physarum* does not appear to have been employed in gravitational biology for the sort of research for which it is most suited as an experimental object – the study of nuclear division.

Among unicellular protozoa, *Paramecium tetraurelia* was flown aboard Salyut 6 (Planel *et al.* 1981, **1982**), the experiment involving measurement of changes in the ionic environment of the cells during flight (**Tixador *et al.* 1981**). A significant observation was increased proliferation in the *Paramecium* cultures (Tixador *et al.* 1984b). Controls indicated that microgravity was the major factor responsible for this response (no stimulation of proliferation in cultures placed on a 1 × g in-flight centrifuge aboard the Spacelab D-1 mission), and, interestingly, exposure to hypergravity on ground centrifuges resulted in a reduced cell growth rate (Planel *et al.* 1986, 1990; Richoilley *et al.* 1988a & b). Stimulation of *Paramecium* proliferation has also been observed in cultures on a fast rotating clinostat (**Hemmersbach-Krause *et al.* 1991; Ayed *et al.* 1992**) (see Section 1.2.2).

An isolated observation, which is potentially interesting for the experimental possibilities it raises, is that hypergravity (1.9 × g) influences the crystal structure of the calcite synthesised by foraminifera (Le Campion & Very 1980). The authors suggest a mechanism operating at the crystal/liquid interface at the cell membrane.

1.3.1 Motile cells

Motile cells can move towards or away from the source of stimulus, thereby exhibiting a *taxis*. In +ve taxis the organism remains in the region of stimulus, homing in on the source; the opposite is a –ve taxis. Responses such as photo- and chemotaxis depend upon the cell sampling the environment through its own movement, and so characterising the stimulus gradient. In the absence of gravitational field gradient over biological distances, gravitaxis results from displacement of the centre of buoyancy and/or centre of propulsion of the cell relative to its centre of gravity. Such displacement results in a torque leading to orientation of the cell within the fluid stream (**Kessler 1986, 1989**, 1992; **Pedley & Kessler 1987**). Kessler demonstrates (theoretically and experimentally) that orientation of cells in a flowing fluid can be entirely physical, requiring no sensory processing but depending on cell shape. Under normal conditions on Earth, even cyclical changes in cell behaviour can be accommodated in a physical model because as cells aggregate the local fluid density changes, creating a convection current which can mix the medium and redisperse the cells (**Kessler 1989**). This analysis does not exclude active signalling processes, but if gravity-dependent aggregation patterns can be achieved entirely through physical interactions - i.e. without cellular sensory perception and/or intercellular signalling, it does add an additional level of complexity to studies of the dynamics of cells in suspension. It also imposes the additional discipline that if such experiments are aimed at studying cell signalling, the purely physical explanation must be rigorously excluded.

This has not been done adequately in instances such as experiments with *Euglena* aboard the TEXUS sounding rocket (**Häder *et al.* 1990**) where absence of aggregation was ascribed to diminished 'cooperativity between the individuals'.

Many motile protozoa exhibit -ve gravitaxes which are believed to act as mechanisms enabling the cells to find the air/water interface, the reaction to the gravity vector supplementing phototaxis and chemotaxis in oxygen gradients (Häder & Vogel 1990; Hemmersbach-Krause *et al.* 1990; Vogel & Häder 1990).

The motion of ciliates lends itself to analysis even in the short periods of microgravity available in drop-tower experiments (**Machemer *et al.* 1992, 1993**). Unfortunately, although undoubtedly technically elegant, these experiments have not yielded any particularly incisive insights. The bias normally shown by *Paramecium* and *Didinium* for swimming vertically (negative gravitaxis) was absent in microgravity; so were gravity-induced changes in velocity of these cells (gravikinesis), which tend to compensate passive sedimentation at 1 × g. The swimming rates of *Paramecium* and *Loxodes* in microgravity were the same as their corresponding rates in the horizontal direction at 1 × g, but in *Didinium* the speed of swimming in microgravity exceeded the horizontal speed at 1 × g. The authors interpreted the data in terms of an electrophysiologically regulated gravisensor, but the evidence is weak and circumstantial.

Longer-term swimming behaviour of a number of ciliates and flagellates has been studied in microgravity, clinostats and hypergravity conditions but with no greater yield of understanding (ciliates: **Hemmersbach-Krause *et al.* 1993**; Hemmersbach-Krause & Briegleb 1994; and flagellates: **Häder *et al.* 1991a & b; Horneck *et al.* 1991; Vogel *et al.* 1993**).

Gravitaxis of *Euglena gracilis* increased at accelerations up to about 3 × g, but decreased, and some of the cells were sedimented, by accelerations of 4.5 to 5 × g. UV irradiation drastically impaired gravitactic orientation, suggesting to the authors that the taxis is an active physiological perception rather than a physical effect such as an asymmetry of the centre of gravity within the cell (**Häder *et al.* 1991b**). However, UV damage to metabolism (e.g. altered distribution of reserve materials) could also shift the centre of buoyancy so this is not a compelling reason to believe in intercellular signalling systems. Gravitaxis was lost within a few minutes of first exposure to real microgravity or onset of rotation on a fast clinostat. After the onset of microgravity, cells persisted in their original swimming direction until they collided with other cells or made random course corrections. The mean velocity was higher under microgravity than at 1 × g (**Vogel *et al.* 1993**).

Despite the varied and considerable investments in these protozoal motility experiments, it seems that the systems are intrinsically too complex to serve as helpful experimental models until sufficient is known for them to be used to test very specific predictions. Being highly adapted and specialised organisms, the protozoa themselves are difficult to understand in terms of any generalised cell biology. The cells are beyond the limit for easy observation of individuals so that most of the experiments have to rely on observing the behaviour of potentially heterogeneous populations, and the experiments have to be designed to distinguish purely physical interactions of small particles in fluid from a variety of (potentially conflicting) tropic responses.

1.3.2 Sessile microorganisms

In sessile organisms response to external stimuli involves tropisms – directed growth towards (+ve) or away (–ve) from the stimulus (see also Section 1.4.1). Amongst microorganisms, this phenomenon is particularly evident in fungi. Gravimorphogenetic responses are known, too. For example, in the apple-scab pathogen, *Venturia inaequalis*, which over-winters in fallen leaves, over 90% of the fruiting structures (ascomes) in the leaves were directed towards the surface that had faced upward during winter (**Gadoury & Machardy 1985**). In a similar phenomenon, Petri dish cultures of the resupinate polypore *Phellinus contiguus*, which normally forms its fruiting bodies (hymenophores) on the lower surfaces of tree branches, produce disorganised masses of tissue when incubated mycelium-side up, but normal hymenophores when inverted and incubated mycelium-side down (**Butler & Wood 1988**).

These fascinating phenomena have never been subjected to microgravity research, not even with clinostats. Indeed, there is an even more remarkable 'lost opportunity' in mycological gravitational biology in that no space experiments seem to have been done with *Phycomyces* sporangiophores. The sporangiophores of *Phycomyces* are essentially individual (though unusually large) hyphal tips. They show a number of interacting tropisms and avoidance growth-responses which have been analyzed in great detail (**Varjú *et al.* 1961**; **Johnson & Gamow 1971**; **Lafay *et al.* 1975**; **Gamow & Böttger 1982**; **Gyure *et al.* 1984**), and as early as 1961, Dennison (**Dennison 1961**) used a centrifuge (with accelerations up to 4.35 × g), conducting some experiments with sporangiophores submerged in a buoyant fluid to distinguish two separate sensory systems; a transient reaction extending over about 5 minutes and the long term, 'normal', gravitropic response. As the direction of the transient response was reversed in the buoyant fluid it was concluded to be a by-product of mechanical forces, probably exerted on the cell wall, arising from the action of gravity on the sporangium (such forces are reversed in direction between experiments in air and in a dense buoyant fluid). This was subsequently shown to be a response to a stretch receptor (**Dennison & Roth 1967**). Later experiments suggested that the gravitropic sensory mechanism may involve displacement of the vacuole changing the distribution of wall synthetic potential (**Dennison & Shropshire 1984**). Although it is doubtful that such a mechanism would have the sensitivity to correct small changes in orientation, to date this remains the only published suggestion for a graviireceptor mechanism in fungi.

Despite the fact that the above-described experiments were contemporaneous with developing space flight opportunities, *Phycomyces* has never been flown. However, there have been a variety of space flight and clinostat experiments with much larger fungal fruit bodies (for reviews see Molitoris 1990; **Moore 1991**; **Kern & Hock 1993**; Moore 1994). Ground based experimentation was recently resumed after a gap of almost 25 years with studies of gravitropic bending in *Coprinus cinereus* stipes (= mushroom stems) (**Kher** ***et al*** 1992; **Greening** ***et al.*** **1993**; **Moore** ***et al.*** **1994a**). Bending is a two-stage process, the first stage being an elastic deformation, the second a rigidification process, and it seems to be different from growth processes concerned with normal extension of the stipe. Control of gravitropic bending is exercised by the apical half of the *Coprinus* stipe and the same tissues are responsible for gravity perception and gravitropic response. The reaction time is about 21 minutes (Moore *et al.* 1994b). Perception of the gravity vector does not involve modulation of Ca^{2+} metabolism, proton channels nor stretch-activated ion channels (**Novak Frazer & Moore 1993**). The perception time for gavitropic bending is 7 minutes. Gravistimulation is not additive and the angle of bending attained depends on gravistimulation dose (implying that sustained stimulation is required for sustained bending). There is no relationship between extent of gravistimulation and rate of response or reaction time (**Hatton & Moore 1992, 1994**). Broadly similar results have been obtained with *Flammulina velutipes* (**Monzer** ***et al.*** **1994**), although the much slower growth and different tissue distributions in this organism result in the kinetics being very different from those observed in *Coprinus*. Experiments with *F. velutipes* were flown on the D-2 Spacelab mission (Hock *et al.* 1993). Although the cultures fruited extremely well and many samples were returned for analysis, only preliminary results have so far been published (Kern & Hock 1994). However, mushroom cultures of a different sort, of the polypore *Polyporus brumalis*, have been flown on a number of Salyut missions (Dudchenko *et al.* 1978), and this species, together with *C. cinereus* and some others, has been the subject of experimentation with clinostats (Gorovoi *et al.* 1989a & b; **Moore 1991**; **Hatton & Moore 1992, 1994**). The indications from both clinostat and space-flown experiments are that the basic form of the mushroom (overall tissue arrangement of stem, cap, gills, hymenium, veil) in agaric and polypore alike is established independently of the gravity vector. Abnormalities of stem growth have been observed in clinostat cultures of *Panus (= Lentinus) tigrinus* and *P. brumalis* (Gorovoi *et al.* 1987; Gorovoi *et al.* 1989a & b), but the morphogenetic event which seems most dependent on gravity is sporulation (in the broadest sense). Cultures of *P. brumalis* on orbiting space craft fail to produce the poroid hymenophore and in clinostat experiments on the ground even karyogamy was rare in similar cultures. *C. cinereus* grown on the clinostat was able to produce apparently normal fruit body primordia which failed to produce spores and then aborted, forming a new flush of primordia on the old (Gorovoi *et al.* 1989a). Taken together with the clear association between observation of gravitropism and the onset of sporulation the implication is that commitment to the pathway incorporating the meiotic division and subsequent spore formation both requires the gravity vector and couples it in some way to fruit body maturation.

An unexpected feature of clinostat experiments with *C. cinereus* fruit bodies is that stipes placed on the clinostat after various gravity exposure times 're-laxed' by 5° immediately after reaching maximum curvature (**Hatton & Moore 1994**). This relaxation process has parallels in the 'tropic reversal' of *Phycomyces* sporangiophores (**Galland & Russo 1985**), 'springback' of plant roots (**Leopold & Wettlaufer 1989**), 'spatial memory' of maize coleoptiles (**Nick *et al.* 1990b**) and 'autotropisms' of seedlings (**Heathcote 1987**; **Chapman *et al.* 1994**). Such similarities suggest that in all these otherwise very different cases, gravitropic bending has an initial, reversible, phase of plastic bending which is followed by a 'fixation' phase if the tropic stimulus has been maintained but which relaxes if the tropic stimulus has been removed.

Other fungal experiments with a bearing on morphogenesis have concerned the persistence of diurnal cycles of conidium formation in *Neurospora crassa*. The conidation pattern generally persisted in microgravity but there were strain differences. One strain had a relatively robust rhythm though its period increased by several hours and alterations in morphology occurred. Another strain gradually lost its rhythm (**Sulzman *et al.* 1984**; Reiss-Bubenheim 1990). This suggests that sensitivity to the gravitational field may have a genetic component which could account for cases where different responses are observed in experiments with the same organism. The unicellular alga *Chlamydomonas reinhardtii* also exhibits a diurnal cycle, but of photo-accumulation (tendency to accumulate in the light alternating with no accumulation). The rhythm is enhanced in microgravity (**Mergenhagen 1986;** Mergenhagen & Mergenhagen 1986), and this is taken to imply that the rhythm is driven by endogenous factors rather than by environmental cues. In *Chlorella*, clinostat treatment causes alterations in the chemistry of membrane lipids (Baraboi *et al.* 1991).

1.4 Plant growth and development

1.4.1 Introduction

In so far as the clinostat was first developed to study plant gravitropism it can be said that microgravity experimental work originated with plant gravitational biology. Inevitably, there is an enormous literature on the topic, of which only a fraction can be reviewed here. This section will attempt to provide a context within which the more recent work on gravitropism, especially that involving space flights, can be emphasised, but the section will begin with a consideration of broader aspects of plant developmental biology.

Being sessile organisms, plant response to external stimuli involves orientation and growth towards (+ve) or away (–ve) from the stimulus, known as a tropism. This is generally considered to result from the differential growth of parts of the plant (**Barlow 1992a & b**) in response to differential distribution of growth hormones – the classic Cholodny–Went theory of the early 1930s (Went & Thimann 1937). Higher plants, mosses, filamentous algae (and fungi too, though they are in a different Kingdom to the plants) all exhibit several types of tropism including response to light (phototropism), chemicals (chemotropism), physical surfaces (thigmotropism), airflow (anemotropism) and gravity (gravitropism) and even strong magnetic fields (magnetotropism) (**Schwarzacher & Audus 1973**; **Jenkins *et al.* 1986**; **Heathcote & Bircher 1987**; **Moore 1991**; **Perbal & Driss–Ecole 1993**). Plant primary roots are positively gravitropic, whilst shoots are negatively gravitropic although the characteristics of the response are similar in both. If a vertically growing root is re-oriented to the horizontal, i.e. gravistimulated, then there is a minimum period of stimulus required to elicit a response, known as the perception time (Audus 1962). Mechanistically, this is the time period required for the sensor system to react to the stimulus. There follows a lag period between initial stimulation and the first signs of response, termed the reaction time. This is the collective time it takes for gravity perception to generate a physiological signal, the signal to travel (if appropriate) from the site of perception to the site of response and the response then to be implemented. In plant roots the tropic response occurs a few millimetres back from the root tip in an actively elongating region, but gravity is believed to be sensed in the root cap since removing the cap virtually abolishes the response (Audus 1962; **Moore & Evans 1986**). This implies that a signal must travel from the tip to the response zone.

Very little research has been done on modulation of the gravitropic reaction. This occurs to ensure that secondary branches of the main axis (whether shoot or root) grow outwards, i.e. inclined away from the vertical (= plagiotropism) and is arguably an important component of mathematical models of plant gravitropism (Stockus, 1992, 1994, 1995). The 1 × g centrifuge on board Spacelab was used during the IML–1 mission to expose roots of lentil seedlings grown in space to varied gravitational stimuli (**Perbal & Driss–**

Ecole 1994). The kinetics of root gravitropism in orbital and ground experiments were different. Root tips exposed to 1 × g in orbit could overshoot the direction of the acceleration vector, ground controls did not, suggesting that a gravity dependent compensation (or inhibition) of root curvature is a component of the normal root response.

1.4.2 Plant growth in microgravity

Levine & Krikorian (1992) grew plantlets of day lily (*Hemerocallis*) (clones regenerated from cell suspensions before flight), and clonal populations of *Haplopappus gracilis* during a 5-day mission aboard the Shuttle Discovery. Statistically significant differences in shoot growth were found (the controls growing more than plantlets subjected to space flight conditions), and between the different plant populations (day lily gaining more shoot material than any of the *Haplopappus* populations and *Haplopappus* seedling clones outperforming the tissue culture-derived *Haplopappus* lines). There were also differences between different growth chambers, implying environmental heterogeneities within the culture apparatus. On the other hand, the fresh weights and hormone contents (abscisic acid and indole-3-acetic acid) of seedlings of *Zea mays* grown for five days during a Shuttle flight were, with minor exceptions, not statistically different from seedlings grown in normal gravity. The tissues of the Shuttle-grown plants appeared normal and the seedlings differed only in the lack of orientation of roots and shoots (**Schulze *et al.* 1992**).

The diversity of these recent results echoes earlier studies. **Cowles *et al.* (1984)** found stem and root growth and fresh and dry weights to be reduced in comparison with ground controls in mung bean, oat and pine seedlings grown on Shuttle and Spacelab missions, and coleoptile growth was reduced, compared with ground controls, in lettuce, cress and *Arabidopsis thaliana* seedlings grown on Salyut missions (Merkys *et al.* 1981, 1984). In these Salyut experiments, more coleoptile growth occurred in microgravity than in an on-board centrifuge, though all flight-grown material was considerably shorter than ground-based controls. Paradoxically, cucumber seedlings grown on an in-flight centrifuge on Cosmos 1667 had longer roots and coleoptiles than those grown in microgravity, and lettuce seedlings grown in space on the same flight were larger than those grown on the ground (Sitnik *et al.* 1984).

Diversity is to be expected, given that the organisms have evolved independently within the Earth's gravitational field and their evolutionary origins have led them to respond to gravitational stresses in different ways. Also, some of the physical effects of microgravity, especially lack of gravity-driven convection and the consequent relative increase in importance of concentration-driven effects and surface-surface interactions, will inevitably influence fluid distributions and gas exchange on which plant growth depends.

Further, these influences will almost certainly differ between plants depending on their sizes and the geometrical structures of their different tissues. However, the diversity of experimental results coupled with the relatively small number of (mostly) short duration experiments which have been completed combine to make generalised interpretations extremely difficult. Longer duration plant growth experiments have been few in number and often suffered from technical difficulties (**Halstead & Dutcher 1987**). *Arabidopsis thaliana* has been successfully cultivated through a complete life cycle (Merkys & Laurinavichius 1983; Merkys *et al.* 1984) aboard Salyut-7. Seeds produced in orbit were subsequently germinated on Earth. They showed reduced germination, increased frequency of embryonic lethality, generally poorer seedling growth and greatly reduced ability to develop into healthy mature plants. Similar results were obtained with seeds produced on plants flown aboard Cosmos 1129 (Parfenov & Abramova 1981). Second generation seeds produced on Earth by the progeny of the space-flown plants were apparently normal. Seemingly, the adverse effects of microgravity on plant development and fecundity are transient, but would they be cumulative in continuously space flown crops? Significantly perhaps, endosperm and embryo nutrition in wheat normally depends upon sedimentation of vesicles within the embryo sac (**Huber & Grabe 1987**).

1.4.3 Plant cell proliferation and differentiation

In lentil roots, the normal gravireaction leads to an increase of the mitotic activity in the whole proximal part of the meristematic zone and also in the lower half of its distal part. Moreover, gravistimulation brings about an increase of nuclear DNA synthesis of cortical cells in the upper half of the distal part of the meristematic zone. The cell cycle of cortical cells is thus changed during the gravitropic curvature. In the roots of *Zea mays* grown aboard the biosatellite Cosmos 1667, the mitotic activity of cortical cells and their differentiation occurred faster than in the ground controls (**Darbelley 1986, 1988**) although other workers reported a decrease in the number of dividing cells in meristems of maize root tips on the same flight (Barmicheva *et al.* 1989). Roots of lentil seedlings grown for 25 h on board the Spacelab D1 Mission showed an increased mitotic index in microgravity in comparison with the flight controls, due either to a shortening of the interphase in roots grown in microgravity or to a lengthening of mitosis. Cell length and mitotic activity were reduced when roots were first grown in microgravity and then subjected to centrifugal acceleration (**Darbelley *et al.* 1989**). Clinostat treatment of similar seedlings of lentil caused cell elongation to occur closer to the tip than in the vertical roots, but the mitotic index was not modified, nor were the frequencies of the G1, S and G2 phases of the cell cycle (**Legue *et al.* 1992**). Similarly, spore germination, direction of growth, and cell division in protonemata of the fern *Adiantum* were not affected by growth on a 3- dimensional clinostat (**Kasahara *et al.* 1994**) although clinostated protonemata were shorter than the controls. Such differences between true microgravity and its clinostat simulation raise the question as to whether adverse effects on mitosis might depend on density differentials being abolished in true microgravity, rather than on absence of a directional gravity vector.

Earlier work suggested that microgravity might cause a block during the later stages of mitosis (**Halstead & Dutcher 1987**), but the data are clearly not unanimous in indicating this specifically. Nevertheless, there are frequent reports of disturbances to the progress of mitosis and to increases in the frequency of chromosomal abnormalities (often with sufficient controls to eliminate radiation as direct cause of the latter). Thus, it may well be that plant cell proliferation requires positional cues which are absent in microgravity; the response to this absence varying between species. In this context, it should be remembered that plant developmental biology depends on the exact positional placement of the new cell wall which separates the daughters produced from each dividing cell. That, in turn, depends on the positioning and orientation of the mitotic division spindle. Perhaps it is these intracellular positioning processes which are gravity-dependent for their correct execution.

There was no increase in protein content of cells of sunflower (Bara 1977) or corn (Tairbekov & Rozov 1978) grown on clinostats, though the former showed enhanced growth and water content. Although growth of cells of *Haplopappus gracilis* cultivated *in vitro* for 35 days on a 2 rpm clinostat had no effect, on a 50 rpm clinostat an increase in wet weight was observed as compared to the control with relatively the same amount of dry weight gain (Klimchuk 1984). Root cap cells of *Arabidopsis* grown on board the Salyut 6 orbital space station had increased vacuolization compared with the ground grown control (Tarasenko *et al.* 1982).

Microgravity may also influence the progress of plant cell differentiation (Kordyum 1994; Laurinavichius, Kentstaviciene *et al.* 1994). Aniseed cell cultures differentiated more rapidly during the D-1 Shuttle flight (**Theimer *et al.* 1986**) and there are other reports in the literature of accelerated development which has been suggested to parallel reports of accelerated aging in animals reared in space (**Halstead & Dutcher 1987**). On the other hand, experience with protoplast regeneration in a number of more recent missions suggests that this aspect of plant cell differentiation is severely retarded in microgravity. Alterations in wall synthesis in microgravity contributed to reduced protoplast regeneration in material flown on the 14 day flight of Biocosmos 9 (Cosmos 2044) (Iversen *et al.* 1992; Klimchuk *et al.* 1992; **Rasmussen *et al.* 1992**). Regeneration of flight rapeseed (*Brassica napus*) was only 56% of the ground control, and with carrot (*Daucus carota*) regeneration in orbit was 82% of the ground control. Peroxidase activity and the amount of protein was lower in flight samples than in the ground controls; the number of different isoenzymes was also decreased in the flight samples. A 54% decrease in the production of cellulose was found in rapeseed, and a 71% decrease in carrot. Hemicellulose production was also decreased in the flight samples compared to the ground controls. Fluorescence analysis showed that cell viability was unaffected by space flight.

One general effect of the stay on board the space vehicle was a retardation of the regeneration process. Callus cultures obtained from the flight samples grew very slowly compared to callus regenerated from the ground controls, and two years after the Biocosmos flight there was no further growth in the samples exposed to microgravity, although callus cultures from the ground controls continued to grow well and in related experiments gave rise to regenerated plants. There was a decrease in the calcium content of cell cultures subjected to space flight compared to the ground controls and a variety of (minor) ultrastructural changes, including folding of the plasmalemma, more cell inclusions and microbodies and greater heterogeneity in mitochondria in the flight samples. The most consistent ultrastructural observation, though, was a 2 to 3-fold decrease in wall thickness in the flight samples, which seems to correlate with the changes in wall constituents referred to above.

Potentially the most interesting microgravity-induced developmental alteration in plants appears to be the least well examined. This is that regeneration of the root cap seems not to occur in orbit (**Moore *et al.* 1987c**). Podlutsky (1992) reported normal root cap development in roots regenerating from callus cultures of *Arabidopsis thaliana* grown on a clinostat, whereas, in similar cultures exposed to 8 days of microgravity in space (flight not specified, but possibly Salyut-6), the root cap was abnormal, lacking the gravireceptor cells normally located in this tissue. This clear implication that a specific and well-defined developmental process needs the normal unilateral gravity vector deserves more extensive investigation.

1.4.4 Plant gravitropism

It is over 100 years since Noll (1892) first suggested that plants might sense gravity by means of a mechanism which depends on sedimentation of microscopic intracellular particles. It is now widely assumed (Wilkins 1984), but by no means fully proved or completely accepted (**Poff & Martin 1989**), that the mechanism of graviperception in plants is the sedimentation of starch-containing amyloplasts (statoliths) in specific cells (statocytes) in the columella (central core) of the root cap and in various locations in shoots (e.g. near vascular bundles or the inner cortical or endodermis layers of the stem). So, while there is discussion about the mechanism (or mechanisms – see Firn 1992) by which plants eventually sense statolith movement (**Björkman 1988**; **Poff & Martin 1989**; Trewavas 1992), there is no debate over the basic premise that gravity perception is the responsibility of particular specialised plant cells and that it depends upon redistribution of intracellular organelles (**Perbal & Driss-Ecole 1994**).

The large size and relative density of the amyloplasts allow them to sediment rapidly. The rate of sedimentation (determined directly and theoretically) is comparable with the perception time (**Sack *et al.* 1984, 1985; Moore & Evans 1986**). Statolith sedimentation polarises the cell, through interaction either with the cytoskeleton, organelle membranes or the plasma membrane (the potential mechanism is discussed below). The physiologically important outcome is differential enhancement of cell elongation in only one side of the growing zone leading to a tropic curvature (downwards in roots, upwards in stems). The Cholodny–Went theory ascribed this specifically to lateral transport of auxin and tropic curvatures can be mimicked experimentally by lateral application of auxin. **Migliaccio & Rayle (1989)** demonstrated that applied auxin gradients as small as 1:1.3 produced >60° curvatures in hypocotyls of sunflower (*Helianthus annuus*) seedlings. Asymmetric auxin distributions (**Harrison & Pickard 1989**), redistribution of auxin-regulated RNA (**McClure & Guilfoyle 1989**) and movement of calcium towards the lower side of the organ (**Lee & Evans 1985a & b; Moore 1985**) have all been recorded as responses to reorientation in plant organs which comply with the Cholodny–Went theory. However, asymmetric (i.e. polarised) distribution of many metabolites have been demonstrated in tropically-responding plant organs, including concentration gradients of potassium and phosphorus (**Goswami & Audus 1976**), Ca^{2+} and H^{+} (**Wright & Rayle 1983; Gehring *et al.* 1990**), Na, Cl, and Mg (**Moore *et al.* 1987a**), gibberellins (**Hestnes 1979; Rood *et al.* 1987**) as well as auxin (Audus 1962; **Jackson & Barlow 1981; Moore, Evans 1986**; Briggs 1992; Evans 1992). Even ethylene has been implicated as a modifier of tropic curvature in *Mimosa* (**Roblin & Perault 1985**) though not in tomato (**Harrison & Pickard 1986**). Extrusion of ionic species induces polarised electrical current flows across reoriented plant organs to which particular importance has been attached because they occur so early in the gravitropic response sequence (**Behrens *et al.* 1985**; Lühring *et al.* 1986; **Iwabuchi *et al.* 1989; Collings *et al.* 1992**).

The great variety of rapid metabolic reactions to reorientation makes it difficult to understand the exact sequence of events in the gravitropic response, despite over a century's research. The debate about how gravitropic bending is effected continues to rage. Even its claimed dependence on a concentration gradient of a hormone is not universally accepted. Differential growth of soybean hypocotyls occurred without free auxin redistribution (**Rorabaugh & Salisbury 1989**), which is interpreted as evidence for differential change in *sensitivity* to auxin (**Evans 1991**). Bruinsma (1992) discusses altered distribution of *inhibitors* of auxin activity as a cause of plant phototropism. This controversy about the way in which the gravitropic response is regulated is beyond the scope of the present review. We will attempt now to concentrate attention on aspects of plant cell physiology which seem to react directly to change in the gravity vector.

1.4.5 Graviperception in plants

It is important to start, perhaps, with the recognition that the perception system may be complex and not necessarily identical even between different phases of the same organism. **Jenkins *et al.* (1986)** isolated several mutant strains of the moss *Physcomitrella patens* which had partial or complete impairment in gravitropism of the caulonema (main stem of the moss sporophyte). Complementation analysis resolved the mutations into at least three genes, but none of the mutants was altered in gametophore gravitropism, suggesting that the gravitropic response of caulonemal filaments may require some gene products that are not required for the response of the multicellular gametophores. These observations issue a warning against seeking a single graviperception mechanism, yet in the absence of detailed genetic analysis in any other system there is no way of knowing whether different observations actually relate to the same mechanism.

The gravisensing cells of plants are very specialised, but they do not contain any unique cell structures. Thus the gravisensing cells are still essentially using general cell structures to sense gravity. Amyloplasts are the most obvious candidates as statoliths (Björkman 1992), but calculations by Audus (1962) showed that structures as small as mitochondria could, in theory at least, act as a very weak gravisensing system. Since structures such as the nucleus, endoplasmic reticulum, golgi complex and other inclusion bodies are sufficiently large and dense these, too, could form the basis of a gravisensing system in an otherwise non specialised cell. In both space-flown and clinostat treated roots of lentil (*Lens culinaris*), as well as the expected change in distribution of amyloplasts when gravistimulated, the shape and position of the nucleus was also altered systematically (**Perbal *et al.* 1987; Perbal & Driss-Ecole 1989; Lorenzi & Perbal 1990a**). In the account which follows, systematic physiological changes in response to change in the gravitational field will be documented.

1.4.5.1 Cell walls

Among the changes which occur most rapidly in plant cells under the simulated hypogravity conditions of the clinostat are alterations in structure and chemistry of the cell wall matrix. Clinostat grown sunflower (*Helianthus annuus*) hypocotyls had increased elongation (28%) and weight (18%) compared to controls. Cellulase activity was also increased but β-1,3 glucanase was not (**Bara & Gordon 1972**). When [U-^{14}C]-proline was fed to shoots of intact marigold (*Tagetes patula*) grown on 15 rpm clinostats they showed a 2-to 9-fold increased incorporation of ^{14}C into salt-extractable material from the cell walls in comparison to controls (**Oputa & Mazelis 1977**).

Growth of marigold (*Tagetes patula*) under experimental hypogravity is characterised by increased peroxidase activity and decreased lignin content (Siegel *et al.* 1978). The specific activity of peroxidase was also elevated significantly in marigold seedlings grown on clinostats or with buoyancy compensators to reduce the mechanical stresses caused by gravity (**Waber *et al.* 1975**) so it is possible that this response arises intracellularly and is independent of the mechanical effects of gravity. Elevation of peroxidase activity of the cell wall, but not of the protoplast, is a short-term response to clinostat treatment (Siegel & Siegel 1983); long-term effects include changes in cell wall composition of Ca, lignin, and protein-bound hemicellulose (possibly cell wall glycoprotein). Lignin contents were also reduced in mung bean seedlings exposed to 194 h of microgravity on the Shuttle STS-3 flight, though this was coupled with reduced phenylalanine ammonia-lyase and peroxidase activities (**Cowles *et aly.* 1984**).

An early change in ultrastructure of protonema cells of the moss *Funaria hygrometrica* grown on a 2 rpm clinostat is that the fibrils of the cell walls became thinner and loosened. Populations of peroxisomes in protonema cells increased (Nedukha 1986) and there was an increase in the deposition of callose (amorphous β-1,3 glucan) in the cell walls (Nedukha *et al.* 1988) and a considerable increase in polygalacturonase and pectin esterase activities (Nedukha & Trutneva 1988). Cell wall lysis has also been observed in cells of the root cap of *Arabidopsis thaliana* grown in orbit (Tarasenko *et al.* 1982).

Cellulase activity (both endo-1,4-β-glucanase and exo-1,4-β-glucanase) increased in the periplasmic space and protonema cell walls of *Funaria hygrometrica* grown in a horizontal clinostat (2 rpm) for 30 days (Nedukha 1992). A decrease of total cellulose content, its crystalline form, and pectic substances as well as an increase of hemicellulose content was observed in the clinostated material compared to controls.

No significant changes were detected in the cell wall polysaccharides of the peripheral cell layers on the upper and lower side of sunflower hypocotyls which had undergone gravicurvature. Neither the relative rates of synthesis of the main polysaccharides nor the turnover of recently incorporated major polysaccharides were influenced significantly by short term changes in growth rate of the peripheral cell layers (**Carrington & Firn 1985**). Although β-D-glucan content increased in proportion in lower halves of gravistimulated oat stems, in concert with an increase in glucan synthase activity in lower halves, the asymmetry arose too slowly to account for initiation of the graviresponse (**Gibeaut *et al.* 1990**). These authors postulated that asymmetrical changes in osmotic solutes act as the driving factor for growth promotion in the graviresponse, while wall extensibility acts as a limiting factor during growth. A similar conclusion, that gravitropic curvature is not driven by new cell wall synthesis but, rather, by changes in cell wall plasticity and osmotic potential, was reached by **Lu *et al.* (1992)**.

Overall effects of such structural changes may be small, or negligible, however. **Masuda *et al.* (1994)** studied seedlings of several species of plant growing on a three-dimensional clinostat (with two rotational axes) and in a centrifuge (30-135 × g). Growth rate and mechanical properties of the cell wall were little affected by either treatment, although the normal gravitropic response of the organs was markedly altered. Far from being a mere enclosure, the plant cell wall is an important organelle which essentially brings the immediate region of the environment under cellular control. There is a constant exchange of information between the cell wall and the cell (see discussion in **Trewavas & Gilroy 1991**), so it is inevitable that changes in the cell wall induced by change in gravitational acceleration will be complex and may well be compensatory.

1.4.5.2 Membrane chemistry

There are also indications of changes in membrane structure. Palladina *et al.* (1984) recorded a considerable decrease in the protein content of plasma membrane preparations isolated from the roots of 5-day old pea seedlings grown on a 2 rpm clinostat. Zhad'Ko *et al.* (1988) report a 20% decrease in the chemiluminescence of pea seedlings in the first 15 to 30 minutes of clinostating, suggesting that lipid peroxidation is one of the first responses of seedlings to clinostating. Similar changes in lipid peroxidation were recorded in *Chlorella* cells, tissue cultures of *Haplopappus* and pea seedlings subjected to slow clinostating, and in tissue cultures of *Haplopappus* and wheat seedlings after space flight (Baraboi *et al.* 1991; Zhad'Ko *et al.* 1994).

Polulyakh (1988) exposed 5 to 6 day-old pea seedlings to a 2 rpm clinostat for 24-48 h and found that phosphatidic acids and phosphatidylinositol decreased in the root cell membranes. Phosphatidylcholine and phosphatidylethanolamine increased during the first, and decreased during the second day of clinostating. The reverse applied to phosphatidylethanolamine and diphosphatidylglycerol. C14:1, C18:0, C18:2, C18:3, and the unsaturation index of cell membrane lipids consistently increased during both days, whereas C14:0, C16:0, and C18:1 decreased. On the other hand, the range of fatty acid changes was wider than that of phospholipids. Clinostat treatment thus prompts major changes in the chemical composition of membranes which are related to an increase in microviscosity (of lipid bilayer of vesicles prepared from root plasmalemma) from 0.05–0.07 to 0.51–0.56 Pa s^{-1} during the first 24 h, though this decreased to 0.02–0.05 Pa s^{-1} after a further 24 h treatment (Polulyakh & Volovik 1989). The decrease in microviscosity during the second day of clinostating was associated with an increase of short- and branched-chain fatty acids, of double bonds in a cis-configuration, and of polarity of the hydrophilic portion of the molecule. These changes increased intermolecular distances in the lipid bilayer, increasing its fluidity and permeability. We see here evidence for adaptation of the membrane to maintain normal fluidity under the hypogravity conditions provided by the clinostat.

Membrane alterations and/or changes in the physical properties of the fluids in microgravity presumably account for the significant increases in fusion products in electrofusion experiments with mesophyll protoplasts of *Nicotiana tabacum* performed aboard the TEXUS 17 sounding rocket (**Mehrle *et al.* 1989; Hampp *et al.* 1990**). Microgravity during the TEXUS 21 and TEXUS 25 sounding rocket flights similarly increased the yield of fusion products between *Digitalis lanata* and *Digitalis purpurea* about 8-fold in comparison to terrestrial conditions (Baumann *et al.* 1990). Interestingly, very similar results have been reported for electrofusion of mammalian (mouse hybridoma) cells. During the flight of TEXUS 18, both fusion frequency and hybrid viability were increased by about 2-fold in the microgravity-exposed samples (**Schnettler *et al.* 1989**).

The most frequently reported membrane phenomenon is a change in ion channels and ion distributions. Tairbekov *et al.* (1980) reported only slight differences between ATPase (mitochondrial and cytoplasmic) activities of maize cells formed on the clinostat and control cells whereas these activities were about 50% higher in cells formed on a $2 \times g$ centrifuge. Increased gravity intensified ATP decomposition and at $2 \times g$ there was a relatively low rate of oxidative phosphorylation. This discrepancy between synthesis and degradation of ATP may lead to structural changes in cell organelles; it was not observed in control or clinostat grown cells. In later work, however, Tairbekov *et al.* (1984) reported the ATP content of clinostat grown cultures of developing maize sprouts to be only 63% of the control. This relatively low ATP content in cells grown on a 2.4 rpm horizontal clinostat was associated with an elevated rate of respiration as measured by CO_2 evolution. These authors pointed out that aerodynamic effects on even slowly-rotating clinostats may increase respiratory activity by eliminating dead zones close to the epidermis and consequently improving gas exchange.

Changes in activity of plasma membrane-bound ATPases are commonly reported. In plasma membrane preparations isolated from the roots of 5-day old pea seedlings grown on a 2 rpm clinostat the activity of Ca^{2+}-ATPase decreased, that of Mg-ATPase increased, and K^+ stimulated the Mg^{2+} effect on the ATPase activity (Palladina *et al.* 1984). Electron cytochemical analysis (transmission electron microscopy using the pyroantimonate precipitation method) showed that in control roots the reaction product from Mg^{2+}-activated ATPase was mostly localized on the plasmalemma, in plasmodesmata, on nuclear chromatin, endoplasmic reticulum, Golgi vesicles; reaction product of the Ca^{2+}-ATPase was localized on the plasmalemma of statocytes. After clinostating, (2 rpm) the Mg^{2+}-ATPase reaction product was found as large granules on the plasmalemma, and as a fine precipitate on vesicles and cisternae of the dictyosomes, internal membrane system of plastids, nuclear membranes, chromatin, and nuclei (Palladina *et al.* 1984; Kordyum *et al.* 1984). The intensive reaction for Ca^{2+}-ATPase activity characteristically limited to the plasmalemma of statocytes of pea roots was greatly reduced in roots grown on a 2 rpm clinostat (Podlutsky 1986). Indeed, Belyavskaya *et al.* (1988) state that no Ca^{2+}-ATPase activity was detected along the plasmalemma in clinostat grown material. Instead it was found in the nuclear envelope, plastid membrane reticulum, granular endoplasmic reticulum, and Golgi membranes, and dictyosomal vesicles. Similar differences in localization of Ca^{2+}-ATPase activity, detected by TEM cytochemistry, were noted between normally grown protonema cells of the moss *Funaria hygrometrica* and those grown for 20 days on a 2 rpm clinostat.

1.4.5.3 The calcium connection

Many of the above observations suggest some disturbance in plasmalemma structure of cells grown on the clinostat which might affect the distribution of Ca^{2+} ions within and around the cell. Sitnik *et al.* (1989) used the pyroantimonate precipitation method for detecting free and weakly- bound calcium in cells of 7-day pea seedling roots grown stationary (control) or on a 2 rpm clinostat. Deposits of calcium pyroantimonate were localized inside organelles and in the periplasmic space in the control. In clinostat grown material, deposits of calcium pyroantimonate were found in the cytoplasm as well. Indeed, chains of deposits were found between amyloplasts in some cases, an interesting observation in view of the localization of calcium in amyloplasts of root cap cells (**Chandra *et al.* 1982**) and the observation that the region of cytoplasm surrounding sedimenting amyloplasts in maize roots stains intensely with toluidine blue, indicating a transient increase in polyanions in the vicinity of the amyloplast as sedimentation commences (**Moore 1986b**). Changes in calcium ion distribution have also been noted in protonema cells of the moss *Funaria hygrometrica* grown for 20 days on a 2 rpm clinostat (Nedukha 1987, 1989a & b). Cultivation of protonema on a clinostat led to a considerable intensification of precipitation of the reaction product in membranes, vacuoles, periplasmic space, and cell walls. At the same time, analysis of pectinase localization using TEM cytochemistry showed that a high reaction intensity in cell walls in comparison to that in the control was distinctive of cells cultivated on a clinostat. Increase of Ca^{2+} content in the periplasmic space of cells grown on the clinostat was suggested to result from hydrolysis of pectins (implied by a decrease in cell wall thickness and loosening of cellulose fibrils) with consequent transfer of calcium from the bound state into the free one (Nedukha 1989a, b). Using a fluorescent calcium probe, Sitnik *et al.* (1989) found that the calcium distribution gradient in cells of *F. hygrometrica* changed considerably during clinostating, and it is suggested that gravity may participate in redistribution or activation of calcium ion channels in the plasmalemma.

1.4.5.4 Coleoptiles, stems and leaves

Slocum & Roux (1983) showed that an accumulation of Ca^{2+} in the upper halves of horizontal, gravistimulated oat (*Avena sativa*) coleoptiles occurs within 10 min of their being placed horizontal, about 20–30 min before large-scale visible bending. Gravitropic curvature was well developed by 3 h and was accompanied by further redistribution of Ca^{2+} to tissues along the upper coleoptile half, centred around the bend. Ultrastructural studies indicated that large amounts of Ca^{2+} were accumulated in cell walls at the upper side of the organ in the region of maximal bending, active transport of Ca^{2+} from cells into walls being performed by a Ca^{2+}–dependent ATPase localized in the plasma membrane (Roux *et al.* 1983). Antagonists of calmodulin (**Biro *et al.* 1982**) and calcium chelators (**Daye *et al.* 1984**) blocks gravitropism in oat and maize (Medvedev & Shtonda 1989) coleoptiles without inhibiting growth. These observations suggest that differential growth resulting in the establishment of gravitropic curvature may be the consequence of Ca^{2+} antagonism of auxin–mediated cell wall loosening and elongation growth at the upper side of aerial plant organs.

Migliaccio & Galston (1987) showed that inhibitors of IAA movement blocked development of both IAA and Ca^{2+} asymmetries, but substances known to interfere with normal Ca^{2+} transport did not significantly alter either IAA or Ca^{2+} asymmetries in gravitropically stimulated pea epicotyls (the part of the seedling stem between the cotyledon and the first leaf). These authors concluded that Ca^{2+} was bound to the cell walls, and that auxin movement is primary and Ca^{2+} movement secondary in stem gravitropism. The hypothesis being that apoplastic Ca^{2+} changed during graviresponse because it was displaced by H^+ secreted through auxin-induced proton release, Ca^{2+} being released from the wall adjacent to the region of high IAA concentration, proton secretion, and growth. In this interpretation, therefore, the asymmetric distribution of Ca^{2+} is a *consequence* of growth stimulation, not a critical step in the early phase of the graviresponse.
Clinostat treatment of plants induces specific changes in the pattern of soluble leaf proteins of *Vicia faba* (broad bean), eight polypeptides increased in amount, whereas four proteins decreased or disappeared. Using immunoblotting, it was shown that the amount of β–tubulin was not affected, but the pattern of polypeptides recognized by calmodulin antibodies was changed by clinostating (**Schulz *et al.* 1992**). The authors interpret these changes as resulting from altered protein turnover as a stress reaction to the rotating gravity vector.

1.4.5.5 Root responses

The pattern of localization of Ca^{2+} in gravistimulated corn roots was quite different from that seen in gravistimulated coleoptiles. In horizontally positioned roots, Ca^{2+}, initially localized within the stele, became localised within the vacuoles of cortical cells along the upper root surface. There was little staining in the walls and, in the root cap, where graviperception is thought to occur, there was little staining of any cellular organelles (**Dauwalder *et al.* 1985**). Nevertheless, calcium ions are firmly implicated in gravity signal perception and transduction (**Poovaiah *et al.* 1987**). Belyavskaya (1992) claims that clinostating and true microgravity exert their disorienting effects by causing unregulated increases of Ca^{2+} in the statocytes of pea roots.
Gravistimulation under normal circumstances induces polar movement of Ca^{2+} across the root tip from the upper side to the lower side in primary roots of maize (*Zea mays*) and pea (*Pisum sativum*). Treatment of maize or pea roots with inhibitors of auxin transport prevents both gravitropism and gravity-induced movement of Ca^{2+} across the root tip (**Lee *et al.* 1984**). **Moore (1986a)** used the pyroantimonate precipitation procedure to localize Ca^{2+} in maize roots. Within 5 min of horizontal reorientation, staining became associated with the portion of the cell wall adjacent to the distal end of the cell. This asymmetrical staining persisted throughout the onset of gravicurvature. No staining of lateral cell walls of columella cells was observed at any stage of gravicurvature, suggesting that a lateral flow of Ca^{2+} through the

columella tissue of horizontally-oriented roots does not occur. Nevertheless, root cap mucilage of *Zea mays* contains effector(s) that induce gravitropic-like curvature (**Fondren & Moore 1987; Moore & Fondren 1986, 1988; Marcum & Moore 1990**). Seemingly, the transverse movement of gravitropic effectors, including Ca^{2+}, to the lower sides of tips of horizontally-oriented roots occurs through the apoplast in the outermost layers of the root. It most strongly occurs in the cytoplasm, cell walls and mucilage of epidermal cells (**Moore *et al.* 1989**).

Björkman & Cleland (1991) used calcium-specific microelectrodes to measure the Ca^{2+} gradient in maize roots. The calcium activity on the upper side of the root tip remained constant for 10 min after gravistimulation (initial pCa about 2.5 mM), then decreased by about 50%. On the lower side, after a similar lag, the calcium activity increased 1.6-fold. The calcium activity gradient (and gravitropism) was eliminated by applying a mobile calcium-binding site (dinitro-BAPTA) to the root cap. Conversely, preventing the formation of apoplastic pH gradients, using MES buffer or 10 mM fusicoccin, did not inhibit curvature, indicating that the calcium gradient is not a secondary effect of proton-induced calcium desorption by a proton gradient [compare with the **Migliaccio & Galston (1987)** pea epicotyl experiment described above].

Gravistimulation is followed by alterations in the external current symmetry (Lühring *et al.* 1986; **Iwabuchi *et al.* 1989**) in *Lepidium* roots. Upon gravistimulation, the membrane potential of statocytes is temporarily depolarized (lag time = 2 s) from a resting potential of *ca* –118 mV to a potential of *ca* –93 mV. This only occurs in statocytes located on the physically lower flank of the root, those on the corresponding upper flank become weakly hyperpolarized (*ca* –13 mV) (**Sievers *et al.* 1984**). Inhibitors of auxin transport blocked gravitropic curvature but not the change in current density, whilst inhibitors of calmodulin activity blocked both curvature and gravity-induced current. The results suggest that auxin transport is not a component of gravity sensing in the root cap, but that calmodulin plays an intrinsic role in gravity sensing (**Björkman & Leopold 1987**) [compare with the **Migliaccio & Galston (1987)** pea epicotyl experiment described above]. Treatment of maize root tips with Ca^{2+} increased the basipetal movement of auxin (Evans *et al.* 1992). Since gravistimulation causes Ca^{2+} movement toward the lower side of the root cap, this observation supports the idea that gravity-induced Ca^{2+} asymmetry is a key step linking gravistimulation to the establishment of auxin asymmetry during root gravitropism.

An important role for calmodulin in root gravitropism was demonstrated by **Stinemetz *et al.* (1987)**, who directly assayed calmodulin in primary roots of maize (whole tissue segments, crude extracts and purified extracts) and found that the time course of the development of gravitropic sensitivity followed the time course of the increase in calmodulin activity in the apical 1 mm of the root. Application of chelators and calcium-ionophores to pea roots (Belyavskaya 1988) and roots of *Zea mays* (**Reddy *et al.* 1987**) resulted in the loss of gravisensitivity. Furthermore, 5-hydroxytryptamine, which is known to promote the hydrolysis of phosphoinositides, sensitized dark-grown maize roots to gravity and increased inositol trisphosphate levels, suggesting roles for both Ca^{2+} and inositol phospholipid turnover in gravity signal transduction. Phosphorylation of two specific polypeptides also seems to be involved in transduction of the gravity signal in maize roots (**Friedmann & Poovaiah 1991**).

Prolonged treatment (2 h) of roots of *Lepidium* with the calcium ionophore A23187 caused a complete loss of graviresponsiveness which was accompanied by loss of statocyte polarity and displacement of the usually distal endoplasmic reticulum (**Wendt & Sievers 1989**). In chelator-treated (50 mM EGTA) roots of *Zea mays*, development of the Golgi apparatus was severely reduced and there was no evidence of a granular secretory product which was encountered in the dictyosomes of control roots (**Bennet *et al.* 1990**). Reduction in amyloplast numbers brought about by EGTA occurred alongside reduced Golgi secretory activity.

1.4.5.6 Organelle distributions

One of the first ultrastructural features observed when protonema cells of the moss *Funaria hygrometrica* are placed on a 2 rpm clinostat is that grana disappear from plastids and their content of starch decreases (Nedukha 1986). Similar decreases in reserve polysaccharides in *Chlorella* grown on the clinostat result from intensified hydrolysis due to an increase of the specific activity of amylase (Popova & Shnyukova 1988, 1989). Soybean seedlings grown on a 1 rpm clinostat or 5 × g centrifuge also exhibited changes in starch metabolism (Brown *et al.* 1994; **Brown & Piastuch 1994**). Starch concentration in the cotyledons was 28% lower in clinorotated plants and 24% higher in centrifuged plants than in control (vertically rotated) plants. The opposite relationship was noted for total lipid concentration. Of six starch metabolic enzyme activities measured, only ADP-glucose pyrophosphorylase was affected by the treatments; being 37% lower in clinorotated plants and 22% higher in those which were centrifuged.

Accelerated chloroplast destruction was noted in tissue cultures of *Haplopappus gracilis* on a 50 rpm clinostat (Lozova *et al.* 1989). On the other hand, mitochondrial volume, assessed by morphometric analysis, increased from 3% to 13% of the cell volume in cultures of *Chlorella vulgaris* grown on the clinostat. The electrophoretic spectrum of succinate dehydrogenase also changed with an increase in the composition of structurally bound forms being observed on the clinostat. The fast and slow electrophoretic variants of the enzyme were most affected by clinostating, those with intermediate rates of migration were present in all samples (Popova *et al.* 1987). Changes in the distribution of succinate and malate dehydrogenases (detected cytochemically with nitrotetrazolium blue) was studied in cells of 5-and 10-day protonema of *Funaria hygrometrica* grown on a 2 rpm clinostat (Nedukha *et al.* 1985).

Morphometric analysis of organelle redistribution after maize roots have been placed horizontally emphasises the correlation of amyloplast sedimentation with graviperception, but also indicates that movement of the nucleus towards the proximal end of the cell (that is, the end of the cell furthest from the root cap) is also correlated with the gravitropic response (**Moore 1986b**). However, the integrity of the statocyte, as well as this polarity in its organelle distribution, seems to be essential for the gravitropic response of the roots of cress (*Lepidium sativum*). Prolonged growth on a 2 rpm clinostat caused damage to the statocytes for as well as redistributing organelles, such treatment caused loss of amyloplast starch, aggregation of lipid droplets and lysis of cell walls (**Hensel & Sievers 1980**). Similar results have been described for pea roots grown on both fast and slow clinostats (Belyavskaya 1983) and cress roots exposed to microgravity on the Bion-10 mission (Laurinavichius *et al.* 1994a & b).

Sievers and co-workers (**Sievers & Volkmann 1972**; Volkmann, Sievers 1979; **Sievers, Heyder-Caspers 1983**) maintained for a number of years that amyloplasts are effective in cress (*Lepidium*) roots because they exert pressure on a complex of rough endoplasmic reticulum consisting of parallel cisternae near the distal (closest to root tip) cell wall. A major reason for this suggestion was the short presentation time (only 12 seconds) in this material, which must mean that the graviperception mechanism is extremely sensitive and cannot depend on sedimentation over large distances.

Endoplasmic reticulum membranes isolated from *Lepidium* roots accumulated Ca^{2+} when incubated with micromolar concentrations of calcium. The accumulation was ATP-dependent and ATP-specific and was not inhibited by protonophore treatment (**Sievers *et al.* 1984**; Sievers 1986). Furthermore, *Lepidium* root gravitropic response was drastically inhibited by treatment with 20 μM cyclopiazonic acid, a known inhibitor of Ca^{2+}-transporting ATPases of the endoplasmic reticulum (**Sievers & Busch 1992**). This mechanism could, therefore, be related to the calcium polarisation which is discussed above. Indeed, **Lawton *et al.* (1986)** claim that their thick section high voltage electron microscopy shows that statoliths in statocytes in the leaf sheath bases from several different grasses sediment onto cushions of tubular endoplasmic reticulum, so the mechanism might may not be limited to the root.

However attractive this idea might be, it is not the end of the story. Audus (1962) had already calculated that the pressure exerted by amyloplasts is extremely small and pressure sensing is, theoretically, a poor basis for a graviperceptive guidance mechanism. **Hensel & Sievers (1981)** showed that even very short periods of contact between these organelles leads to disintegration of the distal ER complex, loss of amyloplast starch, and confluence of lipid droplets, suggesting that contact rather than pressure may be important. Centrifuge experiments (**Wendt *et al.* 1987**) have shown that gravitropic bending of cress roots occurs without contact between amyloplasts and endoplasmic reticulum. Sedimenting amyloplasts showed no consistent contact with any other cellular structure in maize (Moore 1986b), and, indeed, the onset of root gravicurvature in maize does not require a major redistribution of plastids in columella cells (**Moore 1987**).

Furthermore, gravitropic responses still occur in plants free of amyloplast starch because of nutritional or genetical alterations (**Roberts 1984; Moore *et al.* 1987b; Caspar & Pickard 1989; Kiss *et al.* 1989**), although the fact that such mutants *do show* a gravitropic response has been interpreted as evidence that starch is unnecessary for gravity perception (**Caspar & Pickard 1989**). A starchless mutant of *Arabidopsis thaliana* in which amyloplasts did not significantly sediment in either upright or inverted seedlings produced gravitropic curvatures of at least 70% of the wild type (**Caspar & Pickard 1989**). This parallels a much earlier study which showed that physiologically destarched plants can still make gravitropic responses (**Pickard & Thimann 1966**). However, **Kiss *et al.* (1989)** showed that although starch is not required for gravity perception in *Arabidopsis thaliana* roots, it is necessary for full sensitivity; the starchless amyloplasts still function as statoliths. Starch deficiency in a mutant of tobacco (*Nicotiana sylvestris*) greatly reduced gravitropic sensitivity, the mutant bending to an angle of 10° in the time the wild type reached 70° (**Kiss & Sack 1990**).

In lentil (*Lens culinaris*) roots grown in microgravity in orbit during a Spacelab mission, amyloplasts were not randomly distributed as they become on the clinostat, but were more localised in the proximal part of the cells than in $1 \times g$ controls (**Perbal *et al.* 1987; Perbal & Driss-Ecole 1989**). These authors maintain that contact between amyloplasts and the distal complex of endoplasmic reticulum was not required for graviperception, but direct attention to a network of microfibrils, which seems to bind the nucleus to the cell periphery, which may be able to respond to acceleration. **Hensel (1984)** found little evidence for involvement of microtubules in graviperception in statocytes from roots of cress (*Lepidium sativum*). Development of the roots on a horizontal clinostat preserved a youthful state of the microtubular system. Compression of the distal network of microtubules after centrifugation of the roots indicated that microtubules might function in stabilizing the distal complex of endoplasmic reticulum.

An observation which appears to have been given very little attention is that sedimenting amyloplasts in maize roots are associated with a surrounding region of cytoplasm which stains intensely with toluidine blue in sections of resin-embedded material (**Moore 1986b**); the author stresses that this differential staining was unique to the early stages of amyloplast sedimentation, being observable only in material fixed immediately after reorientation. Toluidine blue is a basic dye which reliably stains anions in the presence of resin embedding media. This observation indicates a *transient* increase in polyanions in the vicinity of the amyloplast as sedimentation commences. Perhaps some conformational change in microtubules or microfilaments around the amyloplasts is generated immediately they begin to sediment.

1.4.5.7 Cytoskeleton

Until very recently, the most widely held interpretation of the sorts of results described above has been that gravity perception in plants involved direct interaction between the statoliths and endoplasmic reticulum membranes normally located beneath them (review - Sievers & Hensel 1990). The position of these membranes seems to be determined genetically. When displaced by centrifugation, they relocate to their original position within a few minutes, independently of the direction of gravity in relation to the root axis, or the direction of the applied centrifugation vector (**Wendt & Sievers 1986**). Cytochalasin treatment inhibited relocation of the endoplasmic reticulum, so the cytoskeletal system was seen as being involved simply in locating the endoplasmic reticulum (**Wendt & Sievers 1986**). More recent experiments have certainly changed this view.

Tropic curvature in maize coleoptiles and sunflower hypocotyls is correlated with a change of microtubule orientation from transverse to longitudinal at the slower growing organ flank, whereas the transverse microtubule orientation is maintained (or even augmented) at the faster growing organ flank. These directional changes are confined to the microtubules subjacent to the outer epidermal wall (**Nick *et al.* 1990a**). Application of the antimicrotubular agent, ethyl-*N*-phenylcarbamate, inhibited gravitropism without affecting phototropism, implying a role for the microtubular cytoskeleton in perceiving and transducing statolith movement (**Nick *et al.* 1991**). **Lorenzi & Perbal (1990b)** showed that actin filaments are probably responsible for positioning the nucleus in lentil statocytes in experiments involving space-flown material and clinostat-treated samples. However, it is Sievers and his co-workers who have provided the most definitive results (Sievers *et al.* 1991).

Thin bundles of microfilaments were shown encircling the statoliths as a dense envelope in rhizoids of *Chara* (by labelling actin with rhodamine-conjugated phalloidin), and in root statocytes of *Lepidium sativum*, cytochalasin B treatment caused a 3-fold increase in the rate of statolith sedimentation (**Hensel 1989; Sievers *et al.* 1989; White & Sack 1990**). Statoliths hold a particular position in both these cell types. The distance between the statoliths and the physically lowest point in the cell is up to 25 μm in the *Chara* rhizoid, and up to 4 μm in cress statocytes. Application of cytochalasin to vertical rhizoids or roots causes statoliths to sediment to the lowest point of the cell by displacement of the ER membranes (**Hejnowicz & Sievers 1981; Hensel 1985; Bartnick & Sievers 1988**). Presumably, under normal circumstances statoliths interact with microfilaments to maintain their characteristic position. Experiments during 5 rocket flights (TEXUS 18, 19, 21, 23, 25) showed that within approximately 6 minutes of microgravity exposure the shape of the statolith complex changed and, more importantly, the complex moved approximately 14 μm (*Chara* rhizoids) and 3.6 μm (cress root statocytes) in the opposite direction to the originally-acting gravity vector (**Volkmann *et al.* 1991**). Videomicroscopy of *Chara* rhizoids during rocket flights confirmed that in the weightless state the statoliths moved **away from the tip** (see

Fig. 7), possibly due to relaxation of microfilaments which are normally kept under tension by the weight of the statolith. This idea was further supported by the finding that disorganizing microfilaments with cytochalasin D resulted in sedimentation of statoliths onto the apical cell wall under normal gravity and an absence of basipetal displacement during the microgravity phase of a rocket flight (**Buchen *et al.* 1993**).

The gravity-dependent polarity of cytoplasmic streaming in algal cells (**Wayne *et al.* 1990a; Buchen *et al.* 1991**) has provided a sensitive tool for study of graviperception, even in cells which lack statoliths. Results support the hypothesis that the plasma membrane is the gravireceptor and that Ca^{2+} ions participate in the signal transduction chain that leads to a gravity-induced polarity of cytoplasmic streaming (**Wayne *et al.* 1990b**). The link with work described in the previous paragraph is that Wayne *et al.* (1992) present evidence that the cell-extracellular matrix junction, which includes the cell wall and the outer surface of the plasma membrane, is involved in gravity perception by the statolith-free internodal cells of *Chara*. Gravisensing was sensitive to protease treatment, indicating that proteins in the cell-extracellular matrix junction may be required for gravisensing, and inhibition by the tetrapeptide Arg-Gly-Asp-Ser suggests that the algal gravireceptor may be an integrin-like protein.

In higher fungi, too, there is growing evidence that actin microfilaments are involved directly in the gravity perception component of the gravitropic response of mushroom fruit bodies (Monzer & Haindl 1994; **Monzer, 1995**; Novak Frazer & Moore, 1994). Sedimentation of nuclei within an actin 'cage' has been postulated to account for gravity perception in *Flammulina* (**Monzer, 1995**), and the kinetics of response to a spectrum of inhibitors has been used to deduce that actin microfilaments are specifically involved in perception in *Coprinus* (**Novak Frazer & Moore 1993**, 1994). However, these are all preliminary findings using living material which is extremely difficult to observe. Cells responsible for gravity perception in fungi are not differentiated, as they are in plants. Consequently, the experimental tissue is comprised of a mass of almost indistinguishable cells, and it is difficult to devise experiments which enable observation to focus precisely on the perception event rather than on some later stage in the gravitropic process.

1.5 Animal growth and development

Embryonic development is perhaps the most complex cell biological process, requiring the co-ordination of activities including gene expression, cell division, cell differentiation and cell interactions. Gravity has been cited as being a possible directional cue that could be used for determining pattern formation (**Malacinski *et al.* 1989**). This proposition has been examined experi-

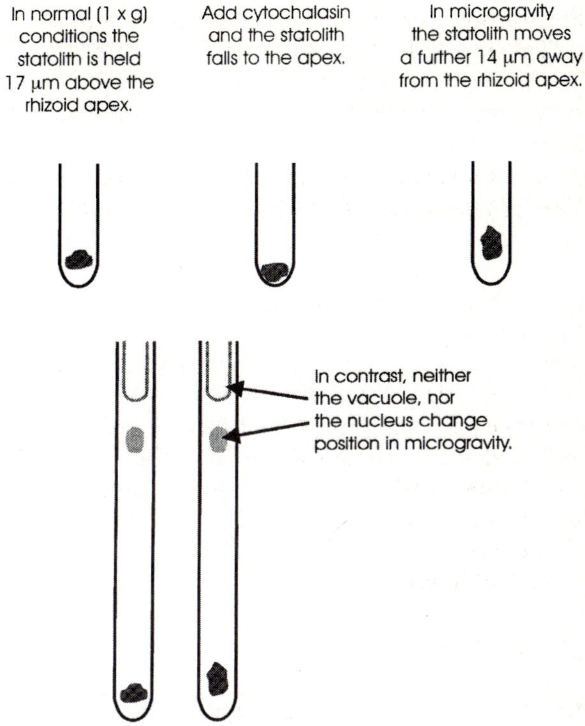

Fig. 7. Summary diagrams of statolith movements in the tips of rhizoids of the alga *Chara*. The rhizoids are growing vertically downwards at the beginning of each experiment depicted here, and this gravitropic response is governed by the statoliths which are normally maintained 17 μm above the apex. When treated with cytochalasin the statoliths fall into the apex, indicating that actin microfilaments are responsible for suspending the statolith complex above the rhizoidal apex. In microgravity (TEXUS sounding rocket) the statolith is pulled back from the apex, suggesting that its normal gravitropic function depends upon its weight exerting a tension stress on the microfilaments of the cytoskeleton. The drawings were made from photographs of video stills in Volkmann *et al.* (1991) and Buchen *et al.* (1993).

mentally in a number of animal systems both in ground-based and space-flown experiments.

Hughes-Fulford (1991) has provided an excellent review of altered cell function in microgravity in which she makes the case for drawing comparisons between physiological alterations seen in single cell prokaryotes and eukaryotes and those observed in animal tissues and even whole animals. It is worth reiterating that many of the physiological changes seen in humans, vertebrate and other multicellular organisms in space flight may originate from dysfunction of basic biological mechanisms caused by microgravity (or other aspects of the orbital environment). Since microgravity seems to affect prokaryotic and eukaryotic cell function at a subcellular and molecular level, research in space offers an opportunity to learn more about basic biological mechanisms which are essential to life. This general justification for microgravity research is a parallel to the observation that as aging in humans shares many of the symptoms seen in astronauts during space flight (reduced cardiac function, loss of bone, reduced immune response, orthostatic hypotension), research in these two areas can only be synergistic; each contributing to the other and both together making more rapid advance than either in isolation (Booth, 1994).

1.5.1 Cell physiology in the whole animal

To a considerable degree, this medical relevance has been a major motive in the design of experiments on animal gravitational biology and has formed the proper context of the research. Human physiology is dealt with in detail in Chapters 2 to 4 and reactions of individual cells are detailed in Section 1.2, so in this introductory section we have chosen to highlight a selection of topics, intending simply to illustrate the wide range of animal physiology which is influenced by change in the gravity vector.

Rats exposed to 3-fold normal gravitational acceleration for 8 months adapted with little chronic stress but at a higher basal metabolic rate. The rats aged faster than controls, the longest survivors living for an average of 520 days on the centrifuge compared to 574 days for controls. The experiment is interpreted as a demonstration of the rate of living theory of aging in mammals (**Economos *et al.* 1982**). Hypergravity acceleration ($1–5 \times g$) did not decrease total fecundity in *Drosophila melanogaster* (**Lints & Le Bourg 1989**), but despite initial indications of a decrease in longevity (**Le Bourg & Lints 1989a**) and in the viability of eggs (**Le Bourg & Lints 1989b**), the effects are not clear and even at $7 \times g$, flies lived for roughly 40 days (**Le Bourg *et al.* 1993**). Thus, the flies remain able to lay viable eggs and to live for a fairly long life, even under sustained hypergravity conditions. Examination of three behavioural traits affected by aging (climbing activity, pattern of movement, spontaneous locomotor activity) suggests that flies age faster in hypergravity (**Le Bourg & Lints 1992a–c**). All of these results also contribute to theories of aging, especially the rate of living theory.

Maternal and foetal plasma prolactin levels were reduced on day 22 of gestation in rats exposed to $3 \times g$ (**Megory & Oyama 1985**). Such an effect on foetuses conceived after adaptation to hypergravity highlights the possibility that gravitational effects on developmental biology might be due to general effects on systemic physiological mechanisms which then impact the developmental processes.

Exposure of chicken embryos to daily 20 min $7 \times g$ acceleration stress had a substantial effect on myocardial differentiation, decreasing Ca^{2+}-stimulated contraction of isolated myocytes, affecting their rhythmic activity and reducing levels of actin, myosin light chain 2, and troponin C in the myocardium (Oganesyan *et al.* 1980).

Articular and growth-plate chondrocytes respond differently to hypergravity. Centrifugation of rabbit growth-plate chondrocytes at 3 × g resulted in a 2-fold increase in incorporation of [^{35}S]sulphate into proteoglycans, but had little effect on [^{3}H]thymidine incorporation into DNA. However, similar treatment of articular chondrocytes caused 1.5-fold increases in both [^{35}S]sulphate incorporation into proteoglycans and [^{3}H]thymidine incorporation into DNA (**Inoue *et al.* 1990**). Space flight cultures (aboard Space Shuttle mission STS-54) of embryonic mouse pre- metatarsals which had initiated chondrogenesis and morphogenetic patterning before flight, resulted in an increase in cartilage rod size and maintenance of rod shape, compared to controls. Older pre-metatarsal tissue, already terminally differentiated to hypertrophied cartilage, maintained rod structure and cartilage phenotype during culture in orbit (**Klement & Spooner 1994**). Cultured mouse embryo lung rudiments continued to grow and branch on the same flight and cultured embryonic mouse pancreas underwent characteristic exocrine acinar tissue and endocrine islet tissue differentiation during space flight (**Spooner *et al.* 1994**).

Osteoporosis was produced in rats by a 19-day space flight on board Cosmos 1129, bones showing no mineralization in the microstructures and a decrease in Ca content (but no change in P content was observed). The adverse effects of the flight continued to develop until the 6th day after return but the bone parameters measured had returned to normal 23 days later (Rogacheva *et al.* 1984).

Chemical measurements and X-ray microtomography indicated a longitudinal gradient of decreasing mineralization toward the distal diaphysis in femurs of rats flown on Cosmos 1887. The techniques identified areas of femoral diaphysis most vulnerable to mineralization deficiencies associated with space flight (**Mechanic *et al.* 1990**). In the same flight, the proliferative zones of rat tibial epiphyseal plates were larger than controls, while calcification zones were reduced (**Duke *et al.* 1990**). These experiments raise another difficulty in literature interpretation as the experimental (flight) animals were exposed to 1 × g for 53.5 h following recovery from a 12.5 day flight. Consequently, as a cell cycles through the growth plate in 2–3 days at 1 × g, the results represent the outcome both of the microgravity exposure and a partial recovery from any effects the exposure may have had. The delay was particularly damaging to experiments involving examination of intestinal tissues with rapid cell turnover (**Sawyer *et al.* 1990**). Although there were no consistent differences between animals in the flight group and those in the control groups, delay in recovering flight animals meant that the only safe conclusion was that any effects of microgravity on jejunal mucosal cells were short-lived. In a subsequent flight (Cosmos 2044), it was established that mucosal cell proliferation in rats was not affected by microgravity conditions associated with space flight. Although the length of villi and depth of crypts were reduced in flight animals this was attributed to changes in the connective tissue core and not to any impairment of proliferation or migration of jejunal mucosal cells (Sawyer *et al.* 1992).

Reductions in human T cell reactivity to mitogens after space flight are well documented and similar results have been obtained using simulations of microgravity (see Section 1.2.2). With rats, the tail suspension model of weightlessness has been used to examine lymphocyte proliferation in response to mitogens including examination of cell surface proteins and assessment of interleukin receptor expression. Significant depression in [^{3}H]thymidine incorporation by mitogen-stimulated peripheral blood lymphocytes was evident after only 1 week of treatment, but little decrease was observed in the proliferation of lymph node lymphocytes and splenocytes. There were no changes in interleukin-2 production or interleukin-2 receptor expression. The tissue specificity of the decrease in mitogen activation was taken to indicate a compartmentalized response to the simulated microgravity treatment (**Nash *et al.* 1991**). Comparison of proliferation of lymphocytes from both spleen and lymph nodes of rats following a 4-day flight aboard the Space Shuttle showed reduced proliferation of lymph node lymphocytes from flight animals, whereas splenocyte proliferation was not depressed, again suggesting tissue-specific response to microgravity (**Nash & Mastro 1992**). Similarly, proliferation of inguinal lymph node lymphocytes from rats flown on the 14-day Cosmos 2044 mission was not significantly different from controls and they produced similar amounts of interleukin-2 (**Nash *et al.* 1992**). These results all support the view that microgravity acts on lymphocytes in a tissue-specific manner.

Differential effects of space flight on the function of natural killer cells has also been observed. The ability of natural killer cells of rats flown on Cosmos 2044 to lyse two different target cell lines differed. Spleen and bone marrow cells obtained from flight rats showed significantly inhibited cytotoxicity for YAC-1 target cells compared with killer cells from control rats, whereas there was no significant difference in cytotoxicity for K-562 target cells (Rykova *et al.* 1992). Thus, experience of microgravity has selective effects on numerous different aspects of the immune response.

Vacek *et al.* (1991) report that the number of erythrocyte and of granulocyte and macrophage progenitors in bone marrow decreased in rats exposed to microgravity during the 14-day flight of Cosmos 2044. On the other hand, on the basis of bone marrow cell differential counts, clonal studies of RBC colony formation, and plasma erythropoietin determinations, **Lange *et al.* (1994)** report a slight increase in granulocytic cells in the bone marrow of Cosmos 2044 flight animals. These authors also report a slight decrease in the percentage of erythroid cells in animals subjected to centrifugation at $2 \times g$.

Reduced secretion of growth hormone and prolactin have been recorded in pituitary cells prepared from rats flown on the 12.5-day mission of Cosmos 1887 and the slightly longer mission of Cosmos 2044. Tail suspension of rats for 14 days resulted in changes similar to those from animals flown in space (Hymer *et al.* 1992).

These few, selected, examples illustrate the variability of response to microgravity exposure, not only between organisms, strains or varieties, but also within an organism there is variability in response between tissues and even between cells within those tissues. The extent of this variability implies the existence of a considerable genetic component in the 'control' of the gravitational response. If this is the case, then the morphogenetic and ontogenetic reaction to changed gravitational acceleration is likely to be even more varied because of the number and complexity of the cell-to-cell and tissue-to-tissue interactions involved in normal animal development.

1.5.2 Morphogenesis and ontogeny

Malacinski & Neff (1984) have described the development of animal systems in terms of a series of overlapping phases: pattern specification; differentiation; growth; and aging; and have reviewed briefly the extent to which microgravity affects these phases. The overall impression gained from this review and from subsequently published work (and including clinostat protocols and orbital experiments alike) is that microgravity alters early pattern specification (dorsal/ventral polarity) but subsequent morphogenesis is much more robust and is not adversely influenced.

Much the same seems to be true for the effects of hypergravity (2–30 × g) exposures. Sensitivity of *Rana temporaria* embryos was increased at the beginning of cleavage into 8 blastomeres and at the gastrula–neurula stage (Parfenov & Oigenblik 1988).

Amphibian and other vertebrate embryology has also been studied using clinostats and orbital experiments (**oung & Tremor 1968**; **Young *et al.* 1970**; Tremor & Souza 1971; **Black & Gerhart 1985**; McLaren 1990). *Xenopus laevis* embryos develop normally on a clinostat, demonstrating that the unilateral gravity vector is not necessary for normal axis formation, presumably because organisation of the egg cytoplasm does not depend solely on gravity-driven processes like differences in density between the components (Ubbels & Brom 1984; **Neff *et al.* 1985**; **Smith & Neff 1986**). The correlation between the point of sperm entry and the orientation of the first division can be disrupted by clinostating, and a similar phenomenon has been observed in microgravity aboard sounding rockets (Ubbels 1992; Ubbels *et al.* 1989, 1990, 1992). Nonetheless, after fertilization in microgravity *Xenopus* embryos do develop bilateral symmetry, presumably cued by the sperm alone (Ubbels *et al.* 1994).

The position of the first cleavage furrow during amphibian egg development can be shifted by placing embryos in novel gravitational fields. Position of the furrow is characterised by the animal/vegetal cleavage ratio (AVCR), which is the ratio of the height of the animal blastomere to the height of the *Xenopus* embryo at the 8 cell stage. Microgravity (and clinostating) increases AVCR, and centrifugation (hypergravity) reduces AVCR. Embryo inversion and D_2O immersion were found to increase AVCR, and cold shock was found to reduce it. Based on the additive or antagonistic effects of combined treatments, it is thought that the primary cause of AVCR changes is an alteration in the distribution of yolk platelets and the rearrangement of microtubule arrays. Embryos with a decreased AVCR exhibited reduced survival in early developmental stages, indicating serious difficulties in cleavage, blastulation and/or gastrulation (**Yokota *et al.* 1992**, 1994).

Some cytoplasmic localisations which occur during normal development on Earth, including clustering of mitochondria in the presumptive dorsal side of the embryo during the first cleavage cycle, are sensitive to disruption by change in orientation (Phillips 1994). Changes in cell morphology associated with clinostat treatment at the 8-cell stage are accompanied by altered distribution of mRNA molecules (detected by *in situ* hybridization), suggesting that gravity dependent differential morphogenesis may result from gravity dependent gene regulation (Neff *et al.* 1994).

Frog eggs were fertilized in microgravity and incubated either on the 1 × g centrifuge or in microgravity during a Space Shuttle flight in 1992. Gastrulae developed in microgravity had a thicker blastocoele roof and formed the blastopore lip nearer the vegetal pole than the 1 × g specimens, but neurulation and subsequent morphogenesis were essentially normal. Living tadpoles returned from the flight differed in some swimming behaviours, but these differences disappeared after one week. Tadpoles from both groups completed metamorphosis (Black *et al.* 1994). In some experiments, despite dramatic differences in early embryogenesis, tadpoles at the feeding stage were largely indistinguishable from controls (**Neff *et al.* 1993**). In other reports (e.g. involving fertilized *Xenopus* eggs flown on Cosmos 2229), later stage flight animals also exhibited a catalogue of deformities (affecting notochords, lungs and thymus glands) when compared to ground controls (Snetkova & Wasserung 1994).

The catalogue of changes caused by real and simulated weightlessness in *Xenopus* and *Rana* includes: (i) the location of the first horizontal cleavage furrow was shifted toward the vegetal pole at the eight-cell stage; (ii) the position of the blastocoele was more centred, and the number of cell layers in the blastocoele roof was increased at the blastula stage; (iii) the dorsal lip appeared closer to the vegetal pole at the gastrula stage; and (iv) head and eye dimensions were enlarged at the hatching tadpole stage, notochords were abnormal and tails elongated, lungs smaller (uninflated?) and thymus significantly reduced. Effects of hypergravity were opposite to those of simulated weightlessness, except that hypergravity reduced the number of primordial germ cells in feeding tadpoles. Abnormalities of early embryos or larvae returned to normal conditions were repaired as the larva developed further and mature animals were essentially normal.

Embryonic development and hatching of Japanese quail proceeded in weightlessness. The main effect seemed to be that the animals were unable to adapt their motor activity to microgravity. Movements, particularly of chicks, were uncoordinated and chaotic. In adults there was an arrest of egg production, dystrophy of ovaries and testes and a decreased testosterone level after 7 days exposure to microgravity (Boda *et al.* 1992; Meleshko *et al.* 1991). **Suda *et al.* (1994)** reported that only one out of ten chick eggs flown in the Space

Shuttle immediately after being fertilized was recovered alive after landing. Viability was better for eggs preincubated for 7 and 10 days on Earth before flight, with all 7-day-old and 9/10 ten-day-old eggs being recovered alive. The high mortality of the 0-day-old eggs is probably related to a need for inner components to separate on the basis of difference in specific gravity to maintain adequate gas exchange during early development (regular turning of bird eggs on Earth is required for the same reason).

Fertilization of eggs of the sea urchin (*Paracentrotus lividus*) occurred normally on a sounding rocket (MASER 4) flight. Although some abnormalities were observed in pluteus larvae developed from the MASER 4 flight samples, these were considered to be artifacts and in a later experiment on MASER 5, larval development was normal to late pluteus stages (Marthy 1990, Marthy *et al.* 1994).

Behavioural changes have been recorded in fish and amphibian larvae which have developed from eggs exposed to microgravity (**Neubert *et al.* 1991**; Rahmann *et al.* 1992; **Slenzka *et al.* 1993**) and have been correlated with ontogenetic effects on the inner ear but also with physiological and biochemical changes affecting membrane function in those parts of the CNS which serve the vestibular system. A related observation is that formation of nerve-associated acetylcholine receptor patches between embryonic spinal neurones and myotomal myocytes in clinostat cell cocultures was significantly inhibited (**Gruener & Hoeger 1990, 1991**). Control cultures indicated that this did not result simply from loss of contact between neurites and myocytes, accelerated diffusion of a putative aggregating factor secreted by neurites, nor from turbulence in the medium. As the authors concluded, these data suggest that embryonic development of the nervous system in space may be markedly different from that normally occurring on earth (see also Section 1.2.4.3).

Eggs of the stick insect, *Carausius morosus* were exposed to space flight conditions on the 7 day Spacelab Mission D1 and flights of the biosatellites Cosmos 1887 and 2044. Nuclear track detectors allowed effects of heavy ions to be separated from microgravity. Microgravity leads to a reduced hatching rate of eggs exposed in the early stages of development, but embryonic development before hatching showed no major difference between flight and ground control specimens either in speed of development or in morphological anomalies (Reitz *et al.* 1989). On the other hand, microgravity unexpectedly enhanced the production of morphogenetic abnormalities caused by hits by heavy ions (**Reitz *et al.* 1990, 1992**).

Results from the last successful flight of the Challenger Shuttle, in early November 1985, indicate that oogenesis and embryonic development of *Drosophila melanogaster* are altered in microgravity (Gonzalez-Jurado *et al.* 1986; **Vernos *et al.* 1989**). Some early stage embryos recovered from the space–flown containers in microgravity showed alterations in the deposition of yolk, and at least 25% of the living embryos recovered from space failed to develop into adults. Studies of late embryos indicated the existence of alterations in the anterior, head and thoracic regions of the animals. The results suggest an interference with the distribution and/or deposition of the mater-

nal components involved in the specification of the anterio-posterior axis of the embryo. Results of **Li & Wang (1992)**, who experimented with *D. melanogaster* during an 8-day flight of a Chinese biosatellite, are in broad agreement with these observations, showing, essentially, freedom from gene mutation in the germ-line but great influence on the embryonic development of the generation bred in space.
An unusual, and isolated, observation is that six days in orbit as a Get-Away-Special in the open bay of the Shuttle Columbia caused a 10-fold increase in hatching of both field-collected and laboratory-reared gypsy moth eggs (**Hayes *et al.* 1991**). This means that, in some way, conditions in orbit greatly shorten diapause in these eggs. The authors suggest the technique may be useful for rearing gypsy moths for pest control programmes. This seems to be a rather unlikely candidate for biotechnology in space. On the other hand, the observation suggests that research on the control of dormancy might be progressed by experiments in orbit and the pest control aspect would be an obvious justification for using microgravity facilities for fundamental research of this sort.

1.6 General Conclusions

Tables 1 and 2 attempt to summarise the most reliable and best-documented experiments which have been conducted in true microgravity (i.e. in parabolic flights, sounding rockets and orbital vehicles).

1.6.1 Mammalian cells

Proliferation. Human T lymphocytes, associated with monocytes as accessory cells showed dramatic changes in the centrifuge, in the clinostat and in space. As free floating cells, mitogenic response is depressed by 90% in microgravity, whereas when attached to a substratum, activation is enhanced by 100% compared to $1 \times g$ controls on the ground and in flight. The duration of phase G1 of the cycle of HeLa cells is reduced in hypergravity, thus resulting in an increase of the proliferation rate. Other systems like Friend cells and WI38 human embryonic lung cells did not show significant changes.

Genetic expression and signal transduction. Human T lymphocytes and monocytes show important changes in the expression of cytokines like interleukins (IL) IL-1, IL-2, interferon-γ and tumor necrosis factor. The data from space experiments (Spacelab, Space Shuttle mid-deck, Biocosmos) have contributed to clarify certain aspects of the mechanism of T cell activation. Epidermoid A431 cells showed changes of genetic expression of proto-oncogenes c-*fos* and c-*jun* in the clinostat and in sounding rockets. Membrane function, in particular the binding of ligands as first messengers of a signal, is not changed in most of the cell systems in microgravity.

Morphology/motility. Free cells (lymphocytes in particular) are able to move and form aggregates in microgravity. Thus cell–cell contacts and communications do take place in microgravity. Dramatic morphological and ultrastructural changes have not been detected in cells cultured in microgravity.

1.6.2 Animals, plants and fungi: multicellular systems

The conventional approach in reviews of the influence of gravity on living organisms is to consider animals and plants quite separately (the other great Kingdom of eukaryotes, the fungi, are rarely brought into the discussion). There is a conceptual convenience in dealing with similar organisms in the same section, but with a topic like 'response to gravity' distinctions drawn between animals, plants and fungi must, at least in part, be artificial.

This belief has an evolutionary foundation. The mechanisms which are evident in animals, plants and fungi of today have presumably been adapted through evolution to amplify graviperception and optimise it to the particular demands and requirements imposed by the structure, physiology and behaviour of the organisms concerned. Yet it must also be that these different mechanisms have common origins in the primeval response of the general cellular apparatus to the gravitational field. Any common features which emerge from comparison of work on different organisms *may* reflect that basic cellular response.

It must be highly significant, therefore, that from the wide range of research in the literature some consistently-encountered responses in eukaryotes do emerge.

- In plants, frequent reports of disturbances to mitosis and to increases in the frequency of chromosomal abnormalities (often with sufficient controls to eliminate radiation as direct cause of the latter) show that plant cell proliferation requires positional cues which are absent in microgravity.
- In animals, hypergravity stimulates proliferation of various mammalian cells in culture, lymphocytes in microgravity show reduced activation, and microgravity strongly decreases EGF-induced expression of proto-oncogenes implicated in regulation of proliferation and differentiation.
- In fungi, sporulation seems most dependent on the normal gravity vector; cultures on orbiting space craft produced mushrooms lacking spore-forming tissue, implying that commitment to meiotic division and spore formation both require the gravity vector.

There are also common themes in the effects of microgravity exposure on eukaryote development and morphogenesis.

Arabidopsis thaliana was cultivated through a complete life cycle aboard Salyut-7; seeds produced in orbit (germinated on Earth) showed reduced germi-

Table 1. Single cells flown in sounding rockets, satellites and orbital laboratories

Cell type	Mission (duration of experiment in brackets)	gravitational effect?*
Bacteria		
Escherichia coli	Vostok 4 & 6 (27–30 h), Zond 5 & 7, Biosatellite (2 d), Biocosmos 2044, Salyut 7 (1 d), D-1 (1 d), D-1 (1 h), STS-42	**NO**
Acetobacter xylinum	parabolic flights	YES
Bacillus subtilis	D-1 (3 d)	YES
Proteus vulgaris	Soyuz-12 (2 d)	YES
Salmonella typhimurium	Biosatellite II (2 d)	YES
Staphylococcus aureus	Salyut 7 (1 d)	YES
Slime mould		
Physarum polycephalum	D-1, IML-1 (20 min - 2 h)	**YES**
Protists		
Chlamydomonas reinhardii	D-1 (6 d)	YES
Euglena	TEXUS sounding rocket	YES(?)
Paramecium aurelia	Salyut 7 (4 d), D-1 (5 d)	**YES**
Pelomyxa carolinensis	Biosatellite II (2 d)	NO
Mammalian cells		
Human immune cells (T & B lymphocytes and monocytes	SL-1, D-1, SLS-1, IML-2 (3–4 d)	**YES**
	MASER 3, MASER 4, Maxus 1 (7–12 min)	YES
	STS 37 & STS 43 (24 h)	YES
	Salyut 6 (7 d) (two missions)	YES
Spleen cells, murine	STS 37 & STS 43 (24 h)	YES
Lymph node cells, murine	STS 37 & STS 43 (24 h)	YES
LML929 TNF-sensitive cell line	STS 50 & STS 57 (24–29 h)	YES
AM2 strain hybridoma, murine	D-1 (6 d)	YES
7E3-N hybridoma, murine	IML-1 (4 d)	
G8 & SP2/0-UZ hybridoma, murine	TEXUS (7 min)	YES
B6MP102 bone marrow-derived macrophages	STS 37, STS 43 & STS 50 (24 h)	YES
Jurkat T cells, human	Cosmos 2044 (24 h)	YES
THP-1 myelomonocytic cells, human	Cosmos 2044 (24 h)	YES
Friend leukemia virus-transformed, murine	IML-1 (4 d)	NO
WI 38 Human embryonic lung cells	Skylab (28 d)	NO
HeLa cells, human	Zond 5 & 7	NO
Hamster embryonic kidney cells	IML-1 (7 d)	NO
Chinese hamster ovary cells	Salyut-6 (6 d)	NO
Human embryonic kidney cells	STS-8 (1, 5 d)	NO
Human erythrocytes	STS-51C & STS-26 (3 d)	YES
A431 epidermoid cells, human	MASER 3, MASER 4 (7 min)	**YES**
L8 rat myoblasts	STS-45 (9 d)	YES
Other animal cells in culture		
Sf9 cells (= army worm)	STS-50, STS-54, STS-57	NO

*entries in bold characters indicate significant **and** well documented results

Table 2. Major findings from experiments with single cells, tissues and embryos in parabolic flights, sounding rockets, satellites and orbital laboratories

Cell type	Finding	Remarks
Cell proliferation		
T lymphocytes, human	Depression of activation by 80–95% in free-floating cells.	5 experiments in Spacelab, 4 of them with in-flight 1 × g, in agreement with data from the fast-rotating clinostat.
	Increase of activation by 100% in microcarrier-adherent cells.	1 experiment in Spacelab with in-flight 1 × g.
7E3-N murine hybridoma cells	Increase by 40% of cell counts and DNA synthesis-rate after 4 days in microgravity, no effect after 2 days.	1 experiment in Spacelab with in-flight 1 × g, microgravity effects become important only after prolonged exposure to space flight conditions.
Paramecium aurelia	60% increase of growth rate.	3 experiments, 2 in Salyut 7 without 1 × g control and 1 in Spacelab with 1 × g in-flight control, in agreement with data from the fast-rotating clinostat.
AM2 murine hybridoma cells Friend cells, murine HaK cells, hamster WI38 human embryonic lung cells HeLa cells *E. coli*	No effect detected.	These data show that gravitational effects occur only in certain cells.
Genetic expression and signal transduction		
A431 human epidermoid cells	Depression by 47% and 26% of the expression of *c-fos* gene and by 56% and 51% of the *c-jun* gene induced, respectively, by EGF or TPA. No effect on the induction by calcium ionophore or forskolin.	2 experiments in sounding rockets, in agreement with data from the fast-rotating clinostat.
T lymphocytes, human	Secretion of Il-2 and IFN-γ correlates with the effect observed on the proliferation rate, i.e. it was strongly depressed in resuspended cells and strongly increased in microcarrier-attached cells.	1 experiment in Spacelab with in-flight 1 × g. The data have contributed to clarification of aspects of the mechanism of T cell activation.
T lymphocytes, human	Binding of Con A to the cell membrane was not altered.	2 experiments in sounding rockets.
Monocytes, human	Strong depression of the secretion of IL-1 in resuspended cells in flight.	1 experiment in Spacelab with in-flight 1 × g.

Table 2 (continued)

Cell type	Finding	Remarks
Escherichia coli	Exchange of chromosomal DNA via conjugation was enhanced, whereas the exchange via transduction or transformation was not altered.	1 experiment in Spacelab with in-flight 1 × g.
Friend cells, murine	No change in the expression of haemoglobin.	
HaK cells, hamster	No change in the secretion of tissue plasminogen activator.	
Morphology and motility		
Cells of the immune system, human	Cells display autonomous movements and cell-cell interactions.	1 experiment in sounding rocket, with in-flight video camera.
	Rearrangement of microtubule structure immediately after exposure to microgravity.	1 experiment in sounding rocket.
	Formation of cell aggregates, cytoplasmic alterations to the ultrastructure of lymphocytes, intact monocytes.	1 experiment in Spacelab with in-flight 1 × g.
	Strong interaction of cell membrane with the surface of microcarrier beads.	1 experiment in Spacelab with in-flight 1 × g.
WI38 human embryonic lung cells	Normal migration of cells adherent to a substratum.	Cinematographic recording in automatic instrument in Skylab.
Physarum polycephalum	Temporary increase of the frequency of cytoplasmic streaming oscillations and velocity.	2 experiments in Spacelab with video recording.
Friend cells Human embryonic kidney cells *Pelomyxa carolinensis*	No ultrastructural alterations detected.	
Metabolism and cell behaviour		
Acetobacter xylinum	Altered microcrystalline structure of cellulose synthesised in flight.	Parabolic flights.
Euglena gracilis	Gravitaxis lost, swimming velocity higher in microgravity, absence of aggregation.	TEXUS sounding rocket.
Mouse hybridoma cells	Fusion frequency and hybrid viability increased about 2-fold in microgravity-exposed samples.	Electrofusion experiment on TEXUS sounding rocket.
Whole plants and plant tissues		
Mung bean Oat Pine seedlings	Stem and root growth, and fresh and dry weights reduced in orbital plants.	Shuttle and Spacelab missions.

Table 2 (continued)

Cell type	Finding	Remarks
Lettuce Cress *Arabidopsis thaliana*	Coleoptile growth reduced, compared with ground controls.	Salyut missions.
Cucumber seedlings	Longer roots and coleoptiles formed on the in-flight centrifuge than in microgravity.	Cosmos 1667 with in-flight centrifuge.
Lettuce seedlings	Those grown in space were larger than those grown on the ground.	Cosmos 1667 with in-flight centrifuge.
Hemerocallis(day lily) *Haplopappus gracilis*	Controls grew more than plants subjected to space flight.	Shuttle flight.
Zea mays	Fresh weights and hormone contents of orbital seedlings not significantly different from controls	Shuttle flight.
	Higher mitotic activity and faster differentiation of root cortical cells	Cosmos 1667.
Lentil seedlings	Roots showed increased mitotic index in microgravity in comparison with in-flight controls.	Spacelab D1 Mission.
Aniseed cell cultures	More rapid differentiation.	Spacelab D1 Mission.
Rapeseed (*Brassica napus*) Carrot (*Daucus carota*)	Protoplast regeneration severely retarded in microgravity.	Biocosmos 9 (Cosmos 2044).
Nicotiana tabacum mesophyll protoplasts *Digitalis lanata* and *Digitalis purpurea* protoplast mixtures	Significant increases in production of fusion products.	Electrofusion experiments on TEXUS sounding rocket.
Arabidopsis thaliana *Zea mays*	No regeneration of the root cap in orbit.	Salyut-6 and Shuttle mission
Chara rhizoid Cress statocytes	Statoliths interact with actin microfilaments to maintain position.	TEXUS sounding rocket.
Arabidopsis thaliana	Cultivated through a complete life cycle. Seeds produced in orbit showed reduced germination, increased frequency of embryonic lethality, generally poorer seedling growth and greatly reduced ability to develop into healthy mature plants. Second generation seeds produced on Earth by the progeny of the space-flown plants were apparently normal.	Salyut-7 & Cosmos 1129.

Table 2 (continued)

Cell type	Finding	Remarks
Animal tissues, eggs and embryos		
Drosophila melanogaster	Generation bred in space showed numerous developmental abnormalities but freedom from gene mutation in the germ–line.	Shuttle and Chinese biosatellite experiments.
Carausius morosus (stick insect)	Space flight results in reduced hatching rate of eggs, but embryonic development before hatching showed no major morphological anomalies, though microgravity enhanced the production of morphogenetic abnormalities caused by heavy ions.	Spacelab Mission D1 and flights of the biosatellites Cosmos 1887 and 2044.
Gypsy moth eggs	Diapause shortened in orbit.	Shuttle flight, no on–board control
Sea urchin (*Paracentrotus lividus*) eggs	Fertilization and larval development normal; some abnormalities in one flight (artifacts?).	MASER 4 & 5 sounding rocket flights.
Frog eggs *Xenopus* eggs	Fertilized in microgravity; in many experiments dramatic differences in early embryogenesis observed, but tadpoles at the feeding stage essentially indistinguishable from controls.	Shuttle experiments and Cosmos 2229.
Japanese quail	Embryonic development and hatching proceeded. Hatchling movements uncoordinated and chaotic. Adults reared from space–hatchlings showed arrest of egg production, dystrophy of gonads and a decreased testosterone level.	Experiments on Salyut and Mir.
Domestic hen eggs	Viability very low in eggs launched immediately after being fertilized.	Space Shuttle experiment.
Fish and amphibian larvae	Developed from eggs exposed to microgravity show numerous behavioural changes, correlated with ontogenetic effects on the inner ear but also with physiological and biochemical changes affecting CNS membrane function.	Space Shuttle experiments.

nation, increased frequency of embryonic lethality, and greatly reduced ability to develop into healthy mature plants.

Oogenesis and embryonic development of *Drosophila melanogaster* were altered in orbited cultures; observations show no gene mutation in the germ-line but great influence on the embryonic development of the generation bred in space, and behavioural changes in fish and amphibian larvae developed from eggs exposed to microgravity suggest altered development of the nervous system.

Clinostat and space experiments show that the basic form of the mushroom (overall tissue arrangement of stem, cap, gills, hymenium, veil) in agaric and polypore alike is established independently of the gravity vector, but particular aspects of development (specifically sporulating tissues) are defective when grown in microgravity.

The demonstration that the major embryonic axes in animals, plants and fungi form independently of gravity incidentally emphasises that experiments in microgravity are the only way to demonstrate that particular features are *not* affected by gravity. E.g. study of phototropism in space is the only way of separating the light response from the otherwise ever-present gravitropic reaction. The small lateral movements made by shoot and root tips as they grow (circumnutations - Barlow *et al.* 1994) have been shown to occur under weightless conditions (Brown & Chapman 1984; Brown *et al.* 1990) implying that gravity is not essential for their occurrence.

Despite the inevitable difficulties in conducting this sort of work (Table 3), research using microgravity facilities has revealed quite categorically that the normal unilateral gravity vector is not required for:

- establishing embryonic axes in animals, plants or fungi;
- circumnutation or plumule hook formation in plants;

but that altered gravitational acceleration *does* cause changes in the following basic cellular processes:

- energy metabolism and intermediary metabolism;
- cell proliferation rates;
- nuclear division;
- membrane composition and function (especially of the plasmalemma);
- cell and tissue differentiation;
- early events in ontogeny;
- rate of aging.

These different effects occur in ANIMALS AND PLANTS ALIKE and can also be traced in fungi and microorganisms including prokaryotes. The overall conclusion must be that the fundamental cellular machinery is dependent on the normal unilateral gravity vector for some of its functioning and is consequently disturbed when exposed to microgravity. Some of the potential value of research on this topic is discussed by Booth (1994).

1.6.2. Theories and models

Based either on experimental results or on theoretical considerations, several authors have discussed gravity effects at the cellular level. The effects reported may be due to important changes in metabolism and/or molecular organization occurring within the cell which allow it to adapt to a new gravitational environment. Other environmental factors such as radiation may play a role as well. This idea of adaptation is not new as many other changes in the environment such as temperature, solute concentration, pH or pressure are also followed by alterations in cell behaviour which reflect the process of adaptation. Therefore, it is not surprising that single cells adapt to altered gravitational conditions.
Concerning the primary gravity receptor within the cell, two alternatives have been discussed in detail in the literature (**Block *et al.* 1992**): (a) a system consisting of dense organelles like nuclei, nucleoli, centrioles, dense vacuoles or mitochondria exerting pressure on the cytoskeleton; and (b) a system based on a gravity dependent ion distribution along membranes, e.g. mitochondrial membranes (Schatz *et al.* 1992). Further detailed analyses of gravity sensing in single cells appear in Albrecht–Buehler (1991) and Todd (1989, 1991). In essence, these discussions depend on consideration of three categories of effect (conclusions being dependent on the relative importance attached to each): (i)*indirect effect*, alterations of the gravitational environment cause significant effects on the microenvironment of the cell with consequences for its metabolism; (ii) *non–equilibrium thermodynamic*, the interaction of gravity with a few organelles may not be sufficient to trigger one event, but a series of small changes could be amplified to generate an important effect; (iii) *direct effect*, the direct interaction of gravity with one or more cellular organelles of different density generating a mechanical response in neighbouring structures (e.g. the cytoskeleton) and, consequently, a signal that is transduced in a biological event.

Indirect effect (physicochemical effects). With respect to changes in the physicochemical environment, lack of sedimentation and thermal convection in

Table 3. Key words which emerge from the gravitational biology literature

UNPREDICTABILITY	For the most part, change in the gravity vector produces effects which are unexpected on the basis of prior knowledge.
VARIABILITY	Between organisms. Between strains or varieties. Between tissues. Between cells. All implying genetic 'control' of the gravitational response.
MULTIPLICITY	There is some (but very little) direct evidence for more than one gravitational target.

microgravity may result in gradients of nutrient, oxygen and waste products. Schatz & Linke-Hommes (1989) pointed out that electric potential and solute variations may occur at the cell-solution interface. Microgravity conditions favour the formation of stationary films (boundary layers) around the cells. Consumption of oxygen and nutrients at rates exceeding those which can be supplied by diffusion decreases the amounts available and increases the gradients of these substances; thereby the cell metabolism may become markedly affected. Schatz & Linke-Hommes (1989) calculated that in $1 \times g$ density convection may be sufficient to partially counterbalance this effect. For a model they took a phospholipid membrane with a surface charge density $q_s = -4.824 \times 10^{-6}$ A.s cm^{-2}, in contact with an electrolyte. In this case the positive ions such as sodium, magnesium, and calcium, accumulate near the membrane surface whereas the negative ions are excluded in response to the electrostatic repulsion of the negatively charged phospholipid groups. In other words a concentration gradient is generated by the membrane potential leading to maximal density variations of $\sim 4 \times 10^{-2}$ g cm^{-3} over a range of 1 to 6 nm. In gravity, two configurations are considered, (i) the membrane surface is horizontally oriented, and (ii) the membrane surface is perpendicular. In the first case the convection in the boundary layer can take place along the membrane surface. In this case the movement of ions is not affected by the electric field because the charges move perpendicular to the field lines. In the second case, however, the action of gravity may result in a displacement of the surface layer away from the membrane surface resulting in a change of the surface potential. In gravity, convection continuously supports the supply of fresh electrolytes but in microgravity the supply of these ions is left to diffusion alone. Likewise, a concentration gradient may develop in the close vicinity of the cell surface when the cells are rapidly consuming substrates such as glucose and oxygen. A boundary layer impoverished in glucose and oxygen may develop.
On the other hand, solutal convection due to concentration gradients (Marangoni convection, which is obscured at $1 \times g$ by convection driven by density differences) may develop in cell cultures and thus favour the movement of particles, like cells, in a fluid (**Napolitano 1984; Scriven & Sternling 1960**; Langbein 1986). Movements of free floating cells have been clearly observed in microgravity, so the lack of sedimentation and thermal convection in microgravity may be compensated, at least in part, by Marangoni convection.

Non-equilibrium thermodynamics (bifurcation theory). An interesting view of a direct action of gravity on single cells has been proposed by Mesland (1990, 1992). This endeavour is a completely new way to explain gravity effects in living organisms. Based on the work of Prigogine & Stengers (1984) and Kondepudi & Prigogine (1983), Mesland applies non-linear non-equilibrium thermodynamics to living cells under changing gravity conditions. Prigogine and coworkers predicted gravitational effects in chemical reactions far from equilibrium. Under the conditions of chemical non-equilibrium the three laws of thermodynamics do not apply, and chemical reactions do not proceed linearly. In thermodynamic terms, this is precisely the principal characteristic of biological systems. Biochemical reaction chains catalyzed by enzymes and controlled by complex feedback mechanisms are nonlinear and far from equilibrium. Under these conditions a cell may display quite unexpected behaviour within a given latitude which is often referred to as chaotic. The constituents of a reaction may remain constant, which is, however, very unlikely, they may oscillate with a known phase, frequency and amplitude, or they may fluctuate chaotically. At the point of such a crossroad, the decision as to which way the reaction develops depends on minuscule differences in the reaction condition such as concentration of substrates, products, catalyst, temperature, and pressure. At the crossroad or point of bifurcation the system is extremely sensitive to changes in the environmental conditions. In fact, the presence of gravity alone can force the reaction to take one direction or the other. Mesland suggested that the lack of gravity alone could cause a cell to behave differently compared with normal gravity conditions. This hypothesis implies that for each cell there must be a threshold gravity force at which the system switches from gravity to microgravity behaviour or vice versa.

With respect to bacteria, yeast, and animal cells, no data are available. In the case of lymphocytes, the space experiments are not in conflict with the bifurcation theory. In microgravity, lymphocyte responsiveness of free-floating cells is virtually nil compared with normal gravity. In a lymphocyte proliferation experiment the individual cell has two possibilities: Upon the addition of the mitogen it can remain dormant in the resting phase of the cell cycle or it can enter it and start to proliferate. With respect to an individual cell, there are no intermediate reactions. Because the lymphocyte population used in the space studies was not uniform (it comprised a great number of subpopulations and clones), the sum of all individual cell reactions may have blurred the transition from microgravity to gravity behaviour. The use of transformed lymphocyte cell lines, which are derived from one single clone, could help to elucidate this interesting question.

Direct effect, gravity receptors. Pollard (1965, 1971) assumed that diffusion and sedimentation inside a single cell might be affected by gravity. Calculations show that the process of diffusion, which is often referred to as Brownian movement, is significant enough to counterbalance gravity effects in cells of roughly spherical shape and below 10 μm in diameter. However, in cells exceeding 10 μm in diameter, sedimentation processes involving organelles may occur. This approach to explain gravity effects by sedimentation and diffusion has become obsolete with the finding that the cytoplasm is not a solution of proteins in a water-like liquid nor a suspension of organelles in the cytosol. Rather, almost no free water is available in the cytoplasm, most of the water is adsorbed to proteins with little bulk water left, which makes the cytoplasm very compact. In eukaryotic cells, the complex latticework of the cytoskeleton further reduces the mobility of organelles. Based on Pollard's considerations, direct gravitational effects in bacteria were ruled out by **Kondo (1968)**.

Nace (1983) proposed that cells may sense gravity by means of the cytoskeleton. Calculations show that gravity exerts a considerable torque on the cytoskeleton. The torque imparted on a cell with a diameter of ~6 µm by starch granules and oil vacuoles is 2.5×10^{-13} dyne cm. The force developed by a bundle of 6 microtubules was found to be $\sim 10^{-6}$ to 10^{-5} dyne. If this force is applied to the lever arm of 6 µm this yields a torque of $\sim 5 \times 10^{-9}$ dyn cm. Neither the actual length of the lever arm of the microtubules (presumably smaller than 6 µm) nor the number of microtubules acting on the organelles are known. Although the torque imparted by the starch granule and the oil vacuole appear to be small compared to the torque microtubules can produce, energy is needed to maintain positional homeostasis under gravity conditions. In microgravity, this requirement for energy expenditure is zero. Thus, cells in space may be expected to use less energy than do cells on Earth and this might be accompanied by structural or biochemical changes.

Until very recently, the most widely held interpretation of gravity perception in plants involved direct interaction between the statoliths and endoplasmic reticulum membranes normally located beneath them. Recent experiments in sounding rockets and in orbit have seriously challenged this view. Theoretically, the pressure of a mass on a surface is a poor candidate as a generator of graviperceptive signals. The reason is that any pressure exerted by the mass is at a maximum when the supporting surface (which is presumed to be detecting the pressure) is horizontal. As the detector tilts the magnitude of the pressure decreases, reaching zero when the detector is vertical. In contrast, for a mobile weight coupled to a shear detector, the shear is zero when the mass is stationary, but increases as tilt increases and gravity causes the mass to move relative to its detector. Further, the *direction* of the shear force relative to the detector depends on the direction of tilt. Consequently, pressure causes effects in the detector which decline as the tilt increases; this is independent of the direction of tilt, and pressure changes sign when the detector surface tilts to the other side of the vertical. On the other hand shear force increases with tilting angle, is directional and changes direction when tilting to either side of the horizontal. Experiments tend to greatly exaggerate phenomena under study, but for most plants in Nature the gravity perception system will only ever have to detect small deflections. Orientation of the seed after dispersal is the most significant variable. After germination, if an organ is normally held vertically then the perception system will need to be adapted to recognise and correct only small deviations from the vertical. In other words, amplification of small deviations is likely to be an advantage. This leads to the conclusion that a shear detector would provide the most effective gravity perception as it would produce signals that increase in magnitude on increased tilt away from normal, and would offer directional sensing that could be incorporated into a negative feedback system. Shear detection is used in the graviperceptive (balance) organs employed by most metazoans to control posture and movement. It is a mechanism which is likely to be equally widely-used in plants. Direct involvement of the cytoskeletal system in shear-dependent gravity sensing in plants seems most probable from recent research (see Section 1.4.5.7) and the elegant experiments in orbit and in sounding rockets which demonstrated this are the clearest examples so far of the way in which space biology can contribute to better understanding of a basic aspect of cell biology which is of fundamental importance on Earth.

Acknowledgements. Work conducted in the laboratory of David Moore has been supported by the Nuffield Foundation, The University of Manchester, the British Mycological Society, SERC, and NSERC (Canada). Work conducted in the laboratory of Augusto Cogoli has been supported by the ETH Zürich, the Swiss National Research Foundation, the Italian Space Agency (ASI), the National Aeronautics and Space Administration (NASA), the European Space Agency (ESA), Oerlikon Contraves AG and Bio-Strath AG. We wish to acknowledge the work of all present and former members of the Space Biology Group at the ETH Zürich and, in particular, Birgitt Bechler, Felix K. Gmünder, Juliet Lee, Giovanna Lorenzi, Alex Tschopp, Pia Fuch-Bislin, Myriam Valluchi-Morf and Isabelle Walther. Finally, we would like to thank Sue B. Criswell, Elisabeth Hunzinger, Helen Joller, Peter Joller, Maria Antonietta Meloni, Otfried Müller, Proto Pippia, Luigi Sciola and Alessandra Spano for their fruitful collaboration.

1.7 Bibliography

This report is dedicated to experiments conducted in hypergravity in centrifuges, in simulated hypogravity in clinostats, and in real microgravity in drop towers, parabolic flights, sounding rockets, and orbital laboratories. The cellular systems considered are single cells like bacteria, protozoa, and animal cells, as well as plant, fungal and animal multicellular systems. The cellular functions analyzed are proliferation and response to mitogens, signal transduction and genetic expression, morphology, motility and morphogenesis.

The data presented in this Chapter are from a total of 471 papers (listed in Section 1.7.1), 279 primary papers published in international journals which are known to be peer-reviewed and 192 appearing in less-accessible journals or as monographs, in books or in reports or proceedings. It is important to note that, at the time of writing, most of the data from experiments carried out in the recent Spacelab missions IML-1, SLJ, D-2 and IML-2 have not yet been published in international journals.

Research on plants, cells and developing embryos in orbital space craft has recently been reviewed by **Halstead & Dutcher (1987)**, Dutcher *et al.* (1994), **Gmünder & Cogoli (1988)**, **Malacinski *et al.*, (1989)** and Cogoli & Gmünder (1991). There have been a number of collected works published (e.g. July

1986 issue of *Naturwissenschaften* dealing with the D1 mission; *FASEB Journal* issue of January 1990 devoted to the Cosmos 1887 flight; NASA *Technical Memorandum 4258* of February 1991 describing NASA's space biology accomplishments in 1989–90; the Proceedings of the 4th (1990) European Symposium on *Life Sciences Research in Space*, (the proceedings of the 5th, 1993, Symposium being on the horizon) and the special issue of *Experimental Gerontology Vol.* 26, numbers 2/3 (1991) containing the proceedings of a 1989 conference on *Correlations of Aging and Space Effects on Biosystems.* Experiments on plants grown in space have featured in a Supplement to *Annals of Botany Vol.*54 (1984) and plant gravitropism has more recently been featured in a special issue of *Plant, Cell and Environment* (1992, vol. 15, pp. 761–794) dealing with the Cholodny–Went theory that tropic curvatures in plants result from unequal distributions of auxin. A detailed account of biological experiments carried out aboard Soviet orbital stations between 1970 and 1990 has recently become available in English (Nechitailo & Mashinsky 1993).

1.7.1 References

Abel JH, Haack DW & Price RW. Effects of weightlessness on the nutrition and growth of *Pelomyxa carolinensis.* In: *The Experiments of Biosatellite II.* Ed. by Saunders JF. Washington, DC, NASA, pp. 291–308, 1971.

Al-Ajmi N, Braidman IP & Moore D. Effect of clinostat rotation on differentiation of embryonic bone *in vitro.* In: *Proceedings of the 5th European Symposium on Life Sciences Research in Space.* Ed. by Oser H, Guyenne TD. Noordwijk, The Netherlands, ESA, ESTEC, pp. 73–6, 1994.

Albrecht-Buehler G. Possible mechanisms of indirect gravity sensing by cells. ASGSB Bulletin 4: 25–34, 1991.

Albrecht-Buehler G. The simulation of microgravity conditions on the ground. ASGSB Bulletin 5: 3–10, 1992.

Audus LJ. The mechanism of the perception of gravity by plants. Symp. Soc. Exptl Biol. 16: 197–226, 1962.

Ayed M, Pironneau O, Planel H, Gasset G & Richoilley G. Theoretical and experimental investigations on the fast rotating clinostat. Microgravity Sci. Technol. 5: 98–102, 1992.

Bara M. The effect of horizontal clinostat on growth, water and protein contents in *Helianthus annuus* hypocotyls. In: *Colloques Internationaux du CNRS, no. 261. Acides Nucleiques et Synthese des Proteines chez les Vegetaux.* Ed. by Bogorad L, Weil JH. Paris, France, Editions du CNRS, pp. 629–33, 1977.

Bara M & Gordon SA. Effect of gravity compensation on growth and cell wall-loosening enzymes in *Helianthus annuus* hypocotyls. Physiol. Plant. 27: 277–80, 1972.

Baraboi VA, Zhad'Ko SI, Kordyum EL & Sidorenko PG. [Lipid peroxidation in plants of different organization levels under microgravity stress]. Izv. Akad. Nauk SSSR Ser. Biol. 3: 368–75 [in Russian], 1991.

Barlow PW. A conceptual framework for investigating plant growth movements, with special reference to root gravitropism, utilising a microgravity environment. Microgravity Q. 2: 77–87, 1992a.

Barlow PW. A flowchart of processes responsible for the gravitropism, nutation and other growth movements of roots. Naturwiss. 79: 34–7, 1992b.

Barlow PW, Parker JS & Brain P. Oscillations of axial plant organs. Adv. Space Res. 14 (8): 149–58, 1994.

Barmicheva EM, Grif VF & Tairbekov MG. [The growth dynamics and cell structure of maize root tips under microgravity]. Tsitologiya 31: 1324–8 [in Russian], 1989.

Bartnik E & Sievers A. *In vivo* observations of a spherical aggregate of endoplasmic reticulum and of Golgi vesicles in the tips of fast-growing *Chara* rhizoids. Planta 176: 1–9, 1988.

Baumann T, Kreis W, Mehrle W, Hampp R & Reinhard E. Regeneration and characterization of protoplast-derived cell lines from *Digitalis lanata* EHRH, and *Digitalis purpurea* L. suspension cultures after electrofusion under microgravity conditions. In: *Proceedings of the 4th European Symposium on Life Science Research in Space.* Ed. by David V. Noordwijk, European Space Agency, ESTEC, pp. 405–10, 1990.

Beaure d'Augéres C, Arnoult J, Bureau J, Duie P, Dupuy-Coin AM, Geraud G, Laquerriere F, Masson C, Prestmal M & Bouteille M. Effect of microgravity on mammalian cell polarization at the ultrastructural level. Naturwiss. 73: 407–9, 1986.

Bechler B & Cogoli A. Lymphozyten sind schwerkraftempfindlich. Naturwiss. 73: 400–3, 1986.

Bechler B, Cogoli A, Cogoli-Greuter M, Müller O, Hunzinger E & Criswell SB. Activation of microcarrier-attached lymphocytes in microgravity. Biotech. Bioeng. 40: 991–6, 1992.

Bechler B, Hunzinger E, Müller O & Cogoli A. Culture of Friend leukemia virus transformed cells in microgravity - Spacelab IML-1 Mission. Biol. Cell 79: 45–50, 1993.

Behrens HM, Gradmann D & Sievers A. Membrane potential responses following gravistimulation in roots of *Lepidium sativum*. Planta 163: 463–72, 1985.

Belyavskaya NA. [Cytological aspects of the hypogravitation effect on gravity sensitive root cells]. Ukr. Bot. Zh. 40: 61–6 [in Ukrainian], 1983.

Belyavskaya NA. [Ultrastructure and localization of ionized calcium in pea root statocytes in the presence of modifiers of calcium homeostasis]. Dokl. Akad. Nauk Ukr. SSR, Ser. B: Geol., Khim. Biol. Nauki 7: 68–71 [in Russian], 1988.

Belyavskaya NA. The function of calcium in plant graviperception. Adv. Space Res. 12 (1): 83–91, 1992.

Belyavskaya NA, Kordyum EL & Podlutsky AG. [Peculiarities of the localization of calcium ATPases in pea root cells under clinostatic rotation]. Tsitologiya 30: 84–7 [in Russian], 1988.

Bennet RJ, Breen CM & Bandu VH. A role for calcium in the cellular differentiation of root cap cells: a re-examination of root growth control mechanisms. Environ. Exp. Bot. 30: 515–23, 1990.

Biro RL, Hale CC II, Wiegand OF & Roux SJ. Effects of chlorpromazine on gravitropism in *Avena* coleoptiles. Ann. Bot. 50: 737–45, 1982.

Björkman T. Perception of gravity by plants. Adv. Bot. Res. 15: 1–41, 1988.

Björkman T. Perception of gravity by plants. Adv. Space Res. 12 (1): 195–201, 1992.

Björkman T & Cleland RE. The role of extracellular free-calcium gradients in gravitropic signalling in maize roots. Planta 185: 379–84, 1991.

Björkman T & Leopold AC. Effect of inhibitors of auxin transport and of calmodulin on a gravisensing-dependent current in maize roots. Plant Physiol. 84: 847–50, 1987.

Black SD & Gerhart JC. Experimental control on the site of embryonic axis formation in *Xenopus laevis* eggs centrifuged before cleavage. Dev. Biol. 108: 310–24, 1985.

Black S, Larkin K & Souza K. Amphibian development at microgravity. COSPAR 94 : 287, 1994.

Block I & Briegleb W. Potential sites for the perception of gravity in the acellular slime mold *Physarum polycephalum*. Adv. Space Res. 9 (11): 75–8, 1989.

Block I, Briegleb W & Wohlfarth-Botterman KE. Gravisensitivity of the acellular slime mold *Physarum polycephalum* demonstrated on the fast-rotating clinostat. Eur. J. Cell Biol. 41: 44–50, 1986.

Block I, Briegleb W, Sobick V & Wohlfarth-Botterman KE. Contraction behaviour and protoplasmic streaming in the slime mold *Physarum polycephalum* (Physarum kinetics). In: *Biorack on Spacelab D-1*. Ed. by Longdon N & David V. ESTEC, Noordwijk, ESA Publications Division, pp. 55–67, 1988.

Block I, Wolke A, Briegleb W, Wohlfarth-Botterman KE, Merbold U, Brinkmann E & Brillouet C. Graviresponse of *Physarum* – investigations in actual weightlessness. Cell Biol. Internat. Repts 16: 1097–102, 1992.

Boda K, Sabo V, Jurani M, Guryeva TS, Kocisova J, Kostal L, Laukova A & Dadasheva OA. Embryonic development and behaviour of japanese quail exposed to microgravity. Acta Vet. Brno 61: 99–107, 1992.

Booth FW. Terrestrial applications of bone and muscle research in microgravity. Adv. Space Res. 14 (8): 373–6, 1994.

Bouloc P & D'Ari R. *Escherichia coli* metabolism in space. J. Gen. Micro. 137: 2839–43, 1991.

Briegleb W. Some qualitative and quantitative aspects of the fast-rotating clinostat as a research tool. ASGSB Bulletin 5: 23–30, 1992.

Briegleb W, Block I & Sobick V. Steady compensations of gravity effects in *Physarum polycephalum*. Naturwiss. 73: 422–4, 1986.

Briegleb W, Schatz A & Neubert J. Das Pendant zum Zentrifugenmikroskop: Klinostaten mikroskop. Umschau Wiss. Technik. 76: 621–3, 1976.

Briggs WR. What remains of the Cholodny-Went theory? It's alive and well in maize. Plant, Cell Environ. 15: 763, 1992.

Brown AH & Chapman DK. Circumnutation observed without a significant gravitational force in spaceflight. Science 225: 230–2, 1984.

Brown AH, Chapman DK, Lewis RF & Venditti AL. Circumnutations of sunflower hypocotyls in satellite orbit. Plant Physiol. 94: 233–8, 1990.

Brown AH, Dahl AO & Chapman DK. Limitations of the use of the horizontal clinostat as a gravity compensator. Plant Physiol. 58: 127–30, 1976.

Brown CS & Piastuch WC. Starch metabolism in germinating soybean cotyledons is sensitive to clinorotation and centrifugation. Plant, Cell Environ. 17: 341–4, 1994.

Brown CS, Piastuch WC & Knott WM. Soybean cotyledon starch metabolism is sensitive to altered gravity conditions. Adv. Space Res. 14 (8): 107–10, 1994.

Brown RM Jr, Kudlicka K, Cousins SK & Nagy R. Gravity effects on cellulose assembly. Amer. J. Bot. 79: 1247–58, 1992.

Bruinsma J. What remains of the Cholodny-Went theory? A phantom in phototropism. Plant, Cell Environ. 15: 765–6, 1992.

Buchen B, Braun M, Hejnowicz Z & Sievers A. Statoliths pull on microfilaments. Experiments under microgravity. Protoplasma 172: 38–42, 1993.

Buchen B, Hejnowicz Z, Braun M & Sievers A. Cytoplasmic streaming in *Chara* rhizoids. Studies in a reduced gravitational field during parabolic flights of rockets. Protoplasma 165: 121–6, 1991.

Butler GM & Wood AE. Effects of environmental factors on basidiome development in the resupinate polypore *Phellinus contiguus*. Trans. Br. Mycol. Soc. 90: 75–83, 1988.

Carlile MJ. Taxes and tropisms: diversity, biological significance and evolution. In: *Primitive Sensory and Communication Systems*. Ed. by Carlile MJ. London, Academic Press, pp. 1–28, 1975.

Carmichael C & Taylor GR. Evaluation of crew skin flora under conditions of a full quarantine lunar-exploration mission. Br. J. Dermatol. 97: 187–96, 1977.

Carrington CMS & Firn RD. Polysaccharide synthesis and turnover in the cell walls of growing and non-growing cells of gravistimulated sunflower hypocotyls. J. Plant Physiol. 118: 49–60, 1985.

Caspar T & Pickard BG. Gravitropism in a starchless mutant of *Arabidopsis*: implications for the starch-statolith theory of gravity sensing. Planta 177: 185–97, 1989.

Chandra S, Chabot JF, Morrison GH & Leopold AC. Localization of calcium in amyloplasts of root cap cells using ion microscopy. Science 216: 1221–3, 1982.

Chapes SK, Morrison DR, Guikema JA, Lewis ML & Spooner BS. Cytokine secretion by immune cells in space. J. Leukocyte Biol. 51: 104–10, 1992.

Chapes SK, Morrison DR, Guikema JA, Lewis ML & Spooner BS. Production and action of cytokines in space. Adv. Space Res. 14 (8): 5–9, 1994.

Chapman DK, Johnsson A, Karlsson C, Brown A & Heathcote DG. Gravitropically-stimulated seedlings show autotropism in weightlessness. Physiol. Plant. 90: 157–62, 1994.

Ciferri O, Tiboni O, Di Pasquale G, Orlandoni AM & Marchesi ML. Effects of microgravity on genetic recombination in *Escherichia coli*. Naturwiss. 73: 418–21, 1986.

Ciferri O, Tiboni O, Orlandoni AM & Marchesi ML. The effects of microgravity on genetic recombination in *Escherichia coli*. In: *Biorack on Spacelab D-1*. Ed. by Longdon N, David V. ESTEC, Noordwijk, ESA Publications Division, pp. 29–35, 1988.

Claassen DE & Spooner BS. Effects of microgravity on liposome-reconstituted cardiac gap junction channelling activity. Biochem. Biophys. Res. Commun. 161: 358–63, 1989.

Cogoli A. The activation of T lymphocytes in space - an overview. Biol. Sci. in Space 7: 1–7, 1993a.

Cogoli A. The effect of hypogravity and hypergravity on cells of the immune system. J. Leukocyte Biol. 54: 259–68, 1993b.

Cogoli A. Space flight and the immune system. Vaccine 11: 496–503, 1993c.

Cogoli A, Bechler B, Cogoli-Greuter M, Criswell SB, Joller H, Joller P, Hunzinger E & Müller O. Mitogenic signal transduction in T lymphocytes in microgravity. J. Leukocyte Biol. 53: 569–75, 1993.

Cogoli A, Bechler B & Lorenzi G. Response of cells to microgravity. In: *Fundamentals of Space Biology*. Ed. by Asashima M, Malacinski GM. Tokyo & Berlin, Japan Sci. Soc. Press, Springer, pp. 97–111, 1990.

Cogoli A, Bechler B, Müller O & Hunzinger E. Effect of microgravity on lymphocyte activation. In: *Biorack on Spacelab D-1*. Ed. by Longdon N, David V. ESTEC, Noordwijk, ESA Publications Division, pp. 89–100, 1988.

Cogoli A & Gmünder FK. Gravity effects on single cells: techniques, findings and theory. Adv. Space Biol. Med. 1: 183–248, 1991.

Cogoli A, Iversen TH, Johnsson A, Mesland D & Oser H. Cell Biology. In: *Life Sciences Research in Space*. Ed. by Oser H, Battrick B. Paris, France, European Space Agency, pp. 49–64, 1989.

Cogoli A & Tschopp A. Gravity and living organisms *in vitro*. Trends Pharmacol. Sci. 3: 403–7, 1982.

Cogoli A & Tschopp A. Lymphocyte reactivity during spaceflight. Immunol. Today 6: 1–4, 1985.

Cogoli A, Tschopp A & Fuchs-Bislin P. Cell sensitivity to gravity. Science 225: 228–30, 1984.

Cogoli A, Valluchi-Morf M, Böhringer HR, Vanni MR & Müller M. Effect of gravity on lymphocyte proliferation. Life Sci. Space Res. 17: 219–24, 1979.

Cogoli A, Valluchi-Morf M, Müller M & Briegleb W. The effect of hypogravity on human lymphocyte activation. Aviat., Space & Environ. Med. 51: 29–34, 1980.

Cogoli M. The fast rotating clinostat: a history of its use in gravitational biology and a comparison of ground-based and flight experiment results. ASGSB Bulletin 5: 59–67, 1992.

Cogoli M, Bechler B, Cogoli A, Arena N, Barni N, Pippia P, Sechi G, Valora N & Monti R. Lymphocytes on sounding rockets. Adv. Space Res. 12 (1): 141–4, 1990.

Cogoli-Greuter M, Pippia P, Sciola L & Cogoli A. Lymphocytes on sounding rocket flights. J. Gravitat. Physiol. 1: 90–1, 1994.

Collings DA, White RG & Overall RL. Ionic current changes associated with the gravity-induced bending response in roots of *Zea mays* L. Plant Physiol. 100: 1417–26, 1992.

Cowles JR, Scheld HW, Lemay R & Peterson C. Growth and lignification in seedlings exposed to eight days of microgravity. Ann. Bot. 54: 33–48, 1984.

Croute F, Gaubin Y, Prevost MC, Beaupain R, Pianezzi B & Soleilvahoup JP. Effects of hypergravity on lung carcinoma cells maintained in continuous organotypic culture. Aviat. Space Environ. Med. 61: 1002–6, 1990.

Darbelley N. Changes in mitotic activity and in nuclear DNA content and synthesis during the gravistimulation of lentil roots. Physiol. Veg. 24: 639–46, 1986.

Darbelley N. Effects of gravistimulation and microgravity on proliferation and cell differentiation in primary roots. Bull. Soc. Bot. Fr. Lett. Bot. 135: 229–50, 1988.

Darbelley N, Driss-Ecole D & Perbal G. Elongation and mitotic activity of cortical cells in lentil roots grown in microgravity. Plant Physiol. Biochem. (Paris) 27: 341–8, 1989.

Dauwalder M, Roux SJ & Rabenberg LK. Cellular and subcellular localization of calcium in gravistimulated corn roots. Protoplasma 129: 137–48, 1985.

Davidson JM, Aquino AM, Woodward SC & Wilfinger WW. Sustained microgravity reduces intrinsic wound healing and growth factor responses in the rat. Wound Repair Regen. 2: 75, 1994.

Davies AH, Preston TM & King CA. Lack of gravity effect on the speed of ameboid locomotion in *Naegleria gruberi*. Experientia 37: 708–9, 1981.

Daye S, Biro RL & Roux SJ. Inhibition of gravitropism in oat coleoptiles by the calcium chelator, ethyleneglycol*bis*(β-aminoethyl ether)-N,N′-tetraacetic acid. Physiol. Plant. 61: 449–54, 1984.

Dedolph RR & Dipert MH. The physical basis of gravity stimulus nullification by clinostat rotation. Plant Physiol. 47: 756-64, 1971.

De Groot RP, Rijken PJ, Den Hertog J, Boonstra J, Verkleij AJ, De Laat SW & Kruijer W. Microgravity decreases c-fos induction and serum response element activity. J. Cell Sci. 97: 33–8, 1990.

De Groot RP, Rijken PJ, Den Hertog J, Boonstra J, Verkleij AJ, De Laat SW & Kruijer W. Nuclear responses to protein kinase C signal transduction are sensitive to gravity changes. Exp. Cell Res. 197: 87–90, 1991.

Demets R. Biobox experiments indicate: *in vitro* bone formation is suppressed under microgravity conditions. Microgravity News from ESA 6: 20–3, 1993.

Dennison DS. Tropic response of *Phycomyces* sporangiophores to gravitational and centrifugal stimuli. J. Gen. Physiol. 45: 23–38, 1961.

Dennison DS & Roth CC. *Phycomyces* sporangiophores: fungal stretch receptors. Science 156: 1386–8, 1967.

Dennison DS & Shropshire W Jr. The gravireceptor of *Phycomyces blakesleeanus*: its development following gravity exposure. J. Gen. Physiol. 84: 845–60, 1984.

De Serres FJ. Effects of radiation during spaceflight on microorganisms and plants on the Biosatellite 2 and Gemini 11 missions. Life Sci. Space Res. 7: 62–6, 1969.

De Serres FJ, Miller IR, Smith DB, Kondo S & Bender MA. The Gemini-11 S-4 spaceflight radiation interaction experiment II. Analysis of survival levels and forward-mutation frequencies in *Neurospora crassa*. Radiat. Res. 39: 435–44, 1969.

Dintenfass L. Effect of zero gravity on blood cells and viscosity. Lancet 335: 239–40, 1990.

Dintenfass L, Osman P, Maguire B & Jedrzejczyk H. Experiment on aggregation of red cells under microgravity on STS 51-C. Adv. Space Res. 6 (5): 81–4, 1986.

Dudchenko LG, Nemchinov AV & Zarubina AP. [Higher basidial fungi under conditions of space flight]. In: *[Production of Higher Edible Fungi in the USSR]*. Kiev, Naukova Diemka, pp. 59–63, 1978.

Duke JC. Suppression of morphogenesis in embryonic mouse limbs exposed *in vitro* to excess gravity. Teratology 27: 427–36, 1983.

Duke PJ, Durnova G & Montufar-Solis D. Histomorphometric and electron microscopic analyses of tibial epiphyseal plates from Cosmos 1887 rats. FASEB (Fed. Am. Soc. Exp. Biol.) J. 4: 41–6, 1990.

Dutcher FR, Hess EL & Halstead TW. Progress in plant research in space. Adv. Space Res. 14 (8): 159–71, 1994.

Economos AC, Miquel J, Ballard RC, Blunden M, Lindseth KA, Fleming J, Philpott DE & Oyama J. Effects of simulated increased gravity on the rate of aging of rats: implications for the rate of living theory of aging. Arch. Gerontol. Geriatr. 1: 349–63, 1982.

Ekberg DR, Silver EC, Bushay JL & Daniels EW. Nuclear and cellular division in *Pelomyxa carolinensis* during weightlessness. In: *The Experiments of Biosatellite II.* Ed. by Saunders JF. Washington, DC, NASA, pp. 273–90, 1971.

Evans ML. Gravitropism: interaction of sensitivity modulation and effector redistribution. Plant Physiol. 95: 1–5, 1991.

Evans ML. What remains of the Choldny–Went theory? Plant, Cell Environ. 15: 767–8, 1992.

Evans ML, Young LM & Hasenstein KH. The role of calcium in the regulation of hormone transport in gravistimulated roots. Adv. Space Res. 12 (1): 211–8, 1992.

Firn R. What remains of the Cholodny–Went theory? Which one? Plant, Cell Environ. 15: 769–70, 1992.

Fleming SD, Edelman LS & Chapes SK. Effects of corticosterone and microgravity on inflammatory cell production of superoxide. J. Leukocyte Biol. 50: 69–76, 1991.

Fondren WM & Moore R. Collection of gravitropic effectors from mucilage of electrotropically-stimulated roots of *Zea mays* L. Ann. Bot. 59: 657–60, 1987.

Friedmann M & Poovaiah BW. Calcium and protein phosphorylation in the transduction of gravity signal in corn roots. Plant & Cell Physiol. 32: 299–302, 1991.

Gadoury DM & Machardy WE. Negative geotropism in *Venturia inaequalis*. Phytopathology 75: 856–9, 1985.

Galland P & Russo VEA. Tropic reversal in *Phycomyces*. Planta 163: 401–4, 1985.

Gamow RI & Böttger B. Avoidance and rheotropic response in *Phycomyces* sporangiophores: evidence for an 'avoidance gas' mechanism. J. Gen. Physiol. 79: 835–48, 1982.

Gasset G, Tixador R, Eche B, Lapchine L, Moatti N, Toorop P & Woldringh C. Growth and division of *Escherichia coli* under microgravity conditions. Res. Microbiol. 145: 111–20, 1994.

Gaubin Y, Croute F, Pianezzi B, Prevost MC & Soleilvahoup JP. Effects of hypergravity on adherent human cells. Microgravity Sci. Technol. 3: 246–50, 1991.

Gehring C, Williams D, Cody S & Parish R. Phototropism and geotropism in maize coleoptiles are spatially correlated with increases in cytosolic free calcium. Nature 345: 528–30, 1990.

Genty C, Palle S & Alexandre C. First osteoblast culture in space: cellular responses to unloading. J. Bone Miner. Res. 8 (Supplement 1): S367, 1993.

Gibeaut DM, Karuppiah N, Chang SR, Brock TG, Vadlamudi B, Kim D, Ghosheh NS, Rayle DL, Carpita NC & Kaufman PB. Cell wall and enzyme changes during the graviresponse of the leaf-sheath pulvinus of oat (*Avena sativa*). Plant Physiol. 94: 411–6, 1990.

Gmünder FK, Baisch F, Bechler B, Cogoli A, Cogoli M, Joller PW, Maass H, Müller J & Ziegler WH. Effect of head down tilt bedrest (10 days) on lymphocyte reactivity. Acta Physiol. Scand. 144:131–41, 1992.

Gmünder FK & Cogoli A. Cultivation of single cells in space. Appl. Micrograv. Technol. 1: 115–22, 1988.

Gmünder FK & Cogoli A. Effect of space flight on lymphocyte function and immunity. In *Handbook of Physiology*, Sec. 4, *Adaptation to the Environment* (Part 3, *The Gravitational Environment*, Section I: *Microgravity*). Ed. by Fregly MJ & Blatteis CM. Oxford, U.K., OxfordUniversity Press, Chapter 35, in press, 1995.

Gmünder FK, Kiess M, Sonnefeld G, Lee J & Cogoli A. A ground-based model to study the effects of weightlessness on lymphocytes. Biol. Cell 70: 33–8, 1990.

Gmünder FK, Nordau CB, Tschopp A, Huber B & Cogoli A. Dynamic cell culture system: a new cultivation instrument for biological experiments in space. J. Biotechnol. 7: 217–28, 1988.

Gonzalez-Jurado J, Vernos I, Calleja M & Carratala M. Biorack experiment FLY 15: Microgravity and other effects of short spaceflights on *Drosophila melanogaster* embryogenesis and life span. In: *Proceedings of the Norderny Symposium on the Scientific Results of the German Spacelab Mission D-1*. Ed. by Sahm PR, Jansen R & Keller MH. Köln, Germany, DFVLR, pp. 438–45, 1986.

Gorovoi LF, Kasatkina TB & Klyushkina NS. [Role of gravitation in the development of carpophores in hymenomycetes]. Mikol. Fitopatol. 21: 301–7 [in Russian], 1987.

Gorovoi LF, Kasatkina TB & Laurinavichius RS. *[Morphogenesis of Mushrooms in Changed Gravitation Conditions]*. Kiev, N. G. Kholodny Institute of Botany, Academy of Sciences of the Ukrainian SSR, 1989a.

Gorovoi LF, Kasatkina TB & Laurinavichius RS. [Significance of gravity in the development of *Polyporus brumalis* Pers. Fr.]. Dokl. Akad. Nauk Ukr. SSR, Ser. B: Geol., Khim. Biol. Nauki 10: 64–7 [in Russian], 1989b.

Goswami KKA & Audus LJ. Distribution of calcium, potassium and phosphorus in *Helianthus annuus* hypocotyls and *Zea mays* coleoptiles in relation to tropic stimuli and curvatures. Ann. Bot. 40: 49–64, 1976.

Greening JP, Holden J & Moore D. Distribution of mechanical stress is not involved in regulating stipe gravitropism in *Coprinus cinereus*. Mycol. Res. 97: 1001–4, 1993.

Gruener R & Hoeger G. Vector-free gravity disrupts synapse formation in cell culture. American Journal of Physiology 258: 489–94, 1990.

Gruener R & Hoeger G. Vector-averaged gravity alters myocyte and neuron properties in cell culture. Aviat., Space & Environ. Med. 62: 1159–65, 1991.

Gu W-L, Peng H-L, Yang Y-L & Jiao R-S. Effects of microgravity and cosmic radiation on some microorganisms. Acta Phytophysiol. Sin. 15: 397–402, 1989.

Gyure DC, Bottger B & Gamow RI. *Phycomyces*: detailed analysis of the anemogeotropic response. J. Gen. Physiol. 84: 727–38, 1984.

Häder DP & Vogel K. Graviorientation in photosynthetic flagellates. In: *Proceedings of the 4th European Symposium on Life Sciences Research in Space. ESA SP-307*. Ed. by David V. Paris, France, European Space Agency, pp. 521–6, 1990.

Häder DP, Liu S-M & Kreuzberg K. Orientation of the photosynthetic flagellate *Peridinium gatunense* in hypergravity. Curr. Microbiol. 22: 165–72, 1991a.

Häder DP, Reinecke E, Vogel K & Kreuzberg K. Responses of the photosynthetic flagellate *Euglena gracilis* to hypergravity. Eur. Biophys. J. 20: 101–8, 1991b.

Häder DP, Vogel K & Schäfer J. Responses of the photosynthetic flagellate *Euglena gracilis* to microgravity. Microgravity Sci. Technol. 3: 110–6, 1990.

Halstead TW & Dutcher FR. Plants in space. Ann. Rev. Plant Physiol. 38: 317–45, 1987.

Halstead TW, Todd P & Powers JV. Clinostats and centrifuges. ASGSB Bulletin 5: 1–82, 1992.

Hamacher H, Fitton B & Kingdon J. The environment of Earth-orbiting systems. In: *Fluid Sciences and Materials Science in Space. A European Perspective*. Ed. by Walter HU. Berlin, Springer-Verlag, pp. 1–50, 1987.

Hampp R, Naton B, Hoffmann E & Mehrle W. Suspensions of plant cells in microgravity. Microgravity Sci. Technol. 3: 168–72, 1990.

Harrison MA & Pickard BG. Evaluation of ethylene as a mediator of gravitropism by tomato hypocotyls. Plant Physiol. 80: 592–5, 1986.

Harrison MA & Pickard BG. Auxin asymmetry during gravitropism by tomato hypocotyls. Plant Physiol. 89: 652–7, 1989.

Hatton JP & Moore D. Kinetics of stem gravitropism in *Coprinus cinereus*: determination of presentation time and 'dosage-response' relationships using clinostats. FEMS Micro. Lett. 100: 82–6, 1992.

Hatton JP & Moore D. Kinetics of stipe gravitropism in the mushroom fungus *Coprinus cinereus* under the conditions of microgravity simulation provided by clinostat treatment. Microgravity Q. 4: 101–6, 1994.

Hayes DK, Morgan NO, Webb RE & Bell RA. Space travel shortens diapause in gypsy moth eggs. Chronobiol. Int. 8: 75–83, 1991.

Heathcote DG. The role of nodal and internodal responses in gravitropism and autotropism in *Galium aparine* L. Plant, Cell Environ. 10: 701–3, 1987.

Heathcote D & Bircher B. Enhancement of phototropic response to a range of light doses in *Triticum aestivum* coleoptiles in clinostat-simulated microgravity. Planta 170: 249–56, 1987.

Hejnowicz Z & Sievers A. Regulation of the position of statoliths in *Chara* rhizoids. Protoplasma 108: 117–37, 1981.

Hejnowicz Z & Wohlfarth-Botterman KE. Propagated waves induced by gradients of physiological factors within plasmodia of *Physarum polycephalum*. Planta 150: 144–52, 1980.

Hemmersbach-Krause R & Briegleb W. Graviperception and graviresponses of *Paramecium*. In: *Proceedings of the 5th European Symposium on Life Science Research in Space*. Ed. by Oser H & Guyenne TD. Noordwijk, The Netherlands, ESA Publications Division, ESTEC, pp. 37–41, 1994.

Hemmersbach-Krause R, Briegleb W, Häder DP & Plattner H. Gravity effects on *Paramecium* cells an analysis of a possible sensory function of trichocysts and of simulated weightlessness on trichocyst exocytosis. Eur. J. Protistol. 27: 85–92, 1991.

Hemmersbach-Krause R, Briegleb W & Häder DP. Swimming behavior of *Paramecium* – first results with the low-speed centrifuge microscope (NIZEMI). Adv. Space Res. 12 (1): 113–6, 1992.

Hemmersbach-Krause R, Briegleb W, Häder DP, Vogel K, Grothe D & Meyer I. Orientation of *Paramecium* under the conditions of weightlessness. J. Euk. Microbiol. 40: 439–46, 1993.

Hemmersbach-Krause R, Häder DP, Köhler M, Peridigao J & Briegleb W. Cellular functions of *Paramecium* under different gravity conditions. In: *Proceedings of the 4th European Symposium on Life Sciences Research in Space*. Ed. by David V. Paris, France, European Space Agency, pp. 285–90, 1990.

Henry RL, Green PD, Wong PP & Guikema JA. Binding of isolated plant lectin by rhizobia during episodes of reduced gravity obtained by parabolic flight. Plant Physiol. 92: 262–4, 1990.

Hensel W. Microtubules in statocytes from roots of cress, *Lepidium sativum*. Protoplasma 119: 121–34, 1984.

Hensel W. Cytochalasin B affects the structural polarity of statocytes from cress roots (*Lepidium sativum* L.). Protoplasma 129: 178–87, 1985.

Hensel W. Tissue slices from living root caps as a model system in which to study cytodifferentiation of polar cells. Planta 177: 296–303, 1989.

Hensel W & Sievers A. Effects of prolonged omnilateral gravistimulation on the ultrastructure of statocytes and on the graviresponse of roots. Planta 150: 338–46, 1980.

Hensel W & Sievers A. Induction of gravity-dependent plasmatic responses in root statocytes by short time contact between amyloplasts and the distal endoplasmic reticulum complex. Planta 153: 303–7, 1981.

Hestnes A. Distribution of radioactivity from exogenously supplied carbon-14 labelled IAA and tritium labelled gibberellin a-1 in geotropically stimulated *Picea abies* roots. Ann. Bot. 44: 567–74, 1979.

Hock B, Kern VD, Monzer J & Haindl E. Influence of gravity on fruiting body development of fungi (Higher basidiomycetes). In: *Research Program of the German Spacelab Mission D-2*. Ed. by Sahm PR, Keller MH & Schiewe B. Köln, Germany, DLR, pp. 174–7, 1993.

Horneck G. Radiobiological experiments in space: a review. Nucl. Tracks Radiat. Meas. 20: 185–205, 1992.

Horneck G, Bucker H, Reitz G & Requardt H. Microorganisms in the space environment. J. Photochem. Photobiol. 11: 69–80, 1991.

Hotchin J, Lorenz PR & Hemenway CL. Survival of micro-organisms in space. Nature 206: 442–5, 1965.

Huber AG & Grabe DF. Endosperm morphogenesis in wheat: transfer of nutrients from the antipodals to the lower endosperm. Crop Sci. 27: 1248–52, 1987.

Hughes-Fulford M. Altered cell function in microgravity. Exp. Gerontol. 26: 247–56, 1991.

Hymer WC, Grindeland R, Krasnov I, Victorov I, Motter K, Mukherjee P, Shellenberger K & Vasques M. Effects of spaceflight on rat pituitary cell function. J. Appl. Physiol. 73: 151S–7S, 1992.

Inoue H, Hiasa K, Samma Y, Nakamura O, Sakuda M, Iwamoto M, Suzuki F & Kato Y. Stimulation of proteoglycan and DNA syntheses in chondrocytes by centrifugation. J. Dent. Res. 69: 1560–3, 1990.

Iversen TH, Rasmussen O, Gmünder FK, Baggerud C, Kordyum EL, Lozovaya VV & Tairbekov MG. The effect of microgravity on the development of plant protoplasts flown on Biokosmos 9. Adv. Space Res. 12 (1): 123–31, 1992.

Iwabuchi A, Yano M & Shimizu H. Development of extracellular electric pattern around *Lepidium* roots: its possible role in root growth and gravitropism. Protoplasma 148: 94–100, 1989.

Jackson MB & Barlow PW. Root geotropism and the role of growth regulators from the cap: a re-examination. Plant, Cell Environ. 4: 107–23, 1981.

Jenkins GI, Courtice GRM & Cove DJ. Gravitropic responses of wild-type and mutant strains of the moss *Physcomitrella patens*. Plant, Cell Environ. 9: 637–44, 1986.

Jerger DE & Volz PA. DNA analysis of fungal wild types and spaceflight phenotypes. Phytologia 35: 265–70, 1977.

Jessup JM, Brown K, Ishii S, Ford R, Goodwin TJ & Spaulding G. Simulated microgravity does not alter epithelial cell adhesion to matrix and other molecules. Adv. Space Res. 14 (8): 71–6, 1994.

Johnson DL & Gamow RI. The avoidance response in *Phycomyces*. J. Gen. Physiol. 57: 41–9, 1971.

Kasahara H, Takeuchi Y, Yamashita M & Yamada M. Effects of simulated microgravity on the germination and elongation of protonemata of *Adiantum capillus-veneris*. Physiol. Plant. 90: 205–9, 1994.

Kawasaki Y, Mizutani H & Kiriu T. Effect of gravity on the morphogenesis of organisms. The case of a cellular slime mold, *Dictyostelium discoideum*. In: *Biological Sciences in Space*. Ed. by Watanabe S, Mitarai G & Mori S. Tokyo, Myu Research, pp. 285–90, 1987.

Kern VD & Hock B. Fungi in space – literature survey on fungi used for space research. Microgravity Sci. Technol. 6: 194–206, 1993.

Kern VD & Hock B. Gravitropism of fungi – experiments in space. In: *Proceedings of the 5th European Symposium on Life Sciences Research in Space*. Ed. by Oser H & Guyenne TD. Noordwijk, The Netherlands, ESA Publications Division, ESTEC, pp. 49–60, 1994.

Kessler JO. Individual and collective fluid dynamics of swimming cells. J. Fluid Mech. 173: 191–205, 1986.

Kessler JO. Path and pattern – the mutual dynamics of swimming cells and their environment. Comments Theor. Biol. 1: 85–108, 1989.

Kessler JO. The internal dynamics of slowly rotating biological systems. ASGSB Bulletin 5: 11–21, 1992a.

Kessler JO. Theory and experimental results on gravitational effects on monocellular algae. Adv. Space Res. 12 (1): 33–42, 1992b.

Kher K, Greening JP, Hatton JP, Novak-Frazer L & Moore D. Kinetics and mechanics of stem gravitropism in *Coprinus cinereus*. Mycol. Res. 96: 817–24, 1992.

Kiss JZ & Sack FD. Severely reduced gravitropism in dark-grown hypocotyls of a starch-deficient mutant of *Nicotiana sylvestris*. Plant Physiol. 94: 1867–73, 1990.

Kiss JZ, Hertel R & Sack FD. Amyloplasts are necessary for full gravitropic sensitivity in roots of *Arabidopsis thaliana*. Planta 177: 198–206, 1989.

Klement BJ & Spooner BS. Pre-metatarsal skeletal development in tissue-culture at unit gravity and microgravity. J. Exp. Zool. 269: 230–41, 1994.

Klimchuk DA. [*In vitro* growth characteristics of *Haplopappus gracilis* cells under clinostat conditions]. Ukr. Bot. Zh. 41: 63-66 [in Ukrainian], 1984.

Klimchuk DA, Kordyum EL, Danevich LA, Tarnavskaya EB, Tairbekov MG, Iversen TH, Baggerud C & Rasmussen O. Structural and functional organisation of regenerated plant protoplasts exposed to microgravity on Biokosmos 9. Adv. Space Res. 12 (1): 133–40, 1992.

Knight TA. On the direction of the radical and germen during vegetation of seeds. Phil. Trans. Roy. Soc., ser. B 96: 99–108, 1806.

Kondepudi DK & Prigogine I. Sensitivity on nonequilibrium chemical systems to gravitational field. Adv. Space Res. 3 (5): 171–6, 1983.

Kondo S. Possibility and impossibility for genetic effects of weightlessness. Japn. J. Genet. 43: 272–8, 1968.

Kordyum EL. Effects of altered gravity on plant cell processes – results of recent space and clinostatic experiments. Adv. Space Res. 14 (8): 77–85, 1994.

Kordyum EL, Belyavskaya NA, Nedukha EM, Palladina TA & Tarasenko VA. The role of calcium ions in cytological effects of hypogravity. Adv. Space Res. 4 (12): 23–6, 1984.

Kordyum, V.A., Polidova, L.V., Mashinsky, A.L., Manko, V.G., Nechitailo, G.N., Susanov, D.V. & Konshin, V.G. Peculiarities of microorganisms' growth in extreme conditions, Abstracts COSPAR 17th Plenary Meeting, Sao Paulo, Brasil, pp. 254–255, 1974.

Kulesh DA, Anderson LH, Wilson B, Otis EJ, Elgin DM, Barker MJ, Mehm WJ & Kearney GP. Space-shuttle flight (STS-45) of L8 myoblast cells results in the isolation of a nonfusing cell-line variant. J. Cell. Biochem. 55: 530–44, 1994.

Kumei Y, Nakajima T, Sato A, Kamata N & Enomoto S. Reduction of G-1 phase duration and enhancement of *c-myc* gene expression in HeLa cells at hypergravity. J. Cell Sci. 93: 221–6, 1989.

Kumei Y, Sato A, Ozawa K, Nakajima T & Yamashita M. Effects of hypergravity on cultured mammalian cells. In: *Biological Sciences in Space*. Ed. by Watanabe S, Mitarai G & Mori S. Tokyo, Myu Research, pp. 291–305, 1987.

Kumei Y, Whitson PA, Sato A & Cintron NM. Hypergravity signal transduction in HeLa cells concomitant phosphorylation of proteins immunoprecipitated with anti-microtubule-associated protein antibodies. Exp. Cell Res. 192: 492–6, 1991.

Lafay J-F, Matricon J & Bodere C. The avoidance response of *Phycomyces*: distance dependence of the response. Physiol. Veg. 13: 259–63, 1975.

Langbein D. Physical parameters affecting living cells in space. Adv. Space Res. 6: 5–14, 1986.

Lange RD, Gibson LA, Driscoll TB, Alleban Z & Ichiki AT. Effects of microgravity and increased gravity on bone-marrow of rats. Aviat., Space & Environ. Med. 65: 730–73, 1994.

Lapchine L, Moatti N, Richoilley G, Templier J, Gasset G & Tixador R. Antibacterial activity of antibiotics in space conditions. In: *Proceedings of the Norderny Symposium on the Scientific Results of the German Spacelab Mission D-1*. Ed. by Sahm PR, Jansen R & Keller MH. Köln, Germany, DLR, pp. 395–403, 1986.

Lapchine L, Moatti N, Richoilley G, Templier J, Gasset G & Tixador R. The Antibio experiment. In: *Biorack on Spacelab D-1*. Ed. by Longdon N & David V. ESTEC, Noordwijk, ESA Publications Division, pp. 45–51, 1988.

Larsen P. Gravitropism. An introduction. In: *Handbuch der Pflanzenphysiologie*. Ed. by Ruhland W. Springer-Verlag: Berlin, pp. 34–73, 1962.

Laurinavichius RS, Kentstaviciene P, Rupainiene O & Nechitailo GS. *In vitro* plant cell growth in microgravity and on clinostat. Adv. Space Res. 14 (8): 87–96, 1994a.

Laurinavichius RS, Stockus A, Buchen B & Sievers A. Structure of cress root statocytes in microgravity (BION-10 mission). COSPAR 94 : 279, 1994b.

Lawton JR, Juniper BE & Hawes CR. Ultrastructural changes in statocytes in grass nodes after short periods of gravistimulation: do statoliths fall through the vacuole? Evidence from thick section high voltage electron microscopy. J. Exp. Bot. 37: 693–704, 1986.

Le Bourg E & Lints FA. Hypergravity and aging in *Drosophila melanogaster.* 2. Longevity. Gerontology 35: 244–52, 1989a.

Le Bourg E & Lints FA. Hypergravity and aging in *Drosophila melanogaster.* 3. Viability. Gerontology 35: 253–9, 1989b.

Le Bourg E & Lints FA. Hypergravity and aging in *Drosophila melanogaster.* 4. Climbing activity. Gerontology 38: 59–64, 1992a.

Le Bourg E & Lints FA. Hypergravity and aging in *Drosophila melanogaster.* 5. Patterns of movement. Gerontology 38: 65–70, 1992b.

Le Bourg E & Lints FA. Hypergravity and aging in *Drosophila melanogaster.* 6. Spontaneous locomotor activity. Gerontology 38: 71–9, 1992c.

Le Bourg E, Lints FA, Fresquet N & Bullens P. Hypergravity aging and longevity in *Drosophila melanogaster.* Comp. Biochem. Physiol., A Comp. Physiol. 105: 389–96, 1993.

Le Campion J & Very JM (1980): Preliminary study of the gravity increase on foraminiferal biomineralization. Crist., Deform., Dissolution Carbonates, Reun., 291–4. Univ. de Bordeaux III: Talence, Fr.

Lee JS & Evans ML. Polar transport of auxin across gravistimulated roots of maize and its enhancement by calcium. Plant Physiol. 77: 824–7, 1985a.

Lee JS & Evans ML. Polar transport of calcium-45 ion across the elongation zone of gravistimulated roots. Plant & Cell Physiol. 26: 1587–95, 1985b.

Lee JS, Mulkey TJ & Evans ML. Inhibition of polar calcium movement and gravitropism in roots treated with auxin-transport inhibitors. Planta 160: 536–43, 1984.

Legue V, Driss-Ecole D & Perbal G. Cell cycle and cell differentiation in lentil roots grown on a slowly rotating clinostat. Physiol. Plant. 84: 386–92, 1992.

Leopold AC & Wettlaufer SH. Springback in root gravitropism. Plant Physiol. 91: 1247–50, 1989.

Levine HG & Krikorian AD. Shoot growth in aseptically cultivated day lily and *Haplopappus* plantlets after a 5-day spaceflight. Physiol. Plant. 86: 349–59, 1992.

Li XG & Wang GZ. The effects of microgravity on the character of progeny of *Drosophila melanogaster.* Microgravity Sci. Technol. 5: 94–7, 1992.

Limouse M, Manie S, Konstantinova I, Ferrua B & Schaffar L. Inhibition of phorbol ester-induced cell activation in microgravity. Exp. Cell Res. 197: 82–6, 1991.

Lints FA & Le Bourg E. Hypergravity and aging in *Drosophila melanogaster.* 1. Fecundity. Gerontology 35: 235–43, 1989.

Lorenzi G, Bechler B, Cogoli M & Cogoli A. Gravitational effects on mammalian cells. The Physiologist 32: S144–7, 1988.

Lorenzi G, Fuchs-Bislin P & Cogoli A. Effects of hypergravity on 'whole blood' cultures of human lymphocytes. Aviat., Space & Environ. Med. 57: 1131–5, 1986.

Lorenzi G, Gmünder FK & Cogoli A. Cultivation of hamster kidney cells in a dynamic cell culture system in space (Spacelab IML-1 mission). Microgravity Sci. Technol. 6: 34–8, 1993.

Lorenzi G & Perbal G. Actin filaments responsible for the location of the nucleus in the lentil statocyte are sensitive to gravity. Biol. Cell 68: 259–64, 1990a.

Lorenzi G & Perbal G. Root growth and statocyte polarity in lentil seedling roots grown in microgravity or on a slowly rotating clinostat. Physiol. Plant. 78: 532–7, 1990b.

Lozova HI, Sydorenko PH, Holod MH, Dobysh KP, Klymchuk DO & Skorokhod TP. [*In vitro* effect of microgravitation on plastid apparatus of *Haplopappus gracilis* Nutt. A. Gray cells]. Ukr. Bot. Zh. 46: 77–82 [in Ukrainian], 1989.

Lu CR, Kim D & Kaufman PB. Changes in the ultrastructure of cell walls, cellulose synthesis, and glucan synthase activity from gravistimulated pulvini of oat (*Avena sativa*). Int. J. Plant Sci. 153: 164–70, 1992.

Lühring H, Behrens HM & Sievers A. Electrical events in growing *Lepidium* root tips seem to be correlated with gravitropic dynamics. In: *Molecular and Cellular Aspects of Calcium in Plant Development.* Ed. by Trewavas AJ. New York & London, Plenum Press, pp. 225–32, 1986.

Lyon CJ. Choice of rotation rate for the horizontal clinostat. Plant Physiol. 46: 355-8, 1970.

Machemer H, Braeucker R, Takahashi K & Murakami A. Short-term microgravity to isolate graviperception in cells. Microgravity Sci. Technol. 5: 119–23, 1992.

Machemer H, Braucker R, Murakami A & Yoshimura K. Graviperception in unicellular organisms a comparative behavioural study under short-term microgravity. Microgravity Sci. Technol. 5: 221–31, 1993.

Malacinski GM & Neff AW. The influence of gravity on the process of development of animal systems. Adv. Space Res. 4 (12): 315–23, 1984.

Malacinski GM, Neff AW, Alberts JR & Souza KA. Developmental biology in outer space. BioScience 39: 314–20, 1989.

Malouvier A, Holy X, Zrath E, Marie PJ, Caissard JC & Schmitt DA. Receptor-mediated endocytosis in osteoblastic cells under 'gravitational stress'. The Physiologist 35: S37–8, 1992.

Marcum H & Moore R. Influence of electrical fields and asymmetric application of mucilage on curvature of primary roots of *Zea mays*. Amer. J. Bot. 77: 446–52, 1990.

Marthy HJ. Microgravity and development of aquatic animals. In: *Proceedings of the 4th European Symposium on Life Sciences Research in Space*. Ed. by David PR. Paris, France, European Space Agency, pp. 255–6, 1990.

Marthy HJ, Schatt P & Santella L. Fertilization of sea urchin eggs in space and subsequent development under normal conditions. Adv. Space Res. 14 (8): 197–208, 1994.

Masuda Y, Kamisaka S, Yamamoto R, Hoson T & Nishinari K. Changes in the rheological properties of the cell-wall of plant seedlings under simulated microgravity conditions. Biorheology 31: 171–7, 1994.

Mattoni R, Ebersold W, Eiserling F, Keller E & Romig W. Induction of lysogenic bacteria in space environment. In: *The Experiments of Biosatellite II*. Ed. by Saunders JF. Washington, DC, NASA, pp. 309–24, 1971.

McClure BA & Guilfoyle T. Rapid redistribution of auxin-regulated RNA during gravitropism. Science 243: 91–3, 1989.

McLaren A. Developmental biology and microgravity. In: *Proceedings of the 4th European Symposium on Life Sciences Research in Space, Trieste, Italy, June 1990. ESA SP-307*. Ed. by David V. Paris, France, European Space Agency, pp. 239–42, 1990.

Mechanic GL, Arnaud SB, Boyde A, Bromage TG, Buckendahl P, Elliott JC, Katz EP & Durnova GN. Regional distribution of mineral and matrix in the femurs of rats flown on Cosmos 1887 biosatellite. FASEB (Fed. Am. Soc. Exp. Biol.) J. 4: 34–40, 1990.

Medvedev SS & Shtonda IA. On the role of calcium ions in gravitropic reaction. Biol Nauki (Mosc) : 94–7, 1989.

Megory E & Oyama J. Intrauterine fetal response to hypergravity by reduction of plasma prolactin levels in the rat. Am. J. Obstet. Gynecol. 153: 811–2, 1985.

Mehrle W, Hampp R, Naton B & Grothe D. Effects of microgravitation on electrofusion of plant cell protoplasts. Plant Physiol. 89: 1172–7, 1989.

Meleshko GI, Shepelev YY, Guryeva TS, Boda K & Sabo W. Embryonic development of birds in weightlessness. Kosm. Biol. Aviakosm. Med. 25: 37–9, 1991.

Mennigmann HD & Lange M. Growth and differentiation of *Bacillus subtilis* under microgravity. Naturwiss. 73: 415–7, 1986.

Mennigmann HD & Lange M. Growth and differentiation of *Bacillus subtilis* under microgravity. In: *Biorack on Spacelab D1. An Overview of the First Flight of Biorack – an ESA Facility for Life Science Research in Microgravity*. Ed. by Longdon N & David V. ESTEC, Noordwijk, ESA Publications Division, pp. 37–44, 1988.

Mergenhagen D. The circadian rhythm in *Chlamydomonas reinhardii* in a Zeitgeber-free environment. Naturwissenschaften. 73: 410–2, 1986.

Mergenhagen D & Mergenhagen E. The behaviour of the circadian clock of *Chlamydomonas reinhardii* in space. In: *Proceedings of the Norderny Symposium on the Scien-*

tific Results of the German Spacelab Mission D-1. Ed. by Sahm PR, Jansen R & Keller MH. Köln, Germany, DLR, pp. 384–95, 1986.

Merkys AJ & Laurinavichius RS. [Complete cycle of individual development of *Arabidopsis thaliana* (L.) Heynh. plants on board the 'Salyut-7' orbital station]. Dokl. Akad. Nauk SSSR 271: 509–12 [in Russian], 1983.

Merkys AJ, Laurinavichius RS, Rupainene OJ, Shvegzhdene DV & Yaroshius AV. Gravity as an obligatory factor in normal higher plant growth and development. Adv. Space Res. 1 (14): 109–16, 1981.

Merkys AJ, Laurinavichius RS & Shvegzhdene DV. Plant growth, development and embryogenesis during Salyut-7 flight. Adv. Space Res. 4 (10): 55–63, 1984.

Mesland D. Gravity effects on cells. In: *Proceedings of the 4th European Symposium on Life Sciences Research in Space. ESA SP-307.* Ed. by David V. Paris, France, European Space Agency, pp. 221–5, 1990.

Mesland D. Possible actions of gravity on single cells. Adv. Space Res. 12 (1): 15–25, 1992.

Migliaccio F & Galston AW. On the nature and origin of the calcium asymmetry arising during gravitropic response in etiolated pea epicotyls. Plant Physiol. 85: 542–7, 1987.

Migliaccio F & Rayle DL. Effect of asymmetric auxin application on *Helianthus* hypocotyl curvature. Plant Physiol. 91: 466–8, 1989.

Miwa M, Kozawa O, Tokuda H, Kawakubo A, Yoneda M, Oiso Y & Takatsuki K. Effects of hypergravity on proliferation and differentiation of osteoblast-like cells. Bone Miner. 14: 15–26, 1991.

Moatti N, Lapchine L, Gasset G, Richoilley G, Templier J & Tixador R. Preliminary results of the 'Antibio' experiment. Naturwiss. 73: 413–4, 1986.

Molitoris HP. Fungi in space-related research. Ukr. Bot. Zh. 47: 70–7, 1990.

Montgomery PO'B, Van Orden F & Rosenblum E. A relationship between growth and gravity in bacteria. Aerosp. Med. 34: 352–4, 1963.

Montgomery PO'B, Cook JE & Franz R. The effects of prolonged centrifugation on *Amoeba proteus*. Exp. Cell Res. 40: 140–2, 1965.

Montgomery PO'B, Cook JE, Reynolds RC, Paul JS, Hayflick L, Stock D, Schulz WW, Kimsey S, Thirolf RG, Rogers T & Campbell D. The response of single human cells to zero gravity. In Vitro 14: 165–73, 1978.

Monzer J. Actin filaments are involved in cellular graviperception of the basidiomycete *Flammulina velutipes*. Eur. J. Cell Biol. 66: in press, 1995.

Monzer J & Haindl E. Evidence for the involvement of the cytoskeleton and organelles in graviperception of the basidiomycete *Flammulina velutipes*. In: *Proceedings of the 5th European Symposium on Life Sciences Research in Space.* Ed. by Oser H & Guyenne TD. Noordwijk, The Netherlands, ESA Publications Division, ESTEC, pp. 89–93, 1994.

Monzer J, Haindl E, Kern VD & Dressel K. Gravitropism of the basidiomycete *Flammulina velutipes*. Morphological and physiological aspects of the gravi-response. Exp. Mycol. 18: 7–19, 1994.

Moore D. Perception and response to gravity in higher fungi – a critical appraisal. New Phytol. 117: 3–23, 1991.

Moore D. Which way is up? Gravity sensing in basidiomycetes. In: *Advances in Mushroom Biotechnology.* Ed. by Nair MC, Gokulapalan C & Das L. Jodhpur, India, Scientific Publishers, pp. 23–6, 1994.

Moore D, Greening JP, Hatton JP & Novak-Frazer L. Gravitational biology of mushrooms: a flow-chart approach to characterising processes and mechanisms. Microgravity Q. 4: 21–4, 1994a.

Moore D, Novak-Frazer L, Hatton JP, Greening JP, Bourne AM & Robson GD. Gravitropic reactions of the mushroom fruit body of *Coprinus cinereus*. In: *Proceedings of the 5th European Symposium on Life Sciences Research in Space.* Ed. by Oser H & Guyenne TD. Noordwijk, The Netherlands, ESA, ESTEC, pp. 133–8, 1994b.

Moore R. Movement of calcium across tips of primary and lateral roots of *Phaseolus vulgaris*. Amer. J. Bot. 72: 785–7, 1985.

Moore R. Cytochemical localization of calcium in cap cells of primary roots of *Zea mays* L. J. Exp. Bot. 37: 73–9, 1986a.

Moore R. A morphometric analysis of the redistribution of organelles in columella cells of horizontally-oriented roots of *Zea mays*. Ann. Bot. 57: 119–32, 1986b.

Moore R. Root gravitropism in a cultivar of *Zea mays* whose columella cells contain starch-deficient amyloplasts. Ann. Bot. 59: 661–6, 1987.

Moore R, Cameron IL, Hunter KE, Olmos D & Smith NKR. The locations and amounts of endogenous ions and elements in the cap and elongating zone of horizontally oriented roots of *Zea mays* L.: an electron-probe EDS study. Ann. Bot. 59: 667–77, 1987a.

Moore R, Cameron IL & Smith NKR. Movement of endogenous calcium in the elongating zone of graviresponding roots of *Zea mays*. Ann. Bot. 63: 589–93, 1989.

Moore R & Evans ML. How roots perceive and respond to gravity. Amer. J. Bot. 73: 574–87, 1986.

Moore R & Fondren WM. The possible involvement of root-cap mucilage in gravitropism and calcium movement across root tips of *Allium cepa* L. Ann. Bot. 58: 381–7, 1986.

Moore R & Fondren WM. A gradient of endogenous calcium forms in mucilage of graviresponding roots of *Zea mays*. Ann. Bot. 61: 113–6, 1988.

Moore R, Fondren WM & Marcum H. Characterization of root agravitropism induced by genetic, chemical, and developmental constraints. Amer. J. Bot. 74: 329–36, 1987b.

Moore R, McClelen CE, Fondren WM & Wang CL. The influence of microgravity on root-cap regeneration and the structure of columella cells in *Zea mays*. Amer. J. Bot. 74: 216–21, 1987c.

Moos PJ, Fattaey HK & Johnson TC. Cell-proliferation inhibition in reduced gravity. Exp. Cell Res. 213: 458–62, 1994.

Muller HJ. Approximation to a gravity-free situation for the human organism achievable at moderate expense. Science 128: 772, 1958.

Nace GW. Gravity and positional homeostasis of the cell. Adv. Space Res. 3: 159–68, 1983.

Nakajima T. Effects of hypergravity on migration proliferation and function of mouse osteoblastic cell line mc3t3-e1. J. Stomatol. Soc. Jpn 58: 529–44, 1991.

Napolitano LG. Marangoni convection in space microgravity experiments. Science 225: 197–8, 1984.

Nash PV, Bour BA & Mastro AM. Effect of hindlimb suspension simulation of microgravity on *in vitro* immunological responses. Exp. Cell Res. 195: 353–60, 1991.

Nash PV, Konstantinova IV, Fuchs BB, Rakhmilevich AL, Lesnyak AT & Mastro AM. Effect of spaceflight on lymphocyte proliferation and interleukin-2 production. J. Appl. Physiol. 73: 186S–90S, 1992.

Nash PV & Mastro AM. Variable lymphocyte responses in rats after space flight. Exp. Cell Res. 202: 125–31, 1992.

Nechitailo GS & Mashinsky AL. *Space Biology. Studies at Orbital Stations.* Moscow, Mir Publishers, 1993.

Nedukha EM. [Effect of clinostating on the ultrastructure of *Funaria hygrometrica* protonema. Ukr. Bot. Zh. 43: 20–3 [in Ukrainian], 1986.

Nedukha EM. [The electron cytochemical detection of calcium ATPase in the protonema of the moss *Funaria hygrometrica* Hedw. under hypogravitation]. Tsitol. Genet. 21: 3–6 [in Russian], 1987.

Nedukha EM. [Effects of clinostat treatment on the distribution of calcium ions in cells of *Funaria hygrometrica* (Hedw.) protonema]. Dokl. Akad. Nauk Ukr. SSR, Ser. B: Geol., Khim. Biol. Nauki 4: 71–3 [in Russian], 1989a.

Nedukha EM. Long clinostation influence on the localization of free and weakly bound calcium in cell walls of *Funaria hygrometrica* moss protonema cells. Adv. Space Res. 9 (11): 83–6, 1989b.

Nedukha EM. The role of cellulases in the mechanism of changes of cell walls of *Funaria hygrometrica* moss protonema at clinostating. Adv. Space Res. 12 (1): 99–102, 1992.

Nedukha EM & Trutneva IA. [Role of pectinases in the mechanism of changes in moss protonema cell walls under clinostating]. Dokl. Akad. Nauk Ukr. SSR, Ser. B: Geol., Khim. Biol. Nauki 7: 71–4 [in Russian], 1988.

Nedukha EM, Tupyk ND, Uvarova SA & Ovruts'ka II. [Effect of hypogravitation on the activity and multiple molecular dehydrogenase forms in *Funaria hygrometrica* protonema]. Ukr. Bot. Zh. 42: 52–55 [in Ukrainian], 1985.

Nedukha EM, Kordyum EL & Danevich LA. [A study of content of callose in moss protonema cells of *Funaria hygrometrica* under hypogravitation]. Tsitologiya 30: 212–7 [in Russian], 1988.

Neff AW, Malacinski GM & Chung HM. Microgravity simulation as a probe for understanding early *Xenopus* pattern specification. J. Embryol. Exp. Morphol. 89: 259–74, 1985.

Neff AW, Yokota H, Chung H-M, Wakahara M & Malacinski GM. Early amphibian anuran morphogenesis is sensitive to novel gravitational fields. Dev. Biol. 155: 270–4, 1993.

Neff AW, Yokata H & Malacinski GM. Gravitational manipulation of *Xenopus* embryo partitions morphogenetic information in animal quartets. COSPAR 94 : 289, 1994.

Nemec M, Horakova D & Kazdova M. The effect of hypergravitation on metabolic activity in bacterial cells of *Staphylococcus aureus*. Folia Fac. Sci. Nat. Univ. Purkynianae Brun. 24: 57–63, 1983.

Neubert J, Rahmann H, Briegleb W, Slenzka K, Schatz A & Bromeis B. Statex II on Spacelab mission D-2 – an overview of the joint project 'graviperception and neuronal plasticity' and preliminary pre-flight results. Microgravity Q. 1: 173–82, 1991.

Nick P, Bergfeld R, Schäfer E & Schopfer P. Unilateral reorientation of microtubules at the outer epidermal wall during photo- and gravitropic curvature of maize coleoptiles and sunflower hypocotyls. Planta 181: 162–8, 1990a.

Nick P, Sailer H & Schäfer E. On the relation between phototropically and gravitropically induced spatial memory in maize coleoptiles. Planta 181: 385–92, 1990b.

Nick P, Schäfer E, Hertel R & Furuya M. On the putative role of microtubules in gravitropism of maize coleoptiles. Plant & Cell Physiol. 32: 873–80, 1991.

Noll F. ***über heterogene Induktion***. Leipzig, Wilhelm Engelman, 1892.

Nose K & Shibanuma M. Induction of early response genes by hypergravity in cultured mouse osteoblastic cells (MC3T3-E1). Exp. Cell Res. 211: 168–70, 1994.

Novak Frazer L & Moore D. Antagonists and inhibitors of calcium accumulation do not impair gravity perception though they adversely affect the gravitropic responses of *Coprinus cinereus* stipes. Mycol. Res. 97: 1113–8, 1993.

Novak Frazer L & Moore D. Gravity perception and response in *Coprinus cinereus*: the role of calcium and cytoskeletal elements. Abstracts, 5th International Mycological Congress 1: 159, 1994.

Oganesyan SS, Gevorkyan RA, Zaminyan TS & Eloyan MA. [Effect of simulated gravity on the chick embryo myocardium]. Kosm. Biol. Aviakosm. Med. 14: 54–7 [in Russian], 1980.

Oputa CO & Mazelis M. Simulated hypogravity and proline incorporation into salt-extractable macromolecules from cell walls. Phytochemistry 16: 673–5, 1977.

Orlob GB & Lorenz PR. Survival of Tobacco Mosaic Virus in space. Phytopathology 58: 955–6, 1968.

Palladina TO, Kordyum EL & Bilyavs'Ka NO. [Activity and localization of transport ATPase in cells of pea seedling roots during hypogravitation]. Ukr. Bot. Zh. 41: 54–7 [in Ukrainian], 1984.

Parfenov GP & Abramova VM. [Flowering and maturing of *Arabidopsis* seeds in weightlessness: experiment on the biosatellite 'Kosmos-1129']. Dokl. Akad. Nauk SSSR 256: 254–6 [in Russian], 1981.

Parfenov GP & Oigenblik ZA. [Influence of hypergravity on the embryogenesis and survival of amphibians]. Izv. Akad. Nauk SSSR Ser. Biol. 3: 346–51 [in Russian], 1988.

Pedley TJ & Kessler JO. The orientation of spheroidal microorganisms swimming in a flow field. Proc. Roy. Soc. (London), ser. B 231: 47–70, 1987.

Perbal G & Driss-Ecole D. Polarity of statocytes in lentil seedling roots grown in space (Spacelab D1 mission). Physiol. Plant. 75: 518–24, 1989.

Perbal G & Driss-Ecole D. Microgravity and root gravitropism. Acta Bot. Gallica 140: 615–32, 1993.

Perbal G & Driss-Ecole D. Sensitivity to gravistimulus of lentil seedling roots grown in space during the IML-1 mission of Spacelab. Physiol. Plant. 90: 313–8, 1994.

Perbal G, Driss-Ecole D, Rutin J & Sall G. Graviperception of lentil seedling roots grown in space (Spacelab D1 mission). Physiol. Plant. 70: 119–26, 1987.

Pfluger EFW. über den Einfluss der Schwerkraft auf die Teilung der Zellen. Arch. gesamte Physiol. Menschen Tiere 31: 311–8, 1883.

Phillips CR. Gravitational effects on the rearrangement of cytoplasmic components during axial formation in amphibian development. COSPAR 94 : 288, 1994.

Pickard BG & Thimann KV. Geotropic response of wheat coleoptiles in absence of amyloplast starch. J. Gen. Physiol. 49: 1065–86, 1966.

Planel H, Tixador R, Nefedov Y, Gretchko G, Richoilley G, Bassler R & Monrozies E. Spaceflight effects on *Paramecium tetraurelia* flown aboard Salyut 6 in the Cytos I and Cytos M experiments. Adv. Space Res. 1: 95–100, 1981.

Planel H, Tixador R, Nefedov Y, Gretchko G & Richoilley G. Effects of space flight factors at the cellular level: results of the CYTOS experiment. Aviat., Space & Environ. Med. 53: 370–4, 1982.

Planel H, Richoilley G, Tixador R, Templier J, Bs JC & Gasset G. The '*Paramecium* experiment' demonstration of a role of microgravity on cell. In: *Proceedings of the Norderny Symposium on the Scientific Results of the German Spacelab Mission D-1*. Ed. by Sahm PR, Jansen R & Keller MH. Köln, Germany, DLR, pp. 376–81, 1986.

Planel H, Richoilley G, Caratero C, Tixador R, Caratero A & Gasset G. Effects of angular speed in responses of *Paramecium tetraurelia* to hypergravity. Microgravity Sci. Technol. 3: 107–9, 1990.

Podlutsky AG. [Cytochemical localization of calcium ATPase in pea roots under normal conditions and clinostating]. Ukr. Bot. Zh. 43: 82–4 [in Ukrainian], 1986.

Podlutsky AG. Ultrastructural analysis of organization of roots obtained from cell cultures at clinostating and under microgravity. Adv. Space Res. 12 (1): 93–8, 1992.

Poff K & Martin H. The site of gravity perception in roots: a re-examination. Physiol. Plant. 76: 451–5, 1989.

Pollard EC. Theoretical studies on living systems in the absence of mechanical stress. J. Theoret. Biol. 8: 112–23, 1965.

Pollard EC. Physical determinants of receptor mechanisms. In: *Gravity and the Organism*. Ed. by Gordon A & Cohen M. Chicago, U.S.A., University of Chicago Press, pp. 25–34, 1971.

Polulyakh YA. [Phospholipid and fatty acid composition of plasma membranes of pea root cells under clinostatic conditions]. Dokl. Akad. Nauk Ukr. SSR, Ser. B: Geol., Khim. Biol. Nauki 10: 69–72 [in Russian], 1988.

Polulyakh YA & Volovik ZN. [Fluorescence analysis of microviscosity of plant cells under scalarization conditions of the gravitation vector]. Dokl. Akad. Nauk Ukr. SSR, Ser. B: Geol., Khim. Biol. Nauki 12: 61–3 [in Russian], 1989.

Poovaiah BW, McFadden JJ & Reddy ASN. The role of calcium ions in gravity signal perception and transduction. Physiol. Plant. 71: 401–7, 1987.

Popova AF & Shnyukova EI. [Ultrastructure of chloroplasts, fractional composition of amylases, and their specific activity in *Chlorella* cells in a clinostat]. Dokl. Akad. Nauk Ukr. SSR, Ser. B: Geol., Khim. Biol. Nauki 7: 78–82 [in Russian], 1988.

Popova AF & Shnyukova EI. [Ultrastructure of plastids, fractional composition and specific activity of amylases in *Chlorella* cells under conditions of clinostating]. Tsitol. Genet. 23: 11–5 [in Russian], 1989.

Popova AF, Tupyk ND & Ivanenko HF. [Mitochondrial ultrastructure and fraction composition of succinate dehydrogenase from *Chlorella vulgaris* Beijer after clinostat treatment]. Ukr. Bot. Zh. 44: 11–5 [in Ukrainian], 1987.

Prigogine I & Stengers I. *Order out of Chaos. Man's New Dialogue with Nature.* Toronto, London, New York, Sydney, Bantam Books, 1984.

Rahmann H, Slenzka K, Körtje KH & Hilbig R. Synaptic plasticity and gravity: ultrastructural, biochemical and physico-chemical fundamentals. Adv. Space Res. 12 (1): 63–72, 1992.

Rasmussen O, Baggerud C & Iversen TH. Preparatory studies for the use of plant protoplasts in space research. Physiol. Plant. 76: 431-7, 1989.

Rasmussen O, Klimchuk DA, Kordyum EL, Danevich LA, Tarnavskaya EB, Lozovaya VV, Tairbekov MG, Baggerud C & Iversen TH. The effect of exposure to microgravity on the development and structural organisation of plant protoplasts flown on Biokosmos 9. Physiol. Plant. 84: 162–70, 1992.

Reddy ASN, McFadden JJ, Friedmann M & Poovaiah BW. Signal transduction in plants: evidence for the involvement of calcium and turnover of inositol phospholipids. Biochem. Biophys. Res. Commun. 149: 334–9, 1987.

Reiss-Bubenheim D (1990): Characterisation of *Neurospora* circadian rhythms in space (CNCR) Post-flight status report; (STS-32) submitted to Space Life Sciences Payloads Office, NASA Ames Research Centre, Moffet Field, California USA (X-AR-00590 REV).

Reitz G, Bücker H, Facius R, Horneck G, Graul EH, Berger H, Rüther W, Heinrich W, Beaujean R, Enge W, Alpatov AM, Ushakov A, Zachvatkin YA & Mesland DAM. Influence of cosmic radiation and/or microgravity on development of *Carausius morosus.* Adv. Space Res. 9 (10): 161–73, 1989.

Reitz G, Bücker H, Rüther W, Graul EH, Beaujean R, Enge W, Heinrich W, Mesland DAM, Alpatov AM, Ushakov IA & Zachvatkin YA. Effects on ontogenesis of *Carausius morosus* hit by cosmic heavy ions. Nucl. Tracks Radiat. Meas. 17: 145–53, 1990.

Reitz G, Bücker H, Lindberg C, Hiendl OC, Rüther W, Graul EH, Beaujean R, Alpatov AM, Ushakov IA & Zachvatkin YH. Radiation and microgravity effects observed in the insect system *Carausius morosus.* Nucl. Tracks Radiat. Meas. 20: 233–9, 1992.

Richoilley G, Tixador R, Gasset G, Templier J & Planel H. Preliminary results of the 'Paramecium' experiment. Naturwiss. 73: 404–6, 1986.

Richoilley G, Tixador R, Templier J, Bs JC, Gasset G & Planel H. The *Paramecium* experiment. In: *Biorack on Spacelab D-1.* Ed. by Longdon N & David V. ESTEC, Noordwijk, ESA Publications Division, pp. 69–73, 1988a.

Richoilley G, Tixador R, Templier J, Gasset G & Planel H. Effects of microgravity and hypergravity on the cell investigations on *Paramecium tetraurelia.* Acta Astronautica 17: 147–50, 1988b.

Rijken PJ, De Groot RP, Briegleb W, Kruijer W, Verkleij AJ, Boonstra J & De Laat SW. Epidermal growth factor-induced cell rounding is sensitive to simulated microgravity. Aviat., Space & Environ. Med. 62: 32–6, 1991.

Rijken PJ, de Groot RP, Kruijer W, de Laat SW, Verkleij AJ & Boonstra J. Identification of specific gravity sensitive signal transduction pathways in human A431 carcinoma cells. Adv. Space Res. 12 (1): 145–52, 1992.

Rijken PJ, de Groot RP, Van Belzen N, De Laat SW, Boonstra J & Verkleij AJ. Inhibition of EGF-induced signal transduction by microgravity is independent of EGF receptor redistribution in the plasma membrane of human A431 cells. Exp. Cell Res. 204: 373–7, 1993.

Roberts JA. Tropic responses of hypocotyls from normal tomato, *Lycopersicon esculentum*, plants and the gravitropic mutant *lazy-1*. Plant, Cell Environ. 7: 515–20, 1984.

Roblin G & Perault JM. Effects of ethephon, 1-aminocyclopropane-1-carboxylic acid, and inhibitors of ethylene synthesis on the gravitropically induced movement of *Mimosa pudica* pulvinus. Plant Physiol. 77: 922–5, 1985.

Rogacheva IV, Stupakov GP, Volozhin AI, Pavlova MN & Polyakov AN. [Characterization of bones of rats flown on board Cosmos-1129]. Kosm. Biol. Aviakosm. Med. 18: 39–44 [in Russian], 1984.

Rood SB, Kaufman PB, Abe H & Pharis RP. Gibberellins and gravitropism in maize shoots endogenous gibberellin-like substances and movement and metabolism of tritiated gibberellin a-20. Plant Physiol. 83: 645–51, 1987.

Rorabaugh PA & Salisbury FB. Gravitropism in higher plant shoots. VI. Changing sensitivity to auxin in gravistimulated soybean hypocotyls. Plant Physiol. 91: 1329–38, 1989.

Roux SJ, Biro RL & Hale CC II. Calcium movements and the cellular basis of gravitropism. Adv. Space Res. 3 (9): 221–7, 1983.

Rykova MP, Sonnenfeld G, Lesnyak AT, Taylor GR, Meshkov DO, Mandel AD, Medvedev AE, Berry WD, Fuchs BB & Konstantinova IV. Effect of spaceflight on natural killer cell activity. J. Appl. Physiol. 73: 196S–200S, 1992.

Sack FD, Suyemoto MM & Leopold AC. Kinetics of amyloplast sedimentation in gravistimulated maize coleoptiles. Planta 161: 459–64, 1984.

Sack FD, Suyemoto MM & Leopold AC. Amyloplast sedimentation kinetics in gravistimulated maize roots. Planta 165: 295–300, 1985.

Sato A, Nakajima T, Kumei Y, Hongo T & Ozawa K. Gravitational effects on mammalian cells. The Physiologist 35: S43–6, 1992.

Sawyer HR, Moeller CL, Phillips RW & Smirnov KL. Effects of spaceflight on the proliferation of jejunal mucosal cells. FASEB (Fed. Am. Soc. Exp. Biol.) J. 4: 92–4, 1990.

Sawyer HR, Moeller CL, Phillips RW & Smirnov KL. Proliferation of jejunal mucosal cells in rats flown in space. J. Appl. Physiol. 73: 148S–50S, 1992.

Schatz A & Linke-Holmes A. Gravity and the membrane-solution interface: theoretical investigations. Adv. Space Res. 9 (11): 61–4, 1989.

Schatz A, Reitstetter R, Briegleb W & Linke-Holmes A. Gravity effects on biological systems. Adv. Space Res. 12 (1): 51–3, 1992.

Schmitt DA, Malouvier A, Holy X, Marie PJ, Caissard JC & Zrath E. Receptor-ligand binding in osteoblasts in microgravity obtained by parabolic flight. The Physiologist 34: S64–5, 1991.

Schmitt DA, Ohlmann P, Gachet C & Cazenave JP. Platelet shape change and protein phosphorylation induced by ADP and thrombin are not sensitive to short periods of microgravity. J. Cell Sci. 104: 805–10, 1993.

Schnettler R, Gessner P, Zimmermann U, Neil GA, Urnovitz HB & Sammons DW. Increased efficiency of mammalian somatic cell hybrid production under microgravity conditions during ballistic rocket flight. Appl. Micrograv. Technol. 2: 3–9, 1989.

Schulz M, Solscheid B & Schnabl H. Changes in the soluble protein pattern and evidence for stress reactions in leaf tissue of *Vicia faba* L. after clinostat rotation. J. Plant Physiol. 140: 502–7, 1992.

Schulze A, Jensen PJ, Desrosiers M, Buta JG & Bandurski RS. Studies on the growth and indole-3-acetic acid and abscisic acid content of *Zea mays* seedlings grown in microgravity. Plant Physiol. 100: 692–8, 1992.

Schwarzacher J & Audus LJ. Further studies in magnetotropism. J. Exp. Bot. 24: 459–74, 1973.

Scriven LE & Sternling CV. The Marangoni effects. Nature 187: 186–8, 1960.

Siegel SM & Siegel BZ. Short-term and long-term clinostat and vibration-induced biochemical changes in dwarf marigold stems. Adv. Space Res. 3: 241–5, 1983.

Siegel S, Speitel T, Shiraki D & Fukumoto J. Effects of experimental hypogravity on peroxidase and cell wall constituents in the dwarf marigold. Life Sci. Space Res. 16: 105–9, 1978.

Sievers A. Transduction of the gravity stimulus in cress roots: a possible role of an ER-localized Ca2+ pump. In: *Molecular and Cellular Aspects of Calcium in Plant Development*. Ed. by Trewavas AJ. New York & London, Plenum Press, pp. 219–23, 1986.

Sievers A, Behrens HM, Buckhout TJ & Gradmann D. Can a calcium pump in the endoplasmic reticulum of the *Lepidium* root be the trigger for rapid changes in membrane potential after gravistimulation? Z. Pflanzenphysiol. 114: 195–200, 1984.

Sievers A, Buchen B, Volkmann D & Hejnowicz Z. Role of the cytoskeleton in gravity perception. In: *The Cytoskeletal Basis of Plant Growth and Form*. Ed. by Lloyd CW. London, Academic Press, pp. 169–82, 1991.

Sievers A & Busch MB. An inhibitor of the calcium-ATPases in the sarcoplasmic and endoplasmic reticula inhibits transduction of the gravity stimulus in cress roots. Planta 188: 619–22, 1992.

Sievers A & Hejnowicz Z. How well does the clinostat mimic the effect of microgravity on plant cells and organs? ASGSB Bulletin 5: 69–75, 1992.

Sievers A & Hensel W. Gravity perception in plants. In: *Fundamentals of Space Biology*. Ed. by Asashima M & Malacinski GM. Berlin, Heidelberg & New York, Springer Verlag, pp. 43–55, 1990.

Sievers A & Heyder-Caspers L. The effect of centrifugal accelerations on the polarity of statocytes and on the graviperception of cress roots. Planta 157: 61–70, 1983.

Sievers A, Kruse S, Kuo-Huang L-L & Wendt M. Statoliths and microfilaments in plant cells. Planta 179: 275–8, 1989.

Sievers A & Volkmann D. Verursacht differentieller Druck der Amyloplasten auf ein Komplexes Endomembransystem die Geoperzeption in Wurzeln? Planta 102: 160–72, 1972.

Silver IL. The dynamics of a discrete geotropic sensor subject to rotation-induced gravity compensation. J. Theoret. Biol. 61: 353–62, 1976.

Sitnik KM, Demkiv OT, Kordyum EL, Nedukha EM & Danevich LA. Calcium gradient in plant cells with polarized growth in simulated microgravity. Adv. Space Res. 9 (11): 41–4, 1989.

Sitnik KM, Kordyum EL, Nedukha EM, Sidorenko PG & Fomicheva VM. *Rastitel'naya Kletka pri Izmenenii Geofizicheskikh Faktorov*. Kiev [in Russian], Naukova Dumka, 1984.

Slenzka K, Appel R & Rahmann H. Brain creatine kinase activity during ontogeny of the cichlid fish *Oreochromis mossambicus* and the clawed toad *Xenopus laevis*, influence of gravity? Neurochem. Int. 22: 405–11, 1993.

Slocum RD & Roux SJ. Cellular and subcellular localization of calcium in gravistimulated oat coleoptiles and its possible significance in the establishment of tropic curvature. Planta 157: 481–92, 1983.

Smith RC & Neff AW. Organisation of *Xenopus* egg cytoplasm: response to simulated microgravity. J. Exp. Zool. 239: 365–78, 1986.

Snetkova E & Wassersung R. The development of *Xenopus laevis* tadpoles in weightlessness. COSPAR 94 : 289, 1994.

Spooner BS, Guikema JA & Barnes G. Binding of alpha fetoprotein by immobilized monoclonal antibodies during episodes of zero-gravity obtained by parabolic flight. Aviat., Space & Environ. Med. 61: 725–8, 1990.

Spooner BS, Hardman P & Paulsen A. Gravity in mammalian organ development – differentiation of cultured lung and pancreas rudiments during spaceflight. J. Exp. Zool. 269: 212–22, 1994.

Stavrinidis C, Stark H, Eilers D & Hornung E. Microgravity quality provided by different flight opportunities. Microgravity Sci. Technol. 3: 191–203, 1991.

Stinemetz CL, Kuzmanoff KM, Evans ML & Jarrett HW. Correlation between calmodulin activity and gravitropic sensitivity in primary roots of maize. Plant Physiol. 84: 1337–42, 1987.

Stokus A. Modelling of gravitropic response. 1. A systemic model. Eksp. biol. 2: 96–104, 1992.

Stokus A. Modelling of gravitropic response. 2. Competence function. Biologija 1: 73–7, 1994.

Stokus A. Action of gravireceptors: lability hypothesis and model. Adv. Space Res. in press, 1995.

Suda T, Abe E, Shinki T, Katagiri T, Yamaguchi A, Yokose S, Yoshiki S, Horikawa H, Cohen GW, Yasugi S & Naito M. The role of gravity in chick embryogenesis. FEBS Letters 340: 34–8, 1994.

Sulzman FM, Ellman D, Fuller CA, Moore-Ede MC & Wassmer G. *Neurospora* circadian rhythms in space: a re-examination of the endogenous-exogenous question. Science 225: 232–4, 1984.

Tairbekov MG, Guzhova NV, Devyatko AV & Rozov AN. [Gas exchange and energy metabolism in plants under conditions of clinostating]. Fiziol. Rast. (Moscow) 31: 385–8 [in Russian], 1984.

Tairbekov MG, Kabitskii EN & Mailyan ES. [ATPase activity in root cells of corn grown under altered gravitational conditions]. Fiziol. Rast. (Moscow) 27: 883–5 [in Russian], 1980.

Tairbekov MG & Rozov AN. [Effect of altered gravity on cytoplasmic viscosity and protein content in the plant cell]. Kosm. Biol. Aviakosm. Med. 12: 76–8 [in Russian], 1978.

Tlas M, Btkai L, Stöger I, Nagy K, Hiros L, Konstantinova I, Rykova M, Mozgovaya I, Guseva O & Kozharinov V. Results of space experiment program 'Interferon'. Acta Microbiol. Hung. 30: 53–61, 1983.

Tarasenko VA, Kordyum EL & Sitnik KM. [Root cap ultrastructure of *Arabidopsis thaliana* under space flight conditions]. Dokl. Akad. Nauk Ukr. SSR, Ser. B: Geol., Khim. Biol. Nauki 7: 79–81 [in Russian], 1982.

Taylor GR. Recovery of medically important microorganisms from Apollo astronauts. Aerosp. Med. 45: 824–8, 1974a.

Taylor GR. Space microbiology. Ann. Rev. Microbiol. 28: 121–37, 1974b.

Taylor GR & Wooley BC. Evaluations of lunar samples for the presence of viable organisms. Geochim. Cosmochim. Acta 2: 2267–74, 1973.

Taylor GR, Ellis W, Johnson PH, Kropp K & Groves T. Microbial assay of lunar samples. In: *Proceedings of the Second Lunar Science Conference*. Cambridge, Massachusetts, U.S.A., MIT Press, pp. 1939–48, 1971.

Taylor GR, Spizizen J, Foster BG, Volz PA, Bücker H, Simmonds RC, Heimpel AM & Benton EV. A descriptive analysis of the Apollo 16 microbial response to space environment experiment. BioScience 24: 505–11, 1974.

Theimer RR, Kudielka RA & Rösch I. Induction of somatic embryogenesis in anise in microgravity. Naturwiss. 73: 442–3, 1986.

Tixador R, Richoilley G, Templier J, Monrozies E, Moatti JP & Planel H. tude de la teneur intra et extracellulaire des electrolytes dans les cultures de Paramecies realises pendant un vol spatial. Biochim. Biophys. Acta 649: 175–8, 1981.

Tixador R, Richoilley G, Gasset G, Bs JC, Templier J, Moatti N, Lapchine EL, Raffin J & Bost R. Study of antibiotics sensitivity and ultrastructure of bacteria performed aboard the orbital station Salyut 7 (CYTOS 2 experiment). Innov. Tech. Biol. 5: 17–25, 1984a.

Tixador R, Richoilley G, Gasset G & Planel H. Effects of gravity and cosmic rays on cell proliferation kinetics in *Paramecium tetraurelia*. Adv. Space Res. 4: 91–5, 1984b.

Tixador R, Richoilley G, Gasset G, Templier J, Bs JC, Moatti N & Lapchine EL. Study of minimal inhibitory concentration of antibiotics on bacteria cultivated *in vitro* in space (CYTOS 2 experiment). Aviat., Space & Environ. Med. 56: 748–51, 1985.

Tobias CA, Risius J & Yang C-H. Biophysical considerations concerning gravity receptors and effectors including experimental studies on *Phycomyces blakesleeanus*. In: *Life Sciences and Space Research XI*, Proceedings of the Open Meeting of the Working Group on Space Biology of the Fifteenth Plenary Meeting of COSPAR (Committee on Space Research), Madrid, 1972. Ed. by Sneath PHA. Akademie-Verlag: Berlin, pp. 127–40, 1973.

Todd P. Gravity-dependent phenomena at the scale of the single cell. ASGSB Bulletin 2: 95–113, 1989.

Todd P. Gravity-dependent processes and intracellular motion. ASGSB Bulletin 4: 35–9, 1991.

Tremor JW & Souza KA. The influence of clinostat rotation on the fertilized amphibian egg. Space Life Sci. 3: 179–91, 1971.

Trewavas A. What remains of the Cholodny-Went theory? A summing up. Plant, Cell Environ. 15: 793–4, 1992.

Trewavas A & Gilroy S. Signal transduction in plant cells. Trends Genet. 7: 356–61, 1991.

Tschopp A & Cogoli A. Hypergravity promotes cell proliferation. Experientia 39: 1323–9, 1983.

Tschopp A, Cogoli A, Lewis ML & Morrison DR. Bioprocessing in space: human cells attach to beads in microgravity. J. Biotechnol. 1: 287–93, 1984.

Ubbels GA. Developmental biology on unmanned spacecraft. Adv. Space Res. 12 (1): 117–22, 1992.

Ubbels GA, Berendsen W & Narraway J. Fertilization of frog eggs on a sounding rocket in space. Adv. Space Res. 9 (11): 187–97, 1989.

Ubbels GA, Berendsen W, Kerkvliet S & Narraway J. Fertilisation of *Xenopus* eggs in space. In: *Proceedings of the 4th European Symposium on Life Sciences Research in Space*. Ed. by David V. Paris, France, European Space Agency, pp. 249–54, 1990.

Ubbels GA, Berendsen W, Kerkvliet S & Narraway J. Fertilization and development of eggs of the South African Clawed toad, *Xenopus laevis*, on sounding rockets in space. Adv. Space Res. 12 (1): 181–94, 1992.

Ubbels GA & Brom TG. Cytoskeleton and gravity at work in the establishment of dorso-ventral polarity in the eggs of *Xenopus laevis*. Adv. Space Res. 4: 9–18, 1984.

Ubbels GA, Reijnen M, Meijerink J & Narraway J. *Xenopus laevis* embryos can establish their spatial bilateral symmetrical body pattern without gravity. Adv. Space Res. 14 (8): 257–69, 1994.

Vacek A, Michurina TV, Serova LV, Rotkovska D & Bartonickova A. Decrease in the number of progenitors of erythrocytes *bfue cfue* granulocytes and macrophages *gm-cfc* in bone marrow of rats after a 14-day flight on board the Cosmos-2044 biosatellite. Folia Biol. (Prague) 37: 35–41, 1991.

Varjú D, Edgar L & Delbrück M. Interplay between the reactions to light and to gravity in *Phycomyces*. J. Gen. Physiol. 45: 47–58, 1961.

Vernos I, Gonzalez-Jurado J, Calleja M & Marco R. Microgravity effects on the oogenesis and development of embryos of *Drosophila melanogaster* laid in the space shuttle during the BIORACK experiment ESA. Int. J. Dev. Biol. 33: 213–26, 1989.

Vogel K & Häder DP. Simultaneous tracking of flagellates in real time by image analysis. In: *Proceedings of the 4th European Symposium on Life Sciences Research in Space*. Ed. by David V. Paris, France, European Space Agency, pp. 541–5, 1990.

Vogel K, Hemmersbach-Krause R, Kuehnel C & Häder DP. Swimming behavior of the unicellular flagellate *Euglena gracilis* in simulated and real microgravity. Microgravity Sci. Technol. 5: 232–7, 1993.

Volkmann D, Buchen B, Hejnowicz Z, Tewinkel M & Sievers A. Oriented movement of statoliths studied in a reduced gravitational field during parabolic flights of rockets. Planta 185: 153–61, 1991.

Volkmann D & Sievers A. Graviperception in multicellular organs. In: *Encyclopedia of Plant Physiology*. Ed. by Haupt W & Feinlieb M. Berlin, Springer-Verlag, pp. 573–600, 1979.

Waber J, Williams BJ, Dubin J & Siegel SM. Changes in peroxidase activity under simulated hypo-gravity. Physiol. Plant. 34: 18–21, 1975.

Wakley GK & Turner RT. Space flight decreases formation of cancellous bone in rat humeri. J. Bone Miner. Res. 8 (Supplement 1): S152, 1993.

Wayne R, Staves MP & Leopold AC. Gravity-dependent polarity of cytoplasmic streaming in *Nitellopsis*. Protoplasma 155: 43–57, 1990a.

Wayne R, Staves M & Moriyasu Y. Calcium, cytoplasmic streaming, and gravity. Curr. Top. Plant Physiol. 4: 86–92, 1990b.

Wayne R, Staves MP & Leopold AC. The contribution of the extracellular matrix to gravisensing in characean cells. J. Cell Sci. 101: 611–23, 1992.

Wendt M, Kuo-Huang L-L & Sievers A. Gravitropic bending of cress roots without contact between amyloplasts and complexes of endoplasmic reticulum. Planta 172: 321–9, 1987.

Wendt M & Sievers A. The polarity of statocytes and the gravisensitivity of roots are dependent on the concentration of calcium in statocytes. Plant & Cell Physiol. 30: 929–32, 1989.

Wendt M & Sievers S. Restitution of polarity in statocytes from centrifuged roots. Plant, Cell Environ. 9: 17–24, 1986.

Went FW & Thimann KV. *Phytohormones*. New York & London, Macmillan, 1937.

Westelind GK, Wakley GK & Turner RT. Two days of reloading following spaceflight does not restore normal gene expression of bone matrix proteins and growth factors. J. Bone Miner. Res. 8 (Supplement 1): S281, 1993.

White RG & Sack FD. Actin microfilaments in presumptive statocytes of root caps and coleoptiles. Amer. J. Bot. 77: 17–26, 1990.

Wiese C, Bechler B, Lorenzi G & Cogoli A. Cultures of erythroleukemic cells (K-562) on a stratospheric balloon flight. In: *Terrestrial Space Radiation and its Biological Effects*. Ed. by McCormack PD, Swenberg CE & Bücker H. New York & London, Plenum Press, pp. 337–43, 1988.

Wilkins MB. Gravitropism. In: *Advanced Plant Physiology*. Ed. by Wilkins MB. London, Pitman, pp. 163–85, 1984.

Woods KM & Chapes SK. Abrogation of TNF-mediated cytotoxicity by space flight involves protein kinase-C. Exp. Cell Res. 211: 171–4, 1994.

Wright LZ & Rayle DL. Evidence for a relationship between proton excretion and auxin in shoot gravitropism. Plant Physiol. 72: 99–104, 1983.

Yokota H, Neff AW & Malacinski GM. Altering the position of the first horizontal cleavage furrow of the amphibian *Xenopus* egg reduces embryonic survival. Int. J. Dev. Biol. 36: 527–35, 1992.

Yokota H, Neff AW & Malacinski GM. Early development of *Xenopus* embryos is affected by simulated gravity. Adv. Space Res. 14 (8): 249–55, 1994.

Young RS, Deal PH & Souza KA, Whitfield O. Altered gravitational field effects on the fertilized frog egg. Exp. Cell Res. 59: 267–71, 1970.

Young RS & Tremor JW. The effect of weightlessness on the dividing egg of *Rana pipiens*. BioScience 18: 609–15, 1968.

Zhad'Ko SI, Baraboi VA, Sidorenko PG & Karnaukh IM. [Initial responses of pea seedlings to clinostating]. Dokl. Akad. Nauk Ukr. SSR, Ser. B: Geol., Khim. Biol. Nauki 8: 66–8 [in Russian], 1988.

Zhad'Ko SI, Polulyakh YA, Vorobyeva TV & Baraboi VA. Lipid-peroxidation of plants under microgravity and its simulation. Adv. Space Res. 14 (8): 103–6, 1994.

Zhukov-Verezhnikov NN, Volkov MN, Maisky IN, Rybakov NI, Guberniev MA, Podoplelov II, Kulagin AN, Aniskin ED, Rybakova KD, Sharyi NI, Voronkova IP, Saxonov PP, Kopyev VY, Antipov VV, Kozlov VA, Parfyonov GP & Orlovsky VI. Experiments with microorganisms and human cell cultures in the Zond 5 and Zond 7 flights. Life Sci. Space Res. 9: 99–103, 1971.

Chapter 2

Physiology of cardiovascular, respiratory, interstitial, endocrine, immune, and muscular systems

HELMUT C. HINGHOFER-SZALKAY

2.1 Cardiovascular system

2.1.1 General overview

Although the impact of gravity on the functioning of the cardiovascular system is obvious, implying that investigations in weightlessness should be crucial for a more complete understanding of circulatory system physiology, only a relatively small number of scientifically sound investigations have been conducted under the conditions of space flight. The understanding of long-term microgravity effects on the cardiovascular system needs more in-depth investigation. Adaptive mechanisms should establish 'homeostasis' along with functional and possible structural changes which can be seen as symptoms of successful adaptation, but at the same time as 'deconditioning' under the circumstances of gravitational loading.

Fluid displacement in the cranial direction is perceived as an increase of effective blood volume by the body's vascular stretch receptor systems which are, interestingly, localized above the hydrostatic indifferent point (HIP). It induces a neuroendocrine response pattern, parts of which seem still unclear. Cardiovascular control is complex and involves the interaction of intrinsic, autonomic, central, reflex, and humoral mechanisms. Even when blood pressure remains unchanged, as in space flight, this can be the average result of balanced resettings within the concert of subsystems contributing to the pressure in the arterial compartment. The ability to vary blood flow in response to local variations in blood pressure, tissue pressure, and concentrations of oxygen and metabolites; humoral agents, such as cortisol and histamine, and autonomic control, that is sympathetic activity which produces constriction of peripheral vessels together control local perfusion and systemic blood pressure.

Cardiovascular regulation during upright standing (orthostasis) is due to a multiplicity of factors (**Levine *et al.* 1991**). Reflex mechanisms produce immediate responses as a result of local blood pressure changes (baroreceptors), and of changes in the chemical composition of the blood with respect to oxygen, carbon dioxide, and acidity (chemoreceptors). A multitude of controlling mechanisms allows an adequate circulation to be delivered to all parts of the body under various conditions, for example when changing from the horizontal to the vertical position. Specifically, as a result of evolutionary adaptation to the upright posture, the cardiovascular system is able to cope with the additional stress of hydrostatic pressure gradients imposed whenever the body's z axis is perpendicularly aligned to the force of gravity. Under terrestrial conditions, the circulatory system has adapted to counter the stress of the cranio-caudal gravitational gradient. This includes prevention of blood pooling in the leg capacitance vessels, and a requirement to pump blood to the brain. The cardiovascular control mechanisms are usually able to respond to such a sudden re-orientation during which a volume of about 500 ml of blood is displaced in the cranio-caudal direction, e.g. to the distensible veins of the lower limb. The functional hypovolemia is aggravated by the tendency of the increased hydrostatic pressure to elicit further loss of fluid by capillary filtration. If allowed to proceed unimpeded the loss of effective circulating blood volume would produce a fall in cardiac output and blood pressure (Sander-Jensen 1991), and cerebral blood pressure would fall reducing cerbral blood flow towards levels inducing syncope. Under normal conditions the increase in muscle activity associated with the upright posture counteracts the progressive loss of intravascular fluid, and this together with cardiovascular reactions to the decrease in preload maintains circulatory stability at the expense of a slightly lower blood pressure, higher heart rate, and peripheral vasoconstriction. In case of experimental or clinical 'autonomic insufficiency' these mechanisms are unable to compensate fully for the loss of effective blood volume and blood pressure drops to levels incompatible with adequate perfusion of the brain (orthostatic intolerance). Similarly, functional deconditioning, as seen after extended bed rest and space flight, aggravates the problem of effective cardiovascular counterregulation upon orthostatic stress.

Arterial and many venous blood vessels together form a system of communicating tubes where hydrostatic pressure gradients are present in case that external acceleration forces are present. On Earth, a standing human of 180 cm length would experience an arterial pressure difference of 18 kPa from head to toe. In a person lying supine, a pressure difference of some 3 kPa still remains along the chest. In the lung vessels, due to pulmonary dimensions, pressure gradients of approximately the same magnitude are operative with any body position. Microgravity conditions cause a major shift of fluid from the lower body towards the chest and head during the first few days in space. The rate of shift reaches a peak after about 24 h, and the new steady state is attained within 3-4 days. Limb measurements show a rapid up to 30% decrease in circumference over this time, to a level which is then sustained for the rest of the flight.

2.1.2 Ground-based experimental results and microgravity simulation

There is a vast body of results on cardiovascular effects of simulated weightlessness from numerous immobilization, bed-rest, and water immersion studies which have been done world-wide, and several detailed reviews are available. Common models to simulate microgravity effects on the cardiovascular and body fluid systems are head-down tilt (Greenleaf *et al.* 1976; Kollias *et al.* 1976; Nicogossian *et al.* 1979; **Hargens *et al.* 1983; Fortney *et al.* 1991; Gharib *et al.* 1992; Vernikos *et al.* 1991**; Greenleaf 1983), headout water immersion (**Epstein 1978; Epstein *et al.* 1980; Greenleaf 1984a & b; Greenleaf & Kozlowski 1982**; **Greenleaf *et al.* 1983; Norsk & Epstein 1988; Epstein 1992**), and lower body positive pressure (**Kravik *et al.* 1986**; Geelen *et al.* 1989, 1992).

If the 1 × g upright head-to-foot gravity vector is removed, as with bed rest, 'cardiovascular deconditioning' ensues (Blomqvist & Stone 1983): upon return to normal orthostatic conditions, heart rate is elevated, arterial blood pressure variations increase, and pulse pressure is narrowed, altogether favouring presyncopal symptoms (**Harrison 1986; Greenleaf *et al.* 1989**). The underlying physiological mechanisms are based on multiple adaptive readjustments: reduced blood volume (**Fortney *et al.* 1991**; Johnson 1975), varied vascular capacitance (**Convertino *et al.* 1989**), cardiac changes, and/or altered cardiovascular control mechanisms (Blomqvist & Stone 1983; **Convertino *et al.* 1990, 1994; Levine *et al.* 1991**).

After bed rest deconditioning, heart rate increases and cardiac output decreases during cardiovascular stress (**Taylor *et al.* 1949**; Greenleaf *et al.* 1976; Sandler ***et al.* 1988; Beck *et al.* 1992**). Cerebral blood flow is maintained at the expense of leg blood flow, due to increased vascular resistance (**Takenaka *et al.* 1994**). The hypothesis that blood pooling in the lower legs contributes to orthostatic intolerance after simulated and real space flight is not supported by findings of unchanged venous compliance with lower body negative pressure (LBNP) after 14 days bed rest (**Melchior & Fortney 1993**); likewise, another study did not show significant compliance (and filtration rate) changes with 20 days bed rest (**Bonde-Petersen *et al.* 1994**).

Recent studies have demonstrated that after only a few hours of head-down tilt bed rest, orthostatic tolerance is significantly reduced (Butler ***et al.* 1991**), i.e. the cardiovascular system rapidly adapts to the requirements of the new stress level (**Lathers & Charles 1994**). Several studies have found reduced baroreflex sensitivity after simulated (**Convertino *et al.* 1990; Eckberg & Fritsch 1992**) or real space flight using the neck chamber method (**Fritsch *et al.* 1992**) and heart rate spectral analysis (Hughson ***et al.* 1994a & b**). In contrast, from the ratio of heart rate *vs* blood pressure change from baseline to baroreceptor-isolated conditions, an increased baroreflex responsiveness was documented after 15 days head-down tilt bed rest (**Crandall *et al.* 1994**). It is unsettled whether these observations are due to higher reflex gain or if, simply, a steeper portion of the baroreflex curve has been reached during the experiments. Results from another study also failed to explain increased orthostatic heart rate responses by sensitivity changes within baroreceptor or pulmonary reflexes (**Haruna *et al.* 1994**).

As a component of cardiovascular deconditioning, appropriate reflexes to meet orthostatic challenge adequately are weakened. Long-lasting antiorthostatic hypokinesia (bed rest) also results in decreased plasma norepinephrine levels (Davydova *et al.* 1986; Robertson ***et al.* 1994**). This could be part of a

general adaptation of the autonomic nervous system to conditions of decreased gravitational load and hypokinesia, as with space flight; pertinent published flight data are limited, however (Leach *et al.* 1983; Kvetnansky *et al.* 1991; **Hinghofer-Szalkay *et al.* 1993; Robertson *et al.* 1994**).
Water immersion increases central blood volume by about 700 ml with a concomitant rise of 200 ml in heart (mainly atrial) volume (Lange ***et al.* 1974**). Contrary to head-down tilt bed-rest, where central venous pressure and central blood volume return to control values after a transient elevation in about 6 h (**Nixon *et al.* 1979**), this is not the case with water immersion (**Norsk 1992**); rather, they remain elevated even after 12 h, whereas water excretion is back to normal after 7 h, probably due to an adaptive decrease in plasma atrial natriuretic hormone (**Stadeager *et al.* 1992**). Other volume-regulating hormones, plasma renin activity, aldosterone, vasopressin, norepinephrine, are suppressed for up to 12 h, whereas excretion of urodilatin, cGMP, and sodium are elevated (**Stadeager *et al.* 1992; Nakamitsu *et al.* 1994**). The situation with (thermoneutral) water immersion (**Epstein 1978, 1992**) seems to be different from the conditions seen before and immediately after beginning of space flight.

2.1.3 Observations in flight

Weightlessness totally removes density gradients. Under no circumstances can this be simulated on Earth for longer periods than a few seconds, as during parabolic flight. Changes in the human cardiovascular system as a result of space flight are thought to be a function of changes in extracellular fluid volume, changes in the contractile state of venous capacitance vessels after fluid redistribution, and possible functional changes in pressure-regulating, neuro-hormonal, mechanisms. Several indices of cardiovascular function have been assessed during space flight. Blood flow through the legs was found to be increased over pre-flight levels, probably as a consequence of reduced peripheral resistance and increased cardiac output. Similarly, venous compliance in the lower limbs was markedly increased in flight although, having reached a peak ten days after launch, the magnitude of the change declined in later weeks. This increase in distensibility is believed to contribute to the exaggerated response to LBNP seen in space (LBNP = lower body negative pressure, or lower body suction: a procedure which should mimic the hydrostatic effects on the cardiovascular system during passive upright standing in a $1 \times g$ environment).
Average resting heart rate early in flight remains normal (Berry & Catterson 1967; Pourcelot ***et al.* 1986;** Bungo ***et al.* 1987**; Gazenko *et al.* 1988a & b; Fritsch *et al.* 1992) or is slightly elevated (Link 1965; Pourcelot *et al.* 1983). Blood pressure data are inconsistent, systolic and diastolic values often change in different directions; dramatic changes do not occur. It is possible that the great variability of responses indicates an increased lability of blood pressure in early flight. By comparing pre-and post-flight (up to 2 weeks)

heart rate power spectra, baroreflex analysis (neck chamber), and plasma catecholamines after orthostatic and Valsalva stimulation, further evidence for increased sympathetic/reduced parasympathetic influence on post-flight arterial pressure control has been presented (**Fritsch-Yelle *et al.* 1994**).

Resting cardiac contractility is not changed even by long duration space flight (Atkov *et al.* 1987). For short term flights, the value of ingestion of isotonic saline, usually taken as salt tablets and water, is questionable (**Buckey *et al.* 1993**) and with longer flight time the saline is even less effective (**Charles & Lathers 1991**). This suggests that whereas the primary change in the first few days of space flight might be explained as a decrease in vascular volume, some other alteration begins to play a role that cannot be fully offset by vascular refilling alone. In flight echocardiographic data from short-duration flights demonstrate unchanged end-diastolic ventricular volume in one astronaut when first measured two days after launch, while stroke volume decreased; the cardiac volumes were increased on days three and four, and decreased to baseline values on day six (Pourcelot *et al.* 1983). Cardiac output and heart rate were higher than preflight baseline values for all days. End-diastolic ventricular volume was unchanged (n = 1) at 3 h or decreased (n = 1) at 9 h into flight, and as with stroke volume, progressively decreased in those 2 subjects on days 4–6 (Pourcelot *et al.* 1986). Heart rate increased in one and remained unchanged in the other subject. Contrary to a previous flight, cardiac output was generally decreased. In another mission, echocardiograms were obtained beginning 4 h after launch and frequently repeated during a one week mission. Right ventricular dimension was decreased, left ventricular, diastolic, and stroke volume were elevated above preflight supine values on flight day 1, then remained about 15% below baseline for one week. Cardiac index was increased on day 1, thereafter not different from preflight supine. Mean blood pressure and heart rate were 20% increased in space compared to preflight supine (Bungo 1990). On the other hand, tetrapolar impedance rheography combined with electrocardiography and blood pressure monitoring on flight days 2–3 did not detect significant changes in stroke volume, cardiac output, and heart rate (Gazenko *et al.* 1988a & b). Taken together, heart volumes and cardiac output seem to be somewhat elevated early in flight with the volumes decreasing thereafter.

Central venous pressure (CVP) while in space was indirectly measured in humans during the Spacelab 1 and D-1 missions (**Kirsch *et al.* 1984**). CVP was already below preflight supine values just 20 minutes after launch and continued to decline over the seven days of the mission. This was confirmed by an experiment performed later on one crew member by using a non-invasive technique: CVP progressively declined over the first three flight days to a stable level that was below the preflight value for the remaining five days (Charles & Bungo 1987). During the SLS-1 and D-2 flights, continuous direct measurements of CVP revealed increased values during launch where up to $+3.5 \times g$ forces compress the thorax, followed by an immediate decrease below supine baseline pressure after entering microgravity conditions (**Buckey *et al.* 1993**). It seems that the volume shift and the rise in CVP may actually take place before launch occurs due to a semisupine position with legs elevated above the chest for at least two hours (**Kirsch *et al.* 1984; Lathers *et al.* 1989**). Head-down tilt studies have shown a similar biphasic pattern of CVP (**Nixon *et al.* 1979**). However, the engorged neck veins and sinus congestion may often persist well into the flight, an observation which needs to be clarified considering the observation of reduced CVP.

Even during flights up to 1 year in duration, no permanent debilitation of the cardiovascular system occurred. Studies performed on Skylab showed that within 4–6 weeks the cardiovascular system seems to have stabilized at a new level thought to represent an adaptive steady state for microgravity (Johnson *et al.* 1977; Dietlein 1983). Differences in the response in flight to LBNP as compared to preflight testing included largely increased heart rate and an elevated gain of leg volumes, conceivably due to the greater shift of central blood volume into the leg veins, which are virtually empty at the onset of testing. There seems to be a tendency for heart rate to increase slightly in long term flight; stroke volume and cardiac output and blood pressure remained essentially unchanged.

During the Salyut 6 and 7 missions, tetrapolar impedance rheography was used to estimate stroke volume and cardiac output at rest and when stressed by LBNP and exercise; heart rate and blood pressure were also measured. Unlike the Skylab study, LBNP response in the Soviet crew members was not different from preflight (Gazenko *et al.* 1988a & b). Exercise echocardiography conducted on the 237-day flight on Salyut 7 on two crew members revealed that stroke volume decreased by 25–30% (**Atkov *et al.* 1987**). It seems that in long-duration space flight, corresponding to a normal physiological response, changes in heart rate are the primary means by which cardiac output regulation occurs.

2.1.4 Post-flight observations

Acute problems with the cardiovascular system on return to 1 × g have been apparent since the earliest space missions, i.e. dizziness, heart palpitations, sweating, orthostatic intolerance, presyncope with postural challenge, and reduced exercise capacity. Recovery is relatively rapid following flights of 4–10 days, and by 3 days post-flight the response to postural challenge may already be similar to what it was preflight. Orthostatic intolerance is regarded as potentially one of the most serious cardiovascular consequences of manned space flight. Post-flight orthostatic intolerance has been a constant feature throughout man's experience in space, and is directly attributable to the abolition of all hydrostatic pressure gradients in the environment of microgravity. The manifestations of orthostatic intolerance range from a rise in heart rate and a fall in systolic pressure, with a consequent narrowing of pulse pressure to an overt tendency to faint. It has been assumed that it is an increase in central blood volume which was responsible for the microgravity-induced cardiovascular changes by acting upon, and resetting, all the controlling mechanisms. However, recent evidence shows that central venous pressure is decreased below preflight control levels at the very beginning of flight. The complex readjustments leading to this and other resettings are still poorly understood, and much work remains to be done with post-flight investigations.

From ground based simulation studies with central volume expansion (horizontal bed rest, head-down tilt bed rest, water immersion, dry body immersion), the following sequence of physiological events was originally expected on transition to weightlessness. First, volume receptors were expected to be activated by the cephalad fluid shift accompanying the transition from $1 \times g$ to microgravity. The body normally responds with a two-staged response: total peripheral resistance in the circulation as well as heart rate would be decreased to reduce arterial pressure to a normal level. The second stage was to reduce extra volume and to set a new steady state condition by increased renal fluid output, hence reducing blood volume. Measurements in flight revealed decreased central venous pressure from the very beginning of flight, however, which can probably be attributed to adaptation due to sustained legs-up position before launch.

Space flights of even short duration (less than 7 d) profoundly effect the way in which the cardiovascular system responds post-flight to postural challenge, and how blood pressure is maintained in this situation; the most consistent finding is elevated heart rate (Link 1965; Volynkin & Vasilyev 1967; Hoffler & Johnson 1975; Kalinichenko 1977; Johnson *et al.* 1977; **Bungo & Johnson 1983; Vorobyov *et al.* 1983**; Bungo *et al.* **1985; Mulvagh *et al.* 1991**). Heart rate seems to exert the main regulatory effect on blood pressure maintenance immediately post-flight since resting mean arterial pressure is unchanged (Hoffler & Johnson 1975; **Mulvagh *et al.* 1991; Fritsch *et al.* 1992**) or slightly elevated (Johnson *et al.* 1977; **Bungo *et al.* 1985; Bungo *et al.* 1987**). Heart rate is elevated by about 35% in the stand test, where the filling volume of the left heart was reduced, and neither contractility nor total peripheral resistance increased to an extent as seen in $1 \times g$.

The baroreflex response of astronauts has been evaluated immediately post-flight by means of the neck cuff method; the data suggest that a reduction of vagal responsiveness to beat-to-beat changes in arterial pressure occurred (**Fritsch *et al.* 1992**). All subjects were able to maintain adequate blood pressure during orthostatic challenge by dramatically increasing their heart rate. Altered autonomic sensory input occurring as a result of space flight may have resulted in some degree of neural plasticity; the decrement in vagal sensitivity was still apparent 8–10 days post-flight (**Fritsch *et al.* 1992**). When regular in flight exercise regimes are used, deconditioning does not worsen linearly with time. Additional evidence for a larger sympathetic, and lower parasympathetic, influence on post-flight arterial pressure control, employing Valsalva and stand tests, with heart rate spectral analysis, neck chamber baroreflex assessment, and plasma catecholamines was recently presented (**Fritsch-Yelle *et al.* 1994**).

With a reduced blood volume in flight (Fischer *et al.* 1967), redistribution into the lower limbs with reimposition of normal gravity would reduce ve-

nous return; however, neither peripheral resistance nor contractility increases to help maintain arterial pressure which is not always adequately maintained despite a large rise in heart rate. Left ventricular stroke index (2-dimensionally directed M-mode echocardiography) was decreased below pre-flight values, but cardiac output was maintained at a normal level (**Bungo & Johnson 1983; Bungo *et al.* 1985, 1987**; Hoffler & Johnson 1975; **Mulvagh *et al.* 1991**). These volume reductions occurred despite fluid loading prior to return. Although the time course to full recovery of heart volumes is not clear (Henry *et al.* 1977), this may be related more to in flight loss of aerobic conditioning than to the actual effects of space flight. Full recovery of cardiovascular parameters has also been reported following long-duration flights of 8–12 months, although about 4 weeks are required for full reconditioning (Atkov 1985; Bogomolov *et al.* 1988).
The greater increase in leg girth apparent in post-flight stand and LBNP testing suggests that a greater than normal volume of blood may pool below hip level with post-flight orthostatic challenge (Berry & Catterson 1967; Hoffler & Johnson 1975).

2.1.5 Conclusions

Long-term circulatory control is based mainly on the regulation of blood volume, and cardiac output, with a relatively independent element controlling arterial blood pressure. Superimposed on this haemodynamic control system are nervous and hormonal 'fine-tuning' mechanisms; the interconnections are multiple and call for an integrative scientific approach. This applies to cardiovascular physiology in general; in space flight experiments, it means that care must be taken to assure correct study design and data evaluation. Effects may occur due not only to the effect of microgravity but also to the interaction of reduced gravity with some other aspect of space travel, e.g. activity, altered pulmonary pressures, or psycho-physical stress. An important question regarding appropriate baseline reference points is the correct postural position for the collection of reference data. Previous studies have typically relied on data collected during preflight recumbency to provide a baseline for in flight measurements because, as in space flight, the effects of the hydrostatic pressure gradient are removed. Ground based measurements with test subjects in an upright (standing or sitting) position may better reflect cardiac status during the relative central hypovolemia induced by orthostasis and may provide a more appropriate comparison for later in flight measurements when blood volume is reduced.
Available experimental evidence raises a number of questions. Does the baroreflex system reset or change as a function of gravitational unloading? Altered cardiovascular sensory inputs may trigger neural plasticity which results in impaired reflexes. Is neurotransmitter release and/or action altered? Does venous compliance change such that venous return and cardiac output are further compromised? Orthostatic intolerance after microgravity expo-

sure is likely to result from multiple abnormalities due to some of, or a combination of all these mechanisms.

Longitudinal measurement of cardiovascular performance has been possible in long-term Soviet flights without compromising mission goals. In future flights, the experimental conditions should be similar to a regular laboratory situation since the cardiovascular system's interconnection to other subsystems renders its functioning highly susceptible to a multitude of potentially confounding influences. This calls for minimally disturbing techniques of high sensitivity and/or specificity, which do not interfere with normal regulatory processes.

Will the cardiovascular system be able to sufficiently re-adapt when subject to terrestrial gravity after extended periods in microgravity? Studies of in flight changes must be complemented by studies done pre- and post-flight in order to provide the database which makes possible quantitative comparisons and the investigation of dynamic, time-critical adaptive changes. For a deeper understanding, investigation of cardiovascular parameters and of biochemical monitoring of compartmental input-output data is needed for ground-based as well as in flight studies (Norsk & Epstein 1991).

2.2 Respiratory system

2.2.1 General overview

The respiratory system, that is the lungs with support structures like trachea, larynx, respiratory muscles, together with the cardiovascular system, and in many various ways connected to the central nervous system, primarily serves as supplier of oxygen, removes carbon dioxide and, hence, acidity from the body, thereby playing a major role in energy metabolism and acid-base-homeostasis. It is also of major importance to blood volume dynamics by holding approximately the same amount of blood as the complete arterial system does (about 700 ml in an adult). Further, water is evaporated to the environment (about 300 to 600 ml d^{-1} under normal circumstances), and its blood vessels significantly contribute to endocrine functions, as by forming angiotensin II through the endothelial angiotensin converting enzyme. Finally, the lungs are able to remove small blood clots, air bubbles, or fat droplets from the blood, and to stop other particles which are potentially able to create arterial embolism when evading the venous side of the circulation.

The air volume which resides within the lung in exhaled resting position is called the functional residual capacity (FRC). Its magnitude depends essentially on the balance of elastic forces between the lung and chest wall, with gravity being a major determinant of their passive mechanical characteristics. In the upright posture, the weight of the intestines pulls down on the diaphragm, bulges the abdominal walls outwards, and expands the lung, which results in an increased FRC as compared to the supine position. In addition, breathing in the supine posture is mainly by expansion of the diaphragm, whereas in the upright posture, expansion of the rib cage is the major mechanism.

The respiratory system is particularly sensitive to the influence of gravity (West 1991). Postural changes alter chest wall mechanics, the relative displacement of rib cage and abdominal compartments and their compliances, as well as ventilation and perfusion patterns. Weightlessness is a unique situation which is ideally suited to test hypotheses of gravitational effects on the lung.

The pulmonary vasculature is exposed to a pressure difference depending on height and acceleration force (e.g. 3 kPa differential over a 0.3 m vertical distance in 1 × g). Since the pulmonary vessels typically do not react to passive distension (P-type vasculature), this leads to a resistance pattern which itself causes a gradient of blood flow within the lung. In many individuals, there may be very little blood flowing through the lung apex under resting upright conditions in a 1 × g environment. Taking into consideration the significance of the 'siphon effect', it has been argued that this reduction in apical flow is not a matter of raising the blood against gravity but a consequence of increased vascular resistance in the upper parts of the lung (**Badeer 1982**). Evidence has been presented that pulmonary blood flow distribution is only marginally influenced by gravity in the canine lung (**Glenny *et al.* 1991**).

There is a regional distribution of various functions of the lung along the gravity vector (West 1991). The elastic properties of the lung are uniform in all lung regions. However, due to the weight of the lung within the chest, respiratory units in the dependent region are less extended than those of the 'upper' parts, respectively. The transpulmonary pressures differ by several mm Hg between those locations, and since air pressures within the lung are equilibrated, the gradient in distending pressure is due to a corresponding gradient in pleural pressures. As a result, the ventilation (per unit volume) to dependent lung regions is greater (up to a factor of about 2) than that to non-dependent regions. These gradients increase as gravitational forces increase (**Bryan *et al.* 1966**) and retain their orientation along the gravitational vector as posture is changed (**Kaneko *et al.* 1966**). In addition to the direct effect of gravity, however, non-gravitational mechanisms also produce inhomogeneities of ventilation and perfusion. Microgravity experiments offer the best approach to evaluate the relative role played by these gravitational and non-gravitational factors.

Ventilation inequality can be measured by performing vital capacity (VC) single-breath nitrogen (N_2) washouts during which a 150 ml bolus of 79% argon and 21% oxygen is inhaled at residual volume followed by a maximal inspiration of pure oxygen. When performed at 1 × g, the concentrations of N_2 and Ar first increase rapidly in the expired air and then rise much more slowly over most of the VC. This slowly ascending portion of the washout curve is referred to as the alveolar plateau (Phase III) and is characterized by its slope and by the amplitude of the superimposed oscillations of expired gas concentrations related to the heartbeat (cardiogenic oscillations: **Fowler**

& Reid 1961). The alveolar plateau ends at a point where there is a sudden increase in expired gas concentrations. This terminal rise (Phase IV) may be described by its height and volume, ie, the volume from the end of phase III to end expiration. The terminal rise in N_2 and Ar concentrations that is seen at 1 × g is thought to be caused by basal airway closure and cessation of flow from units that have low N_2 concentrations because of their large total lung capacity/residual volume (TLC/RV) ratios, and low Ar concentrations because of their closure in early inspiration when the Ar bolus is inhaled. The magnitude of cardiogenic oscillations also primarily reflects TLC/RV expansion differences but it is indepedent of sequential lung emptying between different lung regions (Engel 1986).

Arterial blood gas tensions (pO_2, pCO_2) in pulmonary veins are directly related to the respective alveolar ventilation/perfusion (V_A/Q) ratio. Because there is both a ventilation and perfusion gradient in the lung increasing from lung top to bottom, the distribution of gas tensions will depend on the relative magnitudes of the two gradients. The gradient in perfusion is generally steeper than that of ventilation, therefore the top of the lung is relatively better ventilated than perfused, whereas the lung base is relatively better perfused than ventilated.

Indications on the gravity dependence of interstitial fluid turnover came from studies in intact lungs *in situ*. Both microvascular and interstitial hydraulic pressures increase towards dependent regions (Miserocchi & Negrini 1990). Fluid filtration is predominant in the less dependent regions of the lung, whereas lymphatic drainage of interstitial fluid prevails in the more dependent ones (Miserocchi & Negrini 1991).

Gravity contributes to aerosol deposition within the lung by deposition on the surface of the air passages, and this mechanism disappears in weightlessness. Filtering by gravitational settling prevails in the periphery of the lung, where air velocities are low, the time spent within a given airway increases, and there is no turbulence in the flow. Persistence of aerosol droplets in the air is also prolonged during weightlessness. Less than 20% of the total deposition of medium-sized particles may be observed after inhalation during weightlessness compared with that in a 1 × g environment (**Engel 1991**).

2.2.2 Ground–based experimental results and microgravity simulation

Gravitational loading of the abdomen in an upright position produces a hydrostatic gradient of pressure that passively lengthens and stretches the ventral abdominal wall. Because the abdominal muscles insert on the sternum and the lower ribs, this passive tension is applied to the lower rib cage and acts to pull it downward. By removing the weight of the abdominal contents, transition to weightlessness removes the tension in the abdominal wall and transabdominal pressure becomes nearly zero (Edyvean *et al.* 1991). As a result, the sternum and the ribs move upward and the ventral rib cage expands. A similar change is observed during immersion of seated subjects (**Reid *et al.* 1986**).

Microgravity during parabolic flight produces a slight decrease in forced expiratory volume (FEV1), FEV1/VC ratio, and maximum expiratory flow rates (Foley & Tomashefski 1969; **Guy *et al.* 1991**). In addition, flow rates were higher at large lung volumes and smaller at low lung volumes in microgravity, although not in all subjects (**Guy *et al.* 1991**). These changes have been attributed to alterations in lung recoil pressure due to vascular engorgement and in local airway stresses that alter location and motion of airway choke points during forced expiration (**Prefaut *et al.* 1976**). Abdominal compliance increases on transition from upright to supine posture (Agostini & Mead 1964; **Konno & Mead 1968; Estenne *et al.* 1985**). On transition to weightlessness, abdominal compliance increases by more than 50% relative to 1 × g but decreases (by 35%) when gravitational load is 1.8 × g (**Edyvean *et al.* 1991**).

Studies in parabolic flight have demonstrated consistent changes in the configuration and pattern of movement of the chest wall as well as in FRC on acute transition from 1 × g to microgravity. Removal of gravity was associated with an inward movement of the abdominal wall, a decrease in FRC, and an increased contribution of abdominal movement during tidal breathing. Early measurements in parabolic flights did not reveal consistent changes in vital capacity (VC) between 1 × g and microgravity in seated normal subjects (Foley & Tomashefski 1969; **Michels & West 1978**). Chest radiographs taken at residual volume (RV) and total lung capacity (TLC) after 10 s of microgravity showed no significant effect (**Michels *et al.* 1979**). Later experiments showed an 8% decrease in slow expiratory VC in microgravity in four subjects (**Paiva *et al.* 1989**).

A microgravity induced decrease in VC was found in seven of nine subjects during parabolic flights (**Guy *et al.* 1991**). The residual volume was unchanged or even diminished, therefore this effect must have been due to a reduction in TLC (**Paiva *et al.* 1989**). During periods of increased gravity, changes in the reverse direction were observed. Increased gravity up to 3 × g does not consistently change VC (Foley & Tomashefski 1969; **Pyszczynski *et al.* 1985; Boutellier *et al.* 1985; Arieli *et al.* 1986; Guy *et al.* 1991**). On the other hand, studies during parabolic flight showed a 50–200 ml increase in FRC during the 1.8 × g period preceding the onset of weightlessness (**Paiva *et al.* 1989; Edyvean *et al.* 1991**). The distension of the ventral abdominal wall that occurs on going from supine to upright elicits tonic contraction of the abdominal muscles (**De Troyer 1983**). This response, which opposes the increase in abdominal volume and the descent of the diaphragm dome, may be elicited in response to hypergravity and may minimize the increase in FRC that would otherwise occur.

Redistribution of blood volume to the thorax in weightlessness has been demonstrated by a reduction in thoracic impedance (**Lathers *et al.* 1989**). Removal of the hydrostatic pressure gradient with microgravity conditions also causes redistribution of extracellular fluid. Intrathoracic blood volume increases, which causes alterations in pulmonary diffusion capacity and cardiac output. Using simultaneous measurements of thoracoabdominal and lung volumes, no thoracic blood volume changes were recorded during microgravity in parabolic flights (**Paiva *et al.* 1989**). This suggests that the increase in central blood volume during short periods of microgravity is smaller than that induced by postural changes, head-out immersion, or inflation of 'anti-gravity' suits at 1 × g. The weightless period during parabolic flights might have been too short to permit most of the blood volume redistribution to take place; however, increases in central venous pressure occur immediately after entry into weightlessness and are stable thereafter throughout a 20 s microgravity period (**Norsk *et al.* 1987**).

Lung density as assessed by radiography has been demonstrated to be higher in microgravity than at 1 × g in the upper lung fields relative to the lower ones, in both posteroanterior and lateral views (**Michels *et al.* 1979**). Distribution of pulmonary blood flow has been assessed during parabolic flights (**Michels & West 1978**) by measuring expired pO_2 and pCO_2 during a constant-flow VC exhalation. Immediately before that, the subjects hyperventilated for several seconds and then held their breath at maximal inspiration for 15 s. The rationale was that during breath holding regional gas exchange is determined by blood flow, and the amplitude of the cardiogenic oscillations during the subsequent expiration reflect topographical differences in blood flow per unit lung volume. There is strong evidence that topographical inequalities of blood flow were reduced under weightless conditions; the cardiogenic oscillations were almost eliminated during short periods of microgravity. Increased acceleration has been shown to exaggerate the normal regional differences in perfusion, i.e., at high levels of gravitational acceleration blood flow is confined at the base of the lung (**Bryan *et al.* 1965; Michels & West 1978**).

Experimental data indicate that horizontal posture does not abolish the gravity dependence distribution of lung fluid balance (Negrini *et al.* 1987). Exposure to microgravity causes a more uniform blood flow distribution and increases pulmonary capillary blood volume. Both facts lead to an increase in capillary surface area and fluid filtration rate into the lung parenchyma; a steady state condition is reached if pulmonary lymphatic drainage is correspondingly increased (**Miserocchi *et al.* 1993**). Lung condition under microgravity might be similar to that of an initial pulmonary oedema, a hypothesis which is still to be tested.

2.2.3 Observations in flight

From experiments on dogs, it has been argued that intrathoracic pressure imbalances in the episodes of high gravitational acceleration during launch and reentry phases of space flight might pose a potentially serious hazard (**Rutishauser *et al.* 1966**). No measurements have been done yet during these difficult mission phases. The distribution of volume changes between the chest wall and the lung depends primarily on the relative compliances of these structures. The more compliant the lung relative to the chest wall, the greater the changes would be in lung volume and the smaller those in chest wall volume. Additional volume added to the respiratory system, as by inflation of fracture splints during a pneumothorax or a balloon in the stomach, is primarily accommodated by the chest wall (**Gilroy *et al.* 1985**; **Kimball *et al.* 1986**). Intrathoracic gas volume is well protected since the increase in chest wall volume is about twice as great as the decrease in lung volume.

There have been no observations suggesting serious dysfunction of the respiratory system during space flight; crew members have not complained of impairment in breathing. There is, however, a feeling of fullness in the head, nasal congestion, and a sensation similar to having a cold. This is most evident during the first few days, but may last for several weeks after launch. The sensation is most likely related to the shift of blood and fluid from the lower limbs to the head, resulting in congestion of the facial sinuses by the increase in blood flow.

A change from upright to supine posture and immersion of seated subjects in water are associated with an inward displacement of the anterior abdominal wall, a cranial displacement of the diaphragm, and a fall in FRC. In weightlessness these changes occur because the weight of the abdominal contents is removed; respiratory inductance plethysmography revealed that an FRC decrease in microgravity is primarily due to an abdominal chest wall volume decrease (**Paiva *et al.* 1989**; **Edyvean *et al.* 1991**). End-expiratory rib cage volume was found either unchanged (**Paiva *et al.* 1989**) or increased in microgravity (**Edyvean *et al.* 1991**). Magnetometric determination of changes in rib cage dimensions in seated subjects during parabolic flight confirmed that microgravity is associated with a consistent motion of the sternum in the cranial direction and an increase in the anteroposterior diameter of the lower rib cage as measured at the level of the 5th intercostal space (**Estenne *et al.* 1992**).

Electromyographic recordings from the scalene muscles, which inflate the upper portion of the rib cage during quiet inspiration at 1 × g (**De Troyer 1983**; **Estenne *et al.* 1985**) have shown a consistent decrease in phasic inspiratory activity in weightlessness (Estenne ***et al.* 1992**). This may be part of a reflex referred to as operational length compensation (Green ***et al.* 1978**). When the operating length of the diaphragm increases, as it presumably does in microgravity, the inspiratory activation of all inspiratory muscles is reduced in order to keep tidal volume constant. This mechanism has been demonstrated during changes from upright to supine posture and water immersion (**Druz & Sharp 1981**; **Reid *et al.* 1986**).

It has been predicted that functional residual capacity (FRC) in the weightless state should be intermediate between that found in the upright and supine posture, that is, decrease by about 10% on transition from 1 × g to microgravity (Agostoni & Mead 1964). Early measurements of FRC in parabolic flights have been inconclusive (von Baumgarten *et al.* 1980; Wetzig & von Baumgarten 1987), and chest radiographs made at FRC in seated subjects after 10 s of microgravity showed no significant differences between 1 × g and microgravity, although the lung tended to become somewhat shorter and wider (**Michels *et al.* 1979**). Recent studies have demonstrated a 200–500 ml decrease in FRC during short periods of microgravity. As expected, there was a concomitant rise in end-expiratory oesophageal pressure (**Paiva *et al.* 1989; Edyvean *et al.* 1991; Prisk *et al.* 1993**). The SLS-l data show that FRC was decreased by 15% as compared to standing pre-flight values (Elliott *et al.* 1994). This reduction may be due to blood volume shifts into the thorax and mechanical factors. Measurements of lung volumes in SLS-1 and SLS-2 showed significant changes; FRC was between pre-flight standing and supine values as expected, but RV was reduced by 18%, possibly the result of maximal expiratory effort in the presence of uniform alveolar expansion (West 1991). In upright posture on the ground, the alveoli at the apex of the lung remain large at full expiration because of the distortion of the lung by its weight.

Measurements of VC during prolonged exposure to weightlessness were obtained during the Skylab III and Skylab IV missions. Of the six astronauts studied, four showed a decrease of 10% in VC compared to the preflight control value (**Sawin *et al.* 1976**). Similarly, VC in Shuttle astronauts (Spacelab) was reduced by about 5% on flight day 2 compared to pre-flight standing. However, this reduction was no longer present on subsequent flight days (Elliott *et al.* 1994; **Guy *et al.* 1994**). These data indicate that gravitational unloading produces a slight and transient reduction in VC. This is qualitatively similar to water immersion (**Hong *et al.* 1969**; **Prefaut *et al.* 1976**), simultaneous inflation of an abdominal bladder (**Regnard *et al.* 1990**), and postural changes (Agostoni & Mead 1964). Resting gas exchange showed reduced tidal volume and increased respiratory frequency in flight; alveolar dead space was reduced below preflight values, consistent with the presumed more uniform topographical distribution of blood flow (West 1991).

Short periods of weightlessness during parabolic flight produce only small and inconsistent changes in tidal volume, breathing frequency, and mean inspiratory flow rate (Paiva *et al.* 1989; Edyvean *et al.* 1991). Correspondingly, changes in posture and head-out water immersion do not alter the temporal pattern of breathing at rest (Reid *et al.* 1986). In contrast, prolonged exposure to microgravity may have a different effect. Preliminary results from SLS-1 indicate a 15% reduction in tidal volume accompanied by an increase in breathing frequency in flight compared to pre-flight (Elliott *et al.* 1994). This was obviously not due to increases in interstitial fluid accumulation and stimulation of J receptors, which would result in rapid shallow breathing (**Prisk *et al.* 1993**). The different effects of short and prolonged exposure to microgravity remain to be evaluated.

Nitrogen and argon oscillations have been shown to decrease as well as N_2 Phase IV height and volume in short periods of microgravity (Michels & West 1978) and in crew members of the SLS-1 mission (Guy *et al.* 1994). In contrast, Ar Phase IV height decreased to 40% of control, and Ar Phase IV volume did not change in microgravity. In addition, the presence of Ar Phase IV never became questionable at $0 \times g$. The observation that both height and volume of N_2 Phase IV and the size of cardiogenic oscillations were substantially reduced in microgravity indicates that TLC/RV ratios became much more uniform throughout the lung. However, the persistence of detectable N_2 and Ar terminal rises in microgravity as well as the unchanged Ar phase IV volume demonstrate that airway closure still occurs in weightlessness.

The more uniform TLC/RV ratios in weightlessness produced a marked reduction in Ar slope, which decreased during exposure to microgravity. During exposure to a prolonged period of microgravity the slope of phase III for N_2 was reduced relative to the preflight standing. During parabolic flights, no such changes were observed. It has been suggested that the slope of the alveolar plateau for N_2 is largely gravity dependent, that is, the sloping of phase III would be due to the fact that apical lung units contribute greater proportions of their initial volume during the course of expiration (Anthonisen *et al.* 1970). The observation that the N_2 slope was only slightly reduced in microgravity despite much more uniform TLC/RV ratios argues against this hypothesis. Rather, it suggests that most of the N_2 slope may be due to diffusion-dependent mechanisms within small lung regions (Engel *et al.* 1986; Paiva & Engel 1987). Single breath nitrogen washout measurements showed that ventilatory inhomogeneity remains when weightless; the presence of a phase IV evidenced regional airway closure in weightlessness. The distribution of ventilation was also measured by multi-breath N_2 washouts using tidal volumes, where no change in ventilatory inhomogeneity between preflight standing and in flight occurred. The inhomogeneity was larger, however, during supine compared with upright on Earth, suggesting that ventilatory inhomogeneity during tidal breathing in the upright posture is not primarily gravitational in origin (West 1991).

Distribution of pulmonary blood flow was assessed during the SLS-1 mission (Prisk *et al.* 1993). The cardiogenic oscillations persisted at 60% of the standing pre-flight value during the SLS-1 mission, suggesting that perfusion inhomogeneities in areas of the lung which are relatively close to each other may persist in weightlessness. In the Shuttle missions SLS-1 and SLS-2, a paradoxical finding was that the increase in cardiac output (18%) and ventricular stroke volume (45%) occurred in spite of a reduction in central venous pressure and fall in circulating blood volume. The explanation is unclear but may be due to changes in the pulmonary circulation which altered the cardiac right-left pressure pattern, or to reduced left ventricular compliance; lung volume was not significantly changed (West 1991).

Pulmonary diffusion capacity (D_L), pulmonary capillary blood volume (V_c), and membrane diffusion capacity (D_m) have been measured in four astronauts during the 9-day SLS-1 mission as well as pre- and post-flight in the standing and supine postures. D_L, V_c, and D_m were significantly increased after 1, 4, and 8 days in microgravity. The changes in V_c with microgravity were of similar magnitude as seen with a change from standing to supine on Earth. Postural changes have no substantial effect on D_m, but microgravity caused it to rise significantly, suggesting that there is considerably more surface available for gas exchange. The increase in D_L observed in space was almost twice as great as that elicited by the standing to supine transition, which is directly related to an increase and a more uniform distribution of capillary blood volume throughout the lung (**Prisk *et al.* 1993**).

In the Shuttle missions SLS-1 and SLS-2, D_L was elevated (carbon monoxide method) by 28% compared with preflight standing. This was interpreted as a result of increased V_c and D_L. The latter was expected because of the presumed blood shift to upper body segments which increases filling of the pulmonary capillaries. When comparing the increase in V_c and D_m during transition from upright to supine, and transition from upright to microgravity, it was seen that the increase in capillary blood volume was comparable in the two situations, whereas the rise in diffusing capacity was much larger in the transition to weightlessness, which was unexpected but might be an effect of uniform filling of the pulmonary capillary bed (West 1991).

The mechanisms responsible for airway closure in weightlessness are unknown. Some airways may be closed while others remain open within small regions of the lung at $1 \times g$. Such intraregional inhomogeneity of airway closure may also be present in microgravity, the site of airway closure being then widely dispersed throughout the lung rather than confined to the lung base (**Engel *et al.* 1975**; **Crawford *et al.* 1989**). Peak expiratory flow rate was reduced early in flight compared with preflight upright, consistent with a significant reduction in maximum expiratory time on flight day 2 (West 1991).

2.2.4 Conclusions

Many questions about the impact of space flight on pulmonary function remain unanswered, some still need to be identified. It is unclear why the transverse diameter of the lower rib cage does not increase at $0 \times g$. One explanation is that this dimension is primarily driven by transabdominal pressure.

The distribution of the increased blood volume between the lung and the chest wall in weightlessness is unknown. The lack of any difference between the reduction in thoracoabdominal volume and that in FRC in microgravity phases of parabolic flights suggests that blood volume shifts are relatively small in magnitude during short periods of microgravity. In addition, if two thirds of the volume change are accommodated by the chest wall, then blood volume shifts in microgravity should only have a minor effect on FRC. Its reduction in microgravity would thus be due primarily, if not exclusively, to mechanical factors, a hypothesis which has been tested during space flight: in flight FRC turned out to be between standing and supine $1 \times g$ values (West 1991).

The available data indicate that topographical differences in lung expansion and ventilation are substantially reduced in weightlessness. However, the maintenance of alveolar plateaus with positive slopes, cardiogenic oscillations, and terminal rises in microgravity provides unequivocal evidence for considerable non-gravitational ventilatory inhomogeneity in the normal lung. Virtually nothing is known about neurophysiological adaptation to the profoundly altered biomechanical situation in which the respiratory system finds itself after changing into weightless conditions. Similarly, much remains to be learned about readaptation after transition to 1 × g conditions, particularly after long lasting space flights.

2.3 Interstitial space and fluid regulation

2.3.1 General overview

The matrix system between capillaries, venules and tissue cells has been termed the interstitial space. It contains specialized fibres, fibrils and molecules which form an intricate network and together determine the biophysical conditions, like tunnel flow with certain hydrodynamic resistances and electrical charge, for the percolation of capillary ultrafiltrate and plasma proteins on their way from the blood stream towards the lymphatic drainage system (**Katz 1980; Aukland & Nikolaysen 1981; Aukland & Reed 1993**). Its physiology seems to be delegated to local mechanisms (fluid holding capacity is regulated via net capillary filtration which depends on pre- to postcapillary resistance ratios), whereas blood volume regulation is mainly under central nervous and hormonal influence. The molecular elements of the interstitial matrix are anchored to collagen fibres and to each other, and are slowly but steadily washed away from fluid percolating into the lymphatic and, finally, blood vessels. Hyaluronic acid is one leading component of the interstitial matrix, its concentration in cutaneous lymph exceeds that in plasma by a factor of 10 to 100 times, and its major biodegradation site is the hepatosplanchnic region and the kidneys (Bentsen *et al.* 1986) which has been supposed to exert a pivotal role in the regulation of interstitial compliance (**Lucas & Floyer 1973**).
Because of the crucial position of this interface between exchange vessels and the metabolically active units of life, its heterogeneity and complexity, its biophysical and biochemical features, and a conceivable responsiveness to various humoral inputs, one might think of it as an organ in its own right. This 'interstitial organ' is also a key player in the distribution of body fluid volumes and fluid balance regulation in humans during and after space flight. The heterogeneity of the interstitial organ must be taken into account when performing studies on a local basis, e.g. using girth measurements (Johnston & Dietlein 1977), applying ultrasound thickness determination (**Kirsch *et al.* 1980**), video anthropometry, or mini-invasive tissue investigations as the wick technique (**Snashall *et al.* 1971**; Hargens 1981; **Wiig & Noddeland 1983**). The situation is different when changes in blood composition are related to interstitial fluid dynamics where net sum effects occur.

Plasma proteins can be seen as indicators of such fluid shifts. During steady-state conditions, about 10 g protein h^{-1} move from the plasma compartment into the interstitial fluids; with steady state conditions, the same amount is shuttled back to the bloodstream via the lymphatics (**Renkin 1986**). The relative amount of protein fractions, compared to blood plasma, differs in lymphatic fluid (**Reichel *et al.* 1976**) and depends on relevant conditions (posture, activity) (**Olszewski *et al.* 1977**). The main barrier for albumin exchange is not the capillary wall but the interstitial compartment; the major route of disappearance is the uptake in skin, muscle, lung and gut (Bent-Hansen 1991). The transcapillary escape rate (during steady-state conditions) of albumin is 4.4–4.9% of the intravascular pool h^{-1}, whereas IgG disappears with 2.6–2.8% h^{-1} and IgM only with 1.9% h^{-1} (**Parving *et al.* 1974**). The molecular mechanisms governing this net flux of protein out of the capillary are in the process of being investigated. Current research shows that besides large regional differences, there is a multitude of transport gates, partly receptor-mediated, as shown for albumin and various globulins (Schnitzer 1993). The regulation of body fluid volumes in microgravity is a major element of cardiovascular stability and proper organ functioning, and could be one of the limiting factors within the frame of long-term exposure to weightless conditions.

2.3.2 Ground-based experimental results and microgravity simulation

Since profound interstitial fluid redistribution occurs with space flight, methods are of interest which can mimic such phenomenon on Earth in order to study physiologic principles of interstitial macromolecular and fluid shift regulation. Changes of interstitial fluid content in the interstitial organ can be induced by local application of mechanical force, as with lower body pressure or suction (usually referred to as LBPP, for 'positive', and LBNP, for 'negative' pressure), massage ('lymphatic drainage'), water immersion, or changes of the positioning of body parts (e.g. legs) or the whole body (postural or orthostatic effects). As a well-known example of a frequently employed technique, LBNP is used as a tool to induce venous pooling, and fluid exchange between blood and extravascular space (Musgrave *et al.* 1969), resulting in up to 10% haemoconcentration (Blomqvist & Stone 1983). While a multitude of investigations has looked into blood volume changes with stimuli which change interstitial hydration, few studies have been undertaken to quantify the differential movement of extracellular mobile macromolecules during and after cardiovascular stimuli which disrupt the steady-state of fluid and protein kinetics.

Rapid net transport of fluid from interstitium to blood may occur in skeletal muscle and skin which are large fluid reservoirs, as indicated by plethysmographic studies performed on the human arm (**Lundvall & Länne 1989**). Interstitial volume is reduced, a condition enhancing cardiovascular reflex responses to stress (**Essfeld *et al.* 1993**). An increase in sympathetic tone fol-

lowing the unloading of vascular stretch receptors (Gauer & Thron 1965; **Tripathi *et al.* 1989**) as induced by LBNP could explain transient effects as a volume- and/or baroreceptor driven reflex. Sympathetic drive increases immediately after onset of LBNP (**Rea & Wallin 1989; Joyner *et al.* 1990**); it remains to be proven if and how neural adaptation and/or the intensive in flight fluid loss from the legs changes this mechanism in space.
Transmural pressure adjustment with LBNP (**Tomaselli *et al.* 1987**) causes sequestration of fluid into extravascular spaces which lasts the whole period, and even continues after termination, of lower body suction. It is unclear whether this haemoconcentration occurs after adaptation to microgravity. Head-down tilt produces capillary pressure changes together with changes of the other Starling forces; they have been directly measured in the lower lip of humans using micropressure recordings from tissue and capillaries combined with blood plasma and interstitial fluid colloid osmometry. The results indicated mainly intravascular (pressure and oncotic) changes as responsible for facial edema in short-term (8 h) head-down tilt (**Parazynski *et al.* 1991**). Interstitial fluid shifts have been investigated in long-term bed-rest studies as well. One of them demonstrated unchanged venous-occlusion filtration in the arms, whereas filtration increased in the legs during the first 10 days, to fully recover at the end (day 20) of bed rest (**Bonde-Petersen *et al.* 1994**). Another showed rapid fluid loss during a 24-hour bed rest from the legs, whereas tissue thickness at the forehead continuously went up, with no changes in plasma volume and body mass (Hildebrandt ***et al.* 1994**). This corresponds to the observation of increased forehead and decreased foot cutaneous microvascular blood flow in -6° head-down position, irrespective in magnitude of bed-rest duration up to 1 week (**Aratow *et al.* 1991**); the largest portion of leg interstitial fluid loss occurs within the first 6 h of microgravity simulation (**Hargens 1983**). In another one week bed rest study, orthostatic tolerance was improved through extravascular volume extension (saline ingestion), the effectiveness of which increased further with additional application of LBNP (**Hyatt & West 1977**). The reason might have been lower body interstitial fluid accumulation which could counteract upright plasma volume loss. Similar observations of a tilt-table study led to the conclusion that upright capillary pressure balance is achieved with attenuated plasma volume loss after repeated orthostasis (**Hinghofer-Szalkay *et al.* 1995**). Likewise, smaller interstitial dehydration of the legs may cause better upright blood pressure maintenance after immersion even with a reduced plasma volume (Vogt 1967).

2.3.3 Observations in flight

Few solid data are available on the magnitude and time course of fluid redistribution and readaptation to post-flight $1 \times g$. There are no data about body fluid shifts in the process of adaptation to microgravity cardiovascular and capillary fluid/protein transport parameter changes in simulated or real microgravity. Direct volumetric changes of leg volume have been performed in a very limited number of studies (head-down tilt and bed rest, none in water immersion), and few investigations of fluid shift dynamics exist. Leg volume changes are much larger in space than with application of any simulation model (**Thornton *et al.* 1992**). In addition, the suitability of countermeasures needs to be tested on the grounds of proper understanding of the physiological changes within the cardiovascular system and interstitial organs.

Experimental data on fluid and electrolyte homeostasis during space flight have been inconsistent and are difficult to interpret (**Norsk 1992; Kirsch *et al.* 1993**). Recent measurements revealed increased urodilatin excretion in spite of relative hypovolemia and increased natriuresis (n = 1) (**Drummer *et al.* 1993**). The concentration of blood plasma proteins seems to remain essentially unchanged in microgravity conditions, and intravascular colloid osmotic pressure is unaltered. The redistribution of body fluids along the body axis is obviously due to a unique shift within tissue pressures, volumes, and lymphatic throughput values; the overall pattern of microvascular fluid exchange in astronauts adapted to the weightless state yet remains unclear.

Microgravity profoundly affects the normal distribution and composition of body fluids within minutes to days, the changes being related to adaptation within the cardiovascular and interstitial systems. The redistribution of fluid from the lower body to the chest and head should trigger the physiological sequence of hormonal events to stimulate the kidneys to increase their excretion of water and electrolytes, but in reality, the daily urine volume decreases during the first days in flight. In flight loss of sodium, potassium, and chloride, which is reversed on return to Earth, is frequently observed, however; so the physiological readjustments create a unique pattern within the frame of sustained weightlessness which obviously is a unique biological situation. As a result, the interstitium runs through a peculiar redistribution process upon commencing space flight (Johnson 1975; **Kirsch *et al.* 1993**).

No data exist on ground substance kinetics with space flight adaptation and post-flight readaptation to 1 × g conditions. One investigation has demonstrated a 20% decrease of superficial tissue volume in lower body segments on the 3rd day in weightlessness (**Kirsch *et al.* 1993**). Another experiment was designed to investigate the dynamics of fluid shifts during space flight and suggested a profoundly altered pattern of fluid transients on the 6th day in weightlessness (**Hinghofer-Szalkay *et al.* 1994**). Findings from the German Mir-92 mission have validated hypotheses on fundamental changes of the dynamics of fluid shifts on the lower body induced by controlled cardiovascular stress (Baisch 1993). It is unclear how much neuronal and/or local changes (pre/postcapillary resistance ratio; number of perfused capillaries; altered interstitial compliance) contribute to these phenomena.

2.2.4 Conclusions

There is evidence that the physiology of transcapillary filtration is profoundly altered with adaptation to microgravity. Changes in the hydration of interstitial gel matrix affects the partitioning of plasma proteins between intra- and extravascular spaces. Adaptation to microgravity may alter the amount and distribution of extravascular macromolecules together with interstitial fluid volume and distribution.
Concentration changes in plasma as well as extravascular proteins could be used for calculations of compartmental distribution and kinetics of these compounds. Blood plasma macromolecule concentration profiles could therefore reveal changes within the dynamics of interstitial volume regulation. The pattern of macromolecular exchange between circulating blood, interstitial space, and lymphatics may be altered due to altered reflex patterns which might emerge due to microgravity conditions.
Increased load on central vascular stretch receptors could trigger an immediate sympathetic activity release, resulting in ameliorated precapillary vascular tone and consequently increased microvascular filtration pressure. Comparison of the results from microgravity simulation, and real space flight experiments has demonstrated a much larger volume loss from the lower legs in real weightlessness the physiological reason of which remains unexplained. In summary, whereas blood volume regulation is under a more central regulatory command, the physiological mechanisms governing interstitial pressures, volumes, and composition during and after space flight are under local control and need to be investigated by use of methods which concentrate upon the local tissue level, as well as by precise monitoring of changes in the corresponding blood biochemistry. Hypotheses should be tested in future flights that due to altered reflex patterns, not only the direction of transcapillary fluid movement with low-level LBNP reverses in the early microgravity adaptation stage, but also the pattern of macromolecular exchange between circulating blood, interstitial space, and lymphatics is altered.

2.4 Endocrine system

2.4.1 General overview

Adaptation of endocrine systems to weightlessness is an important question of biomedical research. 'Static' measurement of hormone concentrations in blood plasma or other body fluids are done without additional challenge (hormonal adaptation to space flight as judged from the transition from one steady-state to another), whereas stimulation techniques are used for 'dynamic' investigations of short-term stimulus-response patterns in order to perform systems studies and compartment analysis (e.g. **Norman 1992**). Magnitude and time course of hormonal changes after certain physiological stress factors reveal the responsiveness of various hormones and allow for systematic analysis of the underlying physiology of endocrine mechanisms. The time course of adaptation differs for various physiological systems; some physiological functions need no more than seconds to hours to adapt, whereas others take weeks, months, or even years to do so.

Metabolic or volume load can be created, for example, by glucose or saline load; the astronauts health and fitness is not jeopardized by these procedures, however. Another option which can easily be employed in space is to decrease baroceptor load by means of LBNP. This causes shifts in various hormones which trigger renal and cardiovascular responses that tend to minimize the loss of body water and promote vasoconstriction by altered peripheral sympathetic outflow (Smith & Ebert 1991; **Tidgren *et al.* 1990**). It is not known how dynamic responses to acute cardiovascular stress are altered with adaptation to space flight. In order to study hormones, particularly peptides, availability of adequate processing and storage devices in flight are required.

Another problem is the pulsatile release of many hormones; appropriate investigations must therefore include particular sampling and mathematical data processing strategies (**Kushler & Brown 1991; Pincus & Keefe 1992; OMeara *et al.* 1993**). Frequent blood sampling may be necessary to investigate time-dependent parameter changes after acute stimuli. Different subsystems may exhibit specific hypogravity sensitivity curves, so that the maximum response of one subsystem can occur at different gravity stress than that of other subsystems.

2.4.2 Ground-based experimental results and microgravity simulation

Simulated microgravity reveals several adaptation patterns which can in part be also observed in real space flight. Head-down tilt can produce complex time courses of hormone concentration, e.g. an early decrease of volume-regulating hormones (vasopressin, renin activity, aldosterone) which is reversed after several hours (**Nixon *et al.* 1979**); atrial natriuretic peptide is driven to an opposite time course (Blomqvist & Stone 1983; Rowell 1986). Four days bed rest resulted in an increased plasma renin activity; aldosterone was found both reduced (Dallman *et al.* 1984) or increased (**Annat *et al.* 1986**); no change in vasopressin and neurophysin-I (a non-hormonal vasopressin precursor from the hypothalamic-pituitary system) was found in the latter study. Another investigation found no changes in vasopressin, cortisol, and corticotropin plasma concentration with one week -6° head-down tilt bed rest (Vernikos **et al. 1991**). These findings are partly in conflict with actual space flight data. Plasma catecholamine concentration and urinary excretion of vanillyl-mandelic acid were found lower than with control conditions in a 2–3 week bed rest study (Chobanian ***et al.* 1974**); on the other hand, a 4-day

bed rest study failed to observe changes in plasma catecholamine and methoxyamine levels or urinary excretion (**Pequignot *et al.* 1985**).
Other bed rest studies resulted in increased plasma insulin (and glucagon) concentrations and reduced glucose tolerance (Greenleaf 1983; **Shangraw *et al.* 1988**). It has long been known that bed rest immobilization largely increases blood glucose levels after standard oral glucose loading (**Blotner 1945; Lutwak & Whedon 1959**) and that higher insulin levels are required to handle identical amounts of glucose load in sedentary humans (**Naughton & Wulff 1967**). Clinical consequences are evident since patients kept in prolonged bed rest display physiologically abnormal glucose tolerance test results.
Water immersion is another tool for assessing hormonal effects of space flight because of similar changes in central haemodynamics and regulation (**Epstein 1992**). Vasopressin excretion decreases with water immersion; plasma concentration with water immersion is suppressed only in case of (relative) dehydration. Besides cardiopulmonary receptors, arterial baroreflex mechanisms seem to be involved in the vasopressin response to immersion. A selective suppression of the renin-aldosterone-axis is another feature of water immersion. Due to the acutely elevated central blood volume, atrial natriuretic peptide release is enhanced and plasma atrial natriuretic peptide concentration is increased. Renal water excretion returns to control after 7 h immersion, corresponding with a similar decrease of plasma natriuretic hormone (**Stadeager *et al.* 1992**). Norepinephrine, renin activity, aldosterone and vasopressin are low for at least half a day; sodium, urodilatin, and cGMP excretion rates are increased over pre-immersion values for several hours (**Stadeager *et al.* 1992; Nakamitsu *et al.* 1994**).
Depletion of electrolytes and trace elements could impair performance due to their vital roles; this question has been addressed in a cold water immersion study (thermal protection, intermittent exercise, no fluid repletion). Parallel to an about 2-3-fold increase in urine flow, the excretion of sodium rose about 3-fold, that of potassium, calcium and magnesium doubled; urinary zinc loss was increased 3 to 4-fold (evening compared with morning conditions) during 6 h immersion (**Deuster *et al.* 1989**). Whether these changes were directly mediated by renal mechanisms or by hormonal agents remains to be studied.

2.4.3 Observations in flight

Space flight causes numerous endocrine systems to respond and eventually to find new steady states. It has originally been expected that due to the absence of gravitational pull, energy requirements should be reduced in space (Grigoriev & Egorov 1992a & b). The caloric requirements are primarily a function of overall workload, and space travellers may have unchanged or even increased energy input due to exhaustive activity in space (**Lane 1992**). Insulin was found decreased after 2 weeks in space and increased after most flights (Leach Huntoon ***et al.* 1989**). Catabolic muscle wasting in the lower body occurs due to low mechanic load; interestingly, administration of growth hormone did not affect the atrophic response (m. soleus) in rats flown 4 days in space (**Jiang *et al.* 1993**) where growth hormone and prolactin activity seem to be reduced during space flight (**Sawchenko *et al.* 1992; Hymer *et al.* 1992;** Popova & Grigoriev 1992).

Stress sensitive hormones (corticotropin, catecholamines, corticoids, thyroxine, growth hormone, insulin) are increased after short term flights. Increased urinary excretion of interleukin-6 also indicated a stress response immediately after entering Earth orbit or reentry into gravity (**Stein & Schluter 1994**). Serum cortisol was increased with longer missions as well, whereas corticotropin and insulin decreased (Johnston & Dietlein 1977; **Vorobyov *et al.* 1983; Leach *et al.* 1983;** Leonard 1985; Noskov *et al.* 1989; Macho *et al.* 1991; Grigoriev *et al.* 1992; Huntoon 1992). High epinephrine concentrations accompanied by relatively low norepinephrine, cortisol and corticotropin levels after long-term space flights (**Vorobyov *et al.* 1983**; Kvetnansky *et al.* 1988; Macho *et al.* 1991) seem to indicate a moderate stress response with the prevalence of an emotional component.

Plasma vasopressin levels depend on hydration and osmotic status (Noskov *et al.* 1986). A 4.5-fold vasopressin increase on the 5th, a 5-fold elevated level on the 6th, and a 7-fold increase on the 20th day in space occurred (Grigoriev *et al.* 1992; **Hinghofer-Szalkay *et al.* 1993**). This might be interpreted as a sign of relative dehydration, possibly connected to a reduced pressure gradient in the low-pressure system and decreased cardiac preload. On the other hand, plasma vasopressin was unchanged in SLS-1 Shuttle astronauts (Huntoon 1992). With 7 months flight duration, renal excretion of vasopressin increased, and vasopressin serum concentration fell in 2 cosmonauts (**Noskov *et al.* 1989**).

Plasma renin, angiotensin, and aldosterone levels increased early in flight in Skylab astronauts (Johnston & Dietlein 1977; Leonard 1985); however, angiotensin was below control one day into the flight of SLS-1 (Huntoon 1992). Aldosterone was above control values by 75% on the 5th and by 20% on the 6th, and was back to control level on the 10th flight day. LBNP-induced hormone changes were investigated in one subject on the 6th day in space and were found to be different from 1 × g control. The response of aldosterone was 10 times higher in flight than pre-flight, whereas all other hormones were diminished. The post-flight LBNP response was increased for plasma renin activity but lowered for atrial natriuretic peptide, compared to before or during flight (Grigoriev *et al.* 1992; **Hinghofer-Szalkay *et al.* 1993**). With seven months flight duration, aldosterone excretion increased, which was paralleled by a decrease of 11-DOC (a precursor of aldosterone), suggesting its mobilization for aldosterone synthesis (Noskov *et al.* 1989). Plasma renin activity was elevated on SLS-2 flight day 8 and on landing day (Huntoon 1994, personal communication).

A 3% hyponatremia and decreased body fluid osmolality was found in Skylab missions (Johnston & Dietlein 1977; Leonard 1985) which was not anticipated and would be difficult to explain. It was hypothesized that atrial natriuretic peptide may contribute to this phenomenon, the adaptive value of which might be a protective role of extracellular volume at the expense of extracellular osmolality. Atrial natriuretic peptide was measured in more recent Shuttle and Russian missions; it consistently decreased in SLS-1 (day 1: 20%, day 8: 60%) and SLS-2 (flight days 1, 8, 12) in the presence of sodium excretion, suggesting that it was not significantly associated with urinary sodium levels during space flight; serum osmolality was not changed (Huntoon 1992; Huntoon 1994, personal communication).

Post-flight, increased serum growth hormone, thyroid hormones, angiotensin, and aldosterone were observed while angiotensin and aldosterone findings were inconsistent (Johnston & Dietlein 1977; Leonard 1985; Noskov *et al.* 1989; Macho *et al.* 1991; Grigoriev *et al.* 1992; Huntoon 1992; **Hinghofer-Szalkay *et al.* 1993**).

2.4.4 Conclusions

Space flight activates the sympatho-adrenal system due to a variety of stress factors. The neuroendocrine responsiveness to stimulation seems to be altered by weightlessness, but this needs to be confirmed by further investigations. Research into energy utilization during space missions must include the development of reliable predictors of energy requirements, particularly for long-term flights. Simulation and space flight data are partly in contradiction; integrated investigations must be done to elucidate the cause of these differences. It seems necessary to do parameter estimations and to identify biologically important frame conditions, like stress level, hydration, and nutritional/metabolic status of the astronauts more thoroughly, and to take into account gender differences as well (Sandler & Winter 1978; **Vernikos *et al.* 1991**).

Conflicting hormone data, for example substantial early in flight rise in serum vasopressin observed in several cosmonauts but not in Shuttle astronauts, may depend on specific conditions (crew preparation, pre-launch hydration and positioning, air humidity, stress factors). These various circumstances need to be defined before results can be correctly interpreted.

2.5 Immune system

2.5.1 General overview

A stable immune status is a prerequisite to maintain health and resilience of astronauts. The immune system interacts in many ways with the whole organism. A large part of the information traffic within this system is mediated by cytokines, hormone-like peptides produced by lymphocytes (lymphokines), monocytes/macrophages (monokines), etc. They include, for example, interleukins, haematopoietic factors (CSFs), interferons, and tumour necrosis factor. Immune reactions are modulated by cytokines, which are released primarily as a result of environmental stimuli. Except for lymphokines, cytokines can be produced by many different cell types, and their biological actions are broad and overlap with those of others. They also regulate each other by interaction, competition, and induction in lymphokine circuits and cascades (**Lord 1990**). This complex network is further complicated by the fact that cytokine actions are influenced by other biological signal compounds, as prostaglandins, leukotrienes, corticosteroids, or neuroendocrine peptides. Ligand–receptor interactions add to the complexity of the system the quantitative regulation which is still far from being understood.

Any impairment of the immune system is potentially fatal, therefore the possibility that it could be adversely affected by space flight was of great concern. An autoimmune response to cardiac muscle may have occurred in some cosmonauts, though these findings were not subsequently confirmed. A marked impairment of T–cell function was noted in crews returning from long duration missions, but normal behaviour was re–established a few days after landing, without any clinical significance. Only minor changes in some Ig elements were observed, usually after intermediate and long–duration flights.

In vitro experiments are useful to analyze leucocyte sensitivity to altered gravitational acceleration. They are, however, subject to the extent to which cells contact one another. Since cell–cell interactions mediated by direct contact or by soluble factors are critical for proper leucocyte function, this may obscure the level at which gravity affects immune responses. As an example, responsiveness of T cells to mitogens would be optimized when cells are in close proximity, and relative concentrations of cytokines are high. Some experimental systems, e.g. to measure T–B cell co–operation, actually require short periods of centrifugation to enhance T–B cell conjugate formation and B cell function. It is conceivable that cells in suspension have less opportunity to interact and would be less responsive to immune stimuli. Gravity may have differential effects on distinct elements of the immune system; it would be of great interest to learn if helper and suppressor T cells exhibit opposite and dose dependent responsiveness to microgravity.

2.5.2 Ground–based experimental results and microgravity simulation

Lymphocyte responsiveness is depressed in simulated and real weightlessness. Decreased adhesion may contributes to the depressed *in vitro* responsiveness (**Gmünder *et al.* 1990**). The experimental protocol used can significantly influence responsiveness of white cells (Cogoli & Gmünder 1991). *In vivo* immune responses to hypergravity conditions were studied in animals: in chickens centrifuged at 7 to 8 × g for durations from one minute to several h, lymphocyte counts increased transiently by 30% (**Burton & Smith 1972**; **Smith *et al.* 1979**). Skin graft rejection in rats under different gravita-

tional load was investigated by VanWyk & Brede (1973). Graft rejection requires T-cell recognition of the antigens and cytokine mediated recruitment of inflammatory cell subsets. Whereas control (1 × g) rats rejected skin grafts in 11 days, rats maintained at 2.5 × g tolerated the grafts for an average of 20 days. In the same animals, serum IgG increased, whereas white blood cell counts were reduced during the first 3 days, and returned to normal after 21 days. In contrast, Scibetta *et al.* (1983) were unable to find changes in the antigen-specific antibody titres, haematocrits, or leucocyte counts between control and experimental rats which were centrifuged at 2.1 or 3.1 × g.
Human lymphocytes in a low gravitational acceleration *in vitro* environment, simulated by a fast rotating clinostat, are about half as active in their response to the mitogenic lectin concanavalin A (Con-A) as 1 × g controls (**Cogoli *et al.* 1980**). Cells cultured for 2–3 days in the clinostat and subsequently at 1 × g also exhibited depressed responses to Con-A, however, suggesting that the depressed responsiveness was due to a nonreversible signal delivered to the lymphocytes under simulated microgravity. Because of unique changes in cell ultrastructure as assessed by electron microscopy, it was argued that a gravity induced shift in organelles influenced the mechanism of lymphocyte growth and differentiation.

2.5.3 Observations in flight

Early indications of an impact of space flight on the immune system include the following observations. The plasma and skin bactericidal activity was affected in dogs after short-term space flight, with phagocytotic competence of neutrophils most affected; vibration and acceleration stress led to immunological changes in these animals which persisted for months or even years (Alekseyeva & Volkova 1963). Thymus and spleen tissue of rats flown in space for 22 days were depleted of resting lymphocytes and blast cells, while splenic neutrophils were increased in number. A decrease in erythropoiesis may be due to hyperoxia which also could have induced hyperactivity in the adrenal glands with subsequent hypoplasia in lymphoid organs (**Durnova *et al.* 1976**). Rat lymphocyte interferon-gamma production (7-day SL-3 mission) was decreased, consistent with results from earlier studies. Con-A induced IL-3 (interleukin) production by the same population of lymphocytes was unaffected, however (**Gould *et al.* 1987**).
Crew members of Apollo 7 to Apollo 13 (n = 21) were tested for the ability of their lymphocytes to respond to phytohaemagglutinin (PHA), with no significant effects (Fischer *et al.* 1972). These results must be taken with caution since the method by which the cells were isolated probably depleted B cell and macrophage populations, which contribute to PHA responses. 15 of 29 Apollo astronauts complained of respiratory tract infections in flight or shortly after returning to Earth (Hawkins & Ziegschmid 1975). Increased responses to the T-cell activator PHA were observed after a 20-day flight (**Mandel & Balish 1977**) which could reflect stimulated immune responses.

Increased post-flight levels of leucocytes, particularly neutrophils, were found in the members of the 9 day ASTP which could reflect subclinical infections (Criswell & Cobb 1977). Lymphocytes from the three astronauts were poorly responsive to PHA, while lymphocytes from two of them responded weakly to Con-A. The apparent decrease in T cell responsiveness to mitogens was supported by additional studies: human lymphocytes flown on board Soyuz 6, 7, and 8 (5 days), Skylab 2, 3, and 4 (28, 59, and 84 days), and Salyut 4 (30 days) were poorly responsive to standard mitogenic signals (**Cogoli & Tschopp 1985**). Lymphocytes from the crew of Salyut 6 (140 days) also displayed decreased PHA reactivity and a reduced number of T cells (**Mandel & Balish 1977**). Proliferation was decreased (Spacelab-1 and D-1) by more than 90% compared to ground control, and 80–90% compared to the on board 1 × g centrifuge (Cogoli & Gmünder 1991).

In the first Spacelab mission, human leucocytes were sealed in culture flasks and placed in a carry-on 37° incubator; ground control cultures were established in parallel. Six hours after liftoff, the incubator was activated and Con-A was injected into three flasks. After 69 hours, ^{3}H-thymidine was injected and the incubation allowed to continue for two hours. After injection of a cryopreservative, the cultures were frozen. The in flight Con-A responses turned out to be decreased by more than 97% as compared to the ground controls. This was probably not due to a decrease in intercellular contact required for T cell activation and might reflect gravitational affects on the function of accessory macrophages present in the culture as well as on T cells (**Cogoli & Tschopp 1985**).

In another experiment flown on board Salyut 6, the induction of interferon-alpha in cultured human lymphocytes was determined (Tals *et al.* 1983). In flight cultures produced 4–8 times as much interferon as ground controls. In contrast, lymphocytes obtained from cosmonauts aboard Salyut 6 produced less interferon post-flight than pre-flight, and there was diminished NK-cell activity.

Mitogen responses to various doses of PHA for several incubation times were evaluated from astronauts of the first 4 Shuttle missions. A significant decrease in PHA responsiveness was noted for each astronaut with a corresponding 2–3 fold increase in neutrophils. The diminished T cell reactivity was roughly correlated with a subjectively evaluated level of stress experienced in flight (**Taylor & Dardano 1983**). Some Salyut 6 cosmonauts displayed a positive allergic reaction to *Streptococcus* antigen post-flight, while pre-flight there was no allergic reactivity. Salyut 7 cosmonauts flown for 140–237 days demonstrated an increased propensity for allergic responses to bacterial antigens. The ability to respond to antigens from common bacteria indicated a relatively recent response to an antigenic challenge. The results could have been due to an attenuation of some element of the immune response followed by susceptibility to, and infection with, endogenous microbiota (Konstantinova 1986).

A coordinated, unspecific series of reactions, the so-called acute-phase response, is a physiological mechanism answering stress situations of various kind, as injury or infection. It serves to limit the extent of injury (e.g. by antiprotease activity) and protects the organism against further stress. It was hypothesized that the stress of space flight might elicit such a response. The hypothesis was tested by measuring urinary interleukin-6 (IL-6) excretion. The result was an increased IL-6 (and cortisol) excretion only on the first in flight day in 2 astronauts, and an increased IL-6 excretion in 2 out of 4. It was concluded that the proposed stress effect does occur only during the first in flight day (**Stein & Schluter 1994**).

2.5.4 Post-flight observations

The post-flight immunological changes found in animals or humans can be summarized as follows. Lymphocyte response to mitogen stimuli decreases (**Taylor & Dardano 1983; Cogoli & Tschopp 1985; Sonnenfeld *et al.* 1990**); total blood lymphocyte count is reduced (**Taylor & Dardano 1983**); blood neutrophil count is increased, while eosinophil count is reduced (**Taylor *et al.* 1986**); a slightly decreased production of the cytokines IL-2 and INF-γ occurs (Konstantinova 1991; **Sonnenfeld *et al.* 1992**); CD4-positive lymphocytes are slightly elevated (Peres *et al.* 1994); number and activity of natural killer (NK) cells is significantly reduced (Konstantinova *et al.* 1994); B cells and monocytes are slightly affected (Taylor 1987); and delayed-type hypersensitivity is blunted (**Taylor & Janney 1992**). Leucocytes were not adversely affected by *in vitro* storage under 7 days in microgravity as judged by various activity/viability indicators (**Meehan *et al.* 1989**). Taken together, these data from American and Russian missions provide evidence that immune function is to some extent impaired with space flight.

2.5.5 Conclusions

Each particular experimental approach has inherent limitations. Although it is likely that changes in the immune system can be expected as humans spend extended time in space, detailed effects of long-term space flight on lymphopoiesis have not yet been addressed. Short-term unspecific stress effects (IL-6 excretion) at the beginning and after ending space flights have been documented recently in a limited number of astronauts. In-depth investigations of immune system adaptations to the conditions of space flight, including effects of radiation, nutrition, low physical load, psychic stress, still need to be investigated in detail.

2.6 Muscular system

2.6.1 General overview

Muscle loading is necessary to maintain the balance of contractile protein synthesis compared with catabolism, and muscle mass. Comprising 40–45% of whole body mass, the muscular system contributes most significantly to body composition and weight. Consequently, inactivation of the weight-bearing muscles as in weightlessness has an impact on their mass and functional properties. Generally, reduced activation due to mechanical unloading of skeletal muscle exerts disuse atrophy, i.e. it decreases expression of protein isoforms comprising the contractile system (**Caiozzo *et al.* 1992; Diffee *et al.* 1991, 1993**). This leads to reduced muscle mass, measurable as decreased cross sectional area.

A reduction in body weight following space flight has been a constant feature of early manned space flight programmes. The magnitude of this loss in missions of less than 6 month duration, averaging 6–7 kg, was independent of flight time, with a rapid fall over the first 3–5 days in microgravity, followed by a lesser decline over the rest of the mission. On return to Earth, there was an equally rapid restoration of much of the lost mass over the first few days, and pre-flight levels were usually regained within 2–3 weeks. The early phase of in flight weight reduction has been attributed to the loss of fluids described above, while the later sustained decline included the effect of bone and muscle loss. On the other hand, a gradual weight gain was observed in some Skylab astronauts and Salyut cosmonauts in whom evidence of muscle and bone breakdown had already been seen; this has been explained by an increase in adipose tissue greater than the loss of lean body mass. In most cases, however, fat and muscle mass change in parallel (Forbes 1987).

Exercise capacity is affected by space flight. As a countermeasure, graded in flight stress protocols were soon employed, e.g. involving a cycle ergometer for Skylab. Workload, heart rate, blood pressure, respiratory gas exchange (oxygen consumption and carbon dioxide production), cardiac output and stroke volume were recorded; results showed significant reduction in exercise capacity post-flight compared with pre-flight. At a heart rate of 160 beats per minute, workload, systolic and diastolic blood pressures, cardiac output, stroke volume, and oxygen consumption were all decreased, although the amount of oxygen required to perform a given amount of work (efficiency) was unchanged.

2.6.2 Ground-based experimental results and microgravity simulation

Removal of mechanical loading leads to muscle atrophy as observed with disuse during casting (**Deitrick *et al.* 1948; Patel *et al.* 1969**; Sargeant ***et al.* 1977**), prolonged bed rest without exercise (Saltin *et al.* 1968; Krupina *et al.*

1982), and space flight. A useful ground-based animal model of the effects of space flight on muscle atrophy is hind limb suspension: the legs cannot be used to support the body weight or even allowed to touch any surface. The effects are similar in degree, kind, and affected muscle groups as those observed in space flight, and similar to changes seen in paralysed muscles (**Baldwin *et al.* 1984; Roy *et al.* 1984**). In response to hind limb suspension, it appears that the muscles which atrophy the most, predominantly slow muscles, do not become more fatiguable (**Winiarski *et al.* 1987; Herbert *et al.* 1988**); in the same group of rats there was no activity loss of mitochondrial enzymes that play a primary role in providing energy (**Hauschka *et al.* 1987**). Such maintenance of oxidative enzyme activities up to a month after hind limb suspension has been observed also in rats that have endured space flights of 7–12 days (**Martin *et al.* 1988**). Therefore, even though the muscle maximum force potential is significantly reduced in space flight, the potential of the muscle to sustain force relative to the reduced maximum force appears to be maintained; maximum force changes with available cross-sectional area.

Various forms of muscle training have been employed as countermeasure against muscle wasting in simulated weightlessness. Magnetic resonance imaging has been proven as a highly useful noninvasive approach to assess muscle disuse and countermeasure efficiency by determining cross-sectional areas of selected limb muscles (**Berry *et al.* 1993**). Torque-velocity measurements indicated a larger effect of 30-day bed rest on lower limb extensor than flexor groups (**Dudley *et al.* 1989**). Besides muscle strength measures, anthropometry, and X-ray tomography, percutaneous muscle biopsy has been used to investigate the effects of prolonged head-down tilt immobilization as an analogue for microgravity (**Buchanan & Convertino 1989**). The results indicated that decreased fibre size was paralleled by reduced aerobic energy supply capacity and ultrastructural disorganization (**Hikida *et al.* 1989**). Other 30-day head-down tilt (-6°) bed rest investigations showed that isokinetic exercise was a more efficient countermeasure than isotonic ergometry training (**Ellis *et al.* 1993; Greenleaf *et al.* 1994**).

2.6.3 Observations in flight

In space, there is a profound functional redistribution within the musculoskeletal system, due to the altered activity and force patterns. Muscle force, velocity of movements, and endurance are affected in some if not all muscle groups by space flight. Early observations of partial disuse catabolism were, for example, an increase in flight urine and faecal calcium excretion with an apparent decrease in intestinal calcium absorption. Faecal calcium continues to increase with flight duration, with a calculated total skeletal loss from excess urine and faecal excretion of 25 g calcium in 84 days. The total measured negative calcium balance was about 1% of the calcium in the average skeleton lost over the three month period (Rambaut *et al.* 1979).

Countermeasures such as running on a treadmill to load the skeleton are usually employed; if the exercise does not transmit appropriate load to parts of the weight-bearing system, then muscle and bone are lost from those sites. A series of pre-flight/post-flight measurements for durations of up to eight months revealed that in contrast to the hypothesized loss of trabecula bone from the vertebral body due to unloading, there is little or no bone loss at this site, apparently validating the effectiveness of the exercise regimen in preventing spinal bone loss as well as minimizing calcaneal losses (Oganov *et al.* 1990). The Russian exercise countermeasure programme was designed to simulate the vertical loads produced on the bones during normal ambulation or running, but it did not have a specific component to exercise the antigravity postural muscles in the back. These muscles hold the body erect while standing or help return the torso to its normal position when twisting movements are performed.

Mechanical load exerted by these muscles on the spine are not transmitted to the vertebral bodies, but to the transverse and spinous bony processes to which the muscles are attached, and consequently if the muscles are not exercised they will not load the bone. Analysis of the 3-dimensional computed tomography data showed that even though the vertebral bodies did not lose bone, the posterior elements lost an average of 8%, and this loss correlated with a 4% decrease in the volume of the muscles attached to the bone (Oganov *et al.* 1990). A decrease in the mass and strength of muscles in the back necessary to perform torsional tasks such as assembly during EVA is of operational concern. After 1 year in orbit, one cosmonaut averaged 10% loss of trabecula bone from the upper lumbar vertebrae while the other gained 5% mineral density as measured using dual photon absorptiometry (Grigoriev *et al.* 1991). The X-ray tomography methods were not the same as used in prior flights, and compartmental analyses could not be performed. The data obtained from these studies are observational rather than experimental; the small number of subjects available for study and the constraints placed by operational concerns limits interpretation. Further design of experiments testing hypotheses about musculoskeletal homeostasis in short term flights and over extended durations needs to be carefully evaluated.

Animal studies from short duration flights show that extensor muscles atrophy more severely than flexors and that those extensor muscles which consist of predominantly slow fibres atrophy more rapidly than extensor muscles that consist of predominantly fast fibres. The mean cross-sectional area of muscle fibres of the slow soleus muscle of rats, for example, was reduced by a third after a seven-day space flight (**Martin *et al.* 1988**). More atrophy was noted in the soleus of rats flown on a 12 day Cosmos biosatellite mission; similar degrees of atrophy were found following Cosmos satellite flights lasting twenty-two days (**Ilyina-Kakueva *et al.* 1976**). Severe muscle atrophy can occur within a few days, in extensors more rapidly than flexors, and in slow contracting fibres more severely than in fast ones.

Muscles that have an antigravity function, such as the calf and quadriceps muscles of the leg, rapidly shrink during space flight while others are barely affected. Hip, back, and neck extensor muscles have not been studied in sufficient detail to quantify the degree that they are affected. Calf muscle maximal torque of two cosmonauts measured at 180° s^{-1} was reduced by more than half of the preflight values when tested on the second post-flight day after a 175-day mission (Kozlovskaya *et al.* 1981). Differences were less severe when the maximal torques were tested at lower velocities of movement, suggesting that the reduction in the torque may be due to a loss of velocity as well as force potential. Torque measurements for thigh muscles showed a similar pattern but the changes were less severe than the calf muscles. The pre-post-flight torque differences in the knee were less than those reported after seven days of dry immersion.

The crew of Soyuz 9 had a significant loss in muscle mass and strength of the lower limbs along with marked dysfunctions in movement control, particularly in maintaining an upright posture (Cherepakhin & Pervushin 1970). After 175 days in space, the EMG signal to torque ratio was doubled in these cosmonauts calf muscles, not so in cosmonauts of another, 140-day mission (Kozlovskaya *et al.* 1981; **Clement & Lestienne 1988**). Similar torque-velocity tests of Apollo astronauts suggest a significant loss of force output at the higher velocities of shortening, particularly among extensor muscles of the leg and trunk (Thornton & Rummel 1974). Changes were less in the flexors of the legs and generally in arm muscles. During the Apollo-Soyuz mission, a decreased level of 'electrical efficiency' (integrated EMG over force) was found as a result of 9 days of weightlessness both on the upper (m. brachioradialis) and lower extremities (m. gastrocnemius). It was concluded that the probable site for the microgravity effect is not neural but the muscle fibres themselves (LaFevers *et al.* 1977).

Muscles from rats have been studied following 2-25 day flights (**Ilyina-Kakueva *et al.* 1976**; Chui & Castleman 1980; Martin *et al.* 1988; Rapcsk *et al.* 1990; **Jiang *et al.* 1993**). It appears that the slow extensor muscles of the leg, back, and neck may become more fatiguable because of muscle atrophy, not because of an excessive loss of the supportive metabolic enzymes. In contrast, arm and hand muscles probably would not fatigue faster during space flight than on ground.

2.6.4 Conclusions

Although data have been accumulated over more than 25 years of manned space flight, little is known about the effects of microgravity on musculoskeletal metabolism. Most of the data have been obtained under sub-optimal experimental conditions and with a variety of confounding factors influencing the results. One of the major limitations is that the amount and kind of exercise has differed from individual to individual and from flight to flight. Scientifically valid conclusions on the quantitative effects of space flight without the complication of exercise training on muscle atrophy and function based on solid scientific data can be drawn only in a very limited context.

In response to space flight, the overall shape of a muscle may change. Therefore, changes in anatomical muscle cross-sectional area as occur in space flight may not reflect the physiological impact. Another factor to consider is the change in the relative distribution and amount of fluid within the legs and arms associated with space flight, thus altering muscle volume and, perhaps, muscle cross-sectional area. If these potentially complicating factors are not adequately considered in estimating functional cross-sectional area, the assessment of actual loss of functional muscle mass can be seriously invalidated. The most accurate approach to assess the percent loss of muscle in space flight is to measure the cross-sectional area of a population of muscle fibres by studying biopsy samples.

Exercise is an obvious countermeasure procedure for fighting disuse atrophy in space. Alternately, electrostimulation could be employed if the amount of actually performed exercise is not able to balance the mechanical unloading effects due to microgravity (Duvoisin *et al.* 1989). It seems that novel ways should be identified which allow the astronaut to exercise for an extended time at the desired level without causing overly cumbersome side-effects like excessive boredom, skin lesions, etc. The most logical approach from a physiological point of view would be to employ artificial gravity. The lack of motivation to activate large parts of the musculoskeletal system in microgravity conditions stems from the absence of any 'biological need' to do so, and therefore this is one of the examples how humans must work against an adaptation process which makes physiological sense in the given situation, that is to lose unnecessary muscle mass in weightlessness, but would have dangerous consequences when re-entering $1 \times g$ conditions, particularly after longer-lasting travels in space.

2.7 Postscript

An overall need for the in flight application of experiment standards as used in ground-based laboratories, regardless of operational constraints, is evident and would be extremely important in any fluid-electrolyte, volume balance, interstitial, and endocrine studies. Lack of adherence to such standards has blurred the patterns caused specifically by weightlessness (without additional confounding factors) and rendered interpretation of data obtained so far very difficult, in some cases impossible.

2.8 References

Agostoni E & Mead J. Statics of the respiratory system. In: *Handbook of Physiology. Respiration.* Ed. by Fenn WO & Rahn H. Washington DC: Amer. Physiol. Soc. sect. 3, vol. I, pp. 387–409, 1964.

Alekseyeva OG & Volkova AP. Influence of space-flight factors on the bactericidal activity of the body. In: *Problems of Space Biology* Vol 1. Ed. by Sisakyan NM. NASA TT F-174, Washington DC, pp. 201–10, 1963.

Annat G, Güell A, Gauquelin G & Vincent M. Plasma vasopressin, neurophysin, renin and aldosterone during a 4-day head-down bed rest with and without exercise. Eur. J. Appl. Physiol. 55: 59–63, 1986.

Anthonisen NR, Robertson PC & Roos WRD. Gravity-dependent sequential emptying of lung regions. J. Appl. Physiol. 28: 589–95, 1970.

Aratow M, Hargens AR, Meyer JU & Arnaud SB. Postural responses of head and foot cutaneous microvascular flow and their sensitivity to bed rest. Aviat. Space Environ. Med. 62: 246–51, 1991.

Arieli R, Boutellier U & Farhi LE. Effect of water immersion on cardiopulmonary physiology at high gravity (+Gz). J. Appl. Physiol. 61: 1686–92, 1986.

Atkov OY. The state of cosmonauts cardiovascular systems during long term orbital flights. Bul Bsesoyuznogo Kardiolog Nauchnogo Tsentra AMNSSR. VIII(2): 97–100, 1985. Translated in: USSR Space Life Sciences Digest. 7: 27–28, 1986. NASA Contractor Report No. 3922(08).

Atkov OY, Bednenko VS & Fomina CA. Ultrasound techniques in space medicine. Aviat. Space Environ. Med. 58: A69–73, 1987.

Aukland K & Nicolaysen G. Interstitial fluid volume: local regulatory mechanisms. Physiol. Rev. 61: 556–643, 1981.

Aukland K & Reed RK. Interstitial-lymphatic mechanisms in the control of extracellular fluid volume. Physiol. Rev. 73: 1–78, 1993.

Badeer HS. Gravitational effects on the distribution of pulmonary blood flow: hemodynamic misconceptions. Respiration 43: 408–13, 1982.

Baisch F. Orthostatic stress by lower body negative pressure and its body fluid distribution kinetic under microgravity. The Physiologist 36: S-135–8, 1993.

Baldwin KM, Roy RR, Sacks RD, Blanco C & Edgerton VR. Relative independence of metabolic and neuromuscular activity. J. Appl. Physiol. 56: 1602–7, 1984.

Beck L, Baisch F, Gaffney FA, Buckey JC, Arbeille P, Patat F, TenHarkel ADJ, Hillebrecht A, Schulz H, Karemaker JM, Meyer M & Blomqvist CG. Cardiovascular response to lower body negative pressure before, during, and after ten days head-down tilt bed-rest. Acta Physiol. Scand. 144 (S604): 43–52, 1992.

Bent-Hansen L. Whole body capillary exchange of albumin. Acta Physiol. Scand. 143 (S 603): 5–10, 1991.

Bentsen KD, Henriksen JH & Laurent TC. Circulating hyaluronate: concentration in different vascular beds in man. Clin. Sci. 71: 161–5, 1986.

Berry CA & Catterson AD. Pre-Gemini medical prediction versus Gemini flight results. In: Gemini Summary Conference. NASA SP-138, pp. 197–218, 1967.

Berry P, Berry I & Manelfe C. Magnetic resonance imaging evaluation of lower limb muscles during bed rest – a microgravity simulation model. Aviat. Space Environ. Med. 64(3): 212–218, 1993.

Blomqvist CG & Stone HL. Cardiovascular adjustments to gravitational stress. In: *Handbook of Physiology. The Cardiovascular System. Peripheral Circulation and Organ Blood Flow.* Bethesda, MD: Amer. Physiol. Soc., sect. 2, vol. 3, chapter 28, 1983,

Blotner H. Effect of prolonged physical inactivity on tolerance of sugar. Arch. Int. Med. 75: 39–44, 1945.

Bogomolov VV, Popova IA, Egorov AD & Kozlovskaya IB. The results of medical research during the 326-day flight of Second Principle Expedition on the orbital complex 'Mir.' In: Second US/USSR Joint Working Group Conference on Space Biology and Medicine. Ed. by Gazenko OG. Washington, DC. September 1988.

Bonde-Petersen F, Suzuki Y, Kawakubo K & Gunji A. Effect of 20 days bed rest upon peripheral capillary filtration rate, venous compliance and blood flow in arms and legs. Acta Physiol. Scand. 150 (S616): 65–69, 1994.

Boutellier U, Arieli R & Farhi LE. Ventilation and CO_2 response during +Gz acceleration. Resp. Physiol. 62: 141–151, 1985.

Bryan AC, Macnamara WD, Simpson J & Wagner HN. Effect of acceleration on the distribution of pulmonary blood flow. J. Appl. Physiol. 20: 1129–32, 1965.

Bryan AC, Milic-Emili J & Pengell D. Effect of gravity on the distribution of pulmonary ventilation. J. Appl. Physiol. 21: 778–84, 1966.

Buchanan P & Convertino VA. A study of the effects of prolonged simulated microgravity on the musculature of the lower extremities in man: an introduction. Aviat. Space Environ. Med. 60: 649–52, 1989.

Buckey JC, Gaffney FA, Lane LD, Levine BD, Watenpaugh DE & Blomquist CG. Central venous pressure in space. New England J. Med. 329: 1853–4, 1993.

Bungo MW, Charles JB & Johnson PC. Cardiovascular deconditioning during space flight and the use of saline as a countermeasure to orthostatic intolerance. Aviat. Space Environ. Med. 56: 985–90, 1985.

Bungo MW, Charles JB, Wolf DA & Seddon MR. Similarity of echocardiographic parameters measured during space flight to preflight supine and standing values (Abstract). Aviat. Space Environ. Med. 58: 490, 1987.

Bungo MW, Goldwater DJ, Popp RL & Sandler H. Echocardiographic evaluation of Space Shuttle crew members. J. Appl. Physiol. 62: 278–83, 1987.

Bungo MW & Johnson PC. Cardiovascular examinations and observations of deconditioning during the Space Shuttle orbital flight test program. Aviat. Space Environ. Med. 54: 1001–4, 1983.

Burton RR & Smith AH. Stress and adaptation responses to repeated acute acceleration. Amer. J. Physiol. 222: 1505, 1972.

Butler GC, Xing H, Northey DR & Hughson RL. Reduced orthostatic tolerance following 4 h head-down tilt. Eur. J. Appl. Physiol. 62: 26–30, 1991.

Caiozzo VJ, Ma E, McCue SA, Herrick RE & Baldwin KM. A new animal model for modulating myosin isoform expression by mechanical activity. J. Appl. Physiol. 73: 1432–40, 1992.

Charles JB & Bungo MW. Non-invasive estimation of central venous pressure using a compact Doppler ultrasound system. In: *Results of the Life Sciences DSO's Conducted Aboard the Space Shuttle, 1981–86.* Space Biomedical Research Institute, NASA Johnson Space Center, Houston, TX. pp. 69–71, 1987.

Charles JB & Lathers CM. Cardiovascular adaptations to space flight. J. Clin. Pharmacol. 31: 1010–23, 1991.

Cherepakhin MA & Pervushin VI. Space flight effect on the neuromuscular system of cosmonauts. Kosmicheskaya Biol. Med. 4: 46–9, 1970.

Chobanian AV, Lille RD, Tercyak A & Blevins P. The metabolic and hemodynamic effects of prolonged bed rest in normal subjects. Circulation 49: 551–9, 1974.

Chui LA & Castleman KR. Morphometric analysis of rat muscle fibers following space flight and hypogravity. Physiologist 23: S76–8, 1980.

Clement G & Lestienne F. Adaptive modifications of postural attitude in conditions of weightlessness. Exp. Brain Res. 72: 381–9, 1988.

Cogoli A & Gmünder FK. Gravity effects on single cells: techniques, findings and theory. Adv. Space Biol. Med. 1: 183–248, 1991.

Cogoli A & Tschopp P. Lymphocyte reactivity during space flight. Immunology Today 6: 1–33, 1985.

Cogoli A, Valluchi-Mor M, Müller M & Briegleb W. Effect of hypogravity on human lymphocyte activation. Aviat. Space Environ. Med. 51: 29–34, 1980.

Convertino VA, Doerr DF, Eckberg DL, Fritsch JM & Vernikos-Danellis J. Head-down bed rest impairs vagal baroreflex responses and provokes orthostatic hypotension. J. Appl. Physiol. 68: 1458–64, 1990.

Convertino VA, Doerr DF, Ludwig DA & Vernikos J. Effect of simulated microgravity on cardiopulmonary baroreflex control of forearm vascular resistance. Amer. J. Physiol. 266 (Regulatory Integrative Comp. Physiol. 35): R 1962–9, 1994.

Convertino VA, Doerr DF & Stein SL. Changes in size and compliance of the calf after 30 days of simulated microgravity. J. Appl. Physiol. 68: 1509–12, 1989.

Crandall CG, Engelke KA, Convertino VA & Raven PB. Aortic baroreflex control of heart rate after 15 days of simulated microgravity exposure. J. Appl. Physiol. 77: 2134–9, 1994.

Crawford ABH, Cotton DJ, Paiva M & Engel LA. Effect of airway closure on ventilation distribution. J. Appl. Physiol. 66: 2511–5, 1989.

Criswel BS & Cobb K. Cellular immune response. In: *Apollo-Soyuz Test Project. Summary Science Report*, Vol. I, NASA SP-412: 257–62, 1977.

Dallman MF, Vernikos J, Keil LC, O'Hara D & Convertino V. Hormonal, fluid and electrolyte responses to 6° anti-orthostatic bed rest in healthy male subjects. In:*Stress: The role of Catecholamines and Neurotransmitters*. Ed. by Usdin E, Kvetnansky R & Axelrod J.Gordon & Breach, pp. 1057–77, 1984.

Davydova NA, Shishkina SK, Korneyeva NV, Suprunova YV & Ushakov AS. Biochemical aspects of some neurohumoral system functions during long-term antiorthostatic hypokinesia. Kosm. Biol. Aviakosm. Med. 20: 91–5, 1986.

Deitrick JE, Whedon GD, Shorr E, Toscani V & Davis VB. Effect of immobilization on metabolic and physiologic functions of normal men. Amer. J. Med. 4: 3–35, 1948.

DeTroyer A. Mechanical role of the abdominal muscles in relation to posture. Respir. Physiol. 53: 341–53, 1983.

Deuster PA, Smith DJ, Smoak BL, Montgomery LC, Singh A & Doubt TJ. Prolonged whole-body cold water immersion: fluid and ion shifts. J. Appl. Physiol. 66: 34–41, 1989.

Dietlein LF. Space flight and the telltale heart. Amer. J. Surgery 145: 703–6, 1983.

Diffee GM, Caiozzo VJ, McCue SA, Herrick SE & Baldwin KM. Activity-induced regulation of myosin isoform distribution: comparison of two contractile activity programs. J. Appl. Physiol. 74: 2509–16, 1993.

Diffee GM, Haddad F, Herrick RE & Baldwin KM. Control of myosin heavy chain expression: Interaction of hypothyroidism and hind limb suspension. Amer. J. Physiol. 30: C1099–1106, 1991.

Drummer C, Heer M, Dressendörfer RA, Strasburger CJ & Gerzer R. Reduced natriuresis during weightlessness. Clin. Investig. 71: 678–86, 1993.

Druz WS & Sharp JT. Activity of respiratory muscles in upright and recumbent humans. J. Appl. Physiol. 51: 1552–61, 1981.

Dudley GA, Duvoisin MR, Convertino VA & Buchanan P. Alterations of the *in vivo* torque-velocity relationship of human skeletal muscle following 30 days exposure to simulated microgravity. Aviat. Space Environ. Med. 60: 659–63, 1989.

Durnova GN, Kaplansk AS & Portugalov VV. Effect of a 22-day space flight on the lymphoid organs of rats. Aviat. Space Environ. Med. 47: 588–91, 1976.

Duvoisin MR, Convertino VA, Buchanan P, Gollnick PD & Dudley GA. Characteristics and preliminary observations of the influence of electromyostimulation on the size and function of human skeletal muscle during 30 days of simulated microgravity. Aviat. Space Environ. Med. 60: 671–8, 1989.

Eckberg DL & Fritsch JM. Influence of ten-day head-down bed rest on human carotid baroreceptor-cardiac reflex function. Acta Physiol. Scand. 144 (S 604): 69–76, 1992.

Edyvean J, Estenne M, Paiva M & Engel LA. Lung and chest wall mechanics in microgravity. J. Appl. Physiol. 71: 1956–66, 1991.

Elliott AR, Prisk GK, Guy HJB & West JB. Lung volumes during sustained microgravity on Spacelab SLS-1. In: *Proceedings of the 5th European Symposium on Life*

Sciences Research in Space. Ed. by Oser H, Guyenne TD. Noordwijk, The Netherlands, ESA SP-366, ESTEC, pp. 301–6, 1994.

Ellis S, Kirby LC & Greenleaf JE. Lower extremity muscle thickness during 30-day 6° head-down bed rest with isotonic and isokinetic exercise training. Aviat. Space Environ. Med. 64: 1011–5, 1993.

Engel LA. The dynamic distribution of gas flow. In: *Handbook of Physiology. The Respiratory System*. Ed. by Mead J & Macklem PT. Washington, DC: Amer. Physiol. Soc. sect. 3, vol. III, pp 575–93, 1986.

Engel LA. Effect of microgravity on the respiratory system. J. Appl. Physiol. 70: 1907–11, 1991.

Engel LA, Grassino A & Anthonisen NR. Demonstration of airway closure in man. J. Appl. Physiol. 38: 1117–25, 1975.

Epstein M. Renal effects of head-out water immersion in man: implications for an understanding of volume homeostasis. Physiol. Rev. 58, 529–81, 1978.

Epstein M, DeNunzio AG & Ramachandran M. Characterization of renal response to prolonged immersion in normal man. J. Appl. Physiol. 49, 184–8, 1980.

Epstein M. Renal effects of head-out water immersion in humans: a 15-year update. Physiol. Rev. 72: 563–621, 1992.

Essfeld D, Baum K, Hoffmann U & Stegemann J. Effects of microgravity on interstitial muscle receptors affecting heart rate and blood pressure during static exercise. Clin. Investig. 71: 704–9, 1993.

Estenne M, Yernault JC & De Troyer A. Rib cage and diaphragm abdomen compliance in humans: effects of age and posture. J. Appl. Physiol. 59: 1842–8, 1985.

Estenne M, Gorini M, Van Muylem A, Ninan V & Paiva M. Rib cage shape and motion in microgravity. J. Appl. Physiol. 73: 946–54, 1992.

Fischer CL, Daniel JC, Levi WC, Kimzey SL, Cobb EK & Ritzmann SE. Effect of the space flight environment on man's immune system: II. Lymphocyte counts and reactivity. Aerospace Med. 43: 1122–5, 1972.

Fischer CL, Johnson PC & Berry CA. Red blood cell mass and plasma volume changes in manned space flight. J. Amer. Med. Assoc. 200: 99–103, 1967.

Foley M & Tomashefski C. Pulmonary function during zero-gravity manoeuvers. Aerosp. Med. 40: 655–7, 1969.

Forbes GB. *Human Body Composition – Growth, Aging, Nutrition and Activity*. Springer, New York, 1987.

Fortney SM, Hyatt KH, Davis JE & Vogel JM. Changes in body fluid compartments during a 28-day bed rest. Aviat. Space Environ. Med. 62: 97–104, 1991.

Fowler KT, Reid J, Cardiac oscillations in expired gas tensions, and regional pulmonary blood flow. J. Appl. Physiol. 16: 863–8, 1961.

Fritsch JM, Charles JB, Bennett BS, Jones MM & Eckberg DL. Short duration space flight impairs human arterial baroreflex function. J. Appl. Physiol. 73: 664–71, 1992.

Fritsch-Yelle JM, Charles JB, Jones MM, Beightol LA & Eckberg DL. Space flight alters autonomic regulation of arterial pressure in humans. J. Appl. Physiol. 77: 1776–83, 1994.

Gauer OH & Thron HL. Postural changes in the circulation. In: Handbook of Physiology. Circulation, sec. 2, vol. iii, chap. 67. Ed. by Hamilton WF & Dow P. Washington D.C., Amer. Physiological Society, pp. 2409–39, 1965.

Gazenko OG, Schulzhenko EB, Grigoriev AI, Atkov OY & Yegorov AD. Review of basic medical results of the Salyut-7–Soyuz-T 8-month manned flight. Acta Astronautica 17: 155–60, 1988a.

Gazenko OG, Schulzhenko EB, Turchaninova VF & Yegorov AD. Central and regional hemodynamics in prolonged space flights. Acta Astronautica 17: 173–9, 1988b.

Geelen G, Arbeille P, Saumet JL, Cottet-Emard J, Patat F & Vincent M. Hemodynamic and hormonal effects of prolonged anti-G suit inflation in humans. J. Appl. Physiol. 72: 977–84, 1992.

Geelen G, Kravik SE, Hadj-Aissa A, Leftheriotis G, Vincent M, Bizollon CA, Sem-Jacobsen CW, Greenleaf JE & Gharib C. Effect of antigravity suit inflation on kidney function and cardiovascular and hormonal responses in humans. J. Appl. Physiol. 66: 782–99, 1989.

Gharib C, Maillet A, Gauquelin G, Allevard AM, Güell A, Cartier R & Arbeille Ph. Results of a 4 week head down tilt with and without LBNP countermeasure. I. Volume regulating hormones. Aviat. Space Environ. Med. 63: 3–8, 1992.

Gilroy RJ Jr, Lavietes MH, Loring SH, Mangura BT & Mead J. Respiratory mechanical aspects of abdominal distension. J. Appl. Physiol. 58: 1997–2003, 1985.

Glenny RW, Lamm WJE, Albert RK & Robertson HT. Gravity is a minor determinant of pulmonary blood flow distribution. J. Appl. Physiol. 71: 620–9, 1991.

Gmünder FK, Kiess M, Sonnefeld G, Lee J & Cogoli A. A ground-based model to study the effects of weightlessness on lymphocytes. Biol. Cell. 70: 33, 1990.

Gould CL, Lyte M, Williams J, Mande AD & Sonnenfeld G. Inhibited interferon-gamma but normal interleukin-3 production from rats flown on the Space Shuttle. Aviat. Space Environ. Med. 58: 983, 1987.

Green M, Mead J & Sears TA,.Muscle activity during chest wall restriction and positive pressure breathing in man. Respir. Physiol. 35: 283–300, 1978.

Greenleaf JE. Bed-rest studies: Fluid and electrolyte responses. In: *Space Physiology*. Ed. by Garcia J, Guerin M & Laverlochere C. Cepadus, Toulouse, pp. 335–48, 1983.

Greenleaf JE. Physiology of fluid and electrolyte responses during inactivity: water immersion and bed rest. Med. Sci. Sports Exerc. 16, 20–5, 1984a.

Greenleaf JE. Physiological responses to prolonged bed rest and fluid immersion in humans. J. Appl. Physiol. 57, 619–33, 1984b.

Greenleaf JE. Energy and thermal regulation during bed rest and space flight. J. Appl. Physiol. 67, 507–16, 1989.

Greenleaf JE, Bernauer E M, Ertl A C, Bulbulian R & Bond M. Isokinetic strength and endurance during 30-day 6° head-down bed rest with isotonic and isokinetic exercise training. Aviat. Space Environ. Med. 65: 45–50, 1994.

Greenleaf JE, Greenleaf CJ, VanDerveer D & Dorchak HJ. Adaptation to prolonged bed rest in man: a compendium of research. NASA TM X-3307, Washington DC 1976.

Greenleaf JE & Kozlowski S. Reduction of peak oxygen uptake after prolonged bed rest. Med. Sci. Sports Exercise 14: 477–80, 1982.

Greenleaf JE, T Morse, PR Barnes, J Silver & LC Keil. Hypervolemia and plasma vasopressin response during water immersion in men. J. Appl. Physiol. 55, 1688–93, 1983.

Grigoriev AI, Bugrov SA, Bogomolov VV, Egorov AD, Kozlovskaya IB, Pestov ID, Polyakov VV & Tarasov IK. Medical results of the Mir year-long mission. Physiologist 34: S44–8, 1991.

Grigoriev AI & Egorov AD. General mechanisms of the effect of weightlessness on the human body. Adv. Space Biol. Med. 2: 1–42, 1992a.

Grigoriev AI & Egorov AD. Physiological aspects of adaptation of main human body systems during and after space flights. Adv. Space Biol. Med. 2: 43–82, 1992b.

Grigoriev AI, Noskov VB, Poliakov V, Sukhanov Y, Gharib C, Gauquelin G, Geelen G, Kvetnansky R & Macho L. Fluid-electrolyte metabolism and hormonal control during the second Soviet-French space flight. Aerosp. Ecol. Med. 26: 36–9, 1992.

Guy HJB, Prisk GK, Elliott AR, Deutschman III RA & West JB. Distribution of pulmonary ventilation during sustained microgravity as determined by single - breath washouts. J. Appl. Physiol. 1994 (in press).

Guy HJB, Prisk GK, Elliott AR & West JB. Maximum expiratory flow-volume curves during short periods of microgravity. J. Appl. Physiol. 70: 2587–96, 1991.

Hargens AR (ed). *Tissue Fluid Pressure and Composition*. Williams & Wilkins, 1981.

Hargens AR. Fluid shifts in vascular and extravascular spaces during and after weightlessness. Med. Sci. Sports Exerc. 15: 421–7, 1983.

Harrison MH. Athletes, astronauts and orthostatic tolerance. Sports Med. 3: 428–35, 1986.

Haruna Y, Suzuki Y, Kawakubo K & Gunji A. Orthostatic tolerance and autonomous nervous functions before and after 20-days bed rest. Acta Physiol. Scand. 150 (S616), 71–81, 1994.

Hauschka EO, Roy RR & Edgerton VR. Size and metabolic properties of single fibers in rat soleus after hind limb suspension. J. Appl. Physiol. 62: 2338–47, 1987.

Hawkins WR & Ziegschmid JF. Clinical aspects of crew health. In: *Biomedical Results of Apollo*. Ed. by Johnston RF, Dietlein LF & Berry CA. NASA SP-368, pp. 43–81, 1975.

Henry WL, Epstein SE, Griffith LH, Goldstein RE & Redwood DR. Effects of prolonged space flight on cardiac function and dynamics. In: *Biomedical Results from Skylab*. Ed. by Johnston RS & Dietlein LF. NASA SP-377, pp. 366–71, 1977.

Herbert ME, Roy RR, Hodgson JA & Edgerton VR. Influence of one week hind limb suspension and intermittent high load exercise on rat muscles. Exp. Neurol. 102: 190–8, 1988.

Hikida RS, Gollnick PD, Dudley GA, Convertino VA & Buchanan P. Structural and metabolic characteristics of human skeletal muscle following 30 days of simulated microgravity. Aviat. Space Environ. Med. 60: 664–70, 1989.

Hildebrandt W, Gunga H-C, Hermann J, Röcker L, Kirsch K & Stegemann J. Enhanced slow caudal fluid shifts in orthostatic intolerance after 24 h bedrest. Eur. J. Appl. Physiol. 69: 61–70, 1994.

Hinghofer-Szalkay H, Noskov V, Jezova D, Sauseng-Fellegger G, Füger GF, Sukhanov Y, König EM, Zambo-Polz C, Barowitsch C, Viehböck F, Macho L, Kvetnansky R & Grigoriev AI. Hormonal changes with lower body negative pressure on the 6th day in microgravity in one cosmonaut. Aviat. Space Environ. Med. 64: 1000–5, 1993.

Hinghofer-Szalkay H, Noskov V, Schmied J, Röhrer R, Viehböck F, König EM, Sauseng-Fellegger G, Zambo-Polz C, Heimel H & Grigoriev AI. Changes of blood/plasma composition with LBNP on ground and in space in one subject. Aviat. Space Environ. Med. 65: 214–9, 1994.

Hinghofer-Szalkay HG, Sauseng-Fellegger G & Greenleaf JE. Plasma volume with alternating tilting: effect of fluid ingestion. J. Appl. Physiol. 78(4): 1–5, 1995.

Hoffler GW & Johnson RL. Apollo flight crew cardiovascular evaluations. In: *Biomedical Results of Apollo*. Ed. by Johnston RF, Dietlein LF & Berry CA. NASA SP-368, pp. 227–64, 1975.

Hong SK, Cerretell P, Cruz JC & Rahn H. Mechanics of respiration during submersion in water. J. Appl. Physiol. 27: 535–8, 1969.

Hughson RL, Yamamoto Y, Blaber AP, Maillet A, Fortrat JO, Pavy-LeTraon A, Marini JF & Gharib C. Effect of 28-day head-down bed rest with countermeasures on heart rate variability during LBNP. Aviat. Space Environ. Med. 65: 293–300, 1994a.

Hughson RL, Yamamoto Y, Maillet A, Fortrat JO, Pavy-LeTraon A, Butler GC, Güell A & Gharib C. Altered autonomic regulation of cardiac function during head-up tilt after 28-day head-down bed-rest with counter -measures. Clin. Physiol. 14: 291-304, 1994b.

Huntoon C. Metabolic. 1992 FASEB, Symposium 321: Human Physiology in microgravity: Spacelab SLS-1. Anaheim, California, April 8, 1992.

Hyatt KH & West DA. Reversal of bedrest-induced orthostatic intolerance by lower body negative pressure and saline. Aviat. Space Environ. Med. 48: 120–4, 1977.

Hymer WC, Grindeland R & Krasnov I. Effects of space flight on rat pituitary cell function. J. Appl. Physiol. 73: 151S–7S, 1992.

Ilyina-Kakueva EI, Portugalov VV & Krivenkova NP. Space flight effects on the skeletal muscle of rats. Aviat. Space Environ. Med. 47: 700–3, 1976.

Jiang B, Roy RR, Navarro C & Edgerton VR. Absence of a growth hormone effect on rat soleus atrophy during a 4-day space flight. J. Appl. Physiol. 74: 527–31, 1993.

Johnson PC. Fluid volume changes induced by space flight. Acta Astronautica 2: 311–7, 1975.

Johnson RL, Hoffler GW, Nicogossian AE, Bergman SA & Jackson MM. Lower body negative pressure: Third manned Skylab mission. In: *Biomedical Results from Skylab*. Ed. by Johnston RS & Dietlein LF. NASA SP-377, pp. 284–312, 1977.

Johnston RS & LF Dietlein (eds). *Biomedical Results from Skylab*. NASA SP-377, 1977.

Joyner MJ, Shepherd JT & Seals DR. Sustained increases in sympathetic outflow during prolonged lower body negative pressure in humans. J. Appl. Physiol. 68: 1004–9, 1990.

Kalinichenko VV. Dynamics of orthostatic stability of cosmonauts following 2- to 63-day missions. Kosmicheskaya Biologiya i Aviakosmicheskaya Meditsina 11: 31–7, 1977.

Kaneko K, Milic-Emili J, Dolovich MB, Dawson A & Bates DV. Regional distribution of ventilation and perfusion as a function of body position. J. Appl. Physiol. 21: 767–77, 1966.

Katz MA. Interstitial space - the forgotten organ. Med. Hypoth. 6: 885–98, 1980.

Kimball WR, Kelly KB & Mead J. Thoracoabdominal blood volume change and its effect on lung and chest wall volumes. J. Appl. Physiol. 61: 953–9, 1986.

Kirsch K, Merke J & Hinghofer-Szalkay H. Fluid volume distribution within superficial shell tissues along body axis during changes of body posture in man. The application of a new miniature plethysmographic method. Pflügers Arch. 383: 195–201, 1980.

Kirsch KA, Röcker L, Gauer OH & Krause R. Venous pressure in man during weightlessness. Science 225: 218–9, 1984.

Kirsch KA, Baartz FJ, Gunga HC & Röcker L. Fluid shifts into and out of superficial tissues under microgravity and terrestrial conditions. Clin. Investig. 71: 687–9, 1993.

Kollias J, VanDerveer D, Dorchak KJ & Greenleaf JE. Physiologic responses to water immersion in man: A compendium of research. NASA TM X-3308, Washington DC 1976.

Konno K & Mead J. Static volume-pressure characteristics of the rib cage and abdomen. J. Appl. Physiol. 24: 544–8, 1968.

Konstantinova I. Results of medical research performed on the 'Salyut-6-Soyuz' Space Station complex. In: *Immunological Research on Salyut-6 Prime Crews*. Ed. by Burovsky NN. 1986.

Konstantinova IV. Immune resistance of man in space flight. Acta Astronautica 23, 123–127, 1991.

Konstantinova IV, Rykova M, Meshkov D, Peres C & Schmitt DA. Study of some mechanisms of antivirus resistance: 'Immunology' experiment during the French-Russian 'Antares' project. In: *Proceedings of the 5th European Symposium on Life Sciences Research in Space*. Ed. by Oser H, Guyenne TD. Noordwijk, The Netherlands, ESA SP-366, ESTEC, pp. 411–2, 1994.

Kozlovskaya IB, Kreidich YV, Oganov VS & Koserenko OP. Pathophysiology of motor functions in prolonged space flights. Acta Astronautica 8: 1059–72, 1981.

Kravik SE, Keil LC, Geelen G, Wade CE, Barnes PE, Spaul WA, Elder CA & Greenleaf JE. Effect of antigravity suit inflation on cardiovascular, PRA, and PVP responses in humans. J. Appl. Physiol. 61, 766–74, 1986.

Krupina TN, Tizul AY, Kuzman MP & Tsyganova NI. Clinico-physiological changes in man during long-term antiorthostatic hypokinesia. Aviat. Space Environ. Med. 16: 40–5, 1982.

Kushler RH & Brown M. A model for the identification of hormone pulses. Statistics in Med. 10: 329–40, 1991.

Kvetnansky R, Davydova NA, Noskov VB, Vigas M, Popova IA, Usakov AC, Macho L & Grogoriev AI. Plasma and urine catecholamine levels in cosmonauts during long-term stay on space station Salyut-7. Acta Astronautica 17: 181–6, 1988.

Kvetnansky R, Noskov VB, Blazicek P, Gharib C, Popova IA & Gauquelin G. Activity of the sympathoadrenal system in cosmonauts during 25-day space flight on station Mir. Acta Astronautica 23: 109–16, 1991.

LaFevers EV, Nicogossian AE, Hursta WN & Baker JT. Electromyographic analysis of skeletal muscle. In: *Apollo-Soyuz Test Project, Medical Report.* NASA SP-411, pp. 53–7, 1977.

Lane HW. Energy requirements for space flight. J. Nutr. 122: 13–8, 1992.

Lange L, S Lange, M Echt & OH Gauer. Heart volume in relation to body posture and immersion in a thermo-neutral bath. Pflügers Arch. 352: 219–26, 1974.

Lathers C & Charles JB. Orthostatic hypotension in patients, bed rest subjects, and astronauts. J. Clin. Pharmacol. 34: 403–17, 1994.

Lathers C, Charles JB, Elton KF, Holt TA, Mukai C, Bennett BS & Bungo MW. Acute hemodynamic responses to weightlessness in humans. J. Clin. Pharmacol. 29: 615–27, 1989.

Leach CS, Altchuler SI & Cintron-Trevino NM. The endocrine and metabolic response to space flight. Med. Sci. Sports Exerc. 15: 432–40, 1983.

Leach Huntoon C, Johnson PC & Cintron NM. Hematology, immunology, endocrinology, and biochemistry. In: *Space Physiology and Medicine*, Ed. by Nicogossian AE, Leach Huntoon C & Pool SL. Lea & Febiger, pp. 222–39, 1989.

Leonard JI. Understanding metabolic alterations in space flight using quantitative models: Fluid and energy balance. 36th Congr. IAF, Pergamon Press, 1985.

Levine BD, Buckey JC, Fritsch JM, Yancy CW, Watenpaugh DE, Snell PG, Lane LD, Eckberg DL & Blomqvist CG. Physical fitness and cardiovascular regulation: mechanisms of orthostatic intolerance. J. Appl. Physiol. 70: 112–22, 1991.

Link MM. Mercury medical operations.In: *Space Medicine in Project Mercury.* NASA SP4003, pp. 135–68, 1965.

Lord BI. The architecture of bone marrow cell populations. Int. J. Cell Cloning 8: 317–31, 1990.

Lucas J & Floyer MA. Renal control of changes in the compliance of the interstitial space: A factor in the aetiology of renoprival hypertension. Clin. Sci. 44: 397–416, 1973.

Lundvall J & Länne T. Large transcapillary hydraulic conductance in skeletal muscle and skin of man revealed by a new technique. Acta Physiol. Scand. 135: 417–8, 1989.

Lutwak L & Whedon GD. The effect of physical conditioning on glucose tolerance. Clin. Res. vol 7, p. 143, 1959.

Macho L, Kvetnansky R, Vigas M, Nemeth S, Popova I, Tigranian RA, Noskov VB, Serova L & Grigoriev IA. Effect of space flights on plasma hormone levels in man and in experimental animal. Acta Astronautica 23: 117–21, 1991.

Mandel AD & Balish E. Effect of space flight on cell-mediated immunity. Aviat. Space Environ. Med. 48: 1051–7, 1977.

Martin TP, Edgerton VR & Grindeland RE. Influence of space flight on rat skeletal muscle. J. Appl. Physiol. 65: 2318–25, 1988.

Meehan R, Taylor G, Lionetti F, Neale L & Curren T. Human mononuclear cell function after 4° C storage during 1-G and microgravity conditions of space flight. Aviat. Space Environ. Med. 60: 644–8, 1989.

Melchior FM & Fortney SM. Orthostatic intolerance during a 13-day bed rest does not result from increased leg compliance. J. Appl Physiol. 74: 285–92, 1993.

Michels MD & West JB. Distribution of pulmonary ventilation and perfusion during short periods of weightlessness. J. Appl. Physiol. 45: 987–98, 1978.

Michels MD, Friedman PJ & West JB. Radiographic comparison of human lung shape during normal gravity and weightlessness. J. Appl. Physiol. 47: 851–7, 1979.

Miserocchi G & Negrini D. Gravity dependence of pleural liquid turnover. In: *Proceedings of the 4th European Symposium on Life Sciences Research in Space*. Ed. by David V. Paris, France, European Space Agency, pp. 397–9, 1990.

Miserocchi G & Negrini D. Pleural lymphatics as regulators of pleural fluids dynamics. News in Physiol. Sci. 6: 153–8, 1991.

Miserocchi G, Venturoli D, Negrini D & DelFabbro M. Model of pleural fluid turnover. J. Appl. Physiol. 75: 1798–1806, 1993.

Mulvagh SL, Charles JB, Riddle JM, Rehbein TL & Bungo MW. Echocardiographic evaluation of the cardiovascular effects of short-duration space flight. J. Clin. Pharmacol. 31: 1024–6, 1991.

Musgrave FS, Zechman FW & Mains RC. Changes in total leg volume during lower body negative pressure. Aerospace Med. 40: 602–6, 1969.

Nakamitsu S, Sagawa S, Miki K, Wada F, Nagaya K, Keil LC, Drummer C, Gerzer R, Greenleaf JE, Hong SK & Shiraki K. Effect of water temperature on diuresis-natriuresis: AVP, ANP, and urodilation during immersion in men. J. Appl. Physiol. 77: 1919–25, 1994.

Naughton J & Wulff J. Effect of physical activity on carbohydrate metabolism. J. Lab. Clin. Med. 70: 996, 1967.

Negrini D, Capelli C, Morini M & Miserocchi G. Gravity-dependent distribution of parietal subpleural interstitial pressure. J. Appl. Physiol. 63: 1912–8, 1987.

Nicogossian AET, Whyte AA, Sandler H, Leach CS & Rambaut PC. Chronological Summaries of United State, European, and Soviet Bedrest Studies. NASA, Washington DC 1979.

Nixon JV, Murray RG, Bryant C, Johnson RL, Mitchell JH, Holland OB, Gomez-Sanchez C, Vergne-Marini P & Blomqvist CG. Early cardiovascular adaptation to simulated zero gravity. J. Appl. Physiol. 4: 541–8, 1979.

Norman J. One-compartment kinetics. British J. Anaesthesia 69: 3857–96, 1992.

Norsk P. Gravitational stress and volume regulation. Clin. Physiol. 12: 505–26, 1992.

Norsk P & Epstein M. Manned space flight and the kidney. Amer. J. Nephrol. 11: 81–97, 1991.

Norsk P, Foldager N, Bonde-Petersen F & Elmann-Larsen B. Central venous pressure in humans during short periods of weightlessness. J. Appl. Physiol. 63: 2433–7, 1987.

Noskov VB, Katkov VE, Afonin BV, Chestukhin VV & Sukhanov YV. Central venous pressure and hormonal regulation of the water balance when altered in antiorthostasis. Human Physiology 12: 341–6, 1986.

Noskov VB, Afonin BV, Sukhanov YV & Kvetnansky R. Hormonal control of fluid and electrolyte metabolism during extended periods of weightlessness. In: *Stress. Neurochemical and Humoral Mechanisms*, vol. 2, Ed. by Van Loon A, Kvetnansky A, McCarty A & Axelrod A.Gordon & Breach, pp. 1031–7, 1989.

Oganov VS, Cann CE, Rakhmanov AS & Ternovoy SK. A computer tomographic investigation of the musculoskeletal system of the spine in humans after long-term space flight. Kosmicheskaya Biologiya i Aviakosmicheskaya Meditsina 24: 20–1, 1990.

Olszewski WL, Engeset A & Sokolowski J. Lymph flow and protein in the normal male leg during lying, getting up, and walking. Lymphology 10: 178–83, 1977.

O'Meara NM, Sturis J, Blackman JD, Roland DC, Vancauter E & Polonsky KS. Analytical problems in detecting rapid insulin secretory pulses in norman humans. Amer. J. Physiol. 264: E231–8, 1993.

Paiva M & Engel LA. Theoretical studies of gas mixing and ventilation distribution in the lung. Physiol. Rev. 67: 750–96, 1987.

Paiva M, Estenne M.& Engel LA. Lung volume, chest wall configuration and pattern of breathing in microgravity. J. Appl. Physiol. 67: 1542–50, 1989.

Parazynski SE, Hargens AR, Tucker B, Aratow M, Styf J & Crenshaw A. Transcapillary fluid shifts in tissues of the head and neck during and after simulated microgravity. J. Appl. Physiol. 71: 2469–75, 1991.

Parving HH, Rossing N, Nielsen SL & Lassen NA. Increased transcapillary escape rate of albumin, IgG, and IgM after plasma volume expansion. Amer. J. Physiol. 227: 245–50, 1974.

Patel AN, Razzak ZA & Dastur DK. Disuse atrophy of human skeletal muscles. Arch. Neurol. 20: 413–21, 1969.

Pequignot JM, Guell A, Gauquelin G, Jarsaillon E, Annat G, Bes A, Peyrin L. & Gharib C. Epinephrine, norepinephrine, and dopamine during a 4-day head-down bed rest. J. Appl. Physiol. 58: 157–63, 1985.

Peres C, Konstantinova I, Rykova M, Meshkov D, Schaffar L, Tkaczuk J, Abdal M, Ohayon E & Schmitt DA. Mitogen-induced proliferation of T-lymphocytes from cosmonauts after short and long duration space flight. In: *Proceedings of the 5th European Symposium on Life Sciences Research in Space*. Ed. by Oser H, Guyenne TD. Noordwijk, The Netherlands, ESA SP-366, ESTEC, pp. 413-4, 1994.

Pincus SM. & Keefe DL. Quantification of hormone pulsatility via an approximate entropy algorithm. Amer. J. Physiol. 262: E741-54, 1992.

Popova IA & Grigoriev AI. The effect of space flight on metabolism: the results of biochemical research in rat experiments on the Kosmos biosatellites. Aviakosmich. 26: 4-10, 1992.

Pourcelot L, Savilov AA, Bystrov VV, Kakurin LI, Kotovskaya AR, Patat F, Pottier JM & Zhernakov AF. Results of echocardiographic examination during seven days flight on board Salyut VII, June 1982. Physiologist 26: S66-9, 1983.

Pourcelot L, Arbeille P, Pottier JM, Patat F, Berson M, Roncin A, Toullec C, Guell A & Gharib C. Cardiovascular examination during STS-SlG mission, June 1985. In: *Proceedings, Second International Conference on Space Physiology.* ESA SP-237, pp. 20-2, 1986.

Prefaut, C, Lupi HE & Anthonisen NR. Human lung mechanics during water immersion. J. Appl. Physiol. 40: 320-33, 1976.

Prisk GK, Guy HJB, Elliott AR, Deutschmann III RA & West JB. Pulmonary diffusing capacity, capillary blood volume, and cardiac output during sustained microgravity. J. Appl. Physiol. 75: 15-26, 1993.

Pyszczynski D, Mink SN & Anthonisen NR. Increased gravitational stress does not alter maximum expiratory flow. J. Appl. Physiol. 59: 28-33, 1985.

Rambaut PC, Leach CS & Whedon GD. A study of metabolic balance in crew members of Skylab IV. Acta Astronautica 6: 1313-22, 1979.

Rapcsk M, Oganov VS, Murashko LM, Szilgyi T & Szöör A. Effect of short-term space flight on the contractile properties of rat skeletal muscles with different functions. Acta Physiol. Hung. 76: 13-20, 1990.

Rea RF & Wallin BG. Sympathetic nerve activity in arm and leg muscles during lower body negative pressure in humans. J. Appl. Physiol. 66: 2778-81, 1989.

Regnard J, Baudrillard P, Salah B, Xuan TD, Cabanes L & Lockhart A. Inflation of antishock trousers increases bronchial response to metacholine in healthy subjects. J. Appl. Physiol. 68: 1528-33, 1990.

Reichel A, Rother U, Werner J & Reichel F. On the transport of various endogenous plasma proteins from blood to peripheral lymph in man. Lymphology 9: 118-21, 1976.

Reid MB, Loring SH, Banzett RB & Mead J. Passive mechanics of upright human chest wall during immersion from hips to neck. J. Appl. Physiol. 60: 1561-70, 1986.

Renkin EM. Some consequences of capillary permeability to macromolecules: Starlings hypothesis reconsidered. Amer. J. Physiol. 250: H706-10, 1986.

Robertson D, Convertino VA & Vernikos J. The sympathetic nervous system and the physiologic consequences of space flight: a hypothesis. Amer. J. Med. Sci. 308: 126-32, 1994.

Rowell LB. ***Human Circulation - Regulation During Physical Stress.*** **Oxford University Press. 1986.**

Roy RR, Sacks RD, Baldwin KM, Short M & Edgerton VR. Interrelationships of contraction time, V_{max}, and myosin ATPase after spinal transection. J. Appl. Physiol. 56: 1594-601, 1984.

Rutishauser WJ, Banchero N, Tsakiris AG, Edmundowicz AC & Wood EH. Pleural pressures at dorsal and ventral sites in supine and prone body positions. J. Appl. Physiol. 21: 1500-10, 1966.

Saltin B, Blomqvist G, Mitchell JH, Johnson RL, Wildenthal K & Chapman CB. Response to exercise after bedrest and after training: a longitudinal study of adaptive changes in oxygen transport and body composition. Circulation 38: VII-1-78, 1968.

Sander-Jensen K. *Heart and endocrine changes during central hypovolemia in man.* Doctoral thesis. Department of Medical Physiology, University of Copenhagen. Laegeforeningens Forlag, 1991.

Sandler H, Popp RL & Harrison DC. The hemodynamic effects of repeated bed rest exposure. Aviat. Space Environ. Med. 59: 1047–54, 1988.

Sandler H & Winter DL. *Physiological Responses of Women to Simulated Weightlessness.* NASA SP-430, Washington DC. 1978.

Sargeant AJ, Davies CTM, Edwards RHT, Maunder C & Young A. Functional and structural changes after disuse of human muscle. Clin. Sci. Mol. Med. 52: 337–42, 1977.

Sawin CF, Nicogossian AE, Rummel JA & Michel EL. Pulmonary function evaluation during the Skylab and Apollo-Soyuz missions. Aviat. Space Environ. Med. 47: 168–72, 1976.

Sawin CF, Nicogossian AE, Schachter AP, Rummel JA & Michel EL. Pulmonary Function Evaluation During and Following Skylab Space Flights. In: *Biomedical Results from Skylab.* Ed. by Johnston RS & Dietlein LF. NASA SP-377, 1977.

Schnitzer JE. Update on the cellular and molecular basis of capillary permeability. Trends Cardiovasc. Med. 3: 124–30, 1993.

Scibetta SM, Caren LD & Oyama J. The Effects of hypergravity on the rate of antibody formation in the rat. Physiologist 26: S 135, 1983.

Shangraw RE, Stuart CA, Prince MJ, Peters EJ & Wolfe RR. Insulin responsiveness of protein metabolism *in vivo* following bedrest in humans. Amer. J. Physiol. 225: E548–58, 1988.

Smith AH, Spanger WL, Carlisle B & Kinder G. Effects of brief exposure of domestic fowl to very intense acceleration fields. Aviat. Space Environ. Med. 50: 126, 1979.

Smith JJ & Ebert TJ. General response to orthostatic stress. In: Circulatory Response to the Upright Posture. Ed. by Smith JJ. CRC Press, pp. 1–46. 1991.

Snashall PD, Lucas J, Guz A & Floyer MA. Measurement of interstitial 'fluid' pressure by means of a cotton wick in man and animals: an analysis of the origin of the pressure. Clin. Sci. 41: 35–53, 1971.

Sonnenfeld G, Mandel AD, Konstantinova IV, Taylor GR, Berry WD, Wellhausen SR, Lesnyak AT & Fuchs BB. Effects of space flight on levels and activity of immune cells. Aviat. Space Environ. Med. 61: 648–53, 1990.

Sonnenfeld G, Mandel AD & Konstantinova IV. Space flight alters immune cell function and distribution. J. Appl. Physiol. 73: 191–5, 1992.

Stadeager C, Johansen LB, Warberg J, Christensen NJ, Foldager N, Bie P & Norsk P. Circulation, kidney function, and volume-regulating hormones during prolonged water immersion in humans. J. Appl. Physiol. 73: 530–8, 1992.

Stein TP & Schluter MD. Excretion of IL-6 by astronauts during space flight. Amer. J. Physiol. 266: E448–52, 1994.

Takenaka K, Suzuki Y, Kawakubo K, Haruna Y, YanagiboriI R, Kashihara H, Igarashi T, Watanabe F, Omata M, Bonde-Petersen F & Gunji A. Cardiovascular effects of 20 days bed rest in healthy young subjects. Acta Physiol. Scand. 150 (S616): 59–63, 1994.

Talas M, Batkai L, Stoger I, Nagy K, Hiros L, Konstantinova I & Kozharinov V. Results of space experiment program 'Interferon'. II. Influence of space flight conditions on the activity of interferon preparations and interferon inducer ('Interferon II'). Acta Microbiologica Hungarica 30: 63–7, 1983.

Taylor GR. Cell anomalies associated with space flight conditions. In: Immunobiology of Proteins and Peptides, Vol. 4. Ed. by Zouhair M. New York, Plenum Press, pp. 259–71, 1987.

Taylor GR & Dardano JR. Human cellular immune responsiveness following space flight. Aviat. Space Environ. Med. 54: 555–9, 1983.

Taylor GR & Janney RP. *In vivo* testing confirms a blunting of the human cell mediated immune mechanism during space flight. J. Leucocyte Biol. 92–132, 1992.

Taylor GR, Neale LS & Dardano JS. Immunological analysis of US space shuttle crew members. Aviat. Space Environ. Med. 57, 213 -7, 1986.

Taylor HL, Henshel A, Brozek J & Keys A. Effects of bed rest on cardiovascular function and work performance. J. Appl. Physiol. 2: 223–39, 1949.

Thornton WE, Hedge V, Coleman E, Uri JJ & Moore TP. Changes in leg volume during microgravity simulation. Aviat. Space Environ. Med. 63: 789–94, 1992.

Thornton WE & Rumrnel JA. Muscular deconditioning and its prevention in space flight. NASA TM X-58154. pp. 403–16, 1974.

Tidgren B, Hjemdahl P, Theodorsson E & Nussberger J. Renal responses to lower body negative pressure in humans. Amer. J. Physiol. 259: F573–9, 1990.

Tomaselli CM, Frey MAB, Kenney RA & Hoffler GW. Hysteresis in response to descending and ascending lower body negative pressure. J. Appl. Physiol. 63: 719–25, 1987.

Tripathi AG, Mack G & Nadel ER. Peripheral vascular reflexes elicited during lower body negative pressure. Aviat. Space Environ. Med. 60: 1187–93, 1989.

Van Wyk AJ & Brede HD. The effects of simulated increased gravity on the immunological system of the rat. In: *21st International Congress on Aviation and Space Medicine*, pp. 115–6, 1973.

Vernikos J, Dallman MF, Keil LC, OHara D & Convertino V. Gender differences in endocrine responses to posture and 7 days of -6° head-down bed rest. Amer. J. Physiol. 265: E153–61, 1993.

Vernikos J, Dallman MF, Van Loon G & Keil LC. Drug effects on orthostatic intolerance induced by bedrest. J. Clin. Pharmacol. 31: 974–84, 1991.

Vogt FB. Tilt table and plasma volume changes with short term deconditioning experiments. Aerospace Med. 38: 564–8, 1967.

Volynkin YM & Vasilyev PV. Some results of medical studies conducted during the flight of the 'Voskhod'. In: *Problems of Space Biology*, Vol. VI. Ed. by Sisakyam NM. Nauka Press, Moscow. NASA TT F-528, 52–66, 1967.

Von Baumgarten RJ, Baldrighi G, Voge H & Thumler R. Physiological response to hyper- and hypogravity during rollercoaster flight. Aviat. Space Environ. Med. 51: 145–54, 1980.

Vorobyov EI, Gazenko OG, Genin AM & Egorov AD. Medical results of Salyut-6 manned space flights. Aviat. Space Environ. Med. 54: S31–40, 1983.

West JB. Space. In: *The Lung: Scientific Foundations*, vol. 2. Ed. by Crystal RG & West JB. New York, Raven Press, pp. 2133–41, 1991.

Wetzig J & von Baumgarten R. Respiratory parameters aboard an aircraft performing parabolic flights. In: *Proceedings of the 3rd ESA European Symposium on Life Sciences Research in Space*, Graz, Austria. Ed. by Hunt H & Oser H, ESA SP-271, ESA, Paris, pp. 47–50, 1987.

Wiig H & Noddeland H. Interstitial fluid pressure in human skin measured by micropuncture and wick-in-needle. Scand. J. Clin. Lab. Invest. 43: 255–60, 1983.

Winiarski AM, Roy RR, Alford EK, Chiang PC & Edgerton VR. Mechanical properties of rat skeletal muscle after hind limb suspension. Exp. Neurol. 96: 650–60, 1987.

Chapter 3
Kidney function and fluid homeostasis

P. Bie, M. H. Bestle & L. B. Johansen

3.1 Introduction

Ever since Yuri Gagarin landed in 1961 having performed the first manned space flight, a number of physiological and medical questions related to weightlessness have been investigated. Before the advent of manned space flight it was hypothesized that the absence of gravitational stress would induce a redistribution of body fluid from the legs headward, due to the absence of hydrostatic pressure gradients (for review see **Gauer & Henry 1963**), which would in turn induce changes in kidney function and fluid homeostasis. Because changes in kidney function and fluid homeostasis are relatively slow processes, space flight is the optimal condition for investigations of the influence of weightlessness on these variables. Since it was already well known that water immersion and bed rest induced increases in renal fluid and electrolyte excretion, these models were used as means of simulating the expected effects of weightlessness. Indeed, observations of puffy faces and thin legs ('bird legs') in the first astronauts combined with an almost immediate loss of body mass (**Norsk & Epstein 1991**) initially seemed to verify the hypothesis that increases in central blood volume through stimulation of volume receptors in the thorax induced an increased renal output of fluid and electrolytes and thus a loss in body mass.

Weightlessness may be a valuable tool for investigation of body fluid regulation, not only with the purpose of development of countermeasures, but also to gain new knowledge about the mechanisms of body fluid regulation by observation of the performance of these mechanisms in the absence of the often complicating factor of gravitational force. This could improve the understanding of the mechanisms of various diseases with unexplained retention of sodium and water (i.e. heart and liver diseases).

In this chapter we intend to review results from simulated weightlessness experiments (water immersion, bed rest, and lower body positive pressure) concerning kidney function and fluid homeostasis, comparing these results with those of experiments performed during space flight and parabolic flights.

3.2 Simulated weightlessness

3.2.1 Water immersion

3.2.1.1 Definition

Water immersion (WI) usually describes a protocol during which a subject is immersed to the neck (head-out) in tap water at 34 to 35°C (thermoneutral) (**Epstein 1992**). Usually the subject is seated or standing in order to maximise the effect of immersion. Rarely have studies been performed during which the subject has been partially immersed (i.e. to the hips) or completely immersed using pressure-breathing apparatus. The fact that the subjects are immersed in water puts a limit to the duration of the experiments which usually lasts only a few hours. Thus, in most cases only the acute effects of WI are studied. However, an improved method according to which subjects are protected by a thin plastic sheet has been developed by Russian scientists (dry immersion) and used for long-term experiments lasting some days (Gogolev *et al.* 1980).

3.2.1.2 Comparison with weightlessness

During WI, the hydrostatic pressure gradient of the water column outside the body counteracts to a large extent the hydrostatic pressure gradient inside the body, thereby causing a centripetal redistribution of blood (**Epstein 1992; Norsk 1992**). A situation resembling a negative pressure breathing effect, however, is also induced due to the external pressure of the water on the thoracic wall. This effect enhances the redistribution of blood to the thorax. Furthermore, the hydrostatic pressure gradients of the neck and head remain intact. These inhomogeneous effects of water immersion should be kept in mind when comparing with those of weightlessness during which all hydrostatic pressure gradients are totally abolished.

3.2.1.3 Cardiovascular effects

Immersion immediately induces an increase in central blood volume as evidenced by increases in cardiac output (17–36%), central venous pressure (7–18 mm Hg), and heart volume (180–250 ml) (**Epstein 1992; Norsk 1989, 1992**). Mean arterial blood pressure is unchanged during immersion whereas systolic arterial pressure and pulse pressure increase 5–10 mm Hg. In addition to an increase in central blood volume, total blood volume is increased as a consequence of the water pressure on the legs and the subsequent change in the Starling forces acting across the capillary walls. Concomitantly, a decrease in plasma protein concentration and thus in colloid osmotic pressure is observed (**Johansen *et al.* 1992**). Collectively, these observations indicate that the low-pressure cardiopulmonary volume receptors and high-pressure arterial receptors are stimulated during WI in seated humans.

3.2.1.4 Renal haemodynamics

Glomerular filtration rate is unchanged during WI (**Coruzzi *et al.* 1986**) but effective renal plasma flow is increased.

3.2.1.5 Endocrine responses

WI is associated with profound changes in the plasma levels of volume-regulating humoral factors such as norepinephrine, renin enzymatic activity, aldosterone, vasopressin, atrial natriuretic peptides, renal prostaglandins, and dopamine (**Epstein 1992**).

Plasma renin activity (PRA) is suppressed during WI to the neck usually reaching nadir values of less than 40% of the prestudy value within 3.5 h and remaining suppressed for at least 12 h of WI (**Stadeager *et al.* 1992**). Plasma aldosterone concentration follows a similar temporal profile as PRA with maximal suppression after 3.5 h. Following immersion, both hormones return to baseline and control levels.

Plasma arginine vasopressin concentration (AVP) is usually decreased during WI. Since changes in AVP are difficult to detect in hydrated subjects, experiments have usually been performed with subjects in a slightly water depleted state (e.g. no water ingested from the evening before the experiment). Thus, base line plasma concentrations are easier to detect (**Norsk & Epstein 1988**). In most studies, an immediate decrease of up to 80% is observed during WI, and this decrease is sustained during the entire immersion period. Recently, it has been shown that AVP concentration is suppressed in water replete subjects (200 ml h^{-1}) during immersion (**Larsen *et al.* 1994**).

During WI, the plasma concentration of norepinephrine is immediately decreased and remains so throughout immersion whereas epinephrine concentrations are unchanged (**Stadeager *et al.* 1992**). Furthermore, a decrease in urinary norepinephrine is observed. Urinary dopamine excretion seems to follow a multiphasic course with an initial decrease in the first 2 h followed by an increase (**Grossman *et al.* 1992**).

In response to WI, atrial natriuretic peptide (ANP) concentration in plasma is increased about threefold within 1 h and remains increased for 4 h until a gradual decline sets in (**Stadeager *et al.* 1992**). This decline is associated with an initial increase followed by a decrease in central venous pressure during WI for 12 h and is indicative of a close relationship between the degree of central volume expansion and the release of ANP. Results of a graded WI study by **Larsen *et al.* (1994)** have confirmed such a relationship. The ANP-related peptide, urodilatin (renal natriuretic peptide), which so far has been found only in urine, also is increased during WI (**Norsk *et al.* 1993**). A peak is reached within 3 h followed by a gradual decline towards preimmersion values after 10 h.

Renal prostaglandin E excretion is enhanced during WI with peak levels being attained during hours 2 and 3. Indomethacin treatment administered before immersion attenuates but does not prevent an increase in renal prostaglandin E excretion (**Epstein 1992**).

3.2.1.6 Renal excretion of fluid and electrolytes

WI to the neck induces a marked increase in renal sodium excretion. In normal subjects an increase is observed within 1 h reaching peak values of 3–4fold increases after 3–5 h (**Stadeager *et al.* 1992**). After 5 h, the natriuresis gradually subsides but sodium excretion remains elevated even after 12 h. In subjects on a low-sodium diet (10 mmol day^{-1}) the increase in sodium excretion is much slower, gradually increasing to a peak of approximately 50 mmol min^{-1} at 6 h (**Epstein 1992**). Indications are that the natriuresis is primarily attributable to a decreased tubular reabsorption of sodium at multiple sites in the nephron.

During the initial 4 h, an increase in potassium excretion is observed. Compared to the increased sodium excretion the increase is small, however, and it subsides after 5 h (**Stadeager *et al.* 1992**)

Urine flow rate is markedly increased in water replete subjects attaining a peak value within 2 h. Thereafter, the flow rate gradually returns towards control level (**Stadeager *et al.* 1992**). Although increased compared to a seated time control, the time course of urine flow rate during WI is stable from the 7th h indicating a new steady state. In *hydrated* subjects, the urine flow rate during the initial 6 h of immersion is characterized by a temporary increase in solute-free water clearance. In water *depleted* subjects, however, the increase in urine flow rate is characterized by an increase in excretion of osmotic substances with no increase in solute-free water (**Epstein 1992**).

Due to the increased renal output of fluid, WI of 12 h duration induces decreases in body weight of 0.9±0.2 kg (**Johansen *et al.* 1992**).

3.2.1.7 Summary

Head-out thermoneutral WI induces increases in central and total blood volume. Possibly through stimulation of volume-receptors on the low-pressure (venous) and high-pressure (arterial) side, changes occur in volume regulatory hormones (ANP, PRA, aldosterone, prostaglandin E, and AVP) and renal sympathetic nervous activity are affected. These mediators act in concert on the kidneys resulting in a profound natriuretic and diuretic response in a complex interaction with systemic and renal haemodynamics.

3.2.2 Bed rest

3.2.2.1 Definition

Bed rest (BR) experiments on healthy subjects usually take place over the course of several days or weeks. During this time, subjects are normally not allowed to get out of bed except for very short periods of time and therefore, for all practical reasons, remain supine. Many BR experiments in which the simulation of weightlessness is of prime importance are performed with the subject tilted -3° to -10° (head-down tilted bed rest, HDT) (**Norsk *et al.* 1993; Gharib *et al.* 1988**). Changes during BR and HDT are usually compared with results obtained while the subjects are ambulatory or with the subjects in the supine position in the hours before and after BR and HDT. Only rarely, have seated or supine time control experiments been performed.

3.2.2.2 Comparison with weightlessness

BR and HDT abolish or reverse the foot-to-head hydrostatic gradient. However the position includes a transversal hydrostatic pressure gradient from front-to-back. This is in contrast to weightlessness where all hydrostatic pressure gradients are abolished. Thus, strictly speaking BR and HDT eliminate or reverse a longitudinal gradient and induce a transverse gradient. The former simulates the expected effects of weightlessness on systemic haemodynamics, the latter, however, is a disturbing but unavoidable element of the model which is often overlooked.

3.2.2.3 Cardiovascular effects

Related to the degree of tilt central and total blood volume is increased within the first hours of HDT compared to the both the supine and seated position as evidenced by increases in CVP of some 8 mm Hg, cardiac output of 15%, and stroke volume of 10% (**Norsk 1992**). Mean arterial pressure (MAP) is slightly decreased initially but returns to base line values after 1 h, whereas pulse pressure (PP) remains increased for at least 12 h (Norsk ***et al.* 1993**). Plasma volume is increased initially during the first hours and then decreases gradually below supine pre-HDT values within the first 24 h (**Greenleaf 1984; Greenleaf *et al.* 1992**).

3.2.2.4 Hormonal changes

Evaluation of the significance of the changes in volume-regulating hormones during BR and HDT is of course dependent on the nature of the control position, i.e. whether control experiments were performed with the subject in the seated or supine position. This is illustrated by the fact that when comparing with control values obtained with the subject in the supine position, no significant changes occur in plasma levels of PRA, AVP, aldosterone and norepinephrine within the first hours of 10–30° HDT (**Gharib *et al.* 1988; Goldsmith *et al.* 1985**).

PRA promptly decreases during both -3° and -10° HDT for a least 12 h when comparing with a seated prestudy period and a time control period (**Gharib *et al.* 1988; Norsk *et al.* 1993**). Compared to supine prestudy values, PRA is increased at 24 h after initiation of HDT (**Volicer *et al.* 1976**) and remains increased during 4 days of HDT (**Annat *et al.* 1986**), whereas no changes are observed during horizontal BR for 2–3 weeks (**Chobanian *et al.* 1974**).

Plasma aldosterone is decreased within the first hour and for at least 12 h compared to the seated position and a time control (**Norsk *et al.* 1993**). Plasma aldosterone follows a pattern similar to that of PRA during long-term HDT when compared to the supine position, i.e. an increase beginning at 24 h (**Annat *et al.* 1986**). During horizontal BR, no changes are observed in plasma aldosterone (**Chobanian *et al.* 1974**).

Norepinephrine concentration in plasma decreases immediately and remains suppressed during 12 h of tilted BR compared to a seated control while ANP is increased for at least 3 h (**Norsk *et al.* 1993**).

3.2.2.5 Renal responses

Because most long-term BR and HDT studies have been performed in order to investigate changes in the cardiovascular system, the renal responses to this procedure have not been studied in detail. Furthermore, most BR and HDT experiments do not include appropriate time-control. Particularly because the intakes of food and fluid are likely to change it is difficult under these conditions to evaluate properly the changes in renal output of fluid and electrolytes (**Norsk 1992**). During a 12 h 3° HDT experiment, sodium excretion was increased by 125 μmol min^{-1} above that of a seated time control. This increase was accompanied by an increase in fractional sodium excretion by the kidneys (**Norsk *et al.* 1993**) suggesting a decrease in tubular reabsorption. In a 28-day BR experiment, sodium excretion was slightly increased during the whole BR period compared to an ambulatory control (**Fortney *et al.* 1991**).

Potassium excretion is also increased compared to a seated control during 12 h of HDT (**Norsk *et al.* 1993**). Also during long-term BR potassium excretion remains elevated (**Fortney *et al.* 1991**).

An initial increase in urine flow rate is seen in hydrated subjects during the first 6 h of HDT compared to a seated time control. Later, urine flow rate is indistinguishable from control (**Norsk *et al.* 1993**).

Total body water gradually decreases during horizontal BR compared to ambulatory prestudy values, reaching an average loss of some 1300 ml after 28 days (**Fortney *et al.* 1991**). Body weight did not decrease despite this decline in body water probably due to changes in the ratio of body fat to muscle mass.

3.2.2.6 Summary

BR and HDT increase central blood volume. This increase is accompanied by a diuresis and natriuresis which are quite significant when compared to those of the seated position. When compared to a seated control, BR and HDT induce suppression of plasma levels of AVP, PRA, aldosterone, and norepinephrine, whereas ANP is transiently increased.

3.2.3 Lower body positive pressure

By utilizing lower body positive pressure (LBPP) induced by an antigravity suit, medical anti-shock trousers (MAST garment), or a pressure chamber, several authors have tried to simulate the cardiovascular and renal effects of a decreased gravitational stress (for review see **Norsk 1992**). In one study in the supine position, LBPP of up to 20 mm Hg induced slight increases in CVP and cardiac output with no increases in mean arterial blood pressure (**Shi *et al.* 1993**), whereas higher levels of LBPP (up to 40 mm Hg) induced increases in systolic, mean and diastolic arterial pressures with no further increase in cardiac output. In another study, stroke volume and cardiac output decreased in supine subjects during LBPP of 40 and 100 mm Hg, whereas stroke volume increased by 14% after application of LBPP during head-up tilt (**Gaffney *et al.* 1981**).
When combining an orthostatic stress (70° head-up tilt) with LBPP of 60 mm Hg around the legs and 30 mm Hg around the abdomen (MAST garment), PRA and plasma aldosterone decrease by 50 and 60%, respectively, and renal sodium excretion and urine flow rate increase by 350 and 680%, respectively (**Geelen *et al.* 1989**). Furthermore, plasma norepinephrine decreases by 50% (**Geelen *et al.* 1992**). By exposing conscious squirrel monkeys to 4 days of 20 mm Hg LBPP (**Kass *et al.* 1982**), a marked increase in urine flow rate, and sodium and potassium excretion was observed within the first 12 h. Thereafter, sodium excretion gradually declined while the diuresis persisted. In humans, **Bennet *et al.* (1982)** observed a natriuresis and diuresis with LBPP of 30 mm Hg for 2 h, whereas **Krishna & Danovitch (1984)** did not observe any differences in urine output during LBPP in the supine position compared with that of a supine control. There is no obvious reason for this discrepancy.

3.2.4 Comparison of models

During weightlessness, all hydrostatic pressure gradients are abolished. WI in its usual form only partially abolishes longitudinal hydrostatic pressure gradients and creates a negative pressure breathing effect, while BR and HDT abolish or reverse all hydrostatic gradients along the long axis of the body. However, a transversal hydrostatic pressure gradient persists. LBPP by pressure chamber does probably not to a significant degree change the hydrostatic gradients.

Both WI and BR/HDT induce increases in central blood volume compared to the seated position during at least the initial 12 h, whereas the effects of LBPP on this variable only seem to last a few minutes. All models induce changes in arterial blood pressures: increases in systolic and pulse pressure during WI, increases in pulse pressure during BR/HDT, and increases in mean, systolic and diastolic pressure during LBPP. WI has the most profound effects on volume-regulatory hormones and renal output, whereas the effects of BR/HDT on these variables are less pronounced. LBPP has marked effects on renal sodium excretion and urine flow rate when applied during the erect posture, which in other contexts is considered to provide an orthostatic stress.

3.2.5 Hypothesis

Based on the results of simulated microgravity experiments (WI, BR/HDT, and LBPP) we would expect to observe the following during weightlessness: An initial increase in central and total blood volume as evidenced by increases in CVP and cardiac output. In response to these changes we would expect to see changes in the volume-regulatory hormones accompanied by an increase in renal excretion of sodium and potassium and an increase in urine flow rate. This increased renal output would be followed by a decrease in total body water and a loss of body weight. During long-term space flights (weeks and months) we would expect to observe compensatory changes (secondary decreases in renal output towards control and changes in the volume-regulatory hormones).

3.3 Weightlessness

3.3.1 Body fluids

Prior to the initial phase of the manned space programme in the 1960s, considerations about the human ability to withstand weightlessness were speculative. Based on investigations of simulated weightlessness it was hypothesized that absence of gravity would induce an increase in central blood volume and that this increase would initiate a diuresis (**Graveline & McCally 1962**). Since then considerable work has been performed in order to substantiate this hypothesis.

The first orbital space flights were of very short duration but demonstrated definitely that it is possible for humans to withstand the influence of short term weightlessness. Nevertheless the astronauts from the very beginning reported different symptoms and therefore a number of potential physiological problems soon became evident.

One of the first observations was a sensation of fullness in the head. Later the astronauts also reported a feeling of congestion in the nose and throat (Thornton *et al.* 1977; Kerwin 1977) and during the longer missions the astronauts observed that their legs became thinner, described as 'bird legs' (Schmitt & Reid 1985). During Skylab missions verification of these observations was attempted. Comparison of photographs of the astronauts taken in flight and preflight revealed engorged head veins and a more edematous look of the face in weightlessness (Thornton *et al.* 1977). These observations supported the above mentioned hypothesis of a fluid shift from the lower part of the body to the cephalic part.

Measurement of leg volumes has been performed on Skylab, Apollo–Soyuz, and the US Space Shuttle (**Thornton *et al.* 1987**). Compared to preflight measurements in supine position it was reported that the leg volumes in average decreased 12.8% corresponding to 1.9 l. These data indicate that a fluid shift of approximately 2 l occurred from the legs to other parts of the body. One could postulate that this volume reduction was due to atrophy of leg muscles, but the volume reduction was already evident in the initial days of weightlessness at a time when significant atrophy of muscles is not expected.

3.3.1.1 Body mass

Reduction in body mass of the astronauts was observed even after some of the first orbital space flights and has occurred in virtually all astronauts and cosmonauts (**Norsk & Epstein 1991; Grigoriev *et al.* 1994**). Reduction of body mass is therefore one of the most consistent findings in space physiological research. During the Gemini flights in the 1960s it was observed that loss of body mass was independent of the duration of the flight when this varied between 5 and 14 days. It was also observed that most of the weight loss was regained within the initial 12 h after return to Earth (Minners *et al.* 1962).

During the Apollo 7 to 17 missions, body weight was estimated systematically before and after flight in all astronauts (Hoffler & Johnson 1975). Measurements were made four times before flight at 30, 15 and 5 days before launch and on launch day. The mean value of the launch day was 0.7 kg decreased from the preflight group mean. This decrease could be due to voluntary water withdrawal by the astronauts on launch day in order to avoid urination urge during launch. Nevertheless, when the preflight measurements were compared to measurements immediately post-flight nearly all astronauts demonstrated a significant decrement. The mean decrement was 3.4 kg equivalent to 4.4% of body weight. Most of the astronauts regained the loss within the first days after return, but some showed decreased body weight for many days. These studies were not controlled with respect to intake and excretions. Therefore, it was not possible to distinguish between decreases due to fluid loss, to metabolic loss or to muscular atrophy.

During Skylab missions the time course of the weight loss was investigated for the first time. Mass measurements in flight were now possible because of development of a mass measuring system using the principle of spring-mass oscillations constrained to linear motion. It was realized that any new knowledge with respect to the reasons for the weight loss could be obtained only under controlled circumstances with regard to intake and excretions. Until then three theories for explaining the losses were formulated (Thornton & Ord 1977): (i) increased excretion of water and electrolytes by the kidneys induced by an increase in central blood volume which occurs as a result of a weightlessness induced shift of body fluids from the lower to the upper part of the body; (ii) fluid loss as a result of heat stress and high physical activity during space missions; (iii) decreased fluid intake and food consumption during space missions, possibly in connection with space motion sickness.

In general, the results of the Skylab investigations revealed that body mass decreased 2–3 kg within the first few days in space. Thereafter a decline continued for the rest of the mission but at a much slower rate. This late decline could be minimized by increasing the amount of food and exercise. In most cases the astronauts regained the lost body mass within the first days after return. It was therefore concluded that fluids had to be involved in the acute loss of body mass (Thornton & Ord 1977). Later missions with the US Space Shuttle and Russian missions seem to have confirmed these conclusions (**Grigoriev *et al.* 1994**). The generally accepted hypotheses for the time being is, therefore, that a decrease in body weight occurs within the first few days of weightlessness mainly as a consequence of fluid loss. The further slow and gradual reduction of body mass observed in some astronauts is probably due to loss of muscle and bone substance.

3.3.1.1 Total body water & plasma volume

The hypotheses of fluid loss during weightlessness gave rise to investigations during which it was attempted to measure changes in the body fluids by the use of a radioactive isotope dilution technique. The first measurements of this kind were done during the Apollo program pre- and postflight. By infusion of ^{125}I-albumin, ^{35}S-sulphate, ^{3}H-water and ^{42}KCl measurements of plasma volume, extracellular fluid, total body water and total body potassium, respectively, were made. The results of these measurements were decreases immediately after flight in plasma volume of 4.4%±1.7% in total body water of 2.4%±0.4% (Leach *et al.* 1975). These results are not impressive compared to equivalent periods of bed rest which are associated with a 10% decrease in plasma volume. The reason for this discrepancy could be that some readaptation to the 1 × g environment has already occurred when post-flight measurements were performed.

Measurements of total body water were also performed during the Skylab programme. Similarly to the Apollo results, only slight decreases of 1.1 to 2.5% postflight values were demonstrated. Furthermore, the high value of 2.5% is affected by a heat stress experienced by the astronauts during launch of the Skylab 2 mission (Leach & Rambaut 1977).

Obviously, measurements in flight were needed to make reliable demonstrations of decrements in body water and plasma volume. This was done on board the US space shuttle. By the use of ^{18}O-water as tracer, total body water was measured in flight and compared to pre- and post-flight values. A decrease of 3.4% as a mean among five astronauts was determined after 1 to 3 days in space (**Leach *et al.* 1991**). Using the same technique during another Space Shuttle mission a decrease in total body water of 3% as a mean of 3 astronauts was observed (Cintron *et al.* 1990). During the SLS-1 mission, the plasma volume was found to be 23% below the pre-flight level after 24 h of flight, but only 10% below at landing and at pre-flight level after 6 days (Huntoon *et al.* 1994).

On this basis it seems reasonable to conclude that some decrement in total body water and plasma volume occurs during weightlessness and that this accounts for at least some of the observed loss of body weight.

3.3.2 Central blood volume

3.3.2.1 Central venous pressure (CVP)

In order to substantiate the hypothesis of fluid shifts measurements of the central venous pressure (CVP) during weightlessness have been attempted. The first measurement of this kind was made by Kirsch *et al.* during the first Spacelab mission, SL-1, and the subsequent German Spacelab mission D-1 (**Kirsch *et al.* 1984**, 1985, 1986). They used a method using continuous recording of central venous pressure changes from an arm vein as a measurement of CVP (**Gauer & Sieker 1956**). Results of the study showed a decrease in CVP in flight of 1–4 cm of water and it was concluded that central blood volume was decreased during weightlessness when compared to supine position on ground. Later it has been argued that this method is not suitable for use during weightlessness because the precondition of the method is that pooling of blood in the peripheral arm vein has occurred so that there is a direct fluid-filled opening into the intrathoracic cavity. It is not known whether this is the case in weightlessness (**Norsk & Epstein 1991**). CVP changes were measured during short term weightlessness induced by parabolic flight ($10^{-2} \times g$) using long intrathoracic catheters connected to a pressure transducer (**Norsk *et al.* 1987**). The results showed increased CVP in

seated subjects (1.8 mm Hg) during the last 5 s of the 30 s periods of weightlessness when compared to supine position on ground. When the subjects were in supine position during the parabolic manoeuvre the CVP turned out to be decreased compared to the 1 × g supine position but increased compared to the 1 × g seated position. The discrepancy in these results could be explained by differences in the hydration state of the subjects. CVP measurements in monkeys during parabolic flight induced weightlessness demonstrated that CVP levels were dependent of the degree of hydration (Latham *et al.* 1994). During the Spacelab D–2 mission CVP was measured in one astronaut during launch and the following three hours. Within the initial hours of weightlessness CVP was at or below the level of the ground based measurements in supine position (Foldager, personal communication). These results are in accordance with CVP measurements in one astronaut during the SLS–1 mission (**Buckey *et al.* 1993**). In this study, CVP was lower than the preflight level in supine as well as in seated position.
However, CVP is dependent of factors other than central blood volume. Recent measurements indicate that the intrathoracic pressure is decreased during short term weightlessness induced by parabolic flight (Videbaek, personal communication). It is, therefore, possible that the increase in CVP, expected as a consequence of a weightlessness induced increment in the central blood volume, actually is blunted by a simultaneous decrease in the intrathoracic pressure.

3.3.2.2 Transthoracic electrical impedance

Transthoracic electrical impedance measurement was introduced by Nyboer in the 1950s as a method for estimation of changes in central blood volume. In the sixties NASA developed a device based on this principle for measurement of cardiovascular parameters in astronauts (**Kubicek *et al.* 1966**). Since then the method has been improved especially by use of computer technology and has become more attractive in recent years. The method is based on the fact that any rapid change in electrical conductance reflects a change in fluid volume, an increase in thoracic fluid volume being associated with a decrease in impedance. Being non–invasive and easy to use the method is suitable for investigations during weightlessness.
Impedance measurements during short term weightlessness induced by parabolic flight revealed a decrease in central blood volume when compared to the supine position on ground (3–4%) but an increase compared to the seated position (**Lathers *et al.* 1989**).

3.3.2.3 Echocardiography

Echocardiography measurements was performed on 4 crew members of the Space Shuttle during the STS51–D mission (**Bungo *et al.* 1987**). Measurements were done every day in flight and compared to pre- and post-flight values in supine position. Cardiac index was greatly accentuated during the first day of weightlessness but returned to the preflight value over a week. This increase could be due to an acute increase in central blood volume followed by a decrease when the body adapted to weightlessness.

3.3.2.4 Summary

It is a generally accepted hypothesis that weightlessness increases the central blood volume in the acute state. This hypothesis has been partly substantiated by echocardiography and measurements of thoracic impedance. CVP measurements have shown conflicting results and indeed recent and apparently reliable investigations have demonstrated that CVP values during weightlessness actually are lower than those obtained by ground based measurements.

3.3.3 Renal responses

3.3.3.1 Water and sodium balance

Loss of body fluids during weightlessness has for many years been believed to be caused by increased renal excretion induced by an increased central blood volume (Gauer–Henry theory) (**Gauer *et al.* 1956**). Several ground-based experiments simulating weightlessness by water immersion and head down tilted bedrest have confirmed this hypothesis by demonstrating increased renal excretion of water and sodium (**Norsk 1992**). Due to considerable methodological problems of measurements of urine flow during space flights definitive data are not available and it is difficult to draw precise conclusions on this matter.

Already during the United States second manned space flight rough calculations of intake and excretions before, during and after flight were made. It was concluded that a moderate diuresis occurred even though the fluid intake exceeded the urine output (Minners *et al.* 1962). Sodium excretion was estimated during the Gemini 7 mission. It was found that sodium excretion in one of the two astronauts was increased during the first 7 days and then returned to the preflight level for the rest of the 14 day mission. The other astronaut demonstrated a slight increase during the last week of the mission. Both astronauts showed a significant sodium retention postflight (**Lutwak *et al.* 1969**).

During Apollo 17, investigations of renal function in 3 astronauts were carried out. These demonstrated that urine volume was decreased in flight compared to preflight values (Leach *et al.* 1975). In two of the astronauts the largest decreases occurred within the initial days of weightlessness. When fluid balance was calculated the surprising result was that two of the three astro-

nauts developed a positive fluid balance in flight despite of weight losses of 1.9 and 4.3 kg. Calculations of sodium intake and excretions during the same mission showed an increased sodium balance in flight. Later it was stated that there was some kind of systematic defect in the sodium measurements in this study since the astronauts did not gain weight during the control period and lost weight in flight (**Leach *et al.* 1974**). It is possible that fluid balance calculations were also influenced by this defect (**Norsk & Epstein 1991**).
During Skylab missions 2 (28 days), 3 (59 days) and 4 (84 days) continuous metabolic monitoring began 21 days before flight and continued throughout the flight until 18 days post-flight (Leach & Rambaut 1977). Pre- and post-flight urine samples were collected for each void, in-flight samples for each day. All data were combined and presented as mean values of six days periods. Results from Skylab 2 should be interpreted with caution. Because of the loss of a heat shield during launch the astronauts were exposed to very high temperatures during the initial six days of the mission. It is likely that the heat stress affected the data of fluid and electrolyte balance (**Norsk & Epstein 1991**). Nevertheless, the Skylab programme taken together showed for eight of the nine astronauts a decrease in urine volume (mean of 400 ml) in the first six day period in flight. Thereafter urine volume was similar to pre-flight values until the end of investigation 18 days after the flight (Leach & Rambaut 1977). Fluid intake decreased simultaneously (mean of 700 ml) within the first six day period in flight. Although the quality of these data is not always convincing they led to the conclusion that a negative water balance was established because of a decrease in water intake and not because of an increase in water excretion. Drinking was *ad libitum* in Skylab but it is possible that space motion sickness and anti-emetic drugs played a role in blunting appetite and thirst. Evaporative water loss was not included in these calculations but was expected to be increased during these flights because of low barometric pressures in Skylab (Leach & Rambaut 1977). Later it was reported that evaporative water loss actually was decreased (by an average 8%), except for the period with heat stress on Skylab 2 (**Leach *et al.* 1978**), and it can, therefore hardly be responsible for the unexpected results of water excretion.
Reports of sodium excretion rates from all Skylab missions showed that values in flight were elevated above the pre-flight mean for all nine crewmen (Leach & Rambaut 1977). Nevertheless, interpretations concerning sodium balance are difficult because of the earlier mentioned heat stress. Another reason is that data are presented as relative changes (%) against an unidentified pre-flight value. Data have later been reviewed and the conclusion is that the Skylab missions taken together demonstrated a negative sodium balance which probably was due to increased renal excretion (**Norsk & Epstein 1991**).
Experiments on board the US space shuttle have not yielded results that convincingly contradict earlier studies. Investigations during STS1-4 missions lead to the conclusion that body fluid was decreased due to decreased water intake (**Leach 1983**).

During the German–Russian Mir 92 mission, experiments were made in one astronaut during a one week space flight. This astronaut did not suffer from space motion sickness and therefore these results are not influenced by the use of anti-emetic drugs. Moreover, cortisol was measured from saliva during the whole mission in order to monitor whether stress could have affected the results. This was not the case in this mission. This experiment contradicts the concept that weight loss during weightlessness is due to renal loss of water and sodium (**Drummer *et al.* 1993**). Although the astronaut lost body weight to an extent of more than 5 kg, both water and sodium excretion were decreased during the entire one week mission compared to pre-flight values. Post-flight results were not interpretable.

3.3.3.2 Potassium balance

There are strong indications of potassium loss during weightlessness. Crew members of the Gemini 7 mission demonstrated negative potassium balance during the flight (Leach *et al.* 1975). During Apollo 12–14 total body potassium was determined before and after flight by gamma spectrometric measurements of total body ^{40}K. A significant decrease (3–10%) was demonstrated for seven of the nine astronauts (Leach *et al.* 1975). During the Apollo 15 to 17 and Skylab missions total body exchangeable potassium was determined and found generally to be decreased after flight even though adequate potassium had been ingested throughout the missions (Leach & Rambaut 1977). It is difficult to get a clear picture of potassium excretion from Space Shuttle flights. From the STS1–4 flights it is reported that serum concentrations had decreased significantly after flight (**Leach 1983**). Russian investigators report that blood potassium was usually decreased after short term space flights, but after flights of more than 30 days the decrease was less apparent (**Grigoriev *et al.* 1994**). Similar to the Apollo derived results, a post-flight decrease in total body potassium has been found in cosmonauts (**Grigoriev *et al.* 1994**). Post-flight load tests of potassium chloride demonstrated an increased potassium excretion compared to pre-flight measurements. This has been interpreted as a reduced capacity of tissues to retain potassium after space flight because of decreases in muscle mass and red blood cell mass (see review by **Grigoriev *et al.* 1994**).

3.3.3.3 Summary

In conclusion it appears that when measured with reasonable accuracy renal water excretion is found often to be decreased in flight in conflict with the hypothesis of central hypervolemia. In contrast, the rate of excretion of sodium in most cases has been found to increase and thus support the hypothesis, although noticeable exceptions to this trend have been published. It is remarkable that no investigation has been in full compliance with the hypothesis predicting increases in water and sodium excretion due to volume redistribution. On the other hand, it seems to be a consistent observation that a negative potassium balance develops during weightlessness. Another consistent finding after space missions is the activation of water and sodium retaining mechanisms. This has indeed been used as an argument for a renal loss of sodium and water occurring during weightlessness.

3.3.4 Endocrine responses

Changes in the body composition of fluid and electrolytes due to the weightless condition are expected to involve different hormonal systems. Several simulation experiments have focused on this issue and the results of these suggest that the renal effects are mediated through suppression of the renin-angiotensin-aldosterone system (RAAS), arginine vasopressin (AVP), and catecholamines (**Norsk 1992**). Furthermore, it has been suggested that an increased release of natriuretic peptides (atrial natriuretic peptide from the atrium of the heart and renal natriuretic peptide (urodilatin) from the kidney) during water immersion is involved in the renal excretion of sodium (**Norsk *et al.* 1993**).
The most appropriate way to examine the changes of the endocrine system during weightlessness would be to obtain serial blood samples each day. However, this is seldom possible during space flights and therefore hormone measurements from urine pools have been widely used. This method has the advantage of being non-invasive and cover long periods. However, few studies, if any, include proper investigations of the relations between rate of secretion, plasma concentration and rate of urinary excretion under controlled conditions. Therefore, it is not possible to make conclusions concerning the secretion rate and plasma level of a hormone on basis of the urine concentration or rate of excretion of that hormone.

3.3.4.1 Arginine vasopressin peptide (AVP)

Measurement of urine concentrations of AVP was done during Skylab missions. Except for the Skylab 2 mission results, which are questionable because of the earlier mentioned heat stress, all astronauts showed a decreased urinary excretion of AVP (Leach & Rambaut 1977). During the post-flight period the AVP level did not return to the pre-flight level. Therefore, these results should be interpreted with caution. Also it should be mentioned that the results are in accordance with theory and with simulation experiments, but in contrast to the renal response of the experiment which showed decreased or unchanged urine excretion.
During Spacelab flights plasma AVP concentrations were measured using radioimmunoassay. All astronauts showed increased AVP levels of approximately 100% in the period from 20 to 200 h after launch (Leach *et al.* 1985).

It was later argued that the increase in flight could be due to space motion sickness (Leach *et al.* 1988). Post-flight plasma levels of AVP were also elevated whereas urine concentrations were unchanged. However, if astronauts are separated into users and non-users of the fluid and salt loading procedure as a countermeasure, differences between the groups emerge. Astronauts who used fluid and salt loading showed a 44% increase of AVP excretion in urine, whereas astronauts who did not use fluid/salt loading showed a decrease of 16% (Leach & Johnson 1985).
During a Russian 150 day Salyut-7 space flight urine concentrations of AVP were found to be reduced in two periods on days 43–45 and 86–88 (**Grigoriev *et al.* 1994**), in accordance with the Skylab results. In another 237 day Salyut-7 space flight renal excretion of AVP was elevated on days 216–219 but plasma AVP concentrations were reduced (Gazenko *et al.* 1988). An average of post-flight plasma AVP values from many Russian prolonged spaceflights shows an increase from 2.7±0.8 pg ml^{-1} pre-flight to 9.3±3.4 pg ml^{-1} post-flight (**Grigoriev *et al.* 1994**). Comparison of pre- and post-flight plasma AVP concentrations in relation to reabsorption of osmotically free water have demonstrated decreased AVP efficacy after flight, suggesting a decreased kidney sensitivity to AVP after long term space flights (**Grigoriev *et al.* 1994**). On the other hand Leach and colleagues (**Leach *et al.* 1983**) have suggested an augmented sensitivity of the renal tubules to AVP. Taken together the available evidence appears inconclusive.

3.3.4.2 Renin–angiotensin–aldosterone system (RAAS)

Measurements of urinary excretion of aldosterone have been performed since the 1960s. During Gemini 7 (**Lutwak *et al.* 1969**) and Skylab flights the urinary excretion rates of aldosterone were increased 100–500% throughout the flights (Leach & Rambaut 1977). During Apollo 16 and 17 missions no changes were observed in the renal excretion of aldosterone (**Leach *et al.* 1974**, 1975). During the Russian Saluyt-7 missions the urinary aldosterone excretions were generally increased (Gazenko *et al.* 1988). It seems that elevation of the urinary excretion of aldosterone in space flights is a rather consistent finding.
Measurements of plasma aldosterone concentration in flight does not give a clear picture because of a large individual variation. From the Skylab missions (excluding Skylab 2 because of the influence of heat), it appears that in most cases plasma aldosterone was decreased during the first two weeks and thereafter increased above pre-flight level for the rest of the mission (**Leach *et al.* 1976**; Leach & Rambaut 1977). Results from the more short term Spacelab and SLS-1 missions show that plasma aldosterone were reduced or unchanged (Leach *et al.* 1986, 1992).
On the long term Russian Salyut-7 and Mir missions different changes of plasma renin activity and plasma aldosterone have been demonstrated (**Grigoriev *et al.* 1994**).

Post-flight plasma aldosterone levels are easier to evaluate. After Skylab, Space Shuttle and most of the Russian flights plasma aldosterone levels have always been elevated (Huntoon *et al.* 1994).
Plasma renin activity (PRA) was measured during Skylab and Spacelab missions. Measurements aboard Skylab in flight showed great variability but most measurements were elevated with a tendency towards the largest increase within the initial weeks in space (**Leach *et al.* 1976**; Leach & Rambaut 1977). In Spacelab it was found that PRA was reduced within the initial 48 h but thereafter elevated (Leach 1992). Post-flight PRA values suffer from so much variability that meaningful interpretations are very difficult (Huntoon *et al.* 1994).

3.3.4.2 Natriuretic peptides

Atrial natriuretic peptide (ANP) discovered in 1981 is excreted from the atrium of the heart as a response to atrial distention (**De Bold *et al.* 1981**). Secretion is thus expected to occur in conjunction with an acute increment of the central blood volume. ANP acts on the kidneys by inhibition of sodium reabsorption in the tubules, thereby increasing renal excretion of sodium and water (**Goetz 1988**). It was later shown that ANP reduces plasma volume by an extrarenal effect probably by increasing the vascular permeability to albumin (**Lockette & Brennaman 1990**). Plasma ANP levels are elevated during experiments simulating weightlessness such as water immersion and head-down tilt bedrest studies (**Norsk 1992**). Therefore, it is expected that ANP is involved in the regulation of fluid and electrolytes during adaptation to weightlessness.
During the Spacelab-2 mission plasma ANP was measured for the first time in space (Leach *et al.* 1988; Cintron *et al.* 1990). Samples were obtained from all four astronauts 3 times preflight, twice in flight and 3 times post-flight. ANP levels were increased in all astronauts on flight day 2 by an average of 36% but decreased on flight day 7 (mean −59%). After flight the concentrations returned to the pre-flight level. The direction of the changes were identical for all crew members.
Animal experiments have been performed to study ANP related phenomena during weightlessness. After the 9 day SLS-1 mission the brain of the flown rats were examined with specific interest in the binding sites of ANP (**Herbute *et al.* 1994**). It was demonstrated that the number of ANP bindings sites were increased in the choroid plexus in the rats who returned after 9 days in space. It was suggested that ANP possibly is involved in the regulation of fluid and electrolyte fluxes in the brain during adaptation to weightlessness through modification of the expression of ANP receptors.
The recently discovered peptide urodilatin (renal natriuretic peptide) is produced in the kidneys and acts on kidney tubules by inhibiting sodium reabsorption and thereby increasing the rates of excretion of sodium and water (**Schulz-Knappe *et al.* 1988**). It has been argued that urodilatin rather than

ANP is responsible for regulation of sodium excretion in physiological conditions (**Goetz 1993**). During a 12 h water immersion experiment excretion of urodilatin in urine increased in conjunction with sodium excretion. From this it was concluded that urodilatin is involved in the renal excretion of sodium during water immersion (**Norsk *et al.* 1993**). Therefore, it was expected that urodilatin might be elevated during space flight and mediate a natriuretic and diuretic response contributing to an in flight weight loss (**Drummer *et al.* 1993**). During the one week German–Russian Mir 92 mission urinary urodilatin excretion was measured in one astronaut. Samples were obtained from every void on days 1, 5 and 6. Urodilatin excretion was elevated on day 1 but identical to preflight level on days 5 and 6. However, no increased renal excretion of sodium and water was demonstrated even though the astronaut lost 5 kg body weight during the mission (**Drummer *et al.* 1993**).

3.3.4.3 Intervention experiments

Knowing that a great deal of noise is associated with data obtained in experiments observing the fluid balance of astronauts passively exposed to weightlessness, some investigators have tried to examine the effect of a standardized stimulus during weightlessness and compared this effect to that obtained in the $1 \times g$ environment. One experiment on the Spacelab D–2 mission was directed at the renal and endocrine response to such a standardized volume stimulus (**Norsk *et al.* 1995**). On the sixth day of flight the 4 astronauts received a 20 min infusion of isotonic saline equivalent to 2% of body weight. In-flight results were compared to identical ground based experiments in supine and seated position. Renal excretions of water, sodium and urodilatin were monitored and plasma concentrations of ANP, AVP, aldosterone, renin, norepinephrine and epinephrine were measured. The experiment showed that the renal response to a saline load in weightlessness is augmented compared to seated position on ground but attenuated when compared to the supine position. The plasma levels of norepinephrine, aldosterone and renin were elevated in flight but not related to changes in sodium and water excretion at least not in the same manner as on the ground. Urodilatin excretion seemed to be better correlated to the in flight sodium excretion than norepinephrine and aldosterone. Especially norepinephrine levels were elevated to an extent similar to orthostatic stress on ground. The authors suggest that central hypovolemia is one of the reasons for elevated hormone levels and attenuated renal responses during weightlessness compared to the $1 \times g$ supine condition. Another possibility is that unexpected input to volume and baroreceptors induced relative sodium and fluid retention and increased sympathetic nerve activity.

3.3.4.4 Summary

It is not possible on basis of the present results to arrive at an arguable conclusion with regard to the significance of changes in the secretion of AVP during weightlessness. Plasma AVP levels have been elevated in several flights but it should be noted that space motion sickness elicits AVP secretion and it is likely that this mechanism can account for some of the results. It has been suggested that the function of AVP should be altered during weightlessness i.e. by an altered sensitivity of the renal tubules to AVP, but this hypothesis is so far unsubstantiated and needs further investigation.
The available results do not give clear picture of the effects of weightlessness on the RAAS and its role in renal function. However, it seems that the renal excretion of aldosterone increases and the plasma concentration decreases in weightlessness. There is no explanation for this apparent inconsistency and it is possible that environmental factors add systemic error to the results. Moreover it should be noted that the activity of the RAAS is to some extent a reflection of sodium intake. If the diet in flight is not identical to pre- and post-flight diets with respect to at least sodium content interpretation is difficult. A consistent finding though is that both plasma and urine concentrations of aldosterone are elevated in the initial post-flight days which is an indication of activation of sodium retaining mechanisms after re-entry.
Excretion of ANP and urodilatin seems to be well correlated to renal excretion of sodium in weightlessness as well as in $1 \times g$ conditions. Whether these peptides are involved to a significant extent in the regulation of fluid and electrolytes during adaptation to weightlessness still needs to be demonstrated.

3.4 Conclusion

The original Gauer-Henry theory of increased diuresis and natriuresis during the initial days of weightlessness fits well with the results of experiments of simulated weightlessness. However, investigations in flight have not been able to confirm the hypothesis. Available results point in the direction that a negative water balance occurs within the first days in flight due to decreased water intake rather than increased renal fluid excretion. Results with regard to sodium excretion in flight are inconclusive due to the scatter of the data. Attempts to investigate the mechanisms of body fluid regulation in weightlessness by measurements of different hormone levels in plasma and urine have given conflicting results.
On the other hand it seems beyond reasonable doubt that the initial days of weightlessness are associated with central hypervolemia and a loss of body mass of 3–5 kg indicating that body fluid content actually decreased. It also

seems to be a consistent finding that, in the first days post-flight, water and sodium retaining mechanisms (AVP and RAAS) are activated.
The lack of consistency of the results may be due to environmental disturbances leading to space motion sickness, stress and difficulties in controlling fluid and dietary intake. Alternatively, the hypothesis is incorrect and unknown factors contribute to body fluid regulation during weightlessness.
In this area of research the target variables are generally few, slowly changing, and hard to measure, as they tend to be obscured by noise. High quality experiments should be given sufficient flight opportunities to allow for a number of experiments suitable for appropriate statistical evaluation. Astronauts for life science missions should preferably have experience and understanding of physiological investigations. Fruitful investigations of the spontaneous time course during weightlessness will probably require very standardised conditions which at present seem less compatible with other constraints. In this situation it might be advantageous to give priority to well designed intervention experiments investigating the dynamic response to given stimuli (e.g. hormone administration, forced fluid shifts, water/salt loads). It remains crucially important to appreciate and encourage investigations requesting standardised and monitored circumstances with respect to diet and fluid intake/excretions. Studies of dynamic responses under optimized conditions by use of a sufficient number of subjects appear to be the only way to improve the reliability of the often confusing results to levels competitive with ground based investigations. Such improvement is necessary before real cross fertilization will occur between high-impact ground based research of human disease and investigations during weightlessness of body fluid regulation and kidney function.

3.5 References

Annat G, Güell A, Gauquelin G, Vincent M, Mayet MH, Bizollon CA, Legros JJ, Pottier JM & Gharib C. Plasma vasopressin, neurophysin, renin and aldosterone during a 4-day head-down bed rest with and without exercise. Eur. J. Appl. Physiol. 55: 59–63, 1986.

Bennet ED, Tighe D & Wegg W. Abolition, by dopamine blockade, of the natriuretic response produced by lower-body positive pressure. Clin. Sci. 63: 361–6, 1982.

Buckey JC, Gaffney FA , Lane LD, Levine BD, Watenpaugh DE & Blomquist CG. Central venous pressure in space (letter). New England J. Med. 328: 1853–4, 1993.

Bungo MW; Goldwater DJ, Popp RL & Sandler H. Echocardiographic evaluation of Space shuttle crew members. J. Appl. Physiol. 62: 278–83, 1987.

Chobanian AV, Lille RD, Tercyak A & Blevins P. The metabolic and hemodynamic effects of prolonged bed rest in normal subjects. Circulation 49: 551–9, 1974.

Cintron NM, Lane HW & Leach CS. Metabolic consequences of fluid shifts induced by microgravity. The Physiologist 33: S16 – S19, 1990.

Coruzzi P, Biggi A, Musiari L, Ravanetti C & Novarini A. Renal hemodynamics and natriuresis during water immersion in normal humans. Pfluegers Arch. 407: 638–42, 1986.

De Bold AJ, Borenstein HB, Veress AT, Sonnenberg H. A rapid and potent natriuretic response to intravenous injection of atrial myocardial extract in rats. Life Sci. 28: 89–94, 1981.

Drummer C, Heer M, Dressendörfer RA, Strasburger CJ & Gerzer R. Reduced natriuresis during weightlessness. Clin. Invest. 71: 678–86, 1993.

Epstein M. Renal effects of head-out water immersion in humans: a 15-year update. Physiol. Rev. 72: 563–621, 1992.

Fortney SM, Hyatt KH, Davis JE & Vogel JM. Changes in body fluid compartments during a 28-day bed rest. Aviat. Space Environ. Med. 62: 97–104, 1991.

Gaffney FA, Thal ER, Taylor WF, Bastian BC, Weigelt JA, Atkins JM & Blomqvist CG. Hemodynamic effects of medical antishock trousers (MAST garment). J. Trauma 21: 931–7, 1981.

Gauer OH & Henry JP. Circulatory basis of fluid volume control. Physiol. Rev. 43: 423–481, 1963.

Gauer OH & Sieker HO. The continuous recording of central venous pressure changes from an arm vein. Circ. Res. 4: 74–8, 1956.

Gauer OH, Henry JP & Sieker OH. Changes in central venous pressure after moderate hemorrhage and transfusion in man. Circ. Res. 4: 79–84, 1956.

Gazenko OG, Schultzhenko EB, Grigoriev EB, Atkov Oyu & Egorov AD. Review of basic medical results of the Saluyt-7-Soyuz-T 8 month manned flight. Acta Astronautica 17: 155–60, 1988.

Geelen, G, Arbeille P, Saumet J-L, Cottet-Emard J-M, Patat F & Vincent M. Hemodynamic and hormonal effects of prolonged anti-G suit inflation in humans. J. Appl. Physiol. 77: 977–84, 1992.

Geelen G, Kravik SE, Hadj-Aissa A, Leftheriotis G, Vincent M, Bizollon CA, Sem-Jacobsen CW, Greenleaf JE & Gharib C. Antigravity suit inflation: kidney function and cardiovascular and hormonal responses in men. J. Appl. Physiol. 66: 792–9, 1989.

Gharib C, Gauquelin G, Pequignot JM, Geelen G, Bizollon CA & Güell A. Early hormonal effects of head-down tilt (-10) in humans. Aviat. Space Environ. Med. 59: 624–9, 1988.

Goetz KL. Physiology and pathophysiology of atrial peptides. Amer. J. Physiol. 254 (Endocrinol. Metab. 17): E1–E15, 1988.

Goetz KL. Is urodilatin (rather than atriopeptin) the primary natriuretic peptide of the ANP family? J. Cardiovasc. Pharmacol. 22, Suppl. 2: S84–5, 1993.

Gogolev KI, Aleksandrova EA & Shulzhenko EB. Comparative assessment of changes during antiorthostatic hypokinesia and immersion in man. Fiziologia Cheloveka 6: 978–83, 1980.

Goldsmith SR, Francis GS & Cohn JN. Effect of head-down tilt on basal plasma norepinephrine and renin activity in humans. J. Appl. Physiol. 59: 1068–71, 1985.

Graveline DE & McCally M. Body fluid distribution: implications for zero gravity. Aerospace Med. 33: 1281–90, 1962.

Greenleaf JE. Physiological responses to prolonged bed rest and fluid immersion in humans. J. Appl. Physiol. 57: 619–33, 1984.

Greenleaf JE, Vernikos J, Wade CE & Barnes PR. Effect of leg exercise training on vascular volumes during 30 days of -6° head-down bed rest. J. Appl. Physiol. 72: 1887–94, 1992.

Grigoriev AI, Morukov BV & Vorobiev DV. Water and electrolyte studies during long-term missions on board the space stations SALYUT and MIR. Clin. Invest. 72: 169–89, 1994.

Grossman E, Goldstone DS, Hoffman A, Wacks IR & Epstein M. Effects of water immersion on sympathoadrenal and dopa-dopamine systems in humans. Amer. J. Physiol. 262: R993–R999, 1992.

Herbute, S., Oliver, J., Davet, J., Viso, M., Ballard, R.W., Gharib, C., Gabrion, and J. ANP binding sites are increased in choroid plexus of SLS-1 rats after 9 days of spaceflight. Aviat. Space & Environ. Med. 65:134-8, 1994.

Hoffler GW & Johnson RL. Apollo flight crew cardiovascular evaluations. In: *Biomedical Results of Apollo.* Ed. by Johnston RS & Dietlein F. NASA special report SP-368, pp. 244-65, 1975.

Huntoon CL, Cintron NM & Whitson PA. Endocrine and biochemical functions. In: *Space Physiology and Medicine.* Ed. by Nicogossian AE, Huntoon CL, Pool SL. Lea & Fabiger, 3rd edn, pp. 334-50, 1994.

Johansen LB, Foldager N, Stadeager C, Kristensen MS, Bie P, Warberg J, Kamegai M & Norsk P. Plasma volume, fluid shifts, and renal responses in humans during 12 h of head-out water immersion. J. Appl. Physiol. 73: 539-44, 1992.

Kass DA & Moore-Ede MC. Renal responses to prolonged central volume expansion in conscious primates. Amer. J. Physiol. 242: F649-F656, 1982.

Kerwin JP. Skylab 2 crew observations, summary. In: *Biomedical Results from Skylab.* Ed. by Johnston RS & Dietlein F. NASA special report SP-377, pp. 27-9, 1977.

Kirsch KA, Röcker L, Gauer OH, Krause R, Leach C, Wicke HJ & Landry R. Venous pressure in man during weightlessness. Science 225: 218-9, 1984.

Kirsch KA, Röcker L, Krause R & Gauer OH. Cardiovascular research in space: problems and results. AGARD Conference Proceedings, No. 377: 8/1-8/9, 1985.

Kirsch KA, Röcker L & Haenel F. Venous pressure in space. In: *Proceedings of the Norderny Symposium on the Scientific Results of the German Spacelab Mission D-1.* Ed. by Sahm PR, Jansen R & Keller MH. Köln, Germany, DFVLR, pp. 500-3, 1986.

Krishna GG & Danovitch GM. Renal response to lower body positive pressure produced by military anti-shock trousers. Amer. J. Nephrol. 4: 13-18, 1984.

Kubicek WG, Karnagis JN, Patterson RP *et al.* Development and evaluation of an impedance cardiac output system. Aerospace Med. 37:1208-12, 1966.

Larsen AS, Johansen LB, Stadeager C, Warberg J, Christensen NJ & Norsk P. Volume-homeostatic mechanisms in humans during graded water immersion. J Appl. Physiol. 77: 2832-9, 1994.

Latham RD, Fanton JW, Vernalis MN, Gaffney FA & Crisman RP. Central hemodynamics in a baboon model during microgravity induced by parabolic flight. Adv. Space Res. 14: 349-58, 1994.

Lathers CM, Charles JB, Olden KF, Holt TA, Mukai C, Bennett BS & Bungo MW. Acute hemodynamic responses to weightlessness in humans. J. Clin. Phamacol. 29: 615-27, 1989.

Leach CS. Medical results from STS 1-4: analysis of body fluids. Aviat. Space Environ. Med. 54, Suppl. 1: S50-S54, 1983.

Leach CS. Changes in renal function and fluid and electrolyte regulation in space flight. *World Space Congress, Washington, DC, August 28 - September 5, 1992.* Book of abstracts, paper no. IAF/IAA-92-0256: p. 81, 1992.

Leach CS, Alexander WC & Johnson PC. Endocrine, electrolyte, and fluid volume changes associated with the Apollo missions. In: *Biomedical Results of Apollo.* Ed. by Johnston RS & Dietlein F. NASA special report SP-368, pp. 163-84, 1975.

Leach CS, Altschuler SI & Cintron-Trevino NM. The endocrine and metabolic responses to space flight. Med. Sci. Sports Exerc. 15: 432-40, 1983.

Leach CS, Inners LD & Charles JB. Changes in total body water during spaceflight. J. Clin. Pharmacol. 31: 1001-6, 1991.

Leach CS & Johnson PC. Fluid and electrolyte control in simulated and actual space flight. Physiologist, 28, Suppl. 6, pp. S34-S37, 1985.

Leach CS, Johnson PC & Cintron NM. The regulation of fluid and electrolyte metabolism in weightlessness. *Proc. of the 2nd International Conference on Space Physiology, Toulouse, France 20-22 November, 1985.* ESA SP-237, pp. 31-6, 1985.

Leach CS, Johnson PC & Cintron NM. The endocrine system in space flight. Acta Astronautica 17: 161-6, 1988.

Leach CS, Johnson PC & Rambaut PC. Metabolic and endocrine studies: the second manned Skylab mission. Aviat. Space Environ. Med. 47: 402–10, 1976.

Leach CS, Leonard JI, Rambaut PC & Johnson PC. Evaporative water loss in man in a gravity-free environment. J. Appl. Physiol. 45: 430–6, 1978.

Leach CS & Rambaut PC. Biomedical responses of the Skylab crewmen: an overview. In: *Biomedical Results from Skylab.* Ed. by Johnston RS & Dietlein F. NASA spec. rep. SP-377, pp. 204–15, 1977.

Leach CS, Rambaut PC, Johnson PC. Adrenocortical responses of the Apollo 17 crew members. Aerospace Med. 45: 529–34, 1974.

Lockette W & Brennaman B. Atrial Natriuretic Factor increases vascular permeability. Aviat. Space Environ. Med. 61:1121–4, 1990.

Lutwak L, Whedon GD, Lahance PA, Reid M & Lipscomb S. Mineral, electrolyte and nitrogen balance studies of the Gemini-VII fourteen-day orbital space flight. J. Clin. Endocrin. 29: 1140–56, 1969.

Minners HA, White SC, Douglas WK, Knoblock EC & Gabriel A. Clinical aeromedical observations. In: *Results of the Second United States Manned Orbital Space Flight.* NASA SP-6, pp. 43–53, May 24, 1962.

Norsk P. Influence of low- and high-pressure baroreflexes on vasopressin release in humans. Acta Endocrinol. 121, Suppl. 1: 1–27, 1989.

Norsk, P. Gravitational stress and volume regulation. Clin. Physiol. 12: 505–26, 1992.

Norsk P, Drummer C, Johansen LB & Gerzer R. Effect of water immersion on renal natriuretic peptide (urodilatin) excretion in humans. J. Appl. Physiol. 74: 2881–5, 1993.

Norsk P, Drummer C, Röcker L, Strollo F, Christensen NJ, Warberg F, Bie P, Stadeger C, Johansen LB, Heer M, Gunga H-C & Gerzer R. Renal and endocrine responses in humans to isotonic saline infusion during microgravity. J. Appl. Physiol. 78: 2253–2259, 1995.

Norsk P & Epstein M. Effects of water immersion on arginine vasopressin release in humans. J. Appl. Physiol. 64: 1–10, 1988.

Norsk P & Epstein M. Manned space flight and the kidney. Amer. J. Nephrol. 11:81–97, 1991.

Norsk P, Foldager N, Bonde-Petersen F, Elmann-Larsen B & Johansen TS. Central venous pressure in humans during short term periods of weightlessness. J. Appl. Physiol. 63: 2433–7, 1987.

Norsk P, Stadeager C, Johansen LB, Warberg J, Bie P, Foldager N & Christensen NJ. Volume-homeostatic mechanisms in humans during a 12-h posture change. J. Appl. Physiol. 75: 349–6, 1993.

Schmitt HH & Reid DJ. *Anecdotal Information on Space Adaptation Syndrome.* NASA Space Biomedical Research Inst., USRA, Div. Space Biomedicine, 1985.

Schulz-Knappe P, Forssmann K, Herbst F, Hock D, Pipkorn R & Forssmann WG. Isolation and structural analysis of 'urodilatin', a new peptide of the cardiodilatin-(ANP)-family, extracted from human urine. Klinische Wochenschrift 66: 752–9, 1988.

Shi X, Crandall CG & Raven PB. Hemodynamic responses to graded lower body positive pressure. Amer. J. Physiol. 265 (Heart Circ. Physiol. 34): H69–H73, 1993.

Stadeager C, Johansen LB, Warberg J, Christensen NJ, Foldager N, Bie P & Norsk P. Circulation, kidney function, and volume-regulating hormones during prolonged water immersion in humans. J. Appl. Physiol. 73: 530–8, 1992.

Thornton WE & Ord J. Physiological mass measurements in Skylab. In: *Biomedical Results from Skylab.* Ed. by Johnston RS & Dietlein F. NASA special report SP-377, pp. 175–82, 1977.

Thornton WE, Hofffler GW & Rummel JA. Anthropometric changes and fluid shifts. In: *Biomedical Results from Skylab.* Ed. by Johnston RS & Dietlein F. NASA special report SP-377, pp. 330–8, 1977.

Thornton WE, Moore TP, & Pool SL. Fluid shifts in weightlessness. Aviat. Space Environ. Med. 58, Suppl. 9: A86–90, 1987.

Volicer L, Jean-Charles R & Chobanian A. Effects of head-down tilt on fluid and electrolyte balance. Aviat. Space Environ. Med. 47: 1065–8, 1976.

Chapter 4

Neurosensory and sensory-motor functions

G. Clément & M. F. Reschke

4.1 General Introduction

Correct transduction and integration of signals from all the sensory systems is essential for maintenance of stable vision, spatial orientation, eye–head coordination, and postural and locomotion control. Central nervous system (CNS) plasticity allows individuals to adapt to altered sensory stimulus conditions. The microgravity environment of orbital flight represents such an altered sensory stimulus state.

Exposure to microgravity rearranges the relationships among signals from vestibular, visual, skin, joint, and muscle receptors. Until some level of adaptation to the novel sensory conditions encountered in the weightless environment of space flight is achieved, astronauts and cosmonauts often experience the following: (i) illusory self and/or surround motions, (ii) space motion sickness, (iii) eye–head coordination impairment, and (iv) equilibrium control disturbance. Many of the same types of disturbances are observed during a readaptation period following return to Earth. The magnitude and recovery time course of these neurosensory, sensory–motor and perceptual disturbances tend to vary as a function of mission duration.

Investigations designed to understand, predict and prevent the various neurosensory, sensory–motor and perceptual disturbances listed above are described in the remaining sections of this chapter. Wherever possible, we have attempted to describe the research, the findings and the conclusions from ESA Member States, United States, Japanese and Russian space programmes.

4.2 Sensory functions

4.2.1 Vestibular apparatus

The vestibular end organs consist of the semicircular canals, which respond to rotation of the head, and the otolith organs which signal the head's position with respect to gravity (tilt). In microgravity, static tilt will no longer give rise to a change in otolith activity. However, linear acceleration as well as angular acceleration continues to be transduced. As a result, the CNS must adapt to the absence of certain signals and reinterpret others. One possibility is that sensory compensation occurs, and information from the other sense organs are used to maintain spatial orientation. Alternatively, signals from the otoliths may be reinterpreted by the brain. On Earth, graviceptor signals from the otoliths can be interpreted either as linear motion (translation) or as tilt with respect to gravity (Fig. 1). Because stimulation from gravity is absent during orbital flight, interpretation of the graviceptor signals as tilt is meaningless. Therefore, during adaptation to weightlessness, the brain reinterprets all graviceptor output to indicate translation, and otolith-mediated responses such as postural reflexes, eye movements, and subjective responses would alter accordingly. This is the so-called Otolith Tilt-Translation Reinterpretation (OTTR) hypothesis.

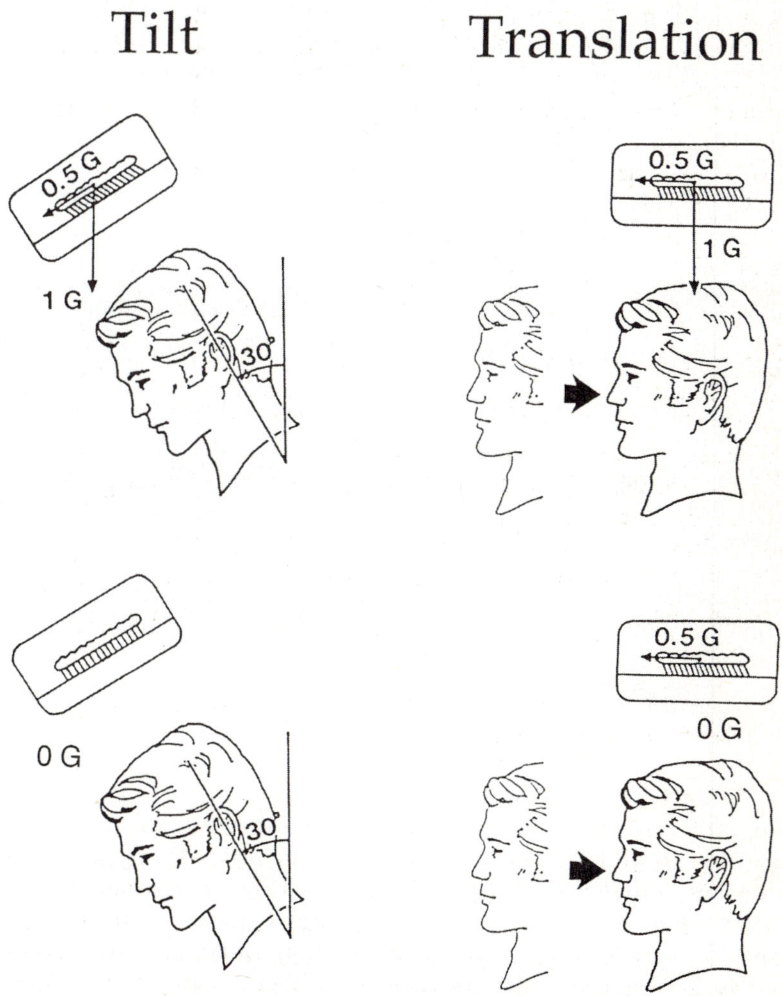

Fig. 1. Displacement of the otoliths during tilt and translation on Earth and in space.

4.2.1.1 Morphological studies

Although it is difficult to measure changes in the vestibular end organs directly, several attempts have been made to determine whether anatomical physiological changes in the vestibular end organs and their primary afferents occur after exposure to microgravity.

The sensory epithelium of the vestibular organ of frogs returned from an 8-day stay aboard the Russian Mir space station remained intact (**Suzuki *et al.* 1993**). The morphology of the vestibular organs of *Xenopus laevis* did not show major alterations following larval development in microgravity aboard the D1-mission (Briegleb et al. 1988; **Hertwig & Hentschel 1989**).

Experiments on the Spacelab Life Sciences-1 (SLS-1) 9-day mission (STS-40) flown in June 1991, examined anatomical changes in the otolith organs of rodents. The hypothesis was that otolith degeneration may occur because of changes in body calcium, carbohydrate and protein metabolism, body fluid redistribution and hormone secretions (Ross 1991). No deleterious effects were seen in the otoconia of the otoliths from flight rodents as compared with the ground controls (**Ross 1993**). This is in contrast to previous data that had shown changes in the structure of the otoconia and otolith membrane in rats after 20 days on board Cosmos 782, suggesting impairment of calcium exchange (Parin & Yemel'yanov 1970; Vinnikov *et al.* 1984). However, it is unclear how many of these changes were due to accelerations during take-off and landing. Changes were also observed in type I (with single large afferent fibre) and type II (with multiple afferent terminals) receptor cells, but were later attributed to the procedure used to remove and fix the vestibular apparatus after flight.

An unexpected change found during the SLS-1 mission, was an increase (by a factor of 12) in the number of synapses in type II hair cells from the in-flight maculae as compared with the control data. These findings suggest that mature utricular hair cells retain synaptic plasticity, permitting adaptation to an altered environment (**Ross 1993**). Consistent with these results are the pilot data showing a decrease in synapse activity in centrifuged rats. If verified, these data suggest that the maculae adapt to changes in gravitational acceleration in either directions by up- or down-regulation of synaptic contacts in an attempt to modulate neural outflow to the CNS (Ross 1992).

4.2.1.2 Electrophysiological studies

As early as 1970, fairly sophisticated physiology was being monitored *in vivo* in frogs in orbiting vehicles (**Bracchi *et al.* 1975**). These experiments showed changes in the vestibular afferents innervating the vestibular end organs. Primary afferent fibres of the vestibular nerve are the first components of the vestibular path, relaying the information originating at the cristae and maculae to the brainstem (within each nerve there is also a system of efferent fibres from the CNS which provides neural feedback to modulate the activity of the peripheral organs). Changes in the vestibular afferents innervating the vestibular end organs were first observed in frogs during a 7-day orbital flight (Gualtierotti 1987). A floating microelectrode continuously monitored changes in the resting activity of single otolith afferents and their response to centrifugal forces. The resting and evoked unitary activity showed significant changes in microgravity compared to the ground control. However, this finding was heavily criticised for the lack of temperature control of the preparation during the flight and failure to demonstrate that recordings were made from the same units throughout the flight.

Recently, a U.S.-Russian collaboration on Cosmos 2044 studied two Rhesus monkeys after a 14-day spaceflight. They recorded extracellular responses from single horizontal semicircular canal afferents pre- and post-flight in two flight monkeys and three control monkeys (**Correia *et al.* 1992**). The spontaneous firing rate of an afferent, characterized by its mean frequency of firing and coefficient of variation, did not change pre- to post-flight. However, the mean gain for nine afferents tested post-flight during passive yaw rotation was twice that for 20 afferents tested during control studies. A repeat experiment aimed at investigating the origin of the peripheral neuron changes flew in late 1992, but results are not yet available (Cohen, personal communication, 1994).

Changes in afferents tested may be reflective of modifications to the end organs themselves or changes in neural components such as hair cells, synapses, or afferent terminals. End organs changes could also be mediated through secondary mechanisms like calcium loss or through CNS pathways. The adaptiveness or maladaptiveness of these changes cannot be determined without combining observations of end organ changes with oculomotor responses.

4.2.1.3 Perceptive studies

After the Space-Sled had been removed from the Spacelab-1 mission, a floatable seat (Body Restraint System, BRS) was used to perform rectilinear accelerations in response to metronome beats. All accelerations were recorded with a three-axis accelerometer fixed to the subject's bite-stick. In addition, a large battery of vestibular tests were conducted on the ground pre- and post-flight, including vestibular threshold detection measurement, luminous line tests, eye counter-rotation and dynamic tilt tests, all three conducted on a special tilt table. Furthermore, rotatory tests, posture platform tests and oscillopsia tests were performed (Kass *et al.* 1984; **Von Baumgarten 1986**). The Spacelab-D1 mission extended these experiments to in flight testing with the use of the European Space Agency's Vestibular Sled.

Sensitivity to linear acceleration was expected to change in space flight if otolith cues were reinterpreted. According to this theory, after adaptation to weightlessness, tilt stimulation of the otoliths would be interpreted by the CNS as a translation, and eye movement reflexes, postural reflexes, and perception of self-motion would alter accordingly – the Otolith Tilt-Translation Reinterpretation (OTTR) hypothesis (**Young *et al.* 1984**; **Parker *et al.* 1985**).

Subjective techniques are used since oculomotor responses are smaller (as well as more difficult to measure) for linear stimulation than for angular motion. The assumption is that proprioceptive and somatosensory thresholds are an order of magnitude higher than vestibular thresholds (**Arrott *et al.* 1990**). Thus, if the accelerating force of vibration is distributed over the body and less than 1 Hz, the otoliths are the primary detection organ and not mechano-receptors in the skin and muscle.

Two measures of sensitivity to linear acceleration are the absolute threshold and time-to-detect. During the Spacelab-1 mission the thresholds for detection of linear oscillatory motion at 0.3 Hz in all three orthogonal axes of the body were determined using the method of limits with a single staircase procedure. Measures obtained on three crew members in flight exhibited thresholds greater (i.e. sensitivity is less) than those obtained preflight. On the ground, threshold were higher in the Z-axis (0.077 m s^{-2}) than in Y and X (0.029 m s^{-2}) (**Benson *et al.* 1986**). Neurophysiological evidence supports a lower gain for Z-axis stimulation as the sensitivity (spikes $s^{-1} \times g$) in neurons signalling motion in Z is 30% less than in neurons signalling motion in X and Y (**Fernandez & Goldberg 1976**). Post-flight, the threshold was still elevated in the X and Y axis in two subjects, whereas it was lowered in all axes in two other crew members who had already undergone a number of vestibular tests including motion stimuli (Benson *et al.* 1986**a** & b). However, the variability also increased during and after the flight, so the changes were not consistent. Threshold determination requires more psychological data than space experiments allow, which may account for the variability. Therefore, the second measure of sensitivity, time-to-detect, may yield clearer results.

Time-to-detect varies inversely with acceleration such that their product is the velocity constant (**Walsh 1961**). After the Spacelab-1 mission, four crew members had to indicate the onset and direction of acceleration using a joystick during step acceleration followed by step deceleration in successive discrete profiles of randomly determined amplitude (ranging from 0.001 to 0.08 $\times$ g) and direction. Post-flight measures of the threshold and velocity constant showed no consistent pattern of change among the four crewmen compared to their preflight baseline other than an increased variability of response (**Arrott & Young 1986**). However, on Spacelab-D1, the average time-to-detect was consistently lower in weightlessness in four other crew members and there were no significant differences between the axes tested (**Arrott et al. 1990**). Post-flight data from Spacelab Life Sciences SLS-1, where four crew members were accelerated sinusoidally at 1 Hz and 0.25 Hz with a peak acceleration of 0.5 $\times$ g showed a lower velocity constant as compared with preflight in the Y-axis, but an increase in the Z-axis (**Young *et al.* 1993**).

The closed-loop assessment test (CLOAT), used before and after Spacelab missions, tested the tilt-translation hypothesis by having six subjects attempt to null their movement on the Sled. Post-flight ability in the performance of this task for longitudinal (Z-axis) acceleration was significantly improved over preflight performance, and decayed gradually during the course of one week (**Arrott & Young 1986; Arrott *et al.* 1990**). However, improved tracking was not observed for lateral (Y-axis) acceleration after longer missions, suggesting a different processing of signals from the sacculus than from the utriculus. The difference in the influence of weightlessness on the sensitivity to low level horizontal accelerations (more sensitive to lateral and less sensitive to longitudinal) is consistent with the conjecture that responses to primarily saccular stimuli appear to be more disturbed by removal of the 1 × g bias than responses mediated by primarily utricular stimuli (**Young *et al.* 1993**).

4.2.1.4 Physiology of semicircular canals/caloric nystagmus

Caloric vestibular testing was performed for the first time during the Spacelab-1 flight. Contrary to general expectation at the time (**Scherer & Clarke 1985**), a clear caloric nystagmus response was elicited in three out of four tests during orbital flight (**Scherer *et al.* 1986**), and during the microgravity phase of parabolic flight (**Clarke *et al.* 1988a**). This finding demonstrated the inadequacy (at least in microgravity) of the widely accepted thermoconvection theory, originally described by Barany (1906).

The caloric experiment of the Spacelab-D1 mission was designed to verify the Spacelab-1 findings and to investigate the thermal vestibular responses using an improved test procedure on three more subjects. Unilateral thermal stimulation was included in the stimulus profile in order to clarify whether hot and/or cold stimuli alone were able to elicit a response (**Von Baumgarten *et al.* 1984**, 1986).

The equipment for stimulation by insufflation of heated or cooled air and electro-oculographic (EOG) measurement of eye movements was identical to that of the Spacelab-1 mission (**Kass *et al.* 1986**). In addition, testing was performed with a special device in order to examine whether the pulsating noise produced by the air system played a role in the elicitation of the nystagmus response. This device consisted of a water-cooled copper plug that was inserted into the external auditory canal to produce a caloric stimulus similar to that achieved by direct water or air irrigation.

The influence of controlled linear acceleration on the ongoing caloric response was studied during sled motion. All three subjects tested exhibited a clear caloric response, with a nystagmus always directed toward the warmer ear. Nystagmus response was also observed with the noise-less ear insert device. Sled runs revealed a cyclical modulation of the caloric nystagmus by concomitant linear oscillation (Scherer & Clarke 1987). Indeed, on flight day one the polarity of the observed nystagmus was reversed by the oscillatory acceleration of the sled (0.2 × g peak): the original direction of caloric nystagmus was maintained during +Gx acceleration, whereas it was reversed during -Gx acceleration. Finally, caloric nystagmus was enhanced when sub-

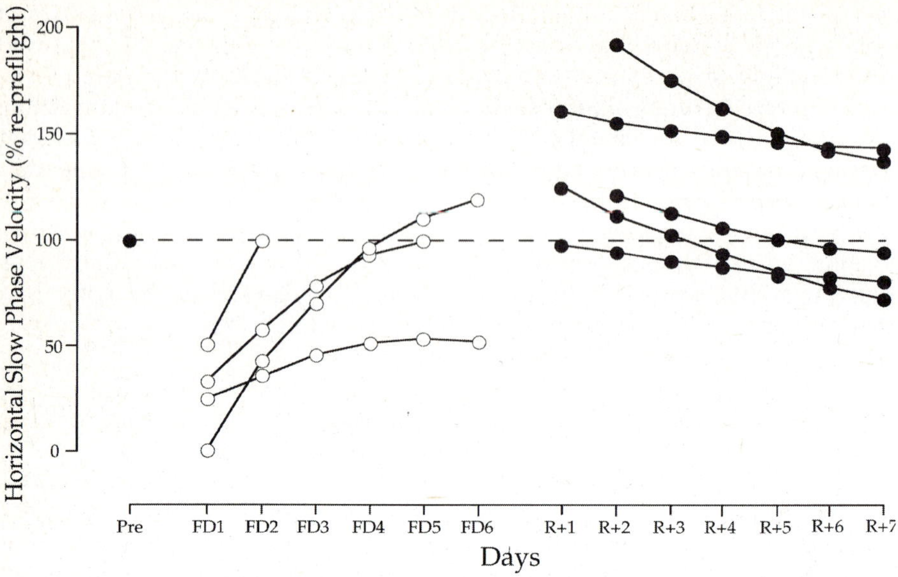

Fig. 2. Results from caloric tests carried out during three spaceflight missions. For each test subject (n = 4 in flight, n = 5 post–flight) the horizontal slow phase velocity of the caloric nystagmus was calculated with respect to that subject's average preflight value. Adapted from **Clarke *et al.* (1993b)**.

jects were released from their seat, thus reducing somatosensory inputs to the central vestibular system (Von Baumgarten *et al.* 1987, **Clarke *et al.* 1988b**).

Caloric testing (air at 15°C insufflated in the left ear) during the German mission aboard Mir in 1992 induced a caloric nystagmus in one crew member on flight day 3 to flight day 6 with a horizontal slow phase velocity about 40% less than that measured before flight (**Clarke *et al.* 1993b**). A readaptation to normal gravity was observed over the 10–day period after landing (Fig. 2).

4.2.2 Vision

Vision has received particular attention because accurate perception of the environment is critical for orientation and adaptation to the space environment. Initially, it was expected that exposure to the space environment would induce changes in visual performance. These changes could result from several factors including altered light conditions, modified receptor physiology, and disturbances of the eye movement control system as a secondary effect of vestibular changes.

The visual environment in space is altered in several ways. First, objects are brighter under solar illumination. The atmosphere of the Earth absorbs at least 15% of the incoming solar radiation. Water vapour, smog, and clouds

Fig. 3. Photograph of the lunar surface taken by Apollo astronauts. Photo NASA.

can increase this absorption considerably. In general, this means that the level of illumination in which astronauts work during daylight is about 25% higher than on Earth. Second, on surfaces such as the moon, there is no atmospheric scattering of light. This causes areas not under direct solar illumination to appear much darker and results in a transformation of normal visual intensity relationships (Fig. 3).

Early anecdotal reports that orbiting astronauts were able to see objects such as trucks and cars moving along highways suggested improved visual acuity in space. However, extensive testing of Gemini-5 astronauts indicated that visual performance was neither degraded nor improved during the eight-day mission (Duntley *et al.* 1966). The reported ability to detect ground vehicles was probably due to dust clouds or shadows associated with traffic significantly enlarging the visual images.

Studies of visual system physiology were undertaken during the Apollo programme. Photographic studies of the retinal vasculature showed a significant decrease in the size of veins for one crew member four hours post-flight, and of both veins and arteries about 3.5 h after flight for another crew member. The degree of constriction of retinal vasculature post-flight in this crew was greater and more persistent than could be accounted for by the vasoconstrictive effect of oxygen alone (Hawkins & Zieglschmid 1975). Such constriction could also be associated with an incompletely understood decrease in central venous pressure in the upper part of the body (Kirsch *et al.* 1986), coupled with an increase in intraocular pressure.

Intraocular pressure rises during flight and drops below preflight levels after landing, as was shown during the Spacelab D-1 mission. Using a hand-applanation tonometer, three crew members measured a mean rise in intraocular pressure of 20 to 25% (Draeger *et al.* 1986). These intraocular tension changes are probably associated with the headward shift of body fluids following orbital insertion (**Mader 1991**) and the subsequent pooling of fluid in the trunk and lower extremities after return to Earth.

More refined visual testing has been performed on several Shuttle flights using a specially-designed visual test apparatus to assess contrast sensitivity, phoria, eye dominance, flicker fusion frequency, stereopsis, and acuity. With the exception of slightly improved stereopsis, no significant changes due to weightlessness were found in 16 crew members on three Shuttle missions (Task & Genco 1986). Another study (Ginsburg & Vanderploeg 1986), which examined near visual acuity and contrast sensitivity in 23 crew members on four flights, reported evidence of reduced contrast sensitivity with little or no change in visual acuity. The spatial frequencies at which the contrast sensitivity changes were noted varied for crew members. The changes were smaller than a factor of two which is too insignificant to impact operational performance. However, if contrast sensitivity continues to change during protracted exposure to weightlessness, as suggested by the work of Lazarev *et al.* (1981), the decrement could become operationally significant.

The effects of weightlessness on vision noted above, may be attributable to intravascular fluid shifts and associated changes in the physical characteristics of the eye. Microgravity is known to produce a headward shift of 700–1400 ml of fluid. Linder & Trick (1986) noted that head-down tilt on Earth produces fluid shifts similar to those observed during space flight. This suggests that the contrast sensitivity changes could result from physiological changes that are a direct consequence of the intraocular pressure changes.

Possible changes during weightlessness in the ability to maintain visual fixation on targets while moving the head and track moving targets have not been fully examined. However, results of several experiments on eye movement control suggest that performance of these tasks might be diminished during the initial period of exposure to weightlessness and immediately after landing. The vestibular nuclei located in the brain stem are part of a system that allows one to fix the gaze on a stationary target during voluntary head motions, as well as, to track moving targets. This system appears to be disturbed during flight, perhaps as a consequence of altered vestibular receptor function due to the absence of gravity. Viéville *et al.* (1986) reported that the amplitudes (eye movement 'gain') of vertical eye movements were diminished during the first three days of weightlessness, when an astronauts moves his head voluntarily while either imagining or fixating a visual target. After four days in orbit, the gains returned to the preflight level, perhaps as a consequence of substituting neck receptor cues for vestibular receptor cues. Grigoriev *et al.* (1986) reported that after 237 days of space flight, the pattern of eye and head movements was significantly altered when cosmonauts moved their heads and eyes to fixate a laterally displaced target. The authors reported that similar disturbances have been observed in patients suffering from cerebellar disorders.

Perceived light flashes in the absence of normal visual stimulation were studied by Hoffman *et al.* (1977) during the Skylab-4 mission. These flashes are caused by heavy ionized cosmic particles passing through retinal cells. During two observation sessions lasting 70 and 55 min, a total of 168 flashes were reported by the pilot. Although no performance disturbance has been associated with these light flashes, it is likely that the flashes mask transient visual stimuli.

4.2.3 Proprioception

Proprioception, used in this context, allows discrimination of position of body parts and discrimination of movement and amplitude of movement of body parts both passively and actively produced. Howard & Templeton (1966) suggested that the behavioral or perceptual responses connected with proprioception arise from stimuli associated with changes in: (i) the length of muscles; (ii) tension, compression, and shear forces (due to the effects of gravity); (iii) the relative movement of body parts; and (iv) muscular contraction. Clearly, gravity or the absence of gravity will modify the stimuli associated with proprioception and impact spatial orientation including knowledge of position in the passive limb, difficulty of target acquisition during voluntary limb movement, modification of tactile sensitivity, and changes in the perception of mass.

Limb position. During the Skylab flights, several crew members reported that, upon waking in the dark, they were unaware of the position of their limbs. Other disturbances of perceived limb position during the Apollo flights have been anecdotally reported (Schmitt & Reid 1985). During the Spacelab-1 mission, one crew member reported 'pins and needles' sensations in his legs late in the flight when exposed to artificial gravity-simulating forces (**Young *et al.* 1984**). Two had difficulty positioning their legs precisely under themselves during 'drop-testing' late in the flight and some hours after landing (Reschke *et al.* 1985, **Wood *et al.* 1985**).

Objective measurement of limb position awareness was conducted during Spacelab-1 by asking three crew members to point at memorized targets in the absence of vision, both in flight and immediately post-flight (**Young *et al.* 1984**). Results demonstrate that pointing accuracy is degraded during and immediately after flight. In an experiment aboard STS-41G (Watt *et al.* 1985), changes in perception of the external world and arm position were studied in two blindfolded subjects. The purpose was to determine how accurately crew members could perform an active pointing task in the absence of vision and to assess changes in performance during the course of the flight and immediately afterward. Both crew members, who were very accurate prior to flight, showed considerable decrement in target acquisition early in flight with one subject showing some improvement during the course of the eight-day flight. Immediately after flight, performance was less than that achieved preflight, but recovery was rapid and complete within a few hours after landing.

External spatial map. Watt *et al.* (1985) suggested that the difficulty in assessing limb position may be the result of abnormal proprioceptive function, and that crews may become disoriented because of the need to establish an egocentric 'down' that would help to develop a correct image of the outside world. An experiment performed on SLS-1 attempted to separate two factors that contribute to the apparent alteration in knowledge of limb and body position in space: a degradation in proprioceptive function and an inaccurate external spatial map (**Young *et al.* 1993**). Four crew members had to point at five remembered target positions on a reference screen, with eyes closed. Target pointing accuracy was demonstrated with a hand-held light pointer recorded both on video and by the observer. Two subjects who were very accurate on the ground demonstrated greatly reduced performance in flight, with pronounced floorwards biasing of their points. Post-flight all subjects made greater errors than preflight. Two subjects made several errors when trying to touch various body parts, and noted that their arms were not exactly where expected when their eyes were opened, both in- and post-flight.
The Monimir experiment performed on the Austrian cosmonaut during his flight aboard Mir station in 1991 studied aiming movements of eyes, head and the arm on visual and acoustic signals. Results showed that variance of time and amplitude of movements, and the number of errors in pointing, increased significantly in weightlessness. Motor patterns for eye-head coordination learned during training and preflight baseline data collection were abandoned in flight in favour of former coordination pattern. Also, arm and head movements, learned under visual control or by passive movement of the arm by another person without visual control and then repeated several times with eyes closed, were changed in flight (Berger *et al.* 1992).
All the above data suggest that in the absence of vision, the maintenance of a stable external spatial map is highly dependent on the presence of the normal gravitational force (**Young *et al.* 1993**).

Internal body scheme. On the other hand, a series of experiments, performed during the Aragatz French-Soviet mission in 1988 aboard Mir (Physalie), suggest that the internal reference system is not easily affected by exposure to microgravity. In one experiment, two cosmonauts were asked to draw ellipses in the air (either parallel or perpendicular to the longitudinal axis of the body) without the aid of vision. Results indicated minimal changes as a function of microgravity suggesting that the egocentric reference system was not disturbed (Fig. 4). The subjects were capable of maintaining a sense of verticality despite disappearance of the main factor contributing to verticality on Earth, gravitational force (Gurfinkel *et al.* 1990, **1993a**).

The results of the head–neck experiment performed aboard Mir (see Section 4.2.4. Neck receptors) led to a contradictory conclusion. In microgravity, when drawing in the air with no vision, bending the head over the trunk caused the cosmonaut's arm movement pattern to be aligned with the head vertical axis, although it was still perceived as 'horizontal' by the subject (Fig. 5). Such sensory–motor mismatch was interpreted as an expression of a body scheme disturbance (Berger *et al.* 1992; Kozlovskaya *et al.* 1994).

Tactile sensitivity. The perception of the orientation of complex tactile stimuli (letters and digits) applied to the palmar surface of the left hand and on the frontal surface of the right thigh was studied in cosmonauts when they were restrained to the deck or free–floating (**Gurfinkel *et al.* 1993b**). The task of interpretation of complex tactile stimuli was not affected by the absence of the gravitational vertical. The number of errors made in flight was often even less than on ground. It was concluded that the body scheme used by the egocentric reference system cannot be easily modified by changing external conditions.

Force feedback. The ability of a cosmonaut to reproduce from memory the different positions of a handle was also investigated on Mir. The temporal parameters of movement and the number of distinguishable positions of a handle were not significantly influenced by microgravity. However, the accuracy of setting the handle to a given position at the beginning of the flight was significantly lower with an error towards a decrease of handle deflection angle. It was assumed that this impairment was due to proprioceptive feedback changes (Lipshits *et al.* 1992).

Mass discrimination. An experiment using five crew members on Spacelab–1 showed that the ability to discriminate between differences in the mass of objects in microgravity was poorer than in normal gravity. Weight discrimination was impaired for two or three days post–flight, and crew members felt their bodies and other objects to be extra heavy post–flight (Ross *et al.* 1984, **1986**). The impairment in flight was partly due to the loss of weight information (a reduction in Z–axis pressure stimulation), and probably also to incomplete adaptation to microgravity. The increase in apparent heaviness of objects reported for static weight judgements post–flight suggests that some central rescaling of the static pressure systems occurs. It was also noted that those subjects who used more rigorously shaking techniques showed better mass discrimination under $0 \times g$. This variable was later investigated system-

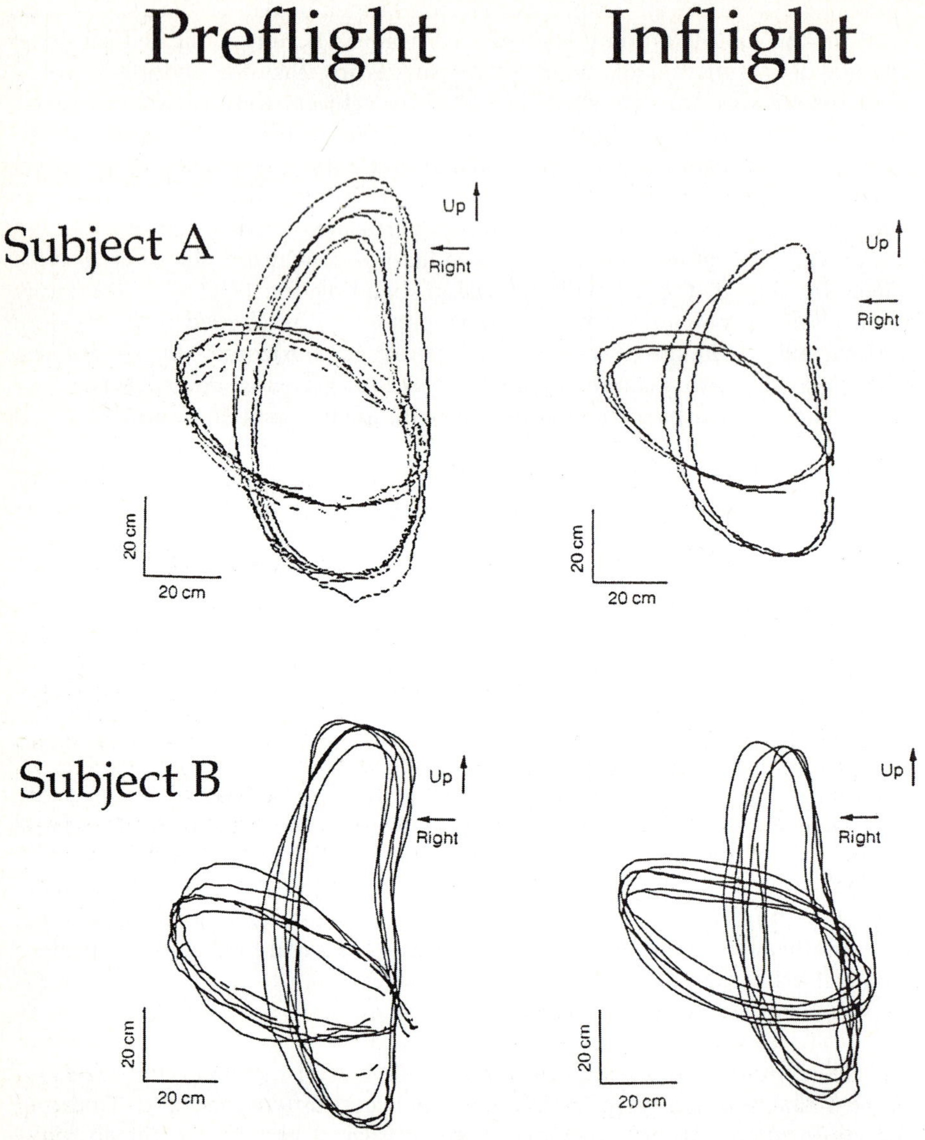

Fig. 4. Trajectory of the fingertip in the frontal plane during ellipses drawing by two cosmonauts before and during a spaceflight. Subjects in the upright position with the eyes closed and with the arm at the side of the trunk were instructed to draw ellipses in the air with a long axis oriented either parallel to the body z-axis or perpendicular to the body z-axis. Fingertip and arm trajectories were recorded by means of two infrared video cameras. No significant differences in the orientation of the ellipse axes were noted for either subject in flight (flight day 19 for subject A, flight day 20 for subject B) compared to preflight values. Adapted from **Gurfinkel *et al.* (1993a).**

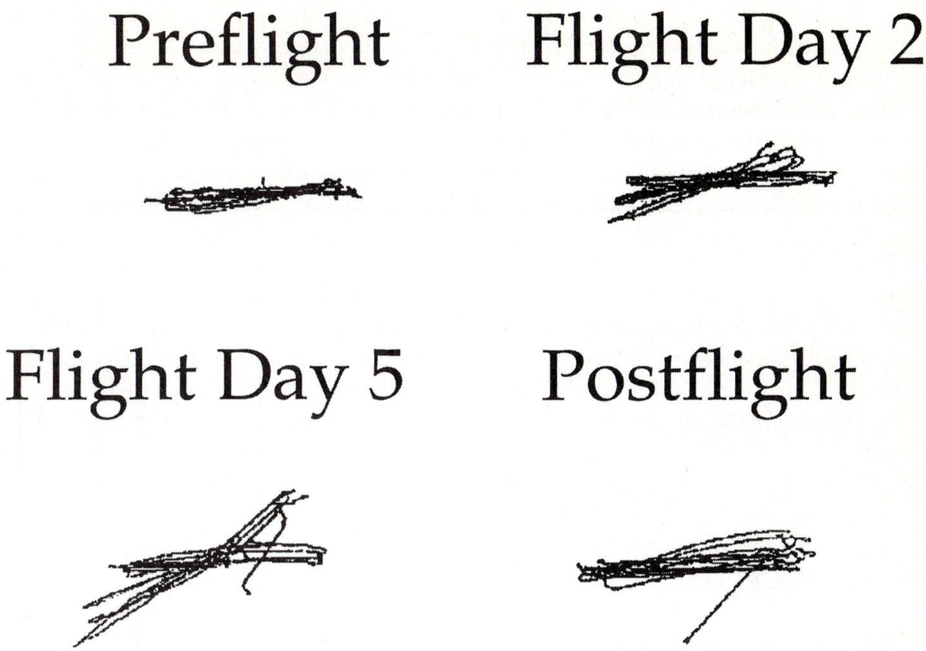

Fig. 5. Trajectory of the arm in the frontal plane during memorized arm movements with eyes closed by one cosmonaut before, during and after a 7-day spaceflight. The subject, lying on the floor of Mir station with his trunk fixed, was asked to track with his outstretched arm the horizontal (side-to-side) displacement of a visual target. Then, he was asked to repeat this arm movement with the eyes closed, first with the head upright, i.e. aligned with the trunk, and then with the head tilted to the left. Head and arms movements were recorded by infrared LEDs and scanner cameras. Preflight, the arm movement was horizontal with both head upright and head tilt. In flight, however, and especially on flight day 5, the arm movement was tilted from horizontal when the head was tilted relative to the trunk, suggesting that the egocentric reference used was the head vertical axis. Adapted from Kozlovskaya *et al.* (1994).

atically during the Spacelab-D1 mission by asking five crew members to make both short and long shaking movements (the former normally giving the higher acceleration) sideways. Video tapes of hand movements while shaking the test objects were made before, during, and after the Spacelab-D1 mission. Movements were faster in flight than preflight, slowed down post-flight and returned to baseline by about the third day (**Ross *et al.* 1986**). It was concluded that errors of weight or mass perception during the maladaptation phase may be due to both inadequate monitoring of command signals and to inappropriate scaling of afferent signals. High acceleration shaking was found to improve mass discrimination but impair weight discrimination. The initial impairment of discrimination post-flight is consistent with an after-effect of impaired sensori-motor coordination (**Ross *et al.* 1987**).

Haptic perception. By touching and tracing the contours of objects, the human haptic system can identify the shape and orientation of these objects. Changes in tactile sensitivity, force feedback as well as a shift in scale of mass discrimination in microgravity could imply a potential change in the way shapes are perceived. The haptic perception was evaluated during an experiment performed on one cosmonaut returning form a short-term space flight (R+0) and two cosmonauts returning form a 6-month flight (R+3). Cosmonauts were asked to move their hands within the limits of an unseen ellipse. The subject's task was to ascertain whether the long axis of each ellipse was oriented in the radial or lateral direction. Haptic performance two days after return from orbit was virtually identical to that measured before the flight. However, it was improved in the cosmonaut tested shortly (24 h) after a short-term space flight, suggesting an increased role of proprioceptive inputs and feedback control of movement in microgravity (Lipshits *et al.* 1994).

4.2.4 Audition

Auditory brain stem response recordings investigated during Shuttle flights (**Thornton *et al.* 1985**) were particularly useful in the study of increased pressure or decreased vascular perfusion in both the labyrinth and lower portions of the CNS. This investigation was aimed at verifying if fluid shift toward the upper body in microgravity could induced such increased pressure, which could play a role in space motion sickness. Auditory evoked potentials were recorded on seven astronauts in flight at various times during the space missions, beginning at 12 h. No significant difference was observed between mean latency values for any potential on the ground or during flight. Slight changes in the morphology of the responses were attributed to in-flight noise and electrical interference aboard the Shuttle. Since the utility of these auditory potentials has primarily been associated with auditory threshold determination in clinic medicine (**Galambos & Hecox 1978**), the absence of significant decrement in flight objectively suggests that an auditory function is not altered in microgravity.
Persons with normal hearing are capable of localizing sound sources in the environment. This faculty depends on binaural hearing. Directional hearing can also be utilized for spatial orientation (**Calhoun *et al.* 1987**). It was assumed that, in the absence of gravity, orientation by acoustic cues would play an important role for spatial orientation. The experiment Audimir which flew aboard Mir during the Austrian–Russian mission in 1991 was designed to trigger an illusion of movement by auditory cues (Persterer *et al.* 1992). The test stimuli included white noise and a few bars of a Viennese waltz and simulated a sound source moving around the subject's head. A second part of the experiment was designed to determine localization accuracy, particularly for sounds presented in the frontal plane. Both eye movements

(EOG) and subjective comments of the subjects were recorded on two cosmonauts. Results showed that the azimuth localization error in microgravity was within the same range as on Earth, i.e. between 1° and 2°. However, a significant downward shift in elevation judgments by approximately 10° was observed. Presentation of a sound moving counter-clockwise induced an illusion of clockwise rotation, which was never observed on ground in the two subjects tested. This sensation was stronger with the waltz than with white noise. Corresponding eye movements were observed, with slow phase directed counter-clockwise. Reversing the direction of rotation of the sound source provoked a reversal of the direction of eye movements but not of the illusion of rotation (Persterer *et al.* 1992), suggesting that spatial orientation was disturbed, but not hearing. No rationale is immediately available to explain this discrepancy.

4.2.5 Taste and olfaction

Several crew members have reported changes in the taste and smell of food during space flight. During the Skylab flights, astronauts asked for more spices and condiments to add taste to the prepared food. Diminished sensitivity to taste and odour could result from the passive nasal congestion reported in conjunction with the headward shift of fluid. Taste, particularly the non-volatile component mediated by the taste buds, may be susceptible to threshold shifts in microgravity because of mechanical factors related to reduced stimulation of the taste buds as a result of changes in convection in weightlessness (Rambaut & Johnson 1989).
Results of an early experiment, performed aboard the 84-day flight of Skylab-4 using paper impregnated with various flavours, suggested that an increased threshold shift did occur, but that these changes were highly individualized (Heidelbaugh *et al.* 1975). In a later experiment aboard STS-41G, olfactory recognition was established by identifying the smell of solutions of lemon, mint, vanilla and distilled water. Taste recognition and detection thresholds were investigated with solutions of sucrose, urea, sodium chloride and citric acid. Preflight responses were compared to those obtained during and immediately after the eight-day flight. No pre- or post-flight subjective changes in taste or aroma or objective changes in olfactory function, taste recognition or taste threshold were observed. Differences in flight length, susceptibility to motion sickness or specific methodology may account for the conflict between the Skylab-4 and STS-41G results (Watt *et al.* 1985).

4.2.6 Summary

There is evidence that the semicircular canal afferent signals function in microgravity, as do otolith afferents. Otoconial formation is probably normal, although an early report showed some morphological changes. Studies on

the morphology and development of otoconial formation in vertebrates, including mammals, in microgravity need to be continued.
Electrophysiological studies in animals showed exaggerated responses within the horizontal semicircular primary afferents at the eighth nerve level. This hypersensitivity might reflect a change in the otolith organs or in the end organs of the semicircular canals (*via* the efferent feedback), or in their afferent responses. Potentially important increases in peripheral type II hair cell synapse development in microgravity needs confirmation. The role of the efferent system in adaptation to weightlessness, and in the apparent change in peripheral time constant remains to be determined.
Perceptive studies showed that the threshold for detection of linear acceleration may be lowered in weightlessness, but the detection is subject to frequent errors and is axis-dependent. After return to Earth, the acceleration levels are strongly over-estimated. The differences between the unloading of the gravity bias on the saccular and on the utricular otolithic membranes have yet to be explored.
Caloric responses exhibit a non-convective component (beating toward the warm ear). However, the mechanism for this non-convective component has not been established.

4.3 Multisensory Interaction

Spatial orientation is the relation established between a body-oriented coordinate system and external reference frames. It results from the integration of sensory signals from the visual, vestibular, tactile, and proprioceptive systems and from a comparison between them and the motor-command or efference-copy signals arising in the brain. A sense of spatial orientation helps to guide behaviour at unconscious or automatic levels.
Perception of self-orientation and self-motion with respect to the environment may be dependent on the ability to detect and integrate information regarding three vectors: gravity/linear acceleration, visual scene polarity, and internal vectors aligned with subject's head or trunk Z-axis. A fourth determinant of perception is the observer's expectations, based on his or her own actions (**Harm & Parker 1993**).
Illusions, reported by 80% to 93% of Russian cosmonauts (**Matsnev *et al.* 1983**; Gorgiladze & Bryanov 1989), are seldom reported by American crews. Although nearly all astronauts adapt to microgravity, individuals apparently differ with respect to the adaptation algorithms they develop to compensate for the sensory disturbances produced by that environment (**Young *et al.* 1986**).

4.3.1 Spatial orientation

Cues used for spatial orientation in initial phases of adaptation to microgravity vary. Some individuals become strongly dependent on vision, or field dependent, orienting with respect to external references such as equipment racks or the floor of the spacecraft (**Clément *et al.* 1987; Clément & Lestienne 1988**; Oman 1988). Other crew members are more 'body oriented' and align their eccentric vertical to lie along their body axis. These people do not become as disoriented when working in unusual attitudes relative to external references or when traversing to places where visual cues for vertical orientation are absent (Young 1985).

Inversion illusion. Dependence on vision and proprioception as a substitute for otolith cues begins on entry into microgravity and can last for many days. This may lead to various illusions of spatial orientation on entering weightlessness, including sudden reversal of orientation, the so-called 'inversion illusion' which has been reported during orbital flight (Oman 1988) and the 0 × g phase of parabolic flight (**Lackner & Graybiel 1979; Lackner & Dizio 1993**). In parabolic flight the illusions tend to abate with increasing number of parabolas (**Clément & André-Deshays 1987**) but some astronauts have reported the inversion illusion for days (Yakovleva *et al.* 1982). Astronauts frequently report what they call the 'downs' which refers to a very strong sensation that the floor is located wherever their feet are positioned. Some astronauts report that this sensation occurs almost immediately, whereas others indicate that it takes two to three days for it to occur, and still others report that this sensation only occurs intermittently (Gibson 1977).

Mittelstaedt (1986a, 1992) has proposed that the inversion illusion may be understood using a model that includes an internal ('ideotropic') orientation vector. This vector may also explain the sensation of the 'downs.' One measure of spatial orientation, the subjective horizontal body position (SHP), is thought to correlate with the inversion illusion and possible motion sickness (Mittelstaedt 1986b). How far the SHP deviates from the true horizontal body position is a measure of the otolithic bias. The bias is determined by summing the afferent discharges according to the direction of the polarization of the receptors producing them. Mittelstaedt developed a model of otolith function (Mittelstaedt 1989) which predicts that individuals whose saccular bias is negative will be inverted in 0 × g since 'up' would be the direction of their feet.

Observations made by astronauts in weightlessness were recently re-examined by **Mittelstaedt & Glasauer (1993a)**. These authors were puzzled by the absence of sensations in the astronauts of trunk tilt and concomitant reflexes when the head is tilted with respect to the trunk, and by the persistence of a perception of 'up' and 'down' in the absence of, as well as in contradiction to, visual cues. They suggest that these observations point out the decisive role of the saccular Z-bias, that is, of a difference of the mean resting discharge of saccular units polarized in the rostral and the caudal direction (**Mittelstaedt & Glasauer 1993b**).

Subjective vertical. One U.S. astronaut reported the perception of a 50–60° pitch forward (at 5° s^{-1}) attitude immediately upon achieving weightlessness (at main engine cut-off) that persisted for approximately 30 s; similar sensations were also associated with the onset of Orbital Manoeuvring System burns throughout the flight. Two astronauts reported the sensation of being at a 10 to 30° pitch up angle when they were parallel to the mid-deck with one of the astronauts reporting that this sensation persisted throughout the entire mission. Many astronauts also report the perception of a 10° to 30° head-down tilt while lying in bed for the first one or two nights following return to Earth (**Harm & Parker 1993**).

Perception of subjective vertical was tested before and after the Spacelab-1 (**Vogel & Kass 1986**) and D1 missions (Kass & Vogel 1986). Eight subjects were titled by means of a tilt table about the roll axis at 15° intervals up to 90°. At each 15° angle, they had to set a luminous line to what was perceived to be the vertical. Errors in setting the vertical for the high tilt of range was greater early post-flight, suggesting that otolith inputs were given a lower weighting factor during processing in the CNS after adaptation to microgravity. However, the results of this test are difficult to interpret, since the necessary control of tilting only the luminous line while the subjects were placed on their backs was not done.

Perception of subjective visual vertical was also investigated in 20 cosmonauts returning from short-term space flights (7–18 days) and 19 cosmonauts returning from long-term space flights (75–365 days) when sitting upright and laterally tilted on one side (Kornilova & Kaspransky 1994). The error between the subjective visual vertical and the gravitational vertical ranged from 10 to 28° post-flight compared to about 5° preflight in both the upright and lateral positions. In the subjects lying on one side, the asymmetry of the error was larger post-flight.

Before and after the German mission aboard Mir in 1992, two cosmonauts were rotated eccentrically in the yaw position, thus provoking a resultant of centrifugal force and gravitational force tilted relative to vertical, and asked to set a luminous line to the subjective vertical (**Wetzig *et al.* 1993**). Compared with a normal group of subjects, the cosmonaut responses were atypical preflight, and early post-flight, since no relationship with the true gravito-inertial vector was observed. It has been suggested that both subjects had suppressed their vestibular information as an effect of vestibular deconditioning which took place before the flight during their training at Star City. An alternative position would suggest that the cosmonauts interpreted the perception of tilt associated with the resultant as an illusion, and simply chose not to report the tilt.

4.3.2 Self- and surround motion illusions

Illusions of self-motion or unusual surround-motion have been addressed in several studies. Several illusions are commonly reported post-flight, including a feeling of heaviness, a sense of disorientation when making sudden pitching or rolling head movements, an inability to move about in the dark, and an illusion of floor motion during vertical body movements (**Graybiel 1980**). Perceived translational self-motion was reported during passive roll stimulation in darkness one to three hours after Shuttle landing (**Parker *et al.* 1985**). Unexpected illusory translation of either the subject or the visual surround was reported by two astronauts who performed voluntary pitch or roll head movements during reentry and after the orbiter had stopped on the runway (**Reschke & Parker 1987**). These observations, as well as similar ones reported by **Young *et al.* (1984)**, support the hypothesis that signals from receptors that respond to linear acceleration are reinterpreted during adaptation to weightlessness (OTTR hypothesis, Section 4.2.1.) (**Parker *et al.* 1985**).
Almost all crew members experience perceptions of self- or surround-motion associated with head movements or combined head and body movements while in orbit, during re-entry, and after landing. Disturbances in orbit appear to increase in intensity as a function of flight day: the longer the flight, the more illusory phenomena reported. Although the data are currently limited to flights of ten days or less, the intensity and duration of re-entry and post-landing perceptions of self- or surround-motion also appears to be a function of mission length, and perhaps of prior space flight experience or the volume of the spacecraft (e.g. mid-deck versus Spacelab).
Even though individual experiences with self- and surround-motion vary, there are three types or categories of disturbances are commonly reported by crew members (Fig. 6): (i) input-output gain disturbances, where the perceived self/surround-motion appears exaggerated either in rate, amplitude, or position following the physical head/body movement; (ii) temporal disturbances, where the perception of self- or surround-motion lags the head/body movement and/or persists after the real physical motion has stopped; and (iii) path disturbances, where angular head and body movements elicit perceptions of linear or combined linear and angular self- or surround-motion (**Harm & Parker 1993**).

4.3.3 Visual illusions

During re-entry and immediately after landing, astronauts frequently report that voluntary head motions produce illusions of motion of the visual surroundings (**Reschke & Parker 1987; Reschke *et al.* 1994; Harm & Parker 1994**). These illusions are probably related to disturbances of the gaze control system.

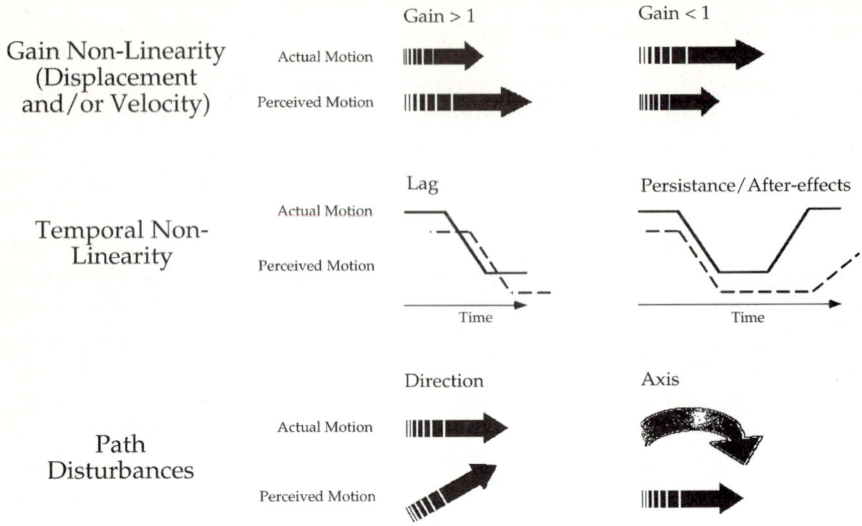

Fig. 6. Motion disturbances commonly reported by astronauts during and after spaceflight.

Oscillopsia. Oscillopsia refers to an apparent displacement of visual targets during passive or voluntary head movement. Apparent oscillation of visual targets was reported by some (n = 5) Skylab astronauts (Graybiel *et al.* 1977), but was attributed to drowsiness. Young and his colleagues (**Young *et al.* 1984**) systematically investigated oscillopsia on the Spacelab–1 mission. Their four subjects reported instability of visual targets during head movement immediately after landing. Bloomberg (personal communication, 1994) has noted that during natural locomotion on a treadmill almost all crew members returning from space flight (n = approximately 25) report substantial illusory movement of a small wall–fixed target.

Oscillopsia is readily understood in terms of miscalibrated compensatory eye movements (inappropriate vestibulo–ocular reflex gain) during head movement in an unusual force environment. The appearance of oscillopsia post–flight indicates that the compensatory eye movements had adjusted to the microgravity condition and were no longer appropriate to normal gravity. This suggests that a carefully designed experiment would reveal oscillopsia during the period immediately following orbital insertion, before astronauts have adapted to weightlessness.

Vection. Visually–induced illusions become stronger in space, since visual cues may not be hindered by constraints from the otolith organs, which in microgravity do not confirm or deny body tilt. This has been confirmed with the rotating dome experiments in Spacelab–1, Spacelab D–1 and SLS–1 (n = 12) where complete rotatory vection, or saturated vection, was generally reached in flight (**Young & Shelhamer 1990**). Results from SLS-1 and IML-1 suggest that visual and tactile cues are weighted differently for each subject such that pre- and post-flight strategies are significantly different and not consistent across subjects (Watt *et al.* 1992). Indeed, beyond the previously reported increased use of visual cues for spatial orientation during and following space flight, the SLS-1 results uncovered interesting variability in individual perceptual styles. Some subjects were strongly tied to a seemingly solid anchor by the localized tactile cues from bungee cord which induced foot pressure; whereas another felt a strong contribution toward self-rotation under the same circumstance. The differences in perceptual styles may be related to the contributions of sensory and body-centred vectors in the normal and adaptive calculation of spatial orientation (**Young *et al.* 1993**).

Linear vertical vection induced by a portable visual stimulator was investigated on one cosmonaut during the Austrian-Russian mission aboard Mir in 1991 (Mueller *et al.* 1992), and on 2 other cosmonauts during subsequent long-term space flights aboard Mir (Mueller *et al.* 1994). Motion of a dot pattern was sinusoidal along the Z-axis (up/down), and the tested subjects were either free-floating or had the feet fixed to the floor. Vection was indicated by means of a joystick; a luminous line was also provided to the subjects for determining the subjective visual vertical. On flight day 6 during the short flight, the downward vection induced by upward optokinetic stimulation started earlier than during ground testing. The tested cosmonaut also showed vertical downbeating spontaneous nystagmus. The phase relationship between vection and visual stimulus was changed during the flight and post-flight, leading to some episodes of inverted vection (in the same direction than visual stimulus). When tested 9 times during a 175-day flight, and 3 times during a 145-day flight, two other cosmonauts showed increased vection onset latencies, in both the free-floating and the restraint position. Furthermore, they reported the illusion of forward body bending and occurrence of pitch vection instead of linear vection (Mueller *et al.* 1994). The observation that linear vection is not saturated in weightlessness is in contrast with results obtained on circular vection (**Young *et al.* 1993**) although it would be expected that linear vection would be more affected than circular vection in weightlessness.

It is interesting to note that frogs born in microgravity showed stronger behavioral response to moving visual scenes when tested after their return on Earth than control animals born on Earth (Wassersug *et al.* 1993b; Izumi-Kurotani *et al.* 1994).

4.3.4 Proprioception

Muscle vibration. During parabolic flight, **Lackner & Dizio (1992)** used a classic technique of muscle vibration to elicit illusory limb movement. Illusions of arm motion were enhanced during $1.8 \times g$ and diminished during the $0 \times g$ portion of parabolic flight. They concluded that muscle spindle receptor output per unit stretch of spindle is influenced by gravitational acceleration. The otoliths are unloaded in microgravity and decrease their descending modulation of alpha and gamma motorneurons, resulting in decreased tonic vibration reflexes. An alternative explanation wherein the increased gravitational load on the arm was equivalent to resistance may also be appropriate.

During orbital flight, the perceptual effect of vibrating (10 s at 70 Hz) the ankle, arm and neck muscles was investigated in four cosmonauts of the French–Russian Aragatz and Antarès missions (Roll *et al.* 1990, 1994) restrained with a back support or supported with elastic cords attached from the waist to the floor. Preflight, stimulation of the anterior tibialis induced a mild sensation of backward tilt while forward tilt was perceived when the soleus was vibrated. In two cosmonauts early in–flight sensations were similar, but of larger magnitude. After 19 to 20 days in flight, the tilting sensation associated with vibration of the anterior tibialis gave way to the sensation that the whole body was raised from the deck of the spacecraft while stimulation of the soleus had no effect (**Roll *et al.* 1993**). Two other cosmonauts reported considerably decreased illusory body tilt during vibration of lower leg muscles when their body was restricted mechanically from flight day 6 (Fig. 7).

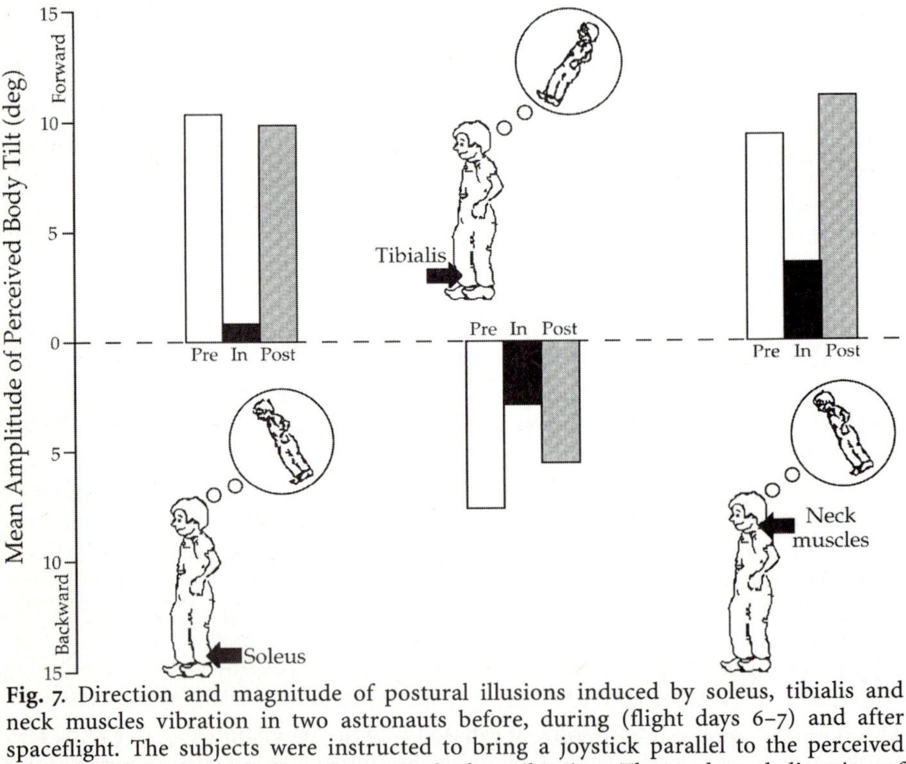

Fig. 7. Direction and magnitude of postural illusions induced by soleus, tibialis and neck muscles vibration in two astronauts before, during (flight days 6–7) and after spaceflight. The subjects were instructed to bring a joystick parallel to the perceived long axis of the body before, during and after vibration. The angle and direction of tilt of the joystick measured from the relative displacement of the joystick were averaged over subjects and trials. Adapted from Roll *et al.* (1994).

Later during the flight (flight days 12–13) vibration gave rise to a sensation of body elevation along the longitudinal axis. Applying artificial forces between the upper trunk and the support surface by means of bungee cords resulted in partial restoration of the Earth (i.e. anterior–posterior) kinaesthetic illusions. A perceived upward motion of a head-fixed target was also reported by the cosmonauts in flight regardless of which leg muscle was vibrated, soleus or tibialis. This visual-proprioceptive illusion was generally weaker in free-floating subjects. During arm vibration, one cosmonaut occasionally reported an illusory whole-body rotation around the visual target. The target was perceived in motion when the vibrated arm was in contact with the spacecraft wall (Roll *et al.* 1994). The authors concluded that muscle proprioception remained intact after prolonged flight, since it was still possible to activate the muscle spindle and generate motor response and kinaesthetic illusions with vibration.

Illusions. Proprioceptive illusions have been noted during and after several Shuttle flights. Watt, as part of his post-flight testing of the Spacelab-1 astronauts, asked his subjects to hop up and down (otolith-spinal reflex evaluation), and received reports that the floor moved up and down like a trampoline under the subject (**Watt *et al.* 1986**). In a follow-up study, Watt and his associates (Watt *et al.* 1985) asked two subjects on STS-41G to perform rhythmical deep knee bends and arm bends. In flight the knee bends were performed with the feet held in foot restraints, and arm bends were accomplished with the subject grasping a hand-hold and alternately flexing and extending the arms. One subject reported that during arm flexions, the wall moved toward him, and during leg bends, there was an illusion of the floor moving up and down (like a trampoline). Illusions of self-motion and aircraft motion were also experienced by subjects performing deep knee bends in the hypergravity phase of parabolic flight. When the knees were bent with the eyes open, there was a sensation that the knees were being bent more rapidly than intended, and that the aircraft had risen upward against the feet. Upon straightening the legs from the bent position, more than the normal amount of effort (to rise against the 1.8 to 2 $\times$ g force) was required, and the sensation, once an upright position was obtained, was that of having risen too far and, frequently, of having momentarily left the deck of the aircraft (**Lackner & Graybiel 1981; Lackner & Dizio 1993**).

Illusions similar to those recorded by Watt were reported by four Spacelab-1 astronauts during drop testing to evaluate changes in the vestibulospinal motor neurons (**Reschke *et al.* 1986**). Three to four hours after landing, subjects reported that they did not have the sensation of falling after being unexpectedly dropped. Rather, they reported that the floor came up to meet their feet. In addition, the subjects frequently did not know where their feet were in relation to the floor and without restraints would have fallen backwards upon landing after the drop. These sensations continued for three days after flight and the instability remained until approximately 24 to 36 h following the first exposure to the drop stimulus. Interestingly, the sensation of not knowing the position of the legs with respect to the trunk was present by the seventh day of flight in the two tested subjects, and the unexpected drop was followed by a backward rotation of the body (**Wood *et al.* 1985**).

The source of proprioceptive illusions is not known, but there are several possibilities. Watt's data (Watt *et al.* 1985) suggest that proprioceptive function is altered in microgravity and is expressed as a sensory deficit. In addition, sensory motor programs could be altered due to the changed mode of locomotion in weightlessness, or the system which compares motor commands with resulting sensory inputs may not, because of the stimulus rearrangement of microgravity and resulting reinterpretation of sensory input to the central nervous system, be capable of correctly distinguishing between self-motion and movement of the world. The latter interpretation fits well with the hypothesis of a tilt-translation reinterpretation (OTTR, Section 4.2.1.), and was proposed by others to account for illusions of self- or surround-motion experienced by returning astronauts (Melvill-Jones 1974; **Young *et al.* 1984**; **Parker *et al.* 1985**; **Reschke *et al.* 1986**; **Reschke & Parker 1987**).

Neck receptors. Among the somatosensory systems projecting to the orientational system of the brain, the position receptors of the cervical column (neck receptors) play an important role. During the Spacelab-D1 mission, the trunk of a crew member was passively bent sidewards or forwards, while keeping his head fixed to the floor of Spacelab (Von Baumgarten *et al.* 1986, 1989), thus stimulating the neck receptors, but not the vestibular system. Surprisingly, the crew member reported an illusory rotation of a head-fixed target cross seen in the monitor of his helmet (Kass *et al.* 1987). It was proposed that temporary disorientation and dizziness in space could possibly be caused by a mismatch between otolith signals and the cervical position receptors (**Von Baumgarten 1987**). However, recently, it was found that one cosmonaut with his head bent over the shoulder was able to reset a luminous line veridically on flight day 3 and flight day 5 (**Hofstetter-Degen *et al.* 1993**). It was suggested that at this time (and also because the cosmonaut's vestibular function had been deconditioned before the mission during training) the missing otolithic input was already totally supplanted by neck receptors.

A 'neck reflex' experiment performed during the Austrian-Russian mission aboard Mir in 1991 studied the influence of the neck receptors on horizontal arm movements. On Earth, stimulation of the neck receptors by assuming various head positions (rotation, side bending) normally causes a slight change in horizontal arm movement. In microgravity, side-bending the head to the right caused the arm's horizontal movement pattern to turn counter-clockwise in a frontal plane. This phenomenon increased on flight day 5 and returned to normal after landing. It was interpreted as an expression of a body scheme disturbance (Berger *et al.* 1992).

4.3.5 Cognitive functions

Mental rotation. The cognitive mechanisms of spatial orientation in microgravity have been investigated recently, on the basis of previous casual observations made by astronauts and cosmonauts. During the STS-51G Shuttle flight, one crew member was asked to report the critical tilt angle of his body with respect to the vehicle, from which he could not mentally reconstruct the original environment (**Clément *et al.* 1987**). The reported angle was about 60° on the first day in flight, but after three days in flight his perception was independent of the respective orientations. It is well known that on Earth, a familiar visual environment, a face or a printed text cannot be recognized or analyzed when it is tilted by more than roughly 60° (**Corballis *et al.* 1978**). Oman (1988) noticed that the inversion illusion (see Section 4.2.1) gives rise to reorientation episodes resembling mental rotation. **Clément *et al.* (1987)** proposed that in the absence of effective gravitational field, the analysis of visual images tilted with respect to the body could be a gravity-dependent process and that weightlessness, by providing a release of this constraint, would facilitate the processing of visual images in any orientation with respect to the body axis.

Verbal descriptions by two astronauts of the Spacelab-D1 mission of visually presented arrays were required under different head positions (straight or tilted). Results showed that spatial assignments are still possible in microgravity. Moreover, different coordinate systems are used in normal gravity and in microgravity: in normal gravity the gravitational vertical was mainly chosen for primary reference, whereas in microgravity it was suggested that retinal information was predominant (**Friederici & Levelt 1987**).

A mental rotation paradigm with pictures of three-dimensional objects (**Shepard & Metzler 1971**) was tested on two cosmonauts during the 26-day orbital flight of the French-Soviet Aragatz mission (Matsakis *et al.* 1990). Responses showed that the average rotation time per degree was shorter in flight than on the ground. This difference seemed to be particularly marked for stimuli calling for roll axis rotations (**Matsakis *et al.* 1993**).

A similar experiment was performed by three cosmonauts during the French-Russian Antarès mission aboard Mir in 1992. In this experiment, the axis of rotation of the 3-D objects was not known in advance by the subject. Also, the repertoire of 3-D objects was different between all experimental sessions to avoid a learning effect. Results showed no significant differences in rotation time about yaw and roll axis versus ground data. During the same flight and during the French-Russian Altar mission aboard Mir in 1993, two cosmonauts were also asked to identify the presence of a symmetry axis in polygonal shapes. Head-on-body position is well known to influence symmetry detection, and it has been proposed that mental rotation process could also be used in a symmetry detection task. Results showed that detection of symmetry was less accurate and required more time in microgravity than on Earth. It has been proposed that these changes are related to the difficulty encountered by the CNS to set an internal vertical reference in the absence of graviceptive otolith information for the detection of symmetry (Léone *et al.* 1994).

It has been suggested recently, on the basis of observations made by using virtual environments, that a poorer ability to mentally rotate the visual environment could be a determinant factor for the apparition of space motion sickness (**Parker & Harm 1992**). Alternatively, the ability of space travellers to arbitrarily set down to the direction of their feet may parallel the course of motion sickness.

Psychometric tests. Tasks which demanded logical reasoning and decision-making functions, memory retrieval functions, and fine manual control were performed in a single subject during the German mission aboard Mir in 1992 (**Manzey *et al.* 1993**). Speed and accuracy of short-term memory retrieval and logical reasoning functions remained unimpaired in flight (flight days 2 and 7), but clear decrements in tracking performance were found, showing difficulties in fine manual control.

A series of psychometric tests involving evaluation of simple reaction time, choice reaction time, mental flexibility, visuospatial perception, spatial working memory and time estimation was performed during the Austrian-Russian mission aboard Mir in 1991 (Cogimir experiment). Overall, there was no significant differences on any variable between in-flight and ground data. A tendency to underestimate longer time periods (6, 8, and 10 s) became obvious on flight days 3 and 6, and continued for two days after the landing (Benke *et al.* 1992). In the U.S. space programme, difficulty in evaluating time periods has also been casually reported by returning Shuttle crew members (Harm, personal communication).

Exposure history. Spatial orientation can be influenced by cognitive factors, and not based primarily under the influence of multiple sensory and motor factors. For example, studies of simulated moon walking under terrestrial conditions have shown that individuals soon take the surface on which they are walking to be the direction of 'down' although it may be tilted 75° from the gravitational vertical (Gurfinkel *et al.* 1981).

Lackner & Dizio (1993) made the interesting observation that although microgravity is free-fall, only a few subjects (astronauts or parabolic flyers) experience a falling sensation when they cannot see visual displacement in relation to their surrounding. Moreover, when subjects in parabolic flight are free-floating and can see their bodies separated from the deck of the aircraft, this usually does not trigger sensations of falling nor fear reactions. This is notable because studies using the 'visual cliff' apparatus have demonstrated that when animals see the support surface visually separated from their feet they exhibit fear and avoidance responses (**Walk & Gibson 1961**). The perception of falling probably depends on a variety of cognitive factors as well as visual and air flow cues, and probably other elements as well. For example, if a person falls or jumps off a ladder, he or she anticipates body displacement and impact. Prior experience is important in affecting behaviour: this means that one can anticipate considerable individual differences in performance. Exposure history and cognitive factors are important in understanding these differences.

4.3.6 Summary

Both static and dynamic visual scene information becomes increasingly dominant for most subjects in weightlessness. Frequently the weighting attached to localized tactile cues is also increased. However, we do not know whether changes in the basic reflexes (vestibulo-ocular, ocular counter-rolling, optokinetic, otolith-ocular) are related to the changes in weighting of visual and tactile cues associated with perception.

The perception of the local vertical is more variable in flight, and pointing experiments generally show a tendency to point low in the dark. It is unclear whether the body centred reference system (body scheme or 'idiotropic vertical'), which contributes to orientation, independent of sensory inputs, undergoes any changes in weightlessness.

An egocentric 'vertical' reference is maintained by everyone in microgravity although there are major intrasubject differences. Some people are strongly 'visual-spatial dependent' and rely on the environment for a reference, while others seem to carry along a reference frame tied to their own trunk ('Z-axis dependent'). It is not fully understood how or why egocentric vertical is maintained, nor why a sensation of free-fall is absent in orbit, or why astronauts do not associate each linear acceleration with a change in body orientation. Also, a relationship between the various orientation perceptual styles and the underlying physiological processes or space motion sickness has not been established.

4.4 Space motion sickness

Exposure to provocative real or apparent motion leads to the progressive cardinal symptoms of terrestrial motion sickness which typically include pallor, increased body warmth, cold sweating, dizziness, drowsiness, nausea and vomiting (**Graybiel *et al.* 1968**). The constellation of signs and symptoms of space motion sickness (SMS), when taken together with the time course of symptom development and movements encountered upon exposure to microgravity, suggests that sickness experienced during space flight is similar to terrestrial motion sickness. There are, however, several important features associated with SMS that may not be part of motion sickness experienced on the Earth (**Money 1970**; Harm 1990).

4.4.1 Symptoms

Among the symptoms experienced as part of SMS which differ from those exhibited during acute provocation on the ground are that flushing is more common than pallor and there is virtually no occurrence of sweating in SMS. Sweating (with the exception of palmar sweat, Oman *et al.* 1990) has been the least frequently reported flight symptom (Homick *et al.* 1984). Nearly universal are malaise, anorexia or loss of appetite, lack of initiative and irritability. In microgravity there are more reports of stomach awareness, vomiting, headache (due perhaps to headward fluid shifts), impaired concentration, lack of motivation, and drowsiness than are typically seen during acute motion sickness on the ground (Homick & Vanderploeg 1989; **Thornton *et al.* 1987b**; Oman *et al.* 1990; **Davis *et al.* 1988**). Vomiting is usually sudden and often without prodromal nausea. The bouts are not frequent, usually separated by one to three hours, with no dry heaves. Bowel sounds, obtained by auscultation, have been found to be decreased or absent in astronauts suffering from SMS (**Thornton *et al.* 1987a**). Gastrointestinal symptoms have their onset from minutes to hours after orbital insertion. Excessive movement early in orbit commonly increases symptoms; however, demanding activities, such as deployment of the Skylab umbrella, may tend to attenuate symptoms. Symptom resolution usually occurs between 30 and 48 h, with a reported range of 12 to 72 h, and recovery is rapid (**Thornton *et al.* 1987b**).

4.4.2 Incidence

The Russian cosmonaut Titov was the first to experience symptoms of SMS, and symptoms have been reported by 48% of the cosmonauts in the Russian space programme (Gorgiladze & Bryanov 1989). There were no reports of motion sickness in the American Mercury and Gemini programmes (Homick 1985) while 35% of Apollo astronauts developed symptoms and the incidence in the Skylab missions was 60% (**Davis *et al.* 1988**). In 24 flights of the Space Shuttle (ending with STS-61C), the incidence of SMS was 67% in first-time flyers. In these flights, there were no statistically significant differences in symptom occurrence between career *vs* non-career astronauts, commanders/pilots *vs* mission specialists, males *vs* females, different age groups, or first-time *vs* repeat flyers. An astronaut's susceptibility or insensitivity to SMS on his or her first flight correctly predicted susceptibility on the second flight in 77% of the cases (**Davis *et al.* 1988**). For STS-27 through STS-38, the incidence of SMS was 70% with 83% of first-time flyers developing symptoms and a 61% occurrence of symptoms in repeat flyers (Beck, personal communication 1991). Symptom recurrence at landing reportedly afflicts 27% of Russian cosmonauts following short duration flights (4 to 14 days) and 92% returning from longer missions (Gorgiladze & Bryanov 1989). No reports of 'readaptation syndrome' were noted in the U.S. Shuttle programme through the mid-1980's (**Thornton *et al.* 1987b**); however, it now appears that this syndrome affects a similar percentage of both U.S. and Russian crews.

4.4.3 Provocative stimuli

Movements that produce changes in orientation, particularly whole body or head movements, are necessary to induce SMS symptoms. A hypersensitivity to angular head motions is commonly reported (Lackner & Graybiel 1979; Jerusalem *et al.* 1985). In particular, crew members report that pitch head movements are initially more provocative than head movements made in other planes (**Watt 1987; Thornton *et al.* 1987b**). However, once sickness has been well established, head movements in any plane are generally minimized by the affected crew member. Indeed, movement of any kind is frequently restricted until the astronaut is on the road to recovery (Oman *et al.* 1990).
Head or body movements made upon transition from microgravity to a gravitational field less than the Earth's, and *vice versa*, may not be as provocative. It is interesting to note that of the 12 Apollo astronauts who walked on the moon, only three reported mild symptoms, such as stomach awareness or loss of appetite, prior to their moon walk. None reported symptoms while in the one-sixth gravity of the lunar surface, and no symptoms were noted upon return to weightlessness (Homick & Miller 1975).
Visual cues are known to play a role in spatial orientation and symptoms can be elicited in orbit during episodes of reorientation (i.e. when the visual scene does not correspond to that expected by the astronaut, requiring an adjustment in perceived 'up' and 'down') (Oman *et al.* 1990). Individual astronauts differ with respect to the adaptation algorithms they develop to compensate for the sensory disturbances encountered following exposure to microgravity. Some astronauts (Type VS or 'visuo-spatial') report that they become predominantly 'visual creatures' in weightlessness. Apparently these astronauts develop a spatial orientation framework by increasing the weighting of visual spatial orientation signals. Type VS astronauts may experience discomfort or symptoms when visual objects that exhibit a consistent polarization with respect to gravity on Earth are seen in unusual orientations in space, i.e., when their feet are not oriented to the deck (**Harm & Parker 1993**). For example, seeing a fellow crew member floating 'upside down' (Oman &. 1990) or viewing the Earth in unexpected orientations through the Shuttle windows may be reported as disturbing by a Type VS astronaut. Other astronauts (Type IZ or 'internal z-axis') appear to compensate for the absence of gravity primarily by increasing the weighting assigned to internally-generated orientation vectors. These astronauts appear to ignore visual polarity information and use the direction of their feet to define 'down' (Harm, personal communication, 1992).

4.4.4 Time course

There are considerable individual differences in susceptibility to SMS, and currently it is not possible to predict with any accuracy those who will have some difficulty with sickness while aloft. Symptom resolution usually occurs between 30 and 48 h; however, the recovery rate, degree of adaptation, and specific symptoms vary widely between individual astronauts.

Although antimotion sickness drugs offer some protection against SMS, some drugs (e.g. scopolamine) may interfere with the adaptation process and symptoms controlled by these drugs are experienced once treatment ceases. This 'state dependency' has been reported by several astronauts who experienced episodes of nausea and vomiting three to four hours following the last dose of a three-day prophylactic regimen of scopolamine (0.4 mg) and dexedrine (5.0 mg) (Davis, personal communication, 1990).

Unless all movement is inhibited, most susceptible astronauts begin to experience space sickness within the first hour of orbital flight. Again, there are wide variations in the latency and intensity of SMS. Several crew members after leaving their seats have vomited only minutes following cut-off of the Shuttle's main engines. However, both the frequency and intensity of symptoms appear to decrease as the flight progresses. By the sixth day of the Skylab flights, all astronauts were free of SMS, and could not be made sick with rotation and programmed head movements (Graybiel et al. 1974).

Adaptation aloft to provocative motion does not convey resistance to motion sickness immediately following flight. Approximately 30% of the Skylab astronauts experienced seasickness while in the command module and aboard the recovery ship. However, within two to three days these same crew members (along with those not susceptible to seasickness) were resistant to the programmed cross-coupled angular accelerations experienced on a rotating litter chair, and this resistance lasted for several weeks (Graybiel *et al.* 1974). This same post-flight resistance has not been seen with astronauts participating in the Shuttle programme. Approximately 10% of the Shuttle astronauts experience motion sickness during re-entry from orbit, or immediately upon landing. Recovery from 'return from orbit' motion sickness is variable, and has the same characteristics as the in flight space sickness.

Symptoms (at least none as catastrophic as vomiting) have never occurred during extravehicular activity (EVA). However, since most space sickness has abated by the third day of flight, mission rules now restrict EVA until the third mission day, and the minimum flight duration is three days to ensure that no astronauts are sick prior to entry and landing (**Davis *et al.* 1988**).

4.4.5 Theories and hypotheses

Two major theories advanced to account for SMS are the fluid shift theory (Barrett & Lokhandwala 1981; **Parker *et al.* 1983**) and the sensory conflict (also known as the neural mismatch, sensory mismatch, or sensory rearrangement) theory (Reason & Brand 1975). Although both theoretical positions have some merit and neither is ideal, the fluid shift theory may have the most difficulty in dealing with the development of motion sickness during space flight. While the fluid shift theory of SMS could be associated with sensory conflict, there are mechanisms whereby the cephalad fluid shift accompanying microgravity could bypass the classic vestibular inputs to induce vomiting.

Briefly, the sensory conflict theory of motion sickness assumes that human orientation in 3-D space, under normal gravitational conditions, is based on at least four sensory inputs to the CNS. The otolith organs provide information about linear accelerations and tilt relative to the gravity vector; angular acceleration information is provided by the semicircular canals; the visual system provides information concerning body orientation with respect to the visual scene or surround; and touch, pressure, and kinaesthetic systems supply information about limb and body position. In normal environments, information from these systems is compatible and complementary, and matches that expected on the basis of previous experience. When the environment is altered in such a way that information from the sensory systems is not compatible and does not match previously stored neural patterns, motion sickness may result.

Sensory conflict explains everything in general, but little in the specific. Shortcomings of the sensory conflict theory include its lack of predictive power, inability to explain those situations where there is conflict but no sickness, failure to include sensory-motor conflict, inability to explain specific mechanisms by which conflict actually gives rise to vomiting, and failure to address the observation that without conflict, there can be no adaptation. The hypotheses outlined below may be helpful in overcoming some of the weaknesses associated with the construct of this theory.

Otolith asymmetry hypothesis. Some investigators (von Baumgarten & Thumler 1979; von Baumgarten *et al.* 1982; Kornilova *et al.* 1983; Gorgiladze *et al.* 1986; **Lackner *et al.* 1987**) have proposed a mechanism complementary to the sensory conflict theory, to explain adaptation to weightlessness, readaptation to 1 × g, and individual differences in SMS susceptibility. They suggest that some individuals possess slight functional imbalances (for example, weight differences) between the right and left otolith receptors that are compensated for by the CNS in 1 × g. A weight imbalance between the otoconia on the left and right macular surfaces is reasonable since there is a continual turn-over of otoconia, and it is unlikely that the two otoliths would ever weigh exactly the same. This compensation is inappropriate in 0 × g, however, since the weight differential is nullified and the compensatory response (either central or peripheral) is no longer correct for the new inertial environment. The result would be a temporary asymmetry producing rotary vertigo, inappropriate eye movements, and postural changes until the imbalance is compensated or adjusted to the new situation. A similar imbalance would be produced upon return to 1 × g, resulting in post-flight vestibular disturbances. Individuals with a greater degree of asymmetry in otolith morphology would thus be more susceptible to SMS. Until recently, no tests were available to evaluate this hypothesis adequately (von Baumgarten *et al.* 1981). However, **Diamond & Markham (1988, 1991)** and **Diamond *et al.* (1990)** have developed a measure of eye torsion (ocular counter-rolling) that should help evaluate in flight otolith asymmetry and hence predict SMS (**Markham & Diamond 1993**). During experiments conducted in parabolic flight, the random ocular torsion which occurs normally while in an upright position was measured in nine former astronauts during parabolic flight. Prior to the flight, the subjects completed SMS history questionnaires. Disconjugate torsional eye movements were measured during both hypo- and hypergravity phases. When ranked by increasing asymmetry during 0 and 1.8 × g the four subjects with the least asymmetry had not experienced SMS symptoms while the five subjects with higher asymmetries had suffered from SMS. These two groups were significantly different on the basis of the asymmetry measures. For the latter group, the latency, severity and duration of SMS was positively correlated with torsional asymmetry scores (**Diamond & Markham 1991**).

Sensory compensation hypothesis. Sensory compensation occurs when the input from one sensory system is attenuated and signals from others are augmented. In the absence of an appropriate graviceptor signal (or perhaps the presence of atypical signals) in microgravity, information from other spatial orientation receptors such as the eyes, the semicircular canals, and the neck position receptors can be used to maintain spatial orientation and movement control (Parker & Parker 1990). Astronauts frequently report that they increase their reliance on visual cues for spatial orientation and motion control (**Young *et al.* 1984; Reschke & Parker 1987**; Reschke *et al.* 1991; **Harm & Parker 1993**). In addition, Reschke *et al.* (1991) found evidence of neck position receptor compensation for maintaining visual target fixation during ocular counter-rolling induced by head tilt performed in orbit.

Otolith tilt-translation reinterpretation hypothesis. The otolith tilt-translation reinterpretation hypothesis (OTTR) was proposed on the basis of data from experiments performed as early as the eighth Shuttle flight (**Parker *et al.* 1985**), and represents a refinement of an otolith reinterpretation hypothesis developed by Young and his colleagues (Young *et al.* 1976; Oman 1982). The rationale for this hypothesis is as follows. First, weightlessness is a form of sensory stimulus rearrangement to which people adapt. Second, because of the fundamental equivalence between linear acceleration and gravity, graviceptors signal both orientation with respect to gravity (tilt) and linear acceleration which is perceived as translation. Third, as a consequence of the absence of gravity during orbital flight, graviceptors do not respond to static pitch or roll in weightlessness; however, they do respond to linear acceleration. Because stimulation from gravity is absent during space flight, interpretation of the graviceptor signals as tilt is meaningless. Therefore, during adaptation to weightlessness, the brain reinterprets all graviceptor output to indicate translation.

4.4.6 Prevention and treatment

The disruptive nature of SMS, occurring as it does during the early, critical stages of a mission, has led to a variety of approaches for the prevention or control of this malady (**Graybiel 1979**). Only limited success has been achieved to date. Research directed toward prevention has proceeded along four broad lines of inquiry: training, preselection/screening, pharmacological treatment and mechanical or electrical devices.

Vestibular training. Vestibular training prior to space flight in the Russian space programme has primarily involved Coriolis (cross-coupled angular) acceleration. However, attempts by the Russian programme to prevent SMS by preselection of individuals with a high tolerance to vestibular stimulation have not been successful (Lapayev & Vorobyev 1986; **Wetzig *et al.* 1993**). Training devices and procedures to pre-adapt astronauts to the sensory stimulus rearrangements of microgravity are currently being pursued by NASA. Two part-task Preflight Adaptation Trainers (PATs), the Tilt-Translation Device (TTD) (Fig. 8) and the Device for Orientation and Motion Environments (DOME), have been developed (**Reschke *et al.* 1988**). Space flight investigations involving the PAT devices have been focused on providing astronauts with demonstrations of and experience with altered sensory stimulus rearrangements which produce perceptual illusions of various combinations of linear and angular self- and/or surround-motion. The findings from one of the recently completed space flight investigations indicate the following: (i) several of the stimulus rearrangement conditions in both PAT devices produce perceptual experiences quite similar to those associated with space flight; (ii) post-flight exposure to the TTD-PAT can re-elicit symptoms identical or similar to SMS symptoms; (iii) post-flight exposure to the TTD-PAT can re-elicit visual disturbances similar to those experienced in-flight; (iv) post-flight exposure to TTD-PAT motion profiles elicits greater perceived linear self-motion than preflight exposure; and (v) asymmetries in perceived tilt angle during exposure to the TTD are common post-flight, whereas they are generally absent preflight (**Harm *et al.* 1991; Harm & Parker 1994**). In general, these results support the OTTR model of sensory adaptation to microgravity and the concept that the PAT devices can simulate the sensory rearrangements present in the weightless environment of orbital flight (Parker *et al.* 1986).

Autonomic response training. The autonomic nervous system initially responds to motion stress with sympathetic activation. Following termination of the stimulus, a decrease in sympathetic activity is noted which may indicate a parasympathetic rebound. If nausea and vomiting are parasympathetic reactions to sympathetic activation, then motion sickness symptoms should

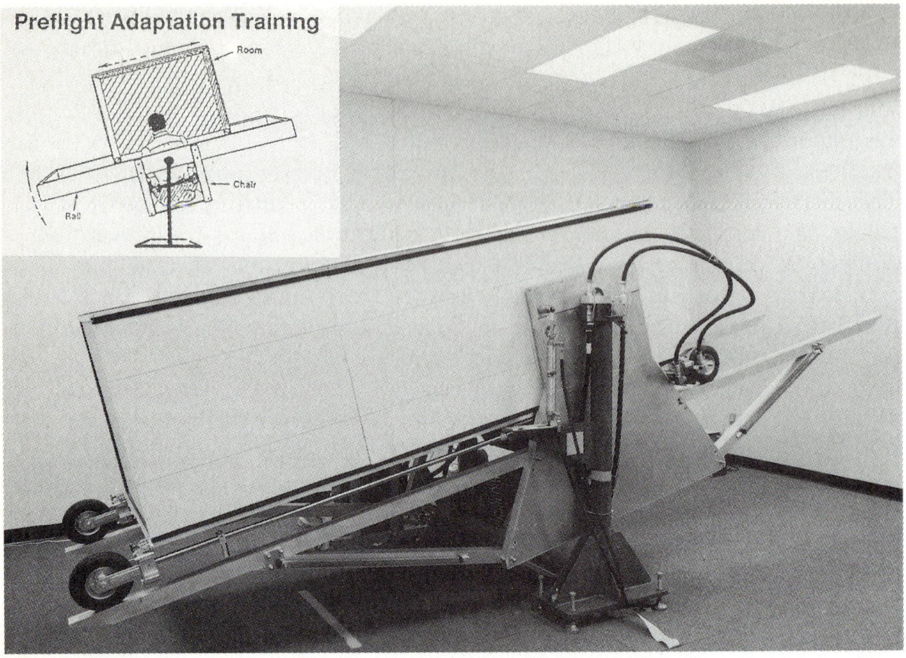

Fig. 8. Tilt-Translation Device (TTD) Preflight Adaptation Trainer (PAT). Subjects placed on the rotation axis are exposed to both body tilt and linear translation of the visual environment. Body pitch is combined with forward-backward linear vection, whereas body roll (see schematic) is combined with side-to-side linear vection. Photo NASA.

be prevented by training an individual to maintain his/her autonomic responses at baseline levels (**Cowings et al.** 1986).

Pharmacological countermeasures. A reported 30% of all Shuttle crew members have received medication for relief of SMS in flight (**Santy & Bungo 1991**). The most detailed reports of the incidence of SMS symptoms and use of antimotion sickness drugs, however, come from the Skylab missions. The variable success of antimotion sickness drugs (scopolamine and d-amphetamine, dimenhydrinate, metoclopramide, naloxone, etc.) administered during flight may be due to changes in drug absorption or metabolism by such factors as dehydration, reduced gastrointestinal motility, alterations in body chemistry related to adaptation to the weightless environment, changes in cabin pressure, and disruption of normal sleep/wake cycles. The concomitant administration of medications for other indications is another confounding factor (Santy & Bungo 1991). The recent success of IM promethazine administration is encouraging; however, whether the effectiveness of the drug is due to the pharmacologic effect of the drug itself or the route of administration and its effect on bioavailability is unknown (**Bagian 1991**).

Mechanical and electrical devices. The Russians have investigated the ability of various mechanical and electrical devices to alleviate the symptoms of motion sickness. The mechanical devices, designed to prevent the complete adaptation of the body to weightlessness, are intended to counteract deconditioning of the body during long term missions, as well as relieve SMS during the first days of flight (**Matsnev *et al.* 1983**). The electrical devices, which pass an electrical current through the body, operate via mechanisms which are as yet unclear (Séverac 1992).

4.4.7 Summary

Based on data from the Russian and U.S. flights it is clear that most (about 67%) space travellers will experience symptoms of SMS. The problem is generally brought on by head movements (and deserves the name space 'motion' sickness). Symptoms are not significantly reduced on a reflight. However, reflights are usually spaced so far apart temporally that no real information exists to adequately answer this question.
The sensory–motor conflict theory of motion sickness needs development and validation to account fully for the occurrence of SMS. The possible influence of otolith organ asymmetry in possibly inappropriate compensation is not understood. A relationship between the occurrence of spatial orientation illusions and SMS has not been established. The generalization of the habituation to weightlessness is not known.
A reliable and validated predictor of space sickness is not available, although measurements of the subjective vertical and of disconjugate eye torsion have correlated with past incidents of SMS. One of the difficulties is that only occurrence of SMS has been evaluated yet, not susceptibility to SMS, since head movements among other activities are neither controlled nor adequately measured in flight. Vestibular training of cosmonauts may affect system sensitivity, and in flight prevention devices have not proven to be successful. Preflight adaptation training (PAT) based on virtual reality and actual body motion needs further study.
The current favourite drug treatment is IM injection of promethazine, rather than the use of scopolamine or other prophylactic medications. The effectiveness and side effects of promethazine are still under investigation.

4.5 Compensatory eye movements

Eye movements serve to hold gaze steady or to shift gaze to an object of interest. A steady image is maintained on the retina during head perturbations by compensatory eye movements produced by the vestibular–optokinetic system (Leigh & Zee 1991). The elimination of gravity alters the inputs to this system. The effect of these alterations on the adequacy of compensatory eye movements and adaptation of the vestibular–optokinetic system to counter these effects has been studied in space, during parabolic flight and on the ground.

4.5.1 Vestibulo–ocular reflex

The vestibulo–ocular reflex (VOR) responds to transient rotation of the head with eye movements in the opposite direction (Fig. 9). This action prevents slippage of images off the retina since the latency of the VOR is less than 16 ms compared with the 70 ms required for retinal processing. During prolonged rotation, acceleration is no longer sensed. In darkness, the initial nystagmus response to rotation produced by the VOR decays over a 30 to 45 s period after which the eyes are stationary. Post–rotational eye movements are also produced by the VOR in response to deceleration (Leigh & Zee 1991).

4.5.1.1 Ground based studies

VOR can be quantitatively described by gain, phase and time constant. VOR gain is the ratio of the amplitude of the movement of the eye to that of the head. For sinusoidal rotation, gain is calculated as the ratio of peak slow–phase eye velocity to peak head velocity. Phase describes the synchronicity of eye and head movements. For the frequency range of natural movement (0.5 to 5.0 Hz), eye movements should be (if the system worked perfectly) of

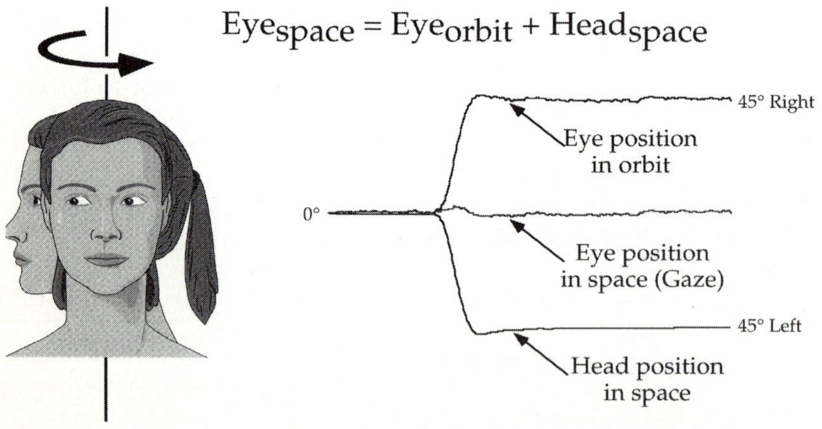

Fig. 9. Vestibulo–ocular reflex. When the head is rapidly rotated about an axis the eyes rotate in the orbit of the skull by a corresponding amount in the opposite direction. Because the rotation of the head in space and the eyes in the orbit are approximately equal and opposite, their sum, i.e. eye position in space or gaze, remains zero. Stable gaze is required to maintain a stable image on the retina.

equal velocity and opposite in direction to head movements resulting in a gain of 1.0 and a phase of 0° (phase shift of 180°). For sustained rotation, gain is computed as the ratio of initial eye velocity to head velocity. In darkness, slow-phase nystagmus velocity decays exponentially and the VOR time constant is defined as the time required for eye velocity to decrease to 37% of its initial value. The nystagmus initially observed in sustained rotation is longer in duration than the signal provided by the vestibular nerve due to velocity storage, a mechanism which prolongs the nystagmic response (Leigh & Zee 1991; **Raphan *et al.* 1979**).

VOR gain varies between subjects and is affected by the distance from the subject to the target, the mental set chosen by the subject (especially in darkness), adaptation to corrective lenses and age. The VOR time constant also exhibits variability between subjects, is shortened by habituation, differs between horizontal and vertical components, and displays vertical directional asymmetry (the time constant for upward slow-phase eye movements is greater than for downward movements) (**Matsuo & Cohen 1984**).

The VOR has been shown to adapt to a number of environmental changes. Habituation to repetitive stimulation in darkness, especially constant velocity rotation or low frequency oscillation, decreases both time constant and gain which may function to reduce motion sickness. Persistent vestibular stimulation also alters the VOR. For constant rotation in darkness, a reverse nystagmus (slow phases in the same direction as rotation) may develop after decay of the initial nystagmus. The brain also adjusts the VOR in response to altered visual stimuli if the VOR cannot maintain image stability, since eye movements resulting from retinal processing occur too late to prevent image slip. If the relationship between eye movements and the visual image is altered by reversing prisms or eyeglasses, the VOR gain and/or phase adapt to the new relationship. Conflict between visual and vestibular cues may provoke motion sickness symptoms during the adaptation period (Reason & Brand 1975).

4.5.1.2 Microgravity investigations

It is believed that the VOR utilizes information provided from both the canals and the otoliths. In weightlessness, otolith signals are modified while canal signals are unaffected perhaps resulting in an inappropriate VOR. The relative contributions of the otoliths and the canals to the VOR is best determined in weightlessness.

4.5.1.3 In flight yaw (horizontal) VOR

On the ground, the direction of the gravity vector sensed by the otoliths does not vary during yaw head oscillations (**Berthoz *et al.* 1986**). Experiments in flight (on about 12 subjects) have relied on voluntary head oscillations at frequencies ranging from 0.25 to 1 Hz (**Thornton *et al.* 1985, 1989**; Viéville *et al.* 1986; Watt *et al.* 1985; **Benson & Viéville 1986**). Passive rotation has also been employed pre- and post-flight on two crew members (**Benson & Viéville 1986**). Head oscillations are performed with eyes open fixating a wall target (where gain is presumably unity) and with eyes open in darkness or eyes closed while imagining a wall-fixed target. Most studies have used conventional electro-oculography to measure eye movements. Few studies have detected pre- or post-flight changes in yaw VOR (**Thornton *et al.* 1989**; Watt *et al.* 1985; **Benson & Viéville 1986**) or the direction of the changes noted has varied between subjects (Grigoriev *et al.* 1991). During active sinusoidal horizontal head movements performed at the beginning of two Russian long-term space flights, an increase of slow phase velocity was observed and an asymmetry appeared (**Kornilova *et al.*** 1992, **1993**). No nystagmic responses, but only sinusoidal compensatory slow eye movements, were observed on flight days 164 and 169, thus leading to decrease of VOR gain.

A single subject aboard STS-51G exhibited decreased VOR gain at 0.25 Hz on his first test, conducted six hours into the mission, which recovered to preflight levels by flight day 7 (Viéville *et al.* 1986). A decrease in VOR gain early in microgravity is consistent with the parabolic flight and centrifuge results of others (Vesterhauge *et al.* 1982; **Lackner & Graybiel 1981**) who have demonstrated decreased VOR gain with decreasing gravity. Since no phase shift accompanied the reduction in VOR in flight, the subject on STS-51G may have suppressed vestibular input to avoid sensory conflict, possibly a learned response from his prior training as a pilot. It is also possible that the subject could not imagine a wall-fixed target in the absence of gravity (Viéville *et al.* 1986).

In humans, parabolic experiments by **DiZio *et al.* (1987, 1988, 1992)** demonstrated that the apparent time constant of post-rotatory nystagmus (PRN) in yaw and pitch was shorter during (but not after) acute exposure to weightlessness. Yaw axis PRN was monitored in flight in one crew member on the SL-1 mission using a hand spun rotating chair; results indicated no change in gain but did suggest a shortened time constant in flight (Oman & Kulbaski 1988). The nystagmus dumping phenomenon (suppression) appeared to be present in flight, suggesting that this phenomenon can be triggered by processes related to active head movements, rather than by gravity *per se* (Oman, personal communication, 1991).

4.5.1.4 In flight pitch (vertical) VOR

In contrast to yaw head movements, pitch head oscillations in $1 \times g$ produce changes in the direction of the gravity vector sensed by the otoliths. The weightless environment offers an ideal way to investigate the contribution of the otoliths to pitch VOR (**Berthoz *et al.* 1986**). A decrease in pitch VOR had been expected early in flight because of the absence of graviceptor signals. However, the gain of pitch VOR would be restored late in flight due to the recalibration of semicircular canal inputs. Finally, pitch VOR gain would be reduced early post-flight due to the negative effects of the otolith-tilt reinterpretation. Indeed, the OTTR would result in reduced perceived pitch amplitude and in increased translational self-motion perception. Consequently, the amplitude of the vertical movement required to maintain gaze fixation would be reduced (Parker & Parker 1990) (Fig. 10). However, as of today, none of the investigations performed has clearly confirmed or denied these expectations.

Investigations of pitch VOR in flight have employed voluntary head oscillations at various frequencies. While pre- and post-flight changes have not been observed in some instances (Watt *et al.* 1985), other investigations have noted alterations in VOR. Two subjects exposed to pitch oscillation at 1 Hz demonstrated significantly increased VOR gain in tests 14 h after landing compared with tests conducted on flight days 5 and 7 and later in the post-flight period (**Berthoz *et al.* 1986**). In these experiments, an increased phase lag was present during the in-flight tests; however, the increase was not significant because of high dispersion of the data.

The subject tested on STS-51G showed a decrease in VOR gain during 0.25 Hz oscillation for the first four flight days after which his gain began returning to preflight levels (**Clément *et al.* 1986**; Viéville *et al.* 1986). The results of these experiments conflict with the increased VOR gain observed during pitch oscillation at 0.25 Hz for the 0 × g portions of parabolic flight (**Clément *et al.* 1992a**). Possible explanations for these conflicting results include: (i) learned suppression of vestibular input by the STS-51G subject (Viéville *et al.* 1986); (ii) occurrence of adaptation before in flight measurements on Spacelab-1; (iii) testing at a frequency (1 Hz) at which the canals were dominant; or (iv) the potential impossibility of imagining a wall-fixed target in 0 × g in the same manner as on Earth (**Berthoz *et al.* 1986**).

No systematic changes in pitch VOR phase or absolute gain across four subjects tested during passive sinusoidal oscillations at 0.05 Hz, 0.2 Hz and 0.8 Hz were noted during the IML-1 Spacelab mission (DiZio *et al.* unpublished observation). However, cross-coupling in eye movements from counter-rolling (roll) to horizontal (yaw) was observed during ramp velocity in the roll plane in microgravity (Reschke *et al.* unpublished observation).

4.5.1.5 Pre- and post-flight yaw and pitch VOR

The horizontal VOR characteristics (gain and dominant time constant) of two rhesus monkeys returning from a two-week flight aboard an unmanned Cosmos satellite did not appear to be significantly changed, as measured 15 h after landing (**Cohen *et al.* 1992**).

Comparison of pre- and post-flight PRN among four Spacelab-1, five D-1, and four SLS-1 astronauts (Oman & Kulbaski 1988; **Oman & Weigl 1989**; **Oman & Balkwill 1993**) showed a residual shortening of the apparent time constant during the first several days after return from week-long flights, but no consistent change in the magnitude of the initial peak slow-phase velocity response. The effects were thus qualitatively similar to those observed by **DiZio *et al.* (1987, 1988, 1992)** in parabolic flight. Responses gradually returned to preflight norms during the first post-flight week. **Oman & Balkwill (1993)** have speculated that as a consequence of the altered graviceptive input in weightlessness, the CNS may reduce the vestibular component driving central velocity storage in favour of visual inputs.

Seven out of 13 cosmonauts (52%) tested on the first day after space flights ranging in duration from 7 to 365 days showed no nystagmic response during active head rotation in yaw or pitch with the eyes closed, but a low amplitude compensatory eye deviation. Three (22%) cosmonauts had a 33% decrease and two (17%) of cosmonauts had a 50% increase in the yaw VOR gain compared to preflight values. During head rotation with the eyes open (without gaze fixation) 62% of the cosmonauts investigated showed an increased amplitude of compensatory deviation and the appearance of nystagmic cycles on it. In this condition, yaw and pitch VOR gain was greater than before the flight (Kornilova & Kaspransky 1994).

4.5.2 Otolith–ocular responses

4.5.2.1 Ground based studies

Compensatory eye movements in response to otolith stimulation are not as easily studied as the VOR. In the case of an angular acceleration signal provided by the canal system, visual compensation is easily predicted because the eye movement needed is equal but opposite to the degree of head rotation regardless of target distance. The desired eye movements needed to compensate for linear acceleration (otolith) signals are less easily approximated. To hold one's gaze on close targets, much bigger compensatory eye movements are needed during translation than if the target is located far away. A simple demonstration of this is looking out of a train window at the nearby trees which go by quickly compared with looking at a house on the horizon.

Since the otoliths respond to both tilt and translation, there are different types of compensatory responses. Otolith-driven eye movements include responses to: head tilt (ocular counter-rolling), linear head motion translation (L-nystagmus), off-vertical axis rotation (OVAR) and analogous centrifuge stimuli. These eye movements must counteract -- or compensate for -- perceived or actual translational (or tilt) changes in head position to maintain ocular fixation on visual targets (Leigh & Zee 1991).

Ocular counter-rolling. Ocular torsion can be elicited from both the semicircular canals and the otolith organs, the saccule and utricle. If the head is tilted laterally, the dynamic component of eye rotation produced from the canals is generally defined as the torsional (or roll) VOR. The static compo-

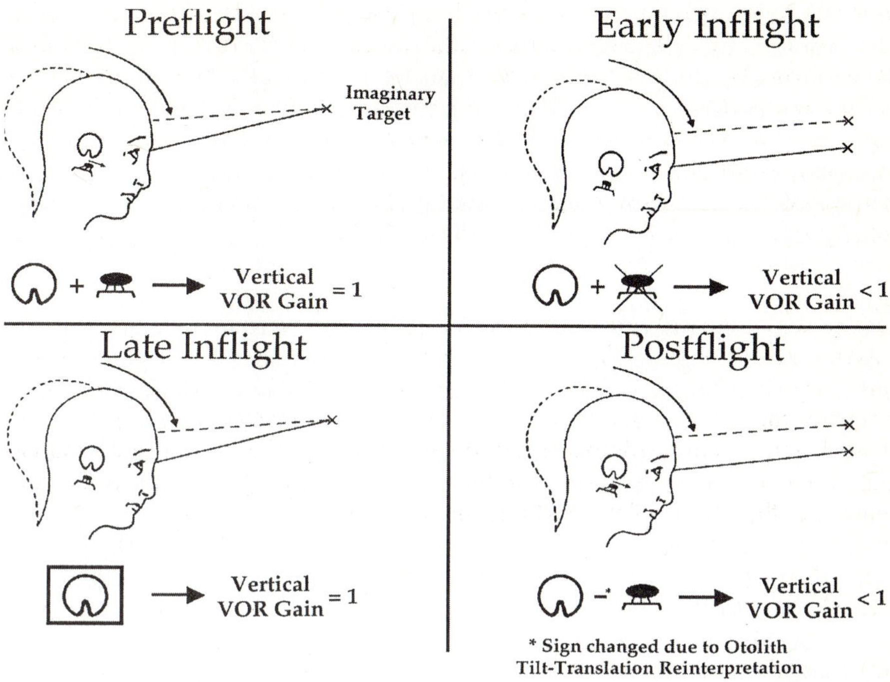

Fig. 10. It is hypothesized that early in flight adaptation to microgravity could result in reduced vertical VOR gain (smaller vertical eye movements would be required to maintain gaze on an imaginary, stationary target). Late in flight the vertical VOR gain would be restored because of the recalibration of canal signals. However, on return to Earth, the OTTR could result in perceived motion after-effect of backward translation during forward pitch. The consequence of translational motion after-effects would be decreased vertical eye movement amplitude. Adapted from Parker & Parker (1990).

nent, or compensatory ocular counter-rolling (OAR), mediated by the otolith organs, holds the eyes in position at the end of the head movement. From the canals, the velocity of the slow phases of the torsional nystagmus is most closely related to the angular velocity of the stimulus, because central processing in canal-ocular pathways is in terms of head and eye velocity. In contrast, the otoliths produce positional changes in OAR, and the central signals related to OAR are in a positional reference scheme (**Diamond & Markham 1988**). Dynamic ocular counter-rolling can be elicited by linear acceleration (**Woellner & Graybiel 1959; Young *et al.* 1984**).

Linear acceleration. Linear acceleration can also induce smooth vertical or horizontal eye movements and nystagmus (the so-called L-nystagmus, or linear VOR) similar to the angular VOR. Nystagmic eye movements during horizontal linear acceleration along the interaural Y-axis were first recorded in man by **Jongkees & Phillipszoon (1962)**. It has been hypothesized that this L-nystagmus is caused by utricular shear. During the vertical stimulation with the head upright, the utricles are only exposed to partial shear due to the anatomic 30° utricular tilt from the horizontal (Frankfurt) plane, thus the vertical sensitivity value is lower than the horizontal value (Young 1967). During centrifuge runs providing 3 × g in the (positive) Z-axis, an up-beating vertical L-nystagmus is also observed (Marcus 1989; **Marcus & Van Holden 1990**; McGrath 1990). Lower levels of linear acceleration (0.03 to 0.34 × g) induced by means of a parallel swing (**Bles & Kapteyn 1973**), as well as the hypergravity phase of parabolic flight also provoke vertical L-nystagmus (**Von Baumgarten *et al.* 1980**; **Clément *et al.* 1989, 1992b**).

Off-vertical axis rotation. During off-vertical axis constant velocity rotation (OVAR), the nystagmus does not decay like it does in the upright position and the post-rotatory nystagmus is much less. Constant-velocity rotation of a subject about his longitudinal axis which has been slightly tilted with respect to gravity produces an unusual and ever changing pattern of stimulation. The effect is equivalent to holding the subject stationary in an upright position, with the ability to rotate an acceleration vector around him at an off-vertical angle of incidence equal to the chair's tilt. Theoretically, it provides adequate stimulation to the otoliths and other graviceptor organs, but probably not to the semicircular canals. The reduction in post-rotatory nystagmus is due to the fact that otolithic information is passed on to the velocity storage integrator which would normally continue the response (Baloh & Honrubia 1990).

Canal-otolith interactions. Evidence that sustained linear accelerations can modify nystagmus elicited by stimulation of the semicircular canals has been reviewed by Benson (1974). Benson notes that Khilov (1936) was one of the first to report that the duration of horizontal after-nystagmus elicited by rotation in the yaw axis was dependent upon the orientation of the head relative to gravity. This effect is manifested by a more rapid decay of after-nystagmus when the acceleration of gravity is in the plane of the horizontal canals, rather than when gravity is perpendicular to these canals (normal position). These findings suggest that the input from the otoliths can modulate an ampullar signal and cause or modify nystagmus. For example, otolith stimulation evokes compensatory reflex eye movements of high velocity when linear motion of the head is combined with concurrent angular motion. Another example of otolith-canal interaction is the reduction in the decay time constant of nystagmus when the plane of the canals is moved out of the plane of rotation during post-rotatory stimulation (e.g. suppression or 'dumping').

4.5.2.2 Microgravity investigations

Ocular counter-rolling. Ocular counter-rolling (OCR) can be induced by static roll tilt or by transient lateral translation. There is surprisingly little objective data about human OCR after space flight, probably because of the difficulty in measuring it. No pre- *vs* post-flight differences in OCR were observed for lateral tilts of ±25° and ±50° following two Gemini flights (**Graybiel *et al.* 1967**). OCR was measured photographically from coloured transparencies of the right eye of four astronauts 18 h following the Spacelab-1 mission (**Vogel & Kass 1986**). The angles of OCR were not given but the authors derived a ratio gain. The gain of OCR was reduced between 28% and 56% for all subjects for left tilt (0° to 90° in 15° increments) after space flight. Changes in gain for right tilt were inconclusive.

A decrease in OCR gain was also present for at least five days after the flight in the German cosmonaut who flew aboard Mir in 1992 (**Hofstetter-Degen *et al.* 1993**). During the flight, there was no static OCR, and the roll VOR during active head roll tilt which occurred and rapidly diminished was most likely caused by the vertical semicircular canals.

An after-image method was used to measure OCR in six astronauts on STS-24 through 26 and STS-28. Unlike the preflight results (recorded in both the upright and supine positions), little or no ocular torsion was observed in flight. Immediately after even relatively brief flights, the after-image method showed a reduction in eye torsion that was slow to return to normal (Reschke *et al.* 1991).

In the Russian space programme, all cosmonauts (n = 28) returning from long duration missions (30 to 175 days) and 14 of 18 returning from short flights (7 to 18 days) exhibited changes in OCR, assessed by an after-image method, when tilted to the right or left of vertical one to two days post-flight. Sixteen of these short flight subjects demonstrated increased OCR while only two demonstrated the decrease predicted by the OTTR hypothesis (one following a long-term and one after a short-term flight). These tests, however, were not conducted immediately after landing. Of the eight subjects tested on R0 following short flights, four showed no change in OCR while the other four demonstrated inappropriate torsional eye movements in the direction of the head movements (i.e. eyeball rotation in the direction of tilt). Only one of the four still exhibited this 'negative reflex' on R+1. Seventeen out of the 28 cosmonauts (62%) returning from long flights demonstrated decreased OCR, and 9 had the 'negative otolith reflex' (Kornilova & Kaspransky 1994). When tested one to two days post-flight, all of the long-term and four of the 18 short-term flight subjects showed greater reflex asymmetry relative to preflight measurements with five cosmonauts demonstrating a change in the direction of their preflight asymmetry. In general, responses returned to normal two to four days following short flights and eight to nine days after long flights (Yakovleva *et al.* 1982). A slight decrease in roll-induced OCR was observed on flight days 1 and 5 which was not in phase with the preflight response (Kornilova *et al.* 1987, **1993**).

Using scleral magnetic research coils to obtain a continuous measure of ocular torsion, OCR was reduced by about 70% for 11 days after space flight in the two rhesus monkeys who flew in the 1992-1993 US-Russian Cosmos Biosatellite Project (Dai *et al.* 1993a & b), suggesting long-lasting alteration in

OCR and otolith function of monkeys as a result of adaptation to microgravity. The authors suggested that differences in species and in performance demands during space flight could account in part for the differences between monkey and human data.

Video-oculography revealed stronger nystagmus activity and an increase in the gain of the torsional (roll) VOR elicited by active head rotation about the roll axis in weightlessness of one crew member. Also, discrepancies between intended and performed head movement were observed in flight: although the subject was convinced to perform pure roll head movement, there were clear components of both roll and yaw motion, indicating an impairment in sensorimotor coordination after prolonged exposure to weightlessness. Interestingly, such a roll/yaw head movement did not elicit a combination of torsional and horizontal eye nystagmus, but rather a combination of torsional and vertical nystagmus (**Clarke et al. 1993a, b** & c).

The gain of roll VOR was reduced in the monkeys tested after the Cosmos flight 2229, although less than the OCR produced by pure otolith stimulation in response to head tilt or during off-vertical axis rotation (Dai *et al.* 1993a).

Linear acceleration. Two subjects exposed to transient lateral acceleration three to five hours after the landing of Spacelab-1 demonstrated OCR with smaller torsional amplitudes than three of the four preflight measures. Torsional amplitude in these subjects steadily increased over most of the post-flight tests. The changes noted were not statistically significant because of high variability in the preflight measurements (**Arrott & Young 1986**). These results are consistent with the OTTR hypothesis prediction of reduced eye torsion immediately post-flight due to reinterpretation of otolith signals as linear translation (**Parker *et al.* 1985**). This hypothesis also predicts a post-flight increase in horizontal eye movements elicited by roll stimulation. Following STS-11,$\pm 15^\circ$ roll stimulation elicited greater horizontal nystagmus in tests conducted 70 to 150 min after landing on three subjects compared to tests conducted preflight or later post-flight; y-axis linear translation also enhanced horizontal eye movements two and three days after landing (**Parker *et al.* 1985**).

Canal-otolith interactions. When testing two rhesus monkeys 15 h after the landing of the Cosmos 2044 mission in 1989, **Cohen *et al.* (1992)** observed significant pre- to post-flight differences in the amplitude of modulation of off-vertical axis rotation (OVAR) and tilt dumping behaviour, confirming the hypothesis of neural modification in the central processing of information from the otolith receptors. One flight monkey lost the ability to shorten its horizontal time constant (to suppress the nystagmus) after tilting out of the plane of the nystagmus (90° roll). The authors suggested that failure to change the horizontal time constant when the head was tilted with respect to the gravity vector indicates that the yaw axis orientation vector had moved with the head, thereby maintaining the dominant time constant. That is, the yaw axis orientation vector had become body- rather than spatially-oriented in microgravity. During nystagmus induced by OVAR, the latency, rising

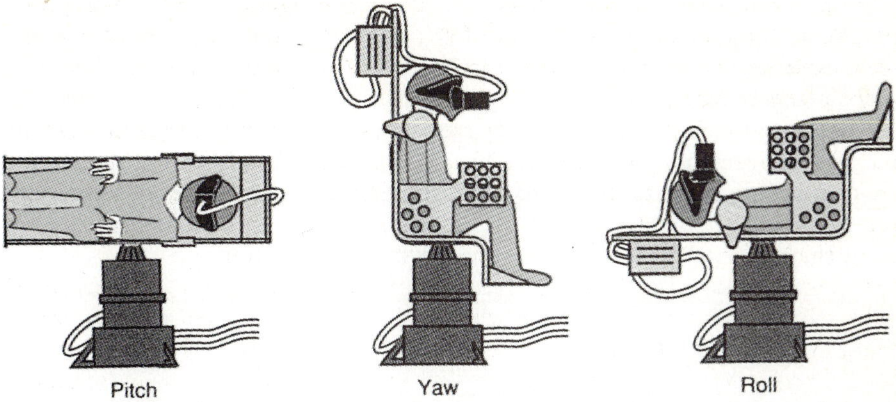

Fig. 11. The Microgravity Vestibular Investigations which flew on the IML-1 Spacelab mission used a rotating chair system to tests the astronaut's visual and vestibular responses to head and body movements. The position of the chair on the rotator could be changed so that a crew member was seated upright and turned around the yaw axis, or laid on one side and moved around the pitch axis, or laid on the back and moved around the roll axis. Note that the head was away from the rotation axis in the pitch and roll orientations, a centripetal acceleration was therefore exerted on the otoliths during rotation.

time constant, steady state eye velocity (depending on the integrity of the velocity storage integrator) and phase of modulation in eye velocity and eye position with respect to head position were similar in both monkeys before and after flight. However, the amplitude of modulation of horizontal eye velocity during OVAR (attributed to the direct otolitho-ocular pathways) was larger after flight. Similar results were observed in the OVAR nystagmus of two astronauts tested 32 h after returning from a 7 day space flight. In addition, the perceptive effects during OVAR at small angles of tilt were much larger post-flight than preflight (Clément *et al.* 1995).

In four crew members tested during passive rotation throughout the SLS-1 mission, head movement from the upright (dumping) had no effect in flight on the time constant of decay for the post-rotatory nystagmus, although the sensation of motion was reduced (**Young *et al.* 1993**).

During the IML-1 space mission four crew members were passively rotated about a yaw axis aligned with their longitudinal body axis, or about pitch or roll axes aligned with their body's centre of gravity (Fig. 11). Therefore in pitch and roll orientations, the head was off-centred about 50 cm. During rotation at constant velocity of $120^{\circ}\ s^{-1}$ the centripetal acceleration exerted at the head level was $0.22 \times g$. In microgravity it was hypothesized that the crew members would experience an inversion illusion due to the presence of this 'steady-state' (60 s) linear acceleration. However, none of the crew members tested reported this illusion, nor showed steady-state, otolith-induced, slow phase eye velocity. In the same pitch or roll orientations, the subjects were submitted to sinusoidal oscillations of the chair at 0.2 Hz, 0.8 Hz, and

1.25 Hz (peak velocity 60° s^{-1}). Thus, a fluctuating, low amplitude linear force, the resultant of radial (0.493 m s^{-2}, peak) and tangential (0.023 m s^{-2}, peak) accelerations rotated about the head during angular acceleration; and might account for the absence of changes in pitch VOR gain and phase in microgravity, or for the presence of cross-coupled eye movements (i.e. oblique eye movements in response to a pure pitch vestibular stimulation). Cross-coupling from torsional to horizontal slow phase eye movements was also observed on flight day 7 in two astronauts during angular acceleration about the roll axis (Reschke *et al.*, unpublished observation).

4.5.3 Optokinetic system

Optokinetic stimulation occurs naturally during sustained self-motion in the light or in motion of the visual world such as in the train tracking example above. In the laboratory the optokinetic nystagmus (OKN) is usually obtained by rotating a large patterned drum around the stationary subject. The subject typically experiences a compelling sensation of self-rotation called circular vection even though the peripheral vestibular signal is absent. During optokinetic stimulation the visual nystagmus is used to track the drum. During this stimulation period the optokinetic system effectively acts as a velocity storage mechanism, for after the lights are turned out, nystagmus continues in the same direction for some seconds (particularly in monkeys) with a declining slow phase velocity; this is optokinetic after-nystagmus (OKAN). The time constant of decay of OKAN is similar to the time constant of the VOR (**Raphan *et al.* 1979**), is a direct reflection of the long time constant of the VOR, and therefore of the storage mechanism. In fact it is the simplest manifestation of the VOR storage mechanism, since OKAN does not contain a contribution due to the deflection of the cupula that occurs during activation of the semicircular canals (**Cohen *et al.* 1977**).

4.5.3.1 Ground based studies

The time constant and direction of OKAN in primates and in humans is strongly affected by head position with respect to gravity (Raphan & Cohen 1988; **Clément & Lathan 1991**). Horizontal OKAN becomes oblique by the addition of a vertical component when subjects are in a 90° static roll position. In the 90° roll position, the time constant of downward OKAN is long. In the upright or erect position, however, the time constant is short and the gain (eye velocity to stimulus) is low. There is also an asymmetry between upward and downward OKN and OKAN. In the upright position, the upward slow phases are faster than downward slow phases. This asymmetry is similar to that observed in vertical vestibular nystagmus (**Matsuo & Cohen 1984**). This OKN asymmetry was found to be reversed when the head is tilted more than 90° (Raphan & Cohen 1988; **Clément & Lathan 1991**). Finally, it has

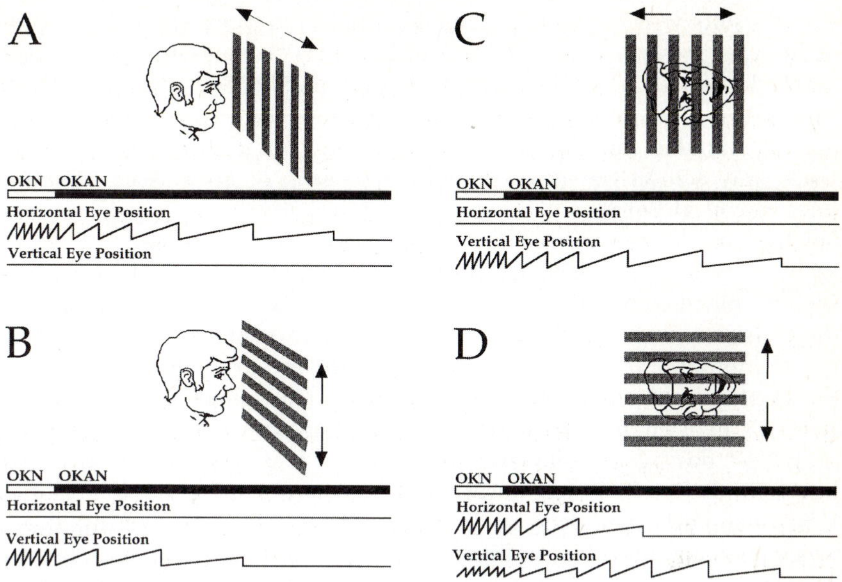

Fig. 12. Changes in OKN and OKAN as a function of head position relative to gravity. On Earth, when subjects are upright a horizontal optokinetic stimulation (A) induces purely horizontal OKN and OKAN: the axis of eye rotation during nystagmus is aligned with gravity. A vertical optokinetic stimulation generates less OKAN (B). However, when the head is tilted 90° (C), thus bringing the axis of eye rotation during OKN and OKAN aligned with gravity, the vertical OKAN is increased. When subjects are on their sides and receive horizontal optokinetic stimulation (D) there is an oblique OKN. The vertical component of OKAN that develops (cross-coupling) is more closely aligned with gravity.

been shown that gravity is important for determining the spatial orientation in slow phases velocity during OKN (**Clément & Lathan 1991**) and vestibular nystagmus (Raphan *et al.* 1992). When subjects are tilted laterally into side down positions (roll tilt) the direction of the OKN slow phases velocity shifts gradually from the axis of stimulation towards the spatial vertical (Fig. 12). During OKAN the response vector shifts further and tends to align with gravity (**Clément & Lathan 1991**). With forwards or backwards tilt (pitch) torsional or roll nystagmus develops in response to the head-horizontal stimulus (**Dai *et al.* 1991**). The appearance of vertical and/or torsional components in response to head-horizontal optokinetic stimulation has been called 'cross-coupling'.

4.5.3.2 Microgravity investigations

Optokinetic nystagmus. Pre- *vs* post-flight changes in optokinetic nystagmus induced by a rotating drum were not observed in tests conducted early in the Shuttle programme on 16 U.S. crew members (**Thornton *et al.* 1985**). These results were perhaps predictable: flight duration was short and the stimulus was presented in such a way that a stable surround was visibly present; additionally, the stimulus was always aligned with the subject's Z-axis. However, two cosmonauts were reportedly unable to track a horizontal optokinetic stimulus moving at 80 stripes min^{-1} in tests following 75 days in space and demonstrated asymmetry at speeds of 40 and 60 stripes min^{-1} (**Matsnev *et al.* 1983**).

Suppression of eye movements during optokinetic stimulation as well as asymmetry and alteration of normal nystagmic components were also observed immediately after Salyut-6 and 7 (Gorgiladze & Maveyev 1990). No in flight changes were noted in two cosmonauts in response to vertical and horizontal optokinetic stimulation ($6^{\circ}\ s^{-1}$) with the exception of a decrease in the amplitude of vertical OKN early in flight (Kornilova *et al.* 1991).

No OKN gain changes were demonstrated for a monocular stimulus moving at $20^{\circ}\ s^{-1}$ during or following the flight of one astronaut aboard STS-51G; however, altered OKN responses were noted for a pattern velocity of $53^{\circ}\ s^{-1}$. Horizontal OKN gain decreased on flight day 1, slowly recovering to preflight levels later in flight. No in flight change in the horizontal beating field was observed. Vertical OKN gain also decreased early in flight and was accompanied by a downward drift of the beating field. The normal OKN asymmetry was reversed for the first three flight days and some downward drift was still evident on flight day 5 although gain and beating field were nearly normal. In tests conducted within an hour of landing, a significant increase in vertical OKN gain was noted along with an upward drift in the beating field. Normal asymmetry was present (**Clément *et al.* 1986**).

No reversal of vertical OKN gain asymmetry was observed in the single cosmonaut tested during the 26 day French-Soviet Aragatz Mission aboard Mir in 1988. However, other characteristics of the OKN response, such as the centre of interest, i.e. the position of the eye in the orbit at the end of the quick phases, which is strongly asymmetrical on the ground, was found to be symmetrical after three weeks in space (**Clément *et al.* 1993**).

The downward drift in the beating field observed in space flight was also evident in parabolic flight tests for both an upward and downward moving stimulus. However, this downward drift has been attributed to the changes in gravito-inertial force levels during parabolic flight (**Clément *et al.* 1989**) and is probably responsible for changes in vertical OKN and VOR gain during parabolic flight (Clément & Berthoz 1988; **Clément *et al.* 1992b**; **Marcus 1993**).

Vertical spontaneous eye movement were also observed during Mir missions in five cosmonauts (Kornilova *et al.* 1991; **Kornilova *et al.* 1993**) and during the IML-1 space mission on four crew members. Interestingly, one astronaut also exhibited a rather strong spontaneous horizontal nystagmus in the dark (Reschke *et al.*, unpublished observation). Examinations performed on 27 cosmonauts returning from long-term space flights and 19 returning from short-term space flights revealed the presence of a spontaneous nystagmus in 31 cosmonauts (24 long flight subjects, 7 short flight subjects). This spontaneous nystagmus had slow phases directed up and left in 9 cosmonauts, down and right in 14 cosmonauts, and purely down in 10 cosmonauts, and changed in direction and intensity when the body was tilted in pitch relative to gravity (Kornilova & Kaspransky 1994).

One possible explanation for the downward drift of the eyes in flight is that, in the absence of gravity, the antigravity tonic influence exerted by the sacculus, which tends to lift upward both the limbs and the eyeball, in order to compensate for the downward pull by gravity is suppressed. Upon return, the adaptation which occurred in space flight would result in a perception of gravity as an upward linear acceleration (opposite in direction to the downward inertial force vector) producing an opposing upward movement of the eyes (**Clément *et al.* 1986**).

The hypothesis of a reorientation in the direction of OKN toward the spatial vertical was investigated in microgravity. On Earth it has been shown that the velocity storage uses gravity for its orientation and the orientation of OKN responses. It is hypothesized that this orientation disappears in microgravity, and that a new reference, such as the long axis (Z-axis) of the body, could be used. However, no cross-coupling from vertical to horizontal eye movements was observed during OKN in one subject in microgravity during the STS-51 G Shuttle mission (**Clément & Berthoz 1990**). No significant change in the OKN gain was observed during the IML-1 space mission in the pooled values from the in-, pre- and post-flight recordings of pitch or yaw OKN in four astronauts (Cohen *et al.*, unpublished observation). However, some episodes of cross-coupling did occur in flight. These findings have been confirmed in an additional 3 astronauts during the STS-60 flight (Reschke *et al.*, unpublished observation).

Optokinetic after-nystagmus. Although not studied extensively in space flight, OKAN was recorded in one subject on STS-51G using a monocular stimulus. The time constant of vertical OKAN with slow-phase up was larger than in normal gravity on flight day 1; this effect disappeared on flight day 2 (**Clément & Berthoz 1990**). In a separate study on the Aragatz mission aboard Mir, the gain of vertical OKAN peak slow-phase velocity was increased in one cosmonaut at the end of the flight relative to preflight measures. These results could suggest a increase of the stored activity related to vertical slow-phase velocity during adaptation to microgravity (**Clément *et al.* 1993**).

No changes in vertical and horizontal OKAN were observed during the IML-1 mission on four astronauts tested. However, after light extinction the presence of gaze resetting to a position near the primary position, and an occasional overshoot beyond the primary position, are indications that optokinetically-induced storage processes, which are gravity sensitive, were affecting gaze stabilization. This was also supported by the development post-flight of strong after-effects which were not present preflight, highly suggestive of vision-evoked destabilization of gaze (Jell *et al.*, unpublished observation).

Cross-coupled OKAN from vertical to horizontal plane was observed after pitch axis optokinetic stimulation on the first day of exposure to microgravity in one subject, but not on later days (**Clément & Berthoz 1990**). Evidence from a shift of the velocity storage from gravitational (spatial) to a body axis after adaptation to microgravity has been obtained in monkeys after the Cosmos 2229 mission (Dai *et al.* 1993b). In one monkey the axis of eye rotation during OKAN following yaw (relative to the head) OKN was only tilted 5° with regard to the spatial vertical before flight when the animal was tilted on its side by 90°. One day after landing the same test yielded an orientation angle of 28° relative to the spatial vertical, a significant shift toward the body axis. At 7 days after recovery, the orientation vector angle had returned to 7° with respect to the spatial vertical, indicating that it was again closely aligned with gravity.

4.5.4 Visual–vestibular interactions

4.5.4.1 Ground based studies

During head rotation it is generally assumed that the outputs of the vestibular and optokinetic system are linearly combined to yield a common slow phase velocity output. Most interaction experiments concerning linear motion have been devoted to motion perception (linear vection), postural control or otolithic contribution to pursuit eye movement. Other experiments have also strongly suggested a synergistic action of acoustic or optokinetic information on otolith–ocular reflex (Dichgans & Brandt 1978).

4.5.4.2 Microgravity investigations

VOR suppression. VOR suppression is tested with eyes open during head oscillations while the subject fixates a head-mounted target. Suppression of horizontal VOR was unchanged pre- and post-flight on the single subject tested during STS-51G while vertical suppression was slightly impaired on flight day 1 (Viéville *et al.* 1986). A slight increase in corrective nystagmus was noted in flight during a VOR suppression task performed on several early Shuttle flights (n = 16) (**Thornton *et al.* 1985**). Gain of VOR suppression in four crew members was significantly lower on R0 following the Spacelab-1 mission than for pre- or later post-flight tests suggesting an increased reliance on vision during weightlessness (**Benson & Viéville 1986**). Interestingly, in a group of eight subjects on STS-6 and 8, **Thornton *et al.* (1989)** reported differences in VOR suppression between those susceptible and not susceptible to SMS. Susceptible subjects had more difficulty suppressing their VOR, but this lack of suppression disappeared when the mean amplitude of VOR error was computed.

Combination of optokinetic and vestibular stimuli. Active head movements may influence OKN during space flight. **Kornilova *et al.*** (1992, **1993**) reported a decrease in horizontal OKN gain as flight length increased and an increase in gain after active head movements. In the vertical plane, active head movements did not affect OKN gain.

Tracking of a moving target was found to be significantly impaired when performed by combined eye and head motion, in 5 astronauts returning from Shuttle missions (Glasauer *et al.* 1994), particularly during vertical tracking. Since no significant changes in the vertical pursuit gain was observed when the head was fixed, it was suggested that the impairment during active head motion was due to the otolithic contribution to the pursuit movement.

The extent to which the slow phase velocity of nystagmus elicited by an optokinetic stimulation with constant velocity could be modulated by sinusoidal angular motion was investigated in parabolic flight (**Clément *et al.* 1992a**) and in orbital flight on four astronauts during the IML-1 Spacelab mission (Clément *et al.*, unpublished observation). Rotation in yaw combined with horizontal optokinetic nystagmus, and rotation in pitch combined with vertical optokinetic nystagmus were investigated with two oscillations frequencies (0.2 and 0.8 Hz) and one velocity of optokinetic stimulus ($40^\circ\ s^{-1}$). Rotation in darkness at the same frequencies, and optokinetic stimulus at 20, 40, and $60^\circ\ s^{-1}$ presented with the rotator in static position, were used as controls. In normal gravity, the interaction of vestibular and optokinetic nystagmus was found to be linear when both responses were elicited either about the subject's yaw or pitch axis. In microgravity, such interaction was still linear when both nystagmus were elicited about the subject's yaw axis. However, a non-linear interaction was observed in the vertical plane, which confirmed previous results obtained in parabolic flight. These results suggest an interaction between the vertical vestibular and the optokinetic systems where the response is modulated as a function of the gravitational state. The observation that major changes occurred for visual-vestibular interaction about the subject's pitch axis when compared to the subject's yaw axis could be attributed to the fact that the synergy between canals and otoliths was disrupted in free fall.

Interaction between passive sinusoidal translational motion along the subject Z-axis and unidirectional vertical optokinetic stimulation at constant velocity has been investigated by Berthoz and his collaborators during the Spacelab-D1 space mission. Although an otolithic modulation was visible on the vertical eye movements on flight day 6 (Von Baumgarten *et al.* 1986), the responses in flight were not found to be significantly different from preflight because of their high variability. Causes for this variability were related to the fact that the optokinetic stimulus covered a very limited part of the subject's visual field ($\pm 10^\circ$), and the linear acceleration component was very low ($0.1 \times g$ peak velocity).

4.5.5 Saccadic eye movements and pursuit tracking

4.5.5.1 Ground based studies

The VOR and optokinetic systems are field-stabilizing reflexes. The eyes can also move alone to fixate a target (saccade) or fixate on a small object of interest that is moving relative to a fixed background and follow it without moving the head (pursuit response). When looking at a moving object, both the head and eyes usually move. The process of turning the head in the direction of the object produces a vestibular stimulus which activates the VOR. Evidence from lesions in humans and from the similarity between visual suppression of the VOR and visual pursuit suggest that the pursuit signal is the signal that cancels the VOR allowing moving objects to be tracked as the head turns (Barnes 1981; **Halmagyi & Gresty 1979**).

4.5.5.2 Microgravity investigations

Eye saccades. In experiments conducted aboard STS-7 and 8, accuracy of horizontal saccades tended to decrease early in flight and increase later in flight, with subjects susceptible to SMS significantly less accurate than non-susceptibles (all inaccuracies were by target undershooting). The six subjects also demonstrated a significant decrease in saccade velocity in flight with non-susceptibles experiencing a greater decrease. All but one of the six subjects exhibited an increase in the latency of saccadic response with higher in flight latencies observed for the SMS susceptible subjects (Thornton *et al.* 1988; Uri *et al.* 1989). In contrast, two cosmonauts on the Aragatz mission showed increased saccade velocity, a trend toward decreased saccade latency and a tendency to overshoot a horizontal target early in flight (one subject) with high accuracy later in flight. The conflicting results could be due to the effects of SMS, which was experienced by three of the six U.S. space flight subjects while neither of the cosmonauts was afflicted (**André-Deshays *et al.*** 1990, **1993**).

Memory-guided ocular saccades in darkness were found to be hypermetric in one cosmonaut in microgravity (**Israel *et al.* 1993**). Direction errors increased in microgravity compared to ground measurements. Since the same subject did not demonstrate any changes in visually-guided saccades in microgravity, the authors concluded that the changes in memory-guided saccades result from alterations of the internal representation of space rather than from the oculomotor system.

Target acquisition performance decreased significantly in 5 astronauts returning from Shuttle missions, as shown by slower velocities of both eye and head movements and changed timing (the head started to move later in post-flight trials) when attempting to acquire horizontal and vertical targets placed at ±68° and ±50° respectively. It was suggested that both delayed head

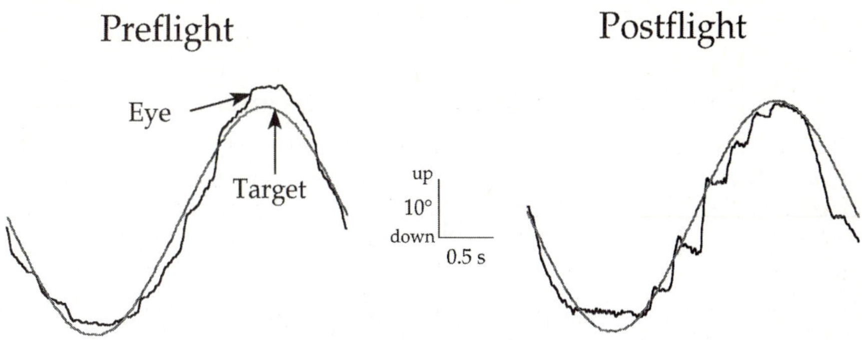

Fig. 13. Example of vertical pursuit tracking of one astronaut before flight and two hours after landing. Frequency of pursuit stimulus was 0.33 Hz, with a peak velocity of $60° \, s^{-1}$. Note the presence of saccades in the post-flight recording.

motion and decreased head velocity point towards a strategy designed to reduce head movements post-flight. The decrease in peak eye velocity might result from a general reduction of gaze shift velocity during readaptation to gravito-inertial forces (Glasauer *et al.* 1994).

Pursuit tracking. Kornilova et al. (1991) have reported changes in pursuit tracking of vertical pulsed movements of a point stimulus manifested on flight days 3 and 5 by decreased eye movement amplitude (undershooting) and the appearance of correction saccades. Pursuit of a vertically or diagonally moving point stimulus also deteriorated in flight while associated saccadic movements were unchanged (**Kornilova *et al.* 1993**). The effects of microgravity on the pursuit function were most pronounced early in flight (flight day 3), after long exposure to weightlessness (flight days 50, 116, and 164) and after landing. Pursuit improved following execution of active head movements in flight indicating that the deficiencies in pursuit function noted in microgravity may be of central origin.
Data collected on two cosmonauts of the French-Soviet Aragatz mission aboard Mir in 1988 showed that horizontal pursuit at 0.25 Hz and 0.5 Hz remained smooth with a gain of 0.98 and minor phase lag (**André-Deshays *et al.* 1993**). In the vertical plane, the eye did not track the target with a pure smooth pursuit eye movement, but the saccadic system contributed to gaze control. Upward tracking was mainly achieved with a succession of saccades, whereas downward tracking was due to combined smooth pursuit and catch-up saccades (Fig. 13). Similar results were observed on astronauts returning from Shuttle missions (Reschke, personal communication, 1992; Glasauer *et al.* 1994) and on cosmonauts returning from Mir missions. In the latter, 22 cosmonauts out of 27 returning from long flights and 4 cosmonauts out of 19 returning from short flights showed saccades while tracking horizontal and particularly vertical moving stimuli (Kornilova & Kaspransky 1994).

4.5.6 Summary

Some evidence supports an initial increase in VOR gain (and reduction in head movement threshold) in weightlessness, followed by a prolonged inhibition after several days in space. The time constant of decay of post-rotational nystagmus appears to be shortened on orbit, reflecting a possible loss of some 'velocity storage'. However, the nature and time of the changes in VOR for rotational head movements in microgravity are not fully known, and we do not know if they are adaptive.

Vertical eye movements, including pitch VOR and vertical pursuit, seem to be more disturbed by weightlessness than are horizontal movements. Spontaneous nystagmus appears and disappears periodically throughout space flight. The details of changes in optokinetic nystagmus thresholds and range remain to be determined definitively. The hypothesized CNS reinterpretation of otolith signals to represent translation rather than tilt (OTTR) has not yet received firm support.

4.6 Posture and locomotion

Disturbances in postural equilibrium and gait upon return from flight have been among the most consistently observed and reported responses associated with space flight. Although some in-flight data have been collected, most postural stability testing has been limited to the comparison between pre-flight and post-flight performance.

4.6.1 Posture studies in flight

4.6.1.1 Motion pattern

Analysis of the behavioural responses of vertebrates to microgravity shows convergent and divergent patterns among organisms. Midwater-dwelling teleost fish make forward somersaults in microgravity, if denied visual cues (**Von Baumgarten *et al.* 1972, 1975**; Scheld *et al.* 1976; **Wetzig 1987**; Takabahashi *et al.* 1990; Griffy *et al.* 1993). Tadpoles of the African clawed-toad (*Xenopus laevis*) similarly make forward somersaults when subjected to microgravity (**Wassersug & Souza 1990; Wassersug 1992**). However, bottom-dwelling fishes and fully-aquatic amphibians either float randomly or make reverse (backward) somersaults when abruptly exposed to microgravity (Wassersug *et al.* 1991).

Adult non-arboreal frogs and salamander larvae rotate along their rostral-caudal axis in response to microgravity. This long axis rotation is similar to their righting reflex when inverted in normal gravity (Wassersug *et al.* 1991; **Wassersug *et al.* 1993**). Arboreal frogs take up an extended-limb gliding/parachuting posture when suspended in air during microgravity (Izumi-Kurotani *et al.* 1992).

A semi-arboreal snake was observed taking a stereotypic defensive posture in microgravity during parabolic flight. Pond turtles in microgravity extend their neck and limbs in an asymmetric fashion identical to their righting response when placed upside-down in normal gravity (**Wassersug & Izumi-Kurotani 1993**).

Birds adopt a flying behaviour (**Oosterveld & Greven 1975**) and mammals, such as hamsters or rats, frequently extend their extremities and back, similar to a flying squirrel in flight, or spiral along their long axis (**Sakuragi *et al.* 1991**).

Two carp, 26 cm in size, intact and otolith-removed, were flown on the Spacelab-J 8-day mission. The behaviour of the fish was recorded by a video system through the window of the water tank together with the EEG cerebellar activity. Both fishes showed slower motion during the flight, compared to pre- and post-flight testing. However, the near absence of motion in the otolith-removed fish was attributable to the tight twisting of the EEG cable which immobilized the carp from the second day in flight (Mori *et al.* 1994). Cable twisting could have been produced by an extensive rotation to the right by the lesioned animal.

Most lower vertebrates react to an abrupt decrease in gravity as if stressed (e.g. the snake). Forward or reverse looping among aquatic vertebrates in microgravity appear to be extensions of escape behaviours, such as rapid ascent or descent respectively. The looping behaviours result, however, when there is an absence of any vestibular feedback to the animals indicating completion of the manoeuvre. Long-axis rotation appears to be the result of the repetitive execution of the righting response (such as that observed in a falling cat) in a situation where it is non-effective. Extension of the extremities would be the consequence of release of the inhibitory influence exerted by the vestibular system on the antigravity muscles, the extensors (**Clément *et al.* 1986, 1989**; Wassersug *et al.* 1993a).

In humans, frame-by-frame analysis of video recordings of on-board activities of crew members have allowed the neutral body position to be characterized (Thornton 1978), as well as prevalent orientations and stereotyped motor acts (**Tafforin *et al.* 1989**; **Tafforin & Lambin 1993**).

4.6.1.2 Responses to voluntary movements

Arm raising and tiptoe raising. In a joint French and Russian experiment aboard Salyut-7, control of upright posture was examined in two cosmonauts during voluntary upward movement of the arm and voluntary raising on tiptoe (**Clément *et al.* 1984, 1985**). Early in flight postural attitude was similar to that on Earth, but as the flight progressed, there was a forward inclination of the body which increased when vision was stabilized. Muscle responses to sudden voluntary perturbations (raising the arm rapidly) indicated a redistribution of tonic activity between extensor and ankle flexor muscles, and a general reduction of extensor tone. In trials conducted in 1 × g, anticipatory activity of postural muscles functions to displace the body's centre of gravity into a new stable position. In flight, results typical of those observed preflight were found on the third flight day. The finding that the sequence of motor patterns were preserved in-flight suggests that terrestrial postural programs continue to operate for a relatively long period of time in weightlessness, independent of how sensory inputs are modified (**Clément *et al.* 1985**).

Bending at the waist. Rapid voluntary pitch movements at the waist (forward and backward) were made while the crew member's feet were fastened to the wall of the spacecraft with velcro bands (Massion *et al.* 1992; **Massion *et al.* 1993**). Kinematic analysis performed on two cosmonauts, in addition to confirming the forward tilt posture reported by **Clément *et al.* (1985)**, showed that upper trunk movements were accompanied by hip and knee movements in the opposite direction, and that there was little difference between in flight measurements and those obtained both pre- and post-flight. The results of electromyographic (EMG) analysis, like that observed during the Salyut-7 flight (**Clément *et al.* 1985**), showed that the early activation of the soleus muscle group observed under terrestrial conditions was replaced in flight by an early activation of the anterior tibialis muscle group. This redistribution of motor program was still in evidence five days following the flight.

Squatting. In flight, simultaneous recordings of EMG activity in the tibialis and soleus muscles while the crew member's feet were fixed to the floor of the spacecraft demonstrated that upright posture was maintained through tonic activity in the flexor muscle (**Clément *et al.* 1984, 1987**). This reported change prompted an investigation on STS-51G into the relationship between conscious appreciation of limb position and body position in space and muscle afferent activity. Two crew members were ask to lower their bodies into a squatting position, pause, and then rise to a fully erect position. By the third day in flight, the subjects reported illusions of floor motion during execution of the deep knee bends. Similar illusions occurred following the Spacelab-1 flight on four crew members tested (**Watt *et al.* 1986**; **Reschke *et al.* 1986**), during STS-41G (n = 2) (Watt *et al.* 1985) and during the hypogravity portions of parabolic flight (**Lackner & DiZio 1993**).

4.6.1.3 Responses to involuntary movements

Support surface translations. Using a foot support platform designed to provide sudden forward translation, postural responses to involuntary body displacements were also investigated during the Salyut-7 flight on two cosmonauts (**Clément *et al.* 1985**). On flight day 2, in response to a sudden plantar flexion, the initial burst of tibialis activity was similar to that observed preflight, but the level of tonic activity in the tibialis was greater than that observed on the ground. By flight day 3, the overall tonic activity remained high, but the initial tibialis burst was similar to that observed following the seventh trial on flight day 2 and the trajectory of the ankle movement showed better stabilization than that observed preflight. On the third day after landing, the tibialis motor response had returned to baseline, but the ankle rotation trajectory suggested postural destabilization. It was suggested that the early tibialis burst resembles the EMG activity of a 'functional stretch reflex' mediated by supraspinal centres (**Melvill-Jones & Watt 1971**), and that the changes in overall EMG amplitude during flight reflect reduced output from the otoliths. These results are consistent with the findings from the Hoffmann reflex (H-reflex) experiment (n = 2) on Spacelab-1 (**Reschke *et al.* 1984, 1986**) and the otolith-spinal reflex measurements made on Spacelab-1 and D-1 missions (n = 4) (**Watt *et al.* 1986**).

Sudden drops (falls). During the Spacelab-1 mission, **Reschke *et al.* (1985)** examined the effects of weightlessness on the H-reflex on two astronauts in flight and on four astronauts post-flight. This monosynaptic spinal reflex can be modulated by a nerve pathway from the otolith receptors. It was hypothesized that exposure to free fall would reduce the necessity for postural reflexes in the major leg muscles, and that postural modifications would reflect a change, not in the peripheral vestibular organs, but more centrally.

EMG from the gastrocnemius muscle and self-motion reports were also recorded and related to the H-reflex data (**Reschke *et al.* 1986**). Early in flight H-reflex amplitude was similar to that recorded preflight, but measurements obtained on flight day 7 did not show a change in potentiation as a function of the drop-to-shock intervals. Immediate post-flight H-reflex response in three of four astronauts tested showed a rebound effect. Increase in post-flight gastrocnemius EMG in response to the sudden and unexpected falls was observed in only one crewman who was tested early post-flight. This effect returned rapidly to baseline. In flight self-motion perception reports suggested that the early in flight drops were perceived like those preflight. Drops later in flight were described as sudden, fast, hard, and translational in nature. Immediately post-flight, the drops were perceived like those late in flight, with the astronauts reporting that they did not feel as though they were falling, but rather that the floor came up to meet them.

In a related Spacelab-1 experiment (**Watt *et al.* 1986**), otolith-spinal reflexes were elicited by sudden, unexpected Earth vertical falls. Like the H-reflex experiment, falls were executed in flight by pulling two subjects to the deck of the Spacelab using elastic cords. EMG activity recorded early in flight from the gastrocnemius-soleus complex during the fall was of lower amplitude than that observed preflight and continued to decline as the flight progressed. Post-flight, the EMG amplitude returned to normal by the first opportunity for measurement. These results agree with the results of the H-reflex experiment showing little or no potentiation of the monosynaptic reflex as a function of a vertical fall late in flight (**Reschke *et al.*** 1985, **1986**).

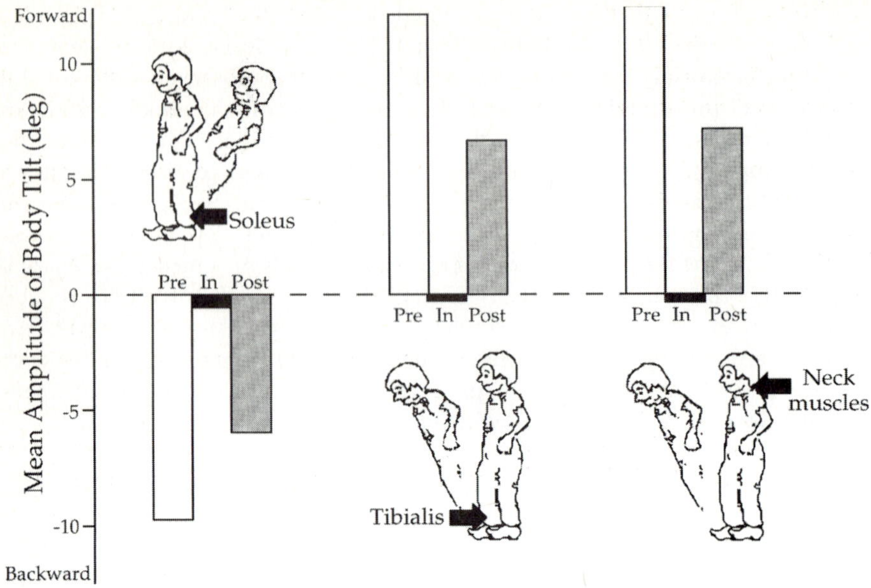

Fig 14. Direction and magnitude of actual body tilt generated by soleus, tibialis and neck muscles vibration in two astronauts before, during (flight days 6–7) and after spaceflight. The angle and direction of tilt given by the relative displacement of a goniometer attached to the subject's ankle were averaged over subjects and trials. Adapted from Roll *et al.* (1994).

Muscle vibration. The role of muscle proprioceptive receptors in control of upright posture was investigated by vibratory stimulation of the soleus and anterior tibialis muscle tendons during the Mir Aragatz and Antarès missions in four cosmonauts (Roll *et al.* 1990; **Roll *et al.* 1993**, 1994). Before flight, vibratory stimulation of the soleus resulted in backward sway about the ankle joint, whereas stimulation of the anterior tibialis resulted in forward sway. During flight, the mean amplitude and the velocity of the vibration-induced postural responses, as assessed from the ankle goniometer, decreased compared with preflight measurements (Fig. 14). The postural responses disappeared after flight day 7, and recovered progressively during the first days post-flight. In addition, the compensatory EMG recorded preflight was altered in flight even though muscle activity concomitant with the vibration was observed in both soleus and anterior tibialis (similar to the classic tonic vibratory response).

With the body restricted mechanically, the illusory body tilt perceived by the astronauts during soleus or tibialis vibration decreased in flight (see Fig. 7). However, vibration applied to the anterior lower leg muscles gave rise to a sensation of body elevation along the longitudinal axis. A perceived upward motion of a head-fixed target was reported by the cosmonauts in flight regardless of which muscle was vibrated, soleus or tibialis. The authors concluded that muscle proprioception remained intact after prolonged flight, since it was still possible to activate the muscle spindle and generate motor response with vibration. However, the characteristics of the response to muscle vibration changed in microgravity, indicating that adaptive sensorimotor responses occur, and that these new responses are appropriate to the environment.

Postural tremor of arm muscle was investigated on one cosmonaut during the Austrian–Russian space mission in 1991 (Gallasch *et al.* 1992) and on two cosmonauts during long-term space flights aboard Mir (Gallasch *et al.* 1994) at rest and in visual feedback tasks using a hand dynamometer. Tremor at rest (microvibration) which was observed on the ground disappeared completely in flight, and increased after the long-term flights but with significant changes compared to the typical pattern. With the stretched arm, the postural tremor was still visible in flight but with a decreased amplitude. These results suggest that the action of the stretch servo system is reduced in microgravity. Isometric tremor exhibited no changes in flight compared to baseline values, probably because the contractile properties of muscle adapt very slowly to the changes in gravito-inertial level.

4.6.2 Pre- and post-flight posture studies

Returning crew members from space missions have reported one or more of four basic unusual sensations associated with posture or locomotion during the first few hours after landing: (i) a sensation of turning or lateral deviation while attempting to walk a straight path (overcompensating for this, many crew members actually walk in a curved path in the opposite direction); (ii) a sudden loss of postural stability, much as though the crew member has been pushed to one side by a 'giant hand', usually experienced while attempting to walk around a corner; (iii) the perception that the pitching and rolling head motions that accompany normal walking are greatly exaggerated; (iv) in a visual environment that is not well structured, some crew members experience a sudden loss of orientation and pitch forward, or fall to the side before position awareness is regained.

Early post-flight measurements of postural ataxia are based on the hypothesis that prolonged exposure to a weightless environment would result in changes in the sensory systems (with the possible exception of vision) necessary for the maintenance of postural stability. It was postulated that these changes would most likely originate at the periphery, and involve modification of input from the receptors serving kinaesthesia, touch, pressure, and otolith functions. Furthermore, as exposure time increased, habituation responses appropriate to the new inertial environment were expected to occur at a central level. Upon return to Earth, postural instability would be manifested as a result of the in flight neural reorganization.

4.6.2.1 Rail tests

Four days after the landing of the Apollo-16 mission (Homick & Miller 1975) and two days after the Skylab-2, 3, and 4 flights (**Homick & Reschke 1977**; Homick *et al.* 1977), metal rails of varying widths were provided for the twelve crew members to stand on in a sharpened Romberg position (feet, heel-to-toe; arms crossed and folded across the chest) with eyes opened or eyes closed. Time before stepping (or time-out) was the performance measure of postural stability. Post-flight decrements in postural stability during the eyes open tests ranged from none to moderate. However, during the eyes closed tests, postural stability was considerably decreased in all crew members tested. The magnitude of the change was greatest during the first post-flight test. Improvement was slow, and appeared to be related to the length of the mission.

Rail tests were repeated with the crew of the Spacelab-1 mission (**Kenyon & Young 1986**). With open eyes, performance on the narrow rail width (1.9 cm) was found to be considerably reduced post-flight, and did not return to pre-flight levels before the last test session, seven days after landing. With the eyes closed, all four crew members tested exhibited a significant decrement in performance immediately post-flight, even while standing on the 5.72 cm wide rail. In at least one case, return to baseline had not occurred by the seventh day post-flight. In addition to the static rail standing task, crewmen were asked to walk on the 1.9 cm rail. All subjects adopted a strategy of speed, trying to complete the test trials as quickly as possible and minimize instability. Post-flight performance was in all cases below that of preflight data, but was only consistently reduced for one subject.

4.6.2.2 Stabilometry

For many of the shorter Soyuz missions (Soyuz-3 through Soyuz-8), postural stability was measured using stabilometry 30 to 40 days before flight and at various times (9, 18, 27 h) after flight. For longer duration flights (Soyuz-9, Soyuz-17), additional repeated observations were collected post-flight. Stabilograms were recorded for periods of one and two minutes during predetermined postural stances, including standing with the head erect eyes open or closed, standing in the Romberg posture, and standing with the head tilted either forward or backward. Primary measures obtained from the stabilogram were the average frequency and amplitude of sway of the derived body centre of gravity in both the sagittal and frontal planes. The post-flight stabilographic data in all assumed postural stances (about 15 subjects tested) were characterized by an increase of sway amplitude primarily in the frontal plane coupled with, in most crew members, a decrease in oscillation frequency (Fig. 15). The magnitude of change was related to the length of flight, with significant changes occurring following the Soyuz-9 flight (Bryanov *et al.* 1976).

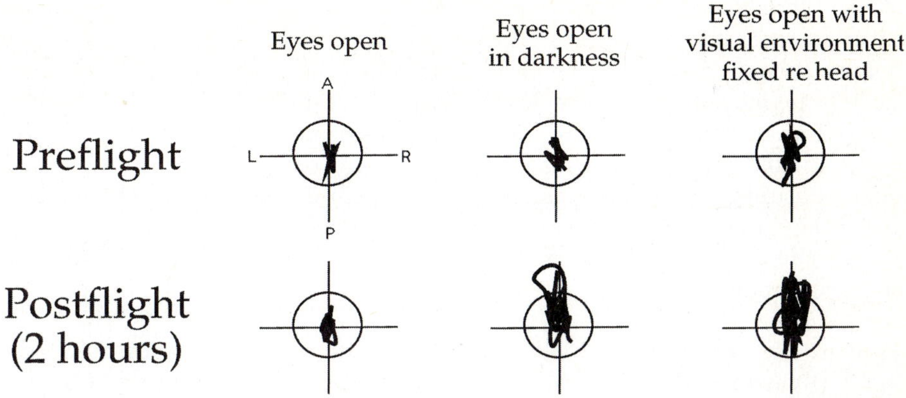

Fig. 15. Displacements of the centre of gravity of one astronaut in the anterior–posterior (A–P) or lateral (L:left, R: right) directions before and after a 7-day space flight. The astronaut is standing upright with the eyes open, or with the eyes open in darkness, or with a visual environment fixed relative to his head (stabilized vision). Post-flight stabilograms show an increase in sway amplitude primarily in the frontal plane, especially when visual inputs are occulted or in conflict with other inputs. Radius of the circle was 5 cm.

In a later study, the prime crew of the Mir–Kvant expedition also participated in postural stability tests using the stabilogram technique. In this study, normal upright posture was perturbed by a calibrated force momentarily applied to the chest. Three cosmonauts participated in this study; two had been in orbit for 151 days and the third for 241 days. Post-flight testing was not initiated until six days after return of the crew. In all but one crew member, less force was required to perturb vertical posture post-flight, and in all crew members the time to recover from the applied perturbation increased post-flight. Overall muscle activity required to maintain upright posture following the perturbation was also increased post-flight. All changes observed on the sixth post-flight day were still present on the eleventh day, but to a lesser degree, and were reportedly similar to those observed following other missions of comparable length (Grigoriev & Yegorov 1990).

4.6.2.3 Moving platform tests

Support base translation. Pre- and post-flight postural stability measurements were made on four crew members from the Spacelab-1 mission using a dynamic posture platform which could be moved parallel to the floor in both predictable and unpredictable patterns, including sinusoids, pseudorandom, and velocity steps (**Anderson *et al.* 1986**). In these studies, the subject attempted to maintain a normal upright stance with eyes either open or closed as the moving platform perturbed his base of support. EMG data obtained from the soleus and anterior tibialis muscles and hip and shoulder displacements relative to the moving platform were recorded throughout the testing period. Post-flight, when the subject's eyes were open and the platform was moved with a backward step function, the subject's response showed an overshoot with the shoulders and an undershoot with the hips relative to his preflight response. Also, the time required to assume a new stable position was greater after flight than before. The EMG data indicated that soleus muscle latency was greater post-flight, and that vision appeared to degrade performance due to head stability problems during platform step function inputs.

Support base rotation. In another dynamic posture platform study on Spacelab-1 (**Kenyon & Young 1986**), the crew member's erect posture was perturbed by pitching the platform base unexpectedly about the ankle joint. EMG activity from the anterior tibialis and gastrocnemius muscles was measured with the eyes open and closed. Post-flight, the early EMG response (first 0.5 s) did not change in latency or amplitude when the platform was pitched. However, the late EMG response (after the first 0.5 s) was found to be higher in amplitude than that obtained preflight.

4.6.2.4 Complex visual, vestibular, and proprioceptive tests

We have mentioned earlier (Section 4.3.3) that frogs hatched in microgravity showed stronger behavioural response to moving visual scenes when tested after their return on Earth than control animals hatched on Earth (Wassersug *et al.* 1993b). Also, vestibular function of six frogs recovered after eight days of space flight showed abnormalities when tested at landing plus two hours: head rotation and body reorientation responses were greater than usual, especially when the animals were exposed to optokinetic stimulus, body rotation, or body tilt (Izumi-Kurotani *et al.* 1994).
After the Spacelab D-1 mission, postural stability of crew members was tested using a tilting room (Von Baumgarten *et al.* 1986). Crew members stood with their feet on an Earth-fixed stabilometer anchored to the floor beneath the tilting room while body sway was measured under conditions of no visual input (eyes closed), conflicting visual-vestibular input (eyes open, room tilted with a sinusoidal motion), normal vision (eyes open, room upright), or reduced somatosensory input (foam rubber placed between the stabilometer and the astronaut's feet). Immediately post-flight (a few hours after landing), two crew members showed an increased reliance on visual feedback for maintenance of upright postural equilibrium; stability was decreased when the room was oscillating and when eyes were closed. By the second day after landing, when three additional subjects were tested, stability under the oscillating room condition was analogous to that observed before flight. On the other hand, postural stability remained impaired for up to five days post-flight when the crew members stood on the foam rubber or closed their eyes.

More recently, a clinical posturography system (Equitest, Neurocom International, Clackamas, OR, USA) has been used to assess the magnitude and recovery time course of post-flight postural instabilities in 23 returning astronauts from six separate Shuttle missions ranging from 5- to 10-day (Paloski *et al.* 1992a & b, **1993**; **Black *et al.* 1994**). This system uses a battery of six sensory organization tests to assess a subject's ability to maintain postural equilibrium under normal and reduced sensory feedback conditions (Paloski *et al.* 1990). The basic paradigm involves measuring hip, shoulder, head, and centre of mass sway over 20 s periods while the subject attempts to maintain a stable upright stance. Sway measurements were made three times under each of the six randomly presented test conditions which included an eyes-open Romberg test, an eyes-closed Romberg test, and four other tests in which vision and/or ankle proprioceptive inputs were selectively eliminated by having the subject close his eyes or by servo-controlling the visual surround and/or support surface to the subject's centre of mass sway. Results showed that overall stability was substantially decreased two to four hours post-flight with all tested crew members demonstrating a post-flight decrease in stability. The recovery time course followed a double exponential path (Paloski *et al.* 1993). An initial rapid improvement in stability during the first eight to ten hours was followed by a more gradual return to preflight stability levels over the next four to eight days. Post-flight postural instability appears to be mediated primarily by alterations in the vestibular (presumably otolithic) feedback loop and, secondarily, by alterations in ankle proprioceptive feedback, at least in some subjects. It also appears that increased reliance on vision may partially compensate for the degraded performance of the other two feedback systems (**Paloski *et al.* 1993**). The increased reliance on vision is demonstrated in Fig. 16, where the ratio of equilibrium scores (inversely proportional to peak sway) obtained with sway-referenced vision to those obtained without vision is plotted for two preflight and five post-flight test sessions. Decreases in this ratio indicate that the subjects are more apt to use visual cues for posture control even though they may be inaccurate.

4.6.2.5 Vestibulo-spinal reflexes

In a number of pre- and post-flight studies using the Hoffmann reflex (H-reflex) and the tendon reflex (T-reflex) techniques, it has been shown that both the alpha and gamma motor systems can be altered by space flight. A generalized hyper-reflexia was reported in the three crew members of the first Skylab mission during the immediate post-flight clinical evaluations (Ross & Hordinsky 1973). In six crew members studied during the Skylab programme, the post-flight T-reflex showed considerable potentiation over preflight baseline values (Baker *et al.* 1977). Before and after the Spacelab-1 mission, the H-reflex showed a potentiation related to the selected drop-to-shock interval (see earlier section of this chapter) for up to seven days post-

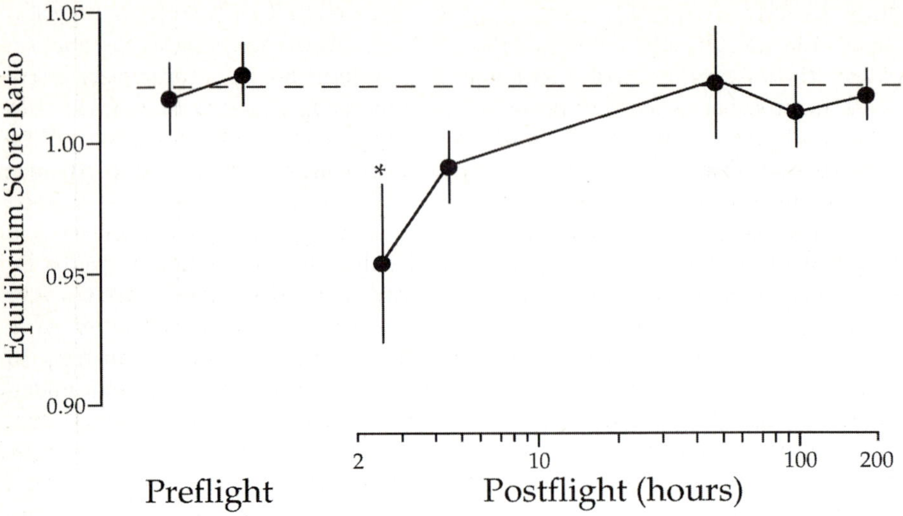

Fig. 16. Ratio of equilibrium scores obtained with sway-referenced vision to those obtained with eyes closed. Each data point on the figure was obtained by averaging the responses of 23 subjects to 4 sensory conditions (mean ± SD; $^*P < 0.05$). Adapted from **Black *et al.* (1994)**.

flight in four crew members (**Reschke *et al.* 1984**). Kozlovskaya *et al.* (1984) demonstrated that two days after landing, the H-reflex in monkeys required a lower stimulus threshold and was potentiated over that observed before flight. More recently, Grigoriev & Yegorov (1990) reported that the T-reflex in crew members who had been in orbit for up to 241 days aboard Mir on the Kvant mission was characterized by a decrease in threshold and a three- to fourfold increase in amplitude over preflight values even on the sixth day post-flight.

Coupled with the gradual decrease in the vestibulo-spinal reflex amplitude observed in-flight (**Reschke *et al.* 1984, 1986; Watt *et al.* 1986**), the post-flight potentiation of the H-reflex and T-reflex suggests response mediation *via* descending vestibular (otolith) pathways, and a reinterpretation of otolith function via adaptation within the central nervous system in response to the stimulus rearrangement of orbital flight.

4.6.3 Pre- and post-flight locomotion studies

General characteristics. Normal gait control depends on the acquisition of pre-programmed patterns of muscle activation and requires the continuous monitoring of external sensory input and internal referent signals. Locomotion pattern generators in the central nervous system are subject to overriding control from higher neural centres (Brooks 1986). Peripheral sensory and internal referent feedback modify patterns of activation emitted by pattern generators to improve ongoing motor performance.

After 30 days in the head-down position, subjects showed an impairment in their posture and gait, evidenced by a decrease in step length, walking velocity and balance stability. By contrast, control subjects who underwent LBNP sessions every day of the bed rest experienced no changes in their gait and balance during the recovery period (**Dupui *et al.* 1992**).

Locomotor behaviour was studied in 14 cosmonauts following missions lasting from 2 to 30 days in the Soyuz spacecraft (Soyuz-9, 12, 13, 14, 15, 16, and 17) using motion picture analysis to determine performance associated with walking, running, standing, long jumps, and high jumps (Bryanov *et al.* 1976). Distinct post-flight performance decrements in gait and jumping behaviour were observed with the duration of the decrements related, in most cases, to the length of the flight. Post-flight gait was modified for 15 to 30 min after two days in space, but was affected for up to two days after flights of six to eight days. This same trend was observed for flights lasting 16 to 18 days. In walking, the cosmonauts place the legs wide apart, with the trunk held to the side of the supporting leg, and the intended path is not maintained. For greater stability, they frequently raise their arms to the side and they walk with small steps of irregular length (Hernandez-Korwo *et al.* 1983).

Head and gaze stability during locomotion. During the performance of various postural and locomotor tasks in 1-g, angular head deviation is maintained with a precision of a few degrees (**Grossman *et al.* 1988**; Berthoz & Pozzo 1988). It has been suggested that postural and gait control mechanisms are designed to stabilize the head during body movement. This concept represents a 'top down' approach to the problem of gait stability and replaces attempts to determine modification of post-flight gait instability using the more classic reactive gait paradigms (Berthoz & Pozzo 1988). The underlying hypothesis is that gait stability is established to maintain head position in space reducing gaze error. Therefore, the maintenance of posture and gait stability is a goal-directed response designed to stabilize the head relative to the Earth's vertical, ensuring gaze stability and the maintenance of visual acuity.

This novel concept was applied to data obtained from the H-reflex experiment flown on Spacelab-1 (**Reschke *et al.* 1984, 1986**). Linear acceleration was provided by a vertical drop of approximately 12 cm. The angle of the head was computed from markers placed on the head. There was approximately twice as much angular deviation of the head three hours after landing than there was before flight (8–10° preflight *vs* 20° post-flight); by the third day post-flight, a strategy had developed which allowed the subject to maintain a stable head position despite the observation that orientation of the trunk and limbs continued to be more variable than that recorded preflight.

In recent experiments, 19 astronauts were asked to walk and run on a motorized treadmill while visually fixating a stationary target positioned in the centre of view (Bloomberg, personal communication, 1994). Tests were conducted 10 to 15 days before launch and two to four hours after landing. Three-dimensional representations of recorded head trajectories during locomotion obtained before and after space flight showed a high repeatability of head motion trajectory during preflight testing. Head motion displayed more variability during locomotion following space flight. There was also considerable variation in the smoothness of the changing head position obtained during post-flight locomotion. These results support the hypothesis that changes in head stability and coordination induced by adaptive modification in 'top down' motor control schemes may be a contributing factor to post-flight locomotor impairment.

4.6.4 Summary

Behavioural responses of vertebrates to sudden exposure to microgravity can be interpreted as manifestations of either escape behaviour or righting reflexes. However, developmental studies are limited. There are critical periods during the development of higher vertebrates which involve the sensory and motor systems; connections between neurons and stabilization of synapses are under the influence of sensory and motor inputs. How the changes in vestibular, proprioceptive and visual systems during development in microgravity affect the postural behaviour of animals and individuals need to be investigated.

Locomotion inside a spacecraft is quickly learned, and this learned behaviour carries over to subsequent flights. Post-flight, most astronauts experience some instability and difficulty in walking, or standing with eyes closed. The recovery is mostly rapid, over several hours. We don't know whether the $1 \times g$ motor control programs in standing and walking are stored or re-learned during return on Earth.

The narrower 'cone of stability' post-flight is associated with rigid standing, and a tendency to round corners wide. Postural instability and illusions seem to be due to an increased reliance upon visual cues, a possible recalibration of the joint angle proprioceptive system, and a reinterpretation of otolith signals. The specific contribution of otolith organ cues (OTTR hypothesis) to posture post-flight is not well understood.

The tendency toward fainting while standing (orthostatic hypotension), which is well documented following return for many astronauts, has not been quantitatively linked to posture.

4.7 References

Anderson DJ, Reschke MF, Homick JL & Werness SAS. Dynamic posture analysis of Spacelab crew members. Exp. Brain Res. 64: 380–91, 1986.

André-Deshays C, Israël I, Berthoz A, Popov K & Lipschitz M. Gaze control and spatial memory in weightlessness. In: *Proceedings of the 4th European Symposium on Life Sciences Research in Space*. Ed. by David V. Paris, France, European Space Agency, pp. 617–21, 1990.

André-Deshays C, Israël I, Charade O, Berthoz A, Popov K & Lipschitz M. Gaze control in microgravity. 1. Saccades, pursuit, eye-head coordination. J. Vestib. Res. 3: 331–44, 1993.

Arrott AP & Young LR. M.I.T./Canadian vestibular experiments on the Spacelab-1 mission: 6. Vestibular reactions to lateral acceleration following ten days of weightlessness. Exp. Brain Res. 64: 347–57, 1986.

Arrott AP, Young LR & Merfeld DP. Perception of linear acceleration in weightlessness. Aviat. Space Environ. Med. 61: 319–26, 1990.

Bagian JP. First intramuscular administration in the U.S. Space Program. J. Clin. Pharmacol. 31: 920, 1991.

Baker JT, Nicogossian AE, Hoffler GW, Johnson RL & Hordinsky J. Changes in the achilles tendon reflexes following Skylab missions. In: *Biomedical Results from Skylab* (NASA SP-377). Ed. by Johnston RS & Dietlein LF. U.S. Government Printing Office, Washington DC, pp. 131–5, 1977.

Baloh RW & Honrubia V. *Clinical neurophysiology of the vestibular system*. FA Davis Company, Philadelphia, 1990.

Barany R. Untersuchungen über den vom Vestibularapparat des Ohres reflektorisch ausgelsten rhythmischen Nystagmus und seine Begleiterscheinungen. Monatschr. Ohrenkeilkde 40: 193–207, 1906.

Barnes GR. Visual-vestibular interaction in the coordination of eye and head movements. In: *Progress in Oculomotor Research*. Ed. by Fuchs AF & Becker W. Elsevier, New York, pp. 299–308, 1981.

Barrett RJ.& Lokhandwala MF. Circulatory and vestibular implications of central angiotensin mechanisms in physiological adaptation to weightlessness. Medical Hypotheses 7: 1415–19,1981.

Benke T, Koserenko O & Gerstenbrand F. Cogimir - How to measure cognitive functions in space. In: *Health from Space Research. Austrian Accomplishments*. Ed. by Austrian Society for Aerospace Medicine. Springer-Verlag, Wien, New York, pp. 55–63, 1992.

Benson AJ. Modification of the response to angular accelerations by linear accelerations. In: *Handbook of Sensory Physiology, Volume VI. Vestibular System, Part 2: Psychophysics, Applied Aspects and General Interpretations*. Ed. by Kornhuber HH. Springer-Verlag, Berlin, pp 281–320, 1974.

Benson AJ, Kass JR & Vogel H. European vestibular experiments on the Spacelab-1 mission: 4. Thresholds of perception of whole-body linear oscillation. Exp. Brain Res. 64: 264–271, 1986a.

Benson AJ. & Viéville T. European vestibular experiments on the Spacelab-1 mission: 6. Yaw axis vestibulo-ocular reflex. Exp.Brain Res. 64: 279–83, 1986.

Benson AJ, Von Baumgarten R, Berthoz A, Brandt T, Brand U, Bruzek W, Dichgans J, Kass J, Probst Th, Scherer H, Viéville T, Vogel H & Wetzig J. Some results of the European vestibular experiments in the Spacelab-1 mission. AGARD Proceedings 377: 1B3–1B14, 1986b.

Berger M, Ferstenbrand F, Kozlovskaya IB, Burlatchkova N, Muigg A, Sokolov A, Babaev B, Grill I, Borisov M, DeCol C, Holzmüller G, Hochmair E & Steinwender G. Eye, head and arm coordination and spinal reflexes in weightlessness - Monimir experiment. In: *Health from Space Research. Austrian Accomplishments*. Ed. by Austrian Society for Aerospace Medicine. Springer-Verlag, Wien, New York, pp 119–135, 1992.

Berthoz A, Brandt T, Dichgans J, Probst T, Bruzek W, Viéville T. European vestibular experiments on the Spacelab-1 mission: 5. Contribution of the otoliths to the vertical vestibulo-ocular reflex. Exp Brain Res. 64: 272–278, 1986

Berthoz A & Pozzo T. Intermittent head stabilization during postural and locomotory tasks in humans. In: *Posture and Gait: Development, Adaptation and Modulation.* Ed. by Amblard B, Berthoz A & Clarac F. Elsevier, Amsterdam, pp. 189–98, 1988.

Black FO, Paloski WH, Doxey-Gasway DD & Reschke M. Vestibular plasticity following space flight: recovery from post-flight postural instability. Acta Otolaryngol. (Stockh.) in press, 1994.

Bles W & Kapteyn TS. Separate recording of the movements of the human eyes during parallel swing tests. Acta Otolaryngol. 75: 6–9, 1973.

Bracchi F, Gualtierotti T, Morabito A & Rocca E. Multiday recordings from the primary neurons of the stratoreceptors of the labyrinth of the bullfrog. Acta Otolaryngol. (Stockh.) suppl. 334:3–27, 1975.

Briegleb WJ, Neubert J, Schatz A & Kruse B. Light microscopic analysis of the gravireceptor in *Xenopus* larvae developed in microgravity. Adv. Space Res. 9:241–244, 1988.

Brooks VB, Locomotion. In: *The Neural Basis of Motor Control.* Oxford University Press, New York, pp 181–194, 1986.

Bryanov II, Yemel'yanov MD, Matveyev AD, Mantsev EI, Tarasov IK, Yakovleva IY, Kakurin LI, Kozerenko OP, Myasnikov VI, Yeremin AV, Pervushin VI, Cherepakhin MA, Purakhin YN, Rudometkin NM & Chekidra IV. Characteristics of statokinetic reactions. In: *Space Flights in the Soyuz Spacecraft: Biomedical Research.* Ed. by Gazenko OG, Kakurin LI & Kuznetsov AG. Leo Kanner Associates, Redwood City, CA. [Translation of *Kosmicheskiye Polety na Korablyakh 'Soyuz' Biomeditsinskiye Issledovaniya.* Nauka Press, Moscow], 1976.

Calhoun GL, Valencia G & Furness TA. Three-dimensional auditory cue simulation for crew station design/evaluation. Proc. Hum. Fac. Soc. 31: 1398–1402, 1987.

Clarke AH, Scherer H & Gundlach P. Caloric stimulation during short episodes of microgravity. Arch. Otorhinolaryngol. 245 (suppl.): 175–9, 1988a.

Clarke AH, Scherer H & Schleibinger J. Body position and caloric nystagmus response. Acta Otolaryngol. (Stockh.) 106: 339–47, 1988b.

Clarke AH, Teiwes W & Scherer H. Evaluation of the torsional VOR in weightlessness. J. Vestib. Res. 3: 207–18, 1993a.

Clarke AH, Teiwes W & Scherer H. Vestibulo-oculomotor testing during the course of a spaceflight mission. Clin. Investig. 71: 740–8, 1993b.

Clarke AH, Teiwes W & Scherer H. Reflex eye movements during head oscillations around each of the three orthogonal axes in weightlessness. *Abstracts, 5th European Symposium on Life Sciences Research in Space.* CNES/ESA, 26 September–1 October 1993, Arcachon, France, 1993.

Clément G & André-Deshays C. Motor activity and visually induced postural reactions during two-g and zero-g phases of parabolic flight. Neurosci. Lett. 79: 113–6, 1987.

Clément G, André-Deshays C & Lathan CE. Effects of gravitoinertial force variations on vertical gaze direction during oculomotor reflexes and visual fixation. Aviat. Space Environ. Med. 60: 1194–8, 1989.

Clément G & Berthoz A. Vestibulo-ocular reflex and optokinetic nystagmus in microgravity. Adv. Oto-Rhino-Laryng. 42: 1–4, 1988.

Clément G, Berthoz A, Cross-coupling between horizontal and vertical eye movements during optokinetic nystagmus and optokinetic after-nystagmus elicited in microgravity. Acta Otolaryngol (Stockh.) 109: 179–187, 1990.

Clément G, Berthoz A & Lestienne F. Adaptive changes in perception of body orientation and mental image rotation in microgravity. Aviat. Space Environ. Med. 58: A159–A163, 1987.

Clément G, Gurfinkel VS, Lestienne F, Lipshits MI & Popov KE. Adaptation of posture control to weightlessness. Exp. Brain Res. 57: 61–72, 1984.
Clément G, Gurfinkel VS, Lestienne F, Lipshits MI & Popov KE. Changes of posture during transient perturbations in microgravity. Aviat. Space Environ. Med. 56: 666–71, 1985.
Clément G & Lathan CE. Effects of static tilt about the roll axis on horizontal and vertical optokinetic nystagmus and optokinetic after-nystagmus in humans. Exp. Brain Res. 84: 335–41, 1991.
Clément G & Lestienne F. Adaptive modifications of postural attitude in conditions of weightlessness. Exp. Brain Res. 72: 381–9, 1988.
Clément G, Petropoulos A, Darlot C & Berthoz A. Eye movements and motion percepion during off-vertical axis rotation (OVAR) at small angles of tilt after space flight. Acta OtoLaryngol. (Stockh.), in press, 1995.
Clément G, Popov KE & Berthoz A. Effects of prolonged weightlessness on human horizontal and vertical optokinetic nystagmus and optokinetic after-nystagmus. Exp. Brain Res. 94: 456–62, 1993.
Clément G, Reschke MF, Verrett CM & Wood SJ. Effects of gravitoinertial force variations on optokinetic nystagmus and on perception of visual stimulus orientation. Aviat. Space Environ. Med. 63: 771–7, 1992b.
Clément G, Viéville T, Lestienne F & Berthoz A. Modifications of gain asymmetry and beating field of vertical optokinetic nystagmus in microgravity. Neurosci. Lett. 63: 271–4, 1986.
Clément G, Wood SJ & Reschke MF. Effects of microgravity on the interaction of vestibular and optokinetic nystagmus in the vertical plane. Aviat. Space Environ. Med. 63: 778–84, 1992a.
Cohen B, Kozlovskaya I, Raphan T, Sirota MG, Yakushin SB & Beloozerova IN. Vestibulo-ocular reflex of rhesus monkey after space flight. J. Appl. Physiol. 73 (Suppl. 2): 121S–31S, 1992.
Cohen B, Matsuo V & Raphan T. Quantitative analysis of the velocity characteristics of optokinetic nystagmus (OKN) and optokinetic after nystagmus (OKAN). J. Physiol. (Lond.) 270: 321–44, 1977.
Corballis MC, Zbrodoff NJ, Shetzer LI & Butler PB. Decisions about identity and orientation of rotated letters and digits. Mem. Cognit. 6: 98–107, 1978.
Correia MJ, Perachio AA & Dickman JD. Changes in monkey horizontal semicircular afferent responses following space flight. J. Appl. Physiol. 73 (Suppl 2), 112S–20S, 1992.
Cowings PS, Suter S, Toscano WB, Kamiya J & Naifeh K. General autonomic components of motion sickness. Psychophysiology 23: 542–51, 1986.
Dai M, Cohen B, McGarvie L, Kozlovskaya I, Sirota M & Raphan T. Reduction of ocular counter-rolling by adaptation to space. In: *Cosmos Biosatellite 2229 Studies, Final Science Report*, NASA, 1993a.
Dai M, Cohen B, Raphan T, McGarvie L, Kozlovskaya I, Reduction of ocular counter-rolling (OCR) after space flight. Soc. Neurosci. Abstr. 19, 1993b.
Dai MJ, Raphan T & Cohen B. Spatial orientation of the vestibular system: dependence of optokinetic after-nystagmus on gravity. J. Neurophysiol. 66: 1422–39, 1991.
Davis JR, Vanderploeg JM, Santy PA, Jennings RT & Stewart DF. Space motion sickness during 24 flights of the Space Shuttle. Aviat. Space Environ. Med. 59: 1185–9, 1988.
Diamond SG & Markham CH. Ocular torsion in upright and tilted positions during hypo- and hypergravity of parabolic flight. Aviat. Space Environ. Med. 59: 1158–62, 1988.
Diamond SG & Markham CH. Prediction of space motion sickness susceptibility by disconjugate eye torsion in parabolic flight. Aviat. Space Environ. Med. 62: 201–5, 1991.

Diamond SG, Markham CH & Money KE. Instability of ocular torsion in zero gravity: possible implications for space motion sickness. Aviat. Space Environ. Med. 61: 899–905, 1990.

Dichgans J & Brandt T. Visual vestibular interaction: effects on self-motion perception and in postural control. In: *Handbook of Sensory Physiology. Vol 8 Perception.* Ed. by Held R, Leibowitz H & Teuber HL, Berlin, Springer-Verlag, pp. 755–804, 1978.

DiZio P & Lackner JR. The effects of gravitoinertial force level and head movements on post-rotational nystagmus and illusory after-rotation. Exp. Brain Res. 70:485–495, 1988.

DiZio P & Lackner JR. Influence of gravito-inertial force level on vestibular and visual velocity storage in yaw and pitch, Vision Research 32: 123–45, 1992.

DiZio P, Lackner JR & Evanoff JN. The influence of gravitoinertial force level on oculomotor and perceptual responses to sudden stop stimulation. Aviat. Space Environ. Med. 58: A224–30, 1987.

Draeger J, Wirt H, Schwartz R & Tomex R. Messung Des Augeninnendrucks Unter MG-Bedingungen. Naturwiss. 73: 450–2, 1986.

Duntley SQ, Austin RW, Taylor JH & Harris JL. Experiment S-8/D-13 visual acuity and astronaut visibility. In: *Gemini Mid-program Conference* (NASA SP-121). National Aeronautics and Space Administration, Washington DC, 1966.

Dupui P, Montoya R, Costes-Salon MC, Séverac A & Güell A. Balance and gait analysis after 30 days -6° bed rest: influence of lower-body-negative-pressure sessions. Aviat. Space Environ. Med. 63: 1004–10, 1992.

Fernandez C & Goldberg JM. Physiology of peripheral neurons innervating otolith organs of the squirrel monkey. III. Response dynamics. J. Neurophysiol. 39: 996–1008, 1976.

Friederici AD & Levelt WJM. Resolving perceptual conflicts: the cognitive mechanisms of spatial orientation. Aviat. Space Environ. Med. 58: A164–9, 1987.

Galambos RG & Hecox K. Clinical applications of the auditory brainstem response. Otolaryngol. Clin. North Am. 11: 709–22, 1978.

Gallasch E, Moser M, Kenner T, Kozlovskaya I, Konev A & Borisov M. Experiment Mikrovib -- Investigation of tremors in microgravity. In: *Health from Space Research. Austrian Accomplishments.* Ed. by Austrian Society for Aerospace Medicine, Springer-Verlag, Wien & New York, pp. 85–105, 1992.

Gallasch E, Kozlovskaya I, Konev A, Rafolt D, Löscher W, Moser M, Hindinger J, Eder H & Kenner T. Results of one short and two long-term flight on human physiological tremor and microvibration. In: *Proceedings of the 5th European Symposium on Life Sciences Research in Space.* Ed. by Oser H, Guyenne TD. Noordwijk, The Netherlands, ESA, ESTEC, pp. 361–6, 1994.

Gibson EG. Skylab 4 crew observations. In: *Biomedical Results from Skylab* (NASA SP-377). Ed. by Johnston RS & Dietlein LF, U.S. Government Printing Office, Washington DC, pp. 22–6, 1977.

Ginsburg AP & Vanderploeg J. Vision in space: near vision acuity and contrast sensitivity. In: *Space Shuttle Medical DSOs: NASA Interim Report*, Johnson Space Centre, Houston, TX, 1986.

Glasauer S, Reschke MF, Berthoz A, Michaud L & Huebner W. The effect of space flight on gaze control strategy. In: *Proceedings of the 5th European Symposium on Life Sciences Research in Space.* Ed. by Oser H, Guyenne TD. Noordwijk, The Netherlands, ESA, ESTEC, pp. 339–44, 1994.

Gorgiladze GI & Bryanov II. Space motion sickness. Kosm. Biol. Aviakosm. Med. 23: 4–14, 1989.

Gorgiladze GI & Maveyev AD. The effect of weightlessness on eye movement responses. In: *Space Biology and Aerospace Medicine: IXth All-Union Conference*, Kaluga, 19–21 June 1990, pp. 51–52, 1990

Gorgiladze GI, Samarin GI & Bryanov II. Interlabyrinthine asymmetry, vestibular dysfunction and space motion sickness. Kosm. Biol. Aviakosm. Med. 20: 19–31, 1986.

Graybiel A. Prevention and treatment of space sickness in shuttle-orbiter missions. Aviat. Space Environ. Med. 50: 171–6, 1979.

Graybiel A. Space motion sickness: Skylab revisited. Aviat. Space Environ. Med. 51: 814–22, 1980.

Graybiel A, Miller EF, Billingham J, Waite R, Berry C & Dietlein LF. Vestibular experiments in Gemini flights V and VII. Aerospace Med. 38: 360–70, 1967.

Graybiel A, Miller EF & Homick JL. Experiment M-131: human vestibular function. 1. Susceptibility to motion-sickness. In: *Proceedings of the Skylab Life Sciences Symposium*, Vol. 1 (TMX-58154), Houston, TX, Johnson Space Centre, pp. 169–98, 1974.

Graybiel A, Miller EF & Homick JL, Experiment M131. Human vestibular function. In: *Biomedical Results from Skylab* (NASA SP-377). Ed. by Johnston RS & Dietlein LF, U.S. Government Printing Office, Washington DC, pp. 74–103, 1977.

Graybiel A, Wood CD, Miller EF & Cramer DB. Diagnostic criteria for grading the severity of acute motion sickness. Aerospace Med. 39: 453–5, 1968.

Griffy C, Watanabe S, Takabayashi A & Satake H. Fish neurovestibular response to changing gravity vectors. Paper presented at the *10th Man in Space Symposium*, Tokyo, April 19–23, 1993.

Grigoriev AI & Yegorov AD. Preliminary medical results of the 180-day flight of prime crew 6 on Space Station MIR. Fourth Meeting of the US/USSR Joint Working Group on Space Biology and Medicine, San Francisco, CA, 16–22 September 1990.

Grigoriev A, Gazenko OG, Kozlovskaya IB. Barmin VA & Kreidich Y. Vestibulo-cerebellar regulation of oculomotor reactions in microgravity conditions. In: *Adaptive Processes in Visual and Oculomotor Systems*. Ed. by Keller EL & Zee DS, Pergamon Press, New York, 1986.

Grigoriev A, Bugrov SA, Bogomolov VV, Egorov AD, Kozlovskaya IB, Pestov, ID, Polyakov VV & Tarasov IK. Medical results of the MIR year-long mission. Physiologist 34: S44–48, 1991.

Grossman GE, Leigh RJ, Abel LA, Lanska DJ & Thurston SE. Frequency and velocity of rotational head perturbations during locomotion. Exp. Brain Res. 70: 470–6, 1988.

Gualtierotti T. Vestibular integrated function and microgravity. In: *Proceedings of the 3rd ESA European Symposium on Life Sciences Research in Space*, Graz, Austria. Ed. by Hunt H & Oser H, ESA SP-271, ESA, Paris, pp. 227–32, 1987.

Gurfinkel VS, Lipshits MI & Popov KE. Stabilization of body position as the main task of postural regulation. Fiziol. Cheloveka 7: 400–10, 1981.

Gurfinkel VS, Lestienne F, Levick YS, Popov KE & Lefort L. The reference systems for tactile perception and for the orientation with respect to body. In: *Proceedings of the 4th European Symposium on Life Science Research in Space*. Ed. by David V. Noordwijk, European Space Agency, ESTEC, pp. 611–15, 1990.

Gurfinkel VS, Lestienne F, Levik YS, Popov KE & Lefort L. Egocentric references and human spatial orientation in microgravity. II. Body-centred coordinates in the task of drawing ellipses with prescribed orientation. Exp. Brain Res. 95: 343–8, 1993a.

Gurfinkel VS, Lestienne F, Levik YS & Popov KE. Egocentric references and human spatial orientation in microgravity. I. Perception of complex tactile stimuli. Exp. Brain Res. 95: 339–42, 1993b.

Halmagyi GM & Gresty MS. Clinical signs of visual-vestibular interaction. J. Neurol. Neurosurg. Psychiatry 42: 934–9, 1979.

Harm DL. Physiology of motion sickness symptoms. In: *Motion and Space Sickness*. Ed. by Crampton GH, CRC Press, Boca Raton, pp. 153–77, 1990.

Harm DL, Bloomberg JJ, Reschke MF, Skinner NC, Liao CA, Michaud LJ & Parker DE. Adaptive modification of gaze control following tilt reinterpretation. Aviat. Space Environ. Med. 62: 477, 1991.

Harm DL & Parker DE. Perceived self-orientation and self-motion in microgravity, after landing and during preflight adaptation training. J. Vestib. Res. 3: 297–305, 1993.

Harm DL & Parker DE. Preflight adaptation training for spatial orientation and space motion sickness. J. Clin. Pharmacol. 34: 618–27, 1994.

Hawkins WR & Zieglschmid JF. Clinical aspects of crew health. In: *Biomedical Results of Apollo* (NASA SP-368). Ed. by Johnston RS, Dietlein LF & Berry CA, U.S. Government Printing Office, Washington DC, pp. 43–81, 1975.

Heidelbaugh ND, Wescott DE, Kare MR, Bourland CT, Smith MC & Rapp R. Taste and aroma testing. In: *Skylab-4 Preliminary Biomedical Report* (JSC 08818), NASA Johnson Space Centre, pp. 2/123 - 2/134, 1975.

Hernandez-Korwo R, Kozlovskaya IB, Kreydich Y, Martinez-Fernandez S, Rakhmanov AS, Fernandez-Pone E & Minenko VA. Effect of seven-day spaceflight on structure and function of human locomotor system. Kosm. Biol. Aviakosm. Med. 17: 37–44, 1983.

Hertwig I & Hentschel J. Vestibular morphology of *Xenopus laevis* (Amphibian, Anura) following larval development in zero gravity (Space shuttle, D-1 Mission). Zool. Jb. Anat. 118:463–72, 1989.

Hoffman RA, Pinsky LS, Osborne WZ & Bailey JV. Visual light flash observations on Skylab 4. In: *Biomedical Results from Skylab* (NASA SP-377). Ed. by Johnston RS & Dietlein LF, U.S. Government Printing Office, Washington DC, pp. 127–30, 1977.

Hofstetter-Degen K, Wetzig J & Von Baumgarten RJ. Oculovestibular interactions under microgravity. Clin. Investig. 71: 749–56, 1993.

Homick JL. Motion sickness: general background and methods. In: *Proceedings of the Space Adaptation Syndrome Drug Workshop*, July 1983. Ed. by Kohl RL, Space Biomedical Research Institute, USRA Division of Space Biomedicine, Houston, TX, pp. 9–19, 1985.

Homick JL & Miller EF II. Apollo flight crew vestibular assessment. In: *Biomedical Results of Apollo* (NASA SP-368). Ed. by Johnston RS, Dietlein LF & Berry CA, U.S. Government Printing Office, Washington DC, pp. 323–40, 1975.

Homick JL & Reschke MF. Postural equilibrium following exposure to weightless space flight. Acta Otolaryngol. 83: 455–64, 1977.

Homick JL, Reschke MF & Miller EF. Effects of prolonged exposure to weightlessness on postural equilibrium. In: *Biomedical Results from Skylab* (NASA SP-377). Ed. by Johnston RS & Dietlein LF, U.S. Government Printing Office, Washington DC, pp. 104–12, 1977.

Homick JL, Reschke MF & Vanderploeg JM. Space adaptation syndrome. Incidence and operational implications for the Space Transportation System Program. In: *Advisory Group for Aerospace Research and Development Conference Proceedings on Motion Sickness: Mechanisms, Prediction, Prevention, and Treatment.* Advisory Group for Aerospace Research and Development, Williamsburg, VA, 1984.

Homick JL & Vanderploeg JM. The neurovestibular system. In: *Space Physiology and Medicine.* Ed. by Nicogossian AE, Huntoon CL & Pool SL, Lea and Febiger, Philadelphia, pp. 154–66, 1989.

Howard IP & Templeton WB. *Human Spatial Orientation.* John Wiley and Sons, London, 1966.

Israel I, André-Deshays C, Charade O, Berthoz A, Popov KE & Lipshits MI. Gaze control in microgravity. 2. Sequences of saccades toward memorized visual targets. J. Vestib. Res. 3: 345–60, 1993.

Izumi-Kurotani A, Wassersug RJ, Yamashita M, Naitoh M & Nagaoka S. Frog behaviour under microgravity. Paper presented at the 9th ISAS Space Utilization Symposium, pp 112–114, 1992.

Izumi-Kurotani A, Yamashita M, Kawasaki Y, Kurotani T, Mogami Y, Okuno M, Oketa A, Shiraishi A, Ueda K, Wassersug RJ & Naitoh T. Behaviour of Japanese tree frogs under microgravity on Mir and in parabolic flight. Adv. Space Res., in press, 1995.

Jongkees LBW & Phillipszoon AT. Nystagmus provoked by linear accelerations. Acta Physiol. Pharmacol. Neerl. 10: 238–47, 1962.

Kass JR, Bruzek W, Probst Th, Thümler R, Viéville T & Vogel H. European vestibular experiments on the Spacelab-1 mission. 2. Experimental equipment and methods. Exp. Brain Res. 64: 247–54, 1986.

Kass JR & Vogel H, Subjective vertical before and after space flight. Adv. Space Res. 6: 171–4, 1986.

Kass JR, Von Baumgarten RJ, Berthoz A, Brandt T, Brand U, Bruzek W, Dichgans J, Probst T, Scherer H, Viéville T, Vogel H & Wetzig J, The European vestibular experiments in Spacelab-1. Adv. Space Res. 4: 3–9, 1984.

Kass JR, Vogel H & Von Baumgarten RJ. Neck receptor stimulation in 0- and 1-G. In: *Proceedings of the 3rd ESA European Symposium on Life Sciences Research in Space*, Ed. by Hunt H & Oser H, ESA SP-271, ESA, Paris, pp. 233–236, 1987.

Kenyon RV & Young LR, MIT/Canadian vestibular experiments on Spacelab-1 mission: 5. Postural responses following exposure to weightlessness. Exp. Brain Res. 64: 335–46, 1986.

Khilov KL. In: *Sb. trudov tsentral'noy nauchnoy psekhofiyiologicheskoy laboratorii po izucheniyu letnogo truda grayhdanskogo vozdushnogo flota SSSR*. [Collection of papers by the central Psychophysiological Laboratory on the study of the activities of the Civil airforce of the USSR]. 1: 5–72, 1936.

Kirsch K, Itaenel F, Rocker L & Wicke HJ. Venous pressure in microgravity. Naturwiss. 73: 447–9, 1986.

Kornilova LN, Bod G & Kaspransky RR. Sensory interaction in weightlessness. Physiologist 30: S85–9, 1987.

Kornilova LN, Goncharenko AM, Bod G, Elkn K, Grigorova V & Manev A. Pathogenesis of sensory disorders in microgravity. Physiologist 34: S36–39, 1991.

Kornilova LN, Grigorova V & Bod G. Vestibular function and sensory interaction in space flight. In: *Abstracts of the XVIIth Brny Society Meeting*, Prague, Czechoslovakia, 1- 5 June 1992, pp. 189–190, 1992.

Kornilova LN, Grigorova V & Bod G. Vestibular function and sensory interaction in space flight. J. Vestib. Res. 3: 219–30, 1993.

Kornilova LN, Kaspransky RR, Vestibular function and sensory interaction after spaceflight. In: *Proceedings of the 5th European Symposium on Life Sciences Research in Space*. Ed. by Oser H, Guyenne TD. Noordwijk, The Netherlands, ESA, ESTEC, pp. 345–350, 1994.

Kornilova LN, Yakovleva IY, Tarasov IK & Gorgiladze GI. Vestibular dysfunction in cosmonauts during adaptation to zero-g and readaptation to 1 g. Physiologist 26: S35–S40, 1983.

Kozlovskaya IB, Babaev BM, Barmin VA, Beloozerova II, Kreidich Y & Sirota MG. The effect of weightlessness on motor and vestibulo-motor reactions. Physiologist 27: S111–S4, 1984.

Kozlovskaya IB, Babaev BM, Borisov ME, Burlachkova NI, Sokolov A, Gerstenbrand F & Berger M. Results of sensorimotor studies performed on board of Mir station. *Paper presented at the 45th Congress of the International Astronautical Federation*, Jerusalem, Israel, October 9–14, 1994.

Jerusalem, I, Lackner JR & Graybiel A. Head movements elicit motion sickness during exposure to microgravity and macrogravity acceleration levels. In: *Vestibular and Visual Control on Posture and Locomotor Equilibrium*. Ed. by Igarashi M & Black FO, Karger, Basel, Switzerland, pp. 170–6, 1985.

Lackner JR & Graybiel A. Parabolic flight: Loss of sense orientation. Science 206: 1105–8, 1979.

Lackner JR & Graybiel A, Variations in gravitoinertial force level affect the gain of the vestibulo-ocular reflex: implications for the etiology of space motion sickness. Aviat. Space Environ. Med. 52: 154–8, 1981.

Lackner JR & Dizio P. Gravitoinertial force level affects the appreciation of limb position during muscle vibration. Brain Res. 592: 175–80, 1992.

Lackner JR & Dizio P. Multisensory, cognitive, and motor influences on human spatial orientation in weightlessness. J. Vestib. Res. 3: 361–72, 1993.

Lackner JR, Graybiel A, Johnson WH & Money KE. Asymmetric otolith function and increased susceptibility to motion sickness during exposure to variations in gravitoinertial acceleration level. Aviat. Space Environ. Med. 58: 652–57, 1987.

Lapayev EV & Vorobyev OA. The problem of vestibular physiology in aerospace medicine and prospects for its solution. In: *Space Biology and Aerospace Medicine*: abstracts of papers delivered at the Eighth All-Union Conference, Kaluga, 25–27 June 1986. Ed. by Gazenko OG, Nauka, Moscow, pp. 85–6, 1986.

Lazarev A, Kovalenok VV, Ivanchenkov A & Avakyan SV. *Atmosphere of Earth from Salyut-6*. [Translation of *Atmosfera Zemli s Salyuta-6*]. Gidrometeoizdat, Leningrad, 1981.

Leigh RJ & Zee DS. *The Neurology of Eye Movements* (2nd edition). FA Davis Company, Philadelphia, 1991.

Léone G, Lipshits MI, Matsakis Y, Gurfinkel V & Berthoz A. Influence of weightlessness upon mental rotation and detection of bilateral symmetry. In: *Proceedings of the 5th European Symposium on Life Sciences Research in Space*. Ed. by Oser H, Guyenne TD. Noordwijk, The Netherlands, ESA, ESTEC, pp. 387–393, 1994.

Linder BJ & Trick GL. Simulation of space flight with whole-body head-down tilt: influence on intraocular pressure and retinocortical processing. *Space Shuttle Medical DSOs*: NASA Interim Report, Johnson Space Centre, Houston, TX, 1986.

Lipshits MI, Gurfinkel EV, Matsakis Y & Lestienne F. Microgravity effects on sensorimotor interaction during operator activity: proprioceptive feedbacks. Anonymous Russian publication, pp. 26–30, 1992.

Lipshits MI, Gurfinkel EV, McIntyre J, Droulez J, Gurfinkel VS & Berthoz A, Influence of weightlessness on haptic perception. In: *Proceedings of the 5th European Symposium on Life Sciences Research in Space*. Ed. by Oser H, Guyenne TD. Noordwijk, The Netherlands, ESA, ESTEC, pp. 367–71, 1994.

Mader TH, Intraocular pressure in microgravity. J. Clin. Pharmacol. 31: 947–50, 1991.

Manzey D, Lorenz B, Schiewe A, Finell G & Thiele G. Behavioral aspects of human adaptation to space: analyses of cognitive and psychomotor performance in space during an 8-day space mission. Clin. Investig. 71: 725–31, 1993.

Marcus JT. Influence of gravito-inertial force on vestibular nystagmus in man. TNO Institute for Perception Report No. IZF 1989–24, 1989.

Marcus JT. Otolith responses in man during parabolic flight. Exp. Brain Res. 96: 328–34, 1993.

Marcus JT & Van Holden MD. Vestibulo-ocular response in man to +Gz hypergravity. Aviat. Space Environ. Med. 61: 631–5, 1990.

Markham CH & Diamond SG. A predictive test for space motion sickness. J. Vestib. Res. 3: 289–96, 1993.

Massion J, Gurfinkel V, Lipshits M, Obadia A & Popov K. Strategy and synergy: two levels of equilibrium control during movement. Effects of microgravity. C.R. Acad. Sci. III 314: 87–92, 1992.

Massion J, Gurfinkel V, Lipshits M, Obadia A & Popov K. Axial synergies under microgravity conditions. J. Vestib. Res: 3: 275–88, 1993.

Matsakis Y, Berthoz A, Lipschits MI & Gurfinkel V. Mental rotation of three-dimensional shapes in microgravity. In: *Proceedings of the 4th European Symposium on Life Science Research in Space*. Ed. by David V. Noordwijk, European Space Agency, ESTEC, pp. 625–9, 1990.

Matsakis Y, Lipshits MI, Gurfinkel V & Berthoz A. Effects of prolonged weightlessness on mental rotation of three-dimensional objects. Exp. Brain Res. 94: 152–62, 1993.

Matsnev EI, Yakovleva IY, Tarasov IK, Alekseev VN, Kornilova LN, Mateev AD & Gorgiladze GI. Space motion sickness: phenomenology, countermeasures, and mechanisms. Aviat. Space Environ. Med. 54: 312–7, 1983.

Matsuo V & Cohen B, Vertical optokinetic nystagmus and vestibular nystagmus in the monkey: up-down asymmetry and effects of gravity. Exp. Brain Res. 53: 197–216, 1984.

McGrath BJ. *Human vestibular response during 3 Gz centrifuge stimulation.* MSc Thesis. Man-Vehicle Laboratory, MIT Cambridge MA, 1990.

Melvill-Jones G. Adaptive neurobiology in space flight. In: *Proceedings of the Skylab Life Sciences Symposium* (JSC-09275), Lyndon B. Johnson Space Centre, Houston, TX, 1974.

Melvill-Jones G & Watt DGD. Muscular control of landing from unexpected falls in man. J. Physiol. 219: 729-41, 1971.

Mittelstaedt H. In search of the causes of the 'inversion illusions'. Paper presented at the *7th International Man in Space Symposium*, Houston, TX, Feb. 10-13, 1986a.

Mittelstaedt H. In-flight and post-flight results on the cessation of inversion illusions and space sickness. In: *Proceedings of the Norderny symposium on the scientific results of the German Spacelab mission D-1.* Ed. by Sahm PR, Jansen R & Keller MH. Köln, Germany, DFVLR, pp. 525-36, 1986b.

Mittelstaedt H. The role of the pitched-up orientation of the otoliths in two recent models of the subjective vertical. Biol. Cybern. 61: 405-16, 1989.

Mittelstaedt H. Somatic versus vestibular gravity reception in man. Ann. N.Y. Acad. Sci. 656: 124-39, 1992.

Mittelstaedt H & Glasauer S. Illusions of verticality in weightlessness. Clin. Investig. 71: 732-9, 1993a.

Mittelstaedt H & Glasauer S. Crucial effects of weightlessness on human orientation. J. Vestib. Res. 3: 307-14, 1993b.

Money KE. Motion sickness. Physiol. Rev. 50: 1-39, 1970.

Mori S, Mitarai G, Takagi S, Takabayashi A, Usui S, Nakamura T, Sakakibara M, Nagatomo M & Von Baumgarten RJ. Space experiment using large-sized fish: in case of carp in Spacelab-J mission. Acta Astronautica 33: 41-7, 1994.

Mueller C, Kornilova L, Wiest G & Deecke L. Optovert - vertical optokinetic stimulation used to examine central orientational processes. In: *Health from Space Research. Austrian Accomplishments.* Ed. by Austrian Society for Aerospace Medicine, Springer-Verlag, Wien & New York, pp. 155-65, 1992.

Mueller C, Kornilova L, Wiest G, Steinhoff N & Deecke L. Results from vertical vection experiments in short and long term space flights. In: *Proceedings of the 5th European Symposium on Life Sciences Research in Space.* Ed. by Oser H, Guyenne TD. Noordwijk, The Netherlands, ESA, ESTEC, pp. 373-6, 1994.

Oman CM. Space motion sickness and vestibular experiments in Spacelab. SAE Technical Paper Series (820833). Society of Automotive Engineers, Warrendale, 1982.

Oman CM. The role of static visual orientation cues in the etiology of space motion sickness. In: *Symposium on Vestibular Organs and Altered Environment.* Ed. by M Igarashi & KG Nute, NASA/SBRI, USRA/DSB, Houston, pp. 25-38, 1988.

Oman CM & Kulbaski M. Spaceflight affects the 1-g postrotatory vestibulo-ocular reflex. Adv. Oto-Rhino-Laryng. 42: 5-8, 1988.

Oman CM, Lichtenberg BK & Money KE. Space motion sickness monitoring experiment: Spacelab 1. In: *Motion and Space Sickness.* Ed. by Crampton GH, CRC Press, Boca Raton, pp. 217-46, 1990.

Oman CM & Weigl H. Post-flight vestibulo-ocular reflex changes in Space Shuttle/ Spacelab D-1 crew. Aviat. Space Environ. Med. 60: 480, 1989.

Oman CM & Balkwill MD. Horizontal angular VOR, nystagmus dumping, and sensation duration in Spacelab SLS-1 crewmembers. J. Vestib. Res. 3: 315-30, 1993.

Oosterveld WJ & Greven AJ. Flight behaviour of pigeons during weightlessness. Acta Otolaryngol. 79: 233-41, 1975.

Paloski WH, Black FO, Reschke MF, Calkins DS & Shupert C. Vestibular ataxia following shuttle flights: effects of microgravity on otolith-mediated sensorimotor control of posture. Am. J. Otology 14: 9-17, 1993.

Paloski WH, Harm DL, Reschke MF, Doxey DD, Skinner NC, Michaud LJ & Parker DE. Postural changes following sensory reinterpretation as an analog to spaceflight. In: *Proceedings of the 4th European Symposium on Life Science Research in Space.* Ed. by David V. Noordwijk, European Space Agency, ESTEC, pp. 175-8, 1990.

Paloski WH, Reschke MF, Black FO, Doxey DD & Harm DL. Recovery of postural equilibrium control following space flight. In: *Sensing and Controlling Motion: Vestibular and Sensorimotor Function.* Ed. by Cohen B, Tomko DL & Guedry F, published in Ann. N.Y. Acad. Sci. 682: 747–54, 1992a.

Paloski WH, Reschke MF, Doxey DD & Black FO. Neurosensory adaptation associated with postural ataxia following spaceflight. In: *Posture and Gait: Control Mechanisms.* Ed. by Woolacott M & Horak F, University of Oregon Press, Eugene, OR, pp. 311–5, 1992b.

Paloski WH, Reschke MF, Doxey DD & Black FO. Spaceflight-induced changes in posture and locomotion, In: *Proceedings of the 14th Congress of the International Society of Biomechanics,* Paris, 1993.

Parin VV & Yemel'yanov MD. *Physiology of the Vestibular Analyzer,* NASA Washington DC, NASA TT-F-616, 1970.

Parker DE & Harm DL. Mental rotation: a key to mitigation of motion sickness in the virtual environment. Presence 1: 329–33, 1992.

Parker DE & Parker KL. Adaptation to the simulated stimulus rearrangement of weightlessness. In: *Motion and Space Sickness.* Ed. by Crampton GH, CRC Press, Boca Raton, pp. 247–62, 1990.

Parker DE, Reschke MF, Arrott AP, Homick JL & Lichtenberg BK. Otolith tilt-translation reinterpretation following prolonged weightlessness: implications for preflight training. Aviat. Space Environ. Med. 56: 601–6, 1985.

Parker DE, Reschke MF, Ouyang L, Arrott AP & Lichtenberg BK. Vestibulo-ocular reflex changes following weightlessness and preflight adaptation training. In *Adaptive Processes in Visual and Oculomotor Systems.* Ed. by Keller EL & Zee DS, Pergamon Press, New York, pp. 103–9, 1986.

Parker DE, Tjernström Ö, Ivarsson A, Gulledge WL & Poston RL. Physiological and behavioral effects of tilt-induced body fluid shifts. Aviat. Space Environ. Med. 54: 402–9, 1983.

Persterer A, Berger M, Koppensteiner C, Mueller C, Nefjodova M & Opitz M. Audimir – directional hearing at microgravity. In: *Health from Space Research. Austrian Accomplishments.* Ed. by Austrian Society for Aerospace Medicine, Springer-Verlag, Wien & New York, pp. 21–38, 1992.

Rambaut PC & Johnson PC. Nutrition. In: *Space Physiology and Medicine.* Ed. by Nicogossian AE, Huntoon CL & Pool SL, Lea and Febiger, Philadelphia, pp. 202–213, 1989.

Raphan T & Cohen B. Organizational principles of velocity storage in three dimensions: the effect of gravity on cross-coupling of optokinetic after-nystagmus. In: *The Representation of Three-Dimensional Space in the Vestibular, Oculomotor and Visual Systems.* Ed. by Cohen B & Henn V, published in Ann. N.Y. Acad. Sci. 545: 74–92, 1988.

Raphan T, Dai M, Cohen B, Spatial orientation of the vestibular system. In: *Sensing and Controlling Motion: Vestibular and Sensorimotor Function.* Ed. by Cohen B, Tomko DL & Guedry F, published in Ann. N.Y. Acad. Sci. 656: 140–57, 1992.

Raphan T, Matsuo V & Cohen B. Velocity storage in the vestibulo-ocular reflex (VOR) arc. Exp. Brain Res. 35: 229–48, 1979.

Reason JT & Brand JJ. *Motion Sickness.* Academic Press, London, 1975.

Reschke MF, Homick JL & Anderson DJ. Development of vestibulospinal reflex measurements as a method for the investigation of statolith function during sustained weightlessness. In: *Vestibular and Visual Control on Posture and Locomotor Equilibrium.* Ed. by Igarashi M & Black FO, Karger, Basel, Switzerland, pp. 151–7, 1985.

Reschke MF, Harm DL, Parker DE & Paloski WH. DSO 459: Otolith tilt-translation reinterpretation. In: *Results of Life Sciences DSOs Conducted Aboard the Space Shuttle,* 1988–1990. NASA Report, pp. 33–50, 1991.

Reschke MF & Parker DE. Effects of prolonged weightlessness on self-motion perception and eye movements evoked by roll and pitch. Aviat. Space Environ. Med. 58: A153–8, 1987.

Reschke MF, Anderson DJ & Homick JL. Vestibulospinal reflexes as a function of microgravity. Science 225: 212–4, 1984.

Reschke MF, Anderson DJ & Homick JL. Vestibulo-spinal response modification as determined with the H reflex during the Spacelab 1 flight. Exp. Brain Res. 64: 367–79, 1986.

Reschke MF, Parker DE, Harm DL & Michaud L. Ground-based training for the stimulus rearrangement encountered during space flight. Acta Otolaryngol. (Stockh.) 460: 87–93, 1988.

Reschke MF, Bloomberg JJ, Harm DL & Paloski WH. Space flight and neurovestibular adaptation. J. Clin. Pharmacol. 34: 609–17, 1994.

Roll JP, Popov K, Gilhodes JC, Roll R, Gurfinkel V, Quoniam C & Charade O. Changes in human proprioceptive functions in weightlessness. In: *Proceedings of the 5th European Symposium on Life Sciences Research in Space.* Ed. by Oser H, Guyenne TD. Noordwijk, The Netherlands, ESA, ESTEC, pp. 351–5, 1994.

Roll JP, Popov K, Gurfinkel V, Lipshits M, André-Deshays C, Gilhodes JC & Quoniam C. Body proprioceptive references in weightlessness as studied by muscle tendon vibration. In: *Proceedings of the 4th European Symposium on Life Science Research in Space.* Ed. by David V. Noordwijk, European Space Agency, ESTEC, pp. 43–48, 1990.

Roll JP, Popov K, Gurfinkel V, Lipshits M, André-Deshays C, Gilhodes JC & Quoniam C. Sensorimotor and perceptual function of muscle proprioception in microgravity. J. Vestib. Res. 3: 259–74, 1993.

Ross CE & Hordinsky JR. Crew health. *Skylab 1/2 R+21 day interim report*, NASA JSC 08706, Houston, pp. 26–59, 1973.

Ross HE, Brodie EE & Benson AJ. Mass discrimination during prolonged weightlessness. Science 225: 219–21, 1984.

Ross HE, Brodie EE & Benson AJ. Mass discrimination in weightlessness and readaptation to Earth's gravity. Exp. Brain Res. 64: 358–66, 1986.

Ross HE, Schwartz E & Emmerson P. Mass discrimination in weightlessness improves with arm movements of higher acceleration. Naturwiss. 73: 453–4, 1986.

Ross HE, Schwartz E & Emmerson P. The nature of sensorimotor adaptation to altered G-levels: evidence from mass discrimination. Aviat. Space Environ. Med. 58, Suppl 9: A148–52, 1987.

Ross MD. A study of the effects of space travel on mammalian gravity receptors (Exp. No. 238), *Spacelab Life Sciences 1*, NASA JSC, Houston, NP 120, 1991.

Ross MD. A study of the effects of space travel on mammalian gravity receptors. Space Life Science-1 *180 day Experimental Reports*, NASA Washington DC, 1992.

Ross MD. Morphological changes in rats vestibular system following weightlessness. J. Vestib. Res. 3: 241–51, 1993.

Sakuragi S, Watanabe S, Mori S & Tanaka M. Posture and behaviour of hamsters under conditions of microgravity. Environ. Med. 35: 147–50, 1991.

Santy PA & Bungo MV. Pharmacologic considerations for shuttle astronauts. J. Clin. Pharmacol. 31: 931–3, 1991.

Scheld HW, Baky A, Boyd JF, Eichler VB, Fuller PM; Hoffman RB, Keefe JR, Kuchnow KP, Oppenheimer JM, Salinas GA & Von Baumgarten RJ. Killifish hatching and orientation (Experiment MA-161), *Apollo-Soyuz Test Project Preliminary Science Report*, NASA TM-X58173, Washington DC, pp. 281–305, 1976.

Scherer H, Brandt U, Clarke AH, Merbold U & Parker R. European vestibular experiments on the Spacelab-1 mission. 3. Caloric nystagmus in microgravity. Exp. Brain Res. 64: 255–63, 1986.

Scherer H & Clarke A. The caloric vestibular reaction in space. Physiological considerations. Acta Otolaryngol. (Stockh.) 100: 328–36, 1985.

Scherer H & Clarke AH. Vestibular response to caloric stimulation during orbital flight. In: *Proceedings of the 3rd ESA European Symposium on Life Sciences Research in Space.* Ed. by Hunt H & Oser H, ESA SP-271, ESA, Paris, pp. 237–9, 1987.

Schmitt HH & Reid DJ. *Anecdotal information on space adaptation syndrome.* NASA, Johnson Space Centre, Houston, 1985.

Séverac A. Electrical vestibular stimulation and space motion sickness. Acta Astronautica 28: 401–8, 1992.

Shepard RN & Metzler J. Mental rotation of three-dimensional objects. Science 171: 701–3, 1971.

Suzuki M, Harada Y & Sekitani T. Vestibular end-organ of the frog after the space flight and postural alteration of the neurectomized frog – its morphological and functional resilience. J. Vestib. Res. 3: 253–8, 1993.

Tafforin C, Thon B, Guell A & Campan R. Astronaut behaviour in orbital flight situation: preliminary ethological analysis. Aviat. Space Environ. Med. 60: 949–56, 1989.

Tafforin C & Lambin M. Analysis of the relationships between sensory-motor alterations and behavioral modifications of the human in space flight (Spacelab-1). Aviat. Space Environ. Med. 64: 146–52, 1993.

Takabahashi A, Watanabe S, Mori S, Tanaka M, Sakuragi S & Takagi S. Posture control of goldfish in microgravity. In: *Proceedings of the 17th International Symposium on Space Technology and Science*, Tokyo, pp 2015–20, 1990.

Task HL & Genco LY. Effects of short-term space flight on several vision functions. In: *Space Shuttle Medical DSOs*, NASA Interim Report. Johnson Space Centre, Houston, TX, 1986.

Thornton WE. Anthropometric changes in weightlessness. In: *Anthropometric Source Book*, vol 1, NASA Ref Pub 1024. National Technical Information Service (NTIS) Springfield, Illinois, 1978.

Thornton WE, Biggers WP, Thomas WG, Pool SL & Thagard NE. Electronystagmography and audio potentials in space flight. Laryngoscope 95: 924–32, 1985.

Thornton WE, Linder BJ, Moore TP & Pool SL. Gastrointestinal motility in space motion sickness. Aviat. Space Environ. Med. 58: A16–21, 1987a.

Thornton WE, Moore TP, Pool SL & Vanderploeg J. Clinical characterization and etiology of space motion sickness. Aviat. Space Environ. Med. 58: A1–8, 1987b.

Thornton WE, Uri JJ, Moore TP & Pool SL. *Eye and head motion during head turns in spaceflight* (NASA TM-100466). National Aeronautics and Space Administration, Washington DC, 1988.

Thornton WE, Uri JJ, Moore TP & Pool SL. Studies of the horizontal vestibulo-ocular reflex in spaceflight. Arch. Otolaryngol. Head Neck Surg. 15: 943–9, 1989.

Uri JJ, Linder BJ, Moore TP, Pool SL & Thornton WE. *Saccadic eye movements during space flight* (NASA TM-100475). National Aeronautics and Space Administration, Washington DC, 1989.

Vesterhauge S, Mansson A, Johansen TS & Zilstorff K. Oculomotoric response to voluntary head rotations during parabolic flights. Physiologist 25: S117–8, 1982.

Viéville T, Clément G, Lestienne F & Berthoz A. Adaptive modifications of the optokinetic vestibulo-ocular reflexes in microgravity. In: *Adaptive Processes in Visual and Oculomotor Systems.* Ed. by Keller EL & Zee DS, Pergamon Press, New York, pp. 111–20, 1986.

Vinnikov YA, Gazenko O, Lychakov D & Pal'mbakh Y. Razvitiye vestibulyarnogo apparata v usloviyakh nevesomosti. Zhurnal oschchey biologii 44:147–63, 1983. Translated as: The development of the vestibular apparatus under conditions of weightlessness. NASA TM-77517, Washington DC, 1984.

Vogel H & Kass JR. European vestibular experiments on the Spacelab-1 mission. 7. Ocular counterrolling measurements pre- and post-flight. Exp. Brain Res. 64: 284–90, 1986.

Von Baumgarten RJ. European vestibular experiments on the Spacelab-1 mission. 1. Overview. Exp. Brain Res. 64: 239–46, 1986.

Von Baumgarten RJ. Orbital weightlessness as a new tool for vestibular research: experiments in two Spacelab missions including experiments on caloric nystagmus. In: *Biological Sciences in Space.* Ed. by Watanabe S, Mitarai G & Mori S, MYU Research, Tokyo, pp. 53–60, 1987.

Von Baumgarten RJ. General remarks on the role of the vestibular system in weightlessness. Acta Oto-Rhino-Laryngol. 244: 135–42, 1987.

Von Baumgarten RJ, Baldrighi G & Shillinger GL. Vestibular behaviour of fish during diminished G-force and weightlessness. Aerospace Med. 43: 626–36, 1972.

Von Baumgarten RJ, Baldrighi G, Vogel H & Thumler R. Physiological responses to hyper- and hypogravity during rollercoaster flight. Aviat. Space Environ. Med. 51: 145–54, 1980.

Von Baumgarten RJ, Benson A, Berthoz A, Bles W, Brandt T, Brenske A, Clarke A, Dichgans J, Eggertsberger R, Jürgens K, Kass J, Krafczyk S, Probst T, Scherer H, Thümler R, Viéville T, Vogel H & Wetzig J. European experiments on the vestibular system during the Spacelab D-1 mission. In: *Proceedings of the Norderny symposium on the scientific results of the German Spacelab mission D-1.* Ed. by Sahm PR, Jansen R & Keller MH. Köln, Germany, DFVLR, pp. 477–90, 1986.

Von Baumgarten RJ, Benson AJ, Brandt T, Brand U, Bruzek W, Dichgans J, Kass J, Probst T, Scherer H, Viéville T, Vogel H & Wetzig J. Effects of rectilinear acceleration and optokinetic and caloric stimulations in space. Science 225: 208–12, 1984.

Von Baumgarten RJ, Boehmer G, Brenske A & Reiser M. A non-thermoconvective mechanism generating caloric nystagmus: Similarities between pressure-induced and caloric nystagmus in the pigeon. In: *The Vestibular System: Neurophysiologic and Clinical Research.* Ed. by Graham MD & Kemink JL, Raven Press, New York, pp. 51–59, 1987.

Von Baumgarten RJ, Kass J, Vogel H & Wetzig J. Influence of proprioceptive information on space orientation on the ground and in orbital weightlessness. Adv. Space Res. 9: 223–30, 1989.

Von Baumgarten RJ, Simmonds RC, Boyd JF & Garriott OK. Effects of prolonged weightlessness on the swimming pattern of fish aboard Skylab 3. Aviat. Space Environ. Med. 46: 902–6, 1975.

Von Baumgarten RJ & Thumler RR. A model for vestibular function in altered gravitational states. In: *Life Sciences and Space Research* (Vol. XVII). Ed. by Holmquist, R, Pergamon Press, Oxford, 1979.

Von Baumgarten RJ, Vogel H & Kass JR. Nausogenic properties of various dynamic and static force environments. Acta Astronaut. 8: 1005–13, 1981.

Von Baumgarten RJ, Wetzig J, Vogel H & Kass JR. Static and dynamic mechanisms of space vestibular malaise. Physiologist 25: S33–6, 1982.

Walk R & Gibson E. A comparative analytical study of visual depth perception. Psychol. Monogr. 75: 2–34, 1961.

Walsh EG. Role of the vestibular apparatus in the perception of motion on a parallel swing. J. Physiol. 155: 506–13, 1961.

Wassersug RJ. The basic mechanisms of ascent and descent by anuran larvae (*Xenopus laevis*). Copeia 3: 890–4, 1992.

Wassersug RJ & Izumi-Kurotani A. The behavioral reactions of a snake and a turtle to abrupt decreases in gravity. Zool. Sci. 10: 505–9, 1993.

Wassersug RJ, Izumi-Kurotani A, Yamashita M & Naitoh T. Motion sickness in amphibians. Behav. and Neural Biol. 60: 42–51, 1993.

Wassersug RJ, Pronych SP, Schofoeld SM & Schmidt GK. On the way of life of aquatic vertebrates and their behavioral responses to reduced gravity. ASGSB Bulletin 5:42, 1991.

Wassersug RJ, Pronych S, Izumi-Kurotani A & Fejtek M. The behavioural responses of vertebrates to microgravity: a comparative approach. Paper no. SB 93-001 In: *Proceedings of the Spacebound '93 Symposium*, Canadian Space Agency, Ottawa, May 16–18, 1993a.

Wassersug RJ & Souza KA. The bronchial diverticula of *Xenopus* larvae. Are they essential for hydrostatic assessment? Naturwiss. 77:443–5, 1990.

Wassersug RJ, Souza KA, Ross M & Black S. Spacelab-J: The frog embryology experiment. Paper presented at the *10th Man in Space Symposium*, Tokyo, April 19–23, 1993b.

Watt DGD. The vestibulo-ocular reflex and its possible roles in space motion sickness. Aviat. Space Environ. Med. 58: A170-4, 1987.

Watt DGD, Landolt JP & Young LR. Effects of long-term weightlessness on roll circular vection. In: *Proceedings of the 7th CASI Conference on Astronautics*, Ottawa, Canada, Nov 1992.

Watt DGD, Money KE, Bondar RL, Thirsk RB, Garneau M & Scully-Power P. Canadian medical experiments on Shuttle flight 41-G. Can. Aeronaut. Space J. 31: 215-26, 1985.

Watt DGD, Money KE & Tomi LM. MIT/Canadian vestibular experiments on the Spacelab-1 mission. 3. Effects of prolonged weightlessness on a human otolith-spinal reflex. Exp. Brain Res. 64: 308-15, 1986.

Watt DGD, Money KE, Tomi LM & Better H. Otolith-spinal reflex testing on Spacelab-1 and D-1. Physiologist 32: S49-S52, 1989.

Wetzig J. Rotation speed of labyrinthectomized fish during short-duration weightlessness. Aviat. Space Environ. Med. 58: A257-61, 1987.

Wetzig J, Hoffstetter-Degen K & Von Baumgarten RJ. Responses to eccentric rotation in two space-bound subjects. Clin. Investig. 71: 757-60, 1993.

Woellner RC & Graybiel A. Counterrolling of the eyes and its dependence on the magnitude of gravitational or inertial force acting laterally on the body. J. Appl. Physiol 14: 632, 1959.

Wood SJ, Reschke MF, Anderson DJ & Homick JL. Postural responses to sudden falls following spaceflight. Aviat. Space Environ. Med. 56: 484, 1985.

Yakovleva IY, Kornilova LN, Tarasov IK & Alekseyev VN. Results of studies of cosmonauts' vestibular function and spatial perception. Kosm. Biol. Aviakosm. Med. 16: 20-6, 1982.

Young LR. Effects of linear acceleration on vestibular nystagmus. In: *Proceedings of the 3rd Symposium on the Role of the Vestibular Organs in Space Exploration*, NASA SP-152, 1967.

Young LR. Perception of the body in space: mechanisms. In: *Handbook of Sensory Physiology. The Nervous System III*. Springer-Verlag, New York, 22: 1023-66, 1985.

Young LR. Space and the vestibular system: what has been learned? Guest editorial. J. Vestib. Res. 3:203-6, 1993.

Young LR, Oman CM, Money KE & Watt DGD. *Scientific and Technical Proposal for Vestibular Experiments in Spacelab.* Man-Vehicle Laboratory, Massachusetts Institute of Technology, Cambridge, MA, 1976.

Young LR, Oman CM, Merfeld D, Watt DGD, Roy S, Deluca C, Balkwill D, Christie J, Groleau N, Jackson DK, Law G, Modestino S & Mayer W. Spatial orientation and posture during and following weightlessness: human experiments on Spacelab-Life-Sciences-1. J. Vestib. Res. 3: 231-40, 1993.

Young LR, Oman CM, Watt DGD, Money KE & Lichtenberg BK. Spatial orientation in weightlessness and readaptation to Earth's gravity. Science 225: 205-8, 1984.

Young LR, Oman CM, Watt DGD, Money KE, Lichtenberg BK, Kenyon RV & Arrott AP. MIT/Canadian vestibular experiments on the Spacelab-1 mission. 1. Sensory adaptation to weightlessness and readaptation to one-g: an overview. Exp. Brain Res. 64: 291-8, 1986.

Young LR & Shelhamer M. Microgravity enhances the relative contribution of visually-induced motion sensation. Aviat. Space Environ. Med. 61: 525-30, 1990.

Chapter 5

Bone and space flight: an overview

JACK J.W.A. VAN LOON, J. PAUL VELDHUIJZEN & ELIZABETH H. BURGER

5.1 Introduction

Even four centuries ago, Galileo Galilei realised that the rigidity of the skeleton of terrestrial animals is related to its load bearing function, and is associated with the size and mass of the animal (Galilei 1974). This phenomenon of differences in mechanical properties also applies to the relatively small variations in the skeleton which occur between individuals of the same species, as well as to variations in bone properties within the same subject. Ward (1836) observed that the trabecular arrangement within the femoral head, the area now known as Ward's triangle, showed patterns comparable to those found in the crossbeam structures of nineteenth century street lights. This very clearly illustrated natural mechanical engineering of the weight bearing properties of bone. Some 50 years later, Wolff postulated his law; 'Das Gezetz der Transformation der Knochen' (Wolff 1892). In this essay he discussed in more detail how the structure of bone reflects its mechanical usage history. The process of bone formation and bone remodelling according to its mechanical history is now generally known as functional adaptation, a term proposed by Roux (1895).

Since the late nineteenth century, many studies have been performed to examine functional adaptation of the skeleton. It has been shown that in people with a sedentary life style the average bone loss is more prominent than in more ambulatory subjects. This demonstrates that not using the skeleton, i.e. not applying loads onto the bones, leads to mechanical inferiority. This loss can be measured as a reduction in bone mineral density, a reduction in trabecular structure or changes in the biochemical composition or arrangement of the organic matrix components. On the other hand, by exceeding the normal loads through weight bearing exercise (**Pruitt *et al.* 1992; Raab *et al.* 1991**), or running (**Aloia *et al.* 1978; Dalsky *et al.* 1988; Yeh *et al.* 1993**) bone mass can increase compared to control or pretreatment conditions.

Bone loss or gain is related to the magnitude, direction and frequency of the stress acting upon the skeleton during application of loads. The resulting strain (ε) is defined as a dimensionless measure of linear deformation, one strain being a 100% deformation in length of a piece of material after applying a load. Such high values are not found in bone biology, where deformations lie in the range of several hundreds of microstrains ($\mu\varepsilon$). Depending on the final shape change of the material, three forms of strain are identified, namely tensile, compressive and shear strains. In all materials, as well as bone, stresses are linearly related to the applied loads. The amount of strain, however, is related to the mechanical properties of the material, the Young's modulus, and does not have to be a linear function of stress.

From various *in vivo* studies (**Lanyon 1984**) it has become clear that strains from 1500–3000 $\mu\varepsilon$ promote bone modelling while strains beneath these values correspond to increased remodelling activity in bone (**Frost 1988**). However, strain values exceeding 6,000 $\mu\varepsilon$ lead to bone fatigue and, eventually, fractures (**Rubin 1984**). So it is the height of the peak strain which renders an anabolic effect on bone mass (**Rubin & Lanyon 1985**), though the frequency of application may also play an important role (**Rubin & Lanyon 1984**).

In free fall the skeleton experiences no weight bearing stress. Thus, the initiation of orbital space flight in the late 1950s created an environment providing very low strains on the skeleton. Since then, several studies have indicated that space flight exposes the human body to various adverse effects as a result of near weightlessness or as a result of accumulating radiation outside the Earth's atmosphere. Most affected are neurovestibular functions (Chapter 4), distribution of various body fluids, cardiovascular function, various haemopoietic parameters (Chapters 1, 2 & 3; **Taylor 1993**) and the musculo-skeletal system (Nicogossian 1989).

How strain affects the various cell populations in the skeleton at the tissue level and which mechanisms are involved, are still unclear. Also, the problem of which cell or cells serve as mechanosensor(s) is not yet solved. However, from a number of *in vitro* studies it has become clear that cells do react to this stimulus.

Isolated bone forming cells, osteoblasts, cultured on a deformable support responded to stretching (shape change) with increased prostaglandin production (**Harell *et al.* 1977**; **Yeh & Rodan 1984**). In high density cultures of isolated cartilage cells, and in developing long bones, shear stress resulting from hydrostatic forces (**Wong & Carter 1990**) are probably responsible for stimulation of mineralization and a reduction in mineral resorption (**Klein Nulend *et al.* 1986, 1987**), changes in cell proliferation (**Veldhuijzen *et al.* 1979**; **van Kampen & Veldhuijzen 1982**) and cAMP production (**Veldhuijzen *et al.* 1979**).

However, it is also possible that, as a result of tissue deformation, fluid streaming is initiated in the system of lacunae and canaliculi in the bone in which the osteocytes are located. These fluid flows in the relative small space between the mineralized matrix and the cells in the lacunae/canaliculi-system will deform the osteocytes and, probably to a lesser extent, the osteoblasts also (**Reich *et al.* 1990**; **Cowin *et al.* 1991**; **Reich & Frangos, 1991**). Recent *in vitro* experiments with isolated osteocytes and osteoblasts subjected to pulsed fluid flow showed an increased prostaglandin production and an altered response to transforming growth factor β (TGF-β) (**Klein Nulend *et al.* 1995**). From these studies it became clear that osteocytes probably serve as mechanosensor in skeletal tissues. In these responses anchorage molecules, such as integrins, and the cytoskeleton are probably involved.

The reduced weight bearing and reduced muscle pull experienced mainly by the long bones and spine under microgravity will result in a reduced strain in the skeletal tissues. A change in maximal strain and strain rate, resulting in direct shape changes and/or shape changes as a result of altered fluid flow in the bone tissue, may be the signal for cellular responses. Studies of the cellular mechanisms for skeletal responses to microgravity must be based on *in vitro* research of cellular responses to increase strain under Earth 1 × g or hypergravity conditions.

The skeleton has evolved during millions of years in unit gravity on Earth, but in a near weightless, or microgravity, state in space there does not seem to be much need for such a rigid weight bearing structure. Based on data from skeleton unloading studies, it seems likely that there may be signs of bone loss after exposure to weightlessness in space. Although most of the changes in physiological parameters found under space flight conditions reach a level of adaptation at about 6 weeks or sooner (Nicogossian 1989) the adaptation time and mechanical stability of bone after long periods of microgravity are still uncertain. The relatively new era of research on vertebrates in space effectively started with the launch of the dog Laika in the Sputnik-2 on 3 November 1957, followed nearly four years later, on 12 April 1961, by the first human, Yuri Gagarin.

In the following we will try to summarize data about *in vivo* bone growth and metabolism which has been accumulated since the start of space flight, in addition the results of the first *in vitro* experiments under microgravity conditions will also be addressed. We have tried to be as complete as possible, using original papers published in, mostly Western, peer reviewed literature. Less attention is given to the Russian, or former Soviet, literature merely due to the language barrier, although some data from Russian investigators is included. One should keep in mind in this respect, that the Russians have far more experience both in number of flights and total flight hours in space than all other nations combined (**Ballard & Connolly 1990**; Ballard & Mains 1990).

5.2 Data from space flight experimentation

5.2.1 *In vivo* spaceflight results

The main space flights covering a period of nearly 30 years involving *in vivo* bone research are tabulated in essentially chronological order in Table 1, whilst Table 2 provides more detailed information about the particular flights involved. Five main groups of flights can be distinguished in this overview: the Gemini flights, the long duration Skylab missions, the Russian Cosmos series, the very long duration Salyut-7 and Mir missions, and the relatively short flights of the U.S. Space Shuttle.

Table 1. Space flights dealt with in this Chapter.

Flight	Launch date	Flight duration d:h:min	Recovery time
Gemini-4	3 March 1965	4:1:58	not known
Gemini-5	21 August 1965	7:22:59	not known
Gemini-7	4 December 1965	13:18:35	not known
Gemini-6	15 December 1965	1:1:51	not known
Biosatellite-III	June 1969	8:20:0	not known
Apollo-17	7 December 1972	12:13:51	not known
Skylab-II	25 May 1973	28:0:50	not known
Skylab-III	28 July 1973	59:11:9	not known
Skylab-IV	4 December 1973	84:1:16	not known
Cosmos 605	October 1973	22:0:0	2 d
Cosmos 782	25 November 1975	19:12:0	3 d
Cosmos 936	3 August 1977	18:12:0	not known
Cosmos 1129	25 September 1979	18:12:0	7–11 h
Cosmos 1514	14 December 1983	5:0:0	over 6 h
Cosmos 1667	10 July 1985	6:12:0	over 6 h
Cosmos 1887	29 September 1987	12:12:0	42 –55h
Cosmos 2044	15 September 1989	14:0:0	6–10 h monkey: 11–34 d
Salyut-7	1982	years	not known
Spacelab-3 (SL-3) (STS-51B)	26 April 1985	7:0:9	Flight: R+11 h Sim: R+60 h
Spacelab-2 (SL-2) (STS-51F)	29 July 1985	7:22:46	not known
Mir	February 1986	years	not known
STS-41	6 October 1990	4:2:10	not known
STS-42	22 January 1992	8:1:15	fixed in flight
STS-52	22 October 1992	9:20:56	not known
STS-54	January 1993	6 days	6–13 h
STS-56	April 1993		NA

L = launch; R = recovery; STS = Space Transportation System (= U.S. Space Shuttle). The 'recovery time' listed in this Table is the actual time between the landing of the space craft and the post flight handling of the samples. NA = Not applicable. Further details of these flights can be found in Chapter 8.

More than four years after the first man in space, the Gemini missions were launched. One of the first studies on bone loss in relation to space flight emerged, and was reported by **Mack *et al.* (1967)** later corrected by **Vose (1974)**. They reported a reduction in bone density during these flights and showed that mineral loss in the os calcis was 3 to 9% after 4 to 14 days of orbital space flight. Metabolic studies were also performed during these flights. **Lutwak *et al.* (1969)** revealed a clear trend towards an increased urinary calcium excretion in the Gemini 7 flight. In one individual this decrease also persisted during a four day post flight period. In addition, urinary phosphate excretion increased in flight in both Gemini astronauts (**Lutwak *et al.* 1969**).
In the Biosatellite-III non-human primate (*Macaca nemestrina*) study, a ground control monkey was included which stayed in the same hardware as

Table 2. Characteristics of the various flights referred to in this Chapter.

Flight	Species	Sex	Age (d) at launch = L at death = †	Weight (g) at launch = L at death = †	Number of samples*
Gemini–4	human	male	adult		F: 2
Gemini–5	human	male	adult		F: 2
Gemini–6	human	male	adult		F: 2
Gemini–7	human	male	adult		F: 2
Bio-satellite III	monkey	male		6000	F: 1
Apollo 17	human	male	adult		F: 3
Skylab–I I	human	male	adult		F: 3
Skylab–I II	human	male	adult		F: 3
Skylab–I V	human	male	adult		F: 3
Cosmos 605	rat	not known	not known	not known	F: 5
Cosmos 782	rat	male	F: 63 (L) S: 63 (L) V: 63 (L)	all 215 (L)	B: 6 (B) F: 6 (F) S: 6 (S) V: 6
Cosmos 936	rat	male	63 (L)	202±13.9 (L)	F: 10 F 1 × g: 10 S: 10 V: 10
Cosmos 1129	rat	male	F: 133 (†) S: 138 (†) V: 135 (†)	290 (L) F: 349±4 (†) S: 359±2 (†) V: 349±4 (†)	F: 5 S: 7 V: 4
Cosmos 1514	rat	female	83 (L)	295 (L) F: 300 (†) S: 360 (†)	B: not known F: 5–10 S: 5–10 V: 5
Cosmos 1667	rat	male	105 (L)	F: 304±46 (†) S: 334±16 (†)	B: not known F: 7–10 S: 7–10 V: 7
Cosmos 1887	rat monkey	male male	B: 85 (†) F: 105 (†) S: 111 (†) V: 108 (†)	B: 316±8 (†) :L–5 d F: 303±2 (†) R+2.4 d S: 349±8 (†) R+5 d V: 342±8 (†) R+3–4 d monkey: 4000	B: 5 F: 5 S: 5 V: not known monkey: 2
Cosmos 2044	rat monkey	male male	rat: B: 108 (†) F: 123 (†) S: 126 (†) V: 129 (†) monkey: 3.5 years	rat: B: 320±4 (†) F: 338±2 (†) S: 343±7 (†) V: 363±2 (†) monkey: 3800 to 3900	rat: F: 5 S: 5 V: 5 monkey: 2
Salyut–7 & Mir	human	not known	adult		F: 9

Table 2. (continued).

Flight	Species	Sex	Age (d) at launch = L at death = †	Weight (g) at launch = L at death = †	Number of samples*
Spacelab-3	rat	male	Flight 1: 84 (†) Flight 2: 56 (†) Fl. Sim. 1: 84 (†) Fl. Sim. 2: 58 (†)	B: 200±6 F 1: 384±9 (†) F 2: 194±10 (†) F Sim.1: 395±23 (†) F Sim.2: 198±5 (†)	F 1: 5 F 2: 6 F Sim. 1: 5 F Sim. 2: 6
Spacelab-2	human				F: 4
STS-41	rat	male		F: 151±4 (†) F Control: 164±2 (†)	F: 8 F Control: 12
STS-42: IML-1 (**van Loon *et al.* 1995**)	mouse	–	16 and 17.5 days prenatal	–	F: 8 F 1 × g: 8 G: 8 G 2 × g: 8
STS-42: IML-1 (**Duke *et al.* 1993**)	mouse	–	prenatal	–	F: ? F 1 × g: ? G: ? G 2 × g: ?
STS-52	rat	male		F: 260±3 (†) F Control: 265±2(†)	F: 6 F Control: 6
STS-54 (Bikle *et al.* 1994)	rat	male		F: 210–233 (†) F Control: 210–220(†)	F: 6 Control: 6
STS-54 (**Klement & Spooner 1994**)	mouse	–	13.7 days prenatal		F: 3 × 8 G: 3 × 8
STS-56	mouse	–	15.5 days prenatal		F: 2 × 4 G: 2 × 4

* B = basal samples (ground analyses to establish condition prior to flight), F = flight animals, F Sim = flight simulation, S = synchronous (animals on the ground but in flight containers subjected to simulated flight stresses) , V = vivarium (unstressed animals on the ground). In the columns 'species' and 'number of samples' in this table, only the species involved in bone research are listed. The number indicated is based on the maximum number of animals used for at least one of the studies reviewed in this Chapter. The total number of samples used by other studies in the same flight may vary from the number indicated. During Cosmos 605, for example, 10 rats and two boxes filled with amphibians and reptiles were on board, though only 5 rats were used for bone related studies (**Yagodovsky *et al.* 1976**).

the flight animal. In these primates an average bone mineral density loss of 4.5% at various anatomical sites was reported after a near 9 day flight (**Mack 1971**).

Few reports on life sciences experiments in the Apollo series of space flights have been published in peer reviewed papers. A metabolic study on calcium and phosphorus metabolism during Apollo 17 has been reported by **Rambaut *et al.* (1975)**. In this study, astronauts scored all their diet intakes and faecal and urinary excretions. A reduction of about 0.2% in total body calcium and 0.7% in total body phosphorus was found after this 14 day flight. These losses were mainly due to an increased faecal calcium and an increased urinary and faecal phosphorus excretion while under microgravity. The losses were comparable to reductions reported in bed rest studies (**Rambaut *et al.* 1975**). The authors argued that the changes found may be attributed not only to hypogravity or immobilization but also to disturbances in gastrointestinal absorption. However, other metabolic studies have been performed, all indicating increased calcium excretion (**Lutwak *et al.* 1969**; **Johnson *et al.* 1973**; **Whedon *et al.* 1975**). In the study by **Rambaut *et al.* (1975)** the authors also reported a small but significant loss of total body weight in the Apollo 17 crew. This has been confirmed during the 28 day Skylab-II mission (**Whedon *et al.* 1975**). For Apollo flights this could have been caused by the near weightless period in space, by immobilization while being strapped into the seats, or by the relative hypogravity on the lunar surface and/or other systemic influences. For Skylab flights, however, probably only weightlessness contributed to bone loss, since there was ample space to move about in the large Skylab modules. Reduced bone mineral content in Apollo crew was also reported by **Vogel & Whittle (1976a)**.

Apart from two fishes (*Fundulus* spp.) in the Skylab-III mission, there were no vertebrate animals on board these long duration flights. In analogy to the Gemini experiment, bone mineral content measurements were performed on the Skylab crew before and after the mission. **Vogel & Whittle (1976b)** reported a negative trend towards a decreased bone mineral content, especially in the os calcis and radius. In two crew members of the Skylab-IV mission, this bone loss had not returned to baseline values even 97 days after the flight (**Tilton *et al.* 1980**)

The calcium balance in the Skylab-II crew was determined in an extensive study by **Whedon *et al.* (1975)**. A mean pre-flight value of this balance of +71 mg d^{-1} was recorded, while the mean balance during the last 16 days in flight was -50 mg d^{-1}, a negative shift of 121 mg d^{-1}. Increased phosphorus and nitrogen excretion was also noted, indicating a loss of soft tissues, probably muscles (**Whedon *et al.* 1975**).

In the mid-1970s a series of unmanned space flights was initiated. These flights were a collaboration of the Soviet space agency with various other nations including East Germany, Czechoslovakia, France, Hungary, Poland, Rumania and the U.S.A. These missions, generally referred to as Cosmos or Biocosmos flights, involved mainly life sciences experiments. At that time these satellites were the only completely automated space craft available for life sciences research in space. They were capable of staying in orbit for nearly three weeks while maintaining normal ambient environmental conditions. They could be retrieved and the biological samples recovered for analysis on the ground. Much of the early, more detailed, bone studies derive from these flights (**Yagodovsky *et al.* 1976**; **Morey & Bayerlink 1978**; **Turner**

et al. 1985; Popova *et al.* 1987). The rat was the main species involved, although, especially in the later missions, non-human primates were also part of the payload. Results of studies on bone and Rhesus monkeys under near weightlessness have been reported for Cosmos 1877 and 2044 flights by Zerath *et al.* (1990, 1991).
The experimental set-up for the Cosmos programme was quite consistent over the years. Basically there were four groups, as listed below (initials in brackets are the symbols used to identify the sample groups in Tables 3 and 4).

Basal (B). This group of animals was killed at the start of the mission to give baseline values for the various parameters investigated. Especially with young rapidly growing rats this group is important to distinguish between growth and space flight induced effects.

Flight (F). This group actually flew on board the spacecraft. Several days before launch the animals were transferred from standard laboratory cages into the special space flight hardware, cylindrical cages 20.5 cm long and 9.4 cm in diameter. Housed in this hardware they were fed pasty diets and water *ad lib.* During launch this group was subjected to various accelerations, vibrations and noise levels. While in orbit there was a regular light-dark cycle (±2 lux intensity) and normal room temperatures.

Synchronous (S). This group of animals was synchronized as closely as possible with the flight group, following the same procedure as used for flight. One important element was that the launch characteristics experienced by the flight group were simulated for the synchronous animals. Noise levels of up to 110 db and vibrations of 50–70 Hz at an amplitude of 0.4 mm for 10 min were imposed upon the animal holding units. After this, the animals were subjected to a series of accelerations for 10 min with a plateau at $4 \times g$ for 7 min to simulate take-off. Re-entry loads during the return to Earth, were simulated by applying gravitational accelerations for 6 min with a $6 \times g$ plateau for 3 min. The actual impact shock at touch down was $50 \times g$ for 10 ms (**Spengler *et al.* 1983**). The timeline of the synchronous group normally lagged behind the flight group. This was to provide time to adjust the ground simulation set-up according to input of actual flight data. Unexpected changes in, for example, temperatures in the flight module were programmed into the ground module to duplicate the orbital status as closely as possible. The differences between the flight and the synchronous group were, of course, the near weightless condition but also the galactic radiation experienced by the flight group and the difference in start time of the experiment. The effects of space flight, or microgravity, on the various parameters were therefore at best compared to the synchronous group. The ultimate control group, however, housed in an on-board $1 \times g$ centrifuge, was used only twice, during the Cosmos 782 and Cosmos 936 flights. These centrifuges had

Table 3. Overview of studies on bone and space flight: details of experimental subjects and materials.

No.	Author	Flight	Experimental organism	Number of samples*	Bone site and/or cells examined	Remarks
1	**Mack *et al.* 1967**	Gemini 4, 5, 7	human	2	Os calcis, talus, phalanges (hand), carpals.	See **Vose 1974** for corrections.
2	**Vose 1974**	Gemini 4, 5	human	2	Os calcis.	Commander–pilot and pilot of each flight.
3	**Lutwak *et al.* 1969**	Gemini 7	human	2	Endocrines; urine, faeces and sweat.	Samples at 10 d pre–flight, in flight and 4 d post–flight.
4	**Mack 1971**	Biosatellite III	non–human primate, *Macaca nemestrina*	F: 1 Control: 1	17 different anatomical sites.	Control in identical satellite on Earth.
5	**Rambaut *et al.* 1975**	Apollo 17	human	F: 3	Endocrines; urine and faeces.	Flight, pre-flight and post-flight values.
6	**Whedon *et al.* 1975**	Skylab–II	human	F: 3	Endocrines; urine and faeces.	
7	**Vogel & Whittle 1976a & b**	Skylab II, III, IV	human	F: 3	Os calcis (left), right radius, right ulna.	
8	**Tilton *et al.* 1980**	Skylab II, III, IV	human	F: 9 Control 1: 5 Control 2: 3	Os calcis.	
9	**Yagodovsky *et al.* 1976**	Cosmos 605	Wistar	rat F*: 5 S*: 5 V*: 5	Right femur, tibia and humerus.	22 d; also a group at R + 27 d.
10	**Organov *et al.* 1991**	Salyut–7 orbital station	human human	F: total 7 Hypo–kinesia: total 9	Lumbar vertebral spongy, anterior and posterior side.	F: 150–237 d. 5° hypokinesia (head down tilt): a) 120 d no exercise + 250 + exercise b) total 270 d + exercise.
11	**Morey & Bayerlink 1978**	Cosmos 782	Wistar rat	F: 6 S: 6 V: 6 Post Flight: 6	Tibia.	Synchronous started 5 d after flight. Also a 26 d post–flight group was included.
12	**Turner *et al.* 1985**	Cosmos 782 Cosmos 936	Wistar rat	not known	Tibia.	Half of the flight group killed at R + 25 d.

Table 3. (continued).

No.	Author	Flight	Experimental organism	Number of samples*	Bone site and/or cells examined	Remarks
13	**Spengler *et al.* 1983**	Cosmos 936	Wistar rat	B: 6 F: 5 F 1 × g: 5 S: 5 V: 10	Femur, left tibiae.	Synchronous started 4 d later than flight. Also group at R + 25 d.
14	Spector *et al.* 1983	Cosmos 782 Cosmos 936	rat		Tibia.	Details in **Morey 1979**.
15	**Wronski & Morey 1983a & b**	Cosmos 1129	Wistar rat	n = 11 (some groups n = 5)	Left tibia, humerus, rib cage (just after flight only).	Synchronous started 5 d after flight. Also a flight group with + 29 d post–flight recovery period; post-flight all housed in vivarium.
16	Stupakov *et al.* 1989	Cosmos 1129, bed rest	Wistar rat, human (bed rest)	Bed rest: 9	Rat: femoral head. Human: iliac crest biopsy.	Rat; Cosmos. Human; bed rest study.
17	**Simmons *et al.* 1983**	Cosmos 1129	rat	F: 5–7	Alveolar bone, ribs, incisors, dentine. Post flight: R + 7 to 11 h.	Synchronous started 5 d after flight.
18	**Pitts *et al.* 1983**	Cosmos 1129	Wistar rat	F: 5 S: 5 V: 5	Skinned, eviscerated carcasses (mainly bone & muscle).	Flight animals killed 36 h after reentry, S at R + 4 d
19	**Cann & Adachi 1983**	Cosmos 1129	Wistar rat	F: 5 S: 5	Rib cage, faeces.	
20	**Wronski & Morey 1983a & b**	Cosmos 782, 936, 1129	rat		Tibial diaphysis.	
21	**Eurell & Kazarian 1983**	Cosmos 1129	Wistar rat	F: 5–7 S: 5–7 V: 5–7 (number of vertebral bodies)	4th and 5th Lumbar vertebrae (L4 and L5) (cortical + trabecular bone).	direct R+6 d R+29 d
22	**Jee *et al.* 1983**	Cosmos 1129	Wistar rat	F: R+0: 7 S: R+0: 7 V R+0: 7 All: R+6 d: 6 All: R+29 d: 5	Left tibia, humeri.	direct R+6 d R+29 d
23	France *et al.* 1982	Cosmos 1129	Wistar rat	not known	Vertebral centrum.	'direct' R+6 d R+29 d

Table 3. (continued).

No.	Author	Flight	Experimental organism	Number of samples*	Bone site and/or cells examined	Remarks
24	**Vico *et al.* 1987**	Cosmos 1514	Wistar rat	F: 5 S: 5 V: 5	Left femur, left tibia, 3rd thoracic (T3) vertebrae, 3rd lumbar (L3) vertebrae.	Loaded: tibia–femur. Unloaded: thoracic + lumbar vertebrae. During flight the rats were from 13 to 18 d pregnant.
25	Pospisilova *et al.* 1987	Cosmos 1514	Wistar rat	F: 5 S & V not known	Femur and skin from flight offspring.	Flight rats 13 – 18 d pregnant in microgravity. 4 out of 5 gave birth to about 12 young. at 23rd d. Samples from neonatal rats at: R+2–3 h, and R+15, 30, 100 d old.
26	Bakulin *et al.* 1985	Cosmos 1514	rat	F: 5 S: 5 V: 5	Humerus; proximal and epiphysis.	Flight: 5 pregnant rats (18 d pregnant at R, 13 d at launch).
27	Pospisilova *et al.* 1986	Cosmos 1514	Wistar rat	F: 4 S: 7	Femur.	Pregnant 13 – 17 d during flight.
28	**Patterson–Buckendahl *et al.* 1987**	Spacelab–3	rat	Preflight: 6 F: 6 Simulation: 6	3rd Lumbar vertebrae (L3), n = 6; right humeri, n = 6.	Pre-flight: killed at launch. Flight: R+12. Flight Controls: 48 h. after flight.
29	**Wronski *et al.* 1987**	Spacelab–3	Sprague Dawley rat	F1: 5 F2: 6 F Control 1: 5 F Control 2: 6	Right tibia shaft, right proximal. humerus and 4th lumbar vertebrae (L4).	Flight: R+11–17 h. Flight Control: F + 48.
30	**Simmons *et al.* 1986**	Spacelab–3	Sprague–D awley rat	F: 6 F Control: 6	Femur and vertebra.	Flight: R + 12 h. Flight Control: R + 2 d.
31	**Shaw *et al.* 1988**	Spacelab–3	Sprague Dawley rat	PreFlight: 6 F: 6 F Control: 6	Humerus and tibia.	Flight: R+11 h. Flight control: F+48 h.
32	Doty *et al.* 1985	Spacelab–3	rat	not known	Tibiae.	
33	Duke *et al.* 1985	Spacelab–3	rat	F: 6 F Control: 6	Proximal tibia, growth plate.	

Table 3. (continued).

No.	Author	Flight	Experimental organism	Number of samples*	Bone site and/or cells examined	Remarks
34	**Vico *et al.* 1988**	Cosmos 1667	Wistar rat	F: 7 S: 7 Local control: 10	Tibia + femur proximal metaphysis, 8th thoracic vertebrae (T8), 1st lumbar vertebrae (L1).	Including 10 French control rats (= local control)
35	**Vico *et al.* 1991**	Cosmos 1667	Wistar rat	F: 7 S: 7 Suspension: 17	Right tibia.	Cosmos and tail suspension
36	Pospisilova *et al.* 1988	Cosmos 1667	Wistar rat	not known	Femur, skin.	
37	Weingart *et al.* 1988	Cosmos 1514 Cosmos 1667	Wistar rat	F: 10 S: 10 F: 10 S: 10	Jaw, molars, incisors, in bone cement, dentine.	Cosmos 1514: rat 15 d pregnant at launch
38	**Morey–Holton *et al.* 1988**	Spacelab–2	human	F: 4	Endocrines; blood plasma.	Samples: L–24, –18, –3, in flight, R+0, +30, +10 d.
39	**Zernicke *et al.* 1990**	Cosmos 1887	Wistar rat	B: 5 F: 5 S: 5 V: 5	6th lumbar vertebrae (L6).	Vivarium (same cages as flight).
40	Kaplansky *et al.* 1990	Cosmos 1887	Wistar rat	not known	Tibia and 4th–6th lumbar vertebrae (L4–L6).	
41	**Simmons *et al.* 1990**	Cosmos 1887	Wistar rat	B: 6 F: 6 S: 6 V: 6	Calvaria, mandibles, incisor, dentine and vertebrae.	B: L–5 d. F: R+2,4 d. S: R+5 d. V: R+3–4 d.
42	Földes *et al.* 1990	Cosmos 1887	Wistar rat	F: 5 S: 5 V: 5	Femur–tibia knee joint. (Distal part of femur and proximal part and diaphysis of tibia).	Also rats immobilized for 8 weeks used, number not known. Post–flight: R+ 4–8 h.
43	**Doty *et al.* 1990**	Cosmos 1887	Wistar rat	B: 5 F: 5 S: 5 V: 5	Tibia (proximal + some diaphysis and shaft). Tibio–fibular junction.	F: R+2 d .
44	**Duke *et al.* 1990**	Cosmos 1887	Wistar rat	B: 5 F: 5 S: 5 V: 5	Right tibia heads and growth plates.	Post–flight: R+ 53.5 h.
45	**Garetto *et al.* 1990**	Cosmos 1887	Wistar rat	B: 4 F: 4 S: 4 V: 4	Fibroblast like osteoblast precursors in PDL of maxilla first molar.	Post–flight: R+55 h.

Table 3. (continued).

No.	Author	Flight	Experimental organism	Number of samples*	Bone site and/or cells examined	Remarks
46	**Mechanic *et al.* 1990**	Cosmos 1887	Wistar rat	B: 5 F: 5 S: 5 V: 5	Right femur.	Post–flight: R+2 d.
47	Zerath *et al.* 1990	Cosmos 1887	rat monkey	F: n = 5 F:n = 2	Iliac crest biopsies (monkey). Left humerus (rat), 9th thoracic vertebrae (T9).	Synchronous controls for the monkeys were the same individuals, and tests were performed 43 d after the actual flight.
48	**Doty *et al.* 1992**	Cosmos 2044	Wistar rat	F: 5 S: 5 V: 5	Tibia (tendon of mid–metatarsal area).	
49	Zerath *et al.* 1991	Cosmos 2044 Cosmos 2044	rhesus monkey Wistar rat	not known F: 5 S: 5 V: 5	Iliac crest (monkey). Left humeri, 9th thoracic vertebrae (T9).	R+34 d Rat samples: R+11d R + 6–10 h.
50	**Garetto *et al.* 1992**	Cosmos 2044	Wistar rat	B: 5 F: 5 S: 5 V: 4	Fibroblast like osteoblast precursors in PDL of maxilla first molar.	Post–flight: R+8.5–11 h.
51	Kaplansky *et al.* 1991	Cosmos 2044	Wistar rat	F: 5 S: 5 V: 5	Fractured fibula and intact tibia.	Also 5 tail suspended rats. Post–flight: R+ 4–7 h.
52	**Pedrini–Mille *et al.* 1992**	Cosmos 2044	Wistar rat	B: 5 F: 5 S: 5 V: 5	Invertebral disks, L3–4 to L6–S1.	Also 5 tail suspended rats.
53	**Vailas *et al.* 1992**	Cosmos 2044	Wistar rat	B: 5 F: 5 S: 5 V: 5	Humerus, femur (cortical bone).	Post–flight: R3 – 11 h.
54	**Vico *et al.* 1993**	Cosmos 2044	Wistar rat	B: 5 F: 5 S: 5 V: 5	Left proximal tibia + femur, 5th dorsal vertebral body (D5), 2nd lumbar vertebral (L2) body and the femoral fossa trochanteri.	Total of 10 of rats per group, 5 used in this study, 5 used for bone repair studies by Kaplansky *et al.* 1991. Post flight: R+6–10 h.
55	**Arnaud *et al.* 1992**	Cosmos 2044	rat	B: 5 F: 5 S: 5 V: 5	Blood plasma.	Also 5 tail suspended rats.

Table 3. (continued).

No.	Author	Flight	Experimental organism	Number of samples*	Bone site and/or cells examined	Remarks
56	Schneider *et al.* 1992	Mir	human		Pelvis, femoral neck, trochanter.	Only flight subjects. In space for 131–312 d.
57	Schmidt *et al.* 1991	parabolic flight	ROS 17/ 2.8 osteoblast like cells	F: ? G: ?	Cell culture.	Fixed in flight.
58	Malouvier *et al.* 1992	parabolic flight	ROS 17/ 2.8 osteoblast like cells	G-stress: ? Control: ?	Cell culture	Fixed in 'flight'
59	**Backup *et al.* 1994**	STS-41 STS-52	Sprague-Dawley rat	4 d mission: F: 8 Ground: 12 10 d mission: F: 6 Ground: 6	Calvaria, left femur, humeri, radii and ulnae periostea.	Post-flight: R+4–6 h. SPE-1: STS-41. SPE-2: STS-52.
60	van Loon *et al.* 1995	STS-42	isolated fetal mouse long bones	G 1 × g: 8 G 2 × g: 8 F microgravity: 8 F 1 × g: 8	Metatarsal long bones.	Fixed in flight.
61	**Duke *et al.* 1993**	STS-42	isolated fetal mouse cartilage cells	G 1 × g: ? G 2 × g: ? F microgravity: ? F 1 × g: ?	Cartilage cells micromass culture.	Fixed in flight.
62	**Klement & Spooner 1994**	STS-54, STS-56	isolated fetal mouse long bones	STS-54: F: microgravity: Ground: STS-56: F: microgravity: Ground:	Metatarsal long bones.	Fixed in flight.
63	Bikle *et al.* 1994	STS-54	Sprague-Dawley rat	F: 6 G: 6	Proximal tibiae.	

a radius of 32 cm and held rats which were individually housed in 9.4 ×20.5 cm space flight cages.

Vivarium (V) The vivarium group derived from the same pool of animals as the former three groups. This group was not housed in the special, rather small, flight cages, but in standard laboratory cages, which gave them more space and greater freedom. This group is not subjected to any launch related phenomena.

Table 4. Overview of studies on bone and space flight: details of methods and results.

No.	Author	Experimental organism	Technique	Results and discussion
1	**Mack *et al.* 1967**	human	Gemini 4: photo's at: L–9, L–3 d & L–0; R+0; R+16 & R+50 d Gemini 5: L–10, –4, –2 and – 0 d; R+ 0, 1, 10, 58 d. Gemini 7: L–10, –3, –0 d; R+0, 1, 11, 57 d. ^{125}I source, monoenergetic photons.	In this very extensive study all bone sites lost bone, especially in the appendicular skeleton. Some calibration mistakes were made see **Vose 1974.**
2	**Vose 1974**	human	Automatic scanning of röntgenograms.	Correction of **Mack *et al.* 1967** on bone density. Gemini 4: 2.9 and 3.8% loss; experiment errors in Gemini 5: 9.2 and 2.5% loss ± 2.5%.
3	**Lutwak *et al.* 1969**	human	Ca^{2+}, $PO_4$3–.	Great inter–individual variability. Calcium balance less positive in both astronauts. In one subject significant increase in urinary Ca excretion. Both subjects showed increased dermal Ca^{2+} and $PO_4$2– loss. Also changes in other ions (S^{2-}, Na^+, K^+, $Mg^{2+}$and Cl^-).
4	**Mack 1971**	non–human primate, *Macaca nemestrina*	X–ray, bone density by radiography.	Average loss in flight animals of ±4.5%. Major losses in patella, distal end of the ulna, the wrist (= capitate) and the hand phalange 4/2. Also losses in control animal, probably due to immobilization.
5	**Rambaut *et al.* 1975**	human	Atomic absorption spectrometry for Ca. Auto analyzer techniques for P.	Increase in urinary and faecal P excretion during flight. Increase in faecal calcium excretion during flight. 0.2% total body calcium loss, 0.7% total body phosphorus loss. Reported changes could also, besides hypogravity unloading, be the result of disturbances due to gastrointestinal absorption.
6	**Whedon *et al.* 1975**	human	Food intake registration, urinary and faecal Ca & P excretion. Ca, P, N, K, Na, and Mg balance studies.	Urinary Ca increased gradually during flight. Faecal Ca excretion did not change. Calcium balance became negative or less positive compared to preflight values. Increased excretion and negative balance of P and N. Significant loss of K, Na and Mg in flight.
7	**Vogel & Whittle 1976a & b**	human	Photon absorptiometry: ^{125}I source, monoenergetic photons. Samples at launch -30, -15, –5 d; R+0, 1 & 7 d and thereafter.	Data was quite variable. Some measurements also displayed an increase in BMD. Lowest decrease was 7.9%, greatest increase 3.1%. Overall there was a negative trend, especially in the os calcis and right radius, there was however, a slight increase in BMD in the ulna.

Table 4. (continued).

No.	Author	Experimental organism	Technique	Results and discussion
8	**Tilton *et al.* 1980**	human	Photon absorptiometry.	Flight crew significant increase of average bone loss compared to a control group But after 5 years exercise and diet both directly effecting bone the difference between the same groups varied *ad lib* over these 5 years.
9	**Yagodovsky *et al.* 1976**	Wistar rat	Histology, LM, EM, morphometry.	Half of the flight samples showed osteoporosis of the metaphysis, usually combined with a decrease in the mass of the primary spongiosa near the epiphyseal cartilaginous plate. Suggests that growth could have been inhibited during flight. LM and EM of flight bones showed the presence of wide osteocyte lacunae with could have been produced by perilacunar osteolysis. Some of the phenomena were also seen in synchronous samples but not as pronounced as in flight. A 27 d post–flight period at $1 \times g$ was not enough to return to control values.
10	**Organov *et al.* 1991**	human human	Quantitative computer tomography, mineral density and DPA (dual photon absorptiometry). Hypokinesia: 5° head down tilt.	Changes in mineral density of lumbar vertebrae in cosmonauts were multidirectional, their magnitudes did not show any correlation with flight duration. Similar changes seen in hypokinesia study (370 d) and were not related to bed rest time either. Hypermineralization of specific vertebral segment was more often related to inadequate or lacking countermeasures. There was a decrease in mineral density in the proximal femur epiphysis. This was more constant in direction (decrease) and regular. The authors state that the results of bone mineral density measurements with non–invasive methods (CT, DPA) were not directly related to the mechanical bone strength: other methods should be used in parallel.
11	**Morey & Bayerlink 1978**	Wistar rat	Tetracycline injection at L–3 (flight), L–8 (synchronous), R+3 and R+26. Morphometry. Arrest lines.	Periosteal bone formation rat decreased significant by 47%. Arrest lines suggest a complete cessation of bone growth during flight. Defects were corrected in a 26 d post–flight period. No change in bone resorption.

Table 4. (continued).

No.	Author	Experi-mental organism	Technique	Results and discussion
12	**Turner *et al.* 1985**	Wistar rat	Micrographs, UV–LM, SEM, TEM. All rats (both flights) labelled with tetracycline at L–1 and at R+3 d (Cosmos 782) and R+4 d (Cosmos 963).	Defect at periosteal surface of tibia; an arrest line. In arrest line: abnormal collagen orientation; hypomineralization. 'Arrest lines are located at interface of bone formed in space and that from post–flight'.
13	**Spengler *et al.* 1983**	Wistar rat	Femur: mechanical test and geometry. Tibia: morphology of thin sections.	In the femur, significantly reduced torque, stiffness, energy in microgravity group compared to in flight 1 × g. In femur as well as tibia, no significant difference in geometry between microgravity, 1 × g flight and synchronous. In flight 1 × g, substantially enhanced bone strength, possibly by promoting more normal tissue maturation.
14	Spector *et al.* 1983	rat	Tetracycline, arrest–lines, SEM, TEM.	Cessation of periosteal bone formation did not occur in places where intrinsic muscle forces continue to act on the tibia. Percentage change in bone formation rate on various sides ranging from –16 to –49% for flight compared to controls. Arrest lines mineralization defect, related to abnormal organic matrix.
15	**Wronski & Morey 1983a & b**	Wistar rat	Tetracycline labelling at launch –3 d & recovery +5D. UV histology, MERZ grid analysis, Arrest lines.	Significantly less periosteal bone formation in the humeral and tibial diaphysis (44%). No change in periosteal bone formation in the rib. In periosteal bone, formation in tibia during post flight period increased significantly. No change in periosteal bone formation in humerus during post flight period. No differences in medullary area in any group = no gross change in endosteal resorption activity. Relation in weight bearing on bone site: tibia > humerus > rib (also lower formation rate plays a part).
16	Stupakov *et al.* 1989	Wistar rat, human (bed rest)	Biomechanical analysis, acoustic emission.	Significant decrease in stress absorbance capacity of flight samples at 6 and 29 d post flight, no difference directly after flight. Also decrease in bone mechanical parameters of bed rest study (370 d & 120 d). Training partially counteracted the decrease.

Table 4. (continued).

No.	Author	Experimental organism	Technique	Results and discussion
17	**Simmons *et al.* 1983**	rat	At L –3 d tetracycline injection: again after 5 or 27 d. Post–flight; density gradient fractionation; Ca, P, Hy–Pro; sulphur, Histomorphometry.	Differences in response of jaws and ribs probably related to size of their respective osteoprogenitor pool. There was an overall decrease in turnover rates without creating a negative balance. Microgravity did not disrupt the normal coupling of bone formation and bone resorption. nor alter periosteal bone formation rate in ribs and regions of mandibles. In microgravity, formation–calcification rates were impaired at sites in jaw of non–continuous muscle. Under microgravity indications of a delay of maturation of bone mineral and matrix. At 29 d post–flight all values had returned to normal. It is concluded that most of non–weight bones are also at risk in microgravity.
18	**Pitts *et al.* 1983**	Wistar rat	Whole body composition of all kinds of parameters (fat, water, etc.). Mainly musculo-skeletal system (+ other parts).	Reduced fraction of total body water, shift from skin to viscera. Marked (significant) reduction in the fraction of bone minerals. A relative net increase in total muscle mass. Many of the differences could also be due to age difference between the groups.
19	**Cann & Adachi 1983**	Wistar rat	Ca isotope tracer techniques. Ca, Mg, K, Zn, Na content. Ash weight.	No differences in faecal excretion mass. No significant changes in bone resorption or bone formation. A significant increase in Na, K and Zn excretion.
20	**Wronski & Morey 1983a & b**	rat	Tetracycline, histomorphometry.	Decreased periosteal bone formation in all flight tibiae: 44–37% decrease.
21	**Eurell & Kazarian 1983**	Wistar rat	Ca, collagen, keratosulphate, chondroitin sulphate, (histochemistry).	Decreased Ca content in flight (significant), only at R+6 d returned to normal at R+29 d. (no change at d R+0). Significantly higher keratosulphate in flight+0 and flight+6 d compared to all other groups. No significant changes in chondroitin sulphate and collagen content. It is argued that all these changes are indicative for an increased aging process.

Table 4. (continued).

No.	Author	Experimental organism	Technique	Results and discussion
22	**Jee *et al.* 1983**	Wistar rat	Histomorphometry, declomicine (= tetracycline).	29 d post–flight: in flight tibia significant reduction in trabecular bone volume compared to synchronous. 0 d post–flight: flight significant decrease in trabecular bone volume compared to synchronous in humerus. Significantly more fat (in marrow) in flight compared to synchronous just after recovery: returned to normal at 29 d post flight. Reduced number of osteoblasts 0.3 mm from the cartilage metaphyseal junction (significant). No changes in osteoclast numbers. Hypothesized that the effects of space flight are not due to osteoblasts or osteoclasts but are due to incomplete/retarded growth of the calcified cartilage of the growth plate = less mineralized trabeculae = less bone. This would also reflect the retarded growth in length.
23	France *et al.* 1982	Wistar rat	Atomic absorption spectrophotometry, specific ion electrodes, Vit–C, colorimetric analysis. Na^+, K^+, Ba^{2+}, Ca^{2+}, Mg^{2+}, Sr^{2+}, Y^{2+}, Pl^{2+}, Mn^{2+}, F^-, Cl^-, PO_43- content.	Only significant change in calcium content: lower in flight compared to synchronous. Unchanged PO_43- and decreased Ca content suggests incomplete osteoid mineralization.
24	**Vico *et al.* 1987**	Wistar rat	Bone histomorphometry, TRAcP.	The only flight data on skeleton of pregnant rats. Bone mass was not changed. Mean thickness of osteoid seams and endosteal resorption thickness in long bones was unaffected. Osteoclast number: increase in thoracic (significant), and in lumbar (not significant). This could also be due to hormonal changes due to a combined pregnancy and microgravity effect. No changes in osteoblast parameters in any bone site.

Table 4. (continued).

No.	Author	Experimental organism	Technique	Results and discussion
25	Pospisilova *et al.* 1987	Wistar rat	Pepsin digestion; collagen Type I, II, III, effects both in bone and skin (hormonal influence?).	More neonatals of microgravity group died (19%) compared to synchronous (2.5%), vivarium (0%). In the bones of the flight rat offspring there was an increased collagen solubility, a fall in the ratio Hy–Pro –glycoproteins, a slowly increase in Type I collagen and persistent higher values of Type II and III collagens. As is seen by relatively more Type III collagen = matrix is less mature. Concluded a retarded maturation of protein components, persisting until maturity.
26	Bakulin *et al.* 1985	rat	Volume–weight, ash weight, Mg, Ca, Zn, P, K, Na, silicon content, mechanical strength (stress–strain, Young's modulus).	A significant decrease in organic + mineral (Ca) content; decreased Ca/P ratio, mainly due to a significant increase in P content. A significant decrease in mechanical properties; ultimate strength and relative strain.
27	Pospisilova *et al.* 1986	Wistar rat	Collagens Type I, II, III.	Significant reduction in collagen Type I in flight, and significant higher overall amount of peptides in flight samples.
28	**Patterson-Buckendahl *et al.* 1987**	rat	Three point bending: Ca, Pi, Osteocalcin, Mg, collagen, Hy–Pro.	In vertebrae: total bone mass, Ca, inorganic P, Mg, osteocalcin: all significantly down in flight compared to flight simulation. Hy–Pro slightly increased in flight. Increase could be due to stimulation of synthesis or reduced degradation. In humeri: mass, mineral, osteocalcin; significantly down in flight, total Hy–Pro unchanged. All resulted in a mineralization deficit. Bone strength data suggest a change in bone quality related to the mineralization of bone matrix.

Table 4. (continued).

No.	Author	Experi-mental organism	Technique	Results and discussion
29	**Wronski** ***et al.*** **1987**	Sprague Dawley rat	Calceine label: large rats at L–9 and L–2 d. Histomorphometry: using labels for large rats and non-labelled for small rats.	Large flight rats tibial diaphysis: significant decrease in periosteal bone formation rate compared to controls, proximal humerus and lumbar vertebrae: Long bone growth: reduced (not significant), no change in trabecular bone volume, proximal humeral metaphysis: osteoclast surface and number: increased (not significant), lumbar vertebrae: unchanged osteoclast surface & number. Small flight rats vertebrae: trabecular bone vol. 25% decreased (not significant), primary spongiosa of proximal tibial metaphysis: osteoblast parameter:down (not significant). In general only trends were found. The number of samples was too small to detect significant differences.
30	**Simmons** ***et al.*** **1986**	Sprague–Dawley rat	Gradient density fractionation, X–ray diffraction. Ca, Hy–Pro and P, Mg content. Ca/P ratio and ash weight.	Flight femoral diaphysis contained more inorganic P than the flight control. In the thoracic vertebrae the flight samples had a lower ash weight. In general there was a shift of matrix profiles and mineral moieties towards both higher (femurs, vertebrae) and lower (vertebrae) portions. All indicated a decrease in bone growth/turnover in flight samples and a relative increase in skeletal maturation. X–ray diffraction of the vertebrae indicated that space flight was associated with a decrease in apatite crystal size/perfection. This in turn could have caused reduced mechanical strength.
31	**Shaw** ***et al.*** **1988**	Sprague Dawley rat	Mechanical testing (3–point bending tests), bone morphometry (gross), bone density.	Total length of tibia and humerus of microgravity significantly less than flight control: with age correction only humeri remained significant. No significant change in diaphyseal cross–sectional morphologies of tibia nor humeri. Tibia diaphyseal cross–sectional densities down (significant) in flight: down non–significantly in humeri. Microgravity impeded the mechanical maturation (strength, stiffness) of bone: more pronounced in tibia than in humerus.

Table 4. (continued).

No.	Author	Experimental organism	Technique	Results and discussion
32	Doty *et al.* 1985	rat	AlP, AcP, dipeptidyl peptidase–II, LM, EM and histomorphometry.	No changes in AlP, AcP, Dipeptidyl peptidase–II. Tendency to reduced cytoplasmic area = less osteoblastic activity = less matrix synthesis.
33	Duke *et al.* 1985	rat	LM, zone heights morphometry. Freeze substitution for microanalysis.	Although there were some technical problems, flight animals revealed a low Ca, P, K and Na content, while Mg was not affected. The S levels were much less in flight compared to ground. All the zones within these long bones, resting, proliferating, and hypertrophic/calcification zone were reduced in length. It was concluded that even short space flight can alter bone mineralization.
34	**Vico *et al.* 1988**	Wistar rat	Local control rats: injected with tetracycline. Bone histomorphometry, TRAP.	Tibia: trabecular bone volume reduced, trabeculae were decreased in number and thickness. Formation activity at trabecular and endosteal levels reduced. Resorption activity unchanged. Femur: trabecular bone volume: down (not significant). Osteoclastic and osteoid parameters unchanged. No changes in bone mass and bone resorption in thoracic and lumbar vertebrae.
35	**Vico *et al.* 1991**	Wistar rat	Bone histomorphometry, TRAP, tetracycline (only suspended rats).	In both unloading experiments (space flight as well as tail suspension) a clear and significant decrease in bone formation parameters. Difference in osteoclastic activity: no change under microgravity while a significant increase osteoclastic parameters with tail suspension. Under microgravity a significant decrease in tibial metaphysis trabecular and endosteal bone formation.
36	Pospisilov *et al.* 1988	Wistar rat	Collagen Type I and III: pepsin digestion (soluble/insoluble), chromatography.	Total amount of collagen (Type I+III) down (significant) by 35% in microgravity (in femurs). No change in total collagen in skin. Pepsin–soluble collagen (Type I+III)up (significant) in microgravity in femurs; down in skin (significant). Type III pepsin soluble insoluble up (significant) in microgravity femurs: up in skin. All indicates: high turn–over; active remodeling, bone and skin: = systemic factors.

Table 4. (continued).

No.	Author	Experimental organism	Technique	Results and discussion
37	Weingart *et al.* 1988	Wistar rat	Carbonapatite concentration by infra-red spectroscopy.	Significant reduction in carbonapatite only in male rats dentine, cement and bone, no changes in female rats. This could be a result of a combination of a short flight and different hormone levels compared to males rats.
38	**Morey-Holton *et al.* 1988**	human	Bone related hormones (1,25)-, (25)-, (24-25)-Vit D, PTH and total Ca, P and albumin.	Significant increase of early in flight 1,25vitD levels; returned to normal later in flight: could be due to reduced food intake and could transiently have activated bone remodeling. No changes in PTH levels. Calcium, P, albumin, 24(OH)D and 24, 25(OH)2D levels, all normal.
39	**Zernicke *et al.* 1990**	Wistar rat	Mechanical testing: compression Hy-Pro, Ca, P, hydroxy-pyridinoline, collagen cross-links.	Re-entry forces and transport not mimicked for synchronous rats. Flight: body weight significantly down compared to synchronous and vivarium. Weight L6: not significantly reduced in flight compared to synchronous, vivarium & basal. A 47% decrease in compression stiffness in flight: (not significant) compared to synchronous and vivarium. Initial maximal load in flight samples was -18% compared to synchronous (not significant). Average linear load in flight: reduced (not significant) synchronous and vivarium. No change in Ca, P & Hy-Pro concentrations in any group. Cross-links: Flight significantly down compared to vivarium: not significantly down compared to synchronous and basal. 'Combined biomechanical and matrix biochemistry results suggest that the effects found were material as well as structural in nature'. Slowed maturation of trabecular bone in vertebral bodies of rapidly growing rats.

Table 4. (continued).

No.	Author	Experi-mental organism	Technique	Results and discussion
40	Kaplansky *et al.* 1990	Wistar rat	Bone histomorphometry, TRAcP.	Reduction of bone in primary and secondary spongiosa shown as a decreased number and thickness of trabeculae in the flight tibia. Thickness and length of the epiphyseal growth plate was diminished in flight samples. Decrease in number and activity of osteoblasts in tibial metaphysis, while number of osteoclasts increased in primary spongiosa of flight bones. Nearly identical patterns were seen in the vertebrae. No significant changes on compact bone.
41	**Simmons** ***et al.*** **1990**	Wistar rat	Bone density fractionation, X-ray diffraction, Ca, P and Mg content. Ash weight.	In calvaria: mineralization profiles to lower weights = indicates impaired mineralization. Space flight did not much alter the composition of dentine. In microgravity overall mineral composition not changed in calvaria, vertebra, mandible. In microgravity: hydroxyapatite crystals in mandibles significant smaller, in calvarias lower Ca/P and Ca/Mg ratios tendencies = indicative of immature bone; 'microgravity does not adversely affect the maturation of newly formed bone only in weight bearing bones, such effects occur throughout the skeleton'.
42	Földes *et al.* 1990	Wistar rat	Histology, histomorphology.	Density and volume of spongious trabecular of metaphysis reduced (significant). Density in diaphysis and the density of trabeculae reduced (significant). Reduction of bone formation parameters. Increase in osteoclastic index, osteoclast activity, Howships lacunae osteolysis. No significant changes of flight and immobilization, except for osteoid volume: up in immobilization. Some synchronous parameters more closely related to flight rather than vivarium = indication for stress/ immobilization.

Table 4. (continued).

No.	Author	Experimental organism	Technique	Results and discussion
43	**Doty** ***et al.*** **1990**	Wistar rat	EM, AlP, NADPase lipid staining, LM and hisomorphometry. (NADPase is involved in collagen synthesis and secretion, localized in Golgi.)	AlP in endothelial and perivascular cells at the LM level was decreased in flight samples of diaphyseal bone compared to control. Significantly more blood vessel profiles in flight bones compared to synchronous controls (could also be due to re-entry hypergravity and impact). No differences in bone, bone marrow area or periosteal perimeter.
44	**Duke** ***et al.*** **1990**	Wistar rat	EM, LM, histomorphometry, HE, Toluidine Blue, collagen orientation, collagen fibrils.	Significantly larger proliferation zone in flight. Reduced reserve and hypertrophic/calcification zone in flight. Significantly more cells in flight proliferation zone per column. Total number of cells significantly up in flight samples. Collagen fibrils were wider in flight compared to synchronous. All results possibly partly an effect of microgravity and recovery at 1 × g. In general, decreased length of the tibia and an impaired mineralization.
45	**Garetto** ***et al.*** **1990**	Wistar rat	Nuclear morphometry & volume analysis.	Flight: significant shift of A'–A (osteoblast progenitor) cells, –40%, towards C–D (pre-osteoblast) cells, +42%. There was also an increase in fibroblast like cells near the bone surface. A strong recovery of osteoblast precursor cells, which was probably due to 55 h recovery time, has been reported. Very significant increase in adrenal weight of flight animals. This could be a systematic response due to fluid shift or hormonal environment. Also note age difference of rats.
46	**Mechanic** ***et al.*** **1990**	Wistar rat	SEM image analysis, element analysis, biochemistry (powdered bones (SPEX)), collagen content, cross-linkage, osteocalcin (OC), Ca, P content.	Significant reduction of mineral and osteocalcin in distal half of femur diaphysis. Reduced collagen content and evidence of increased synthesis in proximal half of femur diaphysis. Longitudinal gradient of decreasing mineralization towards distal diaphysis (with X-ray microtomography).

Table 4. (continued).

No.	Author	Experimental organism	Technique	Results and discussion
47	Zerath *et al.* 1990	monkey and rat	Calcium label, histomorphometry, osteoblast apposition rate, osteoclasts, calcification rate, trabecular bone volume.	Monkey: significant reduction in calcification rate in flight compared to synchronous. Rat: reduced bone mass in humeral metaphysis and vertebral area. No changes in other parameters investigated.
48	**Doty *et al.* 1992**	Wistar rat	LM/EM, AcP, AlP, TEM, silver stain for collagen.	Non significant reduction of osteoblast activity/differentiation in flight compared to synchronous. No confirmation of results of Cosmos 1887 in the vascular system = recovery period? Randomly organized collagen fibrils in flight tendons compared to synchronous. Collagen fibrils: no change in periodicity in flight compared to synchronous samples.
49	Zerath *et al.* 1991	rhesus monkey	Rhesus monkey. Calceine label injected (a) 1 d before launch, (b) at re–entry, (c) before ground sim, (d) at the end of the simulation (d 34), trichrome staining: histomorphometry.	All results from cancellous bone tissue. Flight: significantly reduced mineral apposition rate. Flight: lightly reduced mineralizing surface. Diminished iliac crest bone mineralization activity.
		Wistar rat	Rat. See the above + osteoclast indexes. (Humerus is weight bearing in the rat while thoracic vertebra is not.)	Proximal humeral metaphysis + 9th vertebral body: humerus: reduced bone volume, reduced osteoblast surface (all slight), increased eroded surface (significant). Vertebrae: reduced bone volume (not significant), reduced osteoblast surface (not significant), increased eroded surface (not significant). General: reduced mineralization activity, reduced bone mass: a bone cell activity dissociation.
50	**Garetto *et al.* 1992**	Wistar rat	Nuclear volume analysis and widths of PDL.	No changes in osteoblast histogenesis data. Suggest that recovery time is important factor in these responses.

Table 4. (continued).

No.	Author	Experimental organism	Technique	Results and discussion
51	Kaplansky *et al.* 1991	Wistar rat	Bilateral fracture of fibula in diaphyseal central portion combined with histomorphometry.	Size of fibula calluses and strength of consolidated bone fragments were lower in flight compared to synchronous; callus in flight was smaller than in synchronous. In flight callus more fibrous compared to synchronous. = significantly more cartilaginous. Number of osteoclasts in flight per surface unit of spongy bone of the callus significantly up compared to synchronous. In tibia: osteoporosis in trabecular bone of proximal metaphysis; reduction of bone in primary and secondary spongiosa. Decrease of epiphyseal cartilage growth plate and primary spongiosa zone in flight compared to synchronous: also increased number of osteoclast in flight compared to synchronous. In general there was an inhibition of fracture healing in flight. Inhibition of growth in flight. Disorder in mineralization (= increased osteoid, & reduced number of osteoblasts). Increase of osteoclasts in flight. Osteoporosis in flight. Suspension model = good model for microgravity = same effects as seen in flight.
52	**Pedrini-Mille *et al.* 1992**	Wistar rat	LM, EM, biochemistry, proteoglycans (PG's), collagen, cross-linking.	No significant difference in body weight between flight, synchronous and tail suspension rats. No significant changes between flight and synchronous in collagen content, PG content, LM, EM, hexuronate content, or release, cross-linking of collagen. Significant difference only in annuli weight: flight less than synchronous, and significantly larger collagen to PG ratio in flight compared to synchronous. (For statistical evaluations, most data compares flight with all three other control groups combined).

Table 4. (continued).

No.	Author	Experimental organism	Technique	Results and discussion
53	**Vailas *et al.* 1992**	Wistar rat	Length, thickness, density, 3-point bending, morphometry, ultrasonic testing, collagen cross-links, Ca, hydroxyproline content.	No significant differences between flight and synchronous in any parameter. It is argued that this was due to the relatively old age of the rats resulting in a more 'quiescent' cortical bone.
54	**Vico *et al.* 1993**	Wistar rat	Bone histomorphometry, enzyme histochemistry (TRAcP).	Various differences between flight & synchronous versus vivarium, probably due to physiological stress. In the flight samples there was a significant decrease in width of primary spongiosa of proximal tibia versus synchronous. Significant increase in number of osteoclasts per mm and percentile active resorption surface in femoral fossa trochantiri (flight versus synchronous). Significant increase in trabecular separation in lumbar vertebral body in flight compared to synchronous.
55	**Arnaud *et al.* 1992**	rat	Plasma parathyroid hormone (PTH), calcitonin, Ca, Mg, PO_4 total protein, creatinine.	There were increased levels of phosphate, creatinine and total protein in flight animals (significant). PTH levels were slightly elevated in flight animals (significant). Flight animals also the lowest calcitonin levels, but these were only statistically different from the vivarium group.
56	Schneider *et al.* 1992	human	Whole body BMD, DEXA + exercise of cosmonauts.	After flight: increase of BMD in head (significant) 4.2±1.2%, loss in pelvis 7.4±2.1% (significant) = 5.0±2.2% femoral neck, 9.1±2.8% trochanter. Also in flight lean body mass down = fat increased, muscle mass decreased (significant). Concluded that exercise of cosmonauts minimized muscle atrophy without inhibiting the loss of bone.
57	Schmidt *et al.* 1991	ROS 17/2.8 osteoblast like cells	Gold labelled EGF binding to cell membrane.	No changes in EGF receptor–ligand interaction under microgravity compared to ground controls.
58	Malouvier *et al.* 1992	ROS 17/2.8 osteoblast like cells	Insulin and transferrin receptor mediated endocytosis.	Significant decrease in receptor binding combined with a significant increase in internalization. No difference noted in coated structures on the cell membrane.

Table 4. (continued).

No.	Author	Experimental organism	Technique	Results and discussion
59	**Backup *et al.* 1994**	Sprague-Dawley rat	Tetracycline and calceine labelling. Bone histomorphometry, total cellular RNA, Northern blot analysis of osteocalcin and collagen Type I.	Radial bone growth was unchanged in humeri during the 4 d flight, but significantly decreased during the 10 d flight group. Osteocalcin (4 and 10 d flight) and precollagen Type-I (10 d flight) mRNA levels were significantly decreased in long bones and calvarias periosteal cells after spaceflight.
60	van Loon *et al.* 1995	*in vitro* long bones	morphometry, ^{32}P and ^{45}Ca incorporation, ^{45}Ca release, collagen, glucose	No significant change in longitudinal growth. Significant decrease of glucose consumption. Significant decrease in mineral formation as measured by length of diaphysis, ^{45}Ca and ^{32}P incorporation. Increase in osteoclastic ^{45}Ca release. All microgravity values compared with in flight $1 \times g$ controls.
61	**Duke *et al.* 1993**	*in vitro* cartilage cells	Alcian Blue, SEM, TEM, LM.	Cultures ground and flight differentiated to mature chondrocytes, cell aggregates and nodules. Many of microgravity samples had smooth cell membranes and lack the typical cartilage matrix compared to controls. Unusual formation of either ruffled cell membrane or crystallized calcium salts on both flight samples (microgravity and $1 \times g$ in-flight control).
62	**Klement & Spooner 1994**	*in vitro* long bones	longitudinal growth, mineral formation??	Metatarsals (13.6 and 15.5 days old) underwent normal growth and differentiation under microgravity compared to control ground $1 \times g$ samples.
63	Bikle *et al.* 1994	Sprague-Dawley rat	mRNA levels of IGF-I, IGF-I receptor, AlP, osteocalcin.	Transiently increase of mRNA levels of IGF-I, IGF-I receptor and alkaline phosphatase. Decreased levels of osteocalcin (OC). AlP and OC levels consistent with maturation. IGF-I and IGF-I receptor may be a response to reduced bone formation.

*Index numbers of the publications are the same as used in Table 3.

Cosmos 605 was the first in this series of flights. After the 22 day mission, **Yagodovsky *et al.* (1976)** found signs of osteopenia in rat long bones. In their microscopical study they also reported an increase in the width of osteocyte lacunae in flight samples which suggests perilacunar osteolysis. These unusually wide lacunae were also noticed in bone biopsies, taken from three cosmonauts who died during a fatal descent in June 1973 (Yagodovsky & Gorokhova, unpublished observations, reported in **Yagodovsky *et al.* 1976**).
The Cosmos 782 biosatellite had a total of 25 rats on board. It was the first flight with 1 × g and 0.6 × g centrifuges on board. Unfortunately, the centrifuge samples were not used for the bone studies discussed below.
Tetracycline was used to determine bone mineral apposition rate (**Morey & Bayerlink 1978**). Rats were injected, 3 days before launch and 3 days after reentry with this bone seeking label, which makes it possible to estimate the bone formation rate during the experimental period. From histomorphological observations it was concluded that bone formation was reduced during flight, however, most of the parameters returned towards control values during a 26 day post flight period. Also arrest lines, consisting of a less mineralized and biochemically inferior matrix, appeared to be more frequently present in bones from flight animals of Cosmos 782 (**Morey & Bayerlink 1978; Turner *et al.* 1985**) and Cosmos 936 (**Turner *et al.* 1985**) compared to controls. The abnormal alignment of the collagen fibres and the reduced crystallite content of arrest lines is expected to reduce tensile and shear strength of the bone (**Turner *et al.* 1985**).
Cosmos 936 accommodated 30 rats in total. Twenty stayed under microgravity, the remaining 10 were placed on the on board 1 × g centrifuges. It was shown that the mechanical quality of bone from microgravity animals had diminished after space flight (**Spengler *et al.* 1983**). While there were differences between the microgravity group and the in flight 1 × g group as well as the synchronous group, no significant differences were found between the synchronous and the in flight 1 × g samples with regard to various mechanical parameters. Most of the parameters analyzed had returned to control values 25 days after the flight (**Spengler *et al.* 1983**). In addition, periosteal bone formation in rat tibia was decreased in space flight samples compared to Earth based controls (Spector *et al.* 1983).
In contrast to the two preceding missions Cosmos 1129 had no on board 1 × g centrifuge. It housed a total of 37 young male rats, divided into 5 groups. After this 18.5 day flight the animals were killed at various times between 7 h to 29 days post flight. In general, lower calcium contents were found in flight samples (France *et al.* 1982; **Eurell & Kazarian 1983; Jee *et al.* 1983**). This was confirmed during the Cosmos 1514 mission (Bakulin *et al.* 1985). Bone histomorphological methods have been applied by **Vico *et al.* (1987)** for Cosmos 1514, and later for Cosmos 1667 (**Vico *et al.* 1988, 1991**), Cosmos 1887 (Vico *et al.* 1990) and Cosmos 2044 (**Vico *et al.* 1993**). From these experiments it is clear that bone formation is diminished during orbital space flight. Some of these light microscope studies also revealed an increased number of bone resorbing cells, osteoclasts (Cosmos 1514 and 2044). The mechanical properties

as well as calcium and collagen Type-I contents were diminished after the 5 day Cosmos 1514 flight (Bakulin *et al.* 1985; Pospisilova *et al.* 1986, 1987). This reduction in rat femur collagen Type-I, and Type-III, content was affirmed during the Cosmos 1667 flight (Pospisilova *et al.* 1988). Since there was no change in skin collagen Type-I and III content it was argued by the authors that there was no systemic influence responsible for the reduction in bone collagen, suggesting a specific effect of space flight on bone.

Several papers have been published on the Cosmos 1887 flight which carried 10 rats and 2 rhesus monkeys (see also **Grindeland *et al.* 1990**). The spacecraft landed in bad weather in Siberia so the recovery time was unusually long, being at least 48 h. Various techniques have been used to study bone tissues from this flight. Mechanical tests (**Zernicke *et al.* 1990**), light microscopical histomorphometry, (Doty 1985; Kaplansky *et al.* 1990; Vico *et al.* 1990; Földes *et al.* 1990; **Garetto *et al.* 1990**; **Doty *et al.* 1992**; **Duke *et al.* 1990**; Zerath *et al.* 1990), and measurements of the content of various organic and inorganic compounds were performed (**Mechanic *et al.* 1990**; **Simmons *et al.* 1990**; **Zernicke *et al.* 1990**). In general the data indicate reduced mechanical, structural and/or biochemical competence of bones of flight animals.

Out of the ten rats flown on Cosmos 2044, five were used for bone related studies (see also **Grindeland *et al.* 1992**). Comparable techniques were applied to those used on Cosmos 1887 animals. After the 14 day flight, similar results were reported as for previous flights. In Cosmos 1887 as well as 2044 both weight bearing and non-weight bearing bones in the rat skeleton revealed signs of osteopenia. A significant increase in calcium content of flight long bones has never been reported.

Numerous cosmonauts have occupied the Russian space station Mir and its predecessor, Salyut-7. The nine cosmonauts examined in a study by **Organov *et al.* (1991)** are only a fraction of the total number of visitors to these space stations. In this study there was, surprisingly, no clear change in vertebral bone mineral density in cosmonauts who lived under near weightless conditions for up to 150 and 237 days. However, reductions of bone mineral density in pelvis, femoral neck and trochanter were reported in the Mir crew after a maximum of 312 days in flight (Schneider *et al.* 1992).

To date, the U.S. has flown two Spacelab missions, launched with the Space Shuttle, which have been completely dedicated to microgravity life sciences, Spacelab-3 (SL-3) and SL-2. An advantage of Shuttle missions over Cosmos flights is that Shuttle flights are manned, and crew members can give attention to the hardware and the animals, thereby promoting the well being of animals in flight. Also, the launch and re-entry characteristics for a typical Shuttle mission are less harsh compared to a Cosmos lift-off. A disadvantage, especially with older orbiters, is the relatively short duration of the flights, typical missions lasting up to about 10 days.

A total of 8 non-human primates and 72 rats were flown on Spacelab-3 and Spacelab-2 missions. About 15 rodents were used for bone related studies. There were no launch and re-entry simulations for the animals, nor was there a 1 × g centrifuge on board. After the 7 day SL-3 mission a clear reduction in bone mechanical strength was reported in rat vertebrae and humeri (**Patterson-Buckendahl *et al.* 1987**; **Shaw *et al.* 1988**) combined with decreased mineralization parameters at various sites in the skeleton (**Simmons *et al.* 1986**; Duke *et al.* 1985; **Patterson-Buckendahl *et al.* 1987**; **Wronski *et al.* 1987**). Bone related hormones were screened in the astronauts (**Morey-Holton *et al.* 1988**). During the SL-2 mission, only a transient increase in 1,25-dihydroxy vitamin D levels was found which could have induced increased bone remodelling, all other parameters remained unchanged. Recently, some cellular parameters of bone formation, collagen Type-I and osteocalcin mRNA levels, have been studied by **Backup *et al.* (1994)**. They reported a decrease in both parameters after a 10 day Shuttle flight in ulnae periostea of the rat, while the osteocalcin mRNA level was also reduced after a 4 day mission.

5.2.2 *In vitro* space flight results

Up to now only a very limited number of successful *in vitro* experiments with bone cells or skeletal tissues is performed under microgravity. Some experiments were performed during parabolic flights, which provides short periods of microgravity (20 s) alternated with short periods (20 s) of hypergravity (1.8 × g). Schmitt *et al.* (1991) studied osteoblast like cells (ROS) receptor binding of gold labelled EGF selectively during the 20 s periods of microgravity. Receptor binding was observed during microgravity but no differences were found with the 1 × g controls. Malouvier *et al.* (1992) exposed the same type of bone cells to a series of successive parabola in which the cells were subjected, during an 18 min period, to increased and decreased gravity (gravitational stress). Under these conditions a decreased binding was found for gold labelled insulin and transferrin, while the internalization process (number of labelled endosomes) was accelerated.
The first experiment with isolated skeletal tissues was performed in 1992 during the IML-1 mission (STS-42) in Biorack with fetal mouse long bones (**van Loon *et al.* 1995**). In this experiment developing bones from the feet of 16 and 17.5 day old mouse embryos were used. During a 4 day culture period under microgravity overall growth was not changed and the growth was comparable to normal laboratory conditions. However, mineralization was strongly reduced leading to a 30% decrease in the length of the mineralized diaphysis, and to a 40-50% reduction in the uptake of calcium and phosphorus. On the other hand, osteoclastic mineral resorption (measured as release of ^{45}Ca from long bones prelabelled *in utero*) was significantly enhanced (+43%). All microgravity results were compared to an in-flight 1 × g reference centrifuge. This experiment was partly repeated in 1993 during a Russian Biocosmos mission (BION-10) and in 1994 during the IML-2 Space Shuttle mission (STS 65). Preliminary results of these experiments confirm the data from the experiment during IML-1 mission. These experiments strongly indicate that bone as a tissue is directly sensitive to microgravity

conditions, and that microgravity stimulates the activity of osteoclasts but inhibits the activity of osteoblasts and calcifying chondrocytes of the growth plate. Such a reaction explains the loss of bone density observed in astronauts.

During the IML-1 mission in another experiment, isolated mouse chondrocytes were exposed to microgravity. Some of the preliminary results show that in all groups significant detachment of the cell layers from the substratum occurred (**Duke *et al.* 1993**). No differences were found between cartilage nodules formed in microgravity and 1 × g controls. Although cell detachment made clear conclusions difficult, the authors claim that indications were found for a delayed and/or defective chondrogenesis under microgravity. **Klement & Spooner (1994)** using mouse metatarsals (from 15.5 day old embryos) studied growth and development during two Shuttle missions in 1993 (STS-54, STS-56). Cartilage phenotype was maintained during spaceflight culture but mineralization did not occur (although 10 mM Na-β-glycerophosphate was present in the medium). It should be noted, however, that culture conditions were very different from those applied by **van Loon *et al.* (1995)**. Production of mRNA was studied by Bikle *et al.* (1994) in an experiment flown on STS-54. They studied insulin like growth factor one (IGF-I), its receptor (IGF-IR), alkaline phosphatase (AlP) and osteocalcin mRNA levels in extracts from pulverized rat proximal tibiae. They found a transient increase in IGF-I, IGF-IR and AlP while osteocalcin was reduced. The authors argued that the increase in AlP and decrease in osteocalcin are consistent with a shift towards decreased maturation, whilst increase in IGF-I and IGF-IR levels may be a compensatory response of bone cells to the decreased maturation.

5.3 General conclusions

The total number of flight individuals used for bone related studies during the first 30 years of space flight, included in this overview, are about 90 rats, 5 non-human primates and 26 humans (see Table 3). Some specific details from the published literature are recorded in Tables 3 and 4 in overview form. From these studies the following general conclusions can be drawn.

(i) During space flight less calcium is absorbed and/or more calcium is excreted, both resulting in net calcium loss. This was measured either in total body or at specific anatomical sites during metabolic studies or various X-ray analyses. An increased mineralization of long bones exposed to space flight conditions has never been reported (**Mack *et al.* 1967**; **Lutwak *et al.* 1969**; **Mack 1971**; **Vose 1974**; **Rambaut *et al.* 1975**; **Whedon *et al.* 1975**; **Vogel & Whittle 1976a & b**; **Yagodovsky *et al.* 1976**; **Morey & Bayerlink 1978**; France *et al.* 1982; **Eurell & Kazarian 1983**;

Pitts *et al.* 1983; Spector *et al.* 1983; Bakulin *et al.* 1985; Duke *et al.* 1985; **Turner *et al.* 1985; Patterson-Buckendahl *et al.* 1987; Mechanic *et al.* 1990; Organov *et al.* 1991**; Schneider *et al.* 1992).

(ii) Bone mechanical properties are diminished after space flight, especially in weight bearing bones (**Spengler *et al.* 1983**; Bakulin *et al.* 1985; **Patterson-Buckendahl *et al.* 1987; Shaw *et al.* 1988**; Stupakov *et al.* 1989; **Zernicke *et al.* 1990**). Although most changes after space flight have been found in the weight bearing skeleton, non-weight bearing bones also seem to be at risk (**Simmons *et al.* 1983, 1990**). This could mean that systemic factors also play a role in the effects of space flight conditions on bone.

(iii) Biochemical changes of the skeletal organic matrix compounds have been reported. Changes in the content of keratosulphate (**Eurell & Kazarian 1983**) reductions of Type-I and/or Type III collagen (Pospisilova *et al.* 1986, 1987, 1988; **Mechanic *et al.* 1990**) and increases in total peptide and peptide soluble collagen content have been reported (Pospisilova *et al.* 1986, 1988). In addition, wider collagen fibres were found in cartilage (**Duke *et al.* 1990**). Reduced osteocalcin content (**Mechanic *et al.* 1990**) and down regulated mRNA levels of osteocalcin and collagen Type-I were also reported (**Backup *et al.* 1994**). The altered calcium/glycosaminoglycan ratio (**Eurell & Kazarian 1983**) and retarded maturation of protein components (Pospisilova *et al.* 1987) are indications of retarded maturation of the bone matrix. Changes in the maturation of skeletal tissues after space flight was also suggested by **Simmons *et al.* (1986, 1990)**.

(iv) Histomorphological studies have shown a reduced trabecular bone volume (**Yagodovsky *et al.* 1976; Jee *et al.* 1983; Wronski *et al.* 1987; Vico *et al.* 1988, 1991, 1993**; Földes *et al.* 1990; Kaplansky *et al.* 1990; Zerath *et al.* 1991) or reduced cortical cross sectional area (**Shaw *et al.* 1988**) both resulting in reduced mechanical properties of bone.

(v) Overall longitudinal growth of long bones is affected under space flight conditions sometimes related to decrease in total body weight. Decreases were reported in lengths of humeri and tibial long bone (Duke *et al.* 1985; **Shaw *et al.* 1988**) as well as reductions in lengths of the different growth plate zones within the proximal tibia of the rat (Duke *et al.* 1985). Reduced total body weights have been reported for humans (**Rambaut *et al.* 1975; Whedon *et al.* 1975**) as well as for rats (**Zernicke *et al.* 1990**).

(vi) There are numerous constraints involved in space flight research. The first problem is the restricted number of flight opportunities, aggravated by the restricted number of samples per flight. The limited number of samples is a particular dilemma in human physiology studies. To increase scientific output, animals or samples are often shared between various research groups. While being favourable with respect to the number of participants to a particular flight, tissue sharing may also imply serious restrictions in the choice of techniques. Favourable experimental procedures for one particular study may be disastrous for others, resulting in reduced scientific output and quality.

Another problem, especially relevant to Cosmos flights, is the long recovery time. This time span is important since cellular processes will continue at unit gravity on the return to Earth. When the recovery time is long compared to the rate of the processes studied, a possible microgravity effect may be obscured. A solution to these problems is the preparation of samples in flight, as was performed during the SLS-2 mission (STS-58, October 1993).

Apart from the flight constraints, there is a great variability in age or total body weight of the rats used for the various space flight studies, sometimes even within the same experiment set-up (see e.g. Cosmos 1887). During the Cosmos flights, rats between 63 and 115 days old were used. This problem of variation in age was addressed by using rats of two different ages during the SLS-3 flight. Particularly in bone, there is a distinct difference in metabolic processes between young and adult animals. Also the 'lag-time' time difference between the flight and synchronous or ground controls can contribute to differences and variations; it has sometimes been more than 5 days (**Morey & Bayerlink 1978**).

It is not clear what the long term effects of space flight will be on the metabolic processes of growth and remodelling within the human skeleton. Although there have been in flight training programmes (Garshnek *et al.* 1988; **Greenleaf *et al.* 1989**) and some pharmacological countermeasures have been proposed (**Stepaniak *et al.* 1986**; see review by Whedon & Heaney 1993) very little is documented in the literature about the final impact of these procedures on bone tissue. Therefore, in spite of all efforts, very little is known about effective countermeasures for the bone loss occurring during space flight (Whedon & Heaney 1993; Pavy-Le Traon *et al.* 1994). It is quite likely that different procedures have to be adapted to counteract the tissue loss in muscle compared to bone tissue. As indicated in the beginning of this chapter, rather than the frequency and endurance of in-flight training cycles and prophylactics (cycle ergonometers, treadmills and 'penguin suits') it may be the level of impact (peak strain value) of the loads which is the most important factor for remaining bone integrity in space.

Microgravity simulation *in vivo* in studies involving bed rest (**Donaldson *et al.* 1970**; **LeBlanc *et al.* 1990**) denervation (Marotti *et al.* 1979; **Weinreb *et al.* 1989**; **Yeh *et al.* 1993**) and tail suspension (**Morey 1979**; **Globus *et al.* 1986**; **Vico *et al.* 1991**; **Simske *et al.* 1992**) do contribute to the knowledge of the impact of decreased mechanical loads upon the vertebrate skeleton. But with the realization of the international space station Alpha at the end of this century, a combined American, European, Japanese, and Russian initiative, more data will become available to further characterize the changes seen in human and animal bone during prolonged periods under near weightless conditions. In preparation for this era, long duration flights have already been undertaken. During, for example, the Euromir-95 mission, a more than four months flight, issues like bone strength, bone structure, various loading regimes and Vitamin K suppletion are under investigation. Also *in vitro* experiments using microgravity simulators like clinostats (**Briegleb 1968**) and free fall machines (Mesland 1990) could be used to understand microgravity responses. In addition, hypergravity facilities are useful to extend our knowledge of gravity accelerations on a cellular and tissue level (van Loon *et al.* 1993). Extensive ground based research combined with prolonged periods in space with the increased number of astronauts will allow sufficient information to accrue the development of countermeasures for musculo-skeleton deconditioning. These countermeasures might be used in future treatment programs for terrestrial immobilization osteoporosis, or metabolic processes following accidents.

Acknowledgements. We wish to thank the following for help during the preparation of this manuscript: Dr H. Oser (ESA, Paris), Dr E. Morey-Holton (Ames Research Center, Moffett Field, California, U.S.A.), Mr A. Kuipers (ESA/ESTEC, Noordwijk).

5.4 References

Aloia JF, Cohn SH, Babu T, Abesamis C, Kalici N & Ellis K. Skeletal mass and body composition in marathon runners. Metabolism 27: 1793–6, 1978.

Arnaud SB, Fung P, Popova IA, Morey-Holton ER, Grindeland RE. Circulating parathyroid hormone and calcitonin in rats after spaceflight. J. Appl. Physiol. 73, 169S-73S, 1992.

Backup P, Westerlind K, Harris S, Spelberg T, Kline B & Turner R. Space flight results in reduced mRNA levels for tissue-specific proteins in musculoskeletal system. Amer. J. Physiol. 226: E567–73, 1994.

Bakulin AV, Ilyan EA, Organov VS & Lebedev VI. The state of bones of pregnant rats during an acute stage of adaptation to weightlessness. In: *Proceedings, 2nd International Conference on Space Physiology* (Toulouse, France, 20–22 Nov. 1985), Ed. Hunt JJ, ESA SP-237. ESA, Paris, pp. 225–9, 1985.

Ballard RW & Connolly JP. U.S./U.S.S.R. joint research in space biology and medicine on Cosmos satellites. FASEB J. 4: 5–9, 1990.

Ballard RW & Mains RC. Animal experiments in space; a brief overview. In: *Fundamentals of Space Biology.* Ed. by Asashima M & Malacinski GM. Japan Scientific Press, Tokyo/Springer-Verlag, Berlin, pp. 21–41, 1990.

Bikle DD, Harris J, Halloran BP & Morey-Holton E. Altered skeletal pattern of gene expression in response to space flight and hindlimb elevation. Amer. J. Physiol. 267: E822-7, 1994.

Briegleb W. Physiologische Langzeiteffekte in rotierenden Systemen. Naturwissenschaftliche Rundschau 21(12), 507–12, 1968.

Cann CE & Adachi RR. Bone resorption and mineral excretion in rats during space flight. Amer. J. Physiol. 22: R327–31, 1983.

Cowin SC, Moss-Salentijn L, Moss ML. Candidates for the mechanosensory system in bone. Adv. Bioengin. 20: 313–16, 1991.

Dalsky GP, Stocke KS, Ehsani AA, Lee WC & Birge SJ. Weight bearing exercise training and lumbar bone mineral content in postmenopausal women. Ann. Intern. Med. 108: 824–8, 1988.

Donaldson CL, Hulley SB, Vogel JM, Hattner RS, Bayers JH & McMillan DE. Effect of prolonged bed rest on bone mineral. Metabolism 19: 1071–84, 1970.

Doty SB. Morphologic and histochemical studies of bone cells from SL-3 rats. Physiologist 28: S225–6, 1985.

Doty SB, Morey-Holton ER, Durnova GN, Kaplansky AS. Cosmos 1887: morphology, histochemistry, and vasculature of the growing rat tibia. FASEB J. 4: 16–23, 1990

Doty SB, Morey-Holton ER, Durnova GN & Kaplansky AS. Morphological studies of bone and tendon. J. Appl. Physiol. 73: 10S–13S, 1992.

Duke J, Janer L, Campbell M & Morrow J. Microprobe analyses of epiphyseal plates from Spacelab 3 rats. Physiologist 28: S217–8, 1985.

Duke PJ, Durnova G, Montufar-Solis D. Histomorphometric and electron microscopic analyses of tibial epiphyseal plates from Cosmos 1887 rats. FASEB J. 4: 41–6, 1990.

Duke PJ, Daane EL, Montufar-Solis D. Studies of chondrogenesis in rotating systems. J. Cell Biochem. 51: 274–282, 1993.

Eurell JA & Kazarian LE. Quantitative histochemistry of rat lumbar vertebrae following space flight. Amer. J. Physiol. 244: R315–8, 1983.

Földes I, Rapcsk M, Szilgyi T & Organov VS. Effects of space flight on bone formation and resorption. Acta Physiol. Hung. 75: 271–85, 1990.

France EP & Oloff CM. Kazarian Bone mineral analysis of rat vertebra following spaceflight COSMOS 1129. Physiologist 25: S147–8, 1982.

Frost HM. Vital biomechanics: proposed general concepts for skeletal adaptations to mechanical usage. Calcif. Tissue Int. 42: 145–56, 1988.

Galilei G. Two new sciences. In: *The Second Day*, translated by Drake S. The University of Wisconsin Press, pp. 109–46, 1974.

Garetto LP, Gonsalves MR, Morey ER, Durnova G & Roberts WE. Preosteoblast production 55 hours after a 12.5-day space flight on Cosmos 1887. FASEB J. 4: 24–38, 1990.

Garetto LP, Morey ER, Durnova GN, Kaplansky AS & Roberts WE. Preosteoblast production in COSMOS 2044 rats: short-term recovery of osteogenic potential. J. Appl. Physiol. 73: 14S–18S, 1992.

Garshnek V. Soviet space flight: the human element. ASGSB Bulletin 1: 67–80, May 1988.

Globus RK, Bikle D, Halloran B & Morey-Holton E. Skeletal response to dietary calcium in a rat model simulating weightlessness. J. Bone Min. Res. 1: 191–7, 1986.

Greenleaf JE, Bulbulian R, Bernauer EM, Haskell WL & Moore T. Exercise-training protocols for astronauts in microgravity. J. Appl. Physiol. 67: 2191–204, 1989.

Grindeland RE, Popova IA, Vasques M & Arnaud SB. Cosmos 1887 mission overview: effects of microgravity on rat body and adrenal weights and plasma constituents. FASEB J. 4: 105–9, 1990.

Grindeland RE, Ballard RW, Conolly JP & Vasques MF. Cosmos 2044 mission: overview. J. Appl. Physiol. 73, 1S–3S, 1992.

Harell A, Deckel S, Binderman I. Biochemical effects of mechanical stress on cultured bone cells. Calcif. Tissue Int. 22S: 202–9, 1977.

Jee WSS, Wronski TJ, Morey ER & Kimmel DB. Effects of spaceflight on trabecular bone in rats. Amer. J. Physiol. 244: R310–14, 1983.

Johnson PC, Leach CS & Rambaut PC. Estimates of fluid and energy balances of Apollo 17. Aerospace Med. 44: 1227–30, 1973.

Kaplansky AS, Durnova GN & Ilyina-Kakueva EI. Histomorphometric analysis of bones of Cosmos-1887 rats. Physiologist 33: S20–2, 1990.

Kaplansky AS, Durnova GN, Burkovskaya TE & Vorotnikova EV. The effect of microgravity on bone fracture healing in rats flown on Cosmos 2044. Physiologist 34: S196–9, 1991.

Klein Nulend J, Veldhuijzen JP, Burger EH. Increased calcification of growth plate cartilage as a result of compressive force in vitro. Arthritis Rheum. 29: 1002–9, 1986.

Klein Nulend J, Veldhuijzen JP, de Jong M, Burger EH. In creased bone formation and decreased bone resorption in fetal mouse calvaria as a result of intermittent compressive force *in vitro*. Bone Mineral 2: 441–8, 1987.

Klein Nulend J, vd Plas A, Semeins CM, Ajubi NE, Frangos JA, Nijweide PJ, Burger EH. Sensitivity of osteocytes to biomechanical stress *in vitro*. FASEB J 9: 441–5, 1995.

Klement BJ & Spooner BS. Pre-metatarsal skeletal development in tissue culture at unit- and microgravity. J. Exp. Zool. 269: 230–41, 1994.

Lanyon LE. Functional strain as a determinant for bone remodeling. Calcif. Tissue Int. 36: S56–61, 1984.

LeBlanc AD, Schneider VS, Evans HJ, Engelbretson DA & Krebs JM. Bone mineral loss and recovery after 17 weeks of bed rest. J. Bone Min. Res. 5: 843–50, 1990.

Lutwak L, Whedon GD, Lachance PA, Reid JM & Lipscomb HS. Mineral, electrolyte and nitrogen balance studies of the Gemini-VII fourteen-day orbital space flight. J. Clin. Endocrinol. 29: 1140–56, 1969.

Mack PB. Bone density changes in a *Macaca nemestrina* monkey during the Biosatellite III project. Aerospace Med. 42: 828–33, 1971.

Mack PB, LaChance PA, Vose GP & Vogt FB. Bone demineralization of foot and hand of Gemini-Titan IV, V and VII astronauts during orbital flight. Amer. J. Roentgenol., Rad. Therapy & Nuclear Med. 100: 503–11, 1967.

Malouvier A, Holy X, Zérath E, Marie PJ, Caissard JC, Schmitt DA. Receptor mediated endocytosis in osteoblastic cells under 'gravitational stress'. Physiologist 35(1): S37–38, 1992.

Marotti G, Delrio N, Marotti F & Fadda M. Quantitative analysis of the bone destroying activity of osteocytes and osteoclasts in experimental disuse osteoporosis. Ital. J. Orthop. Traumatol. 5: 225–40, 1979.

Mechanic GL, Arnaud SB, Boyde A, Bromage TG, Buckendahl P, Elliott JC, Katz EP & Durnova GN. Regional distribution of mineral and matrix in the femurs of rats flown on Cosmos 1887 biosatellite. FASEB J. 4: 34–40, 1990.

Mesland DAM. Gravity effects on cells. In: *Proceedings of the 4th European Symposium on Life Sciences Research in Space. ESA SP-307.* Ed. by David V. Paris, France, European Space Agency, pp. 221–5, 1990.

Morey ER. Spaceflight and bone turnover: Correlation with a new rat model of weightlessness. BioScience 29: 168–72, 1979.

Morey ER & Baylink DJ. Inhibition of bone formation during space flight. Science 201: 1138–41, 1978.

Morey-Holton ER, Schnoes HK, DeLuca HF, Phelps ME, Klein RF, Nissenson H & Arnaud CD. Vitamin D metabolites and bioactive parathyroid hormone levels during Spacelab 2. Aviat. Space Environ. Med. 59: 1038–41, 1988.

Nicogossian AE. Overall physiological response to space flight. In: *Space Physiology and Medicine*, 2nd edn. Ed. by Nicogossian AE, Huntoon CS & Pool SL, Lea & Febiger, Philadelphia, p. 148, 1989.

Organov VS, Rakhmanov AS, Novikov VE, Zatsepin ST, Rodionova SS & Cann C. The state of human bone tissue during space flight. Acta Astronautica 23: 129–33, 1991.

Patterson-Buckendahl PE, Arnaud SB, Mechanic GL, Martin RB, Grindeland RE & Cann CE. Fragility and composition of growing rat bone after one week in space flight. Amer. J. Physiol. 252: R240–6, 1987.

Pavy-Le Traon A, Güell A, Saivin S, Houin G, Soulez-Larivière C & Pujos M. The use of medicaments in space. Therapeutic measures and potential impact of pharmacokinetics due to weightlessness. ESA Journal 18: 33–50, 1994.

Pedrini-Mille A, Maynard JA, Durnova GN, Kaplansky AS, Pedrini VA, Chung CB & Fedler-Troester J. Effects of microgravity on the composition of the intervertebral disk. J. Appl. Physiol. 73: 26S–32S, 1992.

Pitts GC, Ushakov AS, Pace N, Smith AH, Rahlmann DF & Smirnova TA. Effects of weightlessness on body composition in the rat. Amer. J. Physiol. 244: R332–7, 1983.

Popova IA, Afonin BV, Davydova NA & Grigoriev AI. Hormonal regulation in space flights of varying duration. Physiologist 30: S42–4, 1987.

Pospisilova J, Serova LV & Pospisil M. The effects of spaceflight on the distribution of collagen types in bones of pregnant rats. Scr. Med. (Brno) 59: 277–82, 1986.

Pospisilova J, Pospisil M & Serova LV. Deviation in the developmental dynamics of the protein composition of the connective tissue of rats gestated during the flight in the biological satellite Kosmos 1524. Scr. Med. (Brno) 60: 277–88, 1987.

Pospisilova J, Pospisil M & Serova LV. Effect of spaceflight on collagen pepsin solubility and collagen type distribution in femoral bone and skin of rats. Physiologist 31: S32–3, 1988.

Pruitt LS, Jackson RD, Bartels RL & Lehnhard HJ. Weight-training effects on bone mineral density in early post-menopausal women. J. Bone Min. Res. 7: 179–85, 1992.

Raab DM, Crenchaw TD, Kimmel DB & Smith EL. A histomorphometric study of the cortical bone activity during increased weight-bearing exercise. J. Bone Min. Res. 6: 741–9, 1991.

Rambaut PC, Leach CS & Johnson PC. Calcium and phosphorus change of the Apollo 17 crew members. Nutr. Metabol. 18: 62–9, 1975.

Reich KM, CV Gay & JA Frangos. Fluid shear stress as a mediator of osteoblast cyclic adenosine monophosphate production. J. Cell. Physiol. 143, 100–4, 1990.

Reich KM & JA Frangos. Effect of flow on prostaglandin E2 and inositol triphosphate levels in osteoblasts. Amer. J. Physiol. 261: c428–c432, 1991.

Roux W. *Gesammelte Abhandlung über Entwicklungsmechanic der Organismen.* Vols I and II. Engelmann W, Leipzig, 1895.

Rubin CT. Skeletal strain and the functional significance of bone architecture. Calcif. Tissue Int. 36: S11–8, 1984.

Rubin CT & Lanyon LE. Regulation of bone formation by applied dynamic loads. J. Bone Joint Surg. 66: 397–402, 1984.

Rubin CT & Lanyon LE. Regulation of bone mass by mechanical strain magnitude. Calcif. Tissue Int. 37: 411–17, 1985.

Schmitt DA, Malouvier A, Holy X, Marie PJ, Caissard JC, Zérath E. Receptor ligand binding on osteoblasts in microgravity obtained by parabolic flight. Physiologist 34(1), S65–66, 1991.

Schneider V, Organov V, LeBlanc A, Rakhmanov A, Bakulin A, Grigoriev A & Varonin L. Space flight bone loss and change in fat and lean body mass. J. Bone Min. Res. 117: S122 (abstract), 1992.

Shaw SR, Vailas AC, Grindeland RE & Zernicke RF. Effects of a 1-week space flight on morphological and mechanical properties of growing bone. Amer. J. Physiol. 254: R78–R83, 1988.

Simmons DJ, Grynpas MD & Rosenberg GD. Maturation of bone and dentine matrices in rats flown on the Soviet biosatellite Cosmos 1887. FASEB J. 4: 29–33, 1990.

Simmons DJ, Russell JE & Grynpas MD. Bone maturation and quality of bone material in rats flown on the space shuttle 'Spacelab 3 mission'. Bone Min. 1: 485–93, 1986.

Simmons DJ, Russell JE, Winter F, Tran Van P, Vignery A, Baron R, Rosenberg GD & Walker WV. Effect of spaceflight on the non-weight-bearing bones of the rat skeleton. Amer. J. Physiol. 244: R319–R326, 1983.

Simske SJ, Guerra KM, Greenberg AR & Luttges MW. The physical and mechanical effects of suspension-induced osteopenia on mouse long bones. J. Biomechanics 25: 489–99, 1992.

Spector M, Turner RT, Morey-Holton E, Baylink DJ & Bell NH. Arrested bone formation during spaceflight results in a hypomineralized skeletal defect. Physiologist 26: S110– 11, 1983.

Spengler DM, Morey ER, Carter DR, Turner RT & Baylink DJ. Effects of spaceflight on structural and material strength of growing bone. Proc. Soc. Exp. Biol. Med. 174: 224–8, 1983.

Stepaniak PC, Furst JJ & Woodard D. Anabolic steroids as a countermeasure against bone demineralization during space flight. Aviat. Space Environ. Med. 57: 174–8, 1986.

Stupakov GP, Kazeikin VS & Morukov BV. Microgravity-induced changes in human bone strength. Physiologist 32: S41–4, 1989.

Taylor GR. Overview of spaceflight immunology studies. J. Leukocyte Biol. 54: 179–88, 1993.

Tilton FE, Degioanni JJC & Schneider VS. Long-term follow-up of Skylab bone demineralization. Aviat. Space Environ. Med. 11: 1209–13, 1980.

Turner RT, Bell NH, Duvall P, Bobyn JD, Spector M, Morey-Holton E & Baylink DJ. Spaceflight results in formation of defective bone (42215). Proc. Soc. Exp. Biol. Med. 180: 544–9, 1985.

Vailas AC, Vanderby R, Martinez DA, Ashman RB, Ulm MJ, Grindeland RE, Durnova GN & Kaplansky AS. Adaptations of young adult rat cortical bone to 14 days of space flight. J. Appl. Physiol. 73: 4S–9S, 1992.

van Kampen GPJ & Veldhuijzen JP. Aggregated chondrocytes as a model system to study cartilage metabolism. Exp. Cell Res. 140: 440–3, 1982.

van Loon JJWA, van den Bergh LC, Schelling R, Veldhuijzen JP & Huijser RH. Development of a centrifuge for acceleration research in cell and developmental biology. 44th Intern. Astron. Congr., IAF/IAA-93-G.4-166, 39, Graz, Austria, 16–22 October 1993.

van Loon JJWA, Bervoets DJ, Burger EH, Dieudonné SC, Hagen JW, Semeins CM, Zandieh Doulabi B & Veldhuijzen JP. Decreased mineralization and increased calcium release in isolated fetal mouse long bones under near weightlessness. J. Bone Min. Res. 10: 550–7, 1995.

Veldhuijzen JP, Bourret LA & Rodan GA. *In vitro* studies of the effect of intermittent compressive forces on cartilage cell proliferation. J. Cell Physiol. 98: 299–306, 1979.

Vico L, Bourrin S, Genty C, Palle S & Alexandre C. Histomorphometric analyses of cancellous bone from Cosmos 2044 rats. J. Appl. Physiol. 75: 2203–8, 1993.

Vico L, Bourrin JM, Very D, Chappard D & Alexandre C. Bone adaptation to real and simulated microgravity. In: *Proceedings of the 4th European Symposium on Life Sciences Research in Space.* Ed. by David V. Paris, France, European Space Agency, pp. 359–61, May 1990.

Vico L, Chappard D, Alexandre C, Palle S, Minaire P, Riffat G, Novikov VE & Bakulin AV. Effects of weightlessness on bone mass and osteoclast number in pregnant rats after a five-day space flight (Cosmos 1514). Bone 8: 95–103, 1987.

Vico L, Chappard D, Palle S, Bakulin AV & Alexandre C. Trabecular bone remodeling after seven days of weightlessness exposure (Biocosmos 1667). Amer. J. Physiol. 255: R243–7, 1988.

Vico L, Novikov VE, Very JM & Alexandre C. Bone histomorphometric comparison of rat tibial metaphysis after 7 day tail suspension vs. 7 day spaceflight. Aviat. Space Environ. Med. 62: 26–31, 1991.

Vogel JM & Whittle MW. Bone mineral changes: The second manned Skylab mission. Aviat. Space Environ. Med. 47: 396–400, 1976a.

Vogel JM & Whittle MW. Bone mineral content changes in the Skylab astronauts. Amer. J. Roentgenol., Rad. Therapy & Nuclear Med. 126: 1296–7, 1976b.

Vose GP. Review of roentgenographic bone demineralization studies of the Gemini space flights. Amer. J. Roentgenol., Rad. Therapy & Nuclear Med. 121: 1–4, 1974.

Ward FO. *Outlines of Human Osteology*, 3rd edn. Henry Renshaw, London, 1836.

Weingart H, Kleber M, Geissler H, Wachtel E, Hecht K, Denissova L, Sergejev I & Serova L. The influence of a one-week space flight on teeth and jaw bones of Wistar-rats - Cosmos 1514 and Cosmos 1667. Physiologist 31: S34–5, 1988.

Weinreb M, Rodan GA & Thompson DD. Osteopenia in the immobilized rat hind limb is associated with increased bone resorption and decreased bone formation. Bone 10: 187–94, 1989.

Whedon GD & Heaney RP. Effects of physical inactivity, paralysis and weightlessness on bone growth. In: *Bone*, vol. 7. Ed. by Hall B & Keith BK. CRC Press, Inc., Boca Raton, FL. U.S.A., pp. 57–77, 1993.

Whedon GD, Lutwak L, Reid J, Rambaut P, Whittle M, Smith M & Leach C. Mineral and nitrogen balance study, results of metabolic observations on Skylab II 28-day orbital mission. Acta Astronautica 2: 297–309, 1975.

Wolff JD. *Das Gezetz der Transformation der Knochen.* Berlin, A. Hirschwald, 1892.

Wong M & Carter DR. Theoretical stress analysis of organ culture osteogenesis. Bone 11: 127–131, 1990.

Wronski TJ & Morey ER. Alterations in calcium homeostasis and bone during actual and simulated space flight. Med. Sci. Sports Exerc. 15: 410–14, 1983a.

Wronski TJ & Morey ER. Effects of space flight on periosteal bone formation in rats. Amer. J. Physiol. 244: R305–9, 1983b.

Wronski TJ, Morey-Holton ER, Doty SB, Maese AC & Walsh CC. Histomorphometric analysis of rat skeleton following space flight. Amer. J. Physiol. 252: R252–5, 1987.

Yagodovsky VS, Triftanidi LA & Gorokhova GP. Space flight effects on skeletal bones of rats (light and electron microscopic examination). Aviat. Space Environ. Med. 47: 734–8, 1976.

Yeh C & Rodan GA. Tensile forces enhance PgE synthesis in osteoblasts grown on collagen ribbon. Calcif. Tissue Int. 36S: 67–71, 1984.

Yeh JK, Liu CC & Aloia JF. Effects of exercise and immobilization on bone formation and resorption in young rats. Amer. J. Physiol. 264: E182–9, 1993.

Zerath E, Nogues C, Borne M & Soudraine P. Bone effects of 13 days of weightlessness on rat and monkey: some results of Biocosmos 1887 and ground simulation. Physiologist 33: S94–5, 1990.

Zerath E, Holy X, Malouvier A, Caissard J-C & Nogues C. Rat and monkey bone study in the Biocosmos 2044 space experiment. Physiologist 34: S194–5, 1991.

Zernicke RF, Vailas AC, Grindeland RE, Kaplansky AS, Salem GJ & Martinez DA. Space flight effects on biomechanical and biochemical properties of rat vertebrae. Amer. J. Physiol. 258: R1327–32, 1990.

Chapter 6

Radiation biology

Jürgen Kiefer, Karin Schenk–Meuser & Michael Kost

6.1 Introduction

Radiation biology in space is multifaceted. It comprises the measurement of radiation field parameters, study of interaction processes with materials of biological interest (including the development of special equipment suitable for flight conditions) as well as the investigation of radiation-related phenomena in biological systems, extending from molecules to the whole body. Investigations span many orders of magnitude in spatial dimensions and time, and require the cooperation of many disciplines. This is all equally true for Earth-bound radiation biology, of course, but investigations in space pose special problems related to the specific properties of the radiation field which cannot be found on the ground and can only be poorly simulated.

An additional factor to be taken into account is the possible interaction with flight-related conditions, of which microgravity is the most prominent one, that can only be studied in a sensible way during space missions. Results obtained will not only be of value for space radiobiology as a subdiscipline but may also shed light on fundamental processes important for the whole field.

Radiation constitutes the most important hazard for humans during long-term space flights, particularly those outside the Earth's magnetic field (**Adams 1992**). Radiation protection is therefore mandatory to safeguard the well-being of astronauts and to limit the occurrence of later damage, like cancer. Passive shielding is practical only to a certain extent, both because of obvious mass limits and physical reasons related to the very high energies of the particles in galactic cosmic radiation (GCR). In spite of very many investigations, the understanding of heavy particle radiation biology, which constitutes the necessary prerequisite for a rational approach to hazard estimation, is still rather rudimentary.

Not all of the questions which remain can be tackled in space, and it is not even advisable to do so since the experiments are difficult to control and high cost would be involved. Rather, there has to be a careful balance and coordination between ground studies and flight experiments, and the latter should concentrate on specific questions. It has to be pointed out here, and this cannot be overstated, that there is no value in transporting into space sophisticated equipment and complex systems which are difficult to handle even in a ground-based laboratory in the hope of obtaining good 'scientific' results. This does not only apply to radiobiology, of course, but especially in this field one finds the tendency, even among peer groups, to recommend support only for studies with mammalian systems because of their obvious relevance to humans even though this does not properly take into account that experimental conditions are very difficult to control even in a well-equipped space laboratory. Measurements in space do not have value in themselves; they have to be scrutinized critically in exactly the same way as observations made on Earth. Since flight investigations can be repeated rarely, if at all, the methods used should be simple. The importance of appropriate controls, and a sufficient number of them, can hardly be exaggerated. Many studies, particularly the early ones, do not meet these criteria and this is certainly one reason why they do not find their way into refereed journals, a circumstance which has sometimes given life science research in space an attribute of 'esoteric science'.

Radiation protection in space is not only an issue for manned missions; it is also a concern for electronic equipment. Electronic circuit structures are nowadays so small that they are comparable to the dimensions of chromosomes and biomolecules. Approaches to the study of their damage are quite similar, both experimentally and theoretically, and these seemingly very distant fields appear now to converge and may be able to gain substantially from each other.

This short introduction sets the scene for this Chapter. In the first part the measurement of radiation field parameters and related problems will be discussed. Effects in biological systems constitute the next section, and the chapter will conclude with particular radiation protection issues. Although results from space observations are at the centre of interest, supporting ground experiments will also be discussed.

6.2 The space radiation field

6.2.1 Observations in space

In general terms, the space environment presents radiation conditions which cannot be found anywhere else. It is not uniform, however, but depends on parameters like altitude, geographical latitude (in low earth orbits) and the activity of the sun. Table 1 gives a schematic summary of the types of radiations, some of their properties and where they are found.

6.2.1.1 Galactic cosmic rays (GCR)

Heavy charged particles as important components of cosmic radiation were discovered as early as 1912 by Hess which earned him a Nobel prize (**Hess 1912**). Their origin is still not known. With the availability of satellites more

Table 1. Sources and properties of space radiations (modified after McCormack & Swenberg 1991).

Type	Characterization	Location	Secondaries
Trapped electrons	Low LET, not very penetrating, easily shielded, little contribution to total dose, intensity not stable.	3 to 12 Earth radii (6378 km); geosynchronous Earth orbit (GEO) (about 36000 km).	Bremsstrahlung
Trapped protons	Large ranges in energy, LET and contribution to dose, moderately penetrating, small contribution to total dose.	Low Earth orbit (LEO) (up to about 500 km).	neutrons
Solar particle events	Mostly protons, small fraction of α-particles and heavier ions, variable in strength and time of occurrence, anomalously large events may cause very high doses.	Outside the magnetosphere, polar orbit, GEO, moon, interplanetary space.	
Galactic cosmic rays	Protons, α-particles and heavier ions; large ranges in energy and LET, very penetrating, contribution to dose significant, nearly isotropic and stable in time.	Outside the magnetosphere, polar orbit, GEO, moon, interplanetary space.	

detailed knowledge about composition, energies and abundances has been gained. A number of summarizing publications have appeared (Allkofer 1975; **Blake & Baker 1992**, NASA 1975b; Ginsburg & Syrovatskii 1964; McDonald 1965; Mewaldt 1994; Simpson 1983; Warman *et al.* 1972; **Webber & Ormes 1967**). It is now clear that GCR consists of atomic nuclei from hydrogen to iron ranging in energies up to about 10^6 MeV nucleon^{-1}. Details of spectral composition and energy distributions can be found in specialised papers (**Derrickson *et al.* 1992; Fowler *et al.* 1992; Seo *et al.* 1992; Webber *et al.* 1990**).

The spectra plotted against energy nucleon^{-1} are all quite similar in shape, though differing in particle fluxes for the different ions. The total flux is rather small, of the order of about 2 cm^{-2} s^{-1}. Nevertheless, they constitute an important hazard both to biological and electronic systems because of the very high local energy deposition. GCR is very constant in time (it has been estimated that it has not changed appreciably over the last 30000 years (McCormack & Swenberg 1991)) but is modulated by the solar cycle both in energy distribution and total intensity. It is highest at solar minimum and *vice versa*.

For near-Earth orbits, interaction of GCR with the magnetic field and the upper atmosphere have to be taken into account. The first point seems to be reasonably understood as documented by measurements and calculations (Beaujean 1993, **Beaujean *et al.* 1991**; Heinrich 1988, **1994**; **Heinrich & Spill 1979**). Since our planet is a magnetic dipole there is a variation of particle fluxes with geographical latitude, being highest near the poles and lowest near the equator. Less energetic ions are either deflected or trapped in the radiation belts (see below) so that, depending on latitude, 'cut-off energies' can be defined (Heinrich 1988; Smart & Shea 1984). Galactic cosmic particles may also be accelerated by the magnetic fields of planets or solar proton winds (**Evenson *et al.* 1979; Chupp 1990**; **Williams *et al.* 1991**).
Transport and attenuation in the atmosphere are more difficult to understand. There are a number of codes available which, however, yield results still to some extent at variance with measurements (**Badhwar & ONeill 1992**; Gussenhoven *et al.* 1994; Ganapol *et al.* 1989; Letaw *et al.* 1983; Silberberg *et al.* 1976, 1983; Tsao *et al.* 1983; **Wilson & Badavi 1986**; **Wilson & Townsend 1988a**; **Wilson *et al.* 1987a**, 1989). High-energy heavy ions undergo fragmentations and nuclear reactions whose cross sections are not yet sufficiently well known although more and more data are being gathered both from experiments in space (**Allkofer & Heinrich 1974**; Mathews 1983; Silberberg *et al.* 1976, 1983; Townsend *et al.* 1984) and on Earth (see section 6.2.2.1). This knowledge is not only necessary to assess the creation of secondary radiations but also for the design of appropriate shielding within the spacecraft. This latter problem has been addressed by a number of authors and calculations as well as measurements behind different thicknesses of material have been performed, although in most cases with dose as parameter (Haskins *et al.* 1992a & b; Heinrich 1988; Heinrich & Beer 1984; Heinrich *et al.* 1994a & b; Jordan & Stassinopoulos 1989; Shinn *et al.* 1994; Townsend *et al.* 1989, 1990; Wilson 1978).

6.2.1.2 Solar cosmic rays (SCR)

The sun is also an important source of radiation. In contrast to GCR, the intensity is not constant but depends on solar activity. SCR consist mainly of protons, and partly also of helium ions. Although heavier particles have also been detected they do not contribute significantly. Proton energies are considerably lower than in GCR with a maximum of about 200 MeV, the fluxes are in the order of 10^9 cm^{-2} y^{-1} (Stassinopoulos 1987).
A typical characteristic of SCR is their variability in time. They are rather loosely classified as ordinary and anomalously-large events. The latter can be a very important hazard since they build up in very short times with a sudden enormous increase in particle intensity. Within one anomalously-large peak fluxes of 10^{10} particles cm^{-2} have been measured. They result in doses which may even cause acute radiation damage. During the last few years considerable effort has been devoted to the measurement of solar flares

and to finding ways for their prediction and the assessment of possible hazards. A great number of papers provide data on spectral distributions and fluxes (**Appenzeller 1991**; Avdjushin *et al.* 1992; Bazilevskaya *et al.* 1994a & b; Foelsche 1963; Foelsche & Wilson 1969; Garrard & Stone 1994; Golightly *et al.* 1992; **Goswami et al. 1988**; Gussenhoven & Mullen 1992; Haskins *et al.* **1992a** & b; King 1974; McDonald 1965; Miroshnichenko 1992; Miroshnichenko *et al.* 1992; Nealy *et al.* 1990, 1994; Normand 1994; **Rust 1982**, 1992; Sauer *et al.* 1990; **Schäfer 1957**; **Schäfer *et al.* 1972**; Shavers *et al.* 1991; Shea & Smart 1994; Smart & Shea 1989; **Townsend *et al.* 1991**; Vahia & Biswas 1983; Vashenyuk *et al.* 1994; Webber 1966; Wilson *et al.* 1970). These data are very useful but they are necessarily retrospective. Reliable ways for prediction are not in sight because of the essentially stochastic behaviour of solar flares. They are linked to the solar cycle, having a higher probability particularly towards the end of solar maximum but neither the time of occurrence nor the strength can be foreseen. Nevertheless, a number of model approaches have been suggested which, however, seem to be far from reliable (Blizard 1969; Feynman & Gabriel 1990; Heckman *et al.* 1984, 1989, 1992a & b; Hirman 1987; Simonsen *et al.* 1991; Slutz *et al.* 1971; Smart & Shea 1992, 1994; **Wilson 1981**).

6.2.1.3 Trapped radiation

As mentioned above, charged particles may be trapped in the Earth's magnetic field, a process which forms the radiation belts (NASA 1975a), which consist of electrons, protons and a very few heavier ions. The electrons are found in two zones, an inner (1 to 2.8 earth radii) and an outer zone (from 2.8 to about 12 earth radii). Since their penetration power is rather low they do not give cause for concern because they are stopped by the walls of the spacecraft. They may, however, produce Bremsstrahlung which is penetrating and may add to the total radiation burden, though only to a minor degree. Protons are considerably more important. They cannot be associated with clearly defined zones but extend from about one Earth radius to more than 4 Earth radii. Their energies and fluxes vary with altitude and geographical latitude (see Stassinopoulos 1987). To give an example: 30 MeV protons are at maximum at about 1.5 Earth radii (measured over the equator) and display a flux of about 10^4 particles cm^{-2} s^{-1}. Not only do they penetrate the shielding material appreciably, but they also give rise to secondary radiations. In the present context it is important to note that the magnetic dipole is not colinear with the Earth's axis but is offset by about 11° and is also displaced so that the radiation belts extend to rather low heights at certain locations. The most important is in the South Atlantic (South Atlantic Anomaly) where even low Earth orbits cross the extension of the radiation belts. Consequently, the highest dose contribution is received when a spacecraft crosses this area. Recently, some more measurements of the radiation belts have become available (**Badhwar *et al.* 1992**; Santoro *et al.* 1986; Schmidt *et al.* 1994; Stassinopoulos 1987; Tulunay 1987).

As already mentioned, the interaction of primary and trapped radiation with atmospheric atoms may lead to the production of secondary radiation to which neutrons make the most important contribution (Foelsche *et al.* 1974). They play a significant role also in air travel at high altitudes, an issue which is touched upon below.

6.2.1.4 Dosimetry in space

With regard to biological action and hazard estimation, dose is the most important initial parameter. Results are available from the Long Duration Exposure Facility (LDEF) (Beaujean *et al.* 1992; **Benton *et al.* 1992a,b**; Kinard *et al* 1994; Reitz 1992), the Russian Cosmos series (Biocosmos 3: **Benton *et al.* 1978**; Biocosmos 5: Benton *et al.* 1981; Biocosmos 8: Akatov *et al.* 1989; **Reitz *et al.* 1990**; Biocosmos 9: Baican *et al.* 1992c; **Beaujean *et al.* 1991**; **Dudkin *et al.* 1992**; Reitz *et al.* 1992; Zens *et al.* 1992), Salyut (Akatov *et al.* 1984), Skylab, Apollo and ASTP (Schaefer *et al.* 1972; **Benton *et al.* 1977**; Billingham *et al.* 1965), Spacelab (**Benton 1984**; **Bücker *et al.* 1986b**; Reitz *et al.* 1986), CRRES (Stapor & Campbell 1992), Mir (Vana *et al.* 1994) and on Prognoz (**Kovalev *et al.* 1989**) as well as Chinese satellites (Ye *et al.* 1994). In most cases plastic nuclear track detectors or thermoluminescent dosimeters were used which allow only total particle fluences or doses to be assessed, although in the first case information about particle type and energy can be extracted which is important for the estimation of radiation quality and LET. More recently, active devices have been introduced which also allow time resolved measurements. They were used aboard Mir (Bottollier-Depois *et al.* 1994; Bouisset *et al.* 1992; Dachev 1992; Dachev *et al.* 1989a & b, 1992; Petrov 1994,Petrov *et al.* 1994a & b) and also on Shuttle missions (**Badhwar *et al.* 1992**; Golightly *et al.* 1992). These investigations confirmed that the largest contribution to total dose is accumulated when crossing the South Atlantic Anomaly.

Apart from dose, LET is a very important quantity since it determines the biological effectiveness to a large extent (see section 6.3). Distribution functions in LET have been determined and also compared to theoretical computations (Adams *et al.* 1986; Akopova *et al.* 1985; Chen *et al.* 1994a & b; Chenette *et al.* 1994; Heinrich 1980; **Heinrich *et al.* 1989**, 1994a; Rusch & Heinrich 1987; **Silberberg *et al.* 1984**; Wiegel *et al.*1987, 1992). The active dosimeters which were flown on Mir do not give LET values explicitly, but determine directly the equivalent dose (Bottollier-Dupois *et al.* 1994; Petrov *et al.* 1994b) (see also 6.4). New approaches to space dosimetry are discussed by McNulty (1987), Hanser & Dichter (1987), Schott (1987) and Ing & Mortimer(1994).

An overview of the available dose determinations may be gained from a few summary papers (Atwell 1990; **Benton 1984**, 1986; Benton & Henke 1983; Benton & Parnell 1987; Kovalev & Petrov 1992; Mullen *et al.* 1987; **Reitz 1993**). Although a considerable body of data has been gathered it seems that they are not yet sufficient to allow the formulation of a general model allowing predictive dose estimates for any given situation. Although some strategies

have been developed to approach this goal (Armstrong *et al.* 1994; Atwell 1990; **Badhwar & O'Neill 1992**; Cucinotta & Wilson 1985, Getselev *et al.* 1992; Letaw *et al.* 1989a & b).
It was mentioned above that radiation hazards are not only of concern for biosystems but also for electronic equipment. A thorough discussion of this point lies outside the scope of this review, but discussion of the topic can be found in McNulty (1992), Mullen & Ray (1994) and Stapor & Campbell (1992).
Another issue deserving mention is the problem of radiation exposure to air transport crew members at high altitude which was recognised to be significant some time ago (**ICRP 1966**; FAA 1980). Much of the knowledge gained in space experiments can be used to assess this important problem. Again, a detailed discussion is out of place here. A useful summary is contained in a meeting report (**Reitz *et a.* 1993**).

6.2.1.5 Conclusions

Space experiments have provided a wealth of data concerning the extraterrestrial radiation field in terms of composition and quantitative levels. Dose levels have been determined for many flight scenarios and their dependence on external influences like solar activity has been established. They constitute important input parameters for a systematic approach but present models seem to fail to fulfil these criteria so that more work is obviously needed on both dosimetric and theoretical aspects. The introduction of active, time resolving dosimeters is a significant improvement which could lead to a more refined description of radiation field parameters.

6.2.2 Supporting ground based physical studies

The study of biological effects of space radiation and the assessment of risk requires a thorough knowledge of interaction processes, both with shielding material and with biological structures. Methods to be used are quite sophisticated and generally cannot be applied during space missions. Related and complementary ground studies are indispensable, therefore. Since the literature is vast, a choice has to be made. The following issues are reviewed here:
(i) initial interaction of protons and heavy charged particles (fragmentation, nuclear interactions, production of radionuclides and secondary radiations, transport theory);
(ii) stopping power and dosimetry;
(iii) track structure and microdosimetry;
(iv) special methods for space applications;
(v) theoretical models for the space situation;
(vi) effects on microelectronics (not dealt with comprehensively).

6.2.2.1 Initial interaction processes

For a complete understanding of particle transport and particle effects on biosystems, cross sections for the most important processes are the most significant input parameters. With protons, which are the most abundant species, nuclear reactions which may lead to target fragmentation and generation of radionuclides are most relevant. **Asano *et al.* (1989)** provided a data set obtained by computer simulation for energy ranges between 100 and 500 MeV. Nuclear reactions in tissues are treated by Beck & Powell (1976) and **Wilson & Khanderwal (1976a & b)**, target fragmentation is dealt with by Shavers & Poston (1992). It appears that the data base for protons of energies up to about 500 MeV is now quite solid. This is not the case for heavy ions. Consequently, a number of studies have aimed at filling this gap (Bertini 1967; **Enghardt *et al.* 1992**; Guzik *et al.* 1994; **Hirzebruch et al. 1989;** Norbury & Townsend 1986; **Schimmerling 1992; Schimmerling & Curtis 1989; Schimmerling *et al.* 1977, 1979; Shavers *et al.* 1990, 1993**; Silberberg *et al.* 1976; **Skoski *et al.* 1973**; Townsend *et al.*1983).
On the basis of available measurements, theoretical approaches, computer models and tables have been generated (Khandelwal & Wilson 1974; **Muthukrishnan & Gobinath 1983; Townsend & Wilson** 1985, **1986, 1988;** Townsend *et al.* 1983; Wilson & Costner 1975; **Wilson & Khandelwal 1976a, b; Wilson *et al.* 1987b, 1988; Wong *et al.* 1990**). All these can be used to develop a comprehensive transport theory which is now taking shape (Lambiotte *et al.* 1971; Lamkin 1974; Lezniak 1979; **Perry 1988**; Shinn & Wilson 1992; Wilson 1977a; **Wilson *et al.* 1990b**). A very useful and thorough treatment of present approaches has been provided by Wilson *et al.* (1991).

6.2.2.2 Dosimetry and stopping power

The deposition of energy by protons and heavy ions is quite different from that by sparsely ionizing radiations like X- or gamma-rays, both macroscopically and microscopically. The latter point will be treated below (Section 6.2.2.3). Important parameters in this respect are stopping power and range. Although a number of tables are available (Andersen & Ziegler 1977; Hubert *et al.* 1990, Janni 1982; Northcliffe & Schilling 1970), there are still discrepancies for heavier projectiles. For protons and alpha particles with energies from 1 keV to 10 GeV (1 GeV for alpha particles) an authoritative compilation has recently been published by the *International Commission on Radiation Units and Measurements* (ICRU 1993), a similar one for heavier particles has been in preparation for some time and is expected to appear in the near future. The lack of experimental data, particularly for high energy heavy ions, is gradually being reduced (**Bakker & Segrè 1951; Perris & Zarris 1989; Takahashi *et al.* 1983a; Xu *et al.* 1984, 1985**). A thorough discussion of the theoretical background has been published by **Ahlen (1980)**. A particular feature of charged particle interaction is the fact that the doses to tissues are maximal at a certain depth just before the end of the range is reached. 'Depth dose distributions' are consequently of special interest. They have been measured in some instances (**Heinrich 1979; Raju *et al.* 1978a–c**; Wilson 1977b; **Wilson & Khandelwal 1976b; Wilson *et al.* 1984**). But dose is not alone in changing with depth, so does the LET which is also influenced by the generation of secondary radiations. Calculations and measurements of this quantity are important for shielding construction and the assessment of biological damage. This problem has been treated in a number of investigations using different approaches (**Cucinotta *et al.* 1990; Heinrich 1977a** & b, 1980; **Xu *et al.* 1985**).

6.2.2.3 Track structure and microdosimetry

Densely ionizing radiations are biologically more effective than X- or gamma-rays. The reason for this is that the spatial pattern of energy deposition at the microscopical level is quite different, so that lesions in the sensitive structures, be they biomolecules or chromosomes, are induced with higher efficiency and/or prove less amenable to repair. This fact has long been realized and marked the beginning of microdosimetry as a scientific subdiscipline which deals essentially with energy transfer processes at the submicrometre level (Dicello *et al.* 1994; ICRU 1983; Rossi & Zaider 1991; Todd 1992). Special symposia on microdosimetry are held regularly (the eleventh in 1992, Menzel *et al.* 1994); only a few salient features will be mentioned here. Whilst in the beginning experimental and theoretical efforts concentrated on sites with sizes in the region of a micrometre or slightly less, today the emphasis lies on track structure which describes energy deposition events in a detailed way around the path of an ion. Because of experimental limitations they are evaluated theoretically by the so-called 'Monte-Carlo' method in which the fate of the primary and all secondary ions is followed when traversing matter. A necessary prerequisite is the knowledge of all interaction cross sections, but these are available with some degree of confidence only for protons and alpha particles and hardly at all for heavy ions. Consequently, results are reported only for these light projectiles (**Berger 1988; Budd *et al.* 1983; Hamm *et al.* 1984; Leuthold & Burger 1988; Olko & Booz 1990; Rohrig & Colvert 1978; Wilson & Paretzke 1980, 1981; Wilson *et al.* 1988a & b; Zaider *et al.* 1983**). An intermediate approach is to consider the radial dose distribution around an ion track, this being pioneered by **Butts & Katz (1971)** at a time when it could not be determined experimentally. In the meantime measurements have become possible, both for protons (**Mills & Rossi 1980; Wingate & Baum 1976**; see also Kiefer & Kost 1988 for older references) and heavy ions (Braby *et al.* 1992), and along with the measurements, models were developed (**Chatterjee & Schaefer 1976; Chatterjee *et al.* 1973; Curtis 1971, 1978; Kiefer & Straaten 1986**; Krämer & Kraft 1992, 1994). The distribution of delta-electrons, which are emitted when a charged particle interacts with matter, is the main determinant for all of them (**Gupta &**

Prasad 1984). Their energies can now be measured even for heavy ions (Schmidt *et al.* 1987; Schmidt-Böcking *et al.* 1992). The whole problem is discussed in a more conceptual way by Kiefer & Kost (1988), **Kraft *et al.* (1991)** and Paretzke (1980, 1988, 1989). Microdosimetric considerations play an important role for the rational evaluation of risk associated with densely ionising radiations (Curtis 1994a & b; Dicello *et al.* 1994; ICRU 1986; **Rossi 1959**).

6.2.2.4 Special experimental methods

The unique radiation environment in space and the restrictions imposed by mission factors demand, in most cases, modification of experimental methods. In some instances these have been discussed in Section 6.2.1, when they were described together with the results obtained during the missions. In this Section they are discussed in so far as they are related to the determination of physical quantities.

Since heavy charged particles form an important part of GCR but are only present at low abundances, special methodologies had to be found to assess them in a reliable quantitative way, particularly if their action was to be related to biological effects. The BIOSTACK concept (Bücker 1975) proved to be very efficient in this respect. The device consists of a sandwich-like structure where biological objects are placed on plastic track detectors so that, after etching, the relative positions of particle traversal and the sensitive site of the objects can be determined. If several layers are employed it is also possible to extract the physical characteristics of the impinging particle by the determination of residual ranges. Equipment and methods are described by **Facius *et al.* (1990), Noll *et al.* (1988)** and **Rusch *et al.* (1991)**. It has been suggested that AgCl-crystals could be used as an alternative to foil-detectors, being activated by UV-irradiation so that measurements can be performed at desired times; however, actual flight tests of this idea are still required (Schopper *et al.* 1992). Thermoluminescence is also a widely used method for dose evaluation in space missions. Its application to heavy particles is discussed by Vana *et al.* (1994). The use of plastic track detectors does not allow determination of the time dependence of the particle impact since it is a passive device. Active dosimeters are, therefore desirable. They have recently become available and passed flight tests successfully (**Badhwar *et al.* 1992**; Nguyen *et al.* 1990). Other techniques with special emphasis on spectrometric resolution are described by Appleby & Golightly (1992), Budnik *et al.* (1992), Slobodrian *et al.* (1994) and **Wilson *et al.* (1989).**

6.2.2.5 Radiation effects on microelectronics

To safeguard proper function of measuring devices and, even more importantly, of the control circuits in spacecraft possible radiation-induced failures are a matter of serious concern. Ground tests are essential. These problems cannot be dealt with here but the following are a few pertinent recent references: Dreute *et al.* (1994), **Geschke *et al.* (1991)** and particularly Chapter 14 in the treatise of Wilson *et al.* (1991).

6.2.2.6 Theoretical models

Theoretical models dealing with data obtained in space or which are validated in missions are, of course, developed on the ground but tested in space. The borderline between space and ground activities is hence rather arbitrary. Some are already to be found in the previous section, others which involve also ground-based experiments are listed here. Many are concerned with the computation of beam data like doses or LET distributions, as, for example, in Chen *et al.* (1994a & b), Hardy & Leavitt (1994), Heinrich (1980) and Miller (1994). Others attempt to link physical parameters with risk assessment, examples are: Cucinotta *et al.* (1991) and McNulty & Stassinopoulos (1994).

6.3 Radiation biology in space

6.3.1 Introduction

Radiation biology in space is not a discipline in its own right but part of a much wider field covering the interaction and effects of radiations in biological systems. Studies to be discussed here, consequently, must be embedded and related to this topic as a whole and can only be understood against this background. Radiation effects can be registered at all levels of biological systems. They all start from the primary absorption of energy which takes place within about 10^{-12} to 10^{-16} s, although ultimate effects may be seen as late as after many generations, as in the case of genetic damage to the germ cells. Fig. 1 illustrates this in a simplified manner.

The development of radiation damage does not proceed in a linear unidirectional manner but is modifiable in a number of ways. The most important process in this respect is the ability of cells to recognise genetic alterations and to repair them. Whether this actually occurs depends not only on the properties of the system under consideration but also on the molecular nature of the initial lesions. It is determined by the spatial distribution of energy deposition, i. e. radiation quality. This is quite different in space compared to the normal environment on Earth (see section 6.2) so that the experimental objectives are not the same as in conventional radiation biology. Environmental factors may also intervene. Whether microgravity plays a role here is still an open, and interesting, question.

There are a number of endpoints which are studied covering the whole hierarchy of biological systems. The most important are: DNA lesions; chromosomal aberrations; cell inactivation; mutation induction; neoplastic transforma-

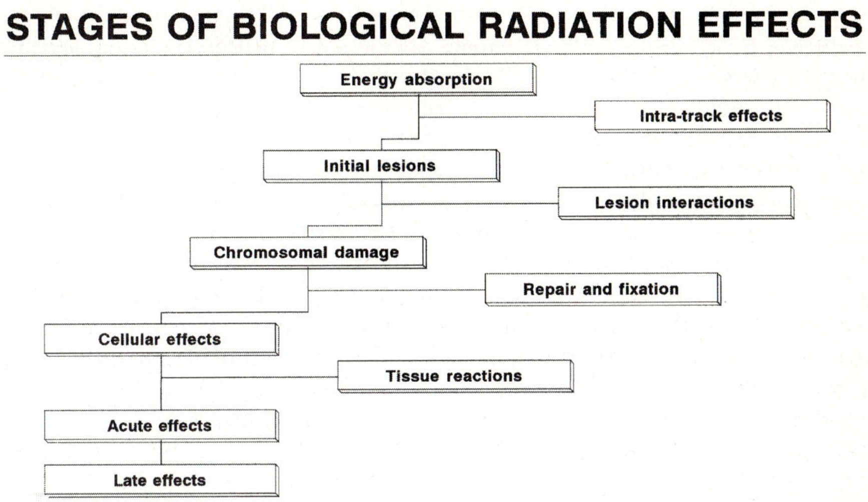

Fig. 1. The development of biological radiation damage and stages of interaction and modifications.

tion (indicating the first step of tumour induction); tissue damage; carcinogenesis in experimental animal and humans; and effects on progeny (genetic and teratogenic). A great many techniques are available, which are under constant development, sometimes rather sophisticated and requiring such careful control of conditions that in most cases they are not applicable in space. This demands that each space experiment is accompanied by many ground studies and an extensive set of controls. More detailed discussion can be found in Alpen (1990), Kiefer (1990) and Tubiana *et al.* (1990).

6.3.2 Radiobiological experiments in space

6.3.2.1 Introduction

The space environment is unique in many respects, not only with regard to its radiation field. Additional flight related factors have to be taken into account which might induce stress responses in biological systems, so changing their radiation response. It is thus hardly surprising that unambiguous results are very difficult to obtain, particularly as repeat experiment can only be performed rarely. A variety of test organisms has been subjected to space conditions with very many different endpoints. Especially in the beginning of the space era it has not always been realized that only extremely carefully controlled conditions allow the conclusion that a particular effect observed can be unequivocally attributed to space radiation and/or its interaction with other environmental factors. There is a number of summary reports about specific missions which contain also radiobiological results (e.g. Bücker 1974; Bücker *et al.* 1984; **Gatuito *et al.* 1978; Gurosky & Ilyin 1978**; Horneck *et al.* 1986; Reitz & Bücker 1992), more detailed papers are reviewed in the following sections Heavy ions are in the centre of interest but their fluences are low so that special experimental techniques had to be developed which allow identification of objects which were hit and assess their biological response.

It has been suggested several times that microgravity might change the repair capacity of cellular systems. The only clear-cut experiments to investigate this problem have to be performed with the aid of an extra radiation source on board and in combination with a 1 × g reference centrifuge. This has been done on some missions but unfortunately without repetitions. Only recently have experimental designs developed to such a stage that clearer answers may be expected. A number of reviews are available (Ballard & Conolly 1990; Bücker 1977; Bücker & Facius 1980; Horneck 1988a & b, **1992**; Horneck *et al.* 1989; McCormack *et al.* 1987; Tobias & Todd 1974).

6.3.2.2 Subcellular systems

Although specific changes at the molecular level are expected from heavy charged particles in space no studies are available as yet, although a few are in preparation. Thin films of bacteriophage T4 were flown on the Soviet ASTP mission, and high yields of mutants (about a factor of 3 compared to non-hit flight controls) were recorded in areas which were traversed by a HZE-particle (Yurov *et al.* 1979, 1983).

6.3.2.3 Chromosomal aberrations

Structural aberrations in chromosomes are a very sensitive and quite specific indicator of radiation damage and may be used for biological dosimetry. They were investigated in a few systems on several missions, also to test the possible interaction of radiation and flight specific factors (see below). There was always an increase in flight controls (i.e. in non-hit objects) compared to those on the ground which may be due to unidentified hits (Grigoriev *et al.* 1978; Gaubin *et al.* 1983) but only small, mostly insignificant differences between hit and non-hit flight samples with single aberrations. Multiple aberrations, on the other hand, which are indicative of heavy particle action were in most cases increased in hit objects (Nevzgodina *et al.* 1984, 1989). Since these studies were performed with lettuce seeds a clear particle-object correlation is quite difficult to establish. It would be advisable to use single cell systems for these kinds of investigation.

6.3.2.4 Unicellular systems

The most studied object here is the spore of *Bacillus subtilis* which was frequently used for the BIOSTACK experiments which will be discussed below.

Other organisms used were *Escherichia coli* (Antipov *et al.* 1969), and *Saccharomyces cerevisiae* (Grigoriev *et al.* 1972; Benevolenskij *et al.* 1982) but no clear cut results were reported. *Chlorella* cells have been flown on several Soviet missions but, again, significant effects of radiation were not found, presumably due to inadequate experimental design (Antipov *et al.* 1969; Vaulina & Anikeeva 1970; Vaulina *et al.* 1971).

6.3.2.5 Plants

Seeds or spores are quite resistant to harsh environmental conditions and plant systems have, therefore, been favourite test objects on many, particularly Soviet flights. A number of parameters were studied. The induction of chromosomal aberrations in lettuce were mentioned above. Other species included *Tradescantia* (Antipov *et al.* 1969), *Arabidopsis* (Gartenbach *et al.* 1994; Grigoriev *et al.* 1978; **Kranz *et al.* 1990**; **Anikeeva *et al.*** 1983, **1990**; Bücker *et al.* 1973, 1984; Bücker 1975), tobacco (Bayonove *et al.* 1984; **Gaubin *et al.* 1990**), rice (Bayonove *et al.* 1992, 1994) and corn (Bücker 1984). Apart from changes at the chromosomal level, germination impairment, mutation induction and anomalous development or tumour formation were demonstrated. Generally, it was found that even without clear interaction with heavy charged particles all parameters indicated an appreciable amount of damage compared to ground controls. Whether this was due to experimental conditions which are obviously more difficult to control during flight, perhaps more so pre-launch and after reentry, or to undetected radiation absorption is not clear but it shows clearly that meaningful results can only be hoped for when 'hit' and 'non-hit' objects can be reliably distinguished. Where this was achieved, e.g. by sandwiching the samples between nuclear detector foils, effects could be recorded but the reproducibility was often poor. In Russian flights significant lethality was reported in *Arabidopsis* (cf. Grigoriev 1982) which was not supported by Kranz (1986, 1987). Lethal mutations seem to be significantly increased but only one experiment has been performed (**Kranz *et al.* 1990**). In tobacco seeds there was a slight loss in germination (Barbier & Dulieu 1978) but careful statistical analysis is lacking. Seedling abnormalities have been variable (Grigoriev *et al.* 1978; Gaubin *et al.* 1983). A very unusual mutation which had not been seen before was reported in one corn plant grown from a seed which received two hits by HZE-particles (Peterson *et al.* 1977).

6.3.2.6 Small animal systems

Because of the special requirements, particularly in unmanned missions, only a few animal systems are suitable for space experiments. These are the cysts of the shrimp *Artemia salina*, eggs of *Tribolium* and eggs of the stick insect *Carausius morosus*. The two latter systems are especially valuable since they

allow investigation of the influence of space factors (including radiation) on development. More recently, the nematode *Caenorhabditis elegans* has been introduced into radiobiological space research which appears to be a promising candidate since so much is already known about its molecular and cell biology. There are plenty of data for *Artemia* but unfortunately they do not give clear-cut answers. Reduction in emergence as a result of particle hits have been reported by Rüther *et al* (1974), Blanquet *et al.* (1977), **Gaubin *et al.* (1979)** and Planel *et al.* (1975) while no changes were found by Grigoriev *et al.* (1977, 1978), Gaubin *et al.* (1983) and Nevzgodina *et al.* (1984). The same general variability is seen with loss of hatching and late lethality (see **Horneck 1992** for a summary of data). There seems to be an increase in larval anomalies but the data base is small.
The effects on development have been studied in two other animal systems, first in *Tribolium confusum*, and later on and more extensively in *Carausius morosus*. In the first case there was no difference in hatching rate between hit and non-hit cells (Graul *et al.* 1975; Bücker *et al.* 1975). Failures of pupal development appeared to increase, as did late lethality and the rate of malformations. In the latter case, however, no in flight controls were available. A large amount of data exist for *Carausius*. Failure to hatch was hardly different between control and hit cells (Graul *et al.* 1975; Bücker *et al.* 1986a & c) and the same was true for late lethality (**Reitz *et al.*** 1989, **1990**). A significant result was achieved when malformations are considered: in three missions (Apollo 17, Spacelab D1 and Cosmos 1887) their rate was clearly higher compared both to in flight and ground controls, which both showed very low levels (**Reitz *et al.*** 1989, **1990**). It appears that the traversal of a heavy charged particle interferes with development in a sensitive and critical way. This certainly has important implications for radioprotection although so far there are no suggestions about the mechanisms involved. It may be hoped that more knowledge can be gained in future, especially by carefully designed experiments with *Caenorhabditis elegans*, both with further flights and accelerator studies on the ground.

6.3.2.7 Mammalian systems

There are not many investigations on mammals as far as radiation action is concerned. The earliest report of tissue damage due to cosmic heavy nuclei is that of **Chase (1954)** and **Chase & Post (1956)** who found that the skin of mice which were flown in balloons at high altitude developed segments of white hair whose frequency could be correlated with the estimated number of particle hits. Damage to hair follicles was also found in animals on board Apollo 17 (Haymaker *et al.* 1975) although an attempt to correlate them with particle traversals failed. In the same study morphologically observable lesions were sought by post-flight electron microscopical analysis but were not found. The total number of particles was quite low (a total of 80 were recorded) due to heavy shielding inside the spacecraft.

The study described was mainly aimed at detecting brain and retinal damage. To this end nuclear detectors were implanted in the head region. This caused inflammations which casts doubt on the reliability of the results obtained. Certainly no clear-cut evidence for particle induced structural lesions was achieved. This has some bearing on the concept of microlesions (see Section 6.3.3).

6.3.2.8 Eye flashes

The astronauts returning from the lunar Apollo 11 missions reported visible impressions at fairly high frequencies which were attributed to passages of heavy charged particles through the eye. This suggestion was later confirmed by accelerator experiments (**Budinger *et al.* 1971**, 1976; **Tobias *et al.* 1971**) and more extended space observations (Budinger *et al.* 1977a & b; Osborne *et al.* 1975). Special helmets were constructed containing time-resolving nuclear detectors to determine the coincidence of particle traversal and visual sensation. This was, however, found only in few cases although the time resolution was about 1 s (Benson & Pinsky 1972; Pinsky *et al.* 1973) which should be sufficient. On the other hand the total number of flashes equalled approximately the number of particles recorded. Further experiments (Hoffman *et al.* 1977) were performed in near-Earth orbits which also showed that the highest number of flashes was seen when areas of high particle fluxes, e.g. polar routes or the South Atlantic Anomaly, were passed. It seems that one is dealing with a real phenomenon, as is also shown by the accelerator studies on the ground, but the exact mechanism is still unclear. A number of possibilities were discussed by Budinger *et al.* (1976) and summarized by **Horneck (1992)**.

6.3.2.9 Single particle correlation studies

The most extensive and elaborate investigations on the action of single HZE-particles on biological systems were performed by applying the BIOSTACK concept (Bücker 1975). Experiments in space were paralleled by measurements on the ground using accelerated heavy ions (see also Section 6.3.3). The object were fixed on nuclear detector foils which were later etched to find the exact location of the particle traversal so that a measured action could be related to the distance to the sensitive site (the impact parameter). The spatial resolution was claimed to be around 0.2 mm which allowed the observation of small objects like bacterial spores. In many space experiments reported above this method was used to distinguish 'hit' and 'non hit' entities. It has, however, greater virtues then just this. In multilayer stacks the direction and even the nature of the passing ion can be assessed (see Section 6.2). It was found in spores of *Bacillus subtilis* that inactivation decreased with the impact parameter but extended to distances up to 4 mm which is difficult to reconcile with standard radiobiological theories. Contributions of other phenomena like thermal spikes or shock waves have been implied as a possible explanation (Facius *et al.* 1978a & b; Horneck *et al.* 1989a & b). It was claimed that for space HZE-particles relative biological effectiveness (RBE) factors of more than 600 can be found (Facius *et al.* 1978a & b, 1979) but this conclusion appears not to stand rigorous statistical analysis.

6.3.2.10 Radiation effects and mission-related factors

As stated in the beginning, a space flight exerts a number of environmental influences on biological systems, particularly microgravity, which may well change their radiation response. This has been discussed in a rather general way in a few reports (Antipov *et al.* 1975; Dons & Fohlmeister 1992; Filatova *et al.* 1983; Horneck 1988b, 1992; Marco *et al.* 1987; Planel *et al.* 1989; Vaulina *et al.* 1984). Since initial damages are effectively repaired in most cells (including those of the human body) the question whether these processes are modified is not only of academic interest. Answers were sought using different approaches. An apparently simple way is to fly pre-irradiated objects and compare the results to those obtained with ground controls. This has to be viewed with a considerable amount of caution since not only is it difficult to control flight factors but handling before and after launch is usually quite different from that under laboratory conditions. Constant temperatures can virtually never be guaranteed. It seems, therefore, hardly surprising that no clear answers were found. Strictly additive (which means independent) and also synergistic effects were found at about equal frequencies (Grigoriev *et al* 1972; Grigoriev 1982). The objects flown were mostly plant seeds which are resting systems and the best suited to test influences on repair *during* the flight. A better way is to irradiate during the flight with the aid of an on-board source. This was done on a number of flights using either β or γ emitters. Not all results obtained can be reviewed here (see **Horneck 1992** for an extensive compilation) but it seems that synergistic action (which suggests but does not prove an impairment of repair capacity) could only be found in active systems like *Tradescantia* plants, *Drosophila*, *Habrabacon* and *Tribolium* (Shank 1974) while *Escherichia coli*, *Salmonella typhimurium* (Saunders 1971) and *Neurospora* spores (Shank 1974) seemed to act in an independent manner. In two Gemini flights (3 and 11) human lymphocytes *in vitro* were exposed to ^{32}P β-irradiation and scored for chromosomal aberrations. In the first (total dose 1.8 Gy, duration 5 h) synergistic action was reported (**Bender *et al.* 1967**) which could not be confirmed by the repeat investigation where, however, total dose and duration were different (**Bender *et al.* 1967**). This experience stresses once more the need for carefully controlled repeat experiments, a postulate which goes without saying on the ground but is very rarely seen in space investigations.

On Spacelab D1 the action of cosmic particles, detected by the BIOSTACK method, was compared under microgravity and in an on-board $1 \times g$ reference centrifuge. Mortality and malformations in *Carausius* which is an active system displayed synergistic interactions. It appears thus that repair processes may indeed be impaired by microgravity although this is very difficult to understand mechanistically. Pross *et al.* (1993, 1994) approached this problem by using a temperature conditional repair mutant of yeast on IML-1 and came to the same conclusion. Unfortunately, technical limitations prevented use of a reference centrifuge so that doubts are left also in this rather straightforward experiment. In summary, there are clear hints that repair is less effective in space flight but obviously more experimentation is warranted. Studies with rats subjected during flight to ^{137}Cs γ irradiation showed no interaction with regard to a number of endpoints (**Gazenko *et al.* 1978**). One might, however, ask whether live animals which are subject to all kinds of stresses, including psychological ones, are really the right objects of study.

6.3.2.11 Conclusions

There is no doubt that space radiation is an important hazard for manned missions. Apart from considerations of human protection, however, there are also interesting fundamental questions, e.g. repair and microgravity, action of single particles, interaction between different components of the radiation field, which can and should be investigated. This can only be done in a sound way if accompanied by very careful ground experiments. By all means the temptation should be resisted to send complex systems into space with the result that their response cannot be interpreted.

6.3.3 Supporting biological ground experiments

6.3.3.1 Introduction

Radiobiological experiments in space are rare and costly. They should, therefore, be reserved for study of specific questions which are special for this environment and cannot be approached on Earth. This implies that fundamental radiobiological issues related to the impact of space radiation have to be tackled on the ground, and the results form the necessary and indispensable prerequisite for successful experimentation on space missions. Since there is hardly a problem of radiobiology which is not touched upon in this context one has to resist the temptation to include the whole wealth of recent research in this review. Obviously a choice has to be made. Since protons and heavy charged particles form the most prominent part of space radiation and create most concern in terms of possible hazards and protection the discussion in this section will concentrate on their action. It is impossible to review all the literature and older papers are not covered comprehensively as they can be found in reviews, e.g. Blakely *et al.* (1984), Kiefer (1986), Kraft (1987), McCormack *et al.* (1987), Swenberg *et al.* (1993a & b), Tobias *et al.*

(1983), Tobias & Grigoriev (1975) and Tobias & Todd (1974). For the sake of completeness, a few research programmes of national and international organisations are also quoted at this point (Antipov *et al.* 1994; Gargir *et al.* 1992; Oser 1992; Ruyters & Gräf 1992; Schimmerling & Sulzman 1994).

6.3.3.2 Molecular changes

Radiation induced molecular alterations are certainly influenced by the spatial distribution of energy deposition (**Magee & Chatterjee 1980**), directly influencing the yield of primary free radicals (**Mozumder 1986**) with consequences for subsequent reactions. Recently this fact has been incorporated into model calculations which are reviewed below. The most important molecule is without doubt DNA, consequently most studies are concerned with DNA lesions. They may be studied at the base level, for example by electron spin resonance (Heck *et al.* 1987; Miller *et al.* 1993; Schaefer *et al.* 1992). Double strand breaks (DSB) are commonly considered to be biologically the most relevant type. While previously it was thought that they are formed with higher efficiency in mammalian cells by densely ionizing radiations (**Blöcher 1988**; Frankenberg-Schwager 1993; **Kampf & Eichhorn 1983; Obutarov *et al.* 1982**; **Tolkendorf & Eichhorn 1983**) in the LET-range around 100 keV mm^{-1} this has recently been doubted (**Jenner *et al.* 1992; Prise *et al.* 1990; Weber & Flentje 1993**) which may also reflect changes in measurement techniques. In yeast in which DSB can be quantified at doses which leave very high surviving fractions, α-particles appear to be clearly more efficient than X-rays (**Frankenberg-Schwager *et al.* 1984, 1990; Löbrich *et al.* 1993**). With heavier ions there is agreement that the induction efficiency per unit energy deposited decreases with LET although a functional dependence has not yet been established (**Akpa *et al.* 1992**; Heilmann & Rink 1993; Kaneko *et al.* 1989; Kraft 1987; Kraft *et al.* 1989; Micke *et al.* 1992 & 1994; **Roots *et al.*** 1989, **1990**; Schäfer *et al.* 1994a & b; Taucher-Scholz *et al.* 1992; Zimmermann *et al.* 1994a & b).

Repair of DNA-DSB has been investigated in several cell types, and it was generally found that its extent (and probably fidelity) decreased with ionization density which is taken to account for the high RBE-values for inactivation, mutation and transformation (**Blöcher** 1988; **Fox & McNally 1990**; Frankenberg 1994; Frankenberg-Schwager 1992; **Frankenberg-Schwager *et al.* 1984; Lett 1992**; Lett & Peters 1992; Lett *et al.* 1989a; **Prise *et al.* 1988; Roots *et al.* 1979, 1980**). A number of theoretical approaches have been developed based on track structure and microdosimetric considerations have been developed to interpret DSB-formation in a quantitative way. So far they are only applicable to comparatively light ions (**Charlton *et al.* 1989**; Chatterjee & Holley 1992a & b; **Holley *et al.* 1990**; Leonteva *et al.* 1984; **Michalik 1991, 1993;** Nikjoo *et al.*1994; Paretzke 1992). DNA lesions can not only lead to inactivation and mutation but they may also change the expression of genetic information which may influence the biological programme of the affected cell in a drastic way, e.g. by selective radiation-induced gene amplification (**Luecke-Huhle *et al.* 1986**; Luecke-Huhle 1993).

6.3.3.3 Cellular endpoints and tissue effects

Inactivation. There is now an impressive body of data for inactivation by charged particles spanning a wide range of energies and atomic numbers. Initial studies used microorganisms like bacteria (Schäfer *et al.* 1978) and yeast (**Schöpfer *et al.* 1980, 1982, 1984**), whilst later on the emphasis shifted understandably to mammalian cells (**Ainsworth *et al.* 1983; Bertsche *et al.* 1983; Bettega *et al.* 1979; Blakely *et al.* 1979; Blomquist *et al.* 1993; Curtis *et al.* 1982; Heyder & Pohlit 1980;** Kraft *et al.* 1983, 1984; **Napolitano *et al.* 1992; Perris *et al.* 1987; Raju *et al.* 1975, 1980, 1991; Roberts & Goodhead 1988; Todd 1965; Tolkendorf & Eichhorn 1983; Wulf *et al.* 1981**). This does not make studies with microbial systems useless, quite on the contrary, particularly in the present context. Reliable methods for studying mammalian cells in space are not yet in sight so that one has still to rely on these so-called 'primitive' organisms and accelerator studies form the indispensable basis for well-founded experimentation. More recently also, multicellular organisms which may be fit to withstand space conditions because they can survive for a long time in a dormant state have been used, for example *Artemia* cysts (**Su *et al.*** 1990) or the nematode *Caenorhabditis elegans* (**Nelson *et al.* 1989, 1994a & b).**

There is also, however, another more fundamental aspect: a characteristic of energy deposition by all charged particles is its pronounced heterogeneity. Objects differing in geometrical dimensions and sensitivity can be used to probe these structures which will ultimately lead to a better understanding of the underlying basic processes in a quantitative manner. A solid theoretical foundation is the only realistic way to a rational hazard estimation.

Because of the structured nature of energy deposition dose is not a good parameter, and particle fluence is preferred. From the functional dependence of various endpoints action cross sections can be derived which have basically only a formal meaning. If they are compared, however, to the dimensions of the cell's sensitive site it may be deduced how many particle traversals are required to cause the effect under consideration.

There are a number of reviews (**Kiefer 1986; Kraft 1987a** & b; **Blakely 1992**) to which the reader is referred for more detailed discussions, only some salient features are mentioned here. Inactivation cross sections rise with LET in the lower range up to a transient plateau whose height depends on ion type. For heavier ions the unequivocal dependence on LET breaks down, each ion has then to be described by a separate curve. At very high LET-values one finds even a decrease which can be understood on the basis of track structure considerations. Since this very high ionization density is found with low ion energy the range of secondary electrons and hence the radial extension of the track (the penumbra) shrinks, causing a decrease in action cross sections (Fig. 2). This behaviour was not only found with inactivation but also

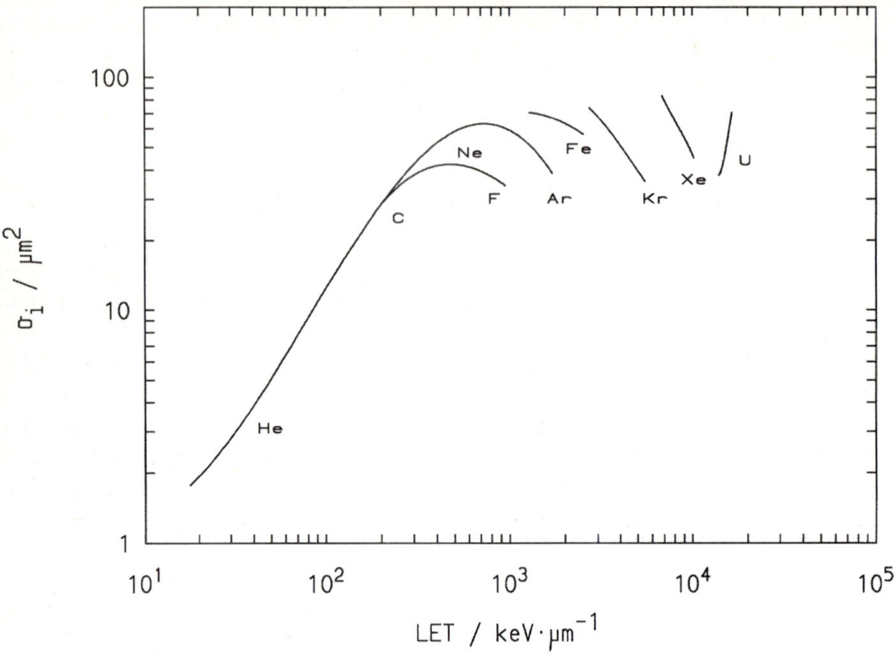

Fig. 2. The dependence of inactivation cross sections of V79 Chinese hamster cells on LET. The different curves indicate different ions. Modified after **Wulf *et al.* 1985.**

for mutation induction and possibly also for other endpoints where the available data do not yet allow a firm conclusion. More important is the fact that in all systems studied the inactivation cross sections were always smaller than the nuclear area (presumed to represent the sensitive site) which means that more than one particle passage was necessary to kill the cell. This has implications for other kinds of damage and for radiation protection. If heavy particles could inactivate by single traversals no surviving mutants or neoplastically transformed cells would be left which would reduce the genetical or tumour hazard. This is obviously not the case, even with uranium ions.

Another question, also pertinent to the space radiation environment, is the possible interaction between different radiation types. While it was formerly assumed that heavy charged particles act via a single hit mechanism so that no interaction was expected, recent studies show that a synergism exists, i.e. cells first exposed to, for example, alpha particles or heavy ions reacted more sensitively to a subsequent X-irradiation and *vice versa* (**Bird *et al.* 1983; McNally *et al.* 1988; Ngo *et al.*** 1980, **1981, 1982**). This was also confirmed at the single cell level (Kost *et al.* 1994, see below). For the space situation this would mean that different components of the radiation field must not be treated separately. Whether this plays a significant role can only be assessed by experiments on satellites which have so far not been performed.

The so far undisputed assumption that in the low LET range there is a unique dependence of inactivation cross sections has recently been called into question by experiments indicating that protons show a maximum killing efficiency at considerably lower LET than heavier ions (**Belli *et al.*. 1989; Belli *et al.* 1992a & b; Folkard *et al.* 1989; Goodhead *et al.* 1992**).

Mutation induction. The first studies on mutation induction with heavy ions were performed in microorganisms (Baltschukat *et al.* 1986; Horneck & Bücker 1983; **Kiefer *et al.* 1983a** & b, reviewed in **Kiefer 1986;** and Horneck *et al.* 1989), and this work continues (**Basha *et al.* 1992**; Horneck et al. 1994; Krasavin *et al.* 1992). Systematic work on mammalian cells was pioneered by **Cox & Masson (1979)** and Thacker *et al.* (1979) employing comparatively light accelerated nuclei. They showed an increase in induction effectiveness with LET up to a peak around 100 keV mm^{-1} followed by a decrease. This was in principle confirmed by **Hei *et al.* (1988a & b)**. Very heavy ions were used by **Kranert *et al.* (1990,** 1992) but the energy at that time was restricted to about 20 MeV $nucleon^{-1}$. Meanwhile, higher energies have become available (Chen *et al.* 1994c; Kiefer **1992**, 1994; Kronenberg 1994; **Tsuboi *et al.* 1992**). The situation is qualitatively similar to inactivation: for low LET values there appears to be a unique relationship but in the higher range this breaks down and each ion has to be represented by a separate curve. It is noteworthy that mutants can even be detected after exposure to uranium ions (**Kranert *et al.* 1990**) which ties in with the observation mentioned above that cells are not killed by single particle passages, even with these very heavy projectiles. A considerable number of studies have been performed on protons (**Belli *et al.* 1991, 1992a**; **Fuhrman Conti *et al.* 1988**; Robertson *et al.* **1987**, 1994; Sideris 1987) and alpha particles (**Barnhardt & Cox 1979; Belli *et al.* 1992b; Iliakis 1984; Thacker *et al.* 1982**), mainly intended to clarify the LET–RBE relationship. The general conclusion is that RBE is higher than for cell inactivation at the peak value around 100–200 keV mm^{-1}. It seems that protons fall out of this common scheme as with inactivation. Here the maximum is found at lower LET–values (**Belli *et al.* 1992b**).

With the advent of powerful techniques of molecular biology like the polymerase chain reaction, it has become possible to characterise radiation-induced mutation also at the molecular level (**Aghamohammadi *et al.* 1992**; Breimer 1987; Hei *et al.* 1994; **Jabberaboansari *et al.* 1991; Thacker 1986**). Such approaches have also been applied recently in heavy ion studies (Chen *et al.* 1994c; Kronenberg 1994; **Kronenberg & Little 1989**). The somewhat surprising result of these investigations, which are still preliminary, is that even with heavy ions there is still a sizeable fraction of mutants which show only small genetic changes and could be classified as point mutations although, as expected, the major proportion consists of deletions of variable extent. More work needs to be done to arrive at a systematic picture.

Mutations may also be studied in multicellular organisms which are able to survive space conditions like the nematode C. *elegans* (Coohill *et al.* 1987; **Nelson *et al.* 1989,** 1994a & b), plants (Mei *et al.* 1994a & b; Pickert *et al.* 1992) or the fruitfly *Drosophila* (Fritz-Niggli *et al.* 1987).

Neoplastic transformations. Fewer data are available for transformation with heavy ion exposure but they are now gradually being collected. Protons (**Bettega *et al.* 1979, 1981, 1986, 1988**) and alpha particles (**Bettega *et al.* 1992;** Durante *et al.* 1992; **Lloyd *et al.* 1979; Robertson *et al.* 1983**) have been studied most, the results being summarised by Curtis (1978). Heavy ions were employed by **Hieber *et al.* (1988,** 1989, 1993), **Miller *et al.* (1990,** only intermediate LET) and **Yang *et al.* (1985,** 1986, 1989). They all used rodent systems for the *in vitro* assay, a technique with human cells would be very valuable but appears not yet to be available in a reliable form. There is a special problem with transformation by densely ionizing radiation like charged particles concerning the dose-rate dependence. Since the first report by **Hill *et al.* (1984)** that dose protraction with neutrons leads to an increase in transformation effectiveness compared to acute doses, a multitude of studies has been devoted to the clarification of this phenomenon, both experimentally as well as theoretically (**Barendsen 1985; Brenner & Hall 1990; Rossi & Kellerer 1986**). The present picture is still not very clear, it appears that the enhancing effect is restricted to a rather small LET range (**Miller *et al.* 1990**) but even the original results have not always been confirmed (**Balzer-Kubitschek & Harrison 1991**). Further clarification is awaited.

Cell progression and cell cycle effects. Irradiated cells are delayed in their progression through the cell cycle, particularly in G2, less so at the G1/S border. The duration is dose dependent, but also depends on ionization density (**Lücke-Huhle 1982; Lücke-Huhle *et al.* 1985; Hieber *et al.* 1982; Raju *et al.* 1980; Sasaki *et al.* 1984**) and may be quite extensive with heavy ions (**Geard 1980; Lücke-Huhle *et al.* 1979; Collyn-d'Hooghe *et al.* 1981;** Scholz *et al.* 1989). This has to be taken into account when other parameters which are expressed in a cycle dependent manner are studied, e.g. chromosome aberrations.

The radiosensitivity depends also on the stage in the cell cycle at which the exposure takes place. This effect is most pronounced with sparsely ionising radiations and vanishes gradually with increasing LET (**Bettega, Conti *et al.* 1982; Bird *et al.* 1980;** Blakely *et al.* 1989).

Chromosomal alterations. Structural chromosomal aberrations, especially dicentrics are a sensitive and comparatively easy indicator of a radiation damage to a cell. It is widely used as a means of biological dosimetry and has, therefore, great importance for practical radiation protection and individual risk assessment. The most used technique is the observation in peripheral blood lymphocytes which are stimulated *in vitro* into the cell cycle by certain chemicals, e.g. phytohaemaglutinine. There is a wealth of studies with this system. In spite of its obvious simplicity there are pitfalls which may be particularly important with heavy particles. Since aberrations can only be scored in mitosis, failure of cells to reach this phase or extensive delay will cause grossly erroneous results (see above). Dose effect relationships have been established for protons (Akhmadieva *et al.* 1984; **Edwards *et al.* 1986;**

Grahn ***et al.*** **1983; Matsubara** ***et al.*** **1990; Rimpl** ***et al.*** **1990; Smith** ***et al.*** **1974; Takatsuji** ***et al.*** **1984, 1989**) and alpha particles (**Bedford & Goodhead 1989; Dufrain** ***et al.*** **1979; Nagasawa** ***et al.*** **1991; Purrott** ***et al.*** **1980; Takatsuji & Sasaki 1984; Takatsuji** ***et al.*** **1989**). In other cell lines also heavier ions have been investigated (**Geard 1986; Geard & Brenner 1990**; Goodwin & Blakely 1992; Goodwin *et al.* 1989; Kraft 1987b; Kraft *et al.* 1989; Ritter & Kraft 1987; Ritter *et al.* 1992; **Skarsgard** ***et al.*** **1967; Suzuki** ***et al.*** **1990;** Yang *et al.* 1994). With alpha particles and heavier ions, but reportedly not with X-rays, a new phenomenon has been observed, chromosomal instability (**Kadhim** ***et al.*** **1992; Evans 1992**; Sabatier *et al.* 1993). It is expressed as the appearance of new aberrations after many cell cycles and interpreted to indicate heritable genomic changes which after some time break up to form chromosomal aberrations.

Apart from dicentrics, other cytogenetic parameters have also been investigated with densely ionizing particles, namely sister chromatid exchanges (SCEs) (**Aghamohammadi** ***et al.*** **1988; Nagasawa** ***et al.*** **1990; Nagasawa & Little 1992**) and micronuclei (**Bettega** ***et al.*** **1980, 1981; Bilbao** ***et al.*** **1989; Brooks** ***et al.*** **1990**). Very heavy projectiles may even cause fragmentation of chromatin (**Cornforth & Goodwin 1991; Kraft 1987a**).

Single ion effects. Particle fluxes in space are quite low which means that most body cells receive only single or at most very few hits during a mission. This fact emphasises the importance of single hit effects (Curtis 1987 a & b). The low particle flux in space demands special techniques to observe cellular effects. They can only be detected if ion passage and the object can be spatially correlated, as can be done with the BIOSTACK (Horneck 1993). This has been carefully tested in accelerator experiments which not only serve the purpose of providing backing for the space studies but also gave interesting insights in the action of single ions and its dependence on the impact parameter under well defined and controlled conditions (Baican *et al.* 1992a–c; **Facius** ***et al.*** **1992**; Facius 1993; Facius *et al.* 1994; Vasilenko *et al.* 1992; **Weisbrod** ***et al.*** **1992**). Generally, it was found that the inactivating action extended further out than expected on the basis of conventional track structure theories (Facius *et al.* 1978b, 1983; Kost & Kiefer 1987; Micke *et al.* 1987; Schäfer *et al.* 1994a). In interaction experiments with alpha particles and X-rays it could be shown that the killing efficiency of the ion is increased at higher impact parameters by a pre-irradiation with sparsely ionizing radiation (Kost *et al.* 1994) which ties in nicely with the above mentioned synergistic effects in conventional survival studies.

Tissue effects. The most studied organ is the eye. Radiation may cause cataract formation (see **Merriam & Worgul (1983)** for clinical implications) whose appearance depends on dose. It is believed that a threshold exists but this is not at all certain. Very high RBE values have been reported for neutrons, so a special interest in protons and heavy ions is also warranted. This problem has been addressed by quite a few investigations (with protons: Cox *et al.*

1993; Kabachenko & Fedorenko 1977; **Lett *et al.* 1987**, 1989a, **1991**; with heavy ions: **Ainsworth & Jose 1982;** Blakely *et al.* 1994; **Brenner *et al.* 1991**; Cox *et al.* 1983, 1987, 1992a & b; **Jose & Ainsworth 1983; Lett *et al.*** 1984, **1985**, 1987, 1989a & b; Medvedovsky *et al.* 1994; **Merriam *et al.* 1984**; Nelson *et al.* 1983; Nelson & Tobias 1983; **Riley *et al.* 1991**; Tao *et al.* 1994; Ulrich 1992; **Worgul *et al.* 1989a & b**; Worgul 1992a & b), retinal damage has been studied by **Krebs *et al.* (1990),** Lett & Williams (1993), **Nelson *et al.* (1981),** Philpott *et al.* (1978, 1980, 1985), Williams & Lett (1992), and cataractogenesis by Worgul (1986), Worgul *et al.* (1989, 1992) and **Worgul *et al.* 1989a & b**). Particular attention has also been paid to the brain and nerve function (Dalrymple *et al.* 1994; Cox & Kraft 1984; Gitsov *et al.* 1992; Hopewell 1994; **Joseph *et al.* 1992a** & b; 1994, **Kraft *et al.* 1979; Miquel *et al.* 1976;** Myers *et al.* 1994; Switzer *et al.* 1994; Vazquez & Worgul 1994; **Woodruff *et al.* 1988**). Other tissues investigated include the spermatogonia (**Alpen & Powers-Risius 1981;** Sapp *et al.* 1992), the alimentary tract (**Alpen *et al.* 1980;** Carr *et al.* 1993; 1994), the skin (Bergtold *et al.* 1983; Cox & Lett 1989; Cox *et al.* 1986; Savchenko 1985; **Tatsuzaki *et al.* 1987**), the respiratory epithelium (Gerzaghi-Howe 1992), and vascular tissue (Griem *et al.* 1994; **Yang & Ainsworth 1982**; Yang & Tobias 1984). General tissue damage was also addressed by a number of investigators (Ainsworth 1992a & b; Ainsworth *et al* 1989; Leith *et al.* 1983; Lett *et al.* 1980, **1986b; Urano *et al.* 1984**). Particularly interesting in the context of space radiation biology are the on-going investigations of a monkey colony exposed in 1966 to protons of different energies about which a number of studies appeared over the years (Cox *et al.* 1983, 1987, 1992a; **Dalrymple *et al.* 1966a–c, 1991;** Lett & Williams 1993; **Lindsay *et al.* 1966; Yochmowitz *et al.* 1985**).
A fairly new development represents the finding that most tissue cells die at the end of their lifespan according to a well coordinated program called *apoptosis.* Radiation interferes with this although in a not yet well known way (Potten 1993; **Rubin *et al.* 1993**). A discussion of these intriguing questions which are also relevant to tumour formation is outside the scope of this review but it should be kept in mind that not only radiation but also other flight related factors could play a role here, so that the problem of interaction discussed in the previous section could appear in a new light.
Since heavy charged particles deposit large amounts of energy when traversing tissue it has been suggested that they may induce 'microlesions' extending over a number of cells around the path (Todd & Walker 1984). Their existence have not been demonatrated unambiguously although a number of investigations have addressed this question (Konariek 1992; Konariek & Worgul 1992; Todd 1989; Worgul *et al.* 1989).

Cancerogenesis and tumour induction. The radiation induced formation of tumours also constitutes one of the major hazards after exposure in space, particularly as it has to be assumed that no threshold dose exists There is quite a solid body of information with sparsely ionising radiation, mainly based on human epidemiological studies (ICRP 1991) but rather little is known about charged particle action. There are a number of animal studies which might help to improve the situation (Alpen *et al.* 1992; Broerse *et al.* 1993; **Burns *et al.* 1972**, 1994; Fry *et al.* 1983, **1985; Mays 1989; Mays *et al.* 1987;** Wood *et al.* 1994) but the obvious question remains how far the results may be applicable to humans (Broerse et al. 1993; Boice & Fraumeni 1984).

General biological effects on body functions. It has been argued for a long time whether :radiation might induce 'unspecific aging'. A number of life-time studies in animals addressed this problem using neutrons in mice (Maisin 1987; **Thomson *et al.* 1981a & b, 1983, 1985a & b, 1986;** see also Ainsworth 1992a). For higher animals the long term monkey colony mentioned above may ultimately yield some information. The general conclusions so far appears to be that radiation-induced life-shortening is overwhelmingly due to tumour induction. Even behaviourial changes are invoked as possible radiation effects and were studied by different techniques although so far with inconclusive results (Hunt *et al.* 1989; Joseph *et al.* **1992a** & b; **Rabin *et al.* 1989**, 1993, 1994).

Repair of radiation effects. The discovery that cells are able to repair damage inflicted by radiation is without doubt one of the most exciting findings in radiobiology. It has, of course, also a bearing on radiation protection. Whether space conditions may influence these processes has been briefly discussed in the previous section. Another problem is their dependence on radiation quality which should be studied on Earth. There is quite a number of studies at different levels of biological organisation (Frankenberg-Schwager 1993; **Frankenberg-Schwager *et al.* 1984, 1990**; **Kiefer 1984; Kiefer & Schneider 1991; Travis *et al.* 1988**) as well as some more general discussions (Fritz-Niggli 1988, 1989).

Theoretical models of biological radiation action. Since the early days of target theory there have been continuing attempts to explain the biological action of radiations in a quantitative way (see Kiefer 1987). Because of the very special nature of the radiation field in space the estimation of hazards to humans necessarily involves a large amount of extrapolation for which these models could be very useful. A number of papers aim at relating the physical properties of the different radiation types with their biological action (Goodhead 1992; Kiefer 1987, 1993; Kiefer & Schneider 1984; Kiefer & Straaten 1989; Kozubek & Horneck 1993; Krämer & Kraft 1994; **Kraft *et al.* 1991; Thomas 1987**). In all of them the spatial pattern of energy deposition plays a central role but the approaches differ. Microdosimetric theories consider the distribution and variation of energy imparted to target volumes within the cell, typically of submicron sizes, to predict cellular endpoints like inactivation, mutation or chromosomal aberrations (Breckow 1993; Bond & Varma 1983; **Fisher & Harty 1982; Hall *et al.* 1978; Kellerer & Chmelevsky 1975; Kellerer & Rossi 1972, 1978; Kliauga *et al.* 1978;** Robertson *et al.* 1994; **Rossi & Zaider 1992;** Takahashi *et al.* 1992). **Katz *et al.* (1971)** have a different concept: they start from the X-ray response and use the calculated mean dose deposited by charged particles to evaluate heavy ion action. This method has been widely used by a number of authors (Cucinotta *et al.* 1991, 1994; Katz 1986; Katz *et al* 1993; Kiefer 1983; **Waligorski *et al.* 1987**; Wilson *et al.* 1993). An intermediate approach is taken by Scholz & Kraft (1992) who follow essentially the ideas of Katz but consider additionally the spatial pattern of energy deposition in the cellular target. It seems presently that a unifying model is not yet in sight.

Radiation and biological evolution. Radiation is one of the driving forces of biological evolution. It is supposed to have played an important part in the formation of prebiotic molecules but it is also a major cause of mutations. These issues have recently been discussed in a few papers, also in relation to space radiation biology (Haynes 1992; Todd 1994; Yang *et al.* 1994).

6.4 Radiation protection issues

6.4.1 Introduction

Radiation protection aims at sparing humans from the hazardous effects of radiation on health and well-being. In principle this can be achieved by physical shielding and setting exposure limits which are considered acceptable as compared to other risks, and which have then to be monitored by suitable devices. This is standard practice on Earth for radiation workers. Chosen limits are based on recommendations by the *International Commission on Radiological Protection* (ICRP) which draws its conclusions on the basis of evaluation of the scientific literature. This implies that revisions have to be made when new insights emerge. The most recent recommendations were published in 1991 (ICRP 60, 1991) which will be referred to in some more detail below.

Table 2 summarises briefly the kind of damage possibly incurred by humans. Acute effects show clear dose thresholds and are hence a matter of concern only when the latter are exceeded. In space this might occur with solar flares but not under normal circumstances, even in outer space. Late effects are as-

Table 2. Radiation effects in humans.

Acute effects	Impairment of organ function; blood forming organs, immune system, alimentary tract, skin, eye. Severity increases with dose Deterministic Threshold doses
Late effects	Germ cell mutations; genetic risk, cancer. Probability increases with dose Stochastic No threshold assumed

sumed to have no thresholds so, theoretically, any dose, however small, has to be considered potentially detrimental. The most important hazard here is the formation of tumours. There is a considerable body of information based on epidemiological data, to a large extent, but not exclusively, from the survivors of Hiroshima and Nagasaki. To obtain usable results large populations are required. This makes it immediately clear that comparable data can never be obtained from space experience so that extrapolations are necessary. They have to proceed in such a way that the differences in the composition of the radiation field have to be carefully evaluated, and then the different contributions weighted according to their biological effectiveness. The latter again involves extrapolations since human data are virtually unavailable. If the mechanism were completely understood this could be done on a more rational basis. This knowledge can only be gained by fundamental research and the formulation of quantitative models which in turn have to be tested by new experiments. The vastly different radiation field is, however, not the only complication in space. Special environmental and flight-related factors, e.g. microgravity, may change the human radiation response. This all shows that radiation protection in space is very complicated and involves a fair amount of sophisticated guesswork.

As described in Section 6.2, dose levels in space are considerably higher than on Earth, sometimes by several orders of magnitude. Physical shielding is feasible only to a certain extent, firstly because of mass limitations, secondly because of the formation of secondary radiations by interaction with the shielding material. A certain radiation background has to be accepted so that an important task for radiation protection lies in the best possible estimate of hazards. In a mixed radiation field the physical quantity 'dose' is not the appropriate parameter to assess biological damage since different radiation types exhibit different effectiveness. In order to account for this a new quantity was introduced, namely 'equivalent dose' (H) (see ICRU 1985) (formerly 'dose equivalent' but changed both in name and definition by ICRP 90). It is obtained by multiplying the dose (D) by a radiation weighting factor (w_R), i.e. $H = w_R D$. If the dose is measured in Gray (Gy) the unit of H is Sievert (Sv). The old unit was 'rem'; 1 Sv = 100 rem). Radiation weighting factors are 1 for X- and γ-rays, for neutrons between 5 and 20 depending on energy, for protons 5 (with energies greater than 2 MeV) and 20 for α-particles and heavy ions. This overestimates the role of heavy particles since it is experimentally known that in many cases, e.g. for mutation induction, the efficiency decreases with LET. To account for this the radiation weighting factor may be replaced by a 'quality factor' (QF) (see ICRU 40, 1986) which depends on LET. The ICRP recommends limits of equivalent doses for occupational exposures. The current value is 20 mSv y^{-1} averaged over 5 y.

If the same system is used in space, namely the application of limits, the situation will become difficult for longer, but not very extended flights. The *National Council on Radiation Protection and Measurements* (NCRP) of the United States published in 1989 its report No. 98 *Guidance on Radiation Received in Space Activities.* Career limits in Sv were recommended based on a fatal cancer risk of 3%: for males: 2 + 0.075(age – 30), and for females: 2 + 0.075 (age – 38). These were based on a lifetime risk factor estimate of 2% per Sv which has, in the meantime, been doubled (ICRP 60; 1991) so, consequently, the above figures have to be reduced accordingly if the approach is to be maintained. Since a large fraction of the equivalent dose is due to heavy particles a reduction in the quality factor would lead to lower values. This consideration explains the great interest in studying the biological action of heavy charged particles.

6.4.2 Space related studies

All the information reviewed in the previous section pertains also to radiation protection, here more specialized publications are discussed. There are already quite a few documentations of dose measurements which are listed in section 6.2.2, for planned missions the expected values have to be estimated. A number of studies have tried to assess radiation burdens for various flight scenarios. They include polar missions (Atwell & Beever 1987), low Earth orbits (Bourrieau & Comet 1987; Curtis *et al.* 1969; Wilson & Cucinotta 1984), the space station (Bücker & Facius 1986; Reitz *et al.* 1987), lunar missions and stations (Nealy *et al.* 1989; Simonsen & Nealy 1991) and Mars missions (Badhwar *et al.* 1994; Letaw *et al.* 1987; **Moore 1992**; Pissarenko 1992, 1994; Simonsen & Nealy 1991; Simonsen *et al.* 1990; **Townsend *et al.* 1992a**) or a combination of them (Adams *et al.* 1992; Grahn 1973; Letaw *et al.* 1987, **1988**; McCormack 1987; **Schäfer 1952**). The important issue of solar particle events which could cause acute radiation effects has been addressed by several authors (**Townsend *et al.* 1991**, 1992a & b; Webber 1963).
A few papers deal with specific shielding problems (Townsend *et al.* 1988; Wrenn *et al.* 1987), whilst a larger number of studies have dealt with the evaluation and application of quality factors or concepts to replace it (Barendsen 1992; Curtis *et al.* 1965, 1992; Edwards 1992; Fry 1992a & b; Grahn 1973; **Sinclair 1985**; Todd 1983; **Townsend *et al.* 1987**; Watt 1989). Protection-directed research needs, and development of methods are discussed or described by Atwell (1994), Conklin & Hagan (1986), Curtis & Wilkinson (1972), **Fry (1991)**, Horneck *et al.* 1989a & b, Kiefer (1994a), **Schimmerling (1992)**, **Shinn *et al.* (1990)**, **Sinclair & Fry (1988)**, Sinclair (1992, 1994), **Wilson *et al.* (1990a)** and Zankl *et al.* (1994). An important question is whether and how animal data can be extrapolated to man, an issue which is touched upon by a few investigators (Ainsworth 1992b; Broerse *et al.* 1993; Hagen 1987; **Lett *et al.* 1986a & b**; McCarthy & Hale 1992). Since at a certain stage it was hoped to overcome the problems linked with human radiation exposures in space with the help of chemical radioprotectors a few studies have investigated this possibility (Grdina *et al.* 1992; Landauer *et al.* 1992; Lowenstein *et al.* 1987; Milas *et al.* 1992; Swenberg 1987).
The majority of publications discuss radiation protection problems in space in a general and conceptual manner (Cassarett & Lett 1983; ESA 1987; Fry

1984, 1986, 1987, 1992a & b; **Fry & Lett 1988; Fry & Nachtwey 1988**; Jacobs 1960; Kovalev 1983; NAS/NRC 1970, 1973; NCRP 1989a & b; Reetz 1965; Reitz *et al.* 1987; Sinclair 1983, 1984, 1986, 1994; Sinclair & Fry 1984; Space Science Board 1970). A spin-off for ordinary terrestrial consumers is the better understanding of radiation hazards connected with extended air travel at high altitudes; for discussion see **Barish (1990), Friedberg *et al.* (1989), Tobias (1952)**, and **Wilson & Townsend (1988b, 1989)**.

6.5 Bibliographic evaluation

This chapter has been based on a compilation of more than 1000 references (though only 530, of which 161 appear in refereed journals, are cited and listed in Section 6.5.1) dealing with the three main topic areas: radiation measurements, biological effects of radiations and radiation protection issues in space. The large number of references illustrates the activity in the field but not necessarily the quality of research. This is obviously very difficult to assess and to provide at least an approximate judgement the references can be assigned to three categories as follows: (i) refereed international journals, (ii) symposia and book contributions, (iii) NASA, ESA and other space 'community' publications. A breakdown (Table 3) reveals clear differences. The proportion of publications in refereed journals is significantly higher for ground-based research, while space results are, to a large extent, presented in special symposia, books and dedicated reports. This is regrettable since even good work does not find its way into the open literature and remains unnoticed within the wider community. Also, this fact nurtures the impression that space research in radiation biology is of lesser quality. This must not necessarily be so but the situation can only be improved if space scientists are willing to stand up to the hard standards of peer review. The very high number of papers in Earth-based radiobiology related to space issues in accepted journals demonstrate that the field is potentially powerful and valid.

Table 3. Bibliographic analysis of the radiation biology literature.

Research field	Publication category			Totals
	Refereed journal	Books & symposia	Space agency publications	
Radiation research in space	68 (28%)	135 (57%)	37 (15%)	240
Ground based radiation research	66 (51%)	46 (35%)	18 (14%)	130
Radiation biology in space	29 (22%)	91 (70%)	11 (8%)	131
Ground based radiation biology	207 (48%)	226 (52%)	3 (1%)	436
Protection issues	22 (24%)	49 (53%)	21 (23%)	92
Totals	392 (38%)	547 (53%)	90 (9%)	1029

Acknowledgements. Special thanks are due to a number of people who helped at different stages of preparation of this review. We want to mention here Gerda Horneck who kindly provided a very helpful compilation of references difficult to find otherwise, Ina Allendörfer for secretarial help, Uwe Stoll and Ingmar Kiefer who served as trouble-shooters.

6.5.1 References

Adams JH jr, Badhwar GD & Letaw JR. Dose equivalent from cosmic rays in deep space. In: *Abstracts, The World Space Congress, Washington, D.C., USA*, p. 568, 1992.

Adams JH jr, Tylka AJ & Stiller B. LET Spectra in low Earth orbit. IEEE Trans. Nuc. Sci., NS-333, 1386, 1986.

Adams JH jr, Cosmic radiation: constraints on space exploration. Nucl. Tracks Rad. Measurements 20, 397–402, 1992.

Aghamohammadi SZ, Goodhead DT & Savage J R K. Induction of sister chromatid exchanges (sce) in Go lymphocytes by plutonium-238 alpha-particles. Int. J. Radiat. Biol. 53: 909–15, 1988.

Aghamohammadi SZ, Morris T, Stevens DL & Thacker J. Rapid screening for deletion mutations in the *hprt* gene using the polymerase chain reaction -- X-ray and alpha-particle mutant spectra. Mutat. Res. 269: 1–7, 1992.

Ahlen SP. Theoretical and experimental aspects of the energy loss of relativistic heavily ionizing particles. Reviews Modern Phys. 52: 121–73, 1980.

Ainsworth EJ. Acute radiation effects in animals: relevance to space radiation scenarios. In: *Abstracts, The World Space Congress, Washington, D.C., USA*, p. 561, 1992a.

Ainsworth EJ. Life span studies in irradiated animals. In: *Abstracts, The World Space Congress, Washington, D.C., USA*, p. 559, 1992b.

Ainsworth EJ & Jose JG. Posterior lens cataracts in mice given single or weekly fractionated doses of 225 MeV 12-C ions or 60-Co gamma radiation. Radiat. Res. 91: 371, 1982 (abstract).

Ainsworth EJ, Afzal SMJ, Crouse DA, Hanson WR & Fry RJM. Tissue responses to low protracted doses of high LET radiations or photons: early and late damage relevant to radio-protective countermeasures. Adv. Space Res. 9, 10 (10): 299–313, 1989.

Ainsworth EJ, Kelly LS, Mahlmann LJ, Schooley JC, Thomas RH, Howard J & Alpen EL. Response of colony-forming units-spleen to heavy charged particles. Radiat. Res. 96: 180–97, 1983.

Akhmadieva AK, Aptikaeva GP, Livanova IA, Antipov AV, Akoev IG & Ganassi EE. Analysis of cytogenetic effects of the secondary radiation resulting from 70 GeV protons of Chinese hamster cells. Adv. Space Res. 4 (10): 227–30, 1984.

Akopova AB, Vikhrov AI, Dudkin VE, Magradze NV & Moiseenko AA Measurement of the spectrum of linear energy losses of cosmic rays by the Cosmos-1129 satellite. Cosmic Research 23: 479–91, 1985 (in Russian).

Akpa TC, Weber KJ, Schneider E, Kiefer J, Frankenberg-Schwager M, Harbich R & Frankenberg D. Heavy ion-induced DNA double-strand breaks in yeast. Int. J. Radiat. Biol. 62: 279–87, 1992.

Akatov YA, Arkhangelsky VV, Kovalev EE, Spurny F & Votochkova I. Absorbed dose measurements on external surface of Kosmos-satellites with glass thermoluminescent detectors. Adv. Space Res. 9 (10): 237–41, 1989.

Akatov YA, Arkhangelsky VV, Aleksandrov AP, Fehér I, Edme S, Szab B, Vgyölgyi J, Szab PO, Csöke A, Rnky M & Farkas B. Thermoluminescent dose measurement onboard SALYUT type orbital stations. Adv. Space Res. 4 (10): 77–81, 1984.

Allkhofer OC & Heinrich W. Attenuation of cosmic ray heavy nuclei fluxes in the upper atmosphere by fragmentation. Nucl. Phys. B B71: 429–38, 1974.

Allkofer OC. *Introduction to Cosmic Radiation.* München, Karl Thieme Verlag, 1975.

Alpen EL. *Radiation Biophysics.* Englewood Cliffs, New Jersey, Prentice Hall International Editions, 1990.

Alpen EL & Powers-Risius P. The relative biological effect of high-Z, high-LET charged particles for spermatogonial killing. Radiat. Res. 88: 132–43, 1981.

Alpen EL, Powers-Risius P & DeGuzman R. The harderian gland model for tumorigenesis. In: *Abstracts, The World Space Congress, Washington, D.C., USA*, p. 558, 1992.

Alpen EL, Powers-Risius P & McDonald M. Survival of intestinal crypt cells after exposure to high Z, high-energy charged particles. Radiat. Res. 83: 677–87, 1980.

Andersen HH & Ziegler JF. *Hydrogen-stopping Powers and Ranges in all Elements. Stopping and Ranges of Ions in Matter.* Ziegler JF, ed. Pergamon Press, New York and London 1977.

Anikeeva ID, Akatov YA, Vaulina EN, Kostina LN, Marenny AM, Portman AI, Rusin SV & Benton EV. Radiobiological experiments with plant seeds aboard the biosatellite Kosmos 1887. Nucl. Tracks Radiat. Meas. 17: 167–71, 1990.

Anikeeva ID, Kostina LN & Vaulina EN. Experiments with air-dried seeds of *Arabidopsis thaliana* (L) Heynh. and *Crepis capillaris* (L). WAllr. Aboard Salyut 6. Life Sci. Space Res. 20 (1): 129–34, 1983.

Antipov VV, Davydov BI, Verigo VV & Svirezhev YM. Combined effect of flight factors. In: *Foundations of Space Biology and Medicine.* Ed by Calvin M & Gazenko OG II, pp. 639–667, Joint USA/USSR Publication, 1975.

Antipov VV, Delone NL, Nikitin MD, Parfyonov GP & Saxonov PP. Some results of radiobiological studies performed in Cosmos-110 biosatellite. Life Sci. Space Res. 6: 207–8, 1969.

Antipov VV, Petrov VM ,Shafirkin AV,Gorlov VV, Fedorenko BS & Chertkov KS. Space radiobiology program in Russia. Adv. Space Res 14 (10): 139–46, 1994.

Appenzeller T. Yohkoh captures a solar flare. Science 254: 793, 1991.

Appleby H & Golightly MJ. Improved radiation environment evaluation technique: application to Space Station Freedom. In: *Abstracts, The World Space Congress, Washington, D.C., USA*, p. 546, 1992.

Armstrong TW, Colborn BL, Harmon BA, Parnell TA, Watts JW jr & Benton EV. Comparison of model predictions with LDEF satellite radiation measurements. Adv. Space Res. 14 (10): 17–20, 1994.

Asano Y, Kariya H & Mori S. Computer simulations of nuclear reactions induced by incident protons in the energy range of 100 to 500 MeV in a human body. Health Phys. 55: 767, 1989.

Atwell W. Astronaut exposure to space radiation: Space Shuttle experience. *SAE Tech. Paper Ser.* 901342, 1990.

Atwell W. Anatomical models for space radiation applications: an overview. Adv. Space Res 14 (10): 415–22, 1994.

Atwell W & Beever ER. Space radiation exposures for manned polar missions: a parametric study. In: *Terrestrial Space Radiation and its Biological Effects.* Ed. by McCormack PD, Swenberg CE & Bücker H., Plenum Press: New York and London, pp. 641–54, 1987.

Avdjushin SJ, Nazarova MN, Perejaslova NK, Petrenko IE & Frolov SG. Solar proton events at the growth phase of the 22nd solar activity cycle. Adv. Space Res. 12 (2/3): 325–29, 1992.

Badhwar GD, Braby LA, Konradi A, Atwell W & Cucinotta FA. Measurements of trapped protons and cosmic rays from recent shuttle flights.Adv. Space Res 14 (10): 67–72, 1994.

Badhwar GD, Konradi A, Hardy AC & Braby LA. Active dosimetric measurements on Shuttle flights. Nucl. Tracks Radiat. Meas. 20: 13–20, 1992.

Badhwar GD & O'Neill PM. An improved model of galactic cosmic radiation for space exploration missions. Nucl. Tracks Radiat. Meas. 20: 403–10, 1992.

Baican B, Heilmann C, Schopper E & Schott JU. Assignment of particles and bio objects in BIOSTACK-experiments with silver halide detectors. In: *Abstracts, The World Space Congress, Washington, D.C., USA*, p. 573, 1992a.

Baican B, Schopper E & Schott JU. E3, E4 Units in LDEF with AgCl-detectors. In: *Abstracts, The World Space Congress, Washington, D.C., USA*, p. 570, 1992b.

Baican B, Schopper E, Wendnagel T, Schott JU & Heilmann C. Experiment 'Seeds' on BIOKOSMOS 9. Dosimetric Part. Adv. Space Res. 12 (2/3): 355–8, 1992c.

Bakker CJ & Segrè E. Stopping power and energy loss for ion pair production for 340MeV protons. Phys. Review 81: 489–92, 1951.

Baltschukat K, Horneck G, Bücker H, Facius R & Schäfer M. Mutation induction in spores of *Bacillus subtilis* by accelerated very heavy ions. Radiat. Env. Biophys. 25: 183–8, 1986.

Balzer-Kubitschek EK & Harrison GH. Lack of dose rate modification (0.0049 vs 0.12 Gy/min) of fission-neutron-induced neoplastic transformation in C3H/10T1/2 cells. Int. J. Radiat. Biol. 59: 1017–26, 1991.

Ballard RW & Connolly JP. U.S./U.S.S.R. joint research in space in biology and medicine on Cosmos biosatellites. FASEB J. 4: 5–9, 1990.

Barbier M & Dulieu HL. Biological study of tobacco seeds flown in the joint Apollo-Soyuz-test-project. Life Sci. Space Res. 16: 143–6, 1978.

Barendsen GW. Do fast neutrons at low dose rate enhance cell transformation *in vitro*? A basic problem of microdosimetry and interpretation. Int. J. Radiat. Biol. 47: 731–4, 1985.

Barendsen GW. RBE for non-stochastic effects. Adv. Space Res. 12 (2/3): 385–92, 1992.

Barish RJ. Health physics concerns in commercial aviation. Health Phys. 59: 199–204, 1990.

Barnhardt BJ & Cox SH. Mutagenicity and cytotoxicity of 4.4-MeV alpha-particles emitted by plutonium 238. Radiat. Res. 80: 542–48, 1979.

Basha SG, Krasavin EA & Kozubek S. Radioprotective action of glycerol and cysteamine on inactivation and mutagenesis in Salmonella tester strains after gamma-irradiation and heavy ion irradiation. Mutat. Res. 269: 237–42, 1992.

Bayonove J, Burg M, Delpooux M & Mir A. Biological changes observed on rice and biological and genetic changes observed on tobacco after spaceflight in the orbital station Salyut-7 (Biobloc III experiments). Adv. Space Res. 4 (10): 97–101, 1984.

Bayonove JF, Mir A & Burg M. Effects of long duration space flight on rice seed (or embryo) irradiations and element microlocalizations. In: *Abstracts, The World Space Congress, Washington, D.C., USA*, p. 570, 1992.

Bayonove JF, Raffi J & Agnel JP. Investigation on rice seeds and embryos after the LDEF, electronic spin resonance identification. Adv. Space Res. 14 (10): 53–60, 1994.

Bazilevskaya GA, Krainev MB, Stozhkov YI, Svirzhevskaya AK, Svirzhevsky NS & Lebedev PN. Stratospheric measurements of cosmic rays in 19th – 22nd solar cycle activity. Adv. Space Res. 14 (10): 779–82, 1994a.

Bazilevskaya GA, Stozhkov YI & Struminsky AB. The influence of interplanetary shocks on solar proton measurements in the stratosphere. Adv. Space Res. 14 (10): 717–20, 1994b.

Beaujean R. Influence of the geomagnetic field and of the solar activity cycle on the cosmic ray energy spectra In: *Biological Effects and Physics of Solar and Galactic Cosmic Radiation.* Ed. by Swenberg CE, Horneck G & Stassinopoulos EG. NATO ASI Series A: Life Sciences 243B, Plenum Press, New York & London, pp. 15–36, 1993.

Beaujean R, Jonathal D & Enge W. Heavy ion measurement on LDEF. Adv. Space Res. 12 (2/3): 359–62, 1992.

Beaujean R, Kopp J & Enge W. Measurements of high-LET-spectra on biocosmos 9. Nucl. Tracks Radiat. Meas. 19: 681–4, 1991.

Beck SM & Powell CA. Proton and deuteron double differential cross sections at angles from 10° to 60° from Be, C, Al, Fe, Cu, Ge, W, and Pb under 558MeV-proton irradiation. NASA TN D-8119, 1976.

Bedford JS & Goodhead DT. Breakage of human interphase chromosomes by alpha particles and X-rays. Int. J. Radiat. Biol. 55: 211–6, 1989.

Belli M, Cera F, Cherubini R, Ianzini F, Moschini G, Sapora O, Simone G, Tabocchini MA & Tiveron P. Mutation induction and RBE-LET relationship of low-energy protons in V79 cells. Int. J. Radiat. Biol. 59: 459–65, 1991.

Belli M, Cera F, Cherubini R, Tabocchini MA & Tiveron P. RBE-LET relationship for survival and mutation induction of V79 cells irradiated with low-energy protons: re-evaluation of the LET values at the LNL facility. Int. J. Radiat. Biol. 61: 145–6, 1992a.

Belli M, Cherubini R, Finotto S, Moschini G, Sapora O, Simone G & Tabocchini MA. RBE-LET relationship for the survival of V79 cells irradiated with low energy protons. Int. J. Radiat. Biol. 55: 93–104, 1989.

Belli M, Goodhead DT, Ianzini F, Simone G & Tabocchini MA. Direct comparison of biological effectiveness of protons and alpha-particles of the same LET. 2. Mutation induction at the hprt locus in V79 cells. Int. J. Radiat. Biol. 61: 625–9, 1992b.

Bender MA, Gooch PC & Kondo S. The Gemini 3 S4 spaceflight radiation interaction experiment. Radiat. Res. 31: 91–111, 1967.

Benevolenskij VN, Marennyi AM *et al.* quoted in: Grigoriev YG, (ed.) *Kosmiceskaja radiobiologija,* Energoizdat, Moscow, pp. 79–117, (German translation), 1982.

Benson RE & Pinsky LE. Visual light flash phenomenon. In: *Apollo 16 Preliminary Science Report,* NASA SP-315, pp. 27/17–27/20, 1972.

Benton EV. Radiation measurements aboard Spacelab. Science 225: 224, 1984.

Benton EV. Summary of radiation dosimetry results on U.S. and Soviet manned spacecraft. Adv. Space Res. 6 (11): 315–28, 1986.

Benton EV, Csige I, Frank AL & Benton ER. Radiation and activation measurements performed on LDEF. In: *Abstracts, The World Space Congress, Washington, D.C., USA*, p. 569, 1992b.

Benton EV, Heinrich W, Parnell TA, Armstrong TW, Derrickson JH, Fishman GJ, Frank AL, Watts JW & Wiegel B. Ionizing radiation exposure of LDEF (pre-recovery estimates). Nucl. Tracks Radiat. Meas. 20: 75–100, 1992a.

Benton EV & Henke RP Radiation exposures during spaceflight and their measurement, Adv. Space Res. 3: 171–187, 1983.

Benton EV, Henke RP, Frank AL, Johnson CS, Cassou RM, Tran MT & Etter E. *Experiment K309 – Space Radiation Dosimetry Aboard Cosmos 1129: U.S. Portion of Experiment.* National Aeronautics and Space Administration, Tech. Memo. TM-81288, National Technical Information Service, Springfield, Virginia, 1981.

Benton EV & Parnell TA Space radiation dosimetry on U.S. and Soviet manned missions. In: *Terrestrial Space Radiation and Its Biological Effects.* Ed. by McCormack PC, Swenberg CE & Bücker H, Plenum Press, New York, pp. 729–794, 1987.

Benton EV, Peterson DD, Bailey JV & Parnell T. High-LET particle exposure of Skylab astronauts. Health Phys. 32: 15–20, 1977.

Benton EV, Peterson DD, Marenny AM & Popov VI. HZE particle radiation studies aboard Kosmos 782. Health Phys. 35: 643–8, 1978.

Berger MJ. Microdosimetric event size distributions in small sites in water vapour irradiated by protons. Phys. Med. Biol. 33: 583–96, 1988.

Bergtold DS, Cox AB, Su CM & Lett JT. Late skin damage in rabbits and monkeys after exposure to particulate radiations. In: *Life Sciences and Space Research 20, 1.* Ed. by Holmquist WR, pp. 221–30, 1983.

Bertini W. *Preliminary data from intranuclear-cascade calculations of 0.75-, 1-, and 2-GeV protons on oxygen, aluminum, and lead, and 1-GeV neutrons on the same elements.* ORNL-TM-1996, U.S. Atomic Energy Commission, 1967.

Bertsche U, Iliakis G & Kraft G. Inactivation of Ehrlich ascites tumor cells by heavy ions. Radiat. Res. 95, 1: 57–67, 1983.

Bettaga D, Calzolari P. & Lombardi LT. Effects of split-dose irradiation on survival and oncogenic transformation induced by 31MeV protons in C3H10T1/2 cells. Int. J. Radiat. Biol. 52: 761–5, 1988.

Bettega D, Birattari C, Bombana M, Conti AMF, Galline E & Tallone LL. Relative Biological effectiveness for protons of energies up to 31 MeV. Radiat. Res. 77: 85–97, 1979.

Bettega D, Birattari C, Candoni B, Copola M, Poli A, Sverzellati PP & Lombardi LT. Energy deposition by proton beams of up to 31 MeV in microscopic volumes. Radiat. Environ. Biophys. 19: 79–89, 1981.

Bettega D, Bombana A, Pelucchi T, Poli A, Lombardi LT & Conti AMF. Multinucleate cells and micronucleus formation in cultured human cells exposed to 12 MeV protons and gamma-rays. Int. J. Radiat. Biol. 37: 1–10, 1980.

Bettega D, Calzolari P, Chiorda GN & Tallonelombardi L. Transformation of C3H10T1/2 cells with 4.3 MeM alpha-particles at low doses -- effects of single and fractionated doses. Radiat. Res. 131: 66–71, 1992.

Bettega D, Calzolari P, Pollara P & Lombardi LT. *In vitro* cell transformations induced by 31 MeV protons. Radiat. Res. 104: 178–81, 1986.

Bettega D, Conti AMF, Gariboldi L, Pelucchi MT, Scaioli E & Lombardi LT. Age response of EUE cells exposed to 31-MeV protons. Radiat. Res. 91: 457–67, 1982.

Bettega D, Dubini S, Fuhrman Conti AM, Pellucchi T & Lombardi LT. Chromosome aberrations induced by protons up to 31 MeV in cultured human cells. Radiat. Environ. Biophys. 19: 91–100, 1981.

Bilbao A, Prosser JS, Edwards AA, Moody JC & Lloyd DC. The induction of micronuclei in human lymphocytes by in vitro irradiation with alpha particles from plutonium-239. Int. J. Radiat. Biol. 56: 287–92, 1989.

Billingham J. Robbins DE, Modisette JL & Higgins PW. Status report on the space radiation effects on the Apollo mission. In: *Second Symposium on Protection Against Radiations in Space*. Ed. by Reetz A. NASA SP-71, pp. 139–56, 1965.

Bird RP, Rohrig N, Colvett D, Geard CR & Marino SA. Inactivation of synchronized Chinese hamster V79 cells with charged- particle track segments. Radiat. Res. 82: 277–89, 1980.

Bird RP, Zaider M, Rossi HH, Hall EJ, Marino SA & Rohrig N. The sequential irradiation of mammalian cells with X rays and charged particles of high LET. Radiat. Res. 93: 444–52, 1983.

Blake J & Baker DN. Long term measurements of the integral galactic cosmic-ray fluxes in geostationary orbits. Nucl. Tracks Rad. Measurements 20: 411–4, 1992.

Blakely EA. Cell inactivation by heavy charged particles. Radiat. Environ. Biophys. 31: 181–96, 1992.

Blakely EA, Chang P, Lommel L, Bjornstad K, Dixon M, Tobias C, Kumar K & Blakely WF. Cell-cycle radiation response: role of intracellular factors. Adv. Space Res. 9 (10): 177–86, 1989.

Blakely EA, Daftari IK, Gillett EL, Lee AC, Lett JT, Castro JR & Char DH. Helium-ion-induced human cataractogenesis. Adv. Space Res. 14 (10): 501–6, 1994.

Blakely EA, Ngo FQH, Curtis SB & Tobias CA. Heavy-ion radiobiology - cellular studies. *Advances in Radiation Biology* 11, 295–390, 1984.

Blakely EA, Tobias CA, Yang TC, Smith KC & Lyman JT. Inactivation of human kidney cells by high-energy monoenergetic heavy-ions. Radiat. Res. 80: 122–60, 1979.

Blanquet Y, Planel H, Grigoriev HG, Gasset G, Benevolensky VA, Delpoux M, Jacqout C, Popov VJ, Pianezzi B & Marennyi AM. Preliminary results of the Biobloc experiment of the Cosmos 782 flight: effects of cosmic rays on brine shrimp eggs and tobacco seeds. Life Sci. Space Res. 15: 165–8, 1977.

Blizard JB. *Long range solar flare prediction.* NASA CR-.61316, 1969.

Blöcher D. DNA double-strand break repair determines the RBE of alpha-particles. Int. J. Radiat. Biol. 54: 761–71, 1988.

Blomquist E, Russell KR, Stenerlow B, Montelius A, Grusell E & Carlsson J. Relative biological effectiveness of intermediate energy protons - comparisons with ^{60}Co gamma-radiation using 2 cell lines. Radiotherapy and Oncology 28: 44–51, 1993.

Boice JD jr & Fraumeni JE jr *Radiation Carcinogenesis.* Raven Press, New York, 1984.

Bond VP & Varma MN. A combined hit theory-microdosimetric explanation of cellular radiobiological action, biological effects of low-level radiation. In: *Procedings of an International Symposium of the International Atomic Energy Agency, IAEA-SM-266*, Vienna, pp. 81–95, 1983.

Bottollier-Depois JF, Nguyen VD, Lebaron-Jacobs L, Siegrist M, Andre-Deshays C, Marsal O, Petrov VM, Kozlova SB, Tognini, M & Avdeev S. Dose rate and LET spectrum aboard Mir station since Antares until Altair mission. In: *Proceedings of the 5th European Symposium on Life Sciences Research in Space*. Ed. by Oser H, Guyenne TD. Noordwijk, The Netherlands, ESA, ESTEC, pp. 217–20, 1994.

Bouisset P, Nguyen VD, Akatov YA, Siegrist M, Parmentier N, Archangelsky VV, Vorojtsov AS, Petrov VM & Kovalev EE. Quality factor and dose equivalent investigations aboard the Soviet space station Mir. Adv. Space Res. 12 (2/3): 363–7, 1992.

Bourrieau J & Comet B. Analysis of the radiative risk during Hermes missions. In: *Terrestrial Space Radiation and its Biological Effects*. Ed by McCormack PD, Swenberg CE & Bücker H. Plenum Press, New York and London, p. 97–116, 1987.

Braby NF, Metting WE, Wilson JW & Toburen LH. Microdosimetric measurements of heavy ion tracks. Adv. Space Res. 12 (2/3): 23–32, 1992.

Breckow J. Concepts of microdosimetry and their applicability to radiation protection problems in manned space missions. In: *Biological Effects and Physics of Solar and Galactic Cosmic Radiation*. Ed. by Swenberg CE, Horneck G & Stassinopoulos EG. NATO ASI Series A: Life Sciences 243A, Plenum Press, New York–London, pp. 269–82, 1993.

Breimer LH. Ionizing radiation induced mutagenesis: molecular genetic changes and role of DNA lesions and repair. In: *Terrestrial Space Radiation and its Biological Effects*. Ed. by McCormack PD, Swenberg CE & Bücker H. Plenum Press: New York & London, pp. 277–300, 1987.

Brenner DJ & Hall EJ. The inverse dose-rate effect for oncogenic transformation by neutrons and charged particles: a plausible interpretation consistent with published data. Int. J. Radiat. Biol. 58: 745–58, 1990.

Brenner DJ, Medvedovsky C, Huang Y, Merriam GR & Worgul BV. Accelerated heavy particles and the lens. 6. RBE studies at low doses. Radiat. Res. 128: 73–81, 1991.

Broerse JJ, Davelaar J & Zurcher C. Tumor induction in animals and the radiation risk for man, In: *Biological Effects and Physics of Solar and Galactic Cosmic Radiation*. Ed. by Swenberg CE, Horneck G & Stassinopoulos EG. NATO ASI Series A: Life Sciences 243A, Plenum Press, New York–London, pp. 161–76, 1993.

Brooks AL, Newton GJ, Shyr LJ, Seiler FA & Scott BR. The combined effects of alpha-particles and X-rays on cell killing and micronuclei induction in lung epithelial cells. Int. J. Radiat. Biol. 58: 799–812, 1990.

Bücker H. The Biostack experiments I and II abroad Apollo 16 and 17. Life Sci. Space Res. 12: 43–50, 1974.

Bücker H. Biostack, a study of the biological effects of HZE galactic cosmic radiation. In: *Biomedical Results of Apollo*. Ed. by Johnston RS, Dietlein LF & Berry CA. NASA SP-368, pp. 343–54, 1975.

Bücker H. Results of radiobiological spaceflight experiments. In: *Proceedings, European Symposium on Life Sciences Research in Space*. ESA SP-130, pp. 263–70, 1977.

Bücker H. Free flyer biostack experiment (A0015). In: *The Long Duration Exposure Facility (LDEF)*. Ed. by Clark L, Kinard WH, Carter DJ jr. & Jones JL jr. NASA SP-473, pp. 139–45, 1984.

Bücker H, Baltschukat K, Beaujean R, Bonting SJ, Delpoux M, Enge W, Facius R, Francois H, Grau EH, Heinrich W, Horneck G, Kranz AR, Pfohl R, Planel H, Portal G, Reitz G, Rüther W, Schäfer M, Schopper E & Schott JU. Advance Biostack, Experiment 1ES 027 on Spacelab-1. Adv. Space Res. 4 (10): 83–90, 1984.

Bücker H & Facius R. Recent radiobiological findings from spaceflight and ground-based studies -- an overview. Life Sci. Space Res. 18: 125–30, 1980.

Bücker H & Facius R. Radiation protection problems for the space station and approaches to their mitigation. Adv. Space Res. 6: 305–14, 1986.

Bücker H, Facius R, Hildebrand D & Horneck G. Results of the *Bacillus subtilis* unit of the Biostack II experiment: physical characteristics and biological effects of individual cosmic HZE particles. Life Sci. Space Res. 13: 161–6, 1975.

Bücker H, Facius R, Horneck G, Reitz G, Graul EH, Berger H, Höffken H, Rüther W, Heinrich W, Beaujean R & Enge E. Embryogenesis and organogenesis of *Carausius morosus* under spaceflight conditions. Adv. Space Res. 6: 115–24, 1986a.

Bücker H, Facius R & Reitz G. Dosimetric mapping inside BIORACK on D-1. Naturwiss. 73: 425, 1986b.

Bücker H & Horneck G. Studies on the effects of cosmic HZE particles on different biological systems in the Biostack experiments I and II flown onboard of Apollo 16 and 17. In: *Radiation Research.* Ed by Nygaard OF, Adler HI & Sinclair WK. Academic Press, New York, pp. 1138–51, 1975.

Bücker H, Horneck G, Allkofer OC, Bartholomä KP, Beaujean R, Cüer P, Enge W, Facius R, Francois H, Graul EH, Henig G, Heinrich W, Kaiser R, Kühn H, Massué JP, Planel H, Portal G, Reinholz E, Rüther W, Scheuermann W, Schmitt R, Schopper E, Schott JU, Soleilhavoup JP & Wollenhaupt H. The Biostack experiment on Apollo 16. Life Sci. Space Res. 11: 295–305, 1973.

Bücker, H, Horneck G, Facius R, Reitz G, Schäfer M, Schott JU, Beaujean R, Enge W, Schopper E, Heinrich W, Beer J, Wiegel B, Pfohl R, Francois H, Portal G, Bonting SL, Graul EH, Rüther W, Kranz AR, Bork U, Koller-Lambert K, Kirchheim B, Starke ME, Planel H & Delpoux M. Radiobiological advanced Biostack experiment. Science 225: 222–4, 1984.

Bücker H, Horneck G, Reitz, G, Graul EH, Berger H, Höffken H, Rüther W, Heinrich W & Beaujean R. Embryogenesis and organogenesis of *Carausius morosus* under space flight conditions. Naturwiss. 73: 433, 1986c.

Budd T, Kwok CS, Marshall M & Lythe S. Microdosimetric properties of alpha-particle tracks measured in a low-pressure cloud chamber. Radiat. Res. 95: 217–30, 1983.

Budinger T, Bichsel H & Tobias CA. Visual phenomena noted by human subjects on exposure to neutrons of energies less than 25 million electron volts. Science 172: 868–70, 1971.

Budinger TF, Tobias CA, Huesman RH, Upham FT, Weiskamp TE & Hoffman RA. Apollo-Soyuz light-flash observations. Life Sci. Space Res. 15: 141–6, 1977a.

Budinger TF, Tobias CA, Huesman RH, Upham FT, Wieskamp TF, Schott JU & Schopper E. Light flash observations (MA 106). In: *Apollo-Soyuz Test Project Summary Science Report.* NASA SP-412, pp. 193–209, 1977b.

Budinger TF, Tobias CA, Schopper E, Schott JU, Huesman RH, Upham FT, Wieskamp TF, Kucala JM, Goulding FS, Landis DA, Walton JT & Walton RE. Quantitative observations of light flash sensations, Experiment MA-106. In: *ASTP Preliminary Science Report.* NASA TM X-58173, pp. 13/1–13/16, 1976.

Budnik E, Morozova N, Pissarenko NF, Sarris E & Tykva R. The function of semiconductor detector response for charged-particle telescopes. In: *Abstracts, The World Space Congress, Washington, D.C., USA*, p. 546, 1992.

Burns FJ, Albert RE, Bennett P & Sinclair IP. Tumor incidence in rat skin after proton irradiation in a sieve pattern. Radiat. Res. 50: 181–90, 1972.

Burns F, Jin Y, Garte S & Hosselet S.. Estimation of risk based on multiple events in radiation carcinogenesis of rat skin. Adv. Space Res. 14 (10): 507–20, 1994.

Butts JJ & Katz R. Theory of RBE for heavy ion bombardment of dry viruses and enzymes. Radiat. Res. 30: 855–71, 1971.

Carr KE, McCullough JS, Brennan P, Hayes TL, Ainsworth EJ & Nelson AC. Heavy ion induced changes in small intestinal parameters. Adv. Space Res. 14 (10): 521–30, 1994.

Carr KE, McCullough JS, Gilmore RStC, Abbas B, Hume SP, Nelson AC, Hayes TL & Ainsworth EJ. Effects of neon ions, neutrons and X rays on small intestine, In: *Biological Effects and Physics of Solar and Galactic Cosmic Radiation.* Ed. by Swenberg CE, Horneck G, Stassinopoulos EG. NATO ASI Series A: Life Sciences 243A, Plenum Press, New York-London, pp. 203–216, 1993.

Casarett GW, Lett JT. Factors involved in the evaluation of space radiation standards – a preface. Life Sci. Space Res. 20 (1): 149–50, 1983.

Charlton DE, Nikjoo H & Humm JL. Calculation of initial yield of SSB and DSB in cell nuclei from electrons, protons and alpha particles. Int. J. Radiat. Biol. 56: 1–19, 1989.

Chase HB. Cutaneous effects of primary cosmic radiation. J. Aviat. Med. 25: 388–91, 1954.

Chase HB & Post YS. Damage and repair in mammalian tissue, exposed to cosmic-ray heavy nuclei. J. Aviat. Med. 27: 533–40, 1956.

Chatterjee A & Holley WR. Biochemical mechanisms and clusters of damage for high-LET radiation. Adv. Space Res. 12 (2/3): 33–43, 1992a.

Chatterjee A & Holley WR. Clusters of DNA damage and repairability. In: *Abstracts, The World Space Congress, Washington, D.C., USA*, p. 550, 1992b.

Chatterjee A & Schaefer HJ. Microdosimetric structure of heavy-ion tracks in tissue. Radiat. Environ. Biophys. 13: 215–27, 1976.

Chatterjee A, Maccabee HD & Tobias CA. Radial cutoff LET and radial cutoff dose calculations for heavy charged particles in water. Radiat. Res. 54: 479–94, 1973.

Chen DJ, Chenette G, Clark R, Garcia-Munoz M, Guzik TG, Pyle KR, Sang Y & Wefel JP. Models of galactic cosmic rays for use in calculating linear energy transfer spectra. Adv. Space Res. 14 (10): 765–70, 1994a.

Chen DJ, Chenette G, Guzik M, Garcia-Munoz M, Pyle K, Sang K & Wefel J. A model of solar energetic particles for use in calculating LET spectra developed from ONR-604 data. Adv. Space Res. 14 (10): 675–80, 1994b.

Chen DJ, Tsuboi K, Nguyen T & Yang TC. Charged particle mutagenesisII: mutagenic effects of high energy charged particles in normal human fibroblasts. Adv. Space Res. 14 (10): 347–54, 1994c.

Chenette DL, Guzik TG & Wefel JP. The LET spectrum and it's uncertainty during the CRRES mission. Adv. Space Res. 14 (10): 809–14, 1994.

Chupp EL. Transient particle acceleration associated with solar flares. Science 250: 229, 1990.

Collyn-d'Hoghe M, Hemon D, Gilet R, Curtis SB, Valleron AJ & Malaise E. Comparative effects of ^{60}Co gamma-rays and neon and helium ions on cycle duration and division probability of emt6 cells. a time laps cinematography study. Int. J. Radiat. Biol. 39: 297–306, 1981.

Conklin JJ & Hagan MP. Research issues for radiation protection for man in prolonged space flight. Adv. in Radiat. Biology 13: 215–84, 1986.

Coohill T, Marshall T, Schubert W & Nelson G. Genetic and developmental responses of radiation sensitive mutants of the nematode *C. elegans* to ultraviolet, high and low LET radiation, In: *Terrestrial Space Radiation and its Biological Effects*. Ed. by McCormack PD, Swenberg CE & Bücker H. Plenum Press: New York and London, pp. 265–76, 1987.

Cornforth MN &Goodwin EH. The dose-dependent fragmentation of chromatin in human fibroblasts by 3.5-MEV alpha particles from Pu-238 – Experimental and theoretical considerations pertaining to single-track effects. Radiat. Res. 127: 64–74, 1991.

Cox AB, Ainsworth EJ, Jose JG, Lee AC & Lett JT. Cataractogenesis from high-LET radiation and the Casarett model. Life Sci. Space Res. 20: 211–20, 1983.

Cox AB, Lee AC & Lett JT. Delayed effects of proton irradiation in the lens and integument: a primate model. In: *Terrestrial Space Radiation and its Biological Effects*. Ed. by McCormack PD, Swenberg CE & Bücker H. Plenum Press: New York & London, pp. 415–22, 1987.

Cox AB, Lee AC, Williams GR & Lett JT. Late cataractogenesis in primates after exposure to particulate radiations. Adv. Space Res. 12 (2/3): 379–84, 1992a.

Cox AB, Lee AC, Williams GR & Lett JT. Late radiogenic cataracts from high-energy particles in selected animal species: a comparative study. In: *Abstracts, The World Space Congress, Washington, D.C., USA*, p. 557, 1992b.

Cox AB, Salmon YL, Lee AC, Lett JT, Williams GR, Broerse JJ, Wagemaker G & Leavitt DD. Progress in the extrapolation of radiation cataractogenesis data across longer-lived mammalian species. In: *Biological Effects and Physics of Solar and Galactic Cosmic Radiation.* Ed. by Swenberg CE, Horneck G & Stassinopoulos EG. NATO ASI Series A: Life Sciences 243A, Plenum Press, New York-London, pp. 177-84, 1993.

Cox AB, Wood DH & Lett JT. Use of primary cell cultures to measure the late effects in the skins of Rhesus monkeys irradiated with protons. Adv. Space Res. 6 (11): 217-22, 1986.

Cox R & Kraft LM. Quantitation of heavy ion damage to the mammalian brain: some preliminary findings. Adv. Space Res. 4 (10): 247-50, 1984.

Cox R & Lett JT. The quantification of wound healing as a method to assess late radiation damage in primate skin exposed to high-energy protons. Adv. Space Res. 9 (10): 125-30, 1989.

Cox R & Masson WK. Mutation and inactivation of cultured mammalian cells exposed to beams of accelerated heavy ions. III. Human diploid fibroblasts. Int. J. Radiat. Biol. 36: 149-60, 1979.

Cucinotta FA, Hajnal F & Wilson JW. Energy deposition at the bone-tissue interface from nuclear fragments produced by high-energy nucleons. Health Phys. 59: 819-25, 1990.

Cucinotta FA, Katz R, Wilson JW, Townsend LW, Nealy JE & Shinn JL. Cellular track model of biological damage to mammalian cell cultures from galactic cosmic rays. NASA TP-3055, 1991.

Cucinotta FA, Katz R, Wilson JW, Townsend LW, Shinn J & Hajnal F. Biological effectiveness of high-energy protons-target fragmentation. Radiat. Res. 127: 130-7, 1991.

Cucinotta FA & Wilson JW. Computer subroutines for estimation of human exposure to radiation in low earth orbit, NASA TM-86324, 1985.

Cucinotta FA, Wilson JW, Shinn JL, Atwell W & Fong M. Dose rate and repair effects on cell damage in earth orbit. Adv. Space Res. 14 (10): 121-4, 1994.

Curtis SB. An analysis of human kidney cell oxygen-enhancement ratios for fast-neutron beams and a prediction for negative pion beams. Radiat. Res: 46, 557-79, 1971.

Curtis SB. Calculated LET distributions of heavy ion beams. Int. J. Radiat. Oncol. Biol. Phys. 3: 87-92, 1978.

Curtis SB. Effects of low and high let radiation on neoplastic transformation in cells and the importance of single track effects in space. In: *Terrestrial Space Radiation and its Biological Effects.* Ed. by McCormack PD, Swenberg CE & Bücker H. Plenum Press, New York and London, pp. 153-62, 1987a.

Curtis SB. Radiation physics related to biology. In: *Terrestrial Space Radiation and its Biological Effects.* Ed. by McCormack PD, Swenberg CE & Bücker H. Plenum Press, New York and London, pp. 301-14, 1987b.

Curtis SB. Importance of dose rate and cell proliferation in the evaluation of biological experimental results. Adv. Space Res. 14 (10): 989-96, 1994a.

Curtis SB. Single-track effects and new directions in GCR risk assessment. Adv. Space Res. 14 (10): 885-96, 1994b.

Curtis SB, Doherty WR & Wilkinson MC. *Study of Radiation Hazards to Man on Extended Near Earth Missions.* NASA CR-1469, 1969.

Curtis SB, Dye DL & Sheldon WR. Fractional cell lethality approach to space radiation hazards. In: *Second Symposium on Protection Against Radiations in Space.* Ed. by Reetz A Jr. NASA SP-71, pp. 219-23, 1965.

Curtis SB, Schilling WA, Tenforde TS, Crabtree K E, Tenforde SD, Lyman T & Howard J. Survival of oxygenated and hypoxic tumor cells in the extended-peak regions of heavy charged-particle beams. Radiat. Res. 90: 292-309, 1982.

Curtis SB, Townsend LW, Wilson JW, Powers-Risius P, Alpen EL & Fry RMJ. Fluence-related risk coefficients using the Harderian gland data as an example. Adv. Space Res. 12 (2/3): 407-16, 1992.

Curtis SB & Wilkinson MC. The heavy particle hazard – what physical data are needed? In: *Proceedings of the National Symposium on Natural and Manmade Radiation in Space.* Ed by Warman EA. NASA TM X-2440, pp. 1007–15, 1972.

Dachev TP. GCR and SCR environment of MIR-space station as observed by Liulin instrument. In: *Abstracts, The World Space Congress, Washington, D.C., USA*, p. 565, 1992.

Dachev TP, Matviichuk YN, Bankov NG, Koleva RT, Vellinov PI, Todorieva L, Semkova JV, Petrov VM, Redko VI, Zil MV & Mitrakas VG. Modeling of the radiation exposure during the flight of the second bulgarian cosmonaut on board the Mir space station. Adv. Space Res. 9 (10): 253–5, 1989b.

Dachev TP, Matviichuk YN, Bankov NG, Semkova JV, Koleva RT, Ivanov YJ, Tomov BT, Petrov, VM, Shurshakov VA, Bengin VV, Machmutov VS, Panova NA, Kostereva, TA, Temny VV, Ponomarev YN & Tykva R. 'Mir' radiation dosimetry results during the solar proton events in September–October 1989. *Adv. Space Res.* 12 (2/3): 321–4, 1992.

Dachev TP, Matviichuk Y, Semkova JV, Koleva RT, Boichev B, Baynov P, Kanchev NA, Lakov P, Ivanov YJ, Tomo PT, Petrov VM, Redko VI, Kojarinov VI & Tykva R. Space radiation dosimetry with active detections for the scientific program of the second Bulgarian cosmonaut on board the Mir space station. Adv. Space Res. 9 (10): 247–51, 1989a.

Dalrymple GV, Leichner PK, Harrison KA, Cox AB, ardy KA, Salmon YL & Mitchell JC. Induction of high grade Astrocytoma (HGA) by protons: molecular mechanisms and RBE considerations. Adv. Space Res. 14 (10): 267–70, 1994.

Dalrymple GV, Lindsay IR, Ghidoni JJ, Hall JD, Mitchell JC, Kundel HL & Morgan IL. Some effects of 138-MeV protons on primates. Radiat. Res. 28: 471–88, 1966a.

Dalrymple GV, Lindsay IR, Ghidoni JJ, Mitchell JC & Morgan IL. Some effects of 400-MeV protons on primates. Radiat. Res. 28: 507–28, 1966c.

Dalrymple GV, Lindsay IR, Hall JD, Mitchell JC, Ghidoni JJ, Kundel HL & Morgan IL. The relative biological effectiveness of 138-MeV protons as compared to Cobalt-60 gamma radiation. Radiat. Res. 28: 489–506, 1966b.

Dalrymple GV, Lindsay IR, Mitchell JC & Hardy KA. A review of the USAF/NASA proton bioeffects project – rationale and acute effects. Radiat. Res. 126: 117–9, 1991.

Derrickson JH, Parnell TA, Austin RW, Selig WJ & Gregory JC. A measurement of the absolute energy spectra of galactic cosmic rays during the 1976–77 solar minimum. Nucl. Tracks Rad. Measurements 20: 415–22, 1992.

Dicello JE, Zaider M & Varma MN. An inductive assessment of radiation risks in space. Adv. Space Res. 14 (10): 899–910, 1994.

Dons RF & Fohlmeister U. Combined injury syndrome in space-related radiation environments. Adv. Space Res. 12 (2/3): 157–63, 1992.

Dreute J, Röcher H, Heinrich W, Harboe-Sorensen R & Adams L. Study of SEUs generated by high energy ions. In: *Proceedings of RADECS 93*, IEEE Report, 1994.

Dudkin YA, Kovalev EE, Ptapov YV, Benton EV, Frank AL, Benton ER, Watts JW, Parnell TA, Schopper E, Baican B, Reitz G, Bücker H, Facius R, Beaujean R & Heilmann C. Cosmic ray LET spectra and doses on board Cosmos-2044 Biosatellite. Nucl. Tracks Radiat. Meas. 20: 149–55, 1992.

Dufrain RJ, Littlefield L, Joiner EE & Frome EL. Human cytogenetic dosimetry: a dose-response relationship for alpha particle radiation of ^{241}Am. Health Phys. 37: 279–90, 1979.

Durante M, Grossi G, Yang TC & Roots R. Genetic changes in mammalian cells transformed by helium ions. Adv. Space Res. 12 (2/3): 137–45, 1992.

Edwards AA. Chromosomal data relevant for Q values. Adv. Space Res. 12 (2/3): 421–30, 1992.

Edwards AA, Lloyd DC, Prosser JS, Finnon P & Moquet JE. Chromosome aberrations induced in human lymphocytes by 8.7 MeV protons and 23.5 MeV helium-3 ions. Int. J. Radiat. Biol. 50: 137–45, 1986.

Enghardt K, Fromm WD, Geissel H, Keller H, Kraft G, Magel A, Manfrass P, Munzenberg G, Nickel F, Pawelke J, Schardt D, Scheidenberger C & Sobiella M. The spatial distribution of positron-emitting nuclei generated by relativistic light ion beams in organic matter. Phys. Med. Biol. 37: 2127–31, 1992.

ESA. *Protection Against Radiation in Space.* ESA Contract No. 6988/86/NL/PP (SC), 1987.

Evans HJ. Radiation biology – alpha-particle after effects. Nature 355: 674–5, 1992.

Evenson P, Garcia-Munoz P, Meyer P, Pyle KR & Simpson, JA A quantitative test of solar modulation theory: the proton, helium and electron spectra from 1965 through 1979. Astrophys. J. 275: L15–L18, 1979.

FAA. *Cosmic radiation exposure during air travel.* FAA-AM-80-2, FAA Advisory Committee on Radiation Biology Aspects of the Supersonic Transport, U.S. Department of Transportation, 1980.

Facius R. Correlation of radiobiological effects with tracks of cosmic heavy ions in mesoscopic test organisms. In: *Abstracts of the 5th European Symposium on Life Sciences Research in Space*, Arcachon, France, 1993.

Facius R, Bücker H, Hildebrand D, Horneck G, Höltz G, Reitz G, Schäfer M & Toth B. Radiobiological results from the *Bacillus subtilis* Biostack experiments within the Apollo and the ASTP spaceflights. Life Sci. Space Res. 16: 151–6, 1978b.

Facius R, Bücker H, Horneck G, Reitz G & Schäfer M. Dosimetric and biological results from the *Bacillus subtilis* Biostack experiment within the Apollo-Soyuz test project. Life Sci. Space Res. 17: 123–8, 1979.

Facius R, Bücker H, Reitz G & Schäfer M. Radial dependence of biological response of spores of *Bacillus subtilis* around tracks of heavy ions, In: *Proceedings of the 6th Symposium on Microdosimetry.* Ed. by Booz J & Ebert HG. EUR 6064 DE-EN-FR, Harwood Academic Publishers Ltd., London, pp. 977–86, 1978a.

Facius R, Reitz G, Bücker H, Nevzgodina LV & Maximova EN. Impact parameter determination for the passage of cosmic heavy ions through mesoscopic biological test organisms. Nucl. Tracks Radiat. Meas. 20: 207–16, 1992.

Facius R, Reitz G, Bücker H, Nevzgodina LV, Maximova EN, Kamisnkaya EV, Vikrov AI, Marenny AM & Akatov YA. Reliability of trajectory identification for cosmic heavy ions and cytogenetic effects of their passage through plant seeds. Nucl. Tracks Radiat. Meas. Int. J. Radiat. Appl. Instrum. Part D 17: 121–32, 1990.

Facius R, Reitz G & Schäfer M. Inactivation of individual bacillus subtilis spores in dependence on their distance to single cosmic heavy ions. Adv. Space Res. 14 (10): 1027–38, 1994.

Facius R, Schäfer M, Baltschukat K & Bücker H. Inactivation probability of heavy ion-irradiated *Bacillus subtilis* spores as a function of the radial distance to the particle's trajectory. Life Sci. Space Res. 20: 85–95, 1983.

Feynman J & Gabriel SB. *A New Model for Calculation and Prediction of Solar Proton Fluences.* AIAA-90-0292, 1990.

Filatova LP, Vaulina EN, Grozdova TY, Prudhommeau C & Proust J. Some results of the effects of space flight factors on *Drosophila melanogaster.* Life Sci. Space Res. 20 (1): 143–8, 1983.

Fisher DR & Harty R. The microdosimetry of lymphocytes irradiated by alpha-particles. Int. J. Radiat. Biol. 41: 315–24, 1982.

Foelsche T. Specific solar flare events and associated radiation doses: space radiation effects. ASTM Special Tech. Publ. No. 363, American Soc. for Testing & Materials, pp. 1–13, 1963.

Foelsche T & Wilson JW. Results of NASA SST-radiation studies including experimental results on solar flare events. Minutes of the Standing Committee on Radiobiology Aspects of the SST, Federal Aviation Administration, 1969.

Foelsche T, Mendell R, Wilson JW & Adams RR. *Measured and Calculated Neutron Spectra and Dose Equivalent Rates at High Altitudes: Relevance to SST Operations and Space Research.* NASA TN D-7715, 1974.

Folkard M, Prise KM, Vojnovic B, Davies S, Roper MJ & Michael BD. The irradiation of V79 mammalian cells by protons with energies below 2 MeV. Part I. Experimental arrangement and measurement of cell survival. Int. J. Radiat. Biol. 56: 221–37, 1989.

Fowler PH, Masheder MRW, Walker RNF & Worley A. Iron fluxes from the Ariel-6 workshop ICRC, Dublin August 1991. Nucl. Tracks Rad. Measurements 20: 423–6, 1992.

Fox JC & McNally NJ. The rejoining of DNA double-strand breaks following irradiation with ^{238}Pu alpha-particles: evidence for a fast component of repair as measured by neutral filter elution. Int. J. Radiat. Biol. 57: 513–21, 1990.

Frankenberg D. Repair of DNA double-strand breaks and its effect on RBE. Adv. Space Res. 14 (10): 235–48, 1994.

Frankenberg-Schwager M. DNA damage, OER and RBE. In: *Abstracts, The World Space Congress, Washington, D.C., USA*, p. 552, 1992.

Frankenberg-Schwager M. Radiation-induced DNA lesions in eukaryotic cells, their repair and biological relevance. In: *Biological Effects and Physics of Solar and Galactic Cosmic Radiation.* Ed. by Swenberg CE, Horneck G & Stassinopoulos EG. NATO ASI Series A: Life Sciences 243A, Plenum Press, New York-London, pp. 1–32, 1993.

Frankenberg-Schwager M, Frankenberg D & Harbich R. Repair of DNA double-strand breaks as a determinant of RBE of alpha particles. Brit. J. Cancer 49: 169–74, 1984.

Frankenberg-Schwager M, Frankenberg D, Harbich R & Adamczyk C. A comparative study of rejoining of DNA double-strand breaks in yeast irradiated with 3.5 MeV alpha-particles or with 30 MeV electrons. Int. J. Radiat. Biol. 57: 1151–68, 1990.

Friedberg W, Faulkner DN, Snyder L, Darden EB jr & O'Brien K. Galactic cosmic radiation exposure and associated health risks for air carrier crewmembers. Aviation, Space & Environ. Med. 60: 1104–8, 1989.

Fritz-Niggli H, Schäppi-Büchi C & Schäppi K. Reactions of genetic systems to heavy ions: acute and late effects. In: *Terrestrial Space Radiation and its Biological Effects.* Ed. by McCormack PD, Swenberg CE & Bücker H. Plenum Press: New York and London, pp. 237–50, 1987.

Fritz-Niggli H. Modifying factors on repair phenomena. Adv. Space Res. 9 (10): 187–95, 1989.

Fritz-Niggli H. The role of repair processes in cellular and genetical response to radiation, In: *Terrestrial Space Radiation and its Biological Effects.* Ed. by McCormack PC, Swenberg CE & Bücker H. Plenum Press, pp. 213–35, 1988.

Fry RJM. Approaches to radiation guidelines for space travel. Adv. Space Res. 4 (10): 121–30, 1984.

Fry RJM. Radiation effects in space. Adv. Space Res. 6 (11): 261–8, 1986.

Fry RJM. Radiation protection guidelines for space missions. in: *Terrestrial Space Radiation and its Biological Effects.* Ed. by McCormack PD, Swenberg CE & Bücker H. Plenum Press, New York & London, pp. 715–28, 1987.

Fry RJM. Radiation carcinogenesis in the whole-body system. Radiat. Res. 126: 157–61, 1991.

Fry RJM. Radiation quality and risk estimation in relation to space missions. Adv. Space Res. 12 (2/3): 403–6, 1992a.

Fry RJM. Deep space missions: radiobiological research needs. In: *Abstracts, The World Space Congress, Washington, D.C., USA*, p. 547, 1992b.

Fry RJM & Lett JT. Radiation hazards in space put in perspective. Nature 335: 305, 1988.

Fry RJM & Nachtwey DS. Radiation protection guidelines for space missions. Health Phys. 55: 159–64, 1988.

Fry RJM, Powers-Risius P, Alpen E & Ainsworth EJ. High-LET radiation carcinogenesis. Radiat. Res. 104: S188–S195, 1985.

Fry RJM, Powers-Risius P, Alpen EL, Ainsworth EJ & Ullrich RL. High-LET radiation carcinogenesis. Life Sci. Space Res. 20 (1): 241–8, 1983.

Fuhrman Conti AM, Francone G, Volonte M & Gallini RE. Induction of 8-azaguanine resistant mutants in human cultured cells exposed to 31 MeV protons. Int. J. Radiat. Biol. 53: 467–76, 1988.

Ganapol BD, Townsend LW & Wilson JW. *Benchmark Solutions for the Galactic Ion Transport Equations: Energy and Spatially Dependent Problems.* NASA TP-2878, 1989.

Gargir G, Bozouklian H & Andre-Deshays C.The space radiation biology program of CNES. In: *Abstracts, The World Space Congress, Washington, D.C., USA*, p. 574, 1992.

Garrard TL & Stone EC. Composition of energetic particles from solar flares. Adv. Space Res. 14 (10): 589–98, 1994.

Gartenbach KE, Pickert M, Zimmermann MW & Kranz AR. Cosmic Ionizing radiation effects in plant seeds after short and long duration exposure flights. Adv. Space Res. 14 (10): 105–8, 1994.

Gatuito OG, Adamovich BA, Grigoriev YuG, Druzhinin YuP, Ilyin EA & Popov VI. Radiobiological experiment aboard the biosatellite COSMOS 690. Aviat. Space and Env. Medicine 49: 42–6, 1978.

Gaubin Y, Delpoux M, Pianezzi B, Gasset G, Heilmann C & Planel H. Investigations of the effects of cosmic rays on *Artemia* cysts and tobacco seeds; results of Exobloc II experiment, flown aboard Biocosmos 1887. Nucl. Tracks Radiat. Meas. 17: 133–43, 1990.

GaubinY, Kovalev EE, Planel H, Nevzgodina LV, Gasset G & Pianezzi B. Development capacity of Artemia cysts and lettuce seeds flown on Cosmos 936 and directly exposed to rays. Aviat. Space Env. Medicine 50: 134–8, 1979.

Gaubin Y, Planel H, Gasset G, Pianezzi B, Clegg J, Kovalev EE, Nevzgodina LV, Maximova EN, Müller AT & Delpoux M. Results of *Artemia* cysts, lettuce and tobacco seeds in the Biobloc 4 experiments flown aboard the Soviet biosatellite Cosmos-1129. Adv. Space Res. 3 (8): 135–40, 1983.

Gazenko OG, Adamovich BA, Grigoriev YG, Druzhinin YP, Ilyin EA & Popov VI. Radiobiological experiment aboard the biosatellite COSMOS 690. Aviat. Space Environ. Med. 49: 42–6, 1978.

Geard CR. Initial changes in cell cycle progression of Chinese hamster V79 cells induced by high-LET charged particles. Radiat. Res. 83: 696–709, 1980.

Geard CR. Charged particle cytogenetics: effects of LET, fluence, and particle separation on chromosome aberrations. Radiat. Res. 104: S112, 1986.

Geard CR & Brenner DJ. Chromosome changes per cell nucleus per charged particle. Radiat. Prot. Dosimetry 31: 285–9, 1990.

Gerzaghi-Howe MT. Effects of radiation on rat respiratory epithelial cells: critical target cell populations and importance of cell interactions. In: *Abstracts, The World Space Congress, Washington, D.C., USA*, p. 558, 1992.

Geschke O, Röcher H, Noll A, Dreute J, Wiegel J, Heinrich W, Harboe-Sorensen R, Adams L & Vetter J. Study of radiation effects of heavy ions in microelectronic devices. Nucl. Tracks Radiat. Meas. 19: 891–7, 1991.

Getselev IV, Ignatiev PP, Kabashova NA, Kontor NN, Moszhukhina AR, Timofeev GA & Khotilovskaya TG. A model of radiation conditions during spacecraft flights in the interplanetary space and in the Earth's magnetosphere. Adv. Space Res. 12 (2/3): 441–4, 1992.

Ginzburg VL & Syrovatskii SI *The Origin of Cosmic Rays.* Translated by Massey HSH and Ed. by Ter Haar D. Macmillan Co., New York & London, 1964.

Gitsov L, Bouneva V & Fedorenko B. The effect of heavy ion radiation on the fine structure of endocrine hypothalamus in rat brain. In: *Abstracts, The World Space Congress, Washington, D.C., USA*, p. 555, 1992.

Golightly M, Hardy AC, Atwell W & Hardy K. Description, analysis and impact of major solar activity during recent U.S. Shuttle missions. Adv. Space Res. 12 (2): 335–8, 1992.

Goodhead DT. Biophysical modelling in radiation protection. In: *Biophysical Modelling of Radiation Effects*. Ed. by Chadwick KH, Moschini G & Varma MN. Adam Hilger, Bristol. Philadelphia, New York, pp. 113–124, 1992.

Goodhead DT, Belli M, Mill AJ, Bance DA, Allen LA, Hall SC, Ianzani F, Simone G, Stevens DL, Stretch A, Tabocchini MA & Wilkinson RE. Direct comparison between protons and alpha-particles of the same LET. 1. Irradiation methods and inactivation of asynchronous V79, Hela and C3H10T1/2 cells. Int. J. Radiat. Biol. 61: 611–24, 1992.

Goodwin E, Blakely EA, Ivery G & Tobias C. Repair and misrepair of heavy-ion-induced chromosomal damage. Adv. Space Res. 9 (10): 83–9, 1989.

Goodwin E & Blakely EA. Heavy ion-induced chromosomal damage and repair. Adv. Space Res. 12 (2/3): 81–9, 1992.

Goswami JN, McGuire RE, Reedy RC, Lal D & Jha R. Solar flare protons and alpha particles during the last three solar cycles. J. Geophys. Res. 93: 7195–205, 1988.

Grahn D. *HZE-particle Effects in Manned Spaceflight*. Space Science Board, National Research Council, National Academy of Sciences, 1973.

Grahn D, Lee CH & Farrington BF. Interpretation of cytogenetic damage induced in the germ line of male mice exposed for over 1 year to ^{239}Pu alpha particels, fission neutrons or ^{60}Co gamma rays. Radiat. Res. 95: 566–83, 1983.

Graul EH, Rüther W, Heinrich W, Allkofer OC, Kaiser R, Pfohl R, Schopper E, Henig GB, Schott JU & Bücker H. Radiobiological results of the Biostack experiment on board Apollo 16 and 17. Life Sci. Space Res. 13: 153–9, 1975.

Grdina DJ, Garnes BA & Nagy B. Protection by WR-2721 and WR-151327 against late effects of gamma rays and neutrons. Adv. Space Res. 12 (2/3): 257–63, 1992.

Griem V, Robotewskyi A & Nagel RH Potential vascular damage from radiation in the space environment. Adv. Space Res. 14 (10): 555–64, 1994.

Grigoriev YG. Die Strahlensensibilität des Organismus unter den Bedingungen des Raumfluges. In: *Kosmiceskaja Radiobiologija,* (German translation). Energoizdat, Moscow, pp. 79–117, 1982.

Grigoriev YG, Benevolensky VP, Druzhinin YP, Shidarov YI, Korogodin VI, Nevzgodina LV, Miller AT & Tsaraphin LS. Influence of Cosmos 368 space flight conditions on radiation effects in yeasts, hydrogen bacteria and seeds of lettuce and pea. Life Sci. Space Res. 10: 113–8, 1972.

Grigoriev YG, Miller AT, Nevzgodina LV, Krustyn AO & Shteine BA. Effects of weightlessness and of artificial gravity on irradiated lettuce seeds. Life Sci. Space Res. 15: 285–9, 1977.

Grigoriev YG, Planel H, Delpoux M, Gaubin-Blanquet Y, Nevzgodina LV, Pianezzi B, Popov VI, Marenny AM, Vinograd YA, Pfohl R, Gasset G & Jacquot C. Radiobiological investigations in Cosmos 782 space flight (Biobloc SF1 experiment). Life Sci. Space Res. 16: 137–42, 1978.

Gupta SK & Prasad MA Dose distribution due to low-energy electrons and delta rays produced along the path of a heavy charged particle in water. Radiat. Res. 76: 1–14, 1984.

Gurosky N & Ilyin Y. Soviet bio-satellites in the Cosmos-series: the main results of the 8-year program. Aviat. Space Env. Med. 49: 1355–6, 1978.

Gussenhoven MS & Mullen EG. Solar particle events as seen on CRRES. In: *Abstracts, The World Space Congress, Washington, D.C., USA*, p. 560, 1992.

Gussenhoven MS, Mullen EG & Brautigam DH,. Near-Earth radiation model deficiencies as seen on CRRES. Adv. Space Res. 14 (10): 927–42, 1994.

Guzik TG, Albergo S, Chen CX, Costa S, Crawford,HJ, Engelage J, Ferrando P,Flores I, Greiner L, Jones FC, Knott CN, Ko S, Lindstrom PJ, Mazotta J, Mitchell JW, Romanski J, Potenza R, Soutoul A, Testard O, Tull CE, Tuve C, Waddington CJ, Webber WR, Wefel JP & Zang X. A program to measure new energetic particle nuclear interaction cross sections. Adv. Space Res 14 (10): 825–30, 1994.

Hagen U. Extrapolation of animal data to man. In: *Terrestrial Space Radiation and its Biological Effects*. Ed. by McCormack PD, Swenberg CE & Bücker H. Plenum Press, New York , London, pp. 465–74, 1987.

Hall EJ, Kellerer AM, Rossi HH & Lam YMP. The relative biological effectiveness of 160 MeV protons. ii. Biological data and their interpretation in terms of microdosimetry. Int. J. Radiat. Oncol. Biol. Phys. 4: 1009–15, 1978.

Hamm RN, Turner JE, Wright HA & Ritchie RH. Calculated ionization distributions in small volumes in liquid water irradiated by protons. Radiat. Res: 97, 16–24, 1984.

Hanser FA & Dichter BK. Space radiation dosimetry. In: *Terrestrial Space Radiation and Its Biological Effects*. Ed. by McCormack PC, Swenberg CE & Bücker H. Plenum Press, New York, pp. 135–52, 1987.

Hardy KA & Leavitt DD. Experimental simulation of proton space radiation environments: a dosimetric perspective. Adv. Space Res. 14 (10): 959–68, 1994.

Haskins PS, McKisson JE, Weisenberger AG, Ely DW, Ballard TA, Dryer CS, Truscott PR, Piercey RB & Ramayya AV. Gamma-ray measurements from the space shuttle during solar flare. Adv. Space Res. 12 (2/3): 331–4, 1992a.

Haskins PS, McKisson JE, Weisenberger AG, Ely DW, Ballard TA, Dryer CS, Truscott PR, Piercey RB, Ramayya AV & Camp DC. Effects of increased shielding on gamma-radiation levels within Spacecraft. Adv. Space Res. 12 (2/3): 461–4, 1992b.

Haymaker W, Look BC, Benton EV & Simmonds RS.The Apollo 17 pocket mouse experiment (BIOCORE). In: *Biomedical Results of Apollo*. Ed. by Johnston RS, Dietlein LF & Berry CA. NASA SP-368, pp. 381–403, 1975.

Haynes RH. Evolutionary significance of dna repair and mutagenesis. In: *Abstracts, The World Space Congress, Washington, D.C., USA*, p. 549, 1992.

Heck T, Hüttermann J & Kraft G. Free radical formation in amino acids induced by heavy ions. In: *Biological Effects and Physics of Solar and Galactic Cosmic Radiation* Ed. by Swenberg CE, Horneck G & Stassinopoulos EG. NATO ASI Series A: Life Sciences 243A, Plenum Press, New York–London, p. 93–6, 1987.

Heckman G, Hirman J, Kunches J & Balch C. The Monitoring and prediction of solar particle events–an experience report. Adv. Space Res. 4 (10): 165–72, 1984.

Heckman G, Kunches JM & Allen JH. Prediction and evaluation of solar particle events based on precursor information. Adv. Space Res. 12 (2/3): 313–20, 1992a.

Heckmann G, Kunches J & Hirman JW. Variations of solar particle events. In: *Abstracts, The World Space Congress, Washington, D.C., USA*, p. 562, 1992b.

Heckmann G, Wagner WJ, Hirman JW & Kunches JM. Strategies for dealing with solar particle events in missions beyond the magnetosphere. Adv. Space Res. 9 (10): 275–80, 1989.

Hei TK, Chen DJ, Brenner DJ & Hall EJ. Mutation induction by charged particles of defined linear energy transfer. Carcinogenesis 9: 1233–6, 1988a.

Hei TK, Komatsu K, Hall EJ & Zaider M. Oncogenic transformation by charged particles of defined LET. Carcinogenesis 9: 747–50, 1988b.

Hei TK, Zhu L, Vannais D & Waldren CA. Molecular analysis of mutagenesis by high LET radiations. Adv. Space Res. 14 (10): 355–62, 1994.

Heilmann J & Rink H. Assessment of heavy ion induced DNA strand breaks in mammalian cells. In: *Biological Effects and Physics of Solar and Galactic Cosmic Radiation*. Ed. by Swenberg CE, Horneck G & Stassinopoulos EG. NATO ASI Series A: Life Sciences 243A, Plenum Press, New York–London, pp. 33–6, 1993.

Heinrich W. Calculation of LET-spectra of heavy cosmic ray nuclei at various absorber depths. Radiat. Effects 34: 143–84, 1977a.

Heinrich W. Calculation of the radiobiological effects of heavy ions on eggs of *Artemia salina* flown in the BIOSTACK experiments. Life Sci. Space Res. 15: 157–63, 1977b.

Heinrich W. Propagation of a high energy heavy ion beam through matter. Rad. Effects 40: 167–71, 1979.

Heinrich W. Predicted LET-spectra of HZE-particles for the free flyer BIOSTACK experiment on the long duration exposure facility mission. Life Sci. Space Res. 8: 143–52, 1980.

Heinrich W. Variation of galactic cosmic radiation by solar modulation, geomagnetic shielding and shielding by material. In: *Terrestrial Space Radiation and Its Biological Effects*. Ed. by McCormack PC, Swenberg CE & Bücker H. Plenum Press, New York, pp. 573–617, 1988.

Heinrich W. Cosmic rays and their interactions with the geomagnetic field and shielding material. Radiat. Phys. Chem. 43: 19–34, 1994.

Heinrich W & Beer J. The radiation situation in space and its modification by geomagnetic field and shielding. Adv. Space Res. 4 (10): 133–42, 1984.

Heinrich W., Benton EV, Wiegel B, Gusch G & Becker E. Comparison of measured cosmic ray spectra with models and predictions. Adv. Space Res. 14 (10): 969–77, 1994a.

Heinrich W., E. V. Benton, B. Wiegel, R. Zens, G. Rusch. Galactic cosmic ray abundances and spectra behind different shielding. Adv. Space Res. 14 (10): 815–24, 1994b.

Heinrich W & Spill A. Geomagnetic shielding of cosmic rays for different satellite orbits. J. Geophys. Res. 84: 4401–11, 1979.

Heinrich W, Wiegel B, Ohrndorf T, Bücker H, Reitz G & Schott JU. LET spectra of cosmic-ray nuclei for near Earth orbits. Radiat. Res. 118: 63–82, 1989.

Hess VF über Beobachtungen der durchdringenden Strahlung bei sieben Freiballon Fahrten. Phys. Z. 13: 1084–91, 1912.

Heyder IR & Pohlit W. Proliferation of lethally damaged mammaliancells irradiated by fast electrons, fast neutrons, alpha-particles and pions. Radiat. Res. 82: 352–63, 1980.

Hieber L, Beck HP & Lücke-Hule C. G2-delay after irradiation with alpha-particles as studied in synchronized cultures and by the bromodeoxyuridine-33258 H technique. Cytometry 2: 175–8, 1982.

Hieber L, Ponsel G, Roos H, Fenn S, Fromke E & Kellerer AM. Absence of a dose-rate effect in the transformation of C3H10T1/2 cells by alpha particles. Int. J. Radiat. Biol. 52: 859–69, 1988.

Hieber L, Ponsel G, Trutschler K, Fenn S & Kellerer AM. Neoplastic transformation of mouse C3H10T1/2 and Syrian hamster embryo cells by heavy ions. Adv. Space Res. 9 (10): 141–9, 1989.

Hieber L, Smida J & Kellerer AM. Neoplastic transformation induced by heavy ions: studies on transformation efficiencies and molecular mechanisms. In: *Biological Effects and Physics of Solar and Galactic Cosmic Radiation*. Ed. by Swenberg CE, Horneck G & Stassinopoulos EG. NATO ASI Series A: Life Sciences 243A, Plenum Press, New York-London, pp. 135–42, 1993.

Hill CK, Han A & Elkind MM. Fission-spectrum neutrons at low dose rate enhance neoplastic transformation in the linear low-dose region (0–10cGy). Int. J. Radiat. Biol. 46: 11–16, 1984.

Hirman JW. Solar proton event forecasts. In: *Terrestrial Space Radiation and Its Biological Effects*. Ed. by McCormack PC, Swenberg CE & Bücker H. Plenum Press, New York, pp. 61–70, 1987.

Hirzebruch SE, Heinrich W, Tolstov KD, Kovalenko, AD & Benton EV. Fragmentation cross sections of ^{16}O between 0.9 and 200 GeV/nucleon. Phys. Rev. C 46: 1487–93, 1989.

Hoffman RA, Pinsky LA, Osborne WZ & Bailey JV. Visual light flash observations on Skylab 4. In: *Biomedical Results from Skylab*. Ed. by Johnston RS & Dietlein LF. NASA SP-377, pp. 127–30, 1977.

Holley WR, Chatterjee A & Magee JL. Production of DNA strand breaks by direct effects of heavy charged particles. Radiat. Res. 121: 161–8, 1990.

Hopewell JW. Models of CNS radiation damage during space flight. Adv. Space Res. 14 (10): 433–42, 1994.

Horneck G. Cosmic ray HZE particle effects in biological systems: results from experiments in space. In: *Terrestrial Space Radiation and Its Biological Effects.* Ed. by McCormack PC, Swenberg CE & Bücker H. Plenum Press, New York, pp. 129–52, Plenum Press, New York, 1988a.

Horneck G. Impact of spaceflight environment on radiation response.In: *Terrestrial Space Radiation and Its Biological Effects.* Ed. by McCormack PC, Swenberg CE & Bücker H. Plenum Press, New York, pp. 707–14, Plenum Press, New York, 1988b.

Horneck G. Radiobiological experiments in space: a review. Nucl. Tracks. Radiat. Meas. 20: 185–205, 1992.

Horneck G. The Biostack concept and its application in space and at accelerators: studies on Bacillus subtilis spores. In: *Biological Effects and Physics of Solar and Galactic Cosmic Radiation.* Ed. by Swenberg CE, Horneck G & Stassinopoulos EG. NATO ASI Series A: Life Sciences 243A, Plenum Press, New York–London, pp. 99–116, 1993.

Horneck G, Brack A, Cadet J, Dose K, Fritz-Niggli H, Kiefer J, Broerse JJ, Raulin F, Reitz G, Roessler K & Schott JU. ERA follow-on scientific study: perspectives of exobiological and radiation biological research by use of free-flying carriers in Earth orbit. ESA contract No. 8116/88/F/BZ (SC), 1989.

Horneck G & Bücker H. Inactivation, mutation induction and repair in *Bacillus subtilis* spores irradiated with heavy ions. Life Sci. Space Res. 10 (1): 79–84, 1983.

Horneck G, Facius R, Enge W, Beaujean R & Bartholomä KP. Microbial studies in the Biostack experiment of the Apollo 16 mission: germination and outgrowth of single *Bacillus subtilis* spores hit by cosmic HZE particles. Life Sci. Space Res. 12: 75–83, 1974.

Horneck G, Greger G & Sahm RP. Bioscience experiments in the german spacelab mission D1: introduction and summary. Adv. Space Res. 6: 85–99, 1986.

Horneck G, KrasavinE & Kozubek S. Mutagenic effects of heavy ions in bacteria. Adv. Space Res. 14 (10): 315–30, 1994.

Horneck G, Schäfer M, Baltschukat K, Weisbrod U, Micke U, Facius R & Bücker H. Cell inactivation, repair and mutation induction in bacteria after heavy ion exposure: results from experiments at accelerators and in space. Adv. Space Res. 9 (10): 105–16, 1989a.

Hubert F, Bimbot R & Gauvin H Range and stopping power tables for 2.5–500 MeV/ nucleon heavy ions in solids. Atom. Nucl. Data Tables: 46, 1–213, 1990.

Hunt WA, Joseph JA & Rabin BM. Behavioral and neurochemical abnormalities after exposure to low doses of high-energy iron particles. Adv. Space Res. 9 (10): 333–6, 1989.

ICRP Task Group Radiobiological aspects of the supersonic transport. Health Phys. 12: 209–226, 1966.

ICRP 60. *Recommendations of the International Commission on Radiological Protection.* Ann. ICRP 21, No. 1–3, 1991.

ICRU 36. *Microdosimetry* International Commission on Radiation Units and Measurements, report 36. Bethesda, Md, 1983.

ICRU 39. *Determination of dose equivalentes resulting from external radiation sourceS.* International Commission on Radiation Units and Measurement Report. Bethesda, Md, 39, 1985.

ICRU 40. *The quality factor in radiation protection.* International Commission on Radiation Units and Measurements, report 40. Bethesda, Md, 1986.

ICRU 49. *Stopping powers and ranges for protons and alpha particles.* International Commission on Radiation Units and Measurements, report 49. Bethesda, Md, 1993.

Iliakis G. The mutagenicity of alpha particles in Ehrlich ascites tumor cells. Radiat. Res. 99: 52–8, 1984.

Ing H & Mortimer A. Space radiation dosimetry using bubble detectors. Adv. Space Res. 14 (10): 73–76, 1994.

Jaberaboansari A, Dunn WC, Preston RJ, Mitra S & Waters LC. Mutations induced by ionizing radiation in a plasmid replicated in human cells. 2. Sequence analysis of alpha-particle-induced point mutations. Radiat. Res. 127: 202–10, 1991.

Jacobs GJ. (ed.) *Proceedings of a Conference on Radiation Problems in Manned Space Flight.* NASA TN D-588, 1960.

Janni JF. Proton range-energy tables 1 KeV–10 GeV-energy loss, range path length, time-of-flight, straggling, multiple scattering and nuclear interaction probability, Part 1. Nuclear Data Tables, 27: 147–339, 1982.

Jenner TJ, Belli M, Goodhead DT, Ianzini F, Simone G & Tabocchini MA. Direct comparison of biological effectiveness of protons and alpha-particles of the same LET. 3. Initial yield of DNA double-strand breaks in V79 cells. Int. J. Radiat. Biol. 61: 631–7, 1992.

Jordan T & Stassinopoulos EG. Effective radiation reduction in space station and missions beyond the magnetosphere. Adv. Space Res. 9 (10): 261–74, 1989.

Jose JG & Ainsworth EJ Cataract production in mice by heavy charged argon, neon, and carbon particles. Radiat. Res. 94: 513–28, 1983.

Joseph JA, Hunt WA, Rabin BM & Dalton TK. Possible accelerated striatal aging induced by Fe-56 heavy-particle irradiation - implications for manned space flights. Radiat. Res. 130: 88–93, 1992a.

Joseph JA, Rabin B, Dalton T, Kandasamy S & Hunt W. 'Accelerated brain aging' induced by putative space radiation (^{56}Fe-particles): implications for motor behaviour changes. In: *Abstracts, The World Space Congress, Washington, D.C., USA*, p. 555, 1992b.

Kabachenko AN, Fedorenko BS. Cataractogenic action of protons with energies 25 and 50 MeV. Kosm. Biol. Med. 11: 59–62, 1977 (in Russian).

Kadhim MA, MacDonald DA, Goodhead DT, Lorimore SA, Marsden SJ & Wright EG. Transmission of chromosomal instability after plutonium alpha-particle irradiation. Nature 355: 738–40, 1992.

Kampf G & Eichhorn K. DNA strand breakage by different radiation qualities and relations to cell killing - further results after the influence of alpha-particles and carbon ions. Stud. Biophys. 93: 17–26, 1983.

Kaneko I, Eguchi-Kasai K, Kosaka T, Nakano K, Maruhashi A & Takahashi T. DNA-lesion and cell death by alpha-particles and nitrogen ions. Adv. Space Res. 9 (10): 73–81, 1989.

Katz R. Biological effects of heavy ions from the standpoint of target theory. Adv. Space Res. 6 (11): 191–8, 1986.

Katz R, Ackerson B, Homayoonfar M & Sharma SC. Inactivation of cells by heavy ion bombardment. Radiat. Res. 47: 402–25, 1971.

Katz R, Cucinotta FA, Wilson JW, Shinn JL & Ngo DM. A model of cell damage in space flight, In: *Biological Effects and Physics of Solar and Galactic Cosmic Radiation.* Ed. by Swenberg CE, Horneck G & Stassinopoulos EG. NATO ASI Series A: Life Sciences 243A, Plenum Press, New York-London, pp. 235–67, 1993.

Kellerer A M & Chmelevsky D. Criteria for the applicability of LET. Rad. Res. 63: 226–34, 1975.

Kellerer AM & Rossi HH. A generalized formulation of dual radiation action. Radiat. Res. 75: 471–88, 1978.

Kellerer AM & Rossi HH. The theory of dual radiation action. Curr. Top. Radiat. Res. Quarterly 8: 85–158, 1972.

Khandelwal GS & Wilson JW Proton tissue dose for the blood forming organ in human geometry: isotropic radiation, NASA TM X-3089, 1974.

Kiefer J. On the interpretation of heay ion survival data. In: *Radiation Protection, Proceedings of the 8th Symposium on Microdosimetry.* Ed by Booz J & Ebert HG, EUR-8395, The Commission of the European Communities, Luxembourg, pp. 729–42, 1983.

Kiefer J. Recovery after exposure to high LET radiation. Brit. J. Cancer 49 Suppl. VI: 151–5, 1984.

Kiefer J. Cellular and subcellular effects of very heavy ions. Int. J. Radiat. Biol. 48: 873–92, 1986.

Kiefer J. Heavy ion effects on cells: an approach to theoretical understanding, In: *Terrestrial Space Radiation and its Biological Effects.* Ed. by McCormack PD, Swenberg CE & Bücker H. Plenum Press, New York and London, pp. 117–28, 1987.

Kiefer J. *Biological Radiation Effects.* Springer Verlag, Berlin & Heidelberg, 1990.

Kiefer J. Heavy ion effects on cells – chromosomal aberrations, mutations and neoplastic transformations. Radiat. Environ. Biophys. 31: 279–88, 1992.

Kiefer J. Theoretical analysis of heavy ion action on cells: model-free approaches, consequences for radiation protection. In: *Biological Effects and Physics of Solar and Galactic Cosmic Radiation.* Ed by Swenberg CE, Horneck G & Stassinopoulos EG. NATO ASI Series A: Life Sciences 243A, Plenum Press, New York–London, pp. 283–90, 1993.

Kiefer J. Issues and open questions in the evaluation of biological radiation hazards in space. Adv. Space Res. 14 (10): 979–988, 1994a.

Kiefer J. Mutation induction in yeast by very heavy ions. Adv. Space Res. 14 (10): 331–338, 1994b.

Kiefer J & Kost M. Fine structures of energy deposition – introductory remarks. In: *Quantitative Mathematical Models in Radiation Biology.* Ed. by Kiefer J. Springer-Verlag, Berlin & Heidelberg, pp. 29–40, 1988.

Kiefer J, Rase S, Schneider E, Straaten H, Kraft G & Liesem H. Heavy ion effects on yeast cells: induction of canavanine-resistant mutants. Int. J. Radiat. Biol. 42: 591–600, 1983a.

Kiefer J, Rase S, Schöpfer F, Schneider E, Weber K & Kraft G. Heavy ion action on yeast cells: inhibition of ribosomal RNA synthesis. Loss of colony-forming ability and induction of mutants. Adv. Space Res. 3: 115–25, 1983b.

Kiefer J & Schneider E. Heavy ion effects on cells – survival of a temperature-conditional repair mutant of yeast. Int. J. Radiat Biol. 59: 1415–23, 1991.

Kiefer J & Schneider E. On the quantitative interpretation of cellular heavy ion action. Adv. Space Res. 4 (10): 199–204, 1984.

Kiefer J & Straaten H. A model of ion track structure based on classical collision dynamics. Phys. Med. Biol. 31: 1201–9, 1986.

Kiefer J & Straaten H. Quantitative interpretation of heavy ions effects: models for the biological effects of heavy ions. Adv. Space Res. 9 (10): 21–29, 1989.

Kiefer J, Stoll U & Schneider E. Mutation induction by heavy ions. Adv. Space Res. 14 (10): 257–266, 1994.

Kinard W, ONeal R, Wilson B, Jones J, Levine A & Calloway R. Overview of the space environmental effects observed on th retrieved long duration exposure facility (LDEF). Adv. Space Res. 14 (10): 7–16, 1994.

King JH. Solar proton fluences for 1977–1983 space missions. J. Spacecr. & Rockets 11: 401–8, 1974.

Kliauga PJ, Colvett RD, Lam YMP & Rossi JJ. The relative biological effectiveness of 160 MeV protons – I. Microdosimeter. Int. J. Radiat. Oncol. Biol. Phys. 4: 1001–8, 1978.

Koniarek JP & Worgul BV. Do heavy ions cause microlesions in cell membranes. Adv. Space Res. 12 (2/3): 417–20, 1992.

Koniarek JP. Can heavy ion irradiation cause permanent microlesions in cell membranes. In: *Abstracts, The World Space Congress, Washington, D.C., USA*, p. 554, 1992.

Kost M, Russmann C & Kiefer J. The cellular inactivation capability of accelerated heavy ions. In: *Proceedings of the 5th European Symposium on Life Sciences Research in Space.* Ed. by Oser H, Guyenne TD. Noordwijk, The Netherlands, ESA, ESTEC, pp. 205–8, 1994.

Kost M & Kiefer J. Biological action of heavy ion irradiation on individual yeast cells. In: *Terrestrial Space Radiation and Its Biological Effects.* Ed. by McCormack PC, Swenberg CE & Bücker H. Plenum Press, New York, pp. 197–204, 1987.

Kovalev EE. Radiation protection during space flight. Aviat. Space Environ. (Med. Suppl. 1) 54: 16–23, 1983.

Kovalev EE, Muratova IA & Petrov VM. Studies of the radiation environment aboard PROGNOZ satellites. Nucl. Tracks Radiat. Meas. 16: 45–8, 1989.

Kovalev EE & Petrov VM Individual doses of cosmonauts over the 30 years of soviet space flights. In: *Abstracts, The World Space Congress, Washington, D.C., USA*, p. 572, 1992.

Kozubek S & Horneck G. Mathematical models of lesion induction and repair in irradiated cells. In: *Biological Effects and Physics of Solar and Galactic Cosmic Radiation.* Ed by Swenberg CE, Horneck G & Stassinopoulos EG. NATO ASI Series A: Life Sciences 243A, Plenum Press, pp. 291–294, 1993.

Kraft G. Radiobiological effects of very heavy ions: Inactivation, induction of chromosome aberrations and strand breaks. Nucl. Sci. Appl. 3: 1–28, 1987a.

Kraft G. Effects of LET, fluence and particle energy on inactivation, chromosomal aberrations and DNA strand breaks. In: *Terrestrial Space Radiation and Its Biological Effects.* Ed. by McCormack PC, Swenberg CE & Bücker H. Plenum Press, New York, pp. 163–84, 1987b.

Kraft G, Blakely EA, Hieber L, Kraft-Weyrather W, Miltenburger HG, Müller W, Schuber M, Tobias CA & Wulf H. HZE effects on mammalian cells. Adv. Space Res. 4 (10): 219–26, 1984.

Kraft G, Kraft-Weyrather W, Meister H, Miltenburger HG, Roots R & Wulf H. The influence of radiation quality on the biological effectiveness of heavy charged particles. In: *Radiation Protection.* Ed by Booz J & Ebert HG. pp. 743–52, 1983.

Kraft G, Kraft-Weyrather W, Ritter S, Scholz M & Stanton J. Cellular and subcellular effect of heavy ions: a comparison of the induction of strand breaks and chromosomal aberrations with the incidence of inactivation and mutation. Adv. Space Res. 9 (10): 59–72, 1989.

Kraft G, Krämer M & Scholz M. LET, track structure and models – a review. Radiat. Environ. Biophys. 31: 161–80, 1991.

Kraft LM, Kelly MA, Johnson JE, Benton EV, Henke RP, Cassuo R, Haymakew M, Philpott DE, Vogel FS & Zeman W. Effects of high-LET neon (20Ne) particle radiation on the brain eyes and other head structures of the pocket mouse: a histological study. Int. J. Radiat. Biol. 35: 33–62, 1979.

Krämer M & Kraft G. Heavy ion track structure calculations. In: *Biophysical Modelling of Radiation Effects.*Ed. by Chadwick KH, Moschini G & Varma MN. Adam Hilger, Bristol, Philadelphia, New York, pp. 61–8, 1992.

KrämerM & Kraft G. Track structure, DNA damage and RBE. Adv. Space Res. 14 (10): 151–60, 1994.

Kranert T, Schneider E & Kiefer J. Mutation induction in V79 chinese hamster cells by very heavy ions. Int. J. Radiat. Biol. 58: 975–87, 1990.

Kranert T, Stoll U, Schneider E & Kiefer J. Mutation induction in mammalian cells by very heavy ions. Adv. Space Res. 12 (2/3): 111–8, 1992.

Kranz AR. Genetic and physiological damage induced by cosmic radiation on dry plant seeds during spaceflight. Adv. Space Res. 6 (12): 135–8, 1986.

Kranz AR. Effects of cosmic heavy ion radiation and space microgravity on plant systems. In: *Proceedings of the 3rd ESA European Symposium on Life Sciences Research in Space*, Graz, Austria. Ed. by Hunt H & Oser H, ESA SP-271, ESA, Paris, pp. 219–23, 1987.

Kranz AR, Bork U, Bücker H & Reitz G. Biological damage induced by ionizing cosmic rays in dry *Arabidopsis* seeds. Nucl. Tracks Radiat. Meas. 17: 155–65, 1990.

Krasavin EA, Kozubek S, Amirtyev KG, Torkarova B & Bonev M. Mutagenic effects of heavy ions in bacteria. Adv. Space Res. 12 (2/3): 103–10, 1992.

Krebs W, Krebs I & Worgul BV. Effect of accelerated iron ions on the retina. Radiat. Res. 123: 213–19, 1990.

Kronenberg A & Little JB. Locus specificity for mutation induction in human cells exposed to accelerated heavy ions. Int. J. Radiat. Biol. 55: 913–24, 1989.

Kronenberg A. Mutation inductions in human lymphoid cells by energetic heavy ions. Adv. Space Res. 14 (10): 339–46, 1994.

Lambiotte JJ jr, Wilson JW & Filippas TA. *Proper 3C: A Nucleon-pion Transport Code.* NASA TM X-2158, 1971.

Lamkin SL. *A Theory for High-energy Nucleon Transport in One Dimension.* M.S. Thesis, Old Dominion University, 1974.

Landauer MR, Davis HD, Kumar KS & Weiss JF. Behavioral toxicity of selected radioprotectors. Adv. Space Res. 12 (2/3): 273–83, 1992.

Leith JT, Ainsworth EJ & Alpen EL. Heavy-ion radiobiology: normal tissue studies. Adv. Radiat. Biol. 10: 192–236, 1983.

Leont'eva GA, Fomenko BS & Antipov AV. Effect of secondary radiation produced by 70 GeV protons on DNA of mammalian cells. Adv. Space Res. 4 (10): 231–5, 1984.

Letaw JR, Silberberg R & Tsao CH. Radiation hazards on space missions outside the magnetosphere. Adv. Space Res. 9 (10): 285–91, 1989a.

Letaw JR, Silberberg R & Tsao CH. Galactic cosmic radiation doses to astronauts outside the magnetosphere. In: *Terrestrial Space Radiation and Its Biological Effects.* Ed. by McCormack PC, Swenberg CE & Bücker H. Plenum Press, New York, pp. 663–674, 1987.

Letaw JR, Silberberg R & Tsao CH. Radiation hazards on space Missions. Nature 330: 709–10, 1988.

Letaw JR, Silberberg R, Tsao CH & Benton EV. Model analysis of space shuttle dosimetry data. Adv. Space Res. 9 (10): 257–60, 1989b.

Letaw J, Tsao CH & Silberberg R Matrix methodss of cosmic ray propagation. In: *Composition and Origin of Cosmic Rays.* Ed. by Shapiro MM, D. Reidel Publ. Co., pp. 337–342, 1983.

Lett JT. Damage to cellular DNA from particulate radiations, the efficacy of its processing and the radiosensitivity of mammalian cells – emphasis on DNA double strand breaks and chromatin breaks. Radiat. Environ. Biophys. 31: 257–77, 1992.

Lett JT, Bergtold DS & Keng PC. Effects of LET infinity on the fate of DNA damage induced in rabbit sensory cells in *situ:* fundamental aspects. Basic Life Sci. 38: 139–45, 1986a.

Lett JT, Cox AB, Bergtold DS & Pickering JE. Late effects from particulate radiations in primate and rabbit tissues. Adv. Space Res. 4 (10): 251–6, 1984.

Lett JT, Cox AB & Bergtold DS. Cellular and tissue responses to heavy ions: basic considerations. Radiat. Environ. Biophys. 25: 1–12, 1986b.

Lett JT, Cox AB, Keng PC, Lee AC, Su CM & Bergtold DS. Late degeneration in rabbit tissues after irradiation by heavy ions. Life Sci. Space Res. 18: 131, 1980.

Lett JT, Cox, AB & Lee AC. Some perspectives on cataractogenesis from heavy charged particles, Radiat. Res. 104: 201–7, 1985.

Lett JT, Cox AB & Lee AC. Selected examples of degenerative late effects caused by particulate radiations in normal tissues. In: *Terrestrial Space Radiation and Its Biological Effects.* Ed. by McCormack PC, Swenberg CE & Bücker H. Plenum Press, New York, pp. 393–414, 1987.

Lett JT, Cox AB & Story MD. The role of repair in the survival of mammalian cells from heavy ion irradiation: approximation to the ideal case of target theory. Adv. Space Res. 9 (10): 99–104, 1989a.

Lett JT, Lee AC & Cox AB. Late cataractogenesis in rhesus monkeys irradiated with protons and radiogenic cataract in other species. Radiat. Res. 126: 147–56, 1991.

Lett JT, Lee AC & Cox AB & Wood DH. Late cataractogenesis caused by particulate radiations and photons in long-lived mammalian species. Adv. Space Res. 9 (10): 325–31, 1989b.

Lett JT & Peters EL. Deoxyribonucleoprotein structure and radiation injury: cellular radiosensitivity is determined by let-dependent DNA damage in hydrated deoxyribonucleoproteins and the extent of its repair. Adv. Space Res. 12 (2/3): 51–8, 1992.

Lett JT & Williams GR. Effects of linear energy transfer on the formation and fate of radiation damage to the photoreceptor cell complement of the rabbit retina: implications for the projected manned mission to mars. In: *Biological Effects and Physics of Solar and Galactic Cosmic Radiation.* Ed by Swenberg CE, Horneck G & Stassinopoulos EG. NATO ASI Series A: Life Sciences 243A, Plenum Press, pp. 185–202, 1993.

Leuthold G & Burger G. Mathematical simulation of proton tracks in water vapor and their microdosimetric analysis. Radiat. Environ. Biophys. 27: 177– 87, 1988.

Lezniak JA. The extension of the concept of the cosmic-ray path-length distribution to nonrelativistic energies. Astrophys. & Space Sci. 63: 279–93, 1979.

Lindsay KR, Dalrymple GV, Ghidoni JJ, Mitchell JC & Morgan IL. Some effects of 55-Mev protons on primates. Radiat. Res. 28: 446–64, 1966.

Lloyd EL, Gemmel MA, Henning CB, Gemmel DS & Zabransky BJ. Transformation of mammalian cells by alpha particles. Int. J. Radiat. Biol. 36: 467–78, 1979.

Löbrich M, Ikpeme S, Haub P, Weber KJ & Kiefer J. DNA double-strand break induction in yeast by X-rays and alpha-particles measured by pulsed-field gel electrophoresis. Int. J. Radiat. Biol.64, 5: 539–46, 1993.

Lowenstein E, Gleeson JL, Hecht E, Factor R, Goldfischer C, Cajigas A & Steinberg JJ. Excision repair is enhanced by WR-2721 radioprotection. In: *Terrestrial Space Radiation and its Biological Effects.* Ed. by McCormack PC, Swenberg CE & Bücker H. Plenum Press, New York, pp. 697–706. 1987.

Lücke-Huhle C. Alpha-irradiation-induced G2 delay: A period of cell recovery. Radiat. Res. 89: 298–308, 1982.

Lücke-Huhle C. On the mechanism and consequences of radiation-induced gene amplification in mammalian cells. In: *Biological Effects and Physics of Solar and Galactic Cosmic Radiation.* Ed. by Swenberg CE, Horneck G & Stassinopoulos EG. NATO ASI Series A: Life Sciences 243A, Plenum Press, p. 143–54, 1993.

Lücke-Huhle C, Blakely EA, Chang PY & Tobias CA. Drastic G2 arrest in mammalian cells after irradiation with heavy ion beams. Radiat. Res. 79: 97–112, 1979.

Lücke-Huhle C, Hieber L & Beck-Bornholdt HP. Cell cycle dependent G2 delay and killing of l929 cells after exposure to ^{241}Am alpha particles. Radiat. Res. 100: 576–84, 1985.

Lücke-Huhle C, Pech M & Herrlich P. Selective gene amplification in mammalian cells after exposure to ^{60}Co gamma rays, ^{241}Am alpha particles, or UV light. Radiat. Res. 106: 345–55, 1986.

Magee JL & Chatterjee A. Radiation chemistry of heavy particle tracks. General considerations. J. Phys. Chem. 84: 3529–36, 1980.

Maisin J. R. Life shortening and causes of death in experimental animals following whole-body exposure to ionizing radiation. In: *Terrestrial Space Radiation and its Biological Effects.* Ed. by McCormack PD, Swenberg CE & Bücker H. Plenum Press, pp. 423–44, 1987.

Marco R, Vernos I, Jurado JG, Carratala M & Calleja M. Analysis of the involvement of the terrestrial space radiation in the microgravity effects on Drosphila Melanogaster development and aging. In: *Terrestrial Space Radiation and its Biological Effects.* Ed. by McCormack PD, Swenberg CE & Bücker H. Plenum Press, pp. 509–16, 1987.

Matsubara S, Ohara H, Hiraoka T, Koike S, Ando K, Yamaguchi H, Kuwabara Y, Hoshina M & Suzuki S. Chromosome aberration frequencies produced by a 70MeV proton beam. Radiat. Res. 123: 182–91, 1990.

Mathews GJ. Complete fragment yields from spallation reactions via a combined time-of-flight and DE-E technique. In: *Composition and Origin of Cosmic Rays.* Ed. by Shapiro MM. Riedel Publishing Co., pp. 317–20, 1983.

Mays CW, Lloyd RD, Taylor GN & Wrenn ME. Cancer incidence and lifespan vs. alpha-particle dose in beagles. Health Phys. 52: 617–24, 1987.

Mays CW. Alpha-particle-induced cancer in humans. Health. Phys. 55: 637–52, 1989.

McCarthy KF & Hale ML. Extrapolation of rodent LD50/30 data to humans. In: *Abstracts, The World Space Congress, Washington, D.C., USA*, p. 559, 1992.

McCormack PD. Radiation hazards in low earth orbit, polar orbit, geosynchronous orbit, and deep space. In: *Terrestrial Space Radiation and its Biological Effects.* Ed. by McCormack PD, Swenberg CE & Bücker H. Plenum Press, pp. 71–96, 1987.

McCormack PD & Swenberg CE. Aerospace medical physics. Encyclop. Appl. Phys. 1: 443–87, 1991.

McCormack PC, Swenberg CE & Bücker H (eds). *Terrestrial Space Radiation and its Biological Effects.* NATO ASI Series, Series A, Life Sciences. Plenum Press: New York and London, 1987.

McDonald FD. Review of galactic and solar cosmic rays. In: *Second Symposium on Protection Against Radiations in Space.* Ed. by Reetz A. NASA SP-71, pp. 19–29, 1965.

McNally NJ, De Ronde J & Folkard M. Interaction between X-ray and alpha-particle damage in V79 cells. Int. J. Radiat. Biol. 53: 917–20, 1988.

McNulty PJ. New directions in space dosimetry. In: *Terrestrial Space Radiation and its Biological Effects.* Ed. by McCormack PC, Swenberg CE & Bücker H. Plenum Press, New York, pp. 819–40, 1987.

McNulty PJ. Hazard from SEUS and other electronic equipment failure caused by energetic cosmic rays. In: *Abstracts, The World Space Congress, Washington, D.C., USA*, p. 566, 1992.

McNulty PJ & Stassinopoulos EG. Uncertainties in radiation effect predictions for the natural radiation environments of space. Adv. Space Res. 14 (10): 947–58, 1994.

Medvedovsky C, Worgul BV, Huang Y, Brenner DJ, Tao F, Miller J, Zeitlin C & Ainsworth J. The influence of dose, dose-rate and particle fragmentation on cataract induction by energetic iron ions. Adv. Space Res. 14 (10): 475–82, 1994.

Mei H, Deng H, Lu Y, Zhuang C, Liu Z, Qiu Q, Qiu Y & Yang TC. Mutagenic effects of heavy ion radiation in plants. Adv. Space Res. 14 (10): 363–72, 1994a.

Mei M, Qiu Y, He Y, Bücker H & Yang TC. Mutational effects of space flight on *Zea mays* seeds. Adv. Space Res. 14 (10): 33–40, 1994b.

Menzel HG, Booz J, Turner JE & Varma MN. Microdosimetry. Radiat. Protection Dosimetry 52: 1-492 1994.

Merriam GR jr. & Worgul BV. Experimental radiation cataract-its clinical relevance. Bull. N. Y. Acad. Med. 59: 372–92, 1983.

Merriam GR jr, Worgul BV, Medvedovsky C, Zaider M & Rossi HH. Accelerated heavy particles and the lens. I. Cataractogenic potential. Radiat. Res. 98: 129–40, 1984.

Mewaldt RA. Galactic cosmic ray composition and energy spectra . Adv. Space Res. 14 (10): 737–48, 1994.

Michalik V. Particle track structure and its correlation with radiobiological endpoint. Phys. Med. Biol. 36: 1001–12, 1991.

Michalik V. Estimation of double-strand break quality based on track-structure calculations. Radiat. Environ. Biophys. 32: 251–8, 1993.

Micke U, Schäfer M, Anton A, Horneck G & Bücker H. Heavy ion induced double strand breaks in bacteria and bacteriophages. Adv. Space Res. 12 (2/3): 59–63, 1992.

Micke U, Schott JU, Horneck G & Bücker H. Heavy ion radiation effects on single spores of *Bacillus subtilis.* In: *Terrestrial Space Radiation and Its Biological Effects.* Ed. by McCormack PC, Swenberg CE & Bücker H. Plenum Press, New York, pp. 193–6, 1987.

Micke U, Horneck G & Kozubek S. Double strand breaks in the DNA of *Bacillus subtilis* cells irradiated by heavy ions. Adv. Space Res. 14 (10): 207–12, 1994.

Milas L, Nishiguchi I, Hunter N, Murray D, Fleck R, Ito H & Travis E. Radiation protection against early and late effects of ionizing irradiation by the prostaglandin inhibitor indomethacin. Adv. Space Res. 12 (2/3): 265–71, 1992.

Miller J. Ground-based simulations of galactic cosmic ray fragmentation and transport. Adv. Space Res. 14 (10): 831–40, 1994.

Miller JH, Frasco DL, Ye M, Swenberg CE, Myers LS jr. & Rupprecht A. Free-radical yields in proton irradiation of oriented DNA. Relationship to energy transfer along DNA chains. In: *Biological Effects and Physics of Solar and Galactic Cosmic Radia-*

tion. Ed by Swenberg CE, Horneck G & Stassinopoulos EG. NATO ASI Series A: Life Sciences 243A, Plenum Press, pp. 85–92, 1993.

Miller RC, Brenner DJ, Randerspehrson G, Marino SA & Hall EJ. The effects of the temporal distribution of dose on oncogenic transformation by neutrons and charged particles of intermediate LET. Radiat. Res. 124: 62–5, 1990.

Mills RE & Rossi HH. Mean energy deposition distribution about proton tracks. Radiat. Res. 84: 434–43, 1980.

Miquel J, Herman MM, Benton EV & Welch G. Effects of high-LET particles (Ar) on the brain of *Drosophila melanogaster*. Int. J. Radiat. Biol. 29: 101–24, 1976.

Miroshnichenko LI. Cyclic variations and sporadic fluctuations of solar cosmic rays. Biofizika SSSR 37: 452–66, 1992.

Miroshnichenko LI, Perez-Peraza J, Gallegos-Cruz A & Vashenyuk EV. Spectrum hardness and solar cosmic rays (SCR) composition in flares with different temperature regimes. In: *Abstracts, The World Space Congress, Washington, D.C., USA*, p. 564, 1992.

Moore FD. Radiation burdens for humans on prolonged exomagnetospheric voyages. FASEB J. 6: 2338–43, 1992.

Mozumder A. Early production of radicals from charged particle tracks in water. Radiat. Res. 104, Suppl. 8: 33–9, 1986.

Mullen EG, Gussenhoven MS, Hardy DA. The space radiation environment at 840 km. In: *Terrestrial Space Radiation and its Biological Effects*. Ed. by McCormack PD, Swenberg CE & Bücker H. Plenum Press, pp. 41–60, 1987.

Mullen EG & Ray KP. Microelectronics effects as seen on CRRES. Adv. Space Res. 14 (10): 797–808, 1994.

Muthukrishnan G & Gopinath DV. Charged-particle transport in one-dimensional systems. Radiat. Res. 95: 1–14, 1983

Myers LS, Carmichael AJ & Pellmar TC. Radiation chemistry of the hippocampal brain slice. Adv. Space Res. 14 (10): 453–6, 1994.

Nagasawa H & Little JB. Induction of sister chromatid exchanges by extremely low doses of alpha-particles. Cancer Res. 52: 6394–6, 1992.

Nagasawa H, Little JB, Inkret WC, Carpenter S, Raju MR, Chen DJ & Strniste GF. Response of X-ray-sensitive CHO mutant cells (XRS-6C) to radiation. 2. Relationship between cell survival and the induction of chromosomal damage with low doses of alpha-particles. Radiat. Res. 126: 280–8, 1991.

Nagasawa H, Robertson J & Little JB. Induction of chromosome aberrations and sister chromatid exchanges by alpha particles in density-inhibited cultures of mouse 10T1/2 and 3T3 cells. Int. J. Radiat. Biol. 57: 35–44, 1990.

Napolitano M, Durante M, Grossi GF, Pugliese M & Gialanella G. Inactivation of C3H10T1/2 cells by monoenergetic high LET alpha-particles. Int. J. Radiat. Biol. 61: 813–20, 1992.

NASA. *The Earth's Trapped Radiation Belts, NASA Space Vehicle Design Criteria (Environment)*. NASA SP-8116, 1975a.

NASA. *Interplanetary Charged Particles, NASA Space Vehicle Design Criteria (Environment)*. NASA SP-8118, 1975b.

NAS/NRC. National Academy of Sciences/National Research Council. Radiation protection guides and constraints for space mission and vehicle design involving nuclear systems. Radiobiological Advisory Panel, Committee on Space Medicine. Ed by Langham WH & Grahn D. National Academy Press, Washington, 1970.

NAS/NRC. National Academy of Sciences/National Research Council. HZE particle effects in manned space flight. In: *Report of the Radiobiological Advisory Panel*. Committee on Space Medicine. Ed. by Grahn D. National Academy Press, Washington, 1973.

NCRP. National Council on Radiation Protection and Measurements. Report No. 98, *Guidance on Radiation Received in Space Activities*. Recommendations of the National Council of radiation protection and measurements, NRCP Publication, Bethesda, MD, 1989a.

NCRP. National Council on Radiation Protection and Measurements. *Influence of Radiation Quality (LET) on Relative Biological Effectiveness*, NCRP draft report of Scientific Committee 40. National Council on Radiation Protection and Measurements, Bethesda, Maryland, 1989b.

Nealy JE, Simonsen LC, Sauer HH, Wilson JW & Townsend LW. *Space Radiation Dose Analysis for Solar Flare of August 1989*. NASA TM-4229, 1990.

Nealy J, Wilson J, Shea MA & Smart DF. Proton flare and magnetic storm effects in the vicinity of the Earth. Adv. Space Res. 14 (10): 655–60, 1994.

Nealy JE, Wilson JW & Townsend LW. *Preliminary Analyses of Space Radiation Protection for Lunar Base Surface Systems*. SAE Tech. Paper Ser. 891487, 1989.

Nelson AC, Hayes TL & Tobias CA. Rapid development of corneal lesions in rats produced by accelerated heavy ions. Adv. Space Res. 3 (8): 195–210, 1983.

Nelson AC, Hayes TL, Tobias CA & Yang TCH. Some indications of structural damage in retina by heavy ion radiation. Scanning Electron Microscopy 4: 79–85, 1981.

Nelson G, Schubert W, Kazarians G, Richards G, Benton EV, Benton ER & Henke R. Radiation effects in nematodies, results from IML-I experiments. Adv. Space Res. 14 (10): 87–92, 1994a.

Nelson G, Schubert WW, Marshall TM, Benton ER & Benton EV. Radiation effects in *Caenorhabditis elegans*, mutagenesis by high and low LET ionizing radiation. Mutat. Res. 212: 181, 1989.

Nelson GA, Schubert WW, Kazarians GA, Richards GF, Benton EV, Benton ER & Henke R. Nematode radiobiology and development in space, results from IML-I. In: *Proceedings of the 5th European Symposium on Life Sciences Research in Space*. Ed. by Oser H, Guyenne TD. Noordwijk, The Netherlands, ESA, ESTEC, pp. 187–92, 1994b.

Nelson AC & Tobias CA. Rapid development of corneal lesions in rats produced by heavy ions. Life Sci. Space Res. 20 (1): 195–210, 1983.

Nevzgodina LV, Gaubin Y, Kovalev EE, Planel H, Clegg J, Gasset G, Maximova EN & Pianezzi B. Changes in developmental capacity of *Artemia* cysts and chromosomal aberrations in lettuce seeds flown aboard Salyut 7 (Biobloc III experiment). Adv. Space Res. 4 (10): 71–6, 1984.

Nevzgodina LV, Maksimova EN & Kaminskaya EV. Effects of prolonged exposure of lettuce seeds to HZE particles on orbital stations. Adv. Space Res. 9 (11): 53–8, 1989.

Ngo FQH, Blakely EA & Tobias CA. Effects of combined low- and high-LET radiations. In: *Biological and Medical Research with Accelerated Heavy Ions at the Bevalac, 1977–1980*. Ed. by Pirruccello MC & Tobias CA. LBL-11220/UC-48, Lawrence Berkeley Laboratory/University of California, Berkeley, pp. 103–12, 1980.

Ngo FQH, Blakely EA & Tobias CA. Sequential exposures of mammalian cells to low- and high-LET radiations I. Lethal effects following X-ray and Neon-ion irradiation. Radiat. Res. 87: 59–78, 1981.

Ngo FQH, Tobias CA, Blakely EA, Holley W, Chang PY, Oleszko O & Lommel L. Asymmetric interactions between effects of high- and low-LET radiations. Radiat. Res. 91: 326–7 (abstract), 1982.

Nguyen VD, Buiset P, Parmentier N, Siegrist N, Akatov YA, Archangelsky VV, Vorojtsov S, Koslova SB, Mitrikas VG, Petrov VM & Kovalev EE. New experimental approach in quality factor and dose equivalent determination during a long-term manned space mission. In: *Proceedings of the 4th European Symposium on Life Sciences Research in Space*. Ed. by David V. Paris, France, European Space Agency, pp. 555–558, 1990.

Nikjoo H, Charlton DE & Goodhead DT. Monte Carlo track srtucturestudies of energy deposition and calculation of initial DSB and RBE. Adv. Space Res 14 (10): 161–80, 1994.

Noll A, Rusch G, Röcher H, Dreute J & Heinrich W.The Siegen automatic measuring system for nuclear track detectors: new developments. Nucl. Tracks Radiat. Meas. 15: 265–8, 1988.

Norbury JW & Townsend LW. *Electromagnetic Dissociation Effects in Galactic Heavy-ion Fragmentation.* NASA TP-2527, 1986.

Normand E. A comparison of two different types of geosynchronous satellite measurements during the 1989 solar proton events. Adv. Space Res. 14 (10): 695–700, 1994.

Northcliffe LC & Schilling RF. *Range and Stopping-power Tables for Heavy Ions.* Nuclear Data, Sect. A, 7, 3–4: 233–463, 1970.

Obaturov GM, Duba VV, Potetnya VI, Kapchigashev SP & Pitkevich VA. Yield of DNA-break production as a function of heavy charged particle and neutron energy. Stud. Biophys. 89: 133–9, 1982.

Olko P & Booz J. Energy deposition by protons and alpha particles in spherical sites of nanometer to micrometer diameter. Radiat. Environ. Biophys. 29: 1–18, 1990.

Osborne WZ, Pinsky LS & Bailey JV. Apollo light flash investigations. In: *Biomedical Results of Apollo.* Ed. by Johnston RS, Dietlein LF & Berry CA. NASA SP-368, pp. 355–65, 1975.

Oser H. ESA's future research in space radiation biology. In: *Abstracts, The World Space Congress, Washington, D.C., USA*, p. 573, 1992.

Paretzke HG. Advances in energy deposition. In: *Advances in Radiation Protection and Dosimetry in Medicine.* Ed. by Thomas RH & Perez-Mendez V, Plenum Press, New York and London, pp. 51–74, 1980.

Paretzke HG. Problems in theoretical track structure research for heavy charged particles. In: *Quantitative Mathematical Models in Radiation Biology.* Ed. by Kiefer J. Springer-Verlag, Berlin & Heidelberg, pp. 49–56, 1988.

Paretzke HG. Physical events in the track structure of heavy ions and their relation to alterations of biomolecules. Adv. Space Res. 9 (10): 15–20, 1989.

Paretzke, HG. The interfaces between physics, chemistry, biological damage of DNA. In: *Abstracts, The World Space Congress, Washington, D.C., USA*, p. 550, 1992.

Perris A, Pialoglou P, Katsanos AA & Sideris EG. Biological Effectiveness of low energy protons. I. survival of chinese hamster cells. Int. J. Radiat Biol 50: 1093–101, 1987.

Perris A & Zarris G. Specific ionization for electrons, protons and alpha particles incident on water. Phys. Med. Biol. 34: 1113–8, 1989.

Perry DJ. On the penetration of fast charged particles. Radiat. Res. 115: 26–43, 1988.

Peterson DD, Benton EV, Tran M, Yang TC, Freeling M, Craise L & Tobias CA. Biological effects of high-LET particles on corn-seed embryos in the Apollo Soyuz Test Project - Biostack III experiment. Life Sci. Space Res. 15: 151–5, 1977.

Petrov V. Overview on experience to date on human exposure to space radiations. Adv. Space Res. 14 (10): 397–409, 1994.

Petrov VM, Benghin VV, Kozlova SB, Nghuen VD & Sigrist M. Radiation Environment on board the Mir Station during the solar proton event of November 2, 1992, according to to the 'NOZICA' device data.In: *Proceedings of the 5th European Symposium on Life Sciences Research in Space.* Ed. by Oser H, Guyenne TD. Noordwijk, The Netherlands, ESA, ESTEC, pp. 213–6, 1994a.

Petrov VM, Mahkmutov VS, Panova NA, Shurshakov VA, Dachev TP, Semkova JV & Matvijchuk YP. Peculiarities of the solar proton events in October 19, 1989 and March 23, 1991 according to the measurements on board the Mir space station. Adv. Space Res 14 (10): 645–50, 1994b.

Philpott DE, Corbett R, Turnbill C, Black S, Dayhoff D, McGourty, Lee R, Harrision G & Savick L. Retinal changes in rats flown on Cosmos 936: a cosmic ray experiment. Aviat. Space Envion. Med. 51: 556–62, 1980.

Philpott DE, Corbett R, Turnbill C, Harrision G, Leafer D, Black S, Sapp W, Klein G & Savick LF. Cosmic ray effects on the eyes of rats flown on Cosmos 782, experiment K-007. Aviat Space Environ. Med. 49: 19–28, 1978.

Philpott DE, Sapp J, Miquel J, Kato K, Corbett R, Stevenson J, Black S, Lindseth KA & Benton EV. The effect of high energy (HZE) particle radiation (40Ar) on aging parameters of mouse hippocampus and retina. Scanning Electron Microscopy III: 1177–85, 1985.

Pickert M., K. E. Gartenbach, A. R. Kranz. Heavy ion induced mutations in genetic effective cells of a higher plant. Adv. Space Res. 12 (2/3): 69–72, 1992.

Pinsky LS, Osborne WZ & Bailey JV. Visual light flash phenomenon. In: *Apollo 17 Preliminary Science Report.* NASA SP-330, pp. 27/1–27/6, 1973.

Pissarenko NF. Radiation situation determining the possibility of a manned flight to Mars and back. Adv. Space Res. 12 (2/3): 435–9, 1992.

Pissarenko NF. Radiation environment due to galactic and solar cosmic rays during the manned mission to Mars in the periods between maximum and minimum solar activity. Adv. Space Res 14 (10): 771–8, 1994.

Planel H, Blanquet Y, Soleilhavoup PJ, Kaiser R & Pianezzi B. Effects of cosmic heavy ions on *Artemia* egg development. In: *Radiation Research.* Ed. by Nygaard OF, Adler HI & Sinclair WK. Academic Press, New York, pp. 1152–63, 1975.

Planel H, Gaubin Y, Pianezzi B & Gasset G. Space environmental factors affecting responses to radiation at the cellular level. Adv. Space Res. 9 (10): 157–60, 1989.

Potten C. Cell death (apoptosis) induced in mammalian tissues by radiation. In: *Book of Abstracts, Fifth European Symposium on Life Sciences Research in Space,* Arcachon, France, 1993.

Prise KM, Davies S & Michael BD. The relationship between radiation-induced DNA double-strand breaks and cell kill in hamster V79 fibroblasts irradiated with 250 KV-X-rays, 2.3MeV Neutrons or ^{238}Pu alpha particles. Int. J. Radiat. Biol. 52: 893–902, 1988.

Prise KM, Folkard M, Davies S & Michael BD. The irradiation of V79 mammalian cells by protons with energies below 2MeV. Part I. Measurement of oxygen enhancement ratios and DNA damage. Int. J. Radiat. Biol. 58: 261–77, 1990.

Pross HD, Kost M & Kiefer J. Repair of radiation induced genetic damage under microgravity, an experimental proposal. In: *Biological Effects and Physics of Solar and Galactic Cosmic Radiation.* Ed. by Swenberg CE, Horneck G & Stassinopoulos EG. NATO ASI Series A: Life Sciences 243A, Plenum Press, pp. 125–8, 1993.

Pross HD, Kost M & Kiefer J. Repair of radiation induced genetic damage under microgravity. In: *Proceedings of the 5th European Symposium on Life Sciences Research in Space.* Ed. by Oser H, Guyenne TD. Noordwijk, The Netherlands, ESA, ESTEC, pp. 193–6, 1994.

Purrott RJ, Edwards AA, Lloyd DC & Stather JW. The induction of chromosome aberrations in human lymphocytes by *in vitro* irradiation with alpha-particles from plutonium-239. Int. J. Radiat. Biol. 38: 277–84, 1980.

Rabin BM, Hunt WA & Joseph JA. An assessment of the behavioural toxicity of high energy iron particles compared to other qualities of radiation. Radiat. Res. 119: 113–22, 1989.

Rabin BM, Hunt WA, Joseph JA, Dalton TK & KandasamySB. An evaluation of the relative behavioral toxicity of heavy particles. In: *Biological Effects and Physics of Solar and Galactic Cosmic Radiation.* Ed by Swenberg CE, Horneck G & Stassinopoulos EG. NATO ASI Series A: Life Sciences 243A, Plenum Press, pp. 227–34, 1993.

Rabin BM, Joseph JA, Hunt WA, Dalton TB, Kandasamy SB, Harris AH & Ludewigt B. Behavioral endpoints for radiation injury. Adv. Space Res. 14 (10): 457–66, 1994.

Raju MR, Amols HI, Bain E, Carpenter SG, Cox RA & Robertson JB. Heavy particle comparative study. 3. OER and RBE. Brit. J. Radiol. 51: 712–19, 1978c.

Raju MR, Amols HI, Dicello JF, Howard J, Lyman JT, Koehler AM & Graves Smathers JB. Heavy particle comparative study. 1. Depth-dose distributions. Brit. J. Radiol. 51: 699–703, 1978a.

Raju MR, Bain E, Carpenter SG, Cox RA & Robertson JB. Heavy particle comparative study. 2. Cell survival versus depth. Brit. J. Radiol. 51: 704–11, 1978b.

Raju MR, Eisen Y, Carpenter S & Inkret WC. Radiobiology of alpha-particles. 3. Cell inactivation by alpha-particle traversals of the cell nucleus. Radiat. Res. 128: pp. 204–9, 1991.

Raju MR & Jett JH. RBE and OER-Variations of mixtures of plutonium alpha particles and X-rays for damage to human kidney cells (T-1). Radiat. Res. 60: 473–81, 1975.

Raju MR, Johnson TS, Tokita N, Carpenter S & Jett J. H. Differences in cell cycle progression delays after exposure to ^{238}Pu alpha-particles compared to X-rays, Res. Radiat. 84: 16–24, 1980.

Reetz A Jr (ed.). *Second Symposium on Protection Against Radiations in Space.* NASA SP-71, 1965.

Reitz G. Preliminary total dose measurements on LDEF. Adv. Space Res. 12 (2/3): 369–73, 1992.

Reitz G. Radiation environment in the stratosphere. Radiat. Protect. Dosimetry 48 (1): 5–29, 1993.

Reitz G & Bücker H. The radiobiological experiments on LDEF. In: *Abstracts, The World Space Congress, Washington, D.C., USA*, p. 568, 1992.

Reitz G., H. Bücker, R. Beaujean, W. Enge, R. Facius, W. Heinrich, T. Ohrendorf and E. Schopper, , Dosimetric mapping inside BIORACK. Adv. Space Res. 6 (12): 107, 1986.

Reitz G, Bücker H, Facius R, Horneck G, Graul EH, Berger H, Rüther W, Heinrich W, Beaujean R, Enge W, Alpatov AM, Ushakov IA, Zachvatkin YA & Mesland DAM. Influence of cosmic radiation and/or microgravity on development of *Carausius morosus*. Adv. Space Res. 9 (10): 161–73, 1989.

Reitz G, Bücker H, Facius R, Schäfer M & Beaujean R. Dosimetric results of Cosmos 2044. Nucl. Tracks Radiat. Meas. 20: 161–5, 1992a.

Reitz G, Bücker H, Rüther W, Graul EH, Beaujean R, Enge W, Heinrich W, Mesland DAM, Alpatov AM, Ushakov IA & Zachvatkin YA. Effects on ontogenesis of *Carausius morosus* hit by cosmic heavy ions. Nucl. Tracks Radiat. Meas. 17: 145–53, 1990.

Reitz G, Facius R & Bücker H. Radiation problems in manned spaceflight - European efforts. In: *Terrestrial Space Radiation and Its Biological Effects.* Ed. by McCormack PC, Swenberg CE & Bücker H. Plenum Press, New York, pp. 619–40, 1987.

Reitz G, Schnuer K & Shaw K. (eds.) *Radiation Exposure of Civil Aircrew.* Rad. Protection Dosimetry, 48: 1–143, 1993.

Riley EF, Lindgren AL, Andersen AL, Miller RC & Ainsworth EJ. Relative cataractogenic effects of X-rays, fission-spectrum neutrons, and Fe-56 particles - a comparison with mitotic effects. Radiat. Res. 125: 298–305, 1991.

Rimpl GR, Schmid E, Braselmann H & Bauchinger M. Chromosome aberrations induced in human lymphocytes by 16.5MeV protons. Int. J. Radiat. Biol. 58: 999–1007, 1990.

Ritter S & Kraft G. Induction of chromosome aberrations in chinese hamster cells after heavy ion irradiation. In: *Terrestrial Space Radiation and Its Biological Effects.* Ed. by McCormack PC, Swenberg CE & Bücker H. Plenum Press, New York, pp. 185–92, 1987.

Ritter S, Kraft-Weyrather W, Scholz m & Kraft G. Induction of chromosome aberrations in mammalian cells after heavy ion exposure. Adv. Space Res. 12 (2/3): 119–25, 1992.

Roberts CJ & Goodhead DT. The effect of ^{238}Pu alpha-particles on the mouse fibroblast cell line C3h10t1/2: characterization of source and RBE for cell survival. Int. J. Radiat. Biol. 52: 871–82, 1988.

Robertson JB, Eaddy JM, Archambeau JO, Coutrakon GB, Miller W, Moyers MF, Siebers DV, Slater JM & Dicello J. Relative biological effectiveness of a mixed energy field of protons up to 200 MeV. Adv. Space Res. 14 (10): 271–6, 1994.

Robertson JB, Koehler A, George JG & Little JB. Oncogenic transformation of mouse Balb/3T3 cells by plutonium-238 alpha particles. Radiat. Res. 96: 261–74, 1983.

Robertson JB, Koehler AM, Weidmann PA & McNulty PJ. High energy proton induced mutations in cultured chinese hamster cells. In: *Terrestrial Space Radiation and Its Biological Effects.* Ed. by McCormack PC, Swenberg CE & Bücker H. Plenum Press, New York, pp. 329–36, 1987.

Rohrig N & Colvert RD. Measurements of W for protons, helium-4-ions and carbon 12 ions in tissue equivalent gas. Radiat. Res. 76: 225–40, 1978.

Roots R, Holley W, Chatterjee A, Irizarry M & Kraft G. The formation of strand breaks in DNA after high-LET irradiation -- a comparison of data from *in vitro* and cellular systems. Int. J. Radiat. Biol. 58: 55–69, 1990.

Roots R, Holley W, Chatterjee A, Rachal E, & Kraft G. The influence of radiation quality on the formation of DNA breaks. Adv. Space Res. 9 (10): 45–55, 1989.

Roots R, Yang TC, Craise L, Blakely EA & Tobias CA. Impaired repair capacity of DNA breaks induced in mammalian cellular DNA by accelerated heavy ions. Radiat. Res. 78: 38–49, 1979.

Roots R, Yang TC, Craise L, Blakely EA & Tobias CA. Rejoining capacity of DNA breaks induced by accelerated carbon and neon ions in the spread Bragg peak. Int. J. Radiat. Biol. 38: 203–10, 1980.

Rossi HH. Specification of radiation quality. Radiat. Res. 10: 522–31, 1959.

Rossi HH & Kellerer A. M. The dose rate dependence of oncogenic transformation by neutrons may be due to variation of response during the cell cycle. Int. J. Rad. Biol. 50: 353–61, 1986.

Rossi HH & Zaider M. Compound dual radiation action. 1. General aspects. Radiat. Res. 132: 178–83, 1992.

Rossi HH & Zaider M. Elements of microdosimetry. Medical Physics 18: 1085–92, 1991.

Rubin LL, Philpott KL & Brook SF. Apoptosis – the cell cycle and cell death. Curr. Biol. 3: 391–4, 1993.

Rusch G & Heinrich W Cosmic ray LET-spectrum measured in the Spacelab 2 mission. In: *Terrestrial Space Radiation and its Biological Effects*. Ed. by McCormack PD, Swenberg CE & Bücker H. Plenum Press, pp. 809–18, 1987.

Rusch G. Winkel E, Noll A & Heinrich W. The Siegen automatic measuring system for track detectors: new developments. Nucl. Tracks Radiat. Meas. 19: 261, 1991.

Rust DM. Solar flares, proton showers and the space shuttle. Science 216: 939–46, 1982.

Rust DM. Solar flares: an overview. Adv. Space Res. 12 (2/3): 289–301, 1992.

Rüther W, Graul EH, Heinrich W, Allkofer OC, Kaiser R & Cüer P. Preliminary results on the action of cosmic heavy ions on the development of eggs of *Artemia salina*. Life Sci. Space Res. 12: 69–74, 1974.

Ruyters G & Gräf P. Future plans in space radiation biology. In: *Abstracts, The World Space Congress, Washington, D.C., USA*, p. 574, 1992.

Sabatier L, Fedorenko B, Hoffschir F, Flüry-Hérard A, Ricoul M, Martins L, Reillaudou M, Gersimenko V & Dutrillaux BV. Radiation-induced chromosomal damages during space missions. In: *Abstracts, 5th European Symposium on Life Sciences Research in Space, Arcachon, France*, abstract RAD-O-04, 1993.

Santoro RT, Alsmiller RG, Barnes JM & Corbin JM. *Shielding of Manned Space Stations Against Van Allen Belt Protons: A Preliminary Scoping Study.* ORNL/TM-10040 (Contract No. IDOD3070-0000), Oak Ridge National Laboratory, 1986.

Sapp WJ, Philpott DE, Williams CS, Williams JW, Kato K, Miquel JM & Serova L. Comparative study of spermatogonial survival after X-ray exposure, high-LET (HZE) irradiation or spaceflight. Adv. Space Res. 12 (2/3): 179–89, 1992.

Sasaki H. Cell killing and division delay in asynchronous and synchronized HeLa cells irradiated with alpha particles or X-rays. Radiat. Res. 99: 311–23, 1984.

Sauer HH, Zwickl RD & Ness MJ. *Summary Data for the Solar Energetic Particle Events of August Through December 1989*. Space Environment Laboratory, National Oceanic and Atmospheric Administration, 1990.

Saunders JF. (ed.). *The Experiments of Biosatellite II*. NASA SP-204, 1971.

Savchenko NY. Skin lesions after exposure to high-energy protons and helium ions. Kosm. Biol. Aviakosm. Med. SSSR, 19: 59–67, 1985.

Schaefer A, Hüttermann J & Kraft G. Direct radiation action of heavy ions on DNA as studied by ESR-Spectroscopy. Adv. Space Res. 12 (2/3): 45–9, 1992.

Schäfer HJ. Exposure hazards from cosmic radiation beyond the stratosphere and in free space. J. Aviation Med. 23: 334–44, 1952.

Schäfer HJ. Cosmic ray dosage during the giant solar flare of February 26. J. Aviation Med. 28: 387–96, 1957.

Schäfer HJ, Benton EV, Henke RP & Sullivan JJ. Nuclear track recordings of the astronauts radiation exposure on the first lunar landing mission Apollo XI. Radiat. Res. 49: 245–71, 1972.

Schäfer M, Bücker H, Facius R, Horneck G, Reitz G & Kraft G. High LET irradiation of Bacillus subtilis spores. In: *Proceedings of the 6th Symposium on Microdosimetry.* Ed. by Booz J & Ebert HG. Harwood Academic Publishers Ltd., London, pp. 1043–9, 1978.

Schäfer M, Facius R & Reitz G. Inactivation of individual B*acillus subtilis* spores in dependence on their distance to single accelerated heavy ions. Adv. Space Res 14 (10): 1039–46, 1994a.

Schäfer M, Schmitz C, & Bücker H. Heavy ion induced DNA double strand breaks in cells of *E. coli.* Adv. Space Res 14 (10): 203–6, 1994b.

Schimmerling W. Radiobiological problems in space - an overview. Radiat. Environ. Biophys. 31: 197–203, 1992.

Schimmerling W & Curtis SB. Biomedical and space-related research with heavy ions at the Bevalac. Radiat. Res. 119: 193–204, 1989.

Schimmerling W, Curtis SB & Vosburgh KG. Velocity spectrometry of 3.5-GeV nitrogen ions. Radiat. Res. 72: 1–17, 1977.

Schimmerling W, Kast JW & Ortendahl D. Measurement of the inclusive neutron production by relativistic neon ions of uranium. Phys. Review Lett. 3: 1985–7, 1979.

Schimmerling W & Sulzmann F. The NASA space radiation health program. Adv. Space Res 14 (10): 133–8, 1994.

Schmidt P, Fellinger J, Hahn T & Hübner K. Investigation of the low-energy component of primary cosmic radiation on the outside of spacecraft. Adv. Space Res 14 (10): 61–6, 1994.

Schmidt S, Kelbach C, Schmidt-Böcking H & Kraft G. Delta-electron emission in heavy ion collision. In: *Terrestrial Space Radiation and its Biological Effects.* Ed. by McCormack PC, Swenberg CE & Bücker H. Plenum Press, New York, pp. 205–10, 1987.

Schmidt-Böcking H, Ramm U, Kraft G, Ullrich J, Berg H, Kelbch C, Olson RE, DuBios R, Hagmann S & Jiazhen J. Delta-electron emission in fast heavy ion atom collisions. Adv. Space Res. 12 (2/3): 7–15, 1992.

Scholz M & Kraft G. A parameter-free track structure model for heavy ion action cross sections, In: *Biophysical Modelling of Radiation Effects.* Ed. by Chadwick KH, Moschini G & Varma MN. Adam Hilger, Bristol. Philadelphia, New York, pp. 185–192, 1992.

Scholz M, Kraft-Weyrather W, Ritter S & Kraft G. Cell cycle delays induced by heavy ion irradiation of synchronous mammalian cells. Adv. Space Res. 9 (10): 91–6, 1989.

Schöpfer F, Kiefer J, Schneider E & Kraft G. Inactivation of dried yeast cells by accelerated heavy ions of very high LET. Radiat. Res 82: 235–43, 1980.

Schöpfer F, Schneider E, Rase S, Kiefer J & Kraft G. Heavy-ion effects on yeast: survival and recovery in vegetative cells of different sensitivity. Radiat. Res. 92: 30–46, 1982.

Schöpfer F, Schneider E, Rase S, Kiefer J, Kraft G & Liesem H. Heavy ion effects on yeast cells: reappearance of the oxygen effect with very high LET. Int. J. Radiat. Biol. 46: 305–16, 1984.

Schopper E, Baican B & Baumgardt H-G. Structure of heavy ion tracks in Ag-Cl detectors. Adv. Space Res. 12 (2/3): 17–21, 1992.

Schott JU. Time resolving detector system for radiobiological investigations of effects of single ions. In: *Terrestrial Space Radiation and its Biological Effects.* Ed. by McCormack PC, Swenberg CE & Bücker H. Plenum Press, New York, pp. 153–63, 1987.

Seo ES, Ormes JF, Streitmatter RE, Stochaj SJ, Jones WV, Stephens SA & Bowen T Measurement of the cosmic ray proton and helium fluxes from 200 MeV nucl^{-1} to 100 GeV nucl^{-1} during the 1987 solar minimum. Nucl. Tracks Rad. Measurements 20: 413–46, 1992.

Shank BB. Results of radiobiological experiments on satellites. In: *Space Radiation Biology and Related Topics*. Ed. by Tobias CA & Todd P. Academic Press, New York, pp. 313–51, 1974.

Shavers MR, Curtis SB, Miller J & Schimmerling W. The fragmentation of 670 MeV neon-20 as a function of depth in water. II. One-generation transport theory. Radiat. Res. 124: 117–30, 1990.

Shavers MR, Frankel K, Miller J, Schimmerling W, Townsend LW & Wilson JW The fragmentation of 670 MeV neon-20 as a function of depth in water. 3. Analytical multigeneration transport theory. Radiat. Res. 136: 1–14, 1993.

Shavers MR & Poston JW. High energy proton fragmentation at the bone-soft tissue interface. In: *Abstracts, The World Space Congress, Washington, D.C., USA*, p. 554, 1992.

Shavers MR, Poston JW, Atwell W, Hardy AC & Wilson JW. *Preliminary Calculation of Solar Cosmic Ray Dose to the Female Breast in Space Missions.* NASA TM-4235, 1991.

Shea MA & Smart DF. Significant proton events of solar cycle 22 and a comparison with events of previous solar cycles. Adv. Space Res 14 (10): 631–8, 1994.

Shinn JL, Nealy JE, Townsend LW, Wilson JW &.Wood JS. Galactic cosmic ray radiation levels in spacecraft on interplanetary missions. Adv. Space Res. 14 (10): 863–72, 1994.

Shinn JL & Wilson JW. *An Efficient Heavy Ion Transport Code: HZETRN.* NASA Tp-3147, 1992.

Shinn JL, Wilson JW & Ngo DM. Risk assessment methodologies for target fragments produced in high-energy nucleon reactions. Health Phys. 59: 141–3, 1990.

Sideris EG. Nature of radiation induced mutations. experimental approaches on the question of intragenic events. In: *Terrestrial Space Radiation and its Biological Effects.* Ed. by McCormack PC, Swenberg CE & Bücker H. Plenum Press, New York, pp. 251–64, 1987.

Silberberg R, Tsao CH & Letaw JR. Improvement of calculations of cross sections and cosmic-ray propagation. composition and origin of cosmic rays. In: *Composition and Origin of Cosmic Rays.* Ed. by Shapiro MM. Riedel Publishing Co., pp. 321–36, 1983.

Silberberg R, Tsao CH & Shapiro MM. Semi-empirical cross sections, and applications to nuclear interactions of cosmic rays. In: *Spallation Nuclear Reactions and their Applications.* Ed. by Shen BSP & Merker M. Riedel Publishing Co., pp. 49–81, 1976.

Silberberg R, Tsao CH, Adams JH & Letaw JR. Radiation doses and LET-distributions of cosmic rays. Radiat. Res. 98: 209–26, 1984.

Simonsen LC & Nealy JE. *Radiation Protection for Human Mission to the Moon and Mars.* NASA TP-3079, 1991.

Simonsen LC, Nealy JE, Sauer HH & Townsend LW. *Solar Flare Protection for Manned Lunar Missions: Analysis of the October 1989 Proton Flare Event.* SAE Tech. Paper Ser. 911351, 1991.

Simonsen LC, Nealy JE, Townsend LW & Wilson JW. *Radiation Exposure for Manned Mars Surface Missions.* NASA TP-2979, 1990.

Simpson JA. Elemental and isotopic composition of the galactic cosmic rays. Ann. Rev. Nucl. Part. Sci. 33: 323–81, 1983.

Sinclair WK. Radiation safety standards: space hazards vs. terrestrial hazards. Life Sci. Space Res. 21: 151–60, 1983.

Sinclair WK. Radiation protection standards in space. Adv. Space Res. 6 (11): 335–43, 1986.

Sinclair WK. Radiation risk estimation and its application to human beings in space. Adv. Space Res. 4 (10): 115–20, 1984.

Sinclair WK. Experimental RBE values of high LET radiations at low doses and the implications for quality factor assignment. Radiat. Prot. Dosim. 13 (1–4): 319–26, 1985.

Sinclair WK. Recent estimates of cancer risk from low-LET ionizing radiation and radiation protection limits. Adv. Space Res. 12 (2/3): 375–8, 1992.

Sinclair WK. Radiation protection issues in galactic cosmic ray risk assessment. Adv. Space Res 14 (10): 879–84, 1994.

Sinclair WK & Fry RJM. General discussion on assessment of radiation risks for space flight. Adv. Space Res. 4 (10): 287–8, 1984.

Sinclair WK & Fry RJM. , Mechanisms of radiation interaction with DNA: potential implications for radiation protection. Radiat. Res. 112: 407–17, 1988.

Skarsgard LD, Kihlman BA, Parker L, Pujara CM & Richardson S. Survival, chromosome abnormalities, and recovery in heavy ion and X-irradiated mammalian cells. Radiat. Res. Suppl. 7: 208–21, 1967.

Skoski L, Merker M & Shen BSP. Absolute cross section for producing ^{11}C from carbon by 270-MeV/u ^{14}N ions. Phys. Lett. 30: 51–4, 1973.

Slobodrian RJ, Potvin L & Rioux C. Spectrometers for fast neutrons from solar flares. Adv. Space Res 14 (10): 725–32, 1994.

Slutz RJ, Gray TB, West ML, Stewart FG & Leftin M. *Solar Activity Prediction.* NASA CR-1939, 1971.

Smart DF & Shea MA. Cosmic exposure factors for shuttle altitudes derived from calculated cut-off rigidities. Adv. Space Res. 4 (10): 161–4, 1984.

Smart DF & Shea MA. Modeling the time-intensity profile of solar flare generated particle fluxes in the inner heliosphere. Adv. Space Res. 12 (2/3): 303–12, 1992.

Smart DF & Shea MA. Geomagnetic cutoffs: a review for space dosimetry applications. Adv. Space Res 14 (10): 787–96, 1994.

Smart DF & Shea MA. PPS-87: a new event oriented solar proton prediction model. Adv. Space Res. 9 (10): 281–4, 1989.

Smith HH, Woodley RG, Maschke A, Combatti NC & McNulty PJ. Relative cytogenetic effectiveness of 28 GeV protons. Radiat. Res. 57: 59–66, 1974.

Space Science Board. *Radiation Protection Guides and Constraints for Space-mission and Vehicle-design Studies Involving Nuclear Systems.* National Academy of Sciences, National Research Council, 1970.

Stapor WJ & Campbell AB. CRRES space results: dosimetry and radiation effects in microelectronics. In: *Abstracts, The World Space Congress, Washington, D.C., USA*, p. 572, 1992.

Stassinopoulos EG. The Earth's trapped and transient space radiation environment. In: *Terrestrial Space Radiation and its Biological Effects.* Ed. by McCormack PC, Swenberg CE & Bücker H. Plenum Press, New York, pp. 5–36, 1987.

Su RZ, Kost M & Kiefer J. The effect of krypton ions on artemia cysts. Radiat. Env. Biophys. 29: 103–7, 1990.

Suzuki K, Suzuki M, Nakano Kaneko I & Watanabe M. Analysis of chromatid damage in G2 phase induced by heavy ions and X-rays. Int. J. Radiat. Biol. 58: 781–90, 1990.

Swenberg CE. DNA and radioprotection. In: *Terrestrial Space Radiation and its Biological Effects.* Ed. by McCormack PC, Swenberg CE & Bücker H. Plenum Press, New York, pp. 675–96, 1987.

Swenberg CE, Horneck G & Stassinopoulos EG. (eds.) *Biological Eeffects and Physics of Solar and Galactic Cosmic Radiation.* NATO ASI Series A Life Sciences, Plenum Press, New York & London, 1993a.

Swenberg CE, Speicher JM & Miller JH. Does the topology of closed supercoiled DNA affect its radiation sensitivity? In: *Biological Effects and Physics of Solar and Galactic Cosmic Radiation.* Ed. by Swenberg CE, Horneck G & Stassinopoulos EG. NATO ASI Series A: Life Sciences 243A, Plenum Press, New York & London, pp. 37–48, 1993b.

SwitzerIII RC, Bogo V & Mickley GA. Histological effects of high energy electron and proton irradiation of rat brain detected with a silver-degeneration stain. Adv. Space Res. 14 (10): 443–452, 1994.

Takahashi T, Awaya Y, Tonuma T, Kumagai H, Izumo K, Nishida M, Hitachi A, Hashizume A, Uchiyama S & Doke T. Stopping power of Ni, Ag, Au and Pb for 7MeV/nucleon O-particles and carbon ions: Z31 deviation from the Bethe formula. Physical Review A27: 1360–4, 1983a.

Takahashi T, Yatagai F & Izumo K. Microdosimetric considerations of effects of heavy ions on E. coli K-12 mutants. Adv. Space Res. 12 (2/3): 65–8, 1992.

Takahashi T, Yatagai F & Kitayama S. Effect of heavy ions on bacterial spores. Life Sci. Space Res. 20 (1): 95–104, 1983b.

Takatsuji T & Sasaki MS. Dose-effect relationship of chromosome aberrations induced by 23MeV alpha particles in human lymphocytes. Int. J. Radiat. Biol. 45: 237–43, 1984.

Takatsuji T, Sasaki MS & Takekoshi H. Effect of static magnetic field on the induction of chromosome aberrations by 4.9MeV protons and 23MeV alpha particles. J. Radiat. Res. 30: 238–46, 1989.

Takatsuji T, Takekoshi H & Sasaki MS. Induction of chromosome aberrations by 4.9MeV protons in human lymphocytes. Int. J. Radiat. Biol. 44: 553–62, 1984.

Tao F, Powers-Risius P, Alpen E, Medvedovsky C, David J, & Worgul BV Radiation effects on late cytopathological parameters in the murine lens relative to particle fluence. Adv. Space Res. 14 (10): 483–92, 1994.

Tatsuzaki H, Inada T, Shimizu T, Arimom T, Satoh S & Akisada M. Early skin reaction following 250 MeV proton peak irradiation. J. Radiat. Res. 28: 150–8, 1987.

Taucher-Scholz G, Stanton JA, Schneider M & Kraft G. Induction of DNA breaks in SV40 by heavy ions. Adv. Space Res. 12 (2/3): 73–80, 1992.

Thacker J. The nature of mutants induced by ionizing radiation in cultured hamster cells. III. Molecular characterization of hprt-deficient mutants induced by gamma-rays of alpha-particles showing that the majority have deletions of all or part of the *hprt* gene. Mutat. Res. 160: 267–75, 1986.

Thacker J, Stretch A & Goodhead DT. The mutagenicity of alpha particles from plutonium-238. Radiat. Res. 92: 343–52, 1982.

Thacker J, Stretch A & Stephens MA. Mutation and inactivation of cultured mammalian cells exposed to beams of accelerated heavy ions. II. Chinese hamster V79 cells. Int. J. Radiat. Biol. 36: 137–48, 1979.

Thomas GE. An approach to modelling radiation damage by fast ionizing particles. Ima J Math Appl Med Biol. 4: 363–77, 1987.

Thomson JF, Williamson FS, Grahn D & Ainsworth EJ. Life shortening in mice exposed to fission neutron and gamma rays. I. Single and short-term fractionated exposures. Radiat. Res. 86: 559–72, 1981a.

Thomson JF, Williamson FS, Grahn D & Ainsworth EJ. Life shortening in mice exposed to fission neutrons and gamma rays. II. Duration-of-life and long-term fractionated exposures. Radiat. Res. 86: pp. 573–579, 1981b.

Thomson JF, Williamson FS & Grahn D. Life shortening in mice exposed to fission neutrons and gamma rays. III: Neutron exposures of 5 and 10 rad. Radiat. Res. 93: 205–9, 1983.

Thomson JF, Williamson FS & Grahn D. Life shortening in mice exposed to fission neutrons and gamma rays. IV. Further studies with fractionated neutron exposures. Radiat. Res. 103: 77–88, 1985a.

Thomson JF, Williamson FS & Grahn D. Life Shortening in mice exposed to fission neutrons gamma rays. V. Further studies with single low doses. Radiat. Res. 104: 420–8, 1985b.

Thomson JF, Williamson FS & Grahn D. Life shortening in mice exposed to fission neutrons and gamma rays. VI. Studies with the white-footed mouse, peromyscus leucopus. Radiat. Res. 108: 176–88, 1986.

Tobias CA. Radiation hazards in high altitude aviation. J. Aviat. Med. 23: 345–72, 1952.

Tobias CA, Blakely EA, Alpen EL, Castro JR, Ainsworth EJ, Curtis SB, Ngo FQH, Rodriguez A, Roots RJ, Tenforde T & Yang TCH. Molecular and cellular radiobiology of heavy ions. Int. J. Radiat. Oncol. Biol. Phys. 8: 2109–20, 1983.

Tobias CA, Budinger TE & Lyman JT. Radiation-induced light flashes observed by human subjects in fast neutron, x-ray, and positive beams. Nature 230: 596–7, 1971.

Tobias CA & Grigoriev YG. Ionizing radiation. In: *Foundation of Space Biology and Medicine*. Ed. by Calvin M & Gazenko OG. Joint USA/USSR Publication, pp. 473–531, 1975.

Tobias CA & Todd P. *Space Radiation Biology and Related Topics*. Academic Press, New York & London, 1974.

Todd P. Heavy ion irradiation of human and Chinese hamster cells *in vitro*. Radiat. Res. 61: 288–97, 1965.

Todd P. Unique biological aspects of radiation hazards -- an overview. Adv. Space Res. 3 (8): 187–94, 1983.

Todd P. Stochastics of HZE-induced microlesions. Adv. Space Res. 9 (10): 31–4, 1989.

Todd P. Multiple cell hits by particle tracks in solid tissues. Adv. Space Res. 12 (2/3): 393–401, 1992.

Todd P. Cosmic radiation and evolution of life on Earth: roles of environment, adaptation and selection. Adv. Space Res. 14 (10): 305–14, 1994.

Todd P & Walker JT. The microlesion concept in HZE particle dosimetry. Adv. Space Res. 4 (10): 187–97, 1984.

Tolkendorf E & Eichhorn K. Effect of ionizing radiation of different linear energy transfer on the induction of cellular death and of chromosome aberrations in cells of the Chinese hamster. Stud. Biophys. 95: 43–56, 1983.

Townsend LW, Cucinotta FA, Shinn JL & Wilson JW. Risk analyses for the solar particle events of August through December 1989. Radiat. Res. 130: 1–6. 1992b.

Townsend LW, Cucinotta FA & Wilson JW. Interplanetary crew exposure estimates for galactic cosmic rays. Radiat. Res. 129: 48–52, 1992a.

Townsend LW, Nealy JE, Wilson JW & Simonsen LC. *Estimates of Galactic Cosmic Ray Shielding Requirements During Solar Minimum.* NASA TM-4167, 1990

Townsend LW, Shinn JL & Wilson JW. Interplanetary crew exposure estimates for the August 1972 and October 1989 solar particle events. Radiat. Res. 126: 108–10, 1991.

Townsend LW & Wilson JW. *Tables of Nuclear Cross Sections for Galactic Cosmic Rays-absorption Cross Sections.* NASA RP-1134, 1985.

Townsend LW & Wilson JW. Energy-dependent parameterization of heavy-ion absorption cross sections. Radiat. Res. 106: 283–7, 1986.

Townsend W & Wilson JW An evaluation of energy-independent heavy ion transport coefficient approximations, Health Phys. 54 (4): 409–12, 1988.

Townsend LW, Wilson JW & Bidasaria HB. *Heavy-ion Total and Absorption Cross Sections Above 25 MeV/u.* NASA TP-2138, 1983.

Townsend LW, Wilson JW & Cucinotta FA. A simple parameterization for quality factor as a function of linear energy transfer. Health Phys. 53: 531–2, 1987.

Townsend LW, Wilson JW & Nealy JE. *Space Radiation Shielding Strategies and Requirements for Deep Space Missions.* SAE Tech. Paper Ser. 891433, 1989

Townsend LW, Wilson JW & Nealy JE. *Preliminary Estimates of Galactic Ray Shielding Requirements for Manned Interplanetary Missions.* NASA TM-101516, 1988.

Townsend LW, Wilson JW, Norbury JW & Bidasaria HB. *An Abrasion-ablation Model Description of Galactic Heavy-ion Fragmentation.* NASA TP-2305, 1984.

Townsend LW, Wilson JW, Shinn JL & Curtis SB. Human exposure to large solar particle events in space. Adv. Space Res. 12 (2/3): 339–48, 1992a.

Travis EL, Curtis SB, Howard J & Fowler JF. Repair but not potentiation observed in mouse lung irradiated with neon ions. Radiat. Res. 112: 500–7, 1988.

Tsao CH, Silberberg R, Adams JH jr & Letaw JR. Cosmic ray transport in the atmosphere-dose and LET-distributions in materials. IEEE Trans. Nuclear Sci. NS-30: 4398–4404, 1983.

Tubiana M, Dutreix J & Wambersie A. *Introduction to Radiobiology.* English translation by Bewley DK. Taylor & Francis, Hampshire, U.K., 1990.

Tulunay YK Ambient electron density distribution at about 500 km altitude at the Earth's ionosphere. In: *Terrestrial Space Radiation and its Biological Effects*. Ed. by McCormack PD, Swenberg CE & Bücker H. Plenum Press, pp. 37–40, 1987.

Tsuboi K, Yang TC & Chen DJ. Charged-particle mutagenesis. 1. Cytotoxic and mutagenic effects of high-LET charged iron particles on human skin fibroblasts. Radiat. Res. 129: 171–6, 1992.

Ullrich RL. High-LET radiation carcinogenesis. In: *Abstracts, The World Space Congress, Washington, D.C., USA*, p. 559, 1992.

Urano M, Verhey LJ, Goitein M, Tepper JE, Suit HD, Mendiondo O, Gragoudas E & Koehler A. Relative biological effectiveness of modulated proton beams in various murine tissues. Int. J. Radiat. Oncol. Biol. Phys. 10: 509–15, 1984.

Vahia MN & Biswas S. Solar energetic particle studies, in: *Composition and Origin of Cosmic Rays,* Ed. by Shapiro MM. Reidel Publishing, Dorcht, Netherlands, p. 155–60, 1983.

Vana N, Schöner W, Fugger M & Akatov Y. DOSIMIR – measurements of absorbed dose and 'average' LET on board of space station Mir. In: *Proceedings of the 5th European Symposium on Life Sciences Research in Space*. Ed. by Oser H, Guyenne TD. Noordwijk, The Netherlands, ESA, ESTEC, pp. 209–12, 1994.

Vashenyuk L, Miroshnichenko I, Sorokin MO, Perez-Peraza J & Gallegos A. Large ground level events (GLE's) of 22nd cycle of solar activity and some peculiarities of relativistic proton acceleration. Adv. Space Res. 14 (10): 711–16, 1994.

Vasilenko A, Sidorenko PG, Pugatch VM, Pavlenko Yu, Vasiliev Yu. The investigation of biological action of single heavy ions on the cellular level. In: *Abstracts, The World Space Congress, Washington, D.C., USA*, p. 548, 1992.

Vaulina EN & Anikeeva ID. The influence of spaceflight on *Chlorella*. Life Sci. Space Res. 8: 12–8, 1970.

Vaulina EN, Anikeeva ID, Gubareva IG & Strauch GA. Survival and mutability of *Chlorella* aboard the Zond vehicles. Life Sci. Space Res. 9: 105–10, 1971.

Vaulina EN, Anikeeva I & Kostina L. Radiosensibility of higher plant seeds after space flight. Adv. Space Res. 4 (10): 103–7, 1984.

Vazquez M, Worgul BV. Neuritogenesis: a model for space radiation effects on the central nervous system. Adv. Space Res. 14 (10): 467–76, 1994.

Waligorski MPR, Sinclair GL & Katz R. Radiosensitivity parameters for neoplastic transformations in C3H10T1/2 cells. Radiat. Res. 111: 424–37, 1987.

Warman EA. (ed) *Proceedings of the National Symposium on Natural and Man-made Radiation in Space.* NASA TM X-2440, 1972.

Watt DE. On absolute biological effectiveness and unified dosimetry. J. Radiol. Prot. 9: 33–49, 1989.

Webber WR. *An Evaluation of the Radiation Hazard Due to Solar-particle Events.* D2-90469, Aero-Space Div., Boeing Co., 1963.

Webber WR. *An Evaluation of Solar-cosmic-ray Events During Solar Minimum.* D2-84274-1, Boeing Co., 1966.

Webber WR. & Ormes JF. Cerénkov-scintillation counter measurements of nuclei heavier than helium in the primary cosmic radiation. 1. Charge composition and energy spectra between 200 MeV/Nucleon and 5 beV/Nucleon. J. Geophys. Res. Space Phys. 71: 5957–76, 1967.

Webber WR, Soutoul A, Ferrando P & Gupta M. The source charge and isotopic abundance of cosmic rays with Z=9–16: a study using new fragmentation cross sections. Astrophys. J. 348: 611–20, 1990.

Weber KJ & Flentje M. Lethality of heavy ion-induced DNA double-strand breaks in mammalian cells. Int. J. Radiat. Biol. 64 (2): 169–78, 1993.

Weisbrod U, Bucker H, Horneck G & Kraft G. Heavy-ion effects on bacteria spores – the impact parameter dependence of the inactivation. Radiat. Res. 129: 250–7, 1992.

Wiegel B, Heinrich W, Benton EV & Frank A. Measurements of LET spectra and comparison to models, Adv. Space Res. 12 (2): 349–53, 1992.

Wiegel B, Ohrndorf T & Heinrich W. Measurements of cosmic ray LET-spectra for the D1 mission using plastic nuclear track detectors. In: *Terrestrial Space Radiation and its Biological Effects*. Ed. by McCormack PD, Swenberg CE & Bücker H. Plenum Press, pp. 795–808, 1987.

Williams DJ, McEntire RW, Krimigis SM, Roelof EC, Jaskulek S, Tossman B, Wilken B, Studemann W, Armstrong TP, Fritz TA, Lanzerotti LJ & Roederer JG. Energetic particles at Venus - Galileo results. Science 253: 1525–8, 1991.

Wilson JW. *Analysis of the Theory of High-energy Ion Transport*. NASA TN D-8381, 1977a.

Wilson JW. Depth-dose relations for heavy ion beams. Virginia J. Sci. 28: 136–8, 1977b.

Wilson JW. Environmental geophysics and SPS shielding. Lawrence Berkeley Laboratory Report LBL-8581, pp. 33–116, 1978.

Wilson JW. Solar radiation monitoring for the high altitude aircraft. Health Phys. 41: 607–17, 1981.

Wilson JW & Badavi FF. Methods of galactic heavy ion transport. Radiat. Res. 108: 231–7, 1986.

Wilson JW, Chun SY, Buck WW & Townsend LW. High energy nucleon data bases. Health Phys. 55: 817–9, 1988a.

Wilson JW & Costner CM. *Nucleon and Heavy-ion Total and Absorption Cross Section for Selected Nuclei*. NASA TN D-8107, 1975.

Wilson JW & Cucinotta F. *Human Exposure in Low Earth Orbit*. NASA TP-2344, 1984.

Wilson JW, Cucinotta FA & Shinn JL. Cell kinetics and track structure. In: *Biological Effects and Physics of Solar and Galactic Cosmic Radiation*. Ed by Swenberg CE, Horneck G & Stassinopoulos EG. NATO ASI Series A: Life Sciences 243A,Plenum Press, pp. 295–338, 1993.

Wilson JW & Khandelwal GS. *Computer Subroutines for the Estimation of Nuclear Reaction Effects in Proton-tissue-dose Calculations*. NASA TM X-3388, 1976a

Wilson JW & Khandelwal GS. Proton-tissue dose buildup factors. Health Phys. 31: 115–8, 1976b.

Wilson JW, Lambiotte JJ jr, Foelsche T & Filippas TA. *Dose Response Functions in the Atmosphere Due to Incident High-energy Protons with Application to Solar Proton Events*. NASA TN D-6010, 1970.

Wilson JW, Lamkin SL, Farhat H, Ganapo BD & Townsend LW. *A Hierarchy of Transport Approximations for High Energy Heavy (HZE) Ions*. NASA TM-4118, 1989.

Wilson JW, Shinn JL & Townsend LW. Nuclear reaction effects in conventional risk assessment for energetic ion exposure. Health Phys. 58: 749–52, 1990a.

Wilson JW & Townsend LW. A benchmark for galactic cosmic-ray transport codes. Radiat. Res. 114: 201–6, 1988a.

Wilson JW & Townsend LW. Radiation safety in commercial air traffic: a need for further study. Health Phys. 55: 1001–3, 1988b.

Wilson JW & Townsend LW. Errata-radiation safety in commercial air traffic: a need for further study. Health Phys. 56: 973–4, 1989.

Wilson JW, Townsend LW & Badavi FF. Galactic HZE propagation through the Earth's atmosphere. Radiat. Res. 109: 173–83, 1987a.

Wilson JW,Townsend LW & Badavi FF. A semi-empirical nuclear fragmentation model. Nuclear Instrum. & Methods Phys. Res. B18: 225–31, 1987b.

Wilson JW, Townsend LW, Bidasaria HB, Schimmerling W, Wong M & Howard J. ^{20}Ne depth-dose relations in water. Health Phys. 46: 1101–11, 1984.

Wilson JW, Townsend LW & Khan F. Evaluation of highly ionizing components in high-energy nucleon radiation fields. Health Phys. 57: 717–24, 1989.

Wilson JW, Townsend LW, Lamkin SL & Ganapol BD. A closed-form solution to HZE propagation. Radiat. Res. 122: 223–8, 1990b.

Wilson JW, Townsend LW, Schimmerling W, Khandelwal GS, Khan F, Nealy JE, Cucinotta FA, Simonsen LC, Shinn JL & Norbury JW. *Transport Methods and Interactions for Space Radiations*. NASA RP-1257, 1991.

Wilson WE, Metting NF & Paretzke HG. Microdosimetric aspects of 0.3- to 20-MeV proton tracks. 1. Crossers. Radiat. Res. 115: 389–402, 1988b.

Wilson WE & Paretzke HG. Calculation of the ionizing frequency distributions in small sites. Radiat. Res. 81: 326–35, 1980.

Wilson WE & Paretzke HG. Calculation of distributions for energy imparted and ionization by fast protons in nanometer sites. Radiat. Res. 87: 521–37, 1981.

Wong M, Schimmerling W, Philips MH, Ludewigt BA, Landis DA, Walton JT & Curtis SB. The multiple coulomb scattering of very heavy charged particles. Med. Phys. 17: 163–71, 1990.

Wood DH, Cox A, Hardy K, Salmon Y, & Trotter R. Head and neck tumors after energetic proton irradiation in rats. Adv. Space Res. 14 (10): 681–4, 1994.

Woodruff KH, Lyman JT & Fabrikant JI. Heavy charged-particle induced lesions in rabbit cerebral cortex. Int. J. Radiat. Oncol. Biol. Phys. 14: 301–9, 1988.

Worgul BV. Cataract analysis and the assessment of radiation risk in space. Adv. Space Res. 6 (11): 285–93, 1986.

Worgul BV. Rationale for studies on integrated biological systems – a biologist's viewpoint. In: *Abstracts, The World Space Congress, Washington, D.C., USA*, p. 553, 1992a.

Worgul BV. Starry eyes: space radiation and the visual system. In: *Abstracts, The World Space Congress, Washington, D.C., USA*, p. 556, 1992b.

Worgul BV, Krebs W & Koniarek JP. Microlesions: theory and reality. Adv. Space Res. 9 (10): 315–23, 1989.

Worgul BV, Medvedovsky C, Brenner DJ, Huang Y & Merriam GR. jr. Low dose/low dose-rate heavy ion catarctogenesis. In: *Abstracts, The World Space Congress, Washington, D.C., USA*, p. 559, 1992.

Worgul BV, Medvedovsky C, Powersrisius P & Alpen E. Accelerated heavy ions and the lens. 4. Biomicroscopic and cytopathological analyses of the lenses of mice irradiated with 600 MeV/u Fe-56 ions. Radiat. Res. 120: 280–83, 1989a.

Worgul BV, Merriam GR, Medvedovsky C & Brenner DJ. Accelerated heavy particles and the lens. III. cataract enhancement by dose fractionation. Radiat. Res. 118: 93, 1989b.

Wrenn GL, Sims AJ, Dyer CS & Truscott PR. Protection of polar platforms from penetrating radiation. In: *Terrestrial Space Radiation and its Biological Effects*. Ed. by McCormack PC, Swenberg CE & Bücker H. Plenum Press, New York, pp. 655–62, 1987.

Wulf H, Kraft-Weyrather W, Miltenburger HG, Blakely EA, Tobias CA & Kraft G. Heavy-ion effects on mammalian cells: inactivation measurements with different cell lines. Radiat. Res. 104 supplement 8: S122–34, 1985.

Wulf H, Miltenberger HG & Kraft G. Inactivation-cross-section of mammalian tissue culture cells irradiated with heavy ions. Radiat. Environ. Biophys. 19 (4): 305, 1981.

Xu YJ, Khandelwal GS & Wilson JW. Low-energy proton stopping power of N_2 O_2, and water vapor, and deviations from Bragg's rule. Phys. Review A29: 3419–22, 1984.

Xu YJ, Khandelwal GS & Wilson JW. Proton stopping cross sections of liquid water. Phys. Review A32: 629–36, 1985.

Yang CH, Craise LM, Durant M & Mei M. Heavy-ion induced genetic changes and evolution processes. Adv. Space Res. 14 (10): 373–82, 1994.

Yang TC, Craise LM, Mei M & Tobias CA. Dose protection studies with low-and high-LET radiations on neoplastic cell transformation *in vitro*. Adv. Space Res. 6 (11): 137–47, 1986.

Yang CH, Craise LM, Mei M & Tobias CA. Neoplastic cell transformation by high-LET radiation: molecular mechanisms. Adv. Space Res. 9 (10): 131–40, 1989.

Yang TC, Craise LM, Mei M & Tobias CA. Neoplastic cell transformation by heavy charged particles. Radiat. Res. 104: S177–87, 1985.

Yang TC & Tobias CA. Effects of heavy ion radiation on the brain vascular system and embryonic development. Adv. Space Res. 4 (10): 239–45, 1984.

Yang VV & Ainsworth EJ. Late effects of heavy charged particles on the fine structure of the mouse coronary artery, Radiat. Res. 91: 135–44, 1982.

Ye ZH, Chen GF, Zhu GW, Shen SZ, Liang JB & Huang HJ. Up-to-date observation of space energetic particles made by the Chinese satellite FENGYUN-1(B). Adv. Space Res. 14 (10): 701–6, 1994.

Yochmowitz MG, Wood DH & Salmon YL. Seventeen-year mortality experience of proton radiation in *Macaca mulatta*. Radiat. Res. 102: 14–34, 1985.

Yurov SS, Akoev IG & Leont'eva GA. Effects of HZE particles and space hadrons on bacteriophages. Adv. Space Res. 3 (8): 51–60, 1983.

Yurov SS, Akoev IG, Akhmidieva AK, Livanova IA, Leont'eva GA, Marennyi AM & Popov VI. Genetic effects of cosmic radiation on bacteriophage $T4Br^+$ (on materials of biological experiment 'Soyuz-Apollo'). Life Sci. Space Res. 17: 129–32, 1979.

Zaider M, Brenner DJ & Wilson WE The applications of track calculations to radiobiology. I. Monte Carlo simulation of proton tracks. Radiat. Res. 95: 231–47, 1983.

Zankl M, Veit R, Petoussi N, Mannweiler E, Wittmann A & Drexler G. Realistic computerized human phantoms. Adv. Space Res. 14 (10): 433–42, 1994.

Zens R, Heinrich W, Wiegel B, Rusch G & Reitz G. Measurement of HZE-particles on BIOCOSMOS 9. In: *Abstracts, The World Space Congress, Washington, D.C., USA*, p. 573, 1992.

Zimmermann H, Schäfer M, Schmitz C & Bücker H. Effects of heavy ions on inactivation and DNA double strand breaks in *Deinococcus radiodurans* R1. Adv. Space Res. 14 (10): 213–6, 1994.

Zimmermann MW, Gartenbach KE & Kranz AR. First radiobiological results of LDEF-1 experiment A0015 with *Arabidopsis* seed embryos and *Sordaria* fungus spores. Adv. Space Res. 14 (10): 47–52, 1994.

Chapter 7
Exobiology

GERDA HORNECK

7.1 Introduction

With the development of space technology, exobiology has been established as a scientific discipline. The term 'exobiology' was introduced by **Lederberg (1960)**, who recognised the emerging capabilities of space exploration for giving new impetus to biology much as the development of astronomy has broadened our understanding of the physical world and the spectral analysis of the stars has proven the universality of the concepts of chemistry. Classical biological research has concentrated on the only example of life so far known, i.e. life on Earth. In contrast, exobiology extends the boundaries of biological investigations beyond the Earth, to other planets, comets, meteorites, and space at large. To date, the over-riding objective of exobiological research has been to attain a better understanding of the principles leading to the emergence of life from inanimate matter, its evolution, and its distribution on Earth and throughout the Universe (Klein 1986). To reach this goal, exobiological research has focused on the different steps of the evolutionary pathways through cosmic history that may be related to the origin, evolution and distribution of life.

Since its introduction more than 30 years ago, exobiology has developed into a broad scientific field, involving the concerted interdisciplinary approach of astrophysics, planetary sciences, organic chemistry, palaeontology in addition to the various sub-disciplines of biology. The bibliography on exobiology in the broadest sense comprises more than 7000 publications, the majority of them appearing in peer reviewed journals (**Roy & Powers 1990**). The International Society for the Study of the Origin of Life (ISSOL) which holds regular meetings and publishes the journal *Origins of Life and Evolution of the Biosphere*, is especially devoted to exobiological topics. Within the Scientific Commission *Life Sciences as Related to Space* of the Committee on Space Research (COSPAR) of the International Commission of Scientific Unions (ICSU), *Planetary Biology and Origins of Life* constitutes a whole subcommission with publications in *Advances in Space Research.* At the International Astronautical Federation (IAF), one subcommission is devoted to the *Search for Extraterrestrial Intelligence* (SETI).

Exobiology has become an integral part of national and international space science programmes. By 1970 the US National Academy of Sciences recommended exobiology and planetary ecology as one of four key research areas in the life sciences programme of NASA, in addition to space biology, human physiology and aerospace medicine, and personnel health and environmental medicine (Glass 1970). NASA reports prepared at a series of science workshops are now available on most topics relevant to the origin and evolution of life throughout the universe. These include the cosmic evolution of the biogenic elements (H, C, N, O, S, P) and the prebiotic chemical evolution of organic molecules (DeVincenzi & Pleasant 1983; Wood & Chang 1985), the beginnings of life on Earth (Hartman *et al.* 1985), the impact of planetary properties and events in space on the evolution of complex life forms (Milne *et al.* 1985), exobiological studies in Earth orbit (DeFrees *et al.* 1989), the wide field of exobiological research in solar system exploration (Carle *et al.* 1992) and the need for planetary protection (National Research Council 1992). The research strategy of NASA's exobiology programme, as defined in these reports, follows the cosmic evolution scheme from the origin of the biogenic elements through the chemical and biological evolution to the rise of complex life and the possible existence of other life forms beyond the Earth (**DeVincenzi 1984**; Klein 1986, 1990).

Within ESA's life sciences programme, exobiology follows a comparably broad approach (Bjurstedt 1979; Horneck 1987; Oser & Battrick 1989). However, concordant with ESA's main responsibility to provide flight opportunities, most exobiological activities concentrate on studies in Earth orbit. Using the peculiar environment of space as a tool, the questions tackled pertain to the relevance of extraterrestrial organic molecules to the emergence of life on Earth or any other planet, to the role of solar UV radiation in evolutionary processes related to life, and to the chances and limits for an interplanetary transfer of life (Horneck *et al.* 1989; Mennigmann 1989). Exobiological aspects are also included in ESA's solar system exploration programme. The Giotto mission to comet Halley has contributed pieces to the puzzle of prebiotic chemistry (Grün 1985). As part of the Cassini mission to the Saturn system, ESA's Huygens probe will investigate the complex organic chemistry of the atmosphere of Titan (Lebreton 1985). Plans for a European contribution to future international Mars exploration initiatives include a step-by-step approach to exobiological research on Mars, where the search for alien life forms, fossils, contemporary forms or precursors, will be preceded by missions gaining more precise knowledge of Martian geology, climate and history (Chicarro *et al.* 1989). The perspectives for a scientific exploration and utilisation of the Moon include exobiological research on organic molecules in dust particles and radioastronomical search for planetary systems and extraterrestrial civilisations (Battrick & Barron 1992).

The state-of-the-art in exobiology at large as well as in several areas related to this field has been progressively compiled over the years in textbooks, surveys, proceedings of relevant meetings as well as popular sciences books. They provide a comprehensive survey of the research concerned with chemical evolution and the origin of life (Miller & Orgel 1974; Dose & Rauchfuss 1975; Fox & Dose 1977; Ponnamperuma 1983; Mason 1991; Brack & Raulin 1991; Greenberg *et al.* 1993), origin and early evolution of life (Kaplan 1978;

Halvorson & Van Holde 1980; Crick 1981; Day 1984; Stanley 1986; Eigen 1987; Trn Thanh Vn *et al.* 1992; Hartmann & Matsuno 1992), the early biosphere as inferred from palaeontology (Schopf 1983; Pflug 1984; Cohen *et al.* 1983; Schidlowski *et al.* 1992; Schopf & Klein 1992) and molecular phylogeny (Patterson 1987), adaptation to environmental extremes (Gregory 1973; Cox 1987; Friedmann 1993), planetary evolution processes related to exobiology (Holland 1984; Levine 1985) and the search for life in the Universe (Goldsmith 1980; Goldsmith & Owen 1980/1984; Horowitz 1986; Marx 1988; Heidmann & Klein 1991; Sagan 1994).
In this report, emphasis will be laid on results in exobiology that have been obtained from space research including radioastronomy, solar system exploration and experiments in Earth orbit, as well as from simulation experiments on the ground. Models and theories will be presented and the currently unknown pieces of information to the puzzle of 'life', its origin, evolution and distribution throughout the Universe, will be identified.

7.2 Evolution of potential precursors of life

7.2.1 In the interstellar medium

Our knowledge of the chemical composition of the interstellar medium has been obtained mainly from radioastronomical observations (for reviews see Winnewisser & Herbst 1987; Mason 1991; Irvine 1992; **Williams 1993**). The observed microwave and radio-frequency lines emitted by molecules in the dense interstellar clouds are compared with the spectra of known species in order to identify the emitting molecule. However, many of the fragile species that can survive in the extreme vacuum of space have not yet been reproduced in the laboratory (Fig. 1).
To date, about eighty molecular types have been identified in interstellar and circumstellar regions from their line emission and absorption characteristics (Winnewisser & Herbst 1987; **Williams 1993**). When hydrogen densities exceed about 1000 atoms cm^{-3}, H_2 molecules may form on dust particles. It is also this dust that prevents UV radiation from leaking into such 'dense' clouds and thereby dissociating the molecules formed. Besides H_2 and He, neither of which emit easily detectable line radiation, there is a small fraction of trace molecules; CO is the most abundant representative of these. A large fraction of the molecules found in the dense clouds of the interstellar medium are of organic nature. Many of them are highly unsaturated, such as polyacetylenes, the small cyclic molecules C_3H_2 and C_3H and the heterogeneously terminated carbon chains C_nH, C_nS and C_nO. The largest molecules, so far discovered, are higher polyacetylenenitriles, H(C C)$_n$CN, with $n = 4$ and 5. It is of interest to note the occurrence of several compounds that may give rise to important biological monomers, including H_2CO and HCOOH as precursors of sugars and lipids; HCN as a precursor of adenine, and together

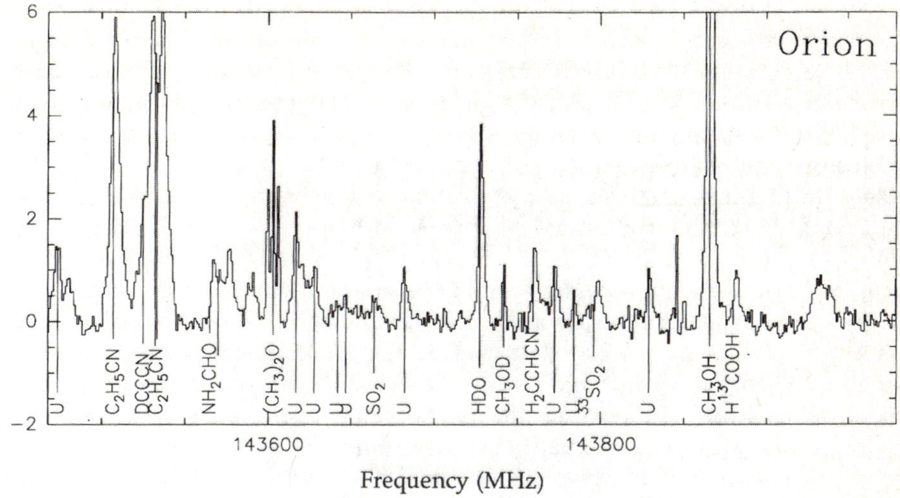

Fig. 1. Spectrum of molecules in gas phase observed in direction of Orion. Most lines have been assigned to known molecules, however several lines remain unidentified (U). The spectrum was acquired from the 30 m millimetre radio telescope, Pic Veleta, Spain. Courtesy Dr. R. Mauersberger, MPI für Radioastronomie, Bonn.

with H(CC)CN, PN, and H_2O as precursors of other nucleic acid components; aldehydes, such as CH_3CHO or H_2CO plus HCN and H_2O as precursors of amino acids and therefore also of proteins (for prebiotic chemistry see Section 7.2.2.5). Extensive search for the smallest amino acid, glycine (NH_2CH_2COOH), has recently led to its detection in the interstellar space (**Snyder *et al.* 1983**, 1994; **Travis 1994**).

The formation processes of interstellar molecules include synthesis on dust grain surfaces and successive evaporation or gas phase ion–molecule reactions. When in the course of prestellar evolution a molecular cloud becomes cooler (10–20°K) and denser (>10^6 cm^{-3}), trace molecules, such as CO, H_2O, CH_3OH, and NH_3, condense onto grain surfaces as ice mantles (**Mauersberger *et al.* 1992**). These ices may be preserved when comets form in the accretion disk of a newly formed star (see Section 7.2.2.2).

Organic molecules in the interstellar medium are mainly formed by gas phase ion–molecule reaction paths which start with the most abundant molecule, H_2, reacting with various ions (Irvine 1992; Williams 1993). Hot or supra thermal reactions (Roessler 1991), photochemical processes and surface reactions on interstellar dust grains (**Greenberg 1976**) also provide satisfying explanations.

From infrared spectroscopy, evidence has been obtained of the existence of small solid particles (grains) in the dense clouds of the interstellar medium (**Greenberg 1976**). According to Greenberg's 'grain model', such interstellar grains consist of a silicate core and a mantle of relatively non-volatile higher molecular weight organic compounds (Greenberg 1983). The steps of their formation proceed in the following manner: (i) silicate particles of a mean radius of 0.05 μm blown from cool stars into space form the seeds on which atoms and molecules of the interstellar gas condense; (ii) ultraviolet light from distant stars photolyzes the condensed mixture thereby creating new molecules and radicals; (iii) grain explosions replenish molecules in the gas phase. In this way, each grain accumulates a complex organic mantle that diffuses and/or regenerates during the passage of the grain through different regions of the interstellar medium (Greenberg 1983). Highly complex organic molecules of this sort, included in composite grain clumps in the interstellar medium, have been proposed as the starting material for biological activity on Earth, or elsewhere in the Galaxy (**Hoyle & Wickramasinghe 1977, 1980**).

In addition to the organic molecules, identified so far, the infrared emission bands give strong support to the existence of even more complex compounds, i.e. polycyclic aromatic hydrocarbons (PAH) (**Allamandola *et al.* 1987a, b, 1989a**) containing 20–40 carbon atoms in the small PAHs, 200–500 carbon atoms in the large PAHs and which form even more complex PAH clusters (PAH hypothesis). The problem in their identification by laboratory spectra lies in the fact that it is not yet possible to produce in the laboratory PAHs in the forms assumed to be present in the interstellar medium. More information is expected from the absorption spectra in the UV range, which can be detected by telescopes deployed in space, e.g. the Hubble Space Telescope, or on a lunar base.

An example how the mere comparison of spectra can lead us astray is the so-called 'bacterial grain model' (**Jabir *et al.* 1983**). From the agreement of interstellar absorption spectra in the infrared with those of desiccated bacteria (see also **Yabushita *et al.* 1986**) it was inferred that one class of the interstellar grains consists of bacterial cells which may be either desiccated or degraded. However, one should bear in mind, that such broad features of molecules on solid bodies may lead to more equivocal interpretations than the sharp lines of molecules in the gas phase. The authors go even further, when suggesting that micro-organisms or even diatoms might exist on comets, the Jovian satellite Europa and in interstellar space (Hoover *et al.* 1986). In the light of all available evidence (see also Section 7.3) the view that life originated in comets of the interstellar medium is untenable (**Bar-Nun *et al.* 1981**).

In laboratory experiments, simulating the conditions of the interstellar medium, volatile inorganic compounds have been photochemically converted to non-volatile organic compounds (**Agarwal *et al.* 1985**). They have been found to consist of a very complex mixture of long, cross-linked carbon chains and probably aromatic carbon compounds (**Mendoza-Gmez & Greenberg 1993**). These compounds are suggested to resemble the organic refractory mantle on interstellar grains.

During the EURECA mission such complex organic residues, produced in the laboratory, were exposed to the full spectrum of solar UV and vacuum-UV light in order to simulate the photolytic processes thought to occur during chemical evolution of interstellar grains (Horneck *et al.* 1989). With respect to the UV flux the exposure of this organic residue to solar UV for about 6 months provided an interstellar equivalent of approximately 5×10^9 years. This time span is in the range of the recycling time of the interstellar medium (Greenberg 1983). Exposure platforms such as the ERA (Exobiology Radiation Assembly) on EURECA have been recommended suitable for *in situ* studies of interstellar organic chemistry processes (Horneck *et al.* 1989).

7.2.2 In the solar system

7.2.2.1 Solar nebula

There is convincing evidence that the solar system began to form some 5 to 4.5×10^9 years ago (for a compilation of the various methods of geochronology and the reliability in dating see Kirsten 1978; Jessberger 1990). According to astronomical models for the formation of the solar system, a dense core region within an interstellar molecular cloud collapsed to form the solar nebula (**Cameron 1988**). These models generally differentiate between an outer region now occupied by the giant outer planets and an inner region corresponding approximately to that of the terrestrial planets (reviewed in Mason 1991). Whereas in the outer reaches of the solar system the molecules from the parental dense cloud were mainly preserved (**Anders & Zinner 1993**), in the hotter central region they were dissociated into free radicals and ions. However, it is still questionable if, and to what extent, the accreting interstellar material survived in molecular form, at least in the outer solar system (Irvine 1992).

Current models concerning the formation of planets from the solar nebula, suggest almost identical initial accretion processes for all bodies (for review see Pollack & Atreya 1992). Through a large number of collisions, small grains aggregate and grow successively to solid bodies of increasing sizes, forming planetesimals. Depending on the distance from the centre, the aggregates have different compositions, because both the temperature and the pressure decrease with distance. In the outer part of the solar nebula, low-temperature material condenses to form so-called 'ices' comprised of oxidised, volatile-rich organic matter. Whilst in the inner regions high-temperature condensates, the so-called 'rocks', prevail that are composed of metallic iron-nickel, SiO_2, MgO, Al_2O_3, and CaO (Mason 1991). In a given region of the solar system, the largest bodies eventually accrete most of the planetesimals that cross their orbit, thereby attaining masses equal to or greater than that of the Earth. It has been suggested that radial mixing of planetesimals during the accretion process could have resulted in substantial exchange of material formed in the different regions of the nebula.

7.2.2.2 Comets

Comets were most likely formed in the outer solar system region. They are widely regarded as being the most undisturbed, primitive remnants of the swarm of planetesimals from which the outer planets of our solar system accreted (Wilkening 1982). From their spectra recorded in the UV, visible, IR and radio range it was inferred that they contain organic material (summarised in **Fechtig & Rahe 1984**). Molecules such as CH, CN, CO, CS, H_2CO, DCO, HCN, and CH_3CN were identified as well as ions in the plasma tail, predominantly CO^+ with contributions of H_2O^+, CO_2+, CH^+ and OH^+ (Table 1). However, almost nothing was known about the abundance of organics and of the composition of the cometary dust until the spacecraft encounters with comet Halley. The detailed investigations of comet Halley during its 1985/1986 apparition by several spacecraft, especially Vega 1 and Vega 2 (**Sagdeev et al. 1986**) and the Giotto mission (**Fechtig & Rahe 1984**) showed that comet Halley is richer in organic material than was predicted (**Kissel *et al.* 1986a, 1986b; Langevin *et al.* 1987; Kissel & Krueger 1987; Moroz *et al.* 1987; Mitchell *et al.* 1992**, for review see also Delsemme 1991; **Huebner & Boice 1992**). Based on these analyses, comet Halley appears to consist of about 14% organic carbon by mass (Delsemme 1991). The chemical composition of Halley's dust particles was inferred from impact-ionisation time-of-flight mass spectrometry as a loosely aggregated mineral core embedded in a fluffy organic ice mantle (**Jessberger *et al.* 1988**). About 30% of the particles were rich in light elements, such as H, C, N, and O (**Kissel *et al.* 1986a, 1986b; Kissel & Krueger 1987**), representing the so-called CHON particles. 35% of the particles were a mixture of CHON particles and silicates (Mg, Si, Fe), and 35% contained no appreciable amount of low atomic number elements except O (**Langevin *et al.* 1987**). The relative mean abundance of the elements in the mineral cores corresponds to that of C1 carbonaceous meteorites. From the decomposition molecules identified the occurrence of HCN and H^3C-CNH has been inferred (**Kissel & Krueger 1987**), which have been suggested as very active reactants in prebiotic chemistry (**Schwartz *et***

Table 1. Comets: considered to be the most pristine celestial bodies in our solar system, bearing witness to the existence of a dynamic organic chemistry from its earliest stages.

Nature	Relic of solar nebula
	Carrier of organic matter and water within our solar system
	Organic compounds so far detected in coma and tail: C, C_2, C_3, CH, CN, CO, CS, CO_2, HCN, CH_3CN (probably)
Aspects of exobiology	Complexity of organic compounds in comet nucleus and mechanisms of their formation
	Mode of transportation of organic matter and water to a planet
Approach	Space missions (ROSETTA)
	Simulation experiments (COSI)

al. 1984). However, the detailed molecular composition of the CHON particles is still mostly unknown. The mass spectra of the organic ice mantles indicate the presence of mainly unsaturated linear and cyclic compounds (**Kissel & Krueger 1987; Mitchell *et al.* 1992**). The presence of PAHs (**Allamandola *et al.* 1987b**) and polymers of formaldehyde (**Mitchell *et al.* 1987; Huebner 1987**) and of HCN (**Krueger *et al.* 1991**) are also suggested. This is an especially important finding because formaldehyde as well as HCN is thought to have played a central role in prebiotic chemistry (see Section 7.2.2.5). These organic molecules, so far detected in cometary coma and tail, point to much more complex 'parent' molecules in the comet nucleus from which they originated. More insight is expected from ESA's planned future comet encounter mission Rosetta with *in-situ* analysis of the nuclear material (Bar-Nun *et al.* 1993).

Laboratory experiments with frozen gases relevant to comet scenarios have demonstrated the formation of complex organic molecules as well as their destruction by energetic particles or γ rays (e.g. **Strazulla 1985; Draganic *et al.* 1985;** Johnson *et al.* 1985) or high energetic photon radiation (e.g. **Roessler *et al.* 1992**) (for review see **Huebner & Boice 1992**). In preparation for the comet rendezvous Rosetta mission (Bar-Nun *et al.* 1993), laboratory studies simulating cometary scenarios have been performed (Kochan *et al.* 1989) including VUV photolysis of potential solid components of comets and grains (**Roessler *et al.* 1992**).

The cometary material, as deduced from comet Halley investigations, bears a remarkable resemblance to the models of chemical and morphological structure of the interstellar dust grains (see Section 7.2.1). These findings are strongly in favour of the formation model proposed by Greenberg (Greenberg 1982, **Greenberg & Hage 1990**). The model suggests that comets are mainly composed of loosely bound aggregates of interstellar dust grains. These grains would consist, originally, of silicate cores surrounded by icy mantles that through subsequent photolysis by UV light in the interstellar environment would have been converted into more refractory compounds rich in organics. Hence, a large fraction of organics in cometary matter would have been originated from the complex refractory organic molecules of the grain mantles. However, one should bear in mind that at least the surface material of comets has been modified since their formation about 4.5×10^9 years ago, by galactic comic rays and by solar radiation. Furthermore, it might be premature, to draw general conclusions from the direct analysis of just one representative of this group of small celestial bodies.

Comets are of special interest to exobiology, because – among all celestial bodies – they contain the largest amount of organic molecules; they are considered the most pristine celestial bodies bearing witness to the existence of a dynamic organic chemistry from the earliest stages of our solar system. They have been suggested to be the major source of the hydrosphere, the atmosphere, and probably also of the organic compounds of the early Earth (**Oro 1961; Irvine *et al.* 1980;** Greenberg 1986; **Chyba *et al.* 1990; Delsemme 1992; Huebner & Boice 1992; Oro *et al.* 1992;** Greenberg & Mendoza-Gomez

1992). Several mechanisms have been proposed that may foster the survival of organic material through a cometary impact with a planet such as grazing incidence encounter, entry at very oblique angles which may result in orbital capture, or preservation of material in the interior of giant comets (**Clark 1988**). It has been argued, that a soft-landing of a cometary nucleus on the surface of a 'suitable' planet may provide all prerequisites for life to originate (**Clark 1988; Krueger & Kissel 1989**).

7.2.2.3 Planets and satellites of the outer solar system

Substantial information on the giant planets, Jupiter, Saturn, Uranus, Neptune and Pluto has been obtained from the Voyager 1 and 2 missions (**Stone & Lane 1979; Smith *et al.* 1981, 1982; Romani & Atreya 1988; Stevens *et al.* 1992; Yelle *et al.* 1993**). They are mainly composed of three groups of material, 'rock', 'ice' and 'gas'. Current models suggest that they were formed very rapidly, within the first 10^7 years before the gases of the solar nebula were dissipated (Pollack & Bodenheimer 1989). Their cores were formed by accretional growth of planetesimals, embedded in the solar nebula. After achieving a critical core mass, they gravitationally attracted the gaseous material of the solar nebula, consisting essentially of H and He. Growth continued until the gas of the solar nebula was dissipated. The atmospheres attained by this process are very thick reducing ones retaining a high fraction of the original solar nebula material.
In addition to H_2 and He, the atmosphere of Jupiter contains CH_4, NH_3, PH_3, AsH_3 (**Lunine 1989**). Small quantities of CO, HCN have been observed in the troposphere, and hydrocarbons of 2 or more C-atoms in the stratosphere. It is assumed that a dynamic atmospheric chemistry takes place in the upper atmosphere, mainly driven by solar UV radiation although solar energy amounts to only 4% of that reaching the Earth. Methane photochemistry is expected to produce high order hydrocarbons, such as C_3H_8, CH_3CCH, C_4H_{10} and even the cyclic C_6H_6 (Pollack & Atreya 1992). Photolysis of NH_3 in the presence of CH_3CCH has been proposed as a feasible pathway for the formation of HCN (**Ferris *et al.* 1992**). The large storms discovered in Jupiter's atmosphere indicate that molecules from the deep interior are carried to the upper layers of the atmosphere by vertical motions, and *vice versa*. Any organic compounds formed in the upper atmosphere when sinking into the deeper layers of the troposphere will be destroyed readily by evaporation, thermal decomposition, or hydrogenation and will thereby be recycled back to the primary ingredients. Thus, the preservation of complex organic compounds on Jupiter seems unlikely. The colourful features observed in Jupiter's atmosphere (**Owen & Terrile 1981**) have been suggested to be a result of one or several of the following chromophores: P_4 which is photolytically produced from PH_3, nitriles, sulphur compounds (**Griffith *et al.* 1992**), or polymers resulting from photochemical reaction pathways of CH_4 and NH_3 (**Ferris *et al.* 1992**).

The atmosphere of Jupiter, as that of any other of the giant planets, has been considered as a natural laboratory for observing organic chemical evolution in a reducing environment (Pollack & Atreya 1992). Here, one of the simplest carbon-containing molecules, CH_4, is converted into more complex hydrocarbons. The Galileo mission which was launched from the Space Shuttle in 1989, will provide further insights into Jupiter's active atmospheric chemistry.

To understand the chemical processes in the Jovian atmosphere simulation experiments in the laboratory (**Khare *et al.* 1978; McDonald *et al.* 1992**) as well as in Earth orbit facilities have been proposed (Horneck *et al.* 1989). The microgravity environment of space provides conditions in which forces to suspend particles can be reduced by as much as 6 orders of magnitude (McKay *et al.* 1986, Fogleman *et al.* 1989). It has been suggested that this allows for the study of gas-particle interactions in time frames much longer than can be achieved on Earth. The radiation environment of space, especially the short-wavelength solar UV radiation, has been suggested to serve as an authentic energy source for studies on the dynamics and mechanisms of the formation of complex organic molecules from simple precursors. Ideal platforms for such studies are the ERA facility flown on EURECA and the Biopan to be attached to a Foton capsule. In addition, a device for keeping the samples at low temperatures, in the range of 10 to 100°K, even when being exposed to solar UV light, has been proposed as an important improvement (Horneck *et al.* 1989).

Of special interest to exobiology is Titan, the largest satellite of Saturn (Table 2). It is large enough to have retained a dense atmosphere mainly composed of N_2 and CH_4, but small enough to allow hydrogen to escape. The temperature is very low, close to 94°K at the surface, so that water is totally frozen out. Hence, Titan possesses a highly reducing atmosphere, where chemical evolution of organic matter occurs in the absence of liquid water (Rolfe & Battrick 1985). Infrared spectrometry on board Voyager 1 (IRIS experiment) detected several hydrocarbons (ethane C_2H_6, acetylene $HCCH$, propane C_3H_8, ethylene $H_2C{=}CH_2$, and diacetylene HCC–CCH) and nitriles (hydrogen cyanide HCN, cyanoacetylene HCC–CN, cyanogen NC=CN) (**Owen**

Table 2. Titan: the largest satellite of Saturn, is considered as a natural laboratory in our solar system, where organic chemical evolution still occurs.

Nature	Moon of Saturn Main constituent of atmosphere is N_2 Organic compounds so far detected: CH_4, C_2H_2, C_2H_4, C_2H_6, C_3H_8, C_4H_2, HCN, C_2N_2, HC_3N, CO, CO_2 Temperature is 93°K
Aspects of exobiology	Origin and evolution of atmosphere Atmospheric organic chemistry Complexity of organic molecules – adenine? Physical state and composition of surface
Approach	Space missions (CASSINI, planned launch 1997)

1982). Traces of CO and CO_2 were also found. These hydrocarbons and nitriles are probably formed from the dissociation of CH_4 and N_2, however the chemistry is not yet known in detail (Owen *et al.* 1992). The following scenario has been proposed. Photochemical reactions will take place in the upper zones of the stratosphere (Yung *et al.* 1984) and – due to the low temperature – the compounds produced will condense as aerosol clouds in the lower stratosphere. Submicron particles formed by polymerisation of acetylene, ethylene, and hydrogen cyanide in the upper stratosphere serve as nucleation centres for condensation processes leading to aerosol formation (**Scattergood *et al.* 1992**; **Toon *et al.* 1992**). At the lower altitudes methane clouds predominate in the upper troposphere, ethane clouds in lower altitudes and possibly lakes or even oceans of a methane/ethane mixture on the surface (Owen *et al.* 1992; **McKay *et al.* 1993**). Since solar UV radiation will not reach the surface of the satellite, organic polymers are suggested to accumulate on the surface.

Hydrocarbons and nitriles, similar to those detected in Titan's atmosphere, have also been produced in ground-based simulation experiments, where N_2/CH_4 gas mixtures have been submitted to electric discharges (Raulin *et al.* 1989), ionising radiation (**Gupta *et al.* 1981**), plasma irradiations (**McDonald *et al.* 1994**) or high-energy UV radiation (**Sagan & Thompson 1984; Scattergood *et al.* 1992**).

Titan has been considered as a natural laboratory in our solar system where organic chemical evolution still occurs (Raulin 1992). It allows for the ongoing study of organic chemical evolution under totally natural conditions over immense time spans. The low temperature and the reducing conditions may favour the preservation of ancient material. More insight into Titan's organic chemistry will be obtained from the Cassini mission that is scheduled to be launched in 1996/1997 for an arrival at Saturn in 2004 (**Raulin *et al.* 1984;** Horneck 1985; Kerridge *et al.* 1992). The Huygens probe contains several instruments which will collect and analyze the gas and aerosols of Titan's atmosphere during the approximately 2.5 h descent. Several questions of exobiological interest will be tackled by this mission, e.g. on the level of chemical complexity reached, such as the occurrence of polymers of HCN (e.g. adenine); on the significance of the different energy sources in the polymerisation reactions; and on the relationship of Titan's atmospheric chemistry to prebiotic chemical pathways on other planets, such as the early Earth. In preparation for the exobiological experiments of the Cassini mission and for their better interpretation, experiments simulating the organic cloud chemistry in Titan's atmosphere are required. Besides laboratory studies (Raulin 1992), experiments in space are necessary that utilise the unique radiation environment as well as the microgravity for extended studies on the behaviour of aerosols and photochemical production and destruction reactions of organic molecules. Feasible facilities will include ERA (Horneck *et al.* 1989), Biopan (Barbier *et al.* 1994) and the Gas Grain Simulation Facility (Fogleman *et al.* 1989).

7.2.2.4 Asteroids and meteorites

The asteroid belt occupies the zone between the orbits of the outer giant planets and those of the terrestrial planets. Telescopic observations from spacecraft as well as IR observations have shown that many asteroids have very dark material on their surface. These dark asteroids, which are very common particularly in the outer part of the main belt, are presumed to be the parent bodies of a special group of meteorites, the carbonaceous chondrites (for survey see Cruikshank & Kerridge 1992). Therefore, most of the information of the chemical composition of the dark asteroids is derived from the analysis of meteorites. Carbonaceous meteorites contain up to 5% organic matter (Hayatsu & Anders 1981; Mullie & Reise 1987). The dominant organic fraction is an insoluble macromolecular component resembling terrestrial kerogen. It consists of a three-dimensional array of aromatic moieties with aliphatic bridging and side chains (Cruikshank & Kerridge 1992). The soluble fraction contains many different classes of organic molecules with high structural diversity. From the Murchison meteorite, 8 of the protein amino acids have been identified among more than 70 amino acids as mixtures of D and L enantiomers in equal proportions (**Lawless & Peterson 1975;** Cronin *et al.* 1988; **Cronin 1976a & b**). However, no clear-cut evidence has been found for the presence of their polymers, the peptides (**Cronin 1976b, Ivanov *et al.* 1984**). In addition to these fundamental building blocks of proteins, some of the purine and pyrimidine bases of the nucleic acids have been identified (**Stoks & Schwartz 1981**) as well as some hydrocarbons and fatty acids (**Deamer 1985, 1986**). Although the carbonaceous chondrites that have been collected and analyzed on Earth may have been subjected to substantial chemical processing during their history (e.g. by exposure to cosmic radiation and solar flares, by local heating of the surface layers and by impact brecciation, which may have altered their organic components) they are considered a potentially important source of organics on the primitive Earth imported during the episode of heavy bombardment.

Contemporary experimental support for such a scenario of an extraterrestrial origin of organics can be found in the meteorites collected on the Earth's surface, micrometeorites found embedded deep within ice and cosmic dust sampled at stratospheric and Earth-orbit level. (for review see Horneck & Brack 1992). Micrometeorites have been extracted from deep-sea sediments (Brownlee 1981), from cryoconite (black sediment) collected from the melt zone of the Greenland ice-cap (**Shimoyama *et al.* 1979; Maurette *et al.* 1987**) and directly from Antarctic old blue ice (Maurette *et al.* 1988; **Maurette *et al.* 1991**). In the latter study a high percentage (30%) of unmelted chondritic micrometeorites from 0.1 to 1 mm in diameter has been repeatedly observed. This indicates that many particles cross the terrestrial atmosphere without drastic modification by thermal treatment. Most of the unmelted micrometeorites are composed of porous aggregates of granular units embedded in an amorphous matrix most likely made of a carbon/oxygen-rich pyrolyzable material.

In the stratosphere, cosmic dust particles of 5 μm to 50 μm in diameter have been recovered by use of high-altitude aircraft (Brownlee 1978). Mineralogical and chemical analysis of these particles revealed a wide diversity of extraterrestrial particles. On the basis of major-element composition related to the mineral phase, the most common particles are referred to as being 'chondritic', which have major and minor element compositions similar to those of carbonaceous chondrites. Unfortunately, they remain poorly characterised. Raman spectra and associated luminescence observations indicate the presence of highly disordered carbons in chondritic particles. Two classes of particles can be distinguished: 'hydrated' particles which comprise minerals that were in contact with liquid water, and 'anhydrous' ones. Comets and asteroids are proposed to be the major sources of cosmic dust; however their relative contributions to cosmic dust are difficult to quantify (Leinert & Grün 1989).

Surfaces returned from space, such as Apollo and Skylab spacecraft windows and thermal blankets and aluminium louvres from Solar Max (Warren *et al.* 1989) provide proof that analyzable particle residues can be returned from low Earth orbits. On the Salyut 7 Space Station in the experiment COMET, dust impacts were collected in October 1985 at an altitude of 350 km when the Earth was crossing the swarm associated with the Giacobini-Zinner comet (**Bibring *et al.* 1988**). Dust was also collected on the NASA Long Duration Exposure Facility LDEF which stayed for nearly 6 years in space (Amari *et al.* 1991). X-ray and ion mass spectrometry analyses showed different classes of grains, identified by their chemical composition. Although space debris of terrestrial origin accounts for the most numerous impacts, many impacting grains have a composition compatible with an extraterrestrial origin. Among these, some have a carbonaceous chondrite-type composition. Several ionic spectra indicate the presence of carbon associated with other light elements. They resemble the spectra of 'CHON-rich' grains detected in the coma of comet Halley (**Langevin *et al.* 1987**). The presence of CHON-rich grains in a cometary swarm indicates that they can survive in the solar UV flux for thousands of years.

Even today, cosmic dust import is not a rare event on Earth. About 10,000 tons of cosmic dust reach Earth's atmosphere each year with a flux of 1 particle $m^{-2}\ y^{-1}$ for 100 μm particles and 1 particle $m^{-2}\ day^{-1}$ for 10 μm particles (**Grün *et al.* 1985**). The collection of cosmic dust in the stratosphere or in Earth orbit constitutes an attractive way for the *direct* measurement of organic molecules of potentially biological importance reaching the Earth's vicinity. Once returned to ground, the samples can be subjected to detailed analysis (mineralogical, elemental, isotopic, molecular, and structural). For this purpose, it is required to reach minimal alteration of the physical and chemical nature of the dust particles during the collection procedure.

7.2.2.5 Terrestrial planets

The formation of the terrestrial planets in the inner part of the solar nebula is best explained by the 'two component accretion model', which was originally developed for the Earth, but may also have some validity for other terrestrial planets (**Wänke 1991**). According to this model, the accretion of the planets started with highly reduced material (component A), which condensed from the solar nebula at elevated temperatures and was therefore devoid of volatiles (**Ringwood 1984; Wänke & Dreibus 1988**). These so-called 'rocks' formed the cores of the planets. The accretion material became progressively more oxidised and richer in moderately volatile elements (component B). Finally, at a late stage, highly volatile elements and molecular species such as water and hydrocarbons were added to the surface. The model assumes a radial mixing of planetesimals in the solar system during the accretion process, caused by outer giant planet orbit perturbations. This would result in substantial exchange of material formed in different regions of the nebula. Hence, substantial amounts of water and organics were imported to the inner planets during this late stage of accretion. It has been suggested that all volatiles of the biosphere, or at least a substantial fraction of them, were brought to the early Earth by comets and other primitive solar system bodies during the period of heavy bombardment (e.g. **Delsemme 1992**, see also Section 7.2.2.2). On the other hand, it has also been argued that collisions between large fast comets and the Earth may have eroded the terrestrial atmosphere including the organic compounds several times in the early history of the Earth (**Melosh & Vickery 1989**). An alternative model suggests that equilibrium partitioning between the core and mantle occurred even during core formation (**Kargel & Lewis 1993**).

Alternatively, it has been suggested that the origin of organics on terrestrial planets is from their production on the surface of the early planet. In model experiments, simulating putative primitive Earth conditions, nearly all the biogenic building blocks of life have been successfully synthesised. They include seventeen of the twenty amino acids that occur in proteins, purine and pyrimidine bases which are the nitrogenous compounds found in nucleotides, and lipid-like materials able to form membranes (for reviews see Dose & Rauchfuss 1975; **Dickerson 1978; Dose 1987**; Brack & Raulin 1991). The first historical experiment was done in 1952, when Miller exposed a mixture of methane (CH_4), ammonia (NH_3), hydrogen (H_2) and water (H_2O) to electric discharges and identified four of the twenty proteinaceous amino acids among the compounds formed (**Miller 1953**). This laboratory synthesis of amino acids is only efficient if a reducing gas mixture containing significant amounts of hydrogen is used. Although the true composition of primitive Earth's atmosphere remains unknown, geochemists now favour a non-reducing atmosphere dominated by CO_2, N_2 and H_2O (**Levine *et al.* 1982**). In non-reducing atmospheres, the production of organic nitrogen-containing compounds appears to be inhibited (**Schlesinger & Miller 1983**), but inorganic compounds of N such as NO_3- and NO_2- are readily synthesised (**Manci-**

nelli & McKay 1988). If the early Earth's atmosphere was dominated by CO_2 and N_2, then the methane-rich atmosphere postulated by Miller would not be a possible route to amino acids on Earth.

7.2.2.6 Conclusions

Our knowledge of the chemical evolution of potential precursors of life is mostly derived from three sources: radioastronomical molecular spectroscopy, comparative planetology and laboratory organic chemistry experiments under simulated extraterrestrial or early Earth conditions. Space missions that were designed to explore our solar system such as Voyager I and II to the giant planets, Vega to comet Halley and Venus, Giotto to Halley and Giacobini–Zinner, as well as collection platforms in Earth orbit, have provided new and sometimes surprising results. They have revealed that organic molecules are frequent in many places of our Galaxy such as in the dense clouds of the interstellar medium, in comets and other small bodies of our solar system, and in the atmospheres of the giant planets and of Titan. But little is known about their molecular structure, complexity and the pathways of their formation and destruction. A fundamental question to be tackled is their role in the chain of events leading to life on Earth or, potentially, on other celestial bodies. Specifically there are uncertainties concerning the nature and extent of similarities in the organic compounds found in the different cosmic sources; and the fate of organic matter during the formation of the planets of our solar system, especially the physical properties and thermal history of the solar nebula and of the bodies formed during the accretion and post-accretion phases.

7.3 Prerequisites for the emergence of life

Exobiological concepts generally assume that life emerges at a certain stage of either cosmic or planetary evolution if the right environmental physical and chemical requirements are provided (for review see Klein 1986; Horneck 1989). On Earth, most of the early geological record has been erased by later events so that we remain ignorant of the true historical facts concerning the origin of life on this planet. On the basis of general commonalities of contemporary terrestrial life, Oro (**Oro *et al.* 1982**) formulated 25 criteria for the emergence and evolution of life in a universal context which he divided in 5 areas: ***stellar*** (single stellar star, planetary system, elemental composition, mass of the star, lifetime of the star), ***planetary*** (mass, orbit, atmosphere, discrete liquid sphere, surface), ***chemical*** (solvents, composition and concentration, energy source, redox potential, pH range), ***protobiological*** (replication of informational molecules, stereospecific catalytic molecules, information transfer molecules, polymerising molecular assembly, interphasic molecular assembly), as well as ***evolutionary*** criteria. Above all, a planetary system with a single central star of sufficient life-time (at least 5×10^9 years) was deemed necessary. This period includes the evolution of 'intelligent' species. For the emergence of life on a planet time-spans of 0.5 to 1×10^9 years seem to be sufficient (Schopf 1983; Schidlowski 1993a & b, see also Section 7.4.1.1). A further important precondition for life to emerge is the availability of an atmosphere (**Owen 1985**). This poses special requirements on the planetary mass and orbit in order to retain a sufficiently dense atmosphere. In addition, temperature and pressure should be in such a range to provide a hydrosphere and a solid surface. Assuming a tolerable temperature range between about -100°C and +100°C, our solar system would provide a habitable zone in an orbit between 0.7 and 2.0 astronomical units (AU) (Goldsmith & Owen 1980/84). Venus, Earth and Mars are situated in this habitable zone. A recent more conservative estimate based on a one-dimensional climate model reduces the width of the habitable zone to the range of 0.95 to 1.37 AU (**Kastings *et al.* 1993**), where climate stability is ensured by feed-back mechanisms.

With regard to the chemical prerequisites, the availability of the so-called 'biogenic' elements CHONSP and relevant 'biogenic' organic compounds are considered to be indispensable, as well as the presence of liquid H_2O (**Oro *et al.* 1982**). The biogenic elements, which make up the bulk of terrestrial biomass, are among the most abundant elements in the Universe (see Mason 1991). Whether the organic starting material relevant to the origin of life comes from *in situ* production on the planet or from importation of extraplanetary sources is still an open and one of the principal questions in exobiology (see Section 7.2). Liquid water is considered as an indispensable prerequisite for the emergence of life. Due to its unique properties, such as the formation of hydrogen bonds, a high polarity and a high dielectric constant, it serves as solvent, as reactant, and as stabiliser of the tertiary structure of complex organic molecules and their organisation into higher aggregates (Lang 1986; **Banin & Mancinelli 1994**). The encounter of abiotically formed organic molecules with liquid water has been considered as one of the decisive steps towards the emergence of life (Horneck & Brack 1992). It was suggested that the special properties of liquid water may generate interesting prebiotic conditions where organic molecules may manage the conflict between hydrophobic groups (e.g. hydrocarbons) and hydrophilic groups (e.g. CHONS) and thereby aggregate to higher complexes (**Brack 1993**). Some authors consider that clays and other minerals with their catalytic surfaces may be involved in the process of selection and concentration of special organic compounds and in their condensation to polymers of increasing complexity (e.g. Cairns-Smith 1982; **Bonner *et al.* 1985**; **Holm 1985**; **Wächtershäuser 1990, 1991, 1994**).

As an energy source effective to drive a prebiotic organic chemistry on a planet, solar (stellar) UV radiation as well as catalytic or latent chemical energy

has been considered as the prevalent form (**Oro *et al.* 1982**). However in the laboratory other energy sources such as electric discharges and ionising radiation are more effective in prebiotic production of organic compounds (e.g. **Miller 1953**).
These criteria, which are mutually interdependent and which have been developed for life as we know it to appear and evolve on our planet, will be used in the following Sections in the search for potential extraterrestrial habitats in our solar system.

7.4 Evolution of life

7.4.1 Early biological evolution

7.4.1.1 Fossil record

Palaeobiology has traced back the history of life on Earth to its very early stages in a geochronological manner (for reviews see Schopf 1983; Schopf & Klein 1992; Schidlowski *et al.* 1992). This is based on the fact that under suitable conditions in their primary sediment, dead organisms are susceptible to fossilisation and thereby leave a morphological fingerprint in sedimentary rocks. The search for fossils spans over a time-period of 3.8 Ga (10^9 years), up to the first evidence of sedimentary rocks. In Precambrian sediments, which are older than 0.6 Ga, microfossils prevail giving evidence of a 'microbial world' for the first billion years of life's history. For this Precambrian microbiota a total of 106 taxa have been described from 236 occurrences in 71 sediment formations (**Schopf 1993**). However, as one goes back in time more than 2.5 Ga to the Archean, only a few sedimentary rocks have survived without alteration. Therefore, it is sometimes difficult, to establish the authenticity of Archean microfossils. According to **Schopf (1993)** microfossils must fulfil the following criteria: (i) they must occur in rocks of known provenance and established Archean age; (ii) they must be demonstrably indigenous to and syngenetic with the primary deposit; (iii) and their biological origin has to be proven. Among the oldest occurrences that satisfy these requirements are the about 3.45 Ga old Fig Tree Group of South Africa possessing non-septate bacterium-like filaments (**Walsh & Lowe 1985**) and the 3.3 to 3.5 Ga old Warrawoona Group of Western Australia with the Apex Chert. In these formations eleven different taxa of cellularly preserved filamentous microbes were discovered which morphologically resemble contemporary oscillatoriacean cyanobacteria (**Schopf 1993**). Also among the oldest are the sedimentary rocks of Towers formation containing sheath-enclosed unicellular aggregates (**Schopf & Packer 1987**). Additional palaeobiological evidence related to microbial life in the Archean includes the stromatolitic deposits, e.g. within the Warrawoona Group, which represent macroscopic fossilised stacks of successively superimposed benthic microbial mats (**Walter *et al.* 1980**).

The frequent occurrence of stromatolites in several Archean formations and the taxonomic diversity of the Archean microfossils give evidence of a highly diversified early Archean microbial life having evolutionary roots at a geological time probably much earlier than 3.5 Ga ago. In this respect it is interesting to note that cell-like morphologies have been identified even in the oldest sedimentary rocks known, that is the 3.8 Ga old Isua metasedimentary suite of West Greenland (**Pflug & Jaeschke-Boyer 1979**). Although their biogenic origin has been questioned - especially because microfossils might not survive the high temperatures and pressures the rocks were subjected to during several episodes of metamorphism (**Bridgewater *et al.* 1981**) - the high content of reduced carbon in the Isua suite (**Pflug 1987**) and its isotopic composition (**Schidlowski 1988**) suggests, but does not prove, the existence of microbial ecosystems already in Isua times (Schidlowski 1993a). Because all common biological pathways of autotrophic carbon fixation discriminate against ^{13}C (**Park & Epstein 1960**), the measurement of the $^{13}C/^{12}C$ fractionation has been used as a means to discriminate between biogenic (organic) carbon and sedimentary carbonate in the deposits (Schidlowski *et al.* 1975). This depletion of ^{13}C is quite conservatively transcribed from the extant biomass through recent marine sediments over billions of years into the Archean period, and with a slight modification even back to the 3.8 Ga old Isua formation (**Schidlowski *et al.* 1979**). This relatively constant $^{13}C/^{12}C$ fractionation of sedimentary organic carbon over 3.8 Ga is indicative of a high degree of evolutionary conservatism in autotrophic carbon fixation, which already existed 3.8 Ga ago (Schidlowski 1993b). Additional evidence of an autotrophic, and probably photosynthetic, carbon fixation in Archean microbiota is derived from the morphological similarity of microfossils with contemporary cyanobacteria (**Schopf 1993**).

To sum up, it is seen that on current available palaeobiological and geochemical data, there is convincing evidence that life has emerged very early on the juvenile Earth - with a degree of certainty earlier than 3.5 Ga ago and probably earlier than 3.8 Ga ago. Autotrophic and probably photosynthetic carbon fixation has been extant since at least 3.8 Ga and therefore must have evolved in much older times than are covered by the terrestrial rock record.

7.4.1.2 Molecular record

Molecular biology reveals a fundamental unity of modern life. All extant organisms are cellular, the genetic information is stored in the DNA, transcribed into RNA, and translated into proteins. All organisms use the same or a very similar genetic code and they use the same amino acids in their proteins. Although there are differences in the transcription and translation machinery, the basic process is very similar in all organisms. All cells use lipid membranes to separate their protoplasm from the environment or from the cell wall. They use the same energy-rich metabolites as well as homologous enzymes to energise their cell membranes. This commonality of the basic biochemical features to all known forms of life on Earth suggests their descent from a universal ancestor.

In order to trace the history of life back from extant forms to the universal ancestor, molecular phylogeny makes use of the fact that at the level of the genotype changes constantly occur which are fixed randomly in time (Kimura 1983). By comparing the genotypic information stored in the sequence of nucleic acids and proteins (**Zuckerkandl & Pauling 1965**), the first unrooted universal phylogenetic tree has been constructed that groups all known organisms in three domains: Bacteria, Archaea and Eucarya (**Woese 1987; Woese *et al.* 1990**) (Fig. 2). The criteria for the selection of a biological macromolecule as a molecular chronometer were defined as follows: no lateral genetic transfer between species, function is as strictly conservative as possible, sufficient number of residues, and rate of change commensurate with the evolutionary distance considered (**Pace *et al.*, 1986**). Among others, the molecules 16S rRNA (**Woese & Olson 1986; Woese 1987; Olson & Woese 1993**), DNA-dependent RNA polymerases (**Pühler *et al.* 1989; Klenk *et al.* 1994; Klenk & Zillig 1994**) and proton pumping ATPases (**Gogarten & Taiz 1992; Kibak *et al.* 1992**) fulfil these requirements and their analysis has provided an increasingly large set of sequences. Most data, obtained even by different methods of sequence comparison (see **Woese 1987; Klenk *et al.* 1994**) and from different molecular markers appear to place the root of the universal tree of life concurrently between the Bacteria on the one side and the Archaea and Eucarya on the other side (Fig. 2) (**Gogarten *et al.* 1989; Iwabe *et al.* 1989; Wheelis *et al.* 1992; Rivera & Lake 1992**).

The nature of the common ancestor of the three domains of life has been discussed on the basis of the many shared characters of extant life and on the results of the molecular analyses. Studies of ancient gene duplications (**Gogarten & Taiz 1992**) particularly suggest that the last common ancestor already possessed a variety of complex enzymatic and regulatory processes, such as the membrane energising ion translocation ATPases already as multi-subunit enzymes, different paralogous elongation factors, two types of ^{met}tRNA, heat shock protein homologues, glutamine synthases, and dehydrogenases for glutamate, malate and lactate (**Gogarten-Boekels *et al.* 1995**). It should concur with (i) the clear distinction in several specific features of Archaea and Bacteria, such as the lipids of the membranes (**DeRosa & Gambacorta 1986**) and the DNA-dependent RNA polymerases (**Zillig *et al.* 1989**); (ii) the striking resemblance in other features of the two domains, such as certain metabolic pathways (**Danson 1988**) and the mechanisms and compo-

▶

Fig. 2. Universal phylogenetic tree, unrooted based on 16S-like rRNA sequences (A) (after Pace *et al.* 1986) and rooted, based on the sequence of 16S/18S rRNA, ATPases and protein elongation factor (B) (after Kandler 1993). For the rooted tree, the non-catalytic subunit of the *Escherichia coli* ATPase has been used as outgroup to root the tree; the Archaea form a sister group to the Eucarya, whereas the Bacteria constitute a second independent lineage coming from the root.

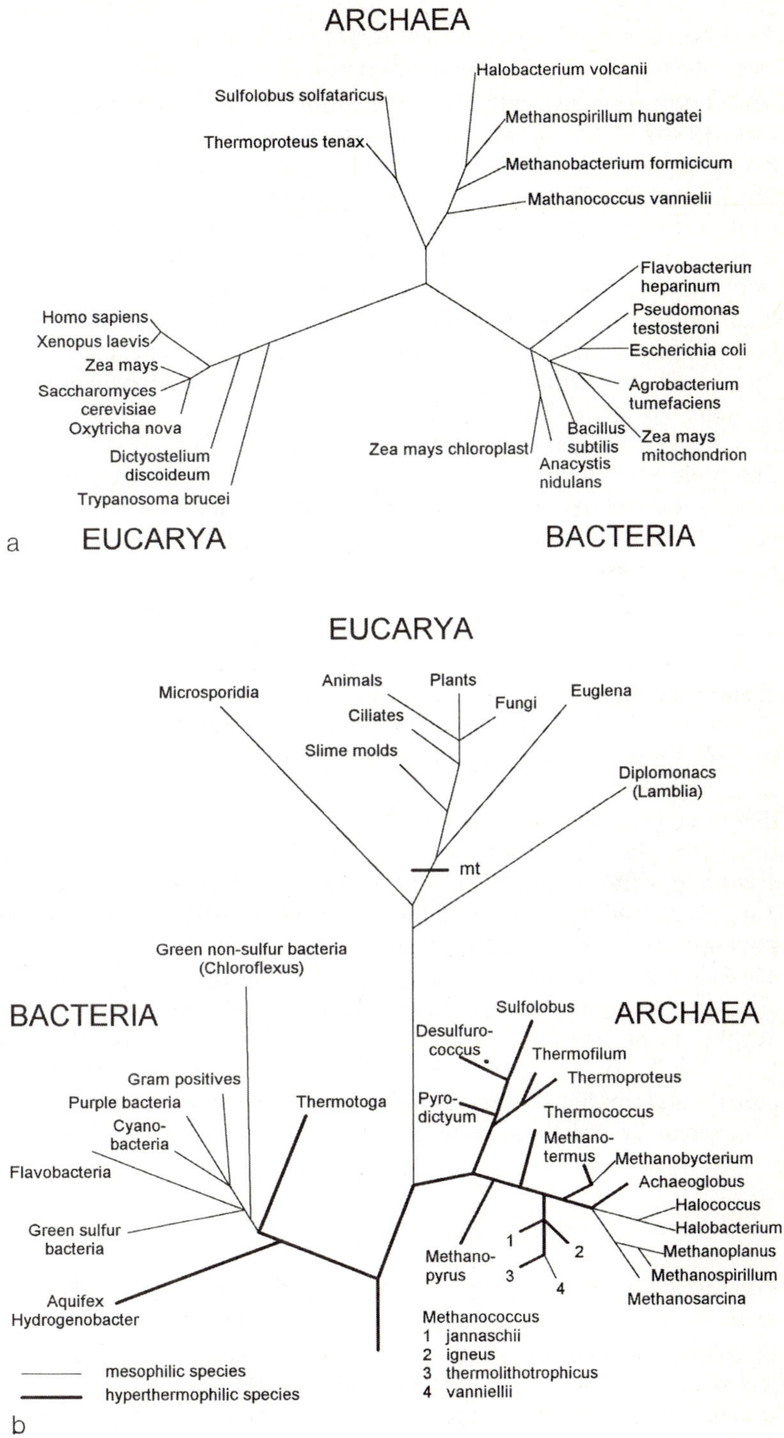
ARCHAEA
Halobacterium volcanii
Sulfolobus solfataricus
Methanospirillum hungatei
Thermoproteus tenax
Methanobacterium formicicum
Mathanococcus vannielii
Flavobacterium heparinum
Homo sapiens
Pseudomonas testosteroni
Xenopus laevis
Escherichia coli
Zea mays
Saccharomyces cerevisiae
Agrobacterium tumefaciens
Oxytricha nova
Bacillus subtilis
Zea mays mitochondrion
Zea mays chloroplast
Dictyostelium discoideum
Anacystis nidulans
Trypanosoma brucei
a
EUCARYA
BACTERIA
EUCARYA
Animals
Plants
Microsporidia
Fungi
Euglena
Ciliates
Slime molds
Diplomonacs (Lamblia)
mt
Green non-sulfur bacteria (Chloroflexus)
BACTERIA
Sulfolobus
ARCHAEA
Desulfurococcus
Thermofilum
Thermoproteus
Gram positives
Purple bacteria
Thermotoga
Pyrodictyum
Thermococcus
Cyanobacteria
Methanotermus
Methanobycterium
Flavobacteria
Achaeoglobus
Halococcus
Green sulfur bacteria
Halobacterium
Methanoplanus
Methanopyrus
Methanospirillum
Aquifex Hydrogenobacter
Methanosarcina
Methanococcus
1 jannaschii
2 igneus
3 thermolithotrophicus
4 vanniellii
mesophilic species
hyperthermophilic species
b

nents involved in gene replication and expression (Zillig *et al.* 1992); and (iii) the entire absence of transitional forms between the domains.

Two alternative pictures of the common ancestor have been suggested: (i) a prokaryotic cell not fundamentally different from present day prokaryotes, that already used DNA, RNA, ribosomes, energy conserving membranes and a variety of sophisticated regulated biochemical pathways **Gogarten & Taiz 1992; Gogarten–Boekels *et al.* 1995**), and (ii) an ancestral gene pool as a population of cells or pre-cells with easy genetic communication among the individuals by horizontal transfer of genes and gene clusters. This ancestral gene pool would already dispose of the main biochemical features common to all today's organisms, such as DNA for the storage of genetic information and the sophisticated protein synthesis machinery (Zillig *et al.* 1992; Kandler 1993). Separation in subpopulations (e.g. pre-bacteria and pre-archaea) would become irreversible, when the diversification of the genetic systems made them incompatible for further genetic communication (Zillig *et al.* 1992).

The molecular record allows also for inferences on the metabolic characteristics of the common ancestor. Thermophilic and hyperthermophilic organisms (**Stetter *et al.* 1990**) occur in numerous phylogenetically highly divergent lineages and could therefore represent an ancient phenotype (**Woese 1987**). In the phylogenetic tree derived from 16S rRNA sequence comparison, hyperthermophiles are among members of both domains, Archaea and Bacteria, that form the deepest branchings (**Woese *et al.* 1990**). They grow either chemolithoautotrophically, employing the redox couples H_2/S^o, H_2/CO_2, H_2/O_2, H_2S/O_2, or H_2/SO_4 as energy sources or organotrophically by sulphur respiration (**Stetter *et al.* 1990**; Stetter 1992; **Segerer *et al.* 1993**). These observations support the assumption of a chemolithoautotrophic hyperthermophilic nature of the common ancestor (Kandler 1993). This organism would depend solely on planetary energy. It is interesting to note that photoautotrophic growth is completely absent among the thermophiles making photoautotrophy a very unlikely feature of the common ancestor (Kandler 1993). However, phylogenetic trees recently reconstructed on the basis of RNA polymerases as molecular marker place the bacterial lineage of hyperthermophiles far away from the root of this domain (**Klenk *et al.* 1994**). Hence, the 'hot pond' appears to be not necessarily the environment of an Archean biosphere. On the other hand, it may have provided refuge for early cellular life to survive global catastrophic environmental changes, such as heavy meteorite impacts (**Gogarten–Boekels *et al.* 1995**).

7.4.1.3 Hypothetical scenario of an early Archean biosphere

The fossil record reveals that microbial autotrophic ecosystems existed on the early Earth already by 3.5 Ga or even 3.8 Ga ago (see Section 7.4.1.1). In order to reconstruct the conditions and events during this epoch or even before, pieces of information have been linked together from different sources, such as palaeontology (e.g. Schopf 1983), molecular phylogeny (Kandler 1993), and

geology (**Maher & Stevenson 1988; Oberbeck & Fogleman 1990; Chyba 1991**). Since on Earth the most ancient rocks known are 3.8 Ga old (**Moorbath *et al.* 1973**) geological data from Moon, Mercury and Mars were added to the picture. The crater counts on these celestial bodies show that the early inner solar system was subjected to an intense heavy bombardment by planetesimals, asteroids, and comets that lasted until approximately 3.8 Ga ago (**Maher & Stevenson 1988**). These impactors would on the one hand, have been delivering the volatiles including water and organic precursors of life (see Section 7.2.2), on the other hand, if sufficiently large and fast, would have eroded the atmosphere (**Melosh & Vickery 1989**). It has been extrapolated from lunar crater records, that during this Hadean period (pre-3.8 Ga ago) the Earth was struck several times by gigantic impacts sufficient to vaporise the entire terrestrial ocean (**Sleep *et al.* 1989; Oberbeck & Fogleman 1990**). As a result of a runaway greenhouse effect, surface temperatures up to 2000°K have been suggested (Chyba 1992) which would have certainly sterilised the Earth. This impact catastrophe scenario implies that, if life did already exist in the Hadean, it may have been extinguished several times, until the end of the heavy bombardment (**Oberbeck & Fogleman 1990**). It is also in agreement with the model of an extreme thermophilic prokaryotic species as last common ancestor, that survived the last nearly complete impact frustration of the Earth and gave rise to the evolution of life on Earth to its present-day variety (**Gogarten-Boekels *et al.* 1995**).
Temperature models of the early Earth (**Kasting & Ackerman 1986**) predict surface temperatures in the range of 100°C in the Hadean that declined to 73°C by 3.5 Ga (**Oberbeck & Mancinelli 1994**). This high temperature would have favoured hyperthermophiles forming the first microbial communities. Whereas this supposition of hyperthermophiles as earliest microbial forms is consistent with the most recent phylogenetic tree (**Woese 1987; Stetter *et al.* 1987**; Kandler 1993), there are controversial views on the metabolic state of those organisms which may have been heterotrophic (**Oberbeck & Mancinelli 1994**), chemolithoautotrophic (Kandler 1993) or photoautotrophic (Schidlowski 1993b). Recently, a theory of a thermophilic chemolithoautotrophic origin of life has been worked out (**Wächtershäuser 1988, 1990, 1991, 1992, 1994**) which is based on pyrite formation as source of reducing power for CO_2-reduction and organic synthesis. The reactions in this primordial biochemistry are homologous to the reductive citric acid cycle. This view of the evolution of biochemical pathways may provide new ideas to the field of prebiotic and early biological evolution.

7.4.2 Possible habitats on other bodies in the solar system

7.4.2.1 Mars

The planets in our solar system that are most similar to Earth are Mars and Venus (Goldsmith & Owen 1980/1984). Mars with a mean distance to the sun of 1.52 AU is located at the outer border of the habitable zone around the sun which is estimated under the premise of the presence of liquid water on the planet's surface at some time during its 4.5 Ga history (**Kasting et al. 1993**, see also Section 7.3) With the exception of the Earth, Mars is by far the most intensively studied of the planets. So far, 13 spacecraft have been sent to Mars (US missions since 1965: Mariner 4, 6, 7, 9; Viking 1 and 2; Soviet missions since 1971: Mars 2, 3, 4, 5, 6, 7; Phobos) that either flew by, landed on, or orbited the planet (summarised in Carr 1981). In 1972, for the first time a spacecraft, Mariner 9, passed over the younger parts of Mars revealing a wide variety of geological processes, indicated by volcanoes, canyons, and channels that resemble dry river beds (Carr 1981). These observations were confirmed by the Viking 1 and 2 extensive mapping of Mars (**Soffen 1976**; **Snyder 1979a, b**; Carr 1981). Three major classes of channels were identified (**Sharp & Malin 1975**): (i) dendritic run-off channels and valley networks over tens to hundreds of kilometres primarily associated with the ancient cratered terrain (3.8 or more Ga old) and which are difficult to reconcile with any origin other than liquid water (**Masursky *et al.* 1977**; Carr 1981, **1987**), (ii) outflow channels suggestive of large-volume flows in cataclysmic events (**Baker & Milton 1974**; **Lucchitta *et al.* 1981**; Carr 1981), and (iii) fretted channels which are steep-walled with smooth flat floors suggestive of erosions by debris flows (**Sharp 1973**; **Squyres 1978**).

Understanding the history of water on Mars appears to be one of the clues to the puzzle on the probability of life on Mars (**Squyres 1989**; **Helfer 1990**; **McKay 1991**; McKay *et al.* 1992) (Table 3). The estimates of the total amount of water that may have existed at the surface of Mars range over two orders of magnitude. A low amount of water ranging from 3.6 to 133 m is suggested from the composition of the contemporary atmosphere, e.g. the D/H ratio (**Yung *et al.* 1988**; **Owen *et al.* 1988**; **de Bergh 1993**), from the impact craters in softened terrains (**Jankowski & Squyres 1993**), and the composition of the

Table 3. Mars: geological evidence that liquid water was once abundant on the Martian surface is the primary motivation for considering the possible origin of life on Mars.

Nature	Terrestrial planet Temperature range is 140 to 295°K Main constituent of atmosphere is CO_2 at a pressure of 590 to 1500 Pa Hydrosphere: permafrost layer, polar caps (rivers, oceans [?] in the past)
Aspects of exobiology	Periods of liquid water on Mars Role of solar UV radiation on surface processes Chance of life to have evolved and expanded on the planet Candidate areas to search for life (relics, contemporary forms, precursors)
Approach	Mission to Mars: phase 1 to collect data on geology and climate and their evolution; phase 2 to search for traces of life

SNC meteorites - a class of meteorites that probably originated from Mars (**Wänke & Dreibus 1988; Gooding 1992**). On the other hand, the geological flow features, such as the channels and valley network (**Pieri 1976, 1980;** Baker 1982), the debris flows (**Carr & Schaber 1977**; **Squyres 1978; Lucchitta 1984),** terrain softenings (**Squyres & Carr 1986**), and fretted terrain (**Sharp 1973)** provide evidence of abundant water at the surface of Mars, at least at some time in the past, assuming a global inventory of water of at least 440 m (reviewed by **Carr 1987**; **Baker *et al.* 1991**; **Carr & Wänke 1992**; Janle 1993). For comparison, the Earth has 2700 m of water on its surface. These uncertainties in the inventory of water on Mars make it difficult to model the Martian climate throughout its history. The climatic conditions required for the formation of the valley networks are especially poorly understood (Carr 1989). Despite of these difficulties, the most plausible explanation is a relatively thick CO_2 atmosphere in the early history of Mars (**Pollack *et al.* 1987**; **Carr 1987**). To allow for mean global surface temperatures above the freezing point of water, an enhanced greenhouse effect is required with an atmospheric pressure ranging between 1 and 5 bar (**Postawko & Kuhn 1986**). Because atmospheric CO_2 would be rapidly fixed as carbonates in the rocks, recharging of the atmosphere with CO_2 by volcanism (**Pollack *et al.* 1987**) or by meteorite impact (**Carr 1989**) has been postulated.

Based on current knowledge and models of the duration of liquid water on Mars (**McKay & Davis 1991**) a scenario of the history of water on the surface of Mars has been suggested which distinguishes four epochs (McKay *et al.* 1992). Epoch I reaching up to the end of the heavy bombardment about 3.8 Ga ago with warm surface and liquid water. Epoch II (about 3.8–3.1 Ga ago) with the temperature fluctuating around the freezing point of water and ice-covered lakes. Epoch III (about 3.1–1.5 Ga ago) with the temperature consistently remaining below freezing and liquid water in porous rocks. Epoch IV (about 1.5 Ga to the present) with pressure at the triple point preventing the presence of liquid water at the surface. For each of these epochs, the probability for indigenous life is discussed taking into consideration the requirements for the emergence of life as defined in Section 7.3 and using terrestrial ecosystems as models for Martian habitats (McKay 1986; Friedmann 1986; **Rothschild 1990**; **Helfer 1990**; Schidlowski 1992; Mancinelli *et al.* 1992; Schwartz *et al.* 1992; **Banin & Mancinelli 1994**; McKay *et al.* 1992).

From a careful analysis of the chemical and physical conditions of early Mars based on a chemical-elemental composition model for the Mars surface (Banin *et al.* 1992) and models of the Martian palaeoclimate (**McElroy *et al.* 1977**; **Anders & Owen 1977**; **Clark & Baird 1979; Carr 1989**; Durham *et al.* 1989) **Banin & Mancinelli (1994)** conclude that the main putative prerequisites for life to arise did exist during epoch I. By analogy with the early Archean biosphere on Earth (Schopf 1983; Kandler 1993; Schidlowski 1993b; **Oberbeck & Mancinelli 1994**, see also Section 7.4.1.3) an early Martian biosphere could be postulated with habitats and microenvironments similar to those on the early Earth (McKay 1986; **Scott *et al.* 1991**; McKay *et al.* 1992). The major uncertainty seems to be whether liquid water was available long

and abundantly enough for life to arise. Therefore, in the future exploration of Mars, focus for exobiological research should be to identify areas of 'palaeontological interest' (Chicarro *et al.* 1989) by (i) getting a better knowledge of the earlier existence of periods of more moderate climatic conditions, characterised by flowing water, higher and constant temperatures and a denser atmosphere, (ii) getting a better understanding of possible cyclic conversion of CO_2, N_2, O_2, H_2O, between soil/carbonate rocks and atmosphere, (iii) identifying suitable sites for exobiological research, e.g. palaeolake deposits (**Scott *et al.* 1991**) or thermal spring deposits (**Walter & Des Marais 1993**), and (iv) then looking for traces of past Martian life, including evidence in chemical (Schwartz & Mancinelli 1989; **Kanavarioti & Mancinelli 1990**; Mancinelli *et al.* 1992; Weckwerth & Schidlowski 1995), isotopic (Rothschild & Des Marais 1989; Schidlowski 1992), morphological sources, including microfossils and stromatolites (McKay 1986, 1992; **Helfer 1990**).
Models suggest that at the end of epoch I atmospheric CO_2 was irreversibly lost due to carbonate formation, and the pressure and temperature then declined (McKay *et al.* 1992). Applying a model of lakes in Antarctica, **McKay & Davis (1991)** have estimated for epoch II a period of 700 +/- 300 My during which ice-covered lakes could have persisted and provided liquid water habitats on early Mars, analogous to ice-covered lakes in Antarctica (McKay 1986; **McKay & Stoker 1989**; **Parker *et al.* 1982**) or cryoconite holes on glaciers (**Wharton *et al.*** 1985, 1989). These terrestrial analogues contain planktonic organisms as well as an abundant benthic community forming microbial mats (**Wharton *et al.* 1983**). Many of these mats develop into stromatolites (Awramik *et al* 1976), organosedimentary structures indicative of phototrophic organisms which are readily preserved on Earth, the oldest dating back more than 3 Ga (**Walter *et al.* 1980**). Such ice-covered lakes might have served as niches for putative life on Mars to retreat by providing both thermal stability against a cooling external environment and enhanced concentrations of CO_2 and N_2 against a thinning atmosphere (**McKay 1991**). Therefore, future Mars missions should search for possible palaeolake deposits on Mars, from orbiters as well as from ground vehicles, which are suggestive of bearing fossil evidence of possible early life on Mars (McKay 1992). Also high resolution studies of the polar cap layers may give insight in the history of the climate on Mars.
The model of **McKay & Davis (1991)** suggests that with gradually decreasing pressure and temperature the ice-covered lakes would eventually have dried out due to lack of melt water supply thereby initiating epoch III where liquid water would be restricted to porous rocks (McKay *et al.* 1992). In order to cope with this dramatic environmental change, the following adaptive steps have been suggested (Friedmann & Koriem 1989): (i) evolutionary selection for psychrophily and xerotolerance, (ii) withdrawal to protected niches, e.g. endolithic habitats. Such cryptoendolithic microbial communities occur on Earth in areas of extreme aridity and frigidity, e.g. dry valleys of Antarctica (**Friedmann & Ocampo 1976**; **Friedmann 1980, 1982**; **Friedmann *et al.* 1987**; **Hirsch *et al.*** 1988; **Siebert & Hirsch 1988**). They form lichen-dominated eco-

systems with psychrophilic nearly exclusively eukaryotic algae and less commonly cyanobacteria as primary producers and psychrotolerant fungi as consumers. They are proposed as a suitable model for the study of processes during the extreme environmental stress on Mars during the cooling-phase of the atmosphere at epoch III (McKay *et al.* 1992). Other potential biotic oases, to which the putative life on Mars might have withdrawn during epoch III, are the polar ice caps and permafrost regions (**Gilichinsky *et al.* 1992, 1993**), submarine hydrothermal vents or other hydrothermal areas in connection with volcanic activities (Ivanov 1989; **Boston *et al.* 1992**), and endoevaporites, i.e. microorganisms that live in evaporites, e.g.. halite or gypsum (**Rothschild 1990**). Such sheltered oases, as first postulated by **Lederberg & Sagan (1962)**, may have provided microhabitats to enable well adapted life to persist on Mars even today.

Prior to their launch in 1975, the pioneering Viking 1 and 2 lander missions to Mars that were designed to address among other scientific issues the question of extant life on Mars (**Klein 1974; Young 1976**), initiated a heyday of microbial studies in simulated Martian environment (**Packer *et al.* 1963; Sagan *et al.* 1968; Hawrylewicz *et al.* 1964**, 1967, 1968; Young *et al.* 1964; **Horowitz *et al.* 1967, 1972**; Lozina-Lozinsky & Bychenkova 1969; Imshenetsky *et al.* 1970, 1973, 1977; Hagen *et al.* 1970; Zaar *et al.* 1970; Lozina-Lozinsky *et al.* 1971; Vishniac & Mainzer 1973; Aksenov *et al.* 1973; **Mitz 1974**). As a result, sophisticated assay systems and techniques for searching for the presence of putative extant biota on Mars were elaborated (**Vishniak 1960**; Levinthal *et al.* 1964; Levin *et al.* 1964; Kok & Varner 1967; Imshenetsky 1969; Klein & Vishniac 1972; Radmer & Kok 1972). Based on the assumption that (i) Martian life, if it exists, will be carbonaceous; (ii) its chemical composition is similar to that of terrestrial life; and (iii) it most likely metabolises simple organic compounds (Horowitz 1964), a life detection instrument package was selected composed of three experiments to detect metabolic activity of potential microbial soil communities (**Klein *et al.* 1972; Klein 1974, 1976, 1992, Young 1976**): (i) the pyrolitic release experiment tested carbon assimilation, i.e. photoautotrophy, as a method of incorporating radioactively labelled carbon dioxide in the presence of sunlight (**Horowitz *et al.* 1976, 1977; Hubbarb 1976**); (ii) the labelled release experiment tested catabolic activity, e.g. respiration, as a metabolic capability to release radioactively labelled carbon from organic nutrient compounds (**Levin & Straat 1976a & b**); and (iii) the gas-exchange experiment tested metabolic production of gaseous by-products in the presence of water and nutrients as produced during respiration (**Oyama *et al.* 1976**). The biology experiments were supported by the lander cameras which would show the presence of any obvious macroscopic form of life (**Huck *et al.* 1972; Levinthal *et al.* 1977**), by the gas chromatography-mass spectrometer (GCMS) experiment capable of detecting organic residues in Martian soil down to ppb for compounds containing 3 or more C and to ppm for compound with 1 or 2 C (**Biemann 1974; Biemann *et al.* 1977**), and by the X-ray fluorescence experiment which analyzed the elemental composition of the Martian surface soil for elements heavier than Mg (**Toulmin *et al.* 1973**).

All three Viking biology experiments gave results indicative of active chemical processes when samples of Martian soil were subjected to incubation under the conditions they normally experienced (summarised in **Horowitz 1977; Klein 1977, 1978, 1979**; Oyama *et al* 1978; McKay 1992). Seven of nine pyrolitic release experiments gave positive results, i.e. after 5 days of incubation a synthesis of organic matter from atmospheric CO and CO_2 appeared to take place at a low rate; this process was thermolabile and inhibited by moisture (**Horowitz *et al.* 1976, 1977**). In all runs of the labelled release experiment, immediately after the first injection of nutrients, radioactive CO_2 was rapidly released, followed by a prolonged slow release of CO_2; this process could hardly be repeated by a second injection and was completely inhibited by heating to 160°C for 3 h (**Levin & Straat 1976a & b, 1977a & b**). In the gas exchange experiment, the soil rapidly released O_2 upon humidification, and CO_2 after addition of nutrients; heating of the sample to 145°C for 3.5 h reduced the activity (**Oyama *et al.* 1977**). Whereas the presence of most essential major biogenic elements C, H, O, N, P, K, Ca, Mg, S on the surface of Mars has been proven, either determined during the Viking missions by GCMS (**Biemann *et al.* 1976, 1977**) or by X-ray fluorescence spectrometry (**Clark *et al.* 1976; Clark & Baird 1979**) – only P was not amenable to the detection – or from SNC meteorites analyses (**Gooding 1992**), so far only N has not been detected (summarised in **Banin & Mancinelli 1994**), no organic carbon was found in the Martian soil by the GCMS experiment (**Biemann *et al.* 1976, 1977; Biemann 1979**).

So far, the mechanisms underlying the results of the biology experiments are not known. A number of hypotheses have been put forward in order to explain the enigma of an active chemistry in the absence of organics or at least some of the phenomena observed, such as suggesting the presence of maghemite and peroxides (**Oyama & Bergdahl 1977; Oyama *et al.* 1977**, 1978), of H_2O_2 produced by UV irradiation of hematite in the presence of traces of H_2O (**Ponnamperuma *et al.* 1977**), or of iron-containing clays without intervention of H_2O_2 (Banin & Rishpon 1979), of higher oxidation states of certain minerals following UV irradiation (Holland *et al.* 1979), of nitrates, partly photolysed by UV radiation, in combination with a sparingly soluble metal carbonate (**Plumb *et al.* 1989**), of at least three different oxidants (McKay 1992), or of micropores in the fine-grain component of Martian soil functioning as gas-traps (**Nussinov *et al.* 1978**). Most models include photochemical reactions that are also responsible for the lack of organics (**Chun *et al.* 1978**, Oro & Holzer 1979) induced by the intense solar UV radiation (**Kuhn & Atreya 1979**).

The results of the Viking experiments have split the scientific community into two groups: those who deny or strongly question the existence of extant life on Mars (Horowitz 1986; **Haynes 1990; McKay & Haynes 1990; McKay *et al.* 1991**) and those who do not rule out the possibility of certain biotic 'oases' on Mars (**Levin & Straat 1977a & b, 1981**; Imshenetsky *et al.* 1980; **Adelman 1986**; Klein 1987). By analogy with terrestrial ecosystems, potential protected niches have been postulated, such as sulphur-rich sub-surface areas for chemoautotrophic communities (**Clark 1979**; **Boston *et al.* 1992**), rocks for endolithic communities (Friedmann, 1986; **Friedmann *et al.* 1987**), permafrost regions (**Gilichinsky 1993**), hydrothermal vents (**Boston *et al.* 1992**) soil (Imshenetsky *et al.* 1981) or evaporite crystals (**Rothschild 1990**). For future exobiology exploration of Mars, a stepwise approach has been recommended to ESA (Chicarro *et al.* 1989). In particular, prior to any 'search for life' experiment, more data are required on Martian geology (palaeolakes, hydrothermal vents, carbonates), climate (hydrosphere) and radiation environment, present state and past evolution as well as organic molecules in sediments.

7.4.2.2 Venus

The current atmospheric environment at the surface of Venus with an average surface temperature of 464°C is too hostile to support life (for criteria see Section 7.3), although the possibility of the production of organic compounds in its atmosphere, e.g. by lightning, has been suggested (**Otroshchenko & Surkov 1974**). Concerning its history, little is known, in spite of 21 probes sent to Venus since 1961 (listed in Cohen & Kasting 1992). This is mainly due to the Venusian high surface temperature and pressure (95 bar), and the permanent thick clouds – primarily composed of aqueous solutions of H_2SO_4 – which obstruct its exploration. Since 1990, however the US spacecraft Magellan has provided a nearly complete global radar map of Venus, revealing valley and channel networks associated with volcanic features (**Head *et al.* 1992**; **Baker *et al.* 1992**; **Komatsu *et al.* 1993**). It would be of special interest to search for ancient surface structures in these radar observations, which could give new information about the archaic surface and climate evolution. Models for Venusian evolution, which take into consideration the very high D/H ratio, i.e. 120 times that of the Earth – determined from spacecraft (Pioneer Venus: **Donahue *et al.* 1982**; **McElroy *et al.* 1982**; International Ultraviolet Explorer: **Bertaux & Clarke 1989**; **Donahue 1989**) or from the Earth (**de Bergh *et al.* 1991; de Bergh 1993**) – assume that Venus possessed a sizeable inventory of water and a cooler climate (100–200°C) in its early history (**Grinspoon 1987**; **Kasting 1988; Kasting *et al.* 1988**; Cohen & Kasting 1992). Uncertainties exist about the amount of water, since these models do not take into consideration the sporadic infall of cometary material (Cohen & Kasting 1992; **de Bergh 1993**). Runaway greenhouse conditions, that still exist today, would then have led to evaporation of the water, its photolysis at higher altitudes and escape of hydrogen (Cohen & Kasting 1992). Today, only traces of water remain in the atmosphere of Venus (**Schubert & Covey 1981**). Hence, Venus may have been considerably more Earth-like in its past, so that early environmental conditions on Venus may have been favourable for the emergence and early evolution of life (Cohen & Kasting 1992). Possible terrestrial model systems might include hyperthermophilic microorganisms occurring in volcanic habitats, such as sulphur-rich boiling springs, mud holes and heated soil (**Segerer *et al.* 1993**).

7.4.2.3 Europa

Europa, a satellite of Jupiter, is another focus for exobiology. More than 95% of the spectroscopically detectable material on its surface is H_2O (Oro *et al.* 1992). Under a crust of H_2O–ice which has been detected by IR spectroscopy from the ground and confirmed during the Voyager flyby (**Smith *et al.* 1979**), an ocean of liquid water might exist. Since liquid water is considered as one of the prerequisites for life to emerge (**Brack 1993**), the possibility of the emergence of life on Europa has been discussed (**Reynolds *et al.* 1983**). In addition to liquid water, prerequisites for habitability are a sufficient supply of energy and the availability of the 'right' substances as precursors of life. Potential energy sources might be solar radiation near the surface, the dissipation of tidal energy in the form of heat, and/or heat transfer from the core to the bottom of the ocean, similar to thermal vents in terrestrial oceans (Oro *et al.* 1992). Although little is known about the chemical composition of Europa, from the spectroscopic similarity of the non-water material of Europa to that of carbonaceous chondrites, carbon compounds are likely to exist on this Jovian satellite. Hence, the major assumed requirements for habitability seem to be fulfilled on Europa. A closer look at Europa is expected from the Galileo mission, which reached the Jovian system in 1994. As follow-on missions, landing probes for *in situ* surface and sub-surface studies have been strongly recommended (Oro *et al.* 1992).

7.4.3 Possible habitats outside the solar system

Estimates of the occurrence and frequency of habitable zones outside our solar system (**Hart 1979**) are based mainly on astronomical concepts of the structure and dynamics of our Galaxy, on planetary atmosphere models (e.g. **Kasting 1988**) and on biological interpretations of the requirements for the emergence and evolution of life (**Oro *et al.* 1982**; Oro 1988, see also Section 7.3). For solar type stars, a habitable belt, ranging from about 0.95 to about 1.5 AU has been suggested (Doyle & McKay 1991; Schwartzman & Volk, 1991). In addition to single main-sequence stars (Whitmire *et al.* 1991), variable stars, giant stars (**Godamski 1961**) and binary systems (Huang 1963) have been examined for supporting a habitable zone (Dole 1964; **Hart 1979**). Within our Galaxy, the orbit of our Sun with a distance of 10^4 parsec from the galactic centre is suggested as especially favourable for supporting life forms and higher civilisations due to the co-rotation circle, an exceptional region of the Galaxy, passing through it (Marochnik & Mukhin 1988). However, at present, there is no unambiguous evidence for the existence of other planetary systems, although several attempts have been made to search for them in our Galaxy (**Black & Suffolk 1973**; **Pendleton & Black 1983**; Doyle 1988; Campbell *et al.* 1988; Paresce 1992). Because observations from the Earth suffer from the limited transparency of the atmosphere and atmospheric turbulence, space-borne telescopes, such as the Hubble Space Telescope and astrometric telescopes mounted on board a space station or on the far side of the Moon (Battrick & Barron 1992) will provide a much higher resolution for the detection of planetary systems beyond our solar system (Tarter 1986; **Tarter *et al.* 1986**; Belton *et al.* 1989).

For estimations of the expected number of habitable planets in our Galaxy, a formula, known as the Drake or Greenbank equation is frequently used, which takes into consideration (i) the rate of star formation, (ii) the fraction of stars which have planetary systems, (iii) the average number of planets per planetary system which fall in a habitable zone, and (iv) the fraction of habitable planets on which life arises (Shkolovsjii & Sagan 1966; **Sagan 1973**; Drake 1974; Morrison 1974; Schwartzman & Rickard 1988). Of all the terms, only the first one is supported by observations. Since interstellar distances are so vast, radio-communication has been deemed the only way of detecting life beyond our solar system (**Cocconi & Morrison 1959**; **Bracewell 1960**; **Drake 1960**; **Schwartz & Townes 1961**; Morrison *et al.* 1977). This requires that additional terms be included in the Drake formula, namely (v) the fraction of planets with life on which intelligence arises, (vi) the fraction of intelligent species that evolve a technological state that enables interplanetary communication, and (vii) the lifetime of such technological civilisation. According to the likely number of advanced civilisations in our Galaxy estimated, scientists can be assigned to the group of 'pessimists' or '$N = 1$ school', which focuses on arguments suggesting the uniqueness of our own civilisation, on the one hand, or to the group of 'optimists' giving numbers as high as 10^6, on the other hand (summarised in Schwartzman & Rickard 1988). It is likely, that in the future, the application of detailed astrophysical and planetary models will lead to a more sound estimation of the occurrence of potential exobiological habitats. So far, radioastronomical search for extraterrestrial intelligence has not given any positive indications (**Drake 1961**; Morrison *et al.* 1977; Marx 1988; Heidmann & Klein 1991).

7.5 Distribution of life throughout the solar system

7.5.1 Interplanetary transfer of life

7.5.1.1 Panspermia revisited

Based on the wide distribution of cosmic dust, the theory of panspermia postulates that microscopic forms of life, e.g. spores, can also be propagated in space driven by radiation pressure from the sun thereby seeding life from one planet to another (**Richter 1865; Arrhenius 1903**, 1908). Since its formulation, this theory has been subjected to several criticisms, with arguments such as (i) it cannot be tested experimentally (**Lederberg 1960**), (ii) it shunts aside the question of the origin of life (Dose 1986a; Allamandola *et al.* 1989b; **Haynes 1990**), and (iii) spores will not survive long-term exposure to the

hostile environment of space, especially vacuum (**Nussinov & Lysenko 1983, 1989, 1992**) and radiation (Shklovskii & Sagan 1966; **Crick & Orgel 1973; Moreno 1988**).
Although it will be difficult to prove that life could be transported through our solar system, estimates of the chances for the different steps of the process to occur can be obtained from measurements, calculations, and experiments. The fundamental phases and requirements of panspermia are: (i) escape process, i.e. removal to space of biological material which has survived being lifted from the surface to high altitudes; (ii) interim state in space, i.e. survival of the biological material over time scales comparable with the interplanetary or interstellar passage; (iii) entry process, i.e. non-destructive deposition of the biological material on another planet (**Horneck & Bücker 1985; Clark 1985; Weber & Greenberg 1985**; Horneck 1988; Allamandola *et al.* 1989b). Recent discoveries have given new support to the idea of panspermia (summarised in Horneck & Brack 1992). These are: (i) the detection of meteorites, some of lunar and some probably of Martian origin (**Kerr 1987**); (ii) the probability of small particles of diameters between 0.5 μm and 1 cm (**Moreno 1988**) or even boulder-sized rocks reaching escape velocities by the impact of large meteorites on a planet, e.g. on Earth (Janle 1984; **Melosh 1984, 1985, 1988**) or Mars (**O'Keefe & Ahrens 1986; Vickery & Melosh 1987**); (iii) the ability of bacterial spores to survive the shock waves of a simulated meteorite impact of 42.5 GPa with low frequency (approximately 10^{-4}) (Horneck & Brack 1992); (iv) the high UV resistance of micro-organisms at the low temperatures of deep space (**Weber & Greenberg 1985**); (v) the high survival of bacterial spores over extended periods in space (**Horneck 1993**; Horneck *et al.* 1994); (vi) the palaeogeochemical evidence of a very early appearance of life on Earth in the form of metabolically advanced microbial prokaryotic ecosystems (see Section 7.4.1.1) leaving not more than approximately 0.4 Ga for the evolution of life from the simple precursor molecules to the level of the prokaryotic, photoautotrophic cell (Schidlowski 1990); (vii) the biochemical evidence of a common ancestor for all life forms on Earth (see Section 7.4.1.2) (Crick 1981; **Crick & Orgel, 1973**); (viii) the likelihood of artificial or directed transport by probes sent to other planets (**Crick & Orgel, 1973; Meotner & Matloff 1979**; McKay & Davis 1989; **McKay & Haynes, 1990; McKay *et al.* 1991**; DeVincenzi 1992; Haynes & McKay 1992, National Research Council 1992).

7.5.1.2 Upper boundary of the biosphere

Distribution of air spora including their identity, behaviour, movements and survival in the troposphere as well as their impact on public health and agriculture are well understood (reviewed in Gregory 1973; Cox 1987; Horneck & Brack 1992). They comprise viruses, bacteria, algae, microfungi, fungal spores, spores of moss and ferns, pollen, minute seeds, and protozoan cysts, up to concentrations of hundreds to thousands m^{-3} (Gregory 1973; Cox 1987).

It has been shown that spores, particularly pigmented species, are especially adapted to survive this inhospitable environment, characterised by low temperature, drought, low pressure, and intense solar radiation (**Mancinelli & Shulls 1978**).

With increasing height, the concentration of air spora rapidly reduces. In a few early experiments, using balloons or meteorological rockets with specially designed analyzers, viable micro-organisms were collected up to 77 km (Rogers & Meier 1936; Bruch 1967; **Imshenetsky *et al.* 1976, 1978**, 1979; Lysenko *et al.* 1980). In most cases, the samplers were pre-sterilised in order to avoid contamination by organisms from lower altitudes (Ramkova *et al.* 1976). They collected predominantly black conidia and spores of fungi. It is assumed that pigmentation offers a selective advantage for the spores, because it protects them against the intense solar UV radiation (**Mancinelli & Shulls 1978**). No viable microorganisms were collected from Earth orbit by use of collectors on board of Gemini 9A and 12 (Hemenway *et al.* 1967).

The natural dynamic forces by which air spora can be transported to the stratosphere or beyond, and what protection mechanisms are required for them to survive their journey are still open questions. Dynamic forces such as gravitation, meteorological drifts, thermal gradients, magnetic or electric fields, and solar radiation pressure, respectively, are considered insufficient to accelerate small particles of the size of a micro-organism or small rock fragments to reach escape velocities (Bücker & Horneck 1969). More promising processes suggested to allow small particles sufficient acceleration are volcanic eruptions, fly-by meteorites or meteorite impacts (Janle 1984; **Melosh 1984, 1985, 1988; Vickery & Melosh 1987**).

Space technology offers a tool to tackle the question of what is the upper boundary of our biosphere, by sampling particles of biological interest in the stratosphere and in Earth orbit by use of satellites. To accomplish this task non-destructive collection devices must be employed, to allow subsequent chemical or biological analysis (Horneck *et al.* 1989; Brownlee *et al.* 1989). Such investigations will provide substantial pieces of information to the mosaic of our understanding of the changes of concentration of air spora with height, and of their circulation over the surface of our planet. This information is also important for assessing the probability of air spora or other minute organisms reaching escape velocities.

7.5.1.3 Responses of microorganisms to space

Interplanetary space, characterised by a high vacuum, an intense radiation climate of solar and galactic origin and extreme temperatures, has been considered to be extremely hostile to all forms of life (details described in Horneck & Brack 1992). In order to study the survival of resistant microbial forms in the upper atmosphere or in free space, microbial samples were exposed *in situ* by use of balloons, rockets or spacecraft and their responses investigated after recovery (reviewed in Horneck 1981; Horneck *et al.* 1989;

Mennigmann 1989; Horneck & Brack 1992). For this purpose, several facilities were developed, such as the exposure device on Gemini, MEED on Apollo, ES029 on SL1, ERA on EURECA, and UV-RAD on Spacelab D2 (survey in Horneck *et al.* 1989). These investigations were supported by studies in the laboratory in which certain parameters of space (high and ultrahigh vacuum, extreme temperature, UV radiation of different wavelengths, ionising radiation) were simulated. The microbial responses (physiological, genetic and biochemical changes) to selected factors applied separately or in combination were determined (reviewed in Horneck *et al.* 1989).

Early experiments in space have demonstrated that certain microorganisms survive short periods of exposure in space – several hours or days – provided they are shielded against the intense solar ultraviolet (UV) radiation (Lorenz *et al.*, 1968, 1969; Bücker *et al.* 1974; **Taylor 1974**; **Taylor *et al.* 1974**; Spizizen *et al.* 1975). To date, the maximum exposure time of microorganisms to the space environment has been the 2107 day duration of the LDEF mission (**Horneck 1993**; Horneck *et al.* 1994). Up to 70% of *Bacillus subtilis* spores survive extended periods of time exposed to the vacuum of space, even if prepared as a monolayer (**Horneck 1993**; Horneck *et al.* 1994). Nevertheless, even in the survivors, the genetic material is affected, indicated by an increased mutation rate, delayed germination, cross linking of DNA and protein, and the requirement of cellular repair processes to restore viability after vacuum exposure, respectively (**Horneck *et al.* 1984a** & b; Dose 1986b). In laboratory experiments, an increased rate of DNA strand breakage was observed (**Dose *et al.* 1991**, 1992). When exposed to vacuum, the spores became dehydrated. With increasing exposure time, free water is removed first, followed by the hydrated water, and finally, even chemically bound water and other volatile molecules may be removed. The deleterious consequences for membranes and macromolecules, such as DNA and proteins, are manifold (**Cox** 1987, **1993**).

Solar UV radiation has been found to be the most deleterious factor of space as tested with dried preparations of viruses, bacterial or fungal spores (Hotchin *et al.* 1967, 1968, Lorenz *et al.* 1968, 1969, Spizizen *et al.* 1975, Bücker *et al.* 1974, **Taylor 1974**, **Taylor et al. 1974**, **Horneck *et al.* 1984a** & b, 1994; **Horneck 1993**). The incidence of the full spectrum of solar UV light (>170 nm) kills 90% of *B. subtilis* spores within seconds. To reach the same effect on Earth, the exposure to sunlight takes approximately 1000 times longer (Horneck & Brack 1992). This difference is mainly attributed to the ozone layer that protects the biosphere from the most harmful fraction of solar UV radiation (<295 nm) (see **Häder 1993**). Action spectra of the solar photons (irradiance data for wavelengths 160 nm to 320 nm from **Labs *et al.* 1987**) for killing bacteriophage T1 (Lorenz *et al.* 1968) or *B. subtilis* spores (**Horneck *et al.* 1984a** & b; **Horneck** 1988, **1993**) correlate closely with the absorption spectrum of DNA. This indicates that the DNA is the critical chromophore for lethality (Fig. 3). Additional proof for UV damage being the main cause of inactivation in free space was obtained with Tobacco Mosaic Virus that, after exposure to space, showed structural changes which are typical of UV induced damage (**Orlob 1968**).

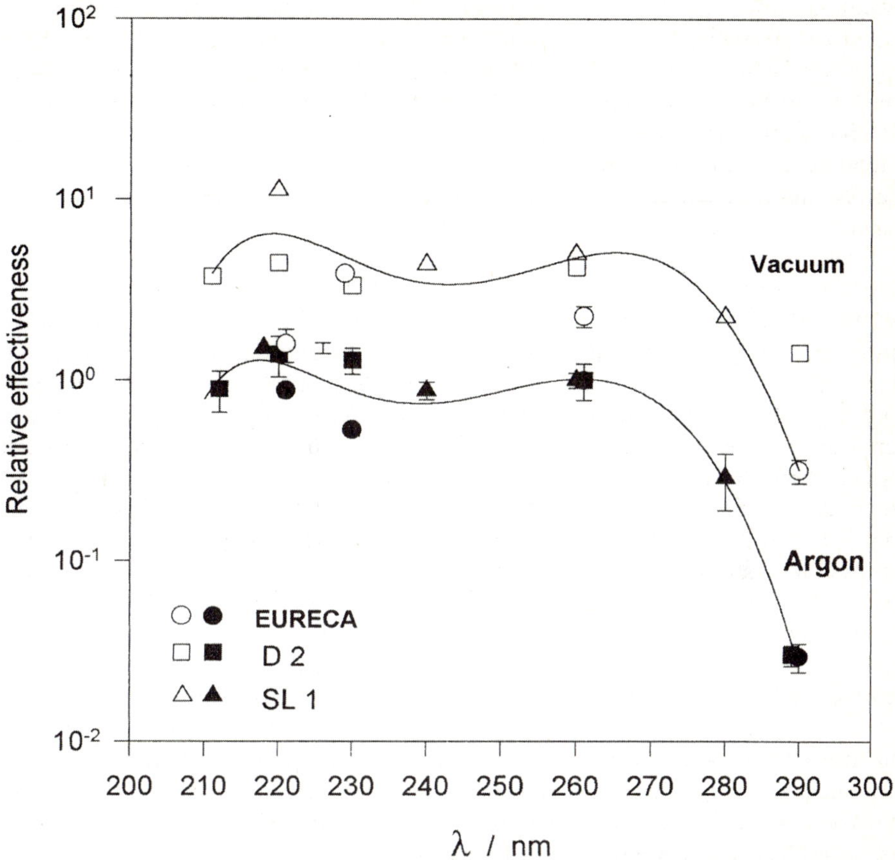

Fig. 3. Action spectrum for lethality induced by extraterrestrial solar UV-radiation in spores of *Bacillus subtilis* after irradiation at atmospheric pressure and in space vacuum (data from space experiments on board Spacelab 1, Spacelab D-2, and EURECA; from Horneck *et al.* 1995).

It should be noted that considerable and relevant information regarding the effects of the far-UV on photobiology and photochemistry of nucleic acids and related compounds under Earth conditions has accumulated during the last 30 years (for reviews see Wang 1976; **Cadet *et al.* 1985**). Various chemical and biochemical assays offering high sensitivity and accuracy are now available for monitoring the formation of the main photoproducts within far-UV irradiated cellular DNA. However, the exact biological role and in particular the mutagenic effects of the four main classes of DNA photoproducts involving cyclobutadipyrimidines, (6-4)pyrimidine-pyrimidone adducts and their Dewar valence isomer (**Taylor & Cohr 1987**) as well as 5,6-dihydro-5-(α-thyminyl)thymine (TDHT) remains to be assessed. Concerning the vacuum-UV light, the other important component of solar electromagnetic radiation, the photobiological and photochemical reactions have recently studied in the laboratory with DNA and microorganisms (**Sontag & Dertinger 1975; Dodonova *et al.* 1982; Ito *et al.* 1983, 1986; Maezawa *et al.* 1984; Hieda *et al.* 1986; Munakata *et al.* 1986; Hieda & Ito 1986; Matsunaga *et al.* 1991; Keller & Horneck 1992; Yamada & Hieda 1992; Mikio 1993**).

If spores are simultaneously exposed to solar UV radiation and space vacuum they respond with an increased sensitivity, to a broad spectrum of solar UV light (>170 nm) as well as to selected wavelengths (**Horneck *et al.* 1984a** & b; **Horneck 1993**) (Fig. 3). Several attempts were made to isolate and identify thymine photoproducts generated within the DNA of resting bacterial spores exposed to UV light in vacuum. In various Earth and space experiments two thymine decomposition products namely, the cis-syn and trans-syn isomers of cyclobutadithymine (ThyThy) were identified (Bücker *et al.* 1979). DNA-protein cross-linkings were found in addition to TDHT, the so-called 'spore photoproduct' (**Horneck *et al.* 1984a** & b; **Lindberg & Horneck 1991**, 1992). From the efficiency of repair processes (photoenzymatic repair, spore photoproduct specific repair) it is concluded, that photoproducts other than cis-syn ThyThy and TDHT seem to be responsible for the UV supersensitivity of spores, if irradiated under vacuum conditions (Bücker & Horneck 1970; **Horneck 1993**).

During the major part of a hypothetical journey through space, if shielded from solar thermal radiation, micro-organisms will be confronted with the 4°K cold emptiness of space. Under these very cold conditions, thermodynamic and chemical reactions are nearly frozen. The photobiological response to solar UV radiation may be completely different from room temperature conditions. Only the latter response (at room temperature) has so far been tested in space. Laboratory experiments under simulated interstellar medium conditions point to a remarkably less damaging effect of UV radiation at these low temperatures. Treating *B. subtilis* spores with 3 simulated parameters of space simultaneously, i.e. UV radiation (>110 nm), vacuum, and low temperature (10°K), results in an unexpectedly high survival, even at very high UV fluences. At this low temperature range, the inactivation cross sections obtained are up to 2–3 orders of magnitude lower than at room temperature (**Weber & Greenberg 1985**). The temperature profile of bacterial spore UV sensitivity shows a maximum at 190°K. From these data it has been estimated that, in the most general environment in space, spores may survive for hundreds of years (**Weber & Greenberg 1985**).

Among the ionising components of radiation in space, the heavy primaries, the so-called HZE particles are the most effective species (Grahn 1973). To understand the ways by which particles of cosmic radiation interact with biological systems, methods have been developed to localise precisely the trajectory of an HZE particle relative to the biological object and to correlate the physical data of the particle relative to the observed biological effects along its path. By use of visual track detectors that were sandwiched between layers of biological objects in a resting state – this concept is known as the 'Biostack' (Bücker & Horneck 1975) – in a variety of test systems including viruses, bacterial spores, plant seeds or brine shrimp cysts, injuries such as somatic mutations in plant seeds, development disturbances and malformations in insect and brine shrimp embryos, or inactivation in bacterial spores were traced back to the traversal of a single HZE particle (recently reviewed by Hagen 1989; **Horneck 1992**). Such HZE particles of cosmic radiation are conjectured to set the ultimate limit on the survival of spores in space since they penetrate even high shielding. The maximum time of a spore to escape a hit by a HZE particle (e.g. iron of LET >100 keV μm^{-1}), has been estimated to amount to 10^5–10^6 years (Horneck 1981; Mennigmann 1989). Laboratory simulation experiments have shown that certain viruses and microorganisms show high survival rates upon exposure to simulated space condition as a combination of low temperature, vacuum and proton irradiation (Koike *et al.* 1992).

To travel from one planet of our solar system to another, e.g. from Mars to Earth by random motion, a mean time of several hundred thousands to millions of years has been estimated for boulder-sized rocks (**Melosh 1988**), but only 2 months for microscopic particles (**Moreno 1988**). These estimates have drawn the following picture for an interstellar transport. The clouds of gaseous and particulate matter between the stars move in a random fashion at speeds of about 10 km s^{-1}. If a bacterial spore is captured in such a cloud, it will be swept along with the gas. Given the distance between neighbouring stars of 0.1–1 parsec, this corresponds to a passage time of 10^5–10^6 years. Thus, for interstellar infection, microorganisms would have to survive at least for 10^5–10^6 years in such a cloud (**Weber & Greenberg 1985**). So far, results from exposure of microorganisms to space up to several years are available (Horneck 1993; Horneck *et al.* 1994). More data on long-term effects are required. Exposure platforms, such as the ERA on EURECA (Horneck *et al.* 1985) and the BIOPAN on Foton are especially suited for this kind of research. Such studies will eventually allow extrapolation to time spans, as required for interplanetary or interstellar transport of life.

7.5.2 Directed transport of life

7.5.2.1 Lunar or planetary base

Our current knowledge of the Moon is essentially derived from past exploration, both automated and human, carried out with the Apollo and Lunar programmes (for overview see Mendell 1985; Hartmann *et al.* 1986; Battrick & Barron 1992). These have demonstrated that the Moon is within reach for human expeditions. Recently, scenarios have been developed leading eventually to the establishment of a manned lunar base (IAA Ad Hoc Committee 'Return to the Moon' 1988; Koelle 1988; Pilcher 1992). Research will concentrate on (i) science of the Moon, including all investigations concerning the Moon as a planetary body, (ii) science from the Moon, using the Moon as a platform for radioastronomical observations, and (iii) science on the Moon, including life sciences activities, such as the development of artificial ecosystems beyond the Earth, the non-destructive collection of cosmic dust particles for organic-chemical or biological analysis, and questions relating to human activities on the Moon (**Duke & Mendell 1988**; Battrick & Barron 1992). A lunar base which will eventually have a permanent human presence has been considered as a first stepping stone on the human path into space (**Koelle 1988**).

The Moon has furnished no evidence of life, or even of organic compounds (Sagan 1972; McKay & Ming 1989). Its environment lacks essential prerequisites for life. Hence, before setting up a lunar base with a permanent human presence, various problems have to be tackled, such as radiation protection (NCRP 1989), health care (**Bluth 1988; Burke 1988**), regenerative life support systems (McElroy *et al.* 1985), and eventually exploitation of lunar resources (White & Hirsch 1985; **Waldron 1988**). The establishment for the first time of an artificial ecosystem on a celestial body beyond the Earth has been considered as one of the most challenging projects (Ming & Henninger 1989; Battrick & Barron 1992).

Following the establishment of a human base on the Moon, further perspectives envisage human exploration of Mars (Manned Mars Mission 1985; Pilcher 1992; Stoker *et al.* 1992). The pros and cons of human presence on Mars, especially in the field of geology and exobiology, and the concern from the viewpoint of planetary protection (see Section 7.5.2.2) have been discussed recently (Horneck 1992). Enterprises, such as human expeditions to Mars, which are mainly promoted by the planetary research community, also bear interest for the life sciences community, especially when human health and well-being is concerned.

7.5.2.2 Planetary protection

The introduction and possible proliferation of terrestrial life forms on other planets by means of orbiters, entry probes or landers could entirely destroy the opportunity to examine the planets in pristine condition. From this concern by the scientific community, the concept of planetary protection has evolved. Its intent is twofold: (i) to protect the planet being explored and to prevent jeopardising search for life studies, including precursors and remnants, and (ii) to protect the Earth from the potential hazards posed by extraterrestrial matter carried on a spacecraft returning from another celestial body (Rummel 1989). Planetary protection issues are bound by an international treaty (UN Doc. A/6621, Dec. 17, 1966) and agreement (UN Gen. Ass. Resol. A/34/68, Dec. 5, 1979). Since 1959, COSPAR has developed planetary protection guidelines that are based on relevant information about the probability of survival and release of organisms either contained in or on exposed surfaces of spacecraft, about the surface and atmosphere characteristics of the planet under consideration, and about the probable distribution and growth of types of organisms involved (Hall 1974). In 1984, this concept of probability of contamination of a planet of biological interest was replaced by a concept of contamination control to be elaborated specifically for certain space-mission/target-planet combinations, such as orbiters, landers, or sample return missions (DeVincenzi & Stabekis 1984; Rummel 1989; Schwehm 1989; McKay & Davis 1989; DeVincenzi & Klein 1989; DeVincenzi 1992; Rummel 1992; Lindberg & Horneck 1995). Information from exobiological experiments on the survival of resistant microorganisms in free space or in simulated planetary environment conditions (e.g. Horneck *et al.* 1971; **Foster *et al.* 1978**; Mancinelli 1989; **Moll & Vestal 1993**; **Horneck 1993**), on shelf-life in enclosed biological systems (e.g. Crowe & Crowe 1992), and on microbial strategies of adaptation to environmental extremes and their ability to build up ecological niches within an otherwise hostile macroenvironment (e.g. Friedmann & Koriem 1989; **Rothschild 1990**; **Gilichinsky *et al.*** 1992, **1993**) has a direct bearing on the planetary protection to be undertaken before any mission becomes reality (National Research Council 1992).

7.5.2.3 Terraforming or ecopoesis

As an extension of the theory of panspermia (**Arrhenius 1903**, see also Section 7.5.1.1), which assumes that life initiates on a planet via random import of e.g. spores, that are driven by the solar radiation pressure through space, the theory of directed panspermia (**Crick & Orgel 1973**; Crick 1981) considers the possibility of emergence of life on Earth *via* deliberate transmission of organisms by intelligent beings on another planet. Their argumentation proceeds on the assumptions that life on Earth appears to be a clone derived from a single organism, that 4 Ga are sufficient for life to develop a technological society, and that Earth-like planets might have existed elsewhere in the Galaxy as much as 6.5 Ga before the formation of our solar system. At the present time the evidence is inadequate to say anything about the probability that directed panspermia has triggered origin of life on Earth; however, we have to realise that with space technology we ourselves have already at our disposal a tool to reach every body in our solar system by spacecraft (Sagan 1980).

Renewed interest in human exploration of Mars (Pilcher 1992) and in the establishment of human settlements on Mars by using indigenous Martian resources (Averner & McElroy 1976) has led to the question of the impact of human activities on Mars (Horneck 1992) and of the possibility of terraforming, i.e. to modify Mars intentionally in such a way to make it more habitable and eventually more Earth-like (**McKay 1982**). To make Mars habitable, in a first step, changes of the environmental parameters of the planet by planetary engineering are suggested, such as the distribution of volatiles, the surface temperature and pressure, the atmospheric composition and opacity, planetary albedo, precipitation and humidity. This would warm the climate sufficiently to allow the presence of liquid water and to increase the thickness of the atmosphere by the release of frozen gases (**McKay *et al.* 1991**). The second step would involve biological engineering, i.e. the production of genetically engineered microorganisms and their implantation as pioneering microbial communities able to proliferate in the newly clement, though still anaerobic, Martian environment (**McKay & Haynes 1990**). This process of establishing an ecosystem, or biosphere, on a lifeless planet has been termed 'ecopoesis' (**McKay & Haynes 1990**; **Haynes 1990**; Haynes & McKay 1992). For terraforming Mars, e.g. for the production of a breathable atmosphere, time scales in the order of 10^4 years have been estimated (**McKay *et al.* 1991**). Such considerations on ecopoesis are based on the assumption that Mars is presently a lifeless planet, which so far has not been conclusively proven. Nevertheless, ecopoesis on Mars might be taken up as a long-term goal for space science and engineering, which will generate a great wealth of new scientific and technical knowledge that is also valuable for the understanding of our own biosphere.

7.6 General conclusions

Exobiological research has allowed scientific objectivity to be applied to questions of the origins and distribution of life in the universe which hitherto have been the province of philosophy and metaphysics. Major research activities that have been tackled by exobiology pertain to:

(i) comparing the overall pattern of chemical evolution of potential precursors of life, in the interstellar medium, and on the planets and small bodies of our solar system;
(ii) tracing the history of life on Earth back to its roots;
(iii) deciphering the environments of the planets in our solar system and of their satellites, throughout their history, with regard to their habitability;
(iv) searching for other planetary systems in our Galaxy and for signals of extraterrestrial civilisations;
(v) testing the impact of space environment on survivability of resistant life forms.

Within the last 20 years, there have been significant advances in the exobiological fields of research mentioned above. Examples are as follows. The organic processes in the interstellar medium and the outer solar system point to a universal organic chemistry to which the origin of life may be directly connected; the concurrence of the fossil and molecular record of life on Earth gives information about the nature of our universal ancestor; comparative planetology point to the kind of planetary environment required for life to emerge and evolve; the high resistance of bacterial spores to the environment of space point to the likelihood of interplanetary transfer of life. With this information, first decisive steps are made towards a better understanding of the principles leading to the emergence of life from inanimate matter, its evolution and distribution on Earth, the solar system, and beyond. This evolutionary approach may eventually sharpen our understanding of the interplay between life on Earth and its environment, from the initial steps throughout its evolution towards perspectives for its future development.

However, many essential pieces of information are still missing. We are ignorant about the premises on the origin of life, we are lacking any historical fact on the initial processes of life on Earth, and we do not know, whether 'life' is a general phenomenon in the overall scenario of cosmic evolution or just an accident confined to a very few isolated planets. To close these gaps, substantial efforts in exobiology are required in various fields of research including theoretical modelling, space exploration, microbial ecology, and simulation experiments in Earth orbit and in the laboratory.

It is important to note the multidisciplinary character of exobiological research which involves scientists from a wide variety of disciplines, such as astronomy, planetary research, organic chemistry, palaeontology and the various subdisciplines of biology including microbial ecology and molecular biology. Pieces of information provided by each discipline have contributed to the conception of the phenomenon of life within the process of cosmic evolution. New techniques that have been developed within the various disciplines are now accessible for exobiology research. Besides space technology and various remote sensing techniques, they include among others: radioastronomical molecular spectroscopy, isotope fractionation analysis, nucleic acid and protein sequencing technology, immunofluorescence techniques for the detection of hitherto uncultured microorganisms, and sensitive assays in organic chemistry and radiation biochemistry. Their application has led to several conceptional breakthroughs, especially in the field of early biological evolution.

Exobiological research requires a concerted action of experts of the various disciplines involved. The Viking missions to Mars are an example of a successful cooperative interdisciplinary enterprise which involved representatives from geology, atmospheric research, organic chemistry, and microbiology. It is due to the comprehensive set of data, including chemical analysis of Martian soil and atmosphere, that the results yielded by the 'search for life' experiments did not lead to a misinterpretation, but were attributed with high probability to non-biological chemical phenomena.

Exobiological research requires extensive pre-flight testing of concepts in simulated environments or with terrestrial analogues. Examples are the exobiology experiments of ERA on the EURECA mission which were defined during several pre-flight simulation runs, studies on simulated cometary material in preparation of the Rosetta mission and microbiology studies in simulated Martian environment preceding the Viking missions. For exobiology experiments in Earth orbit, simultaneous simulation experiments in ground-based laboratories are indispensable. In most cases, post-flight tests are also needed to clarify unexpected flight results. Allowing more sophisticated experimentation, they will provide the information necessary to explain the underlying mechanisms.

Exobiological research requires careful planning of space experiments well in advance of the mission under consideration. The *in situ* experiments on

Mars during the Viking missions, for example, were preceded by a planning period of about 15 years. During this planning period information on the physical and chemical characteristics of the planet or small body under consideration should be acquired. Much of this information can be obtained from non-biology experiments in the field of geology and surface chemistry, atmospheric physics and chemistry, photochemistry and radiation chemistry. There should be a mechanism established to influence the design of such experiments on precursor flights in order to gain the information required for setting up exobiological experiments.

Exobiological research bears the potential of joining the scientific community in an interdisciplinary research programme with space life sciences as the leading discipline. This exobiology programme should include radioastronomy, solar system exploration and experiments in Earth orbit. It should identify targets of interest, e.g. planets, small bodies, space scenarios, and should be adjusted to existing space programmes. In view of upcoming space activities, for the near future, the following topics merit consideration:

Target Mars:

(i) to increase our knowledge of the surface conditions and climate, present–day state and past evolution, necessary to identify potential ecological niches on Mars (Chicarro *et al.* 1989);

(ii) to study the role solar UV radiation plays in evolutionary processes on Mars by simulating Martian UV climate in Earth orbit (Horneck *et al.* 1989);

(iii) to identify and investigate terrestrial analogues of potential Martian habitats (Chicarro *et al.* 1989);

(iv) to develop detection techniques for potential signals of life on Mars, either contemporary, fossil or precursor.

Target Comet:

(i) To identify organic compounds of interest to be searched for during the Rosetta mission (Bar–Nun 1993);

(ii) to study the relevance of organic molecules to the emergence of life on Earth or any other planet by simulating chemical organic chemistry and by constructing an artificial comet in Earth orbit (Horneck & Brack 1992);

(iii) to collect cosmic dust particles non–destructively for organo–chemical analysis (Horneck *et al.* 1989).

Target Moon:

(i) to utilise the lunar environment for exobiological studies, e.g. search for organics (Battrick & Barron 1992);

(ii) to establish – for the first time – an artificial ecosystem on a celestial body beyond the Earth (Battrick & Barron 1992).

Target Earth:

(i) to trace the history of life on Earth by use of palaeontology and molecular biology;

(ii) to learn more about the impact of extraterrestrial events, e.g. meteorite impacts, on the evolution of life;

(iii) to study the influence of oceanic tides and plate tectonics on prebiotic and biological evolution;
(iv) to learn more about the evolutionary events on the early Earth from comparative planetology;
(v) to study the role of the ozone layer in protecting life on Earth from simulation experiments in Earth orbit (Horneck 1994).

Target space:
(i) to determine the chances of interplanetary transfer of life from *in-situ* studies in Earth orbit (Horneck *et al.* 1989);
(ii) to determine the upper boundary of the biosphere by collecting airborne microorganisms up to the stratosphere (Horneck *et al.* 1989).

Whereas in the US a strong exobiology community has been established and supported by NASA (Hartman *et al.* 1985; Milne *et al.* 1985; Wood & Chang 1985; DeFrees *et al.* 1989; Carle *et al.* 1992), the European exobiology activities, although part of ESA's Life Sciences Programme since 1979 (Bjurstedt 1979; Oser & Battrick 1989) are so far restricted to individuals or small research groups. In order to cope with the exobiology tasks during upcoming space activities in Europe, it is therefore necessary to concentrate the scientific potential available. Several excellent scientists in various fields, such as prebiotic organic chemistry, photo-and radiation biology, palaeontology, microbial ecology, and molecular biology, are not yet linked with the ESA space programme, although they have pronounced strong interest in the topic of exobiology during recent scientific meetings.

It is therefore recommended to foster exobiological research within ESA. This could be realised by setting up an interdisciplinary working group constituted of members from both, the Life Sciences within the Microgravity Programme as well as the Space Science Programme. The aim of this interdisciplinary exobiology working group should be:
(i) to activate and concentrate the scientific community, interested in contributing to a universal understanding of the phenomenon of 'life', its origin, evolution and distribution, by use of space research;
(ii) to define an exobiological programme that can be realised within ESA's upcoming space activities;
(iii) to elaborate targeted research activities including the design of facilities and experiments required for a certain mission;
(iv) to develop planetary protection guidelines for ESA's planetary missions.

This study has been supported by an ESA contract. Parts of it are based on a previous ESA Study *ERA-Follow-On Scientific Study, Perspectives of Exobiological and Radiation Biological Research by Use of Free-Flying Carriers in Earth Orbit, Contract No. 8116/88/F/BZ(SC)*, the exobiological part of this study was jointly prepared by A. Brack, K. Dose, J. Cadet, F. Raulin and K. Roessler under the chair of the author. I greatly appreciate the valuable contributions by the team members. I also acknowledge the help of many colleagues who provided me with relevant literature in their field of research, above all J.P. Gogarten, D.P. Häder, J. Heilmann, P. Hirsch, J. Kissel, H.P. Klenk, R. Mauersberger, L. Rothschild, M. Schidlowski, K.H. Schleifer, H.

Wänke. Special thanks go to A. Brack, J.P. Gogarten, P. Janle, J. Kissel, R. Mauersberger and R.L. Mancinelli, who critically reviewed the manuscript or parts of it; their constructive comments have substantially improved the content of the manuscript.

7.7 References

Adelman B. The question of life on Mars. J. Brit. Interplanet. Soc. 39: 256, 1986.

Agarwal VK, Schutte W, Greenberg JM, Ferris JP, Briggs R, Connor S, Van De Bult CPEM & Baas F. Photochemical reactions in interstellar grains, photolysis of CO, NH_3 and H_2O. Origins of Life 16: 21–40, 1985.

Aksenov SI, Babyeva IP & Golubev VI. On the mechanism of adaptation of micro-organisms to conditions of extreme low humidity. Life Sci. Space Res. 11: 55–61, 1973.

Allamandola L, Bar-Nun A, Brock T, Chang S, Davies RE, Greenberg JM, Hochstein L, Horneck G, Huntress W, Miller S, Nealson K & Usher D. *In situ* investigations. In: *Exobiology in Earth Orbit*. Ed. by DeFrees D, Brownlee D, Tarter J, Usher D, Irvine W & Klein H. NASA SP-500, NASA Washington, pp. 89–99, 1989.

Allamandola LJ, Sanford SA & Wopenka B. Interstellar polycyclic aromatic hydrocarbons and carbon in interplanetary dust particles and meteorites. Science 237: 56–59, 1987b.

Allamandola LJ, Tielens AGGM & Barker JR. Infrared emission from interstellar PAHs. In: *Physical Processes in Interstellar Clouds*. Ed. by Morfill E & Scholer M, D. Reidel. Dordrecht, pp. 305–31, 1987a.

Allamandola LJ, Tielens AGGM & Barker JR. Interstellar polycyclic aromatic hydrocarbons: the infrared emission bands, the excitation/emission mechanism, and the astrophysical implications. Astrophys. J. Suppl. Ser. 71: 733–75, 1989.

Amari S, Foot J, Simon C, Swan P, Walker RM, Zinner E, Jessberger EK, Lange G & Stadermann F. SIMS chemical analysis of extended impact features from the trailing edge portion of experiment AO187-2. In: *LDEF – 69 Months in Space Part 1*. Ed. by Levine AS. NASA CP 3134, pp. 503–516, 1991.

Anders E & Owen T. Mars and Earth: origin and abundance of volatiles. Science 198: 453–65, 1977.

Anders E & Zinner E. Interstellar grains in primitive meteorites: diamond, silicon carbide, and graphite. Meteorites 28: 490–514, 1993.

Arrhenius S. Die Verbreitung des Lebens im Weltenraum. Die Umschau 7: 481–5, 1903.

Arrhenius S. *Worlds in the Making*. Harper and Row, New York, 1908.

Averner MM & McElroy RD. *On the Habitability of Mars*. NASA SP-414, NASA, Washington, 1976.

Awramik SM, Margulis L & Barghoorn ES. Evolution processes in the formation of stromatolites, In: *Stromatolites*. Ed. by Walter MR. Elsevier, Amsterdam, pp. 149–62, 1976.

Baker VR. *The Channels of Mars*. University of Texas Press, Austin, 1982.

Baker VR, Komatsu G, Parker TJ, Gulick VC, Kargel JS & Lewis JS. Channels and valleys on Venus: preliminary analysis of Magellan data. J. Geophys. Res. 97: 13421–44, 1992.

Baker VR & Milton DJ. Erosion by catastrophic floods on Mars and Earth. Icarus 23: 27–41, 1974.

Baker VR, Strom RG, Gulick VC, Kargel JS, Komatsu G & Kale FS. Ancient oceans, ice sheets and the hydrological cycle on Mars. Nature 352: 589–94, 1991.

Banin A, Clark BC & Wänke H. Surface chemistry and mineralogy. In: *Mars*. Ed. by Keifer HH, Jakoski BM, Snyder CW & Matthews MS. University of Arizona Press, Tucson, pp. 594–625, 1992.

Banin A & Mancinelli RL. Assessing the chemical and physical limits for the origin and evolution of life on Mars. Icarus, in press.

Banin A & Rishpon J. Experimental simulation of the Viking labeled release (LR) results with iron-absorbed smectic clay minerals. Life Sci. Space Res. 17: 59–64, 1979.

Bar-Nun A, Barucci A, Brussoletti E, Coradini A, Coradini M, Colangeli L, Eberhardt P, Grün E, Hechter M, Keller HU, Kissel J, Klinger J, Langevin Y, Laurance RJ, McDonnell JAM, Milani A, Picardi G, Pillinger C, Schwehm G, Stöffler D & Wänke H. *Rosetta, Comet Rendezvous Mission*, ESA-SCI (97)7. ESA, Paris, 1993.

Bar-Nun A, Lazcano-Aranjo A & Oro J. Could life have evolved in cometary nuclei? Origins of Life 11: 387–94, 1981.

Barbier B, Chabin A, Brack A & Chaput D. UV processing of amino-acids and peptides in Earth orbit. In: *Abstracts, 30th COSPAR Scientific Assembly*, Hamburg, 11–21 July 1994.

Battrick B & Barron C. *Mission to the Moon*. Report of the Lunar Study Steering Group, ESA SP-1150. ESA, Paris, 1992.

Belton M, Betz A, Black J, Cohen M, Encrenau T, Evans N, Feldman PD, Herbst E, Kerridge J, McKay C, Nuth J, Roelling T, Scattergood T, Tarter J, Tielens A & Veverka J. Observational exobiology. In: *Exobiology in Earth Orbit*. Ed. by DeFrees D, Brownlee D, Tarter J, Usher D, Irvine W & Klein H. NASA SP-500, NASA, Washington, pp. 19–75, 1989.

Bertaux JL & Clarke JT. Deuterium content of the Venus atmosphere. Nature 338: 567–8, 1989.

Bibring JP, Borg J, Langevin Y, Rosenbaum B, Salvetat P & Vassent B. The COMET experiment: first results. Lunar & Planetary Sci. 19: 73, 1988.

Biemann K. Test results on the Viking gas chromatograph-mass spectrometer experiment. Origins of Life 5: 417–30, 1974.

Biemann K. The implications and limitations of the findings of the Viking organic analysis experiment. J. Mol. Evol. 14: 65–70, 1979.

Biemann K, Oro J, Toulmin P III, Orgel LE, Nier AO, Anderson DM, Simmonds PG, Flory D, Diaz AV, Rushneck DR & Biller JA. Search for organic and volatile inorganic comounds in two surface samples from the Chryse Planitia region of Mars. Science 194: 72–6, 1976.

Biemann K, Oro J, Toulmin P III, Orgel LE, Nier AO, Anderson DM, Simmonds PG, Flory D, Diaz AV, Rushneck DR, Biller JE & Lafleur AL. The search for organic substances and inorganic volatile compounds in the surface of Mars. J. Geophys. Res. 82: 4641–58, 1977.

Bjurstedt H. *Biology and Medicine in Space*, ESA BR-01, ESA, Paris, 1979.

Black DC & Suffolk GCJ. Concerning the planetary system of Barnard's star. Icarus 19: 353, 1973.

Bluth BJ. Lunar settlements – a socio-economic outlook. Acta Astronaut. 17: 659–67, 1988.

Bonner WA, Hall H, Chow G, Liang Y & Lemmon R. The radiolysis and radioracemization of amino acids on clays. Origin of Life 15: 103–14, 1985.

Boston PJ, Ivanov MV & McKay CP. On the possibility of chemosynthetic ecosystems in the subsurface habitats on Mars. Icarus 95: 300–8, 1992.

Bracewell RN. Communications from superior galactic communities. Nature 186: 670, 1960.

Brack A. Liquid water and the origins of life. Origins of Life 23: 3–10, 1993.

Brack A & Raulin F. *L'évolution Chimique et les Origines de la Vie*. Masson, Paris, 1991.

Bridgewater D, Allaart JH, Schopf JW, Klein C, Walter MR, Barghoorn ES, Strother P, Knoll AH & Gorman BE. Microfossil-like objects from the Archean of Greenland: a cautionary note. Nature 289: 51–3, 1981.

Brownlee DE. Microparticle studies by sampling techniques. In: *Cosmic Dust.* Ed. by McDonnell JAM. John Wiley and Sons, Chichester, p. 295–336, 1978.

Brownlee DE. In: *The Sea.* Ed. by Emiliani C. Wiley, New York, pp. 733, 1981.

Brownlee D, Bunch T, Chang S, Kerridge J & Wolfe J. Cosmic dust collection. In: *Exobiology in Earth Orbit.* Ed. by DeFrees D, Brownlee D, Tarter J, Usher D, Irvine W & Klein H. NASA SP-500, NASA, Washington, pp. 77–87, 1989.

Bruch CW. Microbes in the upper atmosphere and beyond. In: *Airborne Microbes.* Ed. by Gregory PA & Monteith JL. Symp. Soc. Gen. Microbiol., Vol.17, Cambridge University Press, p. 385, 1967.

Bücker H & Horneck G. Discussion of a possible contamination of space with terrestrial life. Life Sci. Space Res. 7: 21–7, 1969.

Bücker H & Horneck G. Survival of microorganisms under simulated space conditions. Life Sci. Space Res. 8: 33–8, 1970.

Bücker H, Horneck G, Wollenhaupt H, Schwager M & Taylor GR. Viability of *Bacillus subtilis* spores exposed to space environment in the M-191 experiment system aboard Apollo 16. Life Sci. Space Res. 12: 209–13, 1974.

Bücker H & Horneck G. Studies on the effects of cosmic HZE-particles on different biological systems in the Biostack I and II flown on board of Apollo 16 and 17. In: *Radiation Research.* Ed. by Nygaard OF, Adler HJ & Sinclair WK. Academic Press, pp. 1138–51, 1975.

Bücker H, Dose K, Horneck G & Thomas C. A special photoproduct of UV-irradiated DNA in vacuo. Life Sci. Space Res. 17: 111–5, 1979.

Burke JD. Development of a lunar infrastructure. *Acta Astronaut.* 17: 669–74, 1988.

Cadet J, Voituriez L, Grand A, Hruska FE, Vigny P & Kan LS. Photosensitized reactions of nucleic acids. Biochimie 67: 277–292, 1985.

Cairns-Smith AG. *Thinking about the Unthinkable: Life Without Carbon.* Cambridge University Press, Cambridge, 1982.

Cameron AGW. Origin of the solar system. Ann. Rev. Astr. Ap. 26: 441, 1988.

Campbell B, Walker GAH & Yang S. A search for planetary mass companions to nearby stars. In: *Bioastronomy – The Next Steps.* Ed. by G. Marx. Kluwer, Dordrecht, pp. 83–90, 1988.

Carle C, Schwartz DE & Huntington JL. *Exobiology in Solar System Exploration.* NASA SP 512, NASA Washington D.C., 1992.

Carr MH. *The Surface of Mars.* Yale University Press, New Haven, 1981.

Carr MH. Water on Mars. Nature 326: 30–5, 1987.

Carr MH. Recharge of the early atmosphere of Mars by impact-induced release of CO_2. Icarus 79: 311–27, 1989.

Carr MH & Schaber GG. Martian permafrost features. J. Geophys. Res. 82: 4039–65, 1977.

Carr MH & Wänke H. Earth and Mars: inventories as clues to accretional histories. Icarus 98: 61–71, 1992.

Chicarro AL, Scoon GEN & Coradini M. *Mission to Mars,* Report of the Mars Exploration Study Team. ESA SP-1117, ESA, Paris, 1989.

Chun S, Pang K, Cutts JA & Ajello JM. Photocatalytic oxidation of organic compounds on Mars. Nature 274: 875–6, 1978.

Chyba CF. Terrestrial mantle siderophiles and the lunar impact record. Icarus 217: 217–33, 1991.

Chyba CF. The violent environment of the origin of life. In: *Frontiers of Life.* Ed. by Trn Thanh Vn J, Trn Thanh Vn K, Mounolou JC, Schneider J, & McKay C. Edition Frontieres, Gif-sur-Yvette, pp. 97–104, 1992.

Chyba DF, Thomas PJ, Brookshaw L & Sagan C. Cometary delivery of organic molecules to the early Earth. Science 249: 366–73, 1990.

Clark BC. Solar-driven chemical energy source for a Martian biota. Origins of Life 9: 241–9, 1979.

Clark BC. Barriers to natural interchange of biologically active material between Earth and Mars. Origins of Life 16: 410–1, 1985.

Clark BC. Primeval procreative comet pond. Origins of Life 18: 209–38, 1988.

Clark BC, Baird AK, Rose HJ, Keil K, Castro AJ, Kelliher WC, Roe CD & Evans PH. Inorganic analyses of Martian surface samples at the Viking landing sites. Science 194: 1283–8, 1976.

Clark BC & Baird AK. Volatiles in Martian regolith. Geophys. Res. Lett. 10: 811–14, 1979.

Cocconi G & Morrison P. Searching for interstellar communications. Nature 184: 844–6, 1959.

Cohen L & Kasting JF. Venus: a search for clues to early biological possibilities. In: *Exobiology in Solar System Exploration.* Ed. by Schwartz DE & Huntington JL. NASA SP 512, Washington DC, pp. 45–65, 1992.

Cohen Y, Castenholz RW & Halvorson HO. *Microbial Mats: Stromatolites.* MBL Lectures in Biology, Vol. 3, AR Liss, New York, 1983.

Cox CS. *The Aerobiological Pathways of Microorganisms.* Wiley, Chichester, 1987.

Cox CS. Role of water molecules in bacteria and viruses. Origins of Life 23: 29–36, 1993.

Crick, FHC. *Life Itself, its Origin and Nature.* Simon and Schuster, New York, 1981.

Crick FHC & Orgel LE. Directed panspermia. Icarus 19: 341–6, 1973.

Cronin JR. Acid-labile amino acid precursors in the Murchison meteorite I: chromatographic fractionation. Origins of Life 7: 337–48, 1976a.

Cronin JR. Acid-labile amino acid precursors in the Murchison meteorite II: a search for peptides and amino acyl amides. Origins of Life 7: 343–8, 1976b.

Cronin JR, Pizarello S & Cruikshank DP. Organic matter in carbonaceous chondrites, planetary satellites, asteroids, and comets. In: *Meteorites and the Early Solar System.* Ed. by Kerridge JF & Matthews MS. Univ. Arizona Press, pp. 819–57, 1988.

Crowe LM & Crowe JH. Anhydrobiosis: a strategy for survival. *Adv. Space Res.* 12 (4): 239–47, 1992.

Cruikshank DP & Kerridge JF. Organic material: asteroids, meteorites and planetary satellites. In: *Exobiology in Solar System Exploration.* Ed. by Carle C, Schwartz DE & Huntington JL. NASA SP 512, Washington D.C., pp. 229–47, 1992.

Danson MJ. Archaebacteria: the comparative enzymology of their central metabolic pathways. Adv. Microbiol. Physiol. 29: 165–231, 1988.

Day W. *Genesis on Planet Earth: The Search for Life's Beginning*, 2nd ed. Yale University Press, New Haven, Conn., 1984.

Deamer DW. Boundary structures are formed by organic components of the Murchison carbonaceous chondrite. Nature 317: 792–4, 1985.

Deamer DW. Role of amphiphilic compounds in the evolution of membrane structure on the early Earth. Origins of Life 17: 3–25, 1986.

de Bergh C. The D/H ratio and the evolution of water in the terrestrial planets. Origins of Life 23: 11–21, 1993.

de Bergh C, Bézard B, Owen T, Crisp D, Maillard JP & Lutz BJ. Deuterium on Venus: observations from Earth. Science 251: 547–9, 1991.

DeFrees D, Brownlee D, Tarter J, Usher D, Irvine W & Klein H. *Exobiology in Earth Orbit*, NASA SP-500, NASA Washington D.C. 1989.

Delsemme AH. Nature and history of the organic compounds in comets: an astrophysical view. In: *Comets in the Post-Halley Era* 1. Ed. by Newburn RL Jr, Neugebauer M & Rahe J. Kluwer, Dordrecht, pp. 377–428, 1991.

Delsemme AH. Cometary origin of carbon, nitrogen and water on the Earth. Origins of Life 21: 279–298, 1992.

DeRosa M & Gambacorta A. Lipid biogenesis in archaebacteria. System. Appl. Microbiol. 7: 278–85, 1986.

DeVincenzi DL. NASA's exobiology program. Origins of Life 14: 793–9, 1984.

DeVincenzi DL. Planetary protection issues and the future exploration of Mars. Adv. Space Res. 12 (4): 121–8, 1992.

DeVincenzi DL & Klein HP. Planetary protection issue for sample return missions. Adv. Space Res. 9 (6): 203–6, 1989.

DeVincenzi DL & Pleasant LG. *First Symposium on Chemical Evolution and Origin and Evolution of Life*. NASA CP 2276, NASA, Washington, 1983.

DeVincenzi DL & Stabekis PD. Revised planetary protection policy for solar system exploration. Adv. Space Res. 4 (12): 291–5, 1984.

Dickerson RE. Chemical evolution and the origin of life. Scientific Amer. 239: 70–86, 1978.

Dodonova NY, Kiseleva MN, Remisova LA & Tsyganenko NM. The vacuum ultraviolet photochemistry of nucleotides. Photochem. Photobiol. 35: 129–32, 1982.

Dole SH. *Habitable Planets for Man*, Blaisdell, New York, 1964.

Donahue TM. Deuterium on Venus. Nature 340: 513–14, 1989.

Donahue TM, Hoffman JH, Hodges RR & Watson AJ. Venus was wet: a measurement of the ratio of D to H. Science 216: 630–3, 1982.

Dose K. Hypotheses on the appearance of life on Earth. Adv. Space Res. 6 (12): 181–6, 1986a.

Dose K. Survival under space vacuum. Adv. Space Res. 6 (12): 307–12, 1986b.

Dose K. Präbiotische Evolution und der Ursprung des Lebens, ein kritischer Rückblick. Chemie in unserer Zeit 21: 177–85, 1987.

Dose K, Bieger-Dose A, Kerz O & Gill M. DNA-strand breaks limit survival in extreme dryness. Origins of Life 21: 177–87, 1991.

Dose K, Bieger-Dose A, Labusch M & Gill M. Survival in extreme dryness and DNA-single-strand breaks. Adv. Space Res. 12 (4): 221–9, 1992.

Dose K & Rauchfu H. *Chemische Evolution und der Ursprung lebender Systeme*. Wissenschaftliche Verlagsgesellschaft, Stuttgart, 1975.

Doyle LR & McKay CP. Exobiological habitats: an overview. In: *Bioastronomy, The Search for Extraterrestrial Life – The Exploration Broadens*. Ed. by Heidmann J & Klein MJ. Springer-Verlag, Berlin, pp. 163–72, 1991.

Doyle LR. Progress in determining the space orientation of stars, In: *Bioastronomy – The Next Steps*. Ed. by Marx, G. Kluwer, Dordrecht, pp. 101–5, 1988.

Draganic ZD, Vujosevic SI, Negron-Mendoza A, Azamar JA & Draganic IG. Radiation chemistry of a multicomponent aqueous system relevant to chemistry of cometary nuclei. J. Mol. Evol. 22: 175–87, 1985.

Drake F. How can we detect radio transmissions from distant planetary systems? Sky Telescope 19: 140, 1960.

Drake FD. Project Ozma. Physics Today 14: 40–6, 1961.

Drake FD. Methods of communication – message content, search strategy, interstellar travel. In: *Interstellar Communications*. Ed. by Ponnamperuma C & Cameron AGW. Houghton Mifflin, Boston, pp. 118–139, 1974.

Duke MB & Mendell WW. Scientific investigations at a lunar base. *Acta Astron.* 17: 675–90, 1988.

Durham R, Schmunk RB & Chamberlain JW. Comparative analysis of the atmospheres of early Earth and Mars. Adv. Space Res. 9 (6): 139–42, 1989.

Eigen M. *Stufen zum Leben*. Piper, München, 1987.

Fechtig H & Rahe J. Comets and ESA's space mission Giotto to Halley's comet. *Naturwiss.* 71: 275–93, 1984.

Ferris JP, Jacobson R & Guillemin C. The photolysis of NH_3 in the presence of substituted acetylenes: a possible source of oligomers and HCN on Jupiter. Icarus 95: 54–59, 1992.

Fogleman G, Huntington JL, Schwartz DE & Fonda ML. Gas-grain simulation facility: fundamental studies of particle formation and interactions. NASA Conference Publ. 10026, 1989.

Foster TL, Winanas L Jr, Casey RC & Kirschner LE. Response of terrestrial microorganisms to simulated Martian environment. Appl. Environ. Microbiol. 35: 730–7, 1978.

Fox SW & Dose K. *Molecular Evolution and the Origin of Life*. Marcel Dekker, New York, 1977.

Friedmann EI. Endolithic microbial life in hot and cold deserts. Origins of Life 10: 223–35, 1980.

Friedmann EI. Endolithic microorganisms in the Antarctic cold desert. Science 215: 1045–53, 1982.

Friedmann EI. The Antarctic cold desert and the search for traces of life on Mars. Adv. Space Res. 6 (12): 265–8, 1986.

Friedmann EI. *Antarctic Microbiology*, Wiley-Liss, New York, 1993.

Friedmann EI & Koriem AM. Life on Mars: how it disappeared (if it was ever there). Adv. Space Res. 9 (6): 167–72, 1989.

Friedmann EI, McKay CP & Nienow JA. The cryptoendolithic microbial environment in the Ross Desert of Antarctica; satellite transmitted continuous nanoclimate data. Polar Biol. 7: 273–87, 1987.

Friedmann EI & Ocampo R. Endolithic blue-green algae in the dry valleys: primary producers and the Antarctic desert ecosystem. Science 193: 1247–9, 1976.

Gilichinsky DA, Soina VS & Petrova MA. Cryoprotective properties of water in the Earth cryolithosphere and its role in exobiology. Origins of Life 23: 65–75, 1993.

Gilichinsky DA, Vorobyova EA, Erokhina LG, Fyordorov-Dayvdov DG & Chaikovskaya NR. Long-term preservation of microbial ecosystems in permafrost. Adv. Space Res. 12 (4): 255–63, 1992.

Glass HB. *Life Sciences in Space*. National Academy of Sciences, Washington, 1970.

Godamski J. Ecospheres of giant stars and ecospheric evolution of Sun's planetary system. *Acta Astronautica* 7: 1–7, 1961.

Gogarten JP, Kibak H, Dittrich P, Taiz L, Bowman EJ, Bowman BJ, Manolson MF, Poole RJ, Date T, Oshima T, Konishi J, Denda K & Yoshida M. Evolution of vacuolar H^+-ATPases: implications for the origin of eukaryotes. Proc. Natl. Acad. Sci USA 86: 6661–5, 1989.

Gogarten JP & Taiz L. Evolution of proton pumping ATPases: rooting the tree of life. Photosynthesis Res. 33: 137–46, 1992.

Gogarten-Boekels M, Hilario E & Gogarten JP. The effects of heavy bombardment on the early evolution – the emergence of the three domains of life. Origins of Life 25: 251–64, 1995.

Goldsmith D. *The Quest for Extraterrestrial Life, a Book of Readings*. University Science Books, Mill Valley, California, 1980.

Goldsmith D & Owen T. *The Search for Life in the Universe*, Benjamin/Cummings, Menlo Park, California, 1980.

Goldsmith D & Owen T. *Auf der Suche nach Leben im Weltall*, Hirzel, Stuttgart, 1984.

Gooding JL. Soil mineralogy and chemistry on Mars: possible clues from salts and clays in SNC meteorites. Icarus 99: 28–41, 1992.

Grahn D. *HZE-Particles in Manned Spaceflight*. National Academy of Sciences, Washington, D.C., 1973.

Greenberg JM. Radical formation, chemical processing and explosion of interstellar grains. Astrophys. Space Sci. 39: 9–18, 1976.

Greenberg JM. What are comets made of? A model based on interstellar dust. In: *Comets*. Ed. by Wilkening LL. University of Arizona Tucson, pp. 131–63, 1982.

Greenberg JM. The largest molecules in space: interstellar dust. In: *Cosmochemistry and the Origin of Life*. Ed. by Ponnamperuma C. Reidel, Dordrecht, pp. 71–112, 1983.

Greenberg JM. The chemical composition of comets and possible contribution to planet composition and evolution. In: *The Galaxy and the Solar System*, Ed. by Smoluchowski R, Bahcall JN & Matthews MS. University of Arizona Press, Tucson, pp. 103–15, 1986.

Greenberg JM & Hage JI. From interstellar dust to comets. Astrophys. J. 361: 260–74, 1990.

Greenberg JM & Mendoza-Gomez C. The seeding of life by comets. Adv. Space Res. 12 (4): 169–80, 1992.

Greenberg JM, Mendoza-Gmez CX & Pironelli V. *The Chemistry of Life's Origin.* Kluwer, Dordrecht, 1993.

Gregory PH. *Microbiology of the Atmosphere.* Leonard Hill, Intertext Publ., Aylesbury, Bucks, 1973.

Griffith CA, Bézard B, Owen T & Gautier D. The tropospheric abundances of NH_3 and PH_3 in Jupiter's great red spot, from Voyager IRIS observations. Icarus 98: 82–93, 1992.

Grinspoon DH. Was Venus wet? Deuterium reconsidered. Science 238: 1702–4, 1987.

Grün E. (Ed.) Comets Halley and Giocobini-Zinner. Adv. Space Res. 5 (12), 1985.

Grün E, Zook HA, Fechtig H & Giese RH. Collisional balance of the meteorite complex. Icarus 62: 244–72, 1985.

Gupta S, Ochiai E & Ponnamperuma C, Organic synthesis in the atmosphere of Titan. Nature 293: 725, 1981.

Häder DP. Risk of enhanced solar ultraviolet radiation for aquatic ecosystems. Progr. Phycol. Res. 9: 1–45, 1993.

Hagen CA, Hawrylewicz EJ, Anderson BT & Cephus ML. Effect of ultraviolet on the survival of bacteria airborne in simulated Martian dust clouds. Life Sci. Space Res. 8: 53–8, 1970.

Hagen U. Radiation biology in space: a critical review. Adv. Space Res. 9 (10): 3–8, 1989.

Hall LB. Ten years of development of the planetary quarantine program of the United States. Life Sci. Space Res. 12: 185–97, 1974.

Halvorson HO & Van Holde KE. *The Origins of Life and Evolution.* AR Liss, New York, 1980.

Hart MH. Habitable zones about main sequence stars. Icarus 37: 351–7, 1979.

Hartman H, Lawless JG & Morrison P. *Search for the Universal Ancestors.* NASA SP-477, 1985.

Hartman H & Matsuno K. *The Origin and Evolution of the Cell.* World Scientific, Singapore, 1992.

Hartmann WK, Phillips RJ & Taylor GJ. *Origin of the Moon.* Lunar Planetary Institute, Houston, 1986.

Hawrylewicz EJ, Hagen CA & Ehrlich R. Survival of microorganisms in a simulated Martian environment. I. *Bacillus subtilis* var. *globigii*. Appl. Microbiol. 12: 215–8, 1964.

Hawrylewicz EJ, Hagen CA, Tolcjacz V, Anderson BT & Ewing M. Probability of growth (pG) of viable microorganisms in Martian environments. Life Sci. Space Res. 6: 146–56, 1968.

Hawrylewicz EJ, Hagen CA, Tolcacz V & Ehrlich R. Effect of reduced barometric pressure on water availability related to microbial growth. Life Sci. Space Res. 5: 174–86, 1967.

Hayatsu R & Anders E. Organic compounds in meteorites and their origins, Topics Curr. Chem. 99: 1, 1981.

Haynes RH. Etablierung von Leben auf dem Mars durch gerichtete Panspermie: Technische und ethische Probleme der ökopoese. Biol. Zent. bl. 109: 193–205, 1990.

Haynes RH & McKay CP. The implantation of life on Mars: feasibility and motivation. Adv. Space Res. 12 (4): 133–40, 1992.

Head JW, Crumpler L, Guest J & Saunders RS. Venus volcanism, classification of volcanic features and structures, associations, and global distribution from Magellan data. J. Geophys. Res. 97: 13153–97, 1992.

Heidmann J & Klein MJ. *Bioastronomy, The Search for Extraterrestrial Life.* Springer, Berlin, 1991.

Helfer HL. Of Martian atmospheres, oceans, fossils. Icarus 89: 228–35, 1990.

Hemenway CL, Hallgreen DS, Coon RE & Bourdillon LA. Technical description of the Gemini S-10 and S-12 micrometeorite experiments. Space Res. 8: 510–20, 1967.

Hieda K & Ito T. Action spectra for inactivation and membrane damage of *Saccharomyces cerevisiae* cells in vacuum by monochromatic synchroton UV radiation (155–250 nm). Photochem. Photobiol. 44: 409–11, 1986.

Hieda K, Hayakawa Y, Ito A, Kobayashi K & Ito T. Wavelength dependence of the formation of single strand breaks and base changes in DNA by ultraviolet radiation above 150 nm. Photochem. Photobiol. 44: 379–83, 1986.

Hilario E & Gogarten JP. Horizontal transfer of ATPase genes – the tree of life becomes a net of life. BioSystems 31: 111–9, 1993.

Hirsch P, Hoffmann B, Gallikowski CA, Mevs U, Siebert J & Sittig M. Diversity and identification of heterotrophs from Antarctic rocks of the McMurdo Dry Valley (Ross Desert). Polarforschung 58: 261–9, 1988.

Holland HD. *The Chemical Evolution of the Atmosphere and Oceans.* Princeton University, Princeton, 1984.

Holland HD, Blackburn TR, Ceasar GP & Quirk RF. X-ray photoelectron spectrometric and gas exchange evidence for surface oxidation of Martian regolith analogues by ultraviolet irradiation. Life Sci. Space Res. 17: 65–76, 1979.

Holm NG. New evidence for a tubular structure of -iron(III) oxide hydroxide akaganéite. Origins of Life 15: 131–9, 1985.

Hoover RB, Hoyle F, Wickramasinghe NC, Hoover MJ & Al-Mufti S. Diatoms on Earth, comets, Europa, and in interstellar space. Earth, Moon, Planets 35: 19–45, 1986.

Horneck G. Survival of microorganisms in space: A review. Adv. Space Res. 1 (14): 39–48, 1981.

Horneck G. Exobiological aspects in view of Titan. In: *The Atmospheres of Saturn and Titan.* Ed. by Rolfe E & Battrick B. ESA SP-241, ESA Paris, pp. 157–159, 1985.

Horneck G. European activities in exobiological research in space. In: *Proceedings of the 3rd ESA European Symposium on Life Sciences Research in Space*, Graz, Austria. Ed. by Hunt H & Oser H. ESA SP-271, ESA, Paris, 1987.

Horneck G. Survival strategies for life in high UV, very low density environment. In: *Bioastronomy – The Next Steps.* Ed. by Marx G. Kluwer Academic Publishers, pp. 201–5, 1988.

Horneck G. Exobiology, In: *Life Sciences Research in Space.* Ed. by H.Oser & Battrick B. ESA SP-1105, ESA, Paris, 1989.

Horneck G. Impact of human expeditions to Mars. Adv. Space Res. 12 (4): 75–140, 1992.

Horneck G. Radiobiological experiments in space: a review. Nucl. Tracks Radiat. Meas. 20: 185–205, 1992.

Horneck G. Responses of *Bacillus subtilis* spores to space environment: results from experiments in space. Origins of Life 23: 37–52, 1993.

Horneck G. Radiation Biology. In: Proceedings, *Symposium on the Scientific Results of the German Spacelab Mission D2*, in press.

Horneck G, Barrett J, Bieger A, Bücker H, Dose K, Greenberg J, Kiefer J, Mennigmann HD, Reitz G & Weber P. Exobiological experiments on the first EURECA mission. Origins of Life 16: 405–6, 1985.

Horneck G & Brack A. Study of the origin, evolution and distribution of life with emphasis on exobiology experiments in Earth orbit, In: *Advances in Space Biology and Medicine*, Vol. 2. Ed. by Bonting SL. JAI Press, Greenwich, CT, pp. 229–62, 1992.

Horneck G, Brack A, Cadet J, Dose K, Fritz-Niggli H, Kiefer J, Broerse JJ, Raulin F, Reitz G, Roessler K & Schott JU. *ERA Follow-on Scientific Study.* ESA Contract No. 8116/88/F/BZ(SC), 1989.

Horneck G & Bücker H. Can microorganisms withstand the multistep trial of interplanetary transfer? Considerations and experimental approaches. Origins of Life 16: 414–5, 1985.

Horneck G, Bücker H, Dose K, Martens KD, Bieger A, Mennigmann HD, Reitz G, Requardt H & Weber P. Microorganisms and biomolecules in space environment. Experiment ES 029 on Spacelab 1. Adv. Space Res. 4 (10): 19–27, 1984b.

Horneck G, Bücker H & Reitz G. Long-term survival of bacterial spores in space. Adv. Space Res. 14 (10): 41–5, 1994.

Horneck G, Bücker H, Reitz G, Requardt H, Dose K, Martens KD, Mennigmann HD & Weber P. Microorganisms in the space environment. Science 225: 226–8, 1984a.

Horneck G, Bücker H & Wollenhaupt H. Survival of bacterial spores under some simulated lunar surface conditions. Life Sci. Space Res. 9: 119–24, 1971.

Horneck G, Eschweiler U, Reitz G, Wehner J, Willimek R & Strauch K. Biological responses to space: results of the experiment 'Exobiological nit' of ERA on EURECA I. Adv. Space Res. 16 (8): 105–18, 1995.

Horowitz NH. The design of Martian biological experiments. Life Sci. Space Res. 2: 133–8, 1964.

Horowitz NH. The search for life on Mars. Scientific Amer. 237: 52–61, 1977.

Horowitz NM. *To Utopia and Back: the Search for Life in the Solar System.* Freeman, New York, 1986.

Horowitz NH, Cameron RE & Hubbarb JS. Microbiology of the dry valleys of Antarctica. Science 176: 242–5, 1972.

Horowitz NH, Hobby GL & Hubbarb JS. The Viking carbon assimilation experiments: interim report. Science 194: 1321–2, 1976.

Horowitz NH, Hobby GL & Hubbarb JS. Viking on Mars: the carbon assimilation experiments. J. Geophys. Res. 82: 4659–662, 1977.

Horowitz NH, Sharp RP & Davies RW. Planetary contamination I: the problem and the agreements. Science 155: 1501–5, 1967.

Hotchin J, Lorenz P, Markusen A & Hemenway C. The survival of microorganisms in space, further rocket and balloon-borne exposure experiments. Life Sci. Space Res. 5: 1–6, 1967.

Hotchin J, Lorenz P & Hemenway CL. The survival of terrestrial microorganisms in space at orbital altitudes during Gemini satellite experiments. Life Sci. Space Res. 6: 108–14, 1968.

Hoyle F & Wickramasinghe NC. Prebiotic molecules in interstellar grain clumps. Nature 266: 241–3, 1977.

Hoyle F & Wickramasinghe NC. Organic grains in space. Astrophys. Space Sci. 69: 511–3, 1980.

Huang SS. Life supporting regions in the vicinity of binary systems, In: *Interstellar Communication.* Ed. by Cameron AGW. W.A. Benjamin, New York, 1963.

Hubbarb JS. The pyrolytic release experiment: measurement of carbon assimilation. Origins of Life 7: 281–92, 1976.

Huck FO, McCall HF, Patterson WR & Taylor GR. The Viking Mars lander camera. Space Sci. Instrum. 1: 189–241, 1972.

Huebner WF. First polymer in space identified in comet Halley. Science 237: 628–30, 1987.

Huebner WF & Boice DC. Comets as possible source of prebiotic molecules, Origins of Life 21: 299–315, 1992.

IAA Ad Hoc Committee 'Return to the Moon'. The case for an international lunar base. *Acta Astronaut.* 17: 463–89, 1988.

Imshenetsky AA. A comparative evaluation of methods for the search for life on Mars. Life Sci. and Space Res. 7: 2–8, 1969.

Imshenetsky AA, Kusjurina LA & Jakshina VM. Multiplication of certain soil microorganisms under simulated Martian conditions. Life Sci. Space Res. 8: 59–61, 1970.

Imshenetsky AA, Kouzyurina LA & Jakshina VM. On the multiplication of xerophilic micro-organisms under simulated Martian conditions. Life Sci. Space Res. 11: 63–6, 1973.

Imshenetsky AA, Lysenko DV & Kazakov GA. Upper boundary of the biosphere, Appl. Environm. Microbiol. 35: 1–5, 1978.

Imshenetsky AA, Lysenko SV, Kazakov GA & Ramkova NV. On micro-organisms of the stratosphere. Life Sci. Space Res. 14: 359–62, 1976.

Imshenetsky AA, Pisarenko NF, Kuziurina LA & Yakshina VM. Physiology of xerophytic microorganisms growing under Martian conditions. Life Sci. Space Res. 15: 47–52, 1977.

Imshenetsky AA, Lysenko SV & Lach SP. Microorganisms of the upper layer of the atmosphere and the protective role of their cell pigments. Life. Sci. Space Res. 17: 105–9, 1979.

Imshenetsky AA, Murzajov BG, Evdokimova MD, Dorofeyeva IK, Kzujurina LA & Jakshina VM. On the possibility of life on Mars. Life Sci. Space Res. 18: 45–51, 1980.

Imshenetsky AA, Murzakov BG, Evdokimova MD & Dorofeyeva IK. Biological studies of Martian soil analogue. Adv. Space Res. 1 (14): 21–6, 1981.

Irvine WM. Chemistry in the cosmos. In: *Frontiers of Life*. Ed. by Trn Thanh Vn J, Trn Thanh Vn K, Mounolou JC, Schneider J, & McKay C. Edition Frontieres, Gif-sur-Yvette, pp. 263–90, 1992.

Irvine WM, Leschine SB & Schloerb FP. Thermal history, chemical composition and relationship of comets to the origin of life. Nature 283: 748–9, 1980.

Ito A, Taniguchi T & Ito T. Wavelength dependence for the inactivation of ATP in the vacuum-ultraviolet region above 140 nm. Photochem. Photobiol. 44: 273–7, 1986.

Ito T, Ito A, Hieda K & Kobayashi K. Wavelength dependence of inactivation and membrane damage to *Saccharomyces cerevisiae* cells by monochromatic synchroton vacuum-UV radiation (145–190 nm). Radiat. Res. 96: 532–48, 1983.

Ivanov IM. Potential for searching for chemolithoautotrophic microorganisms on Mars. USSR *Space Life Sciences Digest* (NASA Contractor Report 3922 (24))21: 38–40, 1989.

Ivanov CP, Stoyanova RZ & Mancheva IN. Some evidence for the possible presence of peptides in two chondrites by use of a sequencing procedure. Origins of Life 14: 61–8, 1984.

Iwabe N, Kuma KI, Hasegawa M, Osawa S & Miyata T. Evolutionary relationships of archaebacteria, eubacteria and eukaryotes inferred from phylogenetic trees of duplicated genes. Proc. Natl. Acad. Sci. USA 86: 9355–9, 1989.

Jabir NL, Hoyle F & Wickramasinghe NC. On the optical properties of bacterial grains I. Astrophys. Space Sci. 91: 327–44, 1983.

Janle P. Besteht die Möglichkeit, da Bakteriensporen auf natürliche Weise von der Erdoberfläche bzw. -atmosphäre in den interplanetaren Raum gelangen, In: *Entstehung und Entwicklung der terrestrischen Planeten*, 4. Ed. by Blum P. DFG Kolloquium über Planetenforschung, Schlo Ringberg, 2–5 Oct., pp. 354–64, 1984.

Janle P. Der Planet Mars. Sterne und Weltraum 32: 102–11 & 262–73, 1993.

Jaskowski DG & Squyres SW. 'Softened' impact craters on Mars: implications for ground ice and the structure of the Martian megaregolith. Icarus 106: 365–79, 1993.

Jessberger R. *Kreationismus, Kritik des modernen Antievolutionismus.* Paul Parey Verlag, Berlin, 1990.

Jessberger EK, Christoforidis A & Kissel J. Aspects of the major element composition of Halley's dust. Nature 332: 691–5, 1988.

Johnson RE, Lanzerotti LI & Brown WL. Sputtering processes: erosion and chemical change. Adv. Space Res. 4: 41–51, 1984.

Kanavarioti A & Mancinelli RL. Could organic matter have been preserved on Mars for 3.5 billion years? Icarus 84: 196–202, 1990.

Kandler O. The early diversification of life. In: *Early Life on Earth, Nobel Symposium 84*. Ed. by Bengtson S. Columbia U.P., New York, pp. 152–60, 1993.

Kaplan RW. *Der Ursprung des Lebens.* Thieme, Stuttgart, 1978.

Kargel JS & Lewis JS. The composition and early evolution of Earth. Icarus 105: 1–25, 1993.

Kasting JF. Runaway and moist greenhouse atmospheres and the evolution of Earth and Venus. Icarus 74: 472–94, 1988.

Kasting JF & Ackerman TP. Climatic consequences of very high carbon dioxide levels in the Earth's early atmosphere. Science 234: 1383–5, 1986.

Kasting JF, Toon OB & Pollack JB. How climate evolved on the terrestrial planets. Scientific Amer. 258: 46–53, 1988.

Kasting JF, Whitmire DP & Reynolds RT. Habitable zones around main sequence stars. Icarus 101: 108–28, 1993.

Keller B & Horneck G. Action spectra in the vacuum-UV and far-UV (122–300 nm) for inactivation of wet and vacuum-dry spores of *Streptomyces griseus* and photoreactivation. J. Photochem Photobiol. B: Biology 16: 61–72, 1992.

Kerr RA. Martian meteorites are arriving. Science 237: 721–3, 1987.

Kerridge SJ, Flury W, Horn LJ, LeBreton JP, Stetson DS, Stoller RL & Tan GH, Cassini: mission to Saturn and Titan. In: *Exobiology in Solar System Exploration*. Ed. by Carle C, Schwartz DE & Huntington JL. NASA SP 512, Washington D.C., pp. 229–247, 1992.

Khare BN, Sagan C, Bandurski EL & Nagy B. Ultraviolet-photoproduced organic solids synthesized under simulated Jovian conditions: molecular analysis. Science 199: 1199–1201, 1978.

Kibak H, Taiz L, Starke T, Bernasconi P & Gogarten JP. Evolution of structure and function of V-ATPases. J. Bioenerg. Biomembranes 24: 415–24, 1992.

Kimura M. *The Neutral Theory of Molecular Evolution*. Cambridge University Press, New York, 1983.

Kirsten T. Time and the solar system, In: *The Origin of the Solar System*. Ed. by Dermott SF. Wiley and Sons, New York, pp. 267–346, 1978.

Kissel J, Brownlee DE, Büchler K., Clark BC, Fechtig H, Grün E, Hornung K, Igenbergs EB, Jessberger EK, Krueger FR, Kuczera H, McDonnell JAM, Morfill GM, Rahe J, Schwehm GH, Sekanina Z, Utterback NG, Völk HJ & Zook HA. Composition of comet Halley dust particles from Giotto observations. Nature 321: 336–7, 1986a.

Kissel J, Sagdeev RZ, Bertaux JL, Angarov VN, Audouze J, Blamont JE, Büchler K, Evlanov EN, Fechtig H, Fomenkova MN, von Hoerner H, Inogamov NA, Khromov VN, Knabe W, Krueger FR, Langevin Y, Leonas VB, Levasseur-Regourd AC, Managadze GG, Podkolzin SN, Shapiro VD, Tabaldyev SR & Zubkov BV. Composition of comet Halley dust particles from Vega observations. Nature 321: 280–2, 1986b.

Kissel J & Krueger FR. The organic component in dust from comet Halley as measured by the PUMA mass spectrometer on board of Vega 1. Nature 326: 755–60, 1987.

Klein HP. Automated life-detection experiments for the Viking mission to Mars. Origins of Life 5: 431–41, 1974.

Klein HP. General constraints of the Viking biology investigation. Origins of Life 7: 273–79, 1976.

Klein HP. The Viking biological investigations: general aspects. J. Geophys. Res. 82: 4677–80, 1977.

Klein HP. The Viking biological investigations: review and status. Origins of Life 9: 157–60, 1978.

Klein HP. The Viking mission and the search for life on Mars. Rev. Geophys. Space Phys. 17: 1655–62, 1979.

Klein HP. Exobiology revisited. Adv. Space Res. 6 (12): 187–92, 1986.

Klein HP. Strategies for the future exploration of Mars. Adv. Space Res. 7: 95–8, 1987.

Klein HP. *The Search for Life's Origin: Progress and Future Directions in Planetary Biology and Chemical Evolution*. National Academic Press, Washington, 1990.

Klein HP. The Viking biology experiments: epilogue and prologue. Origins of Life 21: 255–61, 1992.

Klein HP, Lederberg J & Rich A. Biological experiments: the Viking Mars Lander. Icarus 16: 139–46, 1972.

Klein HP & Vishniac W. Biological instrumentation for the Viking 1975 mission to Mars. Life Sci. Space Res. 10: 201–10, 1972.

Klenk HP, Palm P & Zillig W. DNA-dependent RNA polymerases as phylogenetic marker molecules. System. Appl. Microbiol. 16: 638–47, 1994.

Klenk HP & Zillig W. DNA-dependent RNA-polymerase subunit B as a tool for phylogenetic reconstructions: branching topology of Archaeal domain. J. Molec. Evol. 38: 420–32, 1994.

Kochan H, Feuerbacher B., Joo F, Klinger J, Seboldt W, Bischoff A, Düren H, Stöffler D, Spohn T, Fechtig H, Grün E, Kohl H, Krankowsky D, Rössler K, Thiel K, Schwehm G & Weishaupt U. Comet simulation experiments in the DFVLR space simulators. Adv. Space Res. 9 (3): 113–22, 1989.

Koelle HH. Future Lunar Base. Acta Astronaut. 17: 655–750, 1988.

Koike J, Oshima T, Koike K, Taguchi H, Tanaka R, Nishimura K & Miyaji M. Survival rates of some terrestrial microorganisms under simulated space conditions. Adv. Space Res. 12 (4): 271–4, 1992.

Kok B & Varner JE. Extraterrestrial life detection by means of isotopic oxygen exchange. Life Sci. Space Res. 5: 206–16, 1967.

Komatsu G, Baker VR & Gulick VC. Venusian channels and valleys: distribution and volcanological implications. Icarus 102: 1–25, 1993.

Krueger FR & Kissel J. Biogenesis by cometary grains – thermodynamic aspects of self-organization. Origins of Life 19: 87–93, 1989.

Krueger FR, Korth A & Kissel J. The organic matter of comet Halley as inferred by gas phase and solid phase analyses. Space Sci. Rev. 56: 167–75, 1991.

Kuhn WR & Atreya SK. Solar radiation incident on the Martian surface. J. Mol. Evol. 14: 57–64, 1979.

Labs D, Neckel H, Simon PC & Thuillier G. Ultraviolet solar irradiance measurement from 200 to 358 nm during Spacelab 1 mission. Solar Phys. 107: 203–19, 1987.

Lang EW. Physical-chemical limits for the stability of biomolecules. Adv. Space Res. 6 (12): 251–5, 1986.

Langevin Y, Kissel J, Bertaux JL & Chassefiere E. First statistical analysis of 5000 mass spectra of cometary grains obtained by PUMA I (Vega 1) and PIA (Giotto) impact ionization mass spectrometers in the compressed modes. Astron. Astrophys. 187: 761–6, 1987.

Lawless JG & Peterson E. Amino acids in carbonaceous chondrites. Origins of Life 6: 3–8, 1975.

Lebreton J-P. The Cassini mission to Saturn and Titan – An overview. In: *The Atmospheres of Saturn and Titan.* Ed. by Rolfe E & Battrick B, ESA SP-241, ESA Paris, pp. 225–32, 1985.

Lederberg J. Exobiology: approaches to life beyond the Earth. Science 132: 393–400, 1960.

Lederberg J & Sagan C. Microenvironments for life on Mars. Proc. Natl. Acad. Sci. USA 48: 1473–5. 1962.

Leinert C & Grün E, Interplanetary dust. In: *Physics of the Inner Heliosphere.* Ed. by Schwenn R & Marsh E. Springer Verlag, pp. 207–75, 1989.

Levin GV, Heim AH, Thompson MF, Beem DR & Horowitz NH. 'Gulliver', an experiment for extraterrestrial life detection and analysis. Life Sci. Space Res. 2: 124–32, 1964.

Levin GV & Straat PA. Labeled release – an experiment in radiorespirometry. Origins of Life 7: 293–311, 1976a.

Levin GV & Straat PA. Viking labeled release biology experiment: interim results. Science 194: 1322–9, 1976b.

Levin GV & Straat PA. Recent results from the Viking labeled release experiment on Mars. J. Geophys. Res. 82: 4663–7, 1977a.

Levin GV & Straat PA. Life on Mars? The Viking labeled release experiment. BioSystems 9: 165–74, 1977b.

Levin GV & Straat PA. Search for a non-biological explanation of the Viking labeled release life detection experiment. Icarus 45: 494–516, 1981.

Levine JS. *The Photochemistry of the Atmospheres: Earth, the Other Planets and Comets.* Academic, Orlando, 1985.

Levine JS, Augustsson TR & Nataranjan M. The prebiological paleoatmoshere: stability and composition. Origins of Life 12: 245–59, 1982.
Levinthal E, Hundley L & Lederberg J. Multivator – a biochemical laboratory for Martian experiments. Life Sci. Space Res. 2: 112–23, 1964.
Levinthal EC, Jones KL, Fox P & Sagan C. Lander imaging as a detector of life on Mars, J. Geophys. Res. 82, 4468–78, 1977.
Lindberg C & Horneck G. Action spectra for survival and spore photoproduct formation of *Bacillus subtilis* irradiated with short-wavelength (200–300 nm) UV at atmospheric pressure and *in vacuo*. J. Photochem. Photobiol B: Biol. 11: 69–80, 1991.
Lindberg C & Horneck G. Thymine photoproduct formation and inactivation of intact spores of *Bacillus subtilis* irradiated with short wavelength UV (200–300 nm) at atmospheric pressure and *in vacuo*. Adv. Space Res. 12 (4): 275–9, 1992.
Lindberg C & Horneck G. Planetary protection considerations for Marsnet and Mars sample return missions. Adv. Space Res. 15 (3): 277–80, 1995.
Lorenz PR, Hemenway CL & Hotchin J. The biological effectiveness of solar electromagnetic radiation in space. Life Sci. Space Res. 6: 100–7, 1968.
Lorenz P, Orlob GB & Hemenway CL. Survival of microorganisms in space. Space Life Sci. 1: 491–500, 1969.
Lozina-Lozinsky LK & Bychenkova VN. Resistance of the protozoan *Colpoda maupasi* to Martian conditions of atmospheric pressure and low partial pressure of oxygen. Life Sci. Space Res. 7: 149–155, 1969.
Lozina-Lozinsky LK, Bychenkova VN, Zaar EI, Levin VL & Rumyantseva VM. Some potentialities of living organisms under simulated Martian conditions. Life Sci. Space Res. 9: 159–65, 1971.
Lucchitta BK. Ice and debris in the fretted terrain, Mars. J. Geophys. Res. 89: B409–18, 1984.
Lucchitta BK, Anderson DM & Shoji H. Did ice streams carve Martian outflow channels? Nature 290: 759–63, 1981.
Lunine JI. Origin and evolution of outer solar system atmospheres. Science 245: 141–6, 1989.
Lysenko SV, Kazakov GA & Fedynski AV. Description of an apparatus for studying microorganisms in the upper atmosphere. Life Sci. Space Res. 18: 53–8, 1980.
Maezawa H, Ito T, Hieda K, Kobayashi K, Ito A, Mori T & Suzuki K. Action spectra for inactivation of dry phage T1 after monochromatic (150–254 nm) synchroton irradiation in the presence and absence of photoreactivation and dark repair. Radiat. Res. 98: 227–33, 1984.
Maher KA & Stenvenson DJ. Impact frustration of the origin of life. Nature 331: 612–14, 1988.
Mancinelli RL. Peroxides and the survivability of microorganisms on the surface of Mars. Adv. Space Res. 9 (6): 191–5, 1989.
Mancinelli RL, Marshall JR & White MR. Analyses of exobiological and potential resource materials in the Martian soil. Adv. Space Res. 12 (4): 111–6, 1992.
Mancinelli RL & McKay CP. The evolution of nitrogen cycling. Origins of Life 18: 311–25, 1988.
Mancinelli RL & Shulls WA. Airborne bacteria. Appl. Environ. Microbiol. 35: 1095–101, 1978.
Manned Mars Mission, Working Group papers, Marshall Space flight Center, Huntsville, NASA-TM-8932, 1985.
Maronich LS & Mukhin LM. Belt of life in the Galaxy. In: *Bioastronomy – The Next Steps*. Ed. by Marx G. Kluwer, Dordrecht, pp. 49–59, 1988.
Marx G. *Bioastronomy – The Next Steps*, Kluwer, Dordrecht, 1988.
Mason SF. *Chemical Evolution: Origin of the Elements, Molecules and Living Systems.* Clarendon Press, Oxford, 1991.
Masursky H, Boyce JM, Dial AL, Schaber GG & Strobel ME. Classification and time of formation of Martian channels based on Viking data. J. Geophys, Res. 82: 4016–38, 1977.

Matsunaga T, Hieda K & Nikaido O. Wavelength dependent formation of thymine dimers and (6–4) photoproducts in DNA by monochromatic ultraviolet light ranging from 150 to 365 nm. Photochem. Photobiol. 54: 403–10, 1991.

Mauersberger R, Wilson TL, Mezger PG, Gaume R & Johnston KG. The internal structure of molecular clouds: III. Evidence for molecular depletion in the NGC 2024 condensation. Astron. Astrophys. 256: 640–51, 1992.

Maurette M, Jehanno C, Robin E & Hammer C. Characteristics and mass distribution of extraterrestrial dust from the Greenland ice cap. Nature 328: 699–702, 1987.

Maurette M, Olinger C, Christophe Michel-Levy M, Kurat G, Pourchet M, Brandstätter F & Bourot-Denis M. A collection of diverse micrometeorites recovered from 100 tonnes of Antarctic blue ice. Nature 351: 44–7, 1991.

Maurette M, Pourchet M, Bonny Ph, Jouret C & Siry P. L'Expédition 'Glace Bleue I' de collecte de micrométéorites antarctiques de Janvier–Fevrier 1988: resultats préliminaires et perspectives. *ATP Planétologie*, Besancon, pp. 359–62, 1988.

McDonald GD, Thompson WR, Heinrich M, Khare BN & Sagan C. Chemical investigation of Titan and Triton tholins. Icarus 108: 137–45, 1994.

McDonald GD, Thompson WD & Sagan C. Radiation chemistry in the Jovian stratosphere: laboratory simulations. Icarus 99: 131–42, 1992.

McElroy RD, Klein HP & Averner MM. The evolution of CELSS for lunar bases. In: *Lunar Bases and Activities of the 21st Century.* Ed. by Mendell WW. Lunar Planetary Institute, Houston, pp. 623–33, 1985.

McElroy MB, Kong TY & Yung YL. Photochemistry and evolution of Mars' atmosphere: a Viking perspective. J. Geophys. Res. 82: 4379–88, 1977.

McElroy MB, Prather MJ & Rodriguez JM. Escape of hydrogen from Venus. Science 215: 1614–5, 1982.

McKay CP. Terraforming Mars. J. Brit. Interplanet. Soc. 35: 427–33, 1982.

McKay CP. Exobiology and future Mars missions: the search for Mars' earliest biosphere. Adv. Space Res. 6 (12): 269–85, 1986.

McKay CP. Urey prize lecture: planetary evolution and the origin of life. Icarus 91: 93–100, 1991.

McKay CP. Mars: a reassessment of its interest to biology, In: *Exobiology in Solar System Exploration.* Ed. by Carle C, Schwartz DE & Huntington JL. NASA SP 512, Washington DC, pp. 67–81, 1992.

McKay CP & Davis WL. Planetary protection issues in advance of human exploration of Mars. Adv. Space Res. 9 (6): 197–202, 1989.

McKay CP & Davis WL. The duration of liquid water on early Mars. Icarus 90: 214–21, 1991.

McKay CP, Friedmann EI, Wharton RA & Davis W. History of water on Mars: a biological perspective. Adv. Space Res. 12 (4): 231–8, 1992.

McKay CP & Haynes RH. Should we implant life on Mars? Scientific Amer. 262: 108, 1990.

McKay DS & Ming DW. Mineralogical and chemical properties of the lunar regolith. In: *Lunar Base Agriculture: Soils for Plant Growth.* Ed. by Ming DW & Henninger DL. Amer. Soc. of Agronomy, Crop. Sci. Soc. America, Soil Sci. Soc. America, Madison, pp. 45–68, 1989.

McKay CP, Pollack JB, Lunine JI & Courtin R. Coupled atmosphere-ocean models of Titan's past. Icarus 102: 88–98, 1993.

McKay CP & Stoker CR. The early environment and its evolution on Mars: implications for life. Rev. Geophys. 27: 189–214, 1989.

McKay CP, Stoker CR, Morris J, Conley G & Schwartz D. Space Station gas-grain simulation facility: application to exobiology. Adv. Space Res. 6 (12): 195–206, 1986.

McKay CP, Toon OB & Kasting JF. Making Mars habitable. Nature 352: 489–96, 1991.

Melosh HJ. Impact ejection, spallation, and the origin of meteorites. Icarus 59: 234–60, 1984.

Melosh HJ. Ejection of rock fragments from planetary bodies. Geology 13: 144–8, 1985.

Melosh HJ. The rocky road to Panspermia. Nature 332: 687–8, 1988.

Melosh HJ & Vickery AM. Impact erosion of the primordial atmosphere of Mars. Nature 338: 487–9, 1989.

Mendell WW. *Lunar Bases and Space Activities of the 21st century.* Lunar and Planetary Space Institute, Houston, 1985.

Mendoza-Gmez CX & Greenberg JM. Laboratory simulation of organic grain mantles. Origins of Life 23: 23–8, 1993.

Mennigmann HD. Exobiology: results of spaceflight missions. Adv. Space Res. 9 (6): 3–12, 1989.

Meotner M & Matloff GL. Directed panspermia: a technical and ethical evaluation of seeding nearby solar systems. J. Br. Interplanet. Soc. 32: 419–23, 1979.

Mikio S. Wavelength dependence of the photodecomposition of 2′deoxythymidylyl-(3′–5′–2′-deoxythymindine, dTpdT, irradiated by monochromatic vacuum UV photons above 50 nm. Photochem. Photobiol. 57: 950–7, 1993.

Miller SL. The production of amino acids under possible primitive Earth conditions. Science 117: 528, 1953.

Miller S & Orgel L. *The Origin of Life on the Earth.* Prentice Hall Inc. Englewood Cliffs, New Jersey, 1974.

Milne D, Raup D, Billingham J, Niklaus K & Padian K. *The Evolution of Complex and Higher Organisms*, NASA SP–478 , NASA Washington, 1985.

Ming DW & Henninger DL. *Lunar Base Agriculture: Soils for Plant Growth*, Amer. Soc. of Agronomy, Crop Sci. Soc. America, Soil Sci. Soc. America, Madison, 1989.

Mitchell DL, Lin RP, Anerson KA, Carlson CW, Curtis DW, Korth A, Rème H, Sauvaud JA, d'Uston C & Mendis DA. Evidence for chain molecules enriched in carbon, hydrogen and oxygen in comet Halley. Science 237: 626–8, 1987.

Mitchell DL, Lin RP, Carlson CW, Korth A, Rème H & Mendis DA. The origin of complex organic ions in the coma of comet Halley. Icarus 98: 125–33, 1992.

Mitz MA. Model systems for life processes on Mars. Origins of Life 5: 457–62, 1974.

Moll DM & Vestal JR. Survival of microorganisms in smectite clays: implications for Martian exobiology. Icarus 98: 233–9, 1993.

Moorbath S, O'Nions RK & Pankhurst RJ. Early Archean age for the Isua iron-formation, West Greenland. Nature 245: 138–9, 1973.

Moreno MA. Microorganism transport from Earth to Mars. Nature 336: 209, 1988.

Moroz VI, Combes M, Bibring JP, Coron N, Crovisier J, Encrenaz T, Crifo JF, Sanko N, Grigoriev AV, Bockelée-Morvan D, Gispert R, Nikolsky YV, Emerich C, Lamarre JM, Rocard F, Krasnopolsky VA & Owen T. Detection of parent molecules in comet P/Halley from IKS-Vega experiment. Astron. Astrophys. 187: 513–8, 1987.

Morrison P. Entropy, life and communication (interstellar travel and extra-solar intelligence). In: *Interstellar Communications.* Ed. by Ponnamperuma C & Cameron AGW. Houghton Mifflin, Boston, pp. 168–86, 1974.

Morrison P, Billingham J & Wolfe J. *The Search for Extraterrestrial Intelligence SETI.* NASA SP–419, 1977.

Mullie F & Reisse J. Organic matter in carbonaceous chondrites. *Topics Curr. Chemi.* 139: 83, 1987.

Munakata N, Hieda K, Kobayashi K & Ito T. Action spectra in the ultraviolet wavelength (150-250 nm) for inactivation and mutagenesis of *Bacillus subtilis* spores obtained with synchroton radiation. Photochem. Photobiol. 44: 385–90, 1986.

National Research Council. *Biological Contamination of Mars, Issues and Recommendations*, Space studies Board, National Academy Press, Washington, 1992.

NCRP. *Guidance on Radiation Received in Space Activities.* NCRP Report No. 98, National Council of Radiation Protection and Measurements, Bethesda, 1989.

Nussinov MD, Chernyak YB & Ettinger JL. Model of the fine-grain component of Martian soil based on Viking lander data. Nature 274: 859–61, 1978.

Nussinov MD & Lysenko SV. Cosmic vacuum prevents radiopanspermia. Origins of Life 13: 153–64, 1983.

Nussinov MD & Lysenko SV. Frontiers of the Earth's biosphere and extraterrestrialization. J. Brit. Interplanet. Soc. 42: 431–5, 1989.

Nussinov MD & Lysenko SV. Can spores arrive alive in interstellar space? J. Brit. Interplanet. Soc. 45: 22, 1992.

Oberbeck VR & Fogleman G. Impact constrains on the environment for chemical evolution and the origin of life. Origins of Life 20: 181–95, 1990.

Oberbeck VR & Mancinelli RL. Asteroid impacts, microbes, and the cooling of the atmosphere. Bioscience 44: 173–7, 1994.

O'Keefe JD & Ahrens TJ. Oblique impact: a process for obtaining meteorite samples from other planets. Science 234: 346–9. 1986.

Olson GJ & Woese CR. Ribosomal RNA: a key to phylogeny. *The FASEB J.* 7: 113–23, 1993.

Orlob GB & Lorenz PR. Survival of Tobacco Mosaic Virus in space. Phytopathology 58: 955, 1968.

Oro J. Comets and the formation of biochemical compounds on the primitive Earth. Nature 190: 389–90, 1961.

Oro J. Constraints imposed by cosmic evolution towards the development of life. In: *Bioastronomy – The Next Steps*. Ed. by Marx G. Kluwer, Dordrecht, pp. 161–5, 1988.

Oro J & Holzer G. The effects of ultraviolet light on the degradation of organic compounds: a possible explanation for the absence of organic matter on Mars. Life Sci. Space Res. 17: 77–86, 1979.

Oro J, Mills T & Lazcano A. Comets and the formation of biochemical compounds on the primitive Earth – a review. Origins of Life 21: 267–77, 1992.

Oro J, Rewers K & Odom D. Criteria for the emergence and evolution of life in the solar system. Origins of Life 12: 285–305, 1982.

Oro J, Squyres SW, Reynolds RT & Mills TM. Europa: prospects for an ocean and exobiological implications. In: *Exobiology in Solar System Exploration.* Ed. by Carle C, Schwartz DE & Huntington JL. NASA SP 512, Washington D.C., pp. 229–247, 1992.

Oser H & Battrick B. *Life Sciences Research in Space*, ESA SP–1105, ESA, Paris, 1989.

Otroshchenko VA & Surkov YA. The possibility of organic molecule formation in the Venus atmosphere. Origins of Life 5: 487–90, 1974.

Owen T. The composition and origin of Titan's atmosphere. Planet. Space Sci. 30: 833–8, 1982.

Owen T. Life as a planetary phenomenon. Origins of Life 15: 221–34, 1985.

Owen T, Gautier D, Raulin F & Scattergood T. Titan. In: *Exobiology in Solar System Exploration.* Ed. by Carle C, Schwartz DE & Huntington JL. NASA SP 512, Washington D.C., pp. 127–43, 1992.

Owen T, Maillard JP, De Bergh C & Lutz LH. Deuterium on Mars: the abundance of HDO and the value D/H. Science 240: 1767–70, 1988.

Owen T & Terrile RJ. Colors on Jupiter. J. Geophys. Res. 86: 8797–8814, 1981.

Oyama VI & Berdahl B. The Viking gas exchange experiment results from Chryse and Utopia surface samples. J. Geophys. Res. 82: 4669–76, 1977.

Oyama VI, Berdahl BJ, Carle GC, Lehwalt ME & Ginoza HS. The search for life on Mars: Viking 1976 gas changes as indicators of biological activity. Origins of Life 7: 313–33, 1976.

Oyama VI, Berdahl BJ & Carle GC. Preliminary findings of the Viking gas exchange experiment and a model for Martian surface chemistry. Nature 265: 110–4, 1977.

Oyama VI, Berdahl BJ, Woeller F & Lehwalt M. The chemical activities of the Viking biology experiments and the arguments for the presence of superoxides, γ–Fe_2O_3 and carbon suboxide polymer in the Martian soil. Life Sci. Space Res. 16: 3–8, 1978.

Pace NR, Olson GJ & Woese CR. Ribosomal RNA phylogeny and the primary lines of evolutionary descent. Cell 9: 325–6, 1986.

Packer E, Scher S & Sagan C. Biological contamination of Mars: constraints on the survival of terrestrial microorganisms in simulated Martian environments. Icarus 2: 292–316, 1963.

Paresce F. The search for extrasolar planetary systems. Adv. Space Res. 12 (4): 157–67, 1992.

Park R & Epstein S. Carbon isotope fractionation during photosynthesis. Geochim. Cosmochim. Acta 21: 110–26, 1960.

Parker BC, Simmons GM Jr, Seaburg KG, Cathey DD & Allnutt FTC. Comparative ecology of plankton communities in seven Antarctic oasis lakes. J. Plank. Res. 4: 271–86, 1982.

Patterson C. *Molecules and Morphology in Evolution: Conflict of Compromise*. Cambridge University Press, Cambridge, 1987.

Pendleton YJ & Black DC. Further studies on criteria for the onset of dynamic instability in general three-body systems. Astron. J. 88: 1415–9, 1983.

Pflug HD. *Die Spur des Lebens*. Springer, Berlin, 1984.

Pflug HD. Chemical fossils in early minerals. Topics Curr. Chem. 139: 1–55, 1987.

Pflug HD & Jaeschke-Boyer H. Combined structural and chemical analysis of 3.800-Myr-old microfossils. Nature 280: 483–6, 1979.

Pieri DC. Martian channels: distribution of small channels on the Martian surface. Icarus 27: 25–50, 1976.

Pieri DC. Martian valleys: morphology, distribution and age. Science 210: 895–7, 1980.

Pilcher CB. From a lunar outpost to Mars: science, policy and the U.S. space exploration initiative. Adv. Space Res. 12 (4): 91–7, 1992.

Plumb RC, Tantayanon R, Libby M & Xu WW. Chemical model for Viking biology experiments: implications for the composition of the Martian regolith. Nature 338: 633–5, 1989.

Pollack JB & Atreya SK. Giant planets: clues on current and past organic chemistry in the outer solar system. In: *Exobiology in Solar System Exploration*. Ed. by Carle C, Schwartz DE & Huntington JL. NASA SP 512, Washington D.C., pp. 83–101, 1992.

Pollack JB & Bodenheimer P. Theories of the origin and evolution of the giant planets. In: *Origin and Evolution of Planetary Satellite Atmospheres*. Ed. by Atreya SK, Pollack JB & Mathews MS. University of Arizona, Tucson, pp. 564–602, 1989.

Pollack JB, Kasting JF, Richardson SM & Poliakoff K. The case for a wet, warm climate on early Mars. Icarus 71: 203–24, 1987.

Ponnamperuma C. *Cosmochemistry and the Origin of Life*. Reidel, Boston, 1983.

Ponnamperuma CA, Shimoyama A, Yamada M, Hobo T & Pal R. Possible surface reactions on Mars. Science 197: 455–7, 1977.

Postawko SE & Kuhn WR. Effect of greenhouse gases (CO_2, H_2O, SO_2) on Martian paleoclimates. J. Geophys. Res. 91: D431–8, 1986.

Pühler G, Leffers H, Gropp F, Palm P, Klenk HP, Lottspeich F, Garrett RA & Zillig W. Archaebacterial DNA-dependent RNA polymerases testify to the evolution of the eukaryotic nuclear genome. Proc. Natl. Acad. Sci. USA 86: 4569–73, 1989.

Radmer R & Kok B. An integrated multi-purpose biology instrument utilizing a single detector, the mass spectrometer. Life Sci. Space Res. 10: 211–25, 1972.

Ramkova NV, Nekhorosheva AG, Lysenko SV, Chudnova LB, Kareev NV & Scheglova GV. Sterilization of biological analyser. Life Sci. Space Res. 14: 341–2, 1976.

Raulin F. Titan: A prebiotic planet? In: *Frontiers of Life*. Ed. by Trn Thanh Vn J, Trn Thanh Vn K, Mounolou JC, Schneider J, & McKay C. Edition Frontieres, Gif-sur-Yvette, pp. 327–40, 1992.

Raulin F, Dubouloz N & Frere C. Prebiotic-like organic synthesis in extraterrestrial environments: The case of Titan. Adv. Space Res. 9 (6): 35–47, 1989.

Raulin F, Gautier D & Ip WH, Exobiology and the solar system: The Cassini mission to Titan. Origins of Life 14: 817–24, 1984.

Reynolds R, Squyres SW, Colburn D & McKay CP. On the habitability of Europa. Icarus 56: 246–54, 1983.

Richter H. Zur Darwinschen Lehre. Schmidts Jahrbuch Ges. Med. 126: 243–9, 1865.

Ringwood AE. The Earth's core: its composition, formation and bearing upon the origin of the Earth. Proc. Roy. Soc. A 395: 1–46, 1984.

Rivera MC & Lake JA. Evidence that eukaryotes and eocyte prokaryotes are immediate relatives. Science 257: 74–6, 1992.

Roessler K. Suprathermal chemistry in space. In: *Solid State Astrophysics*. Busoletti E & Strazulla G, North Holland, Amsterdam, pp. 197–266, 1991.

Roessler K, Sauer M & Schulz R. Gaseous products from VUV photolysis of cometary solids. Ann. Geophysicae 10: 226–31, 1992.

Rogers LA & Meier FC. The collection of microorganisms above 36000 feet. Natl. Geogr. Soc. Stratosphere Series 2: 146, 1936.

Rolfe E & Battrick B. *The Atmospheres of Saturn and Titan*. ESA SP–241, ESA Paris, 1985.

Romani PN & Atreya SK. Methane photochemistry and haze production on Neptune. Icarus 74, 424–45, 1988.

Rothschild LJ & Des Marais D. Stable carbon isotope fractionation in the search for life on early Mars. Adv. Space Res. 9 (6): 159–65, 1989.

Rothschild LJ. Earth analogs for Martian life. Microbes in evaporites, a new model system for life on Mars. Icarus 88: 246–60, 1990.

Roy AC & Powers JV. Chemical evolution and the origin of life, cumulative keyword subject index 1970–1986. Origins of Life 20: 425–56, 1990.

Rummel JD. Planetary protection policy overview and application to future missions. Adv. Space Res. 9 (6): 181–4, 1989.

Rummel LD. Planetary protection policy (USA). Adv. Space Res. 12 (4): 129–31, 1992.

Sagan C. The search for indigenous Lunar organic matter. Space Life Sci. 3: 484–9, 1972.

Sagan C. On the detectivity of advanced galactic civilizations. Icarus 19: 350–2, 1973.

Sagan C. *Signale der Erde, unser Planet stellt sich vor.* Droemer Knaur, München, 1980.

Sagan C. *Pale Blue Dot, a Vision of the Human Future in Space*, Random House, New York, 1994.

Sagan C, Levinthal EC & Lederberg J. Contamination of Mars. Science 159: 1191–6, 1968.

Sagan C & Thompson WR. Production and condensation of organic gases in the atmosphere of Titan. Icarus 59: 133–61, 1984.

Sagdeev RZ, Blamont J, Galeev AA, Moroz VI, Shapiro VD, Shevchenko VI & Szegö K. Vega spacecraft encounters with comet Halley. Nature 321: 259–66, 1986.

Scattergood TW, Lau EY & Stone BM. Titan's aerosols. Icarus 99: 98–105, 1992.

Schidlowski M. A 3,800 million-year isotopic record of life from carbon in sedimentary rocks. Nature 333: 313–8, 1988.

Schidlowski M. Life on the early Earth: bridgehead from cosmos or autochthonous phenomenon? In: *From Mantle to Meteorites*. Ed. by Gopalan K, Gaur VK, Somayajulu BLK & Macdougall JD. Indian Acad. Sci, Bangalore, pp. 189–99, 1990.

Schidlowski M. Stable carbon isotopes: possible clues to early life on Mars. Adv. Space Res. 12 (4): 101–10, 1992.

Schidlowski M. The beginnings of life on Earth: evidence from geological record. In: *The Chemistry of Life's Origin*. Ed. by Greenberg JM. Kluwer Academic Publ., pp. 389–414, 1993a.

Schidlowski M. The initiation of biological processes on Earth, summary of empirical evidence. In: *Organic Geochemistry*. Ed. by Engel MH & Macko SA. Plenum Press, New York, pp. 639–55, 1993b.

Schidlowski M, Appel PWU, Eichmann R & Junge CE. Carbon isotope geochemistry of the 3.7 x 10^9 yr old Isua sediments, West Greenland: implications for the Archean carbon and oxygen cycles. Geochim. Cosmochim. Acta 43: 189–99, 1979.

Schidlowski M, Eichmann R & Junge CE. Precambrian sedimentary carbonates: carbon and oxygen isotope geochemistry and implications for the terrestrial oxygen budget. Precambrian Res. 2: 1–69, 1975.

Schidlowski M, Golubic S, Kimberley MM, McKirdy DM & Trudinger PA. *Early Organic Evolution*, Springer, Berlin, 1992.

Schlesinger G & Miller SL. Prebiotic syntheses in atmospheres containing CH_4, CO, and CO_2 I. Amino acids 19: 376–90, 1983.

Schopf JW (ed). *Earth's Earliest Biosphere, its Origin and Evolution.* Princeton University Press, Princeton, New Jersey, 1983.

Schopf JW. Microfossils of the early Archean Apex chert: new evidence of the antiquity of life. Science 260: 640–6, 1993.

Schopf JW & Klein C. *The Proterozoic Biosphere: A Multidisciplinary Study*, Cambridge University Press, Cambridge, 1992.

Schopf JW & Packer BM. Early Archean (3.3-Billion to 3.5-Billion-Year-old) microfossils from Warrawoona Group, Australia. Science 237: 70–3, 1987.

Schubert G & Covey C. The atmosphere of Venus. Scientific Amer. 245: 44–52, 1981.

Schwartz AW, Voet AB & van der Veen M. Recent progress in the prebiotic chemistry of HCN. Origins of Life 14: 91–8, 1984.

Schwartz DE & Mancinelli RL. Bio-markers and the search for extinct life on Mars. Adv. Space Res. 9 (6): 155–8, 1989.

Schwartz DE, Mancinelli RL & Kaneshiro ES. The use of mineral crystals as biomarkers in the search for life on Mars. Adv. Space Res. 12 (4): 117–9, 1992.

Schwartz RM & Townes CH. Interstellar and interplanetary communication by optical masers. Nature 190: 205, 1961.

Schwartzman D & Rickard LJ. Being optimistic about SETI. In: *Bioastronomy -- The Next Steps.* Ed. by Marx G. Kluwer, Dordrecht, pp. 305–12, 1988.

Schwartzman DW & Volk T. Geophysiology and habitable zones around sun-like stars. In: *Bioastronomy, The Search for Extraterrestrial Life - The Exploration Broadens.* Ed. by Heidmann J & Klein MJ. Springer, Berlin, pp. 155–162, 1991.

Schwehm GH. Rosetta - comet nucleus sample return. Adv. Space Res. 9 (6): 185–90, 1989.

Scott DH, Rice JW Jr & Dohm JK. Martian paleolakes and waterways: exobiological implications. Origins of Life 21: 189–98, 1991.

Segerer AH, Burggraf S, Fiala G, Huber G, Huber R, Pley U & Stetter KO. Life in hot springs and hydrothermal vents. Origins of Life 23: 77–90, 1993.

Sharp RP. Mars: fretted and chaotic terrains. J. Geophys. Res. 78: 4073–83, 1973.

Sharp RP & Malin MC. Channels on Mars. Geol. Soc. Amer. Bull. 86: 593–609, 1975.

Shimoyama A, Ponnamperuma C & Yanai K. Amino acids in Yamato carbonaceous chondrite from Antarctica. Nature 282: 394–6, 1979.

Shklovskii IS & Sagan C. *Intelligent Life in the Universe.* Holden Day, New York, 1966.

Siebert J & Hirsch P. Characterization of 15 selected coccal bacteria isolated from Antarctic rock and soil samples from the McMurdo-Dry Valleys (South Victoria Land). Polar Biol. 9: 37–44, 1988.

Sleep NH, Jahnle KJ, Kasting JF & Morowitz HJ. Annihilation of ecosystems by large asteroid impacts on the early Earth. Nature 342: 139–42, 1989.

Smith BA, Soderblom LA, Johnson TV, Ingersoll AP, Collins SA, Shoemaker AM, Hunt GE, Massursky H, Carr MH, Davies ME, Cook II AF, Boyce J, Danielson GE, Owen T, Sagan C, Beebe RF, Veverka J, Strom RG, McCauley JF, Morrison D, Briggs GA & Suomi VE. The Jupiter system through the eyes of Voyager 1. Science 204: 951–72, 1979.

Smith BA, Soderbolm L, Beebe R, Boyce J, Briggs G, Bunker A, Collins SA, Hansen CJ, Johnson TV, Mitchell JL, Terrile RJ, Carr M, Cook II AF, Cuzzi J, Pollack JB, Danielson GE, Ingersoll A, Davies ME, Hunt GE, Masursky H, Shoemaker E, Morrison D, Owen T, Sagan C, Ververka J, Strom R & Suomi VE. Encounter with Saturn: Voyager 1 imaging science results. Science 212: 163–91, 1981.

Smith BA, Soderbolm L, Batson R, Bridges P, Inge J, Masursky H, Shoemaker E, Beebe R, Boyce J, Briggs G, Bunker A, Collins SA, Hansen CJ, Johnson TV, Mitchell JL, Terrile RJ, Cook II AF, Cuzzi J, Pollack JB, Danielson GE, Ingersoll AP, Davies ME, Hunt GE, Morrison D, Owen T, Sagan C, Ververka J., Strom R, Suomi VE, A new look at the Saturn system: the Voyager 2 images, *Science* 215, 504–537, 1982.

Snyder CW. The extended mission of Viking. J. Geophys. Res. 84: 7917–33, 1979a.

Snyder CW. The planet Mars as seen at the end of the Viking mission. J. Geophys. Res. 84: 8487–519, 1979b.

Snyder LE, Hollis JM, Suenram RD, Lovas FJ, Brown LW & Buhl D. An extensive galactic search for conformer II glycine. The Astrophys. J. 268: 123–8, 1983.

Snyder LE, Kuan Y-J, Miao Y & Lovas FJ. Search for biomolecules in SGR B2. In: *Progress in the Search for Extraterrestrial Life (Proceedings of the 1993 Bioastronomy Symposium*). Ed. by Shostak S. LAI Preprint Nr. 1994–5, in press, 1994.

Soffen GA. Scientific results of the Viking missions. Science 194, 1274–6, 1976.

Sontag W & Dertinger H. Energy requirements for damaging DNA molecules. III. The mechanisms of inactivation of bacteriophage X174 by vacuum ultraviolet radiation. Int. J. Radiat. Res. 27: 543–52, 1975.

Spizizen J, Isherwood JE & Taylor GR. Effects of solar ultraviolet radiations on *Bacillus subtilis* spores and T7 bacteriophage. Life Sci. Space Res. 13: 143–9, 1975.

Squyres SW. Martian fretted terrain: flow of erosional debris. Icarus 34: 600–13, 1978.

Squyres SW. Urey prize lecture: water on Mars. Icarus 79: 229–88, 1989.

Squyres SW & Carr MH. Geomorphic evidence for the distribution of ground ice water. Science 231: 249–52, 1986.

Stanley SM. *Earth and Life through Time*. Freeman, New York, 1986.

Stetter KO. Life at the upper temperature border. In: *Frontiers of Life*. Ed. by Trn Thanh Vn J, Trn Thanh Vn K, Mounolou JC, Schneider J, & McKay C. Edition Frontieres, Gif-sur-Yvette, pp. 195–219, 1992.

Stetter KO, Fiala G, Huber R & Segerer A. Hyperthermophilic microorganisms. Microbiol. Rev. 75: 117–24, 1990.

Stetter KO, Lauerer G, Thomm M & Nenner A. Isolation of extremely thermophilic sulfate reducers: evidence for a novel branch of Archaebacteria. Science 236: 822–4, 1987.

Stevens M, Strobel DF & Herbert F. An analysis of the Voyager 2 ultraviolet spectrometer occultation data at Uranus: inferring heat sources and model atmospheres. Icarus 101: 45–63, 1992.

Stoker CR, McKay CP, Haberle RM & Andersen DT. Science strategy for human exploration of Mars. Adv. Space Res. 12 (4): 79–90, 1992.

Stoks PG & Schwartz AW. Nitrogen-heterocyclic compounds in meteorites: significance and mechanisms of formation. Geochim. Cosmochim. Acta 45: 563–9, 1981.

Stone E & Lane AL. Voyager 1 encounter with the Jovian system. Science 204: 945–48, 1979.

Strazulla G. Modification of grains by particle bombardment in the early solar system. Icarus 61: 48–56, 1985.

Tarter J. Survey of Earth orbital telescopes and their potential for exobiology. Adv. Space Res. 6 (12): 219–26, 1986.

Tarter JC, Black DC & Billingham J. Review of methodology available for the detection of extrasolar planetary systems. J. Brit. Interplanet. Soc. 39: 418–24, 1986.

Taylor GR. Space microbiology. Ann. Rev. Microbiol. 28: 121–37, 1974.

Taylor GR, Spizizen J, Foster BG, Volz PA, Bücker H, Simmonds RC, Heimpel AM & Benton EV. A descriptive analysis of the Apollo 16 microbial response to space environment experiment. BioScience 24: 505–11, 1974.

Taylor JS & Cohrs MP. DNA, light, and dewar pyrimidones: the structure and biological significance of TpT3. J. Amer. Chem. Soc. 109: 2834–5, 1987.

Toon OB, McKay CP, Griffith CA & Turco RP. A physical model of Titan's aerosols. Icarus 95: 24–53, 1992.

Toulmin P III, Baird AK, Clark BC, Keil K & Rose HJ Jr. Inorganic chemical investigations by X-ray fluorescence analysis: the Viking Mars lander. Icarus 20: 153–78, 1973.

Trân Thanh Vân J, Trn Thanh Vn K, Mounolou JC, Schneider J & McKay C. *Frontiers of Life*. Edition Frontieres, Gif-sur-Yvette, 1992.

Travis J. Hints of first amino acid outside the solar system. Science, 264: 1668, 1994.

Vickery AM & Melosh HJ. The large crater origin of SNC meteorites. Science 237, 738–43, 1987.

Vishniac WV & Mainzer SE. Antarctica as a Martian model. Life Sci. Space Res. 11, 25–31, 1973.

Vishniak W. Extraterrestrial microbiology. Aerospace Med. 31: 678–80, 1960.

Wächtershäuser G. Pyrite formation, the first energy source for life: a hypothesis. Syst. Appl. Microbiol. 10: 207–10, 1988.

Wächtershäuser G. The case for the chemo-autotrophic origin of life in an iron-sulfur-world. Origins of Life 20: 173–6, 1990.

Wächtershäuser G. Evolution of the first metabolic cycles. Proc. Natl. Acad. Sci. USA 87: 200–4, 1991.

Wächtershäuser G. Groundworks for an evolutionary biochemistry: the iron-sulfur world. Prog. Biophys. Molec. Biol. 58: 85–210, 1992.

Wächtershäuser G. Life in a ligand sphere. Proc. Natl. Acad. Sci. USA 91: 4283–7, 1994.

Waldron RD. Lunar manufacturing: a survey of products and processes. Acta Astron. 17: 691–708, 1988.

Walsh MM & Lowe DR. Filamentous microfossils from 3,500-Myr-old Onverwacht Group, Barberton Mountain Land, South Africa. Nature 314: 530–2, 1985.

Walter MR, Buick R & Dunlop JSR. Stromatolites 3,400–3,500 Myr old from the north Pole area, Western Australia. Nature 284: 443–5, 1980.

Walter MR & Des Marais DJ. Preservation of biological information in thermal spring deposits: developing strategy for the search for fossil life on Mars. Icarus 101: 129–43, 1993.

Wang SY. *Photochemistry and Photobiology*, Vol. 1 and 2, Academic Press, New York, 1976.

Wänke H. Chemistry, accretion, and evolution of Mars. Space Science Rev. 56: 1–8, 1991.

Wänke H & Dreibus G. Chemical composition and accretion history of terrestrial planets. Phil. Trans. Roy. Soc. London A 325: 545–57, 1988.

Warren JL, Allton JH, Dardono CB, Holder JA, Marlow RR, Fook HA, Clanton US & Schultz RA. The detection and observation of meteoroid and space debris impact features on the Solar Max satellite. In: *Proceedings, 19th Lunar Planet Science Conference,* University Press, Houston, pp. 641–57, 1989.

Weber P & Greenberg JM. Can spores survive in interstellar space? Nature 316: 403–7, 1985.

Weckwerth G & Schidlowski M. Phosphorus as a potential quide in the search for extinct life on Mars. Adv. Space Res. 15 (3): 185–91, 1994.

Wharton RA Jr, McKay CP, Mancinelli RL & Simmons GM Jr. Early Martian environments; the Antarctic and other terrestrial analogs. Adv. Space Res. 9 (6): 147–53, 1989.

Wharton RA Jr, McKay CP, Simmons GM Jr & Parker BC. Cryconite holes on glaciers. BioScience 35: 499–503, 1985.

Wharton RA Jr, Parker BC & Simmons GM Jr. Distribution, species composition and morphology of algal mats in Antarctic dry valley lakes. Phycologia 22: 355–65, 1983.

Wheelis ML, Kandler O & Woese CR. On the nature of global classification. Proc. Natl. Acad. Sci. USA 89: 2930–4, 1992.

White DC & Hirsch P. Microbial extraction of hydrogen from lunar dust, In: *Lunar Bases and Activities of the 21st Century.* Ed. by Mendell WW. Lunar Planetary Institute, Houston, pp. 591–602, 1985.

Whitmire DP, Reynolds RT & Kasting JF. Habitable zones for Earth-like planets around main sequence stars. In: *Bioastronomy, The Search for Extraterrestrial Life – The Exploration Broadens.* Ed. by Heidmann J & Klein MJ. Springer, Berlin, pp. 173–8, 1991.

Williams DA. The chemical composition of the interstellar gas. Rev. Modern Astron. 6: 49–67, 1993.

Wilkening LL. *Comets*. University of Arizona Press, Tucson, 1982.
Winnewisser G & Herbst E. Organic and cosmochemistry. In: *Topics in Current Chemistry*. Springer Verlag, Berlin, p. 139, 1987.
Woese CR. Bacterial evolution. Microbiol. Rev. 51: 221–71, 1987.
Woese CR & Olson GJ. Archaebacterial phylogeny: perspectives on the urkingdoms. Syst. Appl. Microbiol. 7: 161–77, 1986.
Woese CR, Kandler O & Wheelis ML. Towards a natural system of organisms: proposal for the domains Archaea, Bacteria and Eucarya. Proc. Natl. Acad. Sci. USA 87: 4576–9, 1990.
Wood JA & Chang S. *Cosmic History of the Biogenic Elements and Compounds*. NASA SP-476, NASA Washington DC, 1985.
Yabushita S, Wada K, Taka T, Inagaki T, Young D & Arakawa ET. A spectroscopic study of the microorganism model of interstellar grains. Astrophys. Space Sci. 124: 377–88, 1986.
Yamada H & Hieda K. Wavelength dependence (150–290 nm) of the formation of the cyclobutane dimer and the (6–4) photoproduct of thymine. Photochem. Phtobiol. 55: 541–48, 1992.
Yelle RV, Herbert F, Sandel BR, Vervack RJ & Wentzel TM. The distribution of hydrocarbons in Neptune's upper atmosphere. Icarus 104: 38–59, 1993.
Young RS. The origin and evolution of the Viking mission to Mars. Origins of Life 7: 271–2, 1976.
Young RS, Deal PH, Bell J & Allen JL. Bacteria under simulated Martian conditions. Life Sci. Space Res. 2: 105–11, 1964.
Yung YL, Allen M & Pinto JP. Photochemistry of the atmosphere of Titan: comparison between model and observations. Astrophys. J. Suppl. 55: 465–506, 1984.
Yung YL, Wen J, Pinto JP, Allen K, Pierce KK & Paulsen S. HDO in the Martian atmosphere: implications for the abundance of crystal water. Icarus 76: 146–59, 1988.
Zaar EI, Zelikson VG, Kitaigorodsky MG, Lozina-Lozinsky LK, Koshelev GV & Rybin MA. The apparatus 'Photostat-I' for simulating Martian environmental conditions. Life Sci. Space Res. 8: 62–7, 1970.
Zillig W, Klenk HP, Palm P, Pühler G, Gropp F, Garrett RA & Leffers H. The polygenetic relations of DNA-dependent RNA polymerases of archaebacteria, eukaryotes and eubacteria. Can. J. Microbiol. 35: 73–80, 1989.
Zillig W, Palm P & Klenk HP. The nature of the common ancestor of the three domains of life and the origin of the Eucarya. In: *Frontiers of Life*. Ed. by Trn Thanh Vn J, Trn Thanh Vn K, Mounolou JC, Schneider J, & McKay C. Edition Frontieres, Gif-sur-Yvette, pp. 181–193, 1992.
Zuckerkandl E & Pauling L. Molecules as documents of evolutionary history. J. Theoret. Biol. 8: 357–66, 1965.

Chapter 8

National and international space life sciences research programmes 1980 to 1993 – and beyond

BRIAN FITTON & DAVID MOORE

8.1 Introduction

In this Chapter we provide short summaries of the national research programmes in the space life sciences of the ESA member states, its associates and major international partners. The purpose is to provide complementary statistical information to the scientific content of earlier review Chapters. This Chapter also presents listings of the life science experiments carried out using the Space Transportation System (STS; the Shuttle), the collaborative research performed with the Russian Mir space station and the Foton recoverable satellite, as well as summaries of the ESA sounding rocket and parabolic flight programme experiments. The main listings cover the period 1980 to 1993. From this basic information it is possible to discern the overall emphasis and trends in the different research areas. Additional information may be found in the ESA Microgravity Database, operated by ESRIN.

In addition to these general data, which are essentially historic, the individual sections covering each of the national programmes summarizes the current activities and some of the plans for the future, where these have been clearly established. In several countries a strategic plan has been established for the future activities in the space life sciences, with specific focus. These plans are outlined where the information has been made available.

8.2 Austria

The research activities in the space life sciences in Austria lie principally in the human physiology area. They have developed slowly over the past decade, during which time Austria has hosted the *European Space Medicine and Biology Meeting* (1986) and the *Third European Symposium on Life Sciences Research in Space* (1987).

The culmination of that development was the AUSTROMIR mission, which took place in October 1991. This comprised ten individual experiments, covering neurophysiology, the cardiovascular and muscular system, genetics, and radiation dosimetry. Further details are given in Table 1.

Austria intends to continue with a modest research programme in the space life sciences, but has chosen to do this outside of the ESA Microgravity Pro-

Table 1. Space life sciences research in Austria.

Topic	Title	Principal Investigator and/or authors	Affiliation
Disturbances in movement coordination following simulation with visual, acoustic and proprioceptive stimuli.	Eye-head-arm coordination and spinal reflexes in weightlessness (MONOMIR).	M. Burger, F. Gersten Brand *et al.;* E. Hochmair, G. Steinwender I.B. Kozlovskaya, A. Sokolov *et al.*	Neurology Dept. And Physics Dept. Univ. Innsbruck IBMP, Moscow
Force-angle velocity relationship of musculature in/against predetermined translatoy movement & correlation with integrated surface EMG.	Development and implementation of the MOTORMIR experiment on the Mir space station.	N. Bachl, R. Baron, K.H. Tschan, M. Mossaheb, W. Bumba, F. Hildebrand I. Kharitonov	Sports & Exercise Dept. Univ. of Vienna IBMP, Moscow
Measurement of attention, psychomotor speed, mental feasibility, time estimation, visuospatial perception, memory, function changes.	A study of cognitive functions in microgravity (COGIMIR).	T. Benk, F. Gerstenbrand B. Koserenko N. Watson	Neurology Dept. Innsbruck Univ. IBMP, Moscow Univ. Brit. Columbia
Accuracy of directional hearing and its role in human orientation in microgravity	Directional hearing in microgravity (AUDIMIR)	A. Persterer, C. Koppensteiner *et al.* M. Berger M. Nefjodova	AKG, Vienna Univ. of Innsbruck IBMP, Moscow
Examine the vertical vection illusion elicited by an optokinetic stimulator system.	Orientational effects from optokinetic stimulation (OPTOVERT).	C. Muller, G. Wiest, L. Deecke L. Kornilova	Neurology Clinic, Univ. Vienna IBMP, Moscow
'PULSTRANS' worktime monitoring, and 'SLEEP' monitoring, of circulatory system.	Monitoring of cardiovascular parameters during AUSTROMIR space flight	M. Moser, E. Gallasch, *et al.* R. Bayevskij, I. Funtowa *et al.*	Physiolog. Inst., Univ. of Graz IBMP, Moscow
Influence of tremor parameters of the changes to muscle loading/ proprioreceptor threshold levels.	The effect of microgravity on tremor (microvibration) .	E. Gallasch, M. Moser *et al.* I.B. Kozlovskaya, M. Borlov	Physiolog. Inst., Univ. of Graz IBMP, Moscow
Changes to dynamics of transcapillary fluid shifts studied by LBNP with sonic velocity measurement technique and measurement of volumetric, osmoregulatory & vasoactive hormone concentrations.	Altered fluid shift dynamics with 6 days in microgravity (body fluids).	H.G. Hinghofer-Szalkay *et al.* J. Schmeid, H. Heimel V.B. Noskov, I. Pestov, A.I. Grogorien	VRSM, Univ. Graz Lab. Messtechnik, Graz IBMP, Moscow

Table 1 (continued).

Topic	Title	Principal Investigator and/or authors	Affiliation
Thermoluminescent dosimeters and track etch foils to measure average LET and absorbed dose.	Radiation measurements inside the soviet space station Mir (DOSIMIR).	N. Vana, W. Schoewer, M. Fugger J. Akatov	ATOMINST., Univ. of Vienna IBMP, Moscow
Investigation of genetic changes in lymphocytes, their possible repair and immunological effects.	Influence of microgravity on immune system and genetic information (MIRGEN).	H. Tuschl, *et al.* M.S. Chajdakow Y.J. Voronkov	Öst. Forsch. Zent. Seibersdorf IBMP, Moscow

gramme. Consequently, Austria does not participate in the Euromir missions. Instead it intends to continue with direct cooperative research ventures with other countries. In particular it will continue to collaborate with Russia in the space life sciences.

8.3 Belgium

There is a small but active programme of space research in the life sciences in Belgium, primarily directed to human physiology. The areas of interest are indicated by the proposals which have been considered for future flight, mainly under the ESA *Columbus Precursor Flights* programme listed in Table 2. The experiment of Dr Schoeters, reflown on the Russian Foton biology spacecraft, will also be a candidate for a future Biopan experiment on Foton. The experiments of Dr Bierkens, Prof. Lapiere, and Prof. Paiva, together with an experiment of Prof. O. Quadens, University of Antwerp, on brain activity during sleep and waking in microgravity, were considered for flight on the Euromir 95 mission, that of Prof. Paiva being accepted. Interest in space life sciences in Belgium is increasing, and further proposals for new activities are expected as prospects for flights improve.

Recent research reports published by the Belgian Science Policy Office, include:

- *The Respiratory System in Microgravity*, by M. Paiva, which covered D-2 and SLS-1 results, using a respiratory inductive plethysmograph, together with data from parabolic aircraft flights. Equipment developed is intended for use on the Euromir 1995 flight. The work has been performed, in concert with the only three other groups involved in such research worldwide, in order to avoid duplication of effort. These are Prof. J.B. West, Univer-

Table 2. Space life sciences research in Belgium: proposals highly recommended for future flight, mainly under the ESA 'Columbus Precursor Flights' programme.

Title	Principal Investigator	Co-Investigators	Proposal no.
Changes in osteogenic marker gene expression and cytokine production	J.G.E.A. Bierkens, VITO, Flemish Inst. for Techn. Research, Mol, Belgium .	G. Schoeters R. van Vlasselner R. van den Heuvel VITO, Mol, Belgium	06
Effects of microgravity on the cytoskeleton, synthesis of extracellular molecules, and production of degenerative enzymes in human fibroblast studies at a pre- and post-tranlational level.	C.M. Lapiere, Lab. de Dermatologie, University of Liege.	J.P. Soleilhavoup, University of Toulouse, France	518
The influence of gravity on the differentiation and maintenance of fast and slow muscle fibres in skeletal muscle.	G. Marechal, Dept. de Physiologie, Univ. de Louvain.	G. Goldspink, Univ. of London. J.M. Gillis, Univ. de Louvain	383
Pulmonary function in microgravity.	M. Paiva, Inst. de Recherche, Interdiscip. Univ. de Bruxelles	L.A. Engel, Westmead Hospital, Australia. G.K. Prisk, H.J. Guy, J.B. West, UCLA, San Diego, U.S.A.	229
Biological dosimetry for cosmic radiation exposure during intra- and extra-vehicular activities in space, using haemopoietic stem cell functions.	G. Schoeters, VITO, Section Biology, Mol, Belgium	A. v.d. Heuvel, J. Bierkens, P. van Vlasselner, VITO. A. Poffijn, K. van Laere, J. Uyttenhove, NPL, Ghent.	81

sity of California, San Diego; Prof. D. Linnarsson, Karolinska Institute, Stockholm (who performed complementary measurements on D-2) and Prof. V. Baranov, Institute of Biomedical Problems, Moscow.

- *Effects of Microgravity on Bone Cell Differentiation*, by G.E.R. Schoeters & J. Bierkens covered the results obtained in the Bion 10 mission, during which the osteogenic activity of MN7 cell cultures was found to be reduced. Interleukin-1 and PTH elicited a larger response.
- *Sleeping and Waking in Microgravity*, by O. Quadens, P. Dequae, R. Olieslagers & K. de Metz. This group has been involved in this type of research since the Spacelab 1 flight. Parabolic aircraft flights have been used to develop techniques and provide basic data in EEG, EMG, and EOG measurements during sleep periods.

8.4 Denmark

Life sciences space research in Denmark is currently concentrated on the DAMEC Research organization, which was established in 1988, primarily to promote space medical and human physiological research. The original formation was due to a nucleus of Danish research workers, who were involved in the preparation of medical experiments for the D-2 Spacelab mission, which took place in April/May, 1993.

There are now five full time research staff in DAMEC, who have focused their research activities on investigations into the relationship between gravitational stress and the mechanisms of renal fluid and electrolyte handling in humans. The work is developed in collaboration with Danish medical university institutes and clinical departments.

This study of fluid volume control during changes in gravitational stress has involved ground based and space experiments, in which the cardiovascular, endocrine and renal variables have been measured. The group, and collaborators, undertook four experiments on the D-2 Spacelab mission dealing with:

- *Peripheral haemodynamics before and after weightlessness during lower body negative pressure in humans*, by O. Henriksen, P. Norsk & A. Gabrielsen.
- *The central venous pressure during weightlessness*, by N. Foldager & P. Norsk.
- *Pulmonary stratification and compartment analysis with reference to micrograyity*, by S. Groth, D. Merhild & F. Jessen.
- *The influence of microgravity on endocrine and renal elements of volume homeostasis in man*, by P. Norsk, P. Bie, J. Warberg, N.J. Christensen & C. Stadeager.

Further experimental programmes, on Euromir 94 and Euromir 95, are:

- *Influence of microgravity on osmo- and volume-regulation in man: dynamic responses to isotonic and hypertonic loads*, by P. Bie, C. Emmeluth, P. Norsk & N.J. Christensen.
- *Central venous pressure during weightlessness*, by N. Foldager & P. Norsk.
- *Influence of microgravity on renal fluid excretion in humans*, by P. Norsk & L.B. Johansen.
- *Central venous pressure during weightlessness*, by N. Foldager & P. Norsk.

Besides the DAMEC group, there is a group at the Institute of Molecular Biology, Aarhus University, under the leadership of Dr O. Rasmussen, which has been involved in space experiments on plant biology. On Bion-9, an experiment was flown on the effects of microgravity on the development of plant protoplasts in collaboration with Dr T.H. Iversen, of Trondheim University. The same group of collaborators carried out an experiment on IML-1, on the effect of microgravity environment on cell wall regeneration, cell divisions, growth, and differentiation of plants from protoplasts. Similar work was also performed during IML-2.

Future research within the DAMEC group is directed towards further study of the mechanisms of fluid volume control in humans. This work will be focused on the distribution of fluid between the interstitial and intravascular compartment, the pulmonary low pressure circulation and total circulatory space, and various low pressure intrapulmonary compartments. There will also be emphasis on the adaptation of total volume to short and long term microgravity and the regulating mechanisms. Ground based experiments have been, and continue to be, an important part of the programme and complement the space experiments.

8.5 France

The French strategies for future research in space biology and life sciences were the subject of a CNES organized seminar at Aix-en-Provence, in April 1991, which revised the 1985 Deauville seminar conclusions. The results of these deliberations are summarised in the following paragraphs.

8.5.1 Space biology

Space biology research was considered to have two basic approaches. The first is directed towards the preparations for the exploration of the solar system and this includes radiobiology, radioprotection, artificial ecosystems, and biology under low gravitational acceleration. The second, and at present the principal approach, is that concerned with the fundamental biological responses to the space environment, including the gravitational responses (e.g. gravitropism), the potential role of gravity on genetic expression, energy transfer and effects of HZE particles, and the origin and development of life on Earth.

Four main research fields were defined, within that framework, namely: (i) exobiology; (ii) radiobiology; (iii) gravitational biology; and (iv) artificial ecosystems. Research objectives were established for each of these fields.

Exobiology. Main objective is understanding of processes leading to origin on life on Earth. Research topics identified are (i) chemistry of interstellar particles (including Antarctic micrometeorites) and planetary surfaces (Mars, Titan, comets), particularly with respect to the presence (or potential for synthesis) of early biochemical compounds (life seed theory); (ii) open exposure of macromolecules to the space environment.

Radiobiology. Principal objectives are characterisation of variations in the radiation environment (both spatial and time), risk assessment and biological implications, definition of preventive measures against radiation risks in space and development of methods for crew protection against space radiation. Subdisciplines identified include (i) physical radiodosimetry, cumulative doses (passive method comparable to Biostack) or on-line monitoring of HZE spectra (active dosimetry); (ii) (bio)chemical characterisation of DNA lesions by HZE particles; (iii) biological and genetic implications of cosmic radiation; cell inactivation, cell mutation and cell transformation (rodents being the preferred study animals).

Gravitational biology. Objectives here include gravity perception and related responses of biological structures at the cellular level and at the organism level. In particular this would include determination and expression of plant and animal bilateral symmetry (e.g. embryo polarity), the effects of varying gravity levels on the organisation of the cytoskeleton and possible effects on cell metabolism (e.g. biosynthesis, respiration). Research objects, or model organisms, identified were: (i) for developmental biology, filamentous algae, moss protonemata, *Drosophila* (fruit fly), marine invertebrates (e.g. ascidia), the nematode *Caenorhabditis elegans*, and certain vertebrates (particularly some amphibia); (ii) for cell biology, both bacterial and animal cell cultures; (iii) in plant science, verification of existing models for cellular reception of gravity (e.g. in statocytes of roots) through experiments with amyloplast-free systems, transformed plants or non-gravitropic mutants.

Artificial ecosystems. The main objective identified here was assessment of the feasibility to construct artificial ecosystems. The recommended programme comprised a long term construction plan for an artificial ecosystem aimed at supporting long duration missions, and a supporting programme for short duration experiments in simplified systems to investigate specific problems (e.g. maintenance and resource requirements). A number of special topics requiring investigation were identified. These were microbial populations, ecophysiology of plants and animals, long term interaction of organisms, population dynamics and modelling, genetic population biology, autarchy of agricultural exploitation management, abiotic factors and potential exploitation of local resources.

8.5.2 Physiology

Physiology research in space was considered to be composed of six major fields: sensorimotor physiology (neuroscience), cardiovascular physiology and respiration, muscle physiology, bone tissue physiology, adaptation to the environment, and space medicine. With the intention that the research experiments be conducted on humans and/or animals (e.g. rodents, monkeys) the research objectives within the fields were identified as follows.

Sensorimotor physiology (neuroscience). Graviperception and motor coordination (e.g. otolithic and muscular stimulation and posture); cognitive processes in microgravity (e.g. recognition of objects and body symmetries (particularly of the retina); concentration problems and electro-physiological patterns); modification of neurosensory structures (e.g. otolithic organs, CNS, synapses); neuronal space adaptation problems (e.g. disorientation, space motion sickness).

Cardiovascular physiology and respiration. Orthostatic intolerance upon return (re-adaptation studies); neuro-vestibular regulation (e.g. of arterial pressure); mechanical regulation in low pressure vascular systems; pharmacological experiments on physiopathology and medication/therapy); long term responses of the cardiovascular system.

Muscle physiology. Mechanisms of muscle atrophy and related functional modifications; role of gravity in organisation, development, differentiation and regeneration of muscle tissues; development of prophylactic and rehabilitation methods.

Bone tissue physiology. Monitoring of bone loss (kinetics, reversibility, localisation, degradation of bone quality); mechanisms of bone loss (tissue level: bone tissue composition, bone restructuring in response to muscle contraction; cellular level: cell biomechanics, hormone influence on mineral metabolism and bone cell characteristics, e.g. proliferation and differentiation); development of bone tissue (e.g. osteoblasts) under microgravity; development of prevention methods against (i) bone degradation on long duration space flights; and (ii) against osteoporosis.
Adaptation to the environment. Confinement effects on small closed groups (psychosocial studies); modification of neuro-endocrine and neurovegetative rhythms (e.g. circadian rhythms); behavioural and neuro-immuno-endocrine responses to stress conditions.

Space medicine: physiological and psychological issues. Medical aspects of crew selection and training; crew health maintenance during space flight; medical aspects of crew post-flight readaptation.

8.5.4 Life science experiment equipment and facilities

Table 3 summarizes the existing and planned facilities for the French research programmes in the life sciences, as of 1992. This list will be modified to take account of changes anticipated in future space programmes.

8.5.5 French life science experiments in U.S. and Russian spacecraft, 1994 & 1995

Table 4 shows the recent, current and near-term missions involving French experiments in the life sciences, and some of the planned missions, while

Table 3. Space life sciences research in France: existing and planned facilities for the French research programmes in the Space Life Sciences.

Facility	Purpose	Status	Flight opportunity
Human physiology & space medicine			
	Echography, diuresis	on-going	Mir 92
PHYSIOLAB	Medical monitoring of astronauts cardiovascular diagnoses; blood pressure, ECG, portable Doppler, spirometer, etc).	predevelopment studies	Mir 95
	Advanced PHYSIOLAB to include measurement of the biochemical parameters of blood and urine.	future development	
Human physiology: neurosciences			
VIMINAL	Sensory-motor (eye/motion) coordination study, perception of the environment, body representation	on-going	Mir 92
VIVA	One eye or two eye optokinetic stimulator	on-going	Spacelab
SUPERPOCKET	Recording & monitoring of physiological signals on the body	on-going	Spacelab
OCULOMETER	Eye movement recording & restitution (pre- and post-flight measurements)	predevelopment studies	
ERGOMETER	(Pre- and post-flight measurements)	predevelopment studies	
COGNILAB	Cognition processes and performance assessment	predevelopment studies	Mir 96
KINESIGRAPH	Motion detection and analysis device	predevelopment studies	Neurolab, SLS-4
Animal physiology as a model			
RHESUS	Monkey holding facility cooperation with NASA	on-going	SLS-3 & 4
Small animal holding facility	For rodents; technology study under way	future developments	
Biology			
IBIS	Instrument for biological investigations in space	on-going	1st flight on Foton, reflight in Shuttle mid-deck
FERTILE	Developmental biology of amphibians	on-going	Mir 96
BIOREACTOR	Cell culture under microgravity	development studies	

Table 4. Space life sciences research in France: recent, current and near-term missions involving French experiments in the life sciences, and some of the planned missions.

Programme	Objectives	Launch	Co-operation
Biology			
Sounding rocket	Fertilization and development of sea urchin eggs. Internalizing of receptors (osteoblasts)	Launched in 1992	ESA
Biorack/IML-1	Resistance of bacteria to antibiotics.	Spacelab/IML-1 launched in 1992	ESA/ U.S.
Antares	Immunology of bacteria to antibiotics. Gravitropic response of plant roots.	Flight on Mir in 1992	Russia
Biocosmos 10	Osteoblasts and DNA lesions.	Launched in 1992	Russia
Altair	Immunology and exposure of biological samples to radiation.	Flight on Mir in 1993	Russia
Biopan	Exposure of samples to radiation (4 experiments).	Flight on Russian capsule in 1993	ESA
Biorack/IML-2	Gravitropic response of plants (lentils). Microenvironment and cellular activation.	Spacelab/IML-2 flights in 1994	ESA/U.S.
IBIS	Cellular biology.	Flight on Russian capsule in 1994 reflight in 1996	Russia
Cassiopea (Fertile)	Developmental biology of amphibians.	Flight on Mir 1996	Russia
Physiology and Space Medicine			
Decubitus	Ground simulation: cardiovascular weakening prophylaxis.	Short duration: 1993 Long duration: 1994	France ESA
Parabolic flight	Muscle, ethology	Campaigns completed in 1992	France
MVI EDO COIS	Vestibular Oculo Reflex (neurophysiology) EDO programme, Viva programme, COIS programme.	Mid-deck in 1992; 1993 Flight IML-1 in 1992 and 1994	U.S.
Biocosmos 10	Support tissues, psychological performance (attention and sleep)	Flight in 1992	Russia
Antares and Post Antares	Neurophysiology, cardiovascular regulation, radiation protection, neurohormonal regulation	Flight on Mir in 1992	Russia
Altair	Neurophysiology, cardiovascular regulation, radiation protection, neurohormonal regulation.	Flight on Mir in 1993	Russia
Spacelab/SLS	Space physiology in rats.	Spacelab/SLS-2 flights in 1993	U.S.

Table 4 (continued).

Programme	Objectives	Launch	Co-operation
Echography	Cardio-vascular regulation.	Spacelab/D2 flight in 1993	ESA/U.S.
Rhesus	Space physiology in primates.	SLS-4 flights in 1998	U.S.
Neurolab	Dedicated mission for all neuroscience disciplines (Kinesiograph).	Spacelab/SLS-4 in 1998	U.S.
Cassiopea (Cognilab)	Cognition processes and performance assessment.	Flight on Mir 1996	Russia

Table 5 summarizes the experiments which have been scheduled for flight in the near future on Russian spacecraft. In addition to such space experiments there is a continuing active programme of short duration microgravity experimentation carried out using sounding rockets and Caravelle-based parabolic flights.

Table 5. Space life sciences research in France: experiments scheduled for flight on Russian spacecraft.

Experiment title	Principal Investigator	Co-investigators
Euromir 94		
Posture and movement in microgravity.	A. Berthoz, LPN, CNRS, Paris	T. Pozzo, Univ. of Bourgogne J. McIntyre, LPN
'STAMP' Experiment – perception of figure symmetry by the two cerebral hemispheres.	A. Berthoz, LPN, CNRS, Paris	D. Viviani, University of Geneva, Switzerland
Spaceflight related orthostatic intolerance: role of autonomic nervous system, water balance, volume regulating hormones, and energy expenditure.	C. Gharib, Lab. de Physiol. de l'Environment, Faculté de Médicine, Lyon	A. Maillat, G. Gauqelin, A.M. Allevard, A. Pavy le Traon, Fac. de Médicine, Lyon. MEDES, Toulouse; R.L. Hughson, J.O. Fortrat, U. of Waterloo, Canada; J.P. Riou, C. Pacchiaudi, S. Normand. INSERM, Lyon.
Changes in mechanical properties of human muscle as result of spaceflight (ground experiment).	F. Goubel, Univ. de Technol. de Compiegne, Compiegne	J.F. Marini, Université d'Aix-Marseille
Bone mass and structure changes and bone remodelling in space.	C. Alexandre, L. Vico, Lab. de Biologie du Tissue Osseux, St. Etienne	
Influence of spaceflight on energy metabolism and its circadian variation.	H. Demaria Pesce, CNRS/INSERM, Inst. de Biologie du College de France, Paris.	S. Daan, Research Group Chronobiology, Haren; H. Visser, Centrum voor Isotopen Onderzoek, Groningen; J. Louis-Sylvestre, Univ. P. et M. Curie, Paris

Table 5 (continued).

Experiment title	Principal Investigator	Co-investigators
Immune changes after space flight	D. Schmitt, Lab d'Immunologie, CHU Rangueil, Toulouse.	
Euromir 95		
Changes in mechanical properties of human muscle during spaceflight.	F. Goubel, U. de Techn. de Compiegne, Compiegne	J.F. Marini, Univ. d'Aix-Marseille
Bone mass and structure measurements during long term spaceflight, using an ultrasound bone densitometer.	C. Alexandre, L. Vico, Lab. de Biologie du Tissue Osseux, St. Etienne	
French experiments on Foton, 1995		
Biobox		
To test whether *in vitro* cultures of osteoblast-like cells are gravity dependent in proliferation and activity	C. Alexandre	Lab. de Biologie du Tissue Osseux, St. Etienne
Biopan		
Survival of organic molecules under space conditions. Catalysis of reactions by solar U.V.	A. Brack	CBM, CNRS, Orléans
Radiation (HZE) induced base modifications in the cell genome.	J. Cadet	SESAM, MDS-CENG, Grenoble
Effects of cosmic radiation and microgravity on dormant shrimp embryos.	A. Hernandorena	Centre d'Etudes et de Recherches, Biarritz
Effects of cosmic radiation on isolated organic macromolecules. The effectiveness of radioprotective drugs.	J.P. Moatti	Université Paul Sabatier, Toulouse.
French experiments on Spacelab IML-2, 1994		
Effect of microgravity on cellular action : role of cytokines.	D. Schmitt	Lab. d'Immunologie, University of Toulouse
Cell microenvironment and signal transduction in microgravity.	P. Bouloc	Université de Paris, 7.
Effect of microgravity on lentil morphogenesis.	G. Perbal	Université de Paris, 6.
The sea urchin larva, a potential model for studying biomineralization and demineralization processes in space.	H.J. Marthy	Laboratoire Arago, Banyuls sur Mer

8.6 Germany

The German Space Agency (DARA) is responsible for defining the general strategic guidelines for space research and for implementing the German space programme by placing industrial contracts and funding of research groups. In addition, the agency represents Germany's space interests in the international arena.

With regard to the life sciences programme, the following goals and objectives have been defined: (i) to achieve fundamental scientific insights into the behaviour of biological systems under space conditions with emphasis on physical, chemical and physiological aspects; (ii) to contribute to the global exploration of the ecosystem of Earth by investigations of artificial ecosystems; (iii) to acquire the medical and technological fundamentals for the maintenance of humans in space; (iv) to contribute to terrestrial application-oriented and basic research, by improvements of existing technologies as well as by the development of novel methods and materials, and (v) to prepare for potential commercial utilization.

Essentially, life sciences research in space is seen as a natural extension of terrestrial research activities. It is therefore developed as an integral part of the terrestrial research fields. Focusing of research is encouraged, as is interdisciplinary cooperation. The national programme objectives are to be complemented by cooperative research programmes within ESA and specific international collaborations.

The programme goals, presented above, are sought whilst taking into account the following consideration: (i) success of theoretical and experimental work in the past; (ii) quality of scientific results gained so far; (iii) experiment proposals of very high scientific quality strongly recommended by peer groups; (iv) avoidance of redundant investigations by cooperative and complementary research; (v) objectives promising visible results in a shorter time frame, trying further to maximize the benefit.

Having established the basic goals, the priorities for future experimental facilities and the flight opportunities could then be defined. The following paragraphs summarize the current programme elements, for biology and for the field of human physiology.

8.6.1 Biology – current German programme

Gravitational biology. Role of gravity for growth, development, reproduction, movement, orientation and other physiological processes; mechanisms of adaptation and compensation; role of gravity for processes regulated by multiple factors; graviperception and signal reduction.

Bioregenerative life support systems. Development of life support systems for basic physiological research, involving: growth unit for higher plants, algal reactor for long term and multi-generation experiments, system for aquatic animals, development of microbial waste management systems and a combined system for investigation of the relationship between the botanical and zoological components, including system-theoretical problems (CEBAS).

Radiation biology. Passive and active dosimetric mapping of cosmic radiation; effects of cosmic radiation on biological material; interaction between cosmic radiation and microgravity.

Exobiology. Life under extreme environmental conditions; origin of life: role of cosmic factors.

Bioprocessing. Bioprocessing techniques: influence of space conditions, especially microgravity, on separation; and on cell fusion techniques; crystallization: growth of high quality monocrystals of biological macromolecules for the subsequent analysis of their structure.

8.6.2 Human physiology – current German programme

Cardiovascular system. Measurement of blood and fluid shift and analysis of the underlying regulatory principles.

Neurophysiology. Graviperception and gravisensitivity; space adaptation syndrome; neural integration and regulation, including neuronal plasticity.

Musculoskeletal system. Training and detraining effects on muscular structure and function; bone decalcification and alterations in connective tissues.

Endocrinology and metabolism. Measurement of hormones in order to obtain an integrative analysis of hormonal regulation; effects of space conditions on the immune system; dose responses and metabolism of medical drugs.

Operational space medicine. Human factors research: circadian rhythms, sleep-wake cycle and human performance; optimization of hygiene and food; isolation and confinement studies, psychological aspects of crew selection and crew training; development of countermeasures against space motion sickness and muscle and bone degradation; development of effective radiation protection in line with results from investigations in radiation biology.

8.6.3 Emphasis and focus in the German programme

Identifying the future focal areas, and looking to the timescale for each of these studies, the scheme shown in Table 6 has been developed. A rough

Table 6. Space life sciences research in Germany: focal points in scientific research areas in the life sciences.

Biology	1992	1997	2002
Gravitational biology	physiological analysis biochemical & biophysical analysis of signal transduction chains		
Bioregenerative life support systems	ground-based research (preparatory phase)		
		– ground-based reference programme for flight experiments – flight experiments with plant & animal systems	
			– flight experiments BLSS
Radiation biology	– passive dosimetric cosmic radiation detection		
		– active dosimetric cosmic radiation detection	
			– analysis of biological effects at morphological physiological and molecular levels
Human physiology			
Cardiovascular system	– measurement of blood and fluid shifts	– analysis of regulatory principles	
Neurophysiology	– analysis of graviperception and gravisensibility	– neuronal regulation and plasticity	
Musculoskeletal system	– training and detraining of muscular structures	– analysis of bone decalcification and alterations of connective tissues	
Endocrinology and metabolism	– measurement of particular hormones	– integrative analysis of hormonal regulation – effect of space conditions on immune system and on responses to medical drugs	
Operational space medicine	– human factors research, e.g. circadian rhythms, sleep-wake cycle, performance – isolation and confinement studies		

idea of the relative emphasis which is given to the different research areas is given by the number of projects funded (in 1993) in each of the various programmes by both DARA and by the DLR (Fig. 1). About 140 flight experiments involving German scientists have been performed in space biology and human physiology since the Apollo 16 flight of 1972; Fig. 2 shows the distribution of these experiments by research area. Many of these experiments were carried out within the flight programme listed in Table 7. Comparison of Figs 1 & 2 shows that research in gravitational biology continues to be a major activity – as indeed it is for ESA life science research as a whole. Radiation biology is also a continuing primary activity, together with research on the cardiovascular system. It is interesting to note also there has been a parallel effort on the more applied science topic of bioprocessing, which is continuing, although at a reduced relative level. A significant new activity is

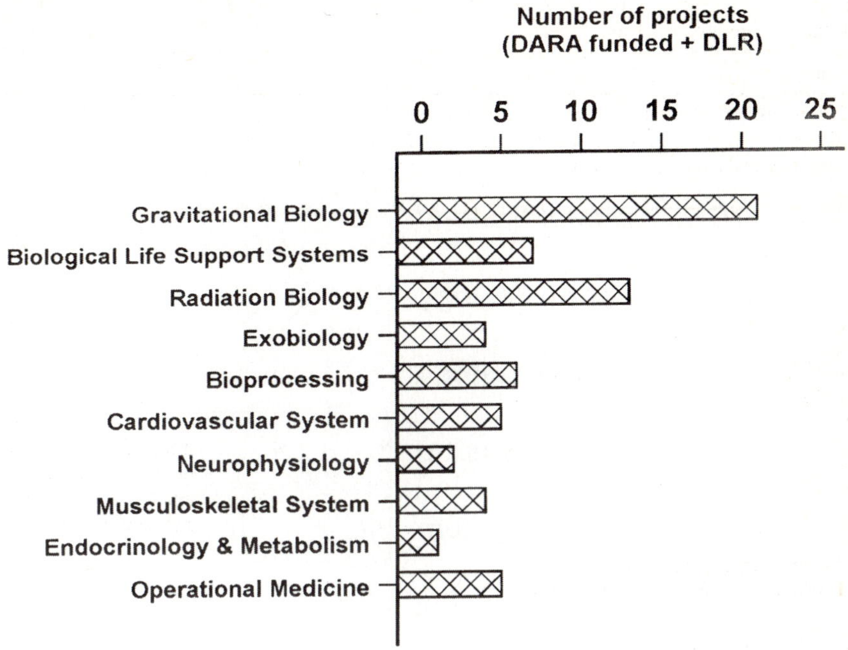

Fig. 1. Numbers of projects in the German space life sciences programme funded by DARA and DLR as of May 1993.

the work on bioregenerative life support systems, which now involves several research groups in Germany.

In addition to these space experiments, there has also been an active programme of microgravity biology research carried out within the TEXUS sounding rocket programme, which started in 1977. The TEXUS 29/30 launches, of 1992/93 contained 4 biology experiments. These are summarized below in order to give an illustration of the type of experimentation undertaken by this means.

- *Gravitaxis and phototaxis in flagellates*, by D. P. Häder (Univiversität Erlangen), third and fourth experiment to observe the movement and behaviour of gravitaxis in the motile unicellular alga *Euglena*, including phototactic reactions in microgravity.
- *Function of cytoskeleton in sedimentation of statoliths in Chara rhizoids*, by A. Sievers (Univiversität Bonn), fourth and fifth experiment to investigate the cytoplasmic streaming under microgravity and to determine the role of cytoskeleton in graviperception of rhizoid cells.
- *Effect of changes in gravitation on energy metabolism of plant cells*, by R. Hampp (Univiversität Tübingen), the effect of short term microgravity conditions on the amount of adenine and pyridine nucleotides in plant protoplasts was investigated (second experiment).
- *Protein pattern in mesophyll protoplasts of Vicia faba*, by H. Schnabl (Univiversität Bonn), this first of a series of experiments investigated the effect

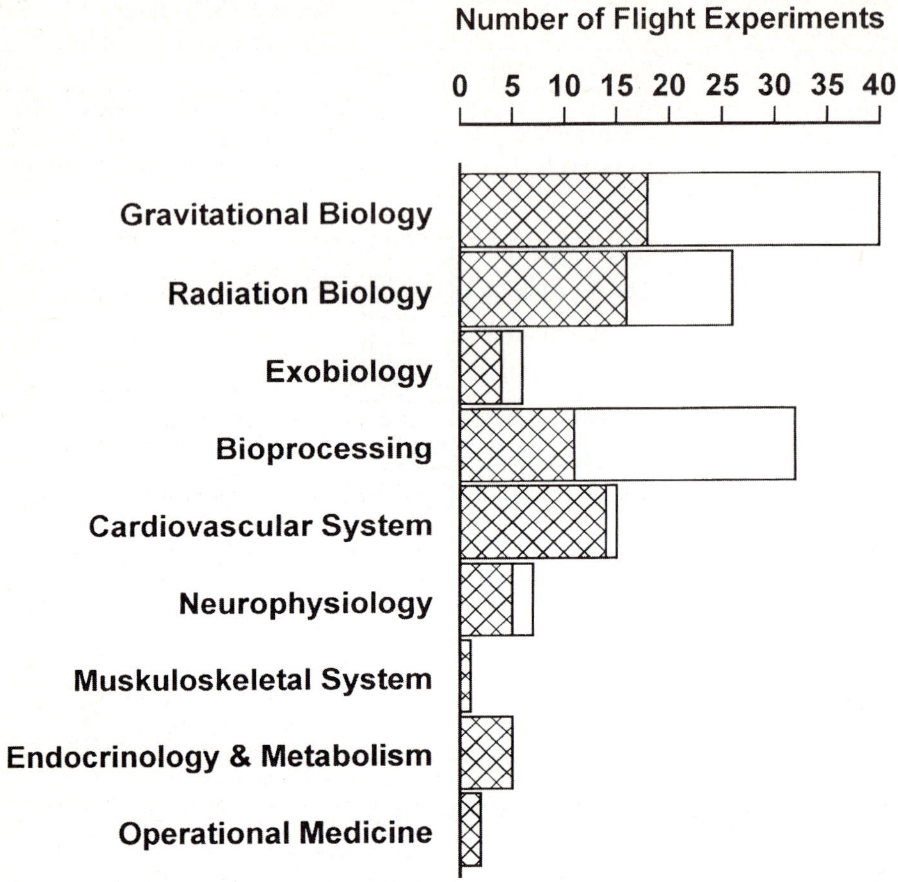

Fig. 2. Distribution of experiments performed by German scientists in space biology and human physiology between different research areas. Data covers 140 flight experiments performed since Apollo 16 (1972) to 1993. Experiments with European Principal Investigators are shown with cross-hatched shading.

of short duration microgravity on the pattern of special proteins in protoplasts as a marker for stress situations.

8.6.4 International collaboration in the German programme

International collaboration continues to be an important part of the overall German microgravity programme, and falls within the strategic guidelines for future planning of missions. The principal collaborations, as they have developed, and current arrangements for future collaborative missions are as follows.

Table 7. Space life sciences research in Germany: some milestones in German space life sciences research activities.

Year	Mission	Remarks
1978	Salyut-6	First experiments in the Intercosmos programme during the first flight of a German crew member on a Soviet space station.
1983	First Spacelab mission	Four life science experiments, as well a being the first Spacelab mission with a German crew member.
1984	LDEF	The Long Duration Exposure Facility launched with German experiments in radiation biology and exobiology (retrieved in 1990).
1985	D1	14 experiments in life sciences during the first German Spacelab mission.
1986	MIKROBA	First launch of a balloon-borne drop capsule.
1987	Biocosmos 8	First life sciences experiment (radiation biology) on a re-entry satellite.
1991	Drop Tower	First experiments on gravitational biology in the Bremen Drop Tower.
1992	IML-1	Participation in the International Microgravity Lab Spacelab mission with 7 German experiments in life sciences and a German crew member.
1992	Mir 92	13 life sciences experiments (mainly on human physiology) during the flight of a German crew member on the permanently manned space station Mir.
1992	EURECA	7 life sciences experiments on an unmanned orbiting platform in the ESA microgravity programme.
1993	D-2	Second German Spacelab mission, with 30 life sciences experiments and two German crew members.
1994	Bion 10	3 biology experiments carried out, in collaboration with Russian and Mir experimenters, on its re-entry satellite
1994	IML-2	4 biology experiments (see Table 23).

In addition to these space experiments, there has also been an active programme of microgravity biology research carried out within the TEXUS sounding rocket programme, which started in 1977; see text.

8.6.4.1 Collaboration with NASA

FSLP (1983); D-1 (1985); IML-2 (January 1992); D-2 (May 1993); IML-2 (1994); Get Away Special (GAS) programme with shuttle (BIOMAUS); middeck experiments under consideration; Spacehab; parabolic flights (KC-135); NASA/DARA-DLR LSPWG; sample sharing agreement; space life sciences training programme for students NSCORT programme.

8.6.4.2 Collaboration with ESA

Parabolic flights; sounding rockets (TEXUS/MAXUS); Bion; IML-1 1992 (Biorack, Biostack); Mir 92 (Vesta); D-2 1993 (Anthrorack); EURECA-1 1992/1993 (ERA); IML-2 1994 (NIZEMI, Biorack, Biostack); Euromir 94 and 95.

8.6.4.3 Collaboration with CIS Institutions

Mir 92 and Mir 92-E(xtension); Euromir 95; Re-entry satellites (Bion); agreement between DARA (DLR Institute of Aerospace Medicine) and ZPK, Moscow on cooperation in human physiology and operational medicine.

8.6.4.4 Collaboration with CNES

CNES/DARA Life Science Working Group; hardware sharing on CIS missions.

8.6.4.5 Collaboration with NASDA

Scientific working group; co-operative use of Drop Tower/Drop Shaft.

8.6.4.6 Collaboration with CSA

CSA/DARA Life Science Working Group.
The current international collaborations, at the scientific level, are listed in Table 8. Life science experimentation on the Russian Space Station Mir has played an important role in the German programme. Table 9 lists experiments carried out on Mir 92, and German experiments on Euromir 94 and Euromir 95.

8.6.5 German experimental facilities

Principal experimental facilities developed to support biological and physiological research in space as part of the overall German microgravity programme are listed in Table 10, with their application and current status.
Besides experimental facilities, Germany has designated a number of centres of excellence to support and encourage specific focal areas of the space life sciences. These are: gravitational biology, AGRAVIS (University of Bonn); radiation biology, SCORT (University of Giessen); bioregenerative life support systems, CEBAS (University of Bochum); human physiology (Institute for Aerospace Medicine, DLR, Cologne).

Table 8. Space life sciences research in Germany: international collaborations, as at August 1993.

Germany		Short title or Field of research	Collaborating country	
Scientist	University		Scientist	University
Space biology				
Machemer	Bochum	Gravitational sensitivity of microbes	Takahashi	Tokyo, Japan
Blüm	Bochum	C.E.B.A.S. – Aquarack Programme	Schreibman Doty Gitelson	New York, U.S.A. New York, U.S.A. Krasnoiarsk, Russia
Kranz	Frankfurt	Free Flyer Biostack, ESA Bion 10	Shevchenko Seitz Hellmann	Moscow, Russia Strasbourg, France Strasbourg, France
Sievers	Bonn	Growth and development of plants	Yamada Masuda Laurinavicius	Tokyo, Japan Osaka, Japan Vilnius, Lithuania,
Hampp	Tübingen	Electrofusion of plant cells	Outlaw Vasil	Tallahassee, U.S.A. Gaines-ville, U.S.A.
Schnabl	Bonn	Membrane physiology and electrofusion of cells	Bornmann	Lund, Sweden
Zimmermann	Würzburg	Electrofusion of cells	Sammons Neill	Arizona, U.S.A Iowa, U.S.A
Horneck	DLR	Radiation biophysics	Kozubek	Brno, CSFR
Space physiology				
Stegemann DSHS College		Respiratory monitoring system	Swanson	Denver, U.S.A.
		O_2 uptake kinetic analysis	Hughson	Waterloo, Canada
von Baumgarten	Mainz	Vestibular physiology	Watanabe	Nagoya, Japan
Baisch	DLR Cologne	Cardio-vascular physiology Fluid shift	Blomquist Baranov	Dallas, U.S.A. IBMP Moscow, Russia
Wegmann	DLR Cologne	Countermeasures	Guell	MEDES Toulouse, France
Gundel	DLR Cologne	Sleep monitoring	Monk	Pittsburg, U.S.A.
Kirsch FU Berlin		Fluid shift and metabolism	Matsui	Nagoya, Japan
		Tissue thickness and compliance	Kupzev	ZPK, Moscow, Russia
Scherer	FU Berlin	Vestibular-ocular-reflex	Kornilova	IBMP Moscow, Russia

Table 9. Space life sciences research in Germany: life science experiments on the Mir space station.

Experiment title	Investigators	Affiliation
Mir 92		
Self tonometry under microgravity.	J. Draeger, R. Schwartz	University Hospital, Eppendorf
Effects of microgravity on interstitial muscle receptors affecting heart rate and blood pressure during static exercise.	D. Essfeld, K. Baum, U. Hoffmann, J. Stegemann	Physiological Institute Deutsche Sporthochschule, Cologne
Dosimetry in the space radiation field.	G.Reitz, R. Beaujean, N. Heckley, G. Obe	DLR, Institut Flugmedizin Univ. Kiel; Univ. G.H., Essen
Sleep and circadian rhythm during a short space mission.	A. Gundel, E.Reucher, M. Vejvodva, J. Zulley, V. Nalishiti	DLR, Inst. Luft- und Raumfahrtmedizin MPI Psychiatrie, Münich ZPK, Star City
Behavioural aspects of human adaptation to space analysis of cognitive and psychomotor performance.	D. Manzey, B. Lorenz, A. Schiewe, G. Finell G. Thiele	DLR, Inst. Luft- und Raumfahrtmedizin, DLR, Astronautenburo
Illusions of verticality in weightlessness	H. Mittelstaedt, S. Glasauer	MPI Verhalten-physiologie, Seewiesen; Neurolog. Klinik, Grosshadern, Münich
Vestibulo-oculomotor testing during space flight.	A.H. Clarke, W. Teiwes, H. Scherer	HNO Klinik, Klinikum Stieglitz, Free Univ., Berlin
Oculovestibular interactions under microgravity	K. Hoffstetter-Degen, J.Wetzig, R. Von Baumgarten	Physiolog. Institut J.G. Univ., Mainz
Responses to eccentric rotation in 2 space subjects.	K. Hoffstetter-Degen, J.Wetzig, R. Von Baumgarten	Physiolog. Institut Univ., Mainz
Reduced natriuresis during weightlessness.	R.Gerzer, C.Drummer, M.Heer, R.A. Dressendorfer, C.J. Strasburger	Klinik Innenstadt, Münich + DLR, Flugmedizin
Fluid shifts into/out of superficial tissues under microgravity and terrestrial conditions.	K.A. Kirsch, F.J. Baartz, H.C. Gunga, L. Rocker	Inst. for Physiology, Free Univ. Berlin
Body fluid distribution in man; space and effect on LBNP treatment.	F.J. Baisch	DLR, Inst. Luft- und Raumfahrtmedizin, Cologne

Table 9 (continued).

Experiment title	Investigators	Affiliation
Euromir 94		
Circadian rhythms and sleep during a 30-day space mission.	A. Gundel, A. Smael, A.M. Lebedjev, I. Ponomariova	DLR, Inst. Luft- und Raumfahrtmedizin. Inst. of Psychology, Moscow; IBMP, Moscow
Fluid and electrolyte balance during weightlessness and possibilities of their regulation.	C. Drummer, R. Gerzer, C. Kirschbaum, C.J. Strassburger	DLR, Inst. Luft- und Raumfahrtmedizin, Univ. Trier, Univ. München
Magnetic resonance spectroscopy imaging of human muscles before and after space flight (ground experiment).	J. Zange, K. Mueller, H. Schuber, A. Haase, V.S. Oganov, B. Shenkmann	DLR, Inst. Luft- und Raumfahrtmedizin, Univ. Würzburg, IBMP, Moscow
Radiation health during prolonged space flight (environmental and personal dosimetry).	G. Reitz, R. Facius, H. Schäfer, J.U. Schott, R. Beaujean, W. Heinrich, C. Heilmann, E.E. Kovalev, V.E. Dudkin.	DLR, Inst. Luft- und Raumfahrtmedizin, Univ. Kiel, Univ. Siegen, Baen, Strasburg, S/C Rad. Safety, Moscow
Chromosomal aberrations in peripheral lymphocytes of astronauts (ground experiment).	G. Obe, N. Heckeley	Univ. Essen, Dept. of Genetics.
Fluid shifts into and out of superficial tissue-stability along the body axis.	K.A. Kirsch, H.C. Gunga, F.J. Baartz, L. Roecker	Inst. for Physiology, Free Univ., Berlin.
Effects of changes in central venous pressure on the erythropoietic system under 1 × g and microgravity.	H.C. Gunga, K. Kirsch, F. Baartz, L. Roecker	Inst. for Physiology, Free Univ., Berlin.
Spatial orientation and space sickness.	H. Mittelstaedt, S. Glasauer	MPI Verhalten-physiol. Seewiesen. Lab. Physiol. Neurosens., Paris
Adaptation of basic vestibulo-oculomotor mechanisms to altered gravity conditions	H. Scherer	Universitaet Klinikum Stieglitz, Free Univ. Berlin
Gastroenteropancreatic peptides during microgravity and possible involvement in space motion sickness.	R.L. Riepl, C. Drummer, F. Fiedler	Klinikum Innenstadt, Univ. of Münich, Faculty of Clinical Medicine, Univ. Heidelberg
Non-invasive stress monitoring in space light by hormone measurements in saliva	C.J. Strassburger	Klinikum Innenstadt, Univ. of München,

Table 9 (continued).

Experiment title	Investigators	Affiliation
Euromir 95 Monitoring of drug metabolism and drug effect during prolonged weightlessness	R. Gerzer, J. Wenzel, C.Drummer J.M. Heim, G. Wolf, V.S. Shashkov, H.B. Lakota, G.P. Stupakov	DLR, Inst. Flugmedizin, Univ. of Münich IBMP, Moscow
Radiation health during prolonged space flight.	G. Reitz, R. Facius, H. Schaeffer, J.U. Schott, R. Beaujean, W. Heinrich, C. Heilmann, E.E. Kovalev, V.E. Dudkin, I.A. Petrov	DLR, Inst. Fugmedizin Univ. Kiel, Univ. Siegen Baen, Strassburg, Res. Ctr, S/C Safety, Moscow IBMP, Moscow
Chromosomal aberrations in peripheral lymphocytes of astronauts.	G. Obe, N. Heckeley	Dept. Genetics, Univ. G.H. Essen
Magnetic resonance spectroscopy and imaging of human muscles and bones, before and after space flight.	J. Zange, K. Mueller, B. Topic, M. Schuber, A. Haase	DLR, Inst. Luft- und Raumfahrtmedizin; Univ. Wurzburg
Differential effects of otolith input on ocular lateropulsion, cyclorotation, perceived visual vertical, straight ahead, & tonic neck reflexes in man.	H. Dieterich, S. Krafczyk, T. Brandt, A. Straube	Neurology Department, Univ. of Münich

8.7 The Netherlands

The Space Research Organisation of the Netherlands (SRON) supports a programme of microgravity life science research, with the emphasis on human physiology, cell and developmental biology, and the crystallization of proteins. Table 11 summarizes the experiments carried out during 1992 within this programme, which involved Russian as well as ESA/NASA space systems, together with ESA sounding rockets.

SRON support is now provided on an individual research topic basis, with fixed start and defined termination dates. Selection of research proposals was based on peer group reviews, apparently without any deliberate attempt to focus on specific life science research areas. At present an in depth re-evaluation of microgravity research projects is taking place. It is likely that this will lead to a focusing in the future on areas of cell biology, cardiovascular,

Table 10. Space life sciences research in Germany: experimental facilities for biological and physiological research in space developed as part of the overall German microgravity programme.

Facility	Research Field	Mission	Status
BIOLABOR BB	Biological basic research		
Incubator 10-20° C with 1 × g reference centrifuge (STATEX)	Animal gravitational biology	D-1/D-2	available
Incubator 20-24° C with 1 × g reference centrifuge (BOTEX)	Plant gravitational biology/ cell cultivation	D-2	available
Incubators 37° C with/ without 1 × g reference centrifuge	Bioprocessing/cell cultivation	D-2	available
Electro cell fusion unit with workbench/ microscope	Bioprocessing/electro cell fusion	D-2	available
Threshold centrifuge (THZ)	Gravitational biology	D-2	available
NIZEMI Slow-rotating centrifuge microscope	Gravitational biology	IML-2	available
BIOMAUS higher plant growth chamber	Gravitational biology	Small payloads programme	in development
CEBAS minimodule closed equilibrated aquatic system	Bioregenerative life support system	Small payloads programme	in development
SIMPLEX Incubator with 1 × g reference-centrifuge	Gravitational Biology	Small payloads programme	in development
MEDEX	Human physiology		available
Body impedance ECG measurement device	Cardiovascular physiology	Mir 92, D-2	available
Pulse and blood pressure measurement device			available
Applied potential tomography		Mir 92	available
Leg volume measurement device			available
Lower body negative pressure device (LBNP)	Operational space medicine		in development
Leg ergometer		Mir 92	available
TONOMETER	Cardiovascular physiology	Mir 92/D-2	available
VOG Mobile vestibular laboratory	Vestibular physiology	Mir 92, Mir 92-E	available
HSD tissue thickness and compliance	Cardiovascular physiology	Mir 92, D-2 Mir 92E	available

Table 11. Space life sciences research in Holland: experiments carried out during 1992.

Project title	Principal Investigator	Mission	Launch date
Effect of microgravity and mechanical stimulation on the *in vitro* mineralization and resorption of fetal mouse long bones.	J.P. Veldhuijzen, ACTA	IML-1	22 January
The role of gravity in the establishment of the dorso-ventral axis in the amphibian embryo.	G.A. Ubbels, Hubrecht Laboratory	IML-1	22 January
Bacterial proliferation under microgravity conditions.	C.L. Woldringh, UvA	IML-1	22 January
Regulation of cell growth and differentiation by microgravity.	A.J. Verkleij, UU; S.W. de Laat, Hubrecht Laboratory	MASER 5	9 April
Plasma membrane fusion in human fibroblasts during short term microgravity.	J.F. Jongkind, EUR	MASER 5	9 April
Photoprocessing of grain mantle analogues and shielding of spores.	M.J. Greenberg, RUL	EURECA	31 July
Vestibular rotation and nystagmus.	W.J.Oosterveld,AMC	ESA parabolic flights	March, April
Control of blood pressure in humans.	J.M. Karemaker, UvA	NASA parabolic flights	April
Effect of microgravity and mechanical stimulation on the *in vitro* mineralization and resorption of fetal mouse long bones.	J.P. Veldhuijzen, ACTA	Bion 10	29 December
Effects of microgravity on plasma membrane-cytoskeleton interactions during cell division in *Chlamydomonas*	H. v.d. Ende, UvA	Bion 10	29 December
Free running circadian rhythmicity during space flight	W.J. Rietveld, RUL	Bion 10	29 December

and vestibular studies. During 1994, the existing programme will involve the reflight of the Bion-10 experiments of Professors Rietveld and van den Ende on the Foton spacecraft. There are also three Dutch experiments on the IML-2 mission; that of Dr G. A. Ubbels, on developmental biology, of Professor S.W. de Laat, on cell growth and differentiation effects in mouse embryonic cells, and of Dr J.P. Veldhuijen, studying *in vitro* mineralization and resorption of fetal mouse bones.

Dutch researchers will also be involved with the Euromir programme. For the Euromir 94 mission, Dr W. Bles, of the TNO Institute for Human Facilities, will be responsible for a ground experiment entitled *Does sickness induced by centrifugation mimic the Space Adaptation Syndrome?* This experiment will involve subjecting the crew to prolonged exposure to $3 \times g$ both before and after flight while recording blood pressure and other relevant parameters, including vestibular tests. Dr G.H. Visser, of the Isotope Research Centre and Dr S. Daan, of the Zoology Laboratory of Groningen University will be co-investigators in the experiment of Dr V. Demaria-Pesce, of CNRS, Paris, entitled *Influence of space flight on energy metabolism and its circadian variation.*

For the Euromir 95 mission, Dr C. Vermeer of the Biochemistry Department, University of Limburg, has been selected to carry out an experiment entitled *Effect of Vitamin K supplements on bone mass during microgravity conditions.* An in-flight bone densitometer will be used, together with urine sample analysis, to observe the effects of administering vitamin K during flight.

In addition to these space experiments, Dutch biologists have also been active in rocket based experimentation, and that activity will continue in the future.

8.8 Italy

The participation of Italian scientists in the microgravity life sciences programmes has been limited in the past, but is now developing as new opportunities become available. The main historical elements of Italian participation in space life sciences research, together with current plans and future developments in both long and short duration missions are given in Table 12.

8.9 Norway

At present there are three groups in Norway active in space life sciences research: (i) Dr T-H. Iversen, Department of Botany, University of Trondheim; Dr A. Johnsson, Department of Physics Department at the same University, and Dr R. Bjerkvig, Department of Anatomy, University of Bergen.

Table 12. Space life sciences research in Italy.

Date	Flight	Title	Principal investigators and collaborators	Affiliation
1983	Space-lab-1	3-D Ballistocardiography in space experiment in Spacelab.	A. Scano, V. Masini, F. Strollo, G. Brazzoduro, F. Prandi, C. Carna, E. Rispoli	University of Rome
1985	D-1	Effects of microgravity on genetic recombination in *Escherichia coli* (experiment on BIORACK).	O. Ciferri, O. Tiboni, A. M. Orlandoni, M. L. Marchesi	University of Pavia
1987	Parabolic flights	Performance of sub-maximal leg exercise in microgravity during parabolic flights.	D. Linnarsson, P.E. di Prampero, J. Stegemann, U. Hoffmann, D. Essfeld, C. Sunberg	Karolinska Institute Stockholm; Univ. of Udine; Univ. of Cologne, Univ. of Stockholm
1988	Parabolic flights	Effects of short term microgravity on the cardio-pulmonary function during exercise.	D. Linnarsson, P.E. di Prampero, G. Antonutto	Karolinska Institute, Stockholm; Univ. of Udine
1989/90	Parabolic flights	Blood lactate during leg exercise in microgravity.	P. Zamparo, C. Campbell, G. Antonutto	University of Udine
1990	MASER 4	Lymphoid cells in space.	A. Cogoli, M. Cogoli, N. Arena, S. Barri, P. Pippia, G. Monaco, L. Sciola, A. Spano, M. A. Meloni, R. Monti	Univ. of Sassari & Univ. of Naples
1992	MAXUS 1B	Lymphoid cells in space.	A. Cogoli, M. Cogoli, N. Arena, P. Pippia, G. Monaco, L. Sciola, A. Spano, M. A. Meloni, R. Monti	Univ. of Sassari & Univ. of Naples
1993	D-2	Effects of spaceflight on pituitary-gonad-adrenal function in humans (experiment for ANTHRORACK).	G. Riondino, F. Strollo, M. More, L. Bollanti, A. Ciarmatori, A. Scano, N. Mangrosse	INCRA University of Rome

Table 12 (continued).

Date	Flight	Title	Principal investigators and collaborators	Affiliation
1993	Parabolic flights	Effects of short duration micro-gravity. Cardiopulmonary function during sub-maximal leg exercise evaluated by means of the photo-acoustic analyzer.	P.E. di Prampero, G. Antonutto, C. Capelli, P. Zamparo, M. Girardis	University of Udine
1993	Parabolic flights	Endocrine changes and motion sickness during parabolic flights.	E. Strollo, G. Giondino, M. More, L. Bollanti, A. Ciarmatori, T. Orsi, W. Mangrosse	INCRA University of Rome
1994	IML-2	Lymphoid cells in space (BIORACK on Spacelab).	A. Cogoli, M. Cogoli, S. Barri, P. Pippia, G. Monaco, L. Sciola, A. Spano, M. A. Meloni	University of Sassari
1994 and 1995	Euro-mir 94 Euro-mir 95	Effects of microgravity on the bio-mechanical and bioenergetic characteristics of human skeletal muscle (experiments on the ground, pre- and post flight only).	P.E. di Prampero, G. Antonutto, C. Capelli, P. Zamparo, D. Linnarsson	University of Udine; Karolinska Institute, Stockholm
1995	Euro-mir 95	Interstitial fluid balance under microgravity, with special reference to pulmonary mechanics.	D. Negrini	University of Milan
1995	MAXUS 2	Lymphoid cells in space.	A. Cogoli, M. Cogoli, P. Pippia, G. Monaco, L. Sciola, A. Spano, M.A. Melano	University of Sassari
1995	MAXUS 2		Prof. Righetti	University of Milan
1996	MASER 7	Effects of microgravity on enzyme catalysis.	Prof. Vanni, Dr Lucarini	University of Firenze

The Trondheim group have had experiments flown on Biokosmos 9 (1989) and the Space Shuttle, (IML-1) in 1992. These are listed in Tables 14 and 19. Further experiments were flown on IML-2 in 1994 (Table 23) . The Trondheim group is also planning experiments for missions after 1995.
Research on IML-1 and 2 is concerned with: (i) studies of gravity perception on higher plants (experiment GTHRES); (ii) the effect of a microgravity environment on regeneration and differentiation of mature plants from protoplasts (experiment PROTO); (iii) the effects of microgravity on higher plants (experiment SHOOTS); root orientation, growth regulation and agravitropic behaviour of genetically transformed roots and regeneration of transgenic protoplasts (experiment TRANSFORM); spontaneous plant growth movements (experiment RANDOM) (see Tables 22 and 24).
The University of Bergen has proposed an experiment entitled *Neural cell development under microgravity conditions* which is now being evaluated for possible flight on Bion 11 in 1996.

8.10 Spain

The Comision Nacional de Ciencia y Tecnologia has established a National Space Plan which is intended to cover the general development of space related activities in Spain. It does not, however, reach down to the level of defining the details of a space life sciences programme. Instead, research activities are funded on an essentially ad hoc basis. Recently, there were only two active groups carrying out life sciences research in Spain. They are those of Professor R. Marco, Department of Biochemistry, Universidad Autnoma de Madrid and Professor J. Miquel, Institute of Neurosciences, Universidad de Alicante. These two groups have collaborated on earlier programmes.
The Madrid group is currently interested in two particular topics, deriving from work in the general field of developmental biology and aging. One is the identification and molecular genetic characterization of the main cytoskeletal and Ca^{++} binding proteins in the oocyte and early embryogenesis and muscle development. The other concerns the molecular genetic characterization of the mitochondrial components and machinery. Within these research interests, microgravity experiments may be performed where changes in gravitational acceleration could lead to a new perspective in the investigation.
Earlier work concerned development and aging, and the role of gravity in these basic processes. Several experiments were carried out. One in the Biorack on the Spacelab D-1 mission (1985), and another on the Bion-9 mission (1989), used *Drosophila melanogaster* fruit flies as experimental animals. The experiments were repeated on IML-1 in 1992, and extended on the Bion-10 mission in 1992. Proposals have been made for experiments using dormant *Artemia* (brine shrimp) cysts, for flight on Biopan and Biostack on IML-2.

Looking further to the future, the group is preparing the following programmes: (i) effect of the microgravity environment and cosmic radiation on the cytoskeletal organization of the oocyte and the early embryo, using *Drosophila* and *Artemia* as test systems; (ii) effects of microgravity and cosmic radiation on the patterns of cell proliferation and movements in the early *Drosophila* and *Artemia* embryo; (iii) study of the involvement of external/environmental forces in the acquisition of the morphological form of the *Artemia* nauplius larva; (iv) use of ground-based simulation equipment, clinostats and low speed centrifuges, to establish whether particular *Drosophila* mutants are more labile to the altered gravitational and/or radiation conditions found in space; (v) understanding the physiological and molecular response to microgravity in *Drosophila melanogaster* males.
In addition, efforts will be made to develop specific hardware in two categories: (i) following the development of the automatic unit for *Drosophila* mobility (DEMIR), to extend it to measure the oxygen consumption; (ii) explore the applications of microwaves in space biology, in staining, histoprocessing and sterilization.

8.11 Sweden

Space research in Sweden is conducted under the auspices of the Swedish National Space Board. The space life sciences research has been primarily undertaken by the Environmental Physiology Laboratory of the Karolinska Institute, in Stockholm. There the interest has focused principally upon the function of the lungs, and upon the influences of long term exposure to microgravity on the control of arterial blood pressure during exercise and on the performance, structure, and biochemistry, of skeletal muscle.
An experiment *Pulmonary perfusion and ventilation in microgravity, under rest and exercise*, was carried out on the Spacelab D-2 mission in May, 1993 with Professor D. Linnarsson of the Karolinska Institute as Principal Investigator. This work will continue, with the experiment of Professor Linnarsson entitled *Effects of short term and long term microgravity on the pulmonary gas exchange, respiratory and cardiovascular control during rest and exercise* being selected for the Euromir 95 mission.
That group was also involved, as co-investigators, on D-2 in the experiment *Determination of segmental fluid content and perfusion* by Dr F. Baisch, of DLR, and as co-investigators on SLS-1 in June, 1991. An active programme of ground based experiments is undertaken to study the effects of simulated weightlessness on the structural, metabolic, and functional properties of skeletal muscle, and cardiovascular control during rest and exercise in preparation for future flights.
Several parabolic flight experiments have also been conducted, the latest being to study the effects on blood circulation and gas exchange during the

transition to microgravity in upright exercising subjects. It is also intended to test on a parabolic flight a gravity independent strength ergometer. This was designed by the group for resistance training of postural muscle, as an exercise countermeasure during long term space flight.

8.12 Switzerland

There is no national Space Agency in Switzerland and no domestic programme set apart to support space activities. Access to ESA programmes is guaranteed by Switzerland's status as a Member State of the Agency and bilateral cooperation with national agencies is also open to Swiss investigators. However they must compete with the rest of the scientific community for the financing of their experiments. An exception to this rule is the ESA Prodex programme, to which Swiss scientists can apply to seek funding for activities that include a major industrial component, and such support is being used also in the life sciences (e.g. the bioreactor project of Dr Cogoli on IML-2).
In the field of cell biology, space activities have been largely confined to the Space Biology Group of Dr A. Cogoli, at the Swiss Federal Institute of Technology in Zürich (ETH-Z). A space biologist from the early days, Dr Cogoli has flown several experiments on the Space Shuttle on Spacelab, on Biokosmos and on Mir, as well as on sounding rockets (MAXUS). His work has included investigating the effects of microgravity on cell differentiation processes (Friend leukaemia cells), cytoskeletal protein expression and organisation (fibroblasts) and cellular physiological responses (lymphocyte activation). Dr Cogoli has also participated in key technical developments for multi-user facilities such as Biorack.
In human physiology, Dr E. Fern (NESTEC Laboratories, Vevey) had an experiment measuring nitrogen turnover in the D-2 Spacelab mission astronauts. Currently, the Physiology Department of the University of Geneva (Drs Ceretelli, Ferretti and Kayser) is involved in experiments on Mir, dealing with respiratory and muscular physiology (using, amongst others, a Respiratory Monitoring System developed by the Swiss industry in cooperation with Dr Buess of ETH-Z), whereas Dr Ruegg (University of Fribourg) will have an experiment on the effects of exposure to microgravity on the preparation and execution of voluntary movements in connection with the Euromir 95 flight. In the field of bone physiology, the groups of Dr Uebelhardt (University of Geneva) and Stüssi (ETH-Z) have experiments on the Euromir 95 flight, and Dr Stüssi's Laboratory of Biomechanics is also developing a Torque Velocity Dynamometer and an instrument to monitor the resistance of long bones to fracture. In space-related sleep research, The Pharmacology Department of the University of Zürich (Drs Borbély and Tobler) has participated in EMSI/ISEMSI and in HYDRA IX experiments, supported by the Directorate of Space Station and Microgravity/Long Term Programme Office.

Finally, in human neurology, the group of Dr Viviani (University of Geneva) is actively cooperating with Dr Berthoz (CNRS, Paris) on the project STAMP which will investigate the perception of figure symmetry by the two cerebral hemispheres on the Euromir 94 flight.

8.13 United Kingdom

There is no U.K. National Programme for the microgravity life sciences, despite the interest of several research groups. Individual research has taken place by way of collaborations.

Dr L. G. Briarty, of the Department of Life Science, University of Nottingham has worked together with Dr E. P. Maher of the University of Edinburgh, on the IML–1 Biorack experiment SHOOTS, flown in 1992.(see Trondheim Group, Sect. 8.9). Proposals have been made to extend this work on a future Spacelab or Mir mission. Dr A. J. Benson, of the RAF Institute of Aviation Medicine, Farnborough, who has been involved with the ESA-SLED project since Spacelab-1 and D-1 missions is preparing to participate in experiments on Neurolab (1997). One experiment, originating from NASA J.S.C., Houston, is concerned with investigating perceptual and vestibular responses to sustained linear acceleration (produced using the turntable flown previously on IML–1). The other concerns oculometer responses to linear oscillation, and is in conjunction with Dr Clarke of the Free University, Berlin.

U.K. scientists have participated in several parabolic flights, with Dr H. Ross of Stirling University involved in studies of human motor skills, Dr D. Grundy of Sheffield University observing the effects of body fluid shifts, and Dr J. Watkins of the University of Wales studying cerebral blood flow.

The Euromir programmes have also attracted interest from U.K. life scientists, with proposals from Dr D. Moore of Manchester University for Euromir 94, to study the gravitational rest state of fungal cells and embryonic rat calvaria cells. Two U.K. proposals for Euromir 95 have been accepted by ESA, both dealing with bone mass changes. That of Professor A. Goodship, of Bristol University, proposes the application of mechanical stimulation to prevent loss of bone mass during long term spaceflight. A study of the effect of venous pressure on bone mineral density in microgravity conditions was proposed by Dr I. D. McCarthy and Dr S. P. F. Hughes, of the Royal Postgraduate School, Hammersmith Hospital. Both of these studies will make use of the bone densitometer to monitor changes in flight.

In addition to these direct experimental activities, the group at Brunel University, under Prof. H. Wolff, have been involved in the design and development of experimental systems for the ESA life science programme, and Prof. Wolff was Chairman of the ESA Microgravity Advisory Committee for several years.

8.14 Hungary

ESA has a formal cooperation agreement with Hungary and also with Poland and Romania, to provide for studies of joint projects, access to laboratories and to databases. For the countries of Eastern Europe, the break-up of the Intercosmos organization has required a reappraisal of their space programmes and the collaborative arrangements. Agreements such as the one with ESA are a consequence of this process.

In Hungary, the new space research programme includes a space life sciences element, although first priority has been given to earth observations. In the life sciences, there is a new agreement with the Institute for Bio-Medical Problems (IBMP), Moscow, which will establish a common research programme.

Current space related research in the life sciences includes studies of the morphological and functional changes of the antigravity muscles in rats, of the adaptation mechanisms in the circulatory systems in humans, and the effects of microgravity on the mineral content of bone, using monkey subjects (with IBMP, Moscow). Studies of the sensorimotor adaptive processes have also used animal subjects.

In biology, there have been *in vitro* studies of animal cells in microgravity, and the interferon producing capacity of human lymphocytes *in vitro* and *in vivo*. Psychological research is also being undertaken into the effects of long term space flight, and radiation dosimetry experiments are included on Euromir 95.

8.15 Space life sciences research programmes of ESA and its partners

Tables 13 and 14 list all experiments which have been conducted on the Shuttle and Spacelab from 1981 onwards. We also present a summary of the experiments in biology and related subjects, carried out jointly with the USSR-CIS, in recent years in Tables 15 and 16. From these can be judged the actual relative emphasis of the ESA Life Sciences Programmes.

8.15.1 U.S.A.

In his address to the U.S. House of Representatives Task Force on Defense, Foreign Policy, and Space, in April 1992, Professor R. H. Moser, on behalf of the National Research Council, U.S. National Academy of Sciences, made this opening statement: 'Exploring the effects of microgravity on the development and maintenance of living systems is of considerable scientific interest. It is

Table 13. Human life sciences (physiology and medicine) experiments on the Space Shuttle (1981-1993).

Mission	Flight	Launch	Experiment location	Experiment topic	Investigator	Affiliation
All STS-	Start STS-1	12-4-81	Mid-deck	Tests & counter measures for motion sickness; prediction/ prevention of motion sickness	J.L. Homick	Med. Sci. Div. (NASA)
All STS-	Start STS-2	12-11-81		Cardiovascular deconditioning countermeasures	M.W. Bungo	Med. Sci. Div. (NASA)
DOD	STS-4	27-6-82	Mid-deck	Trivalent chromium level in the body; changes in chromium levels & carbohydrate metabolism.	K.R. Hauerspre-ger	Mecklenbury High School
DOD	STS-4	27-6-82	Mid-deck	Diet, exercise, zero-g, and lipoprotein profiles; lactic acid levels.	A.M. Kusske	Hill Jun. High School, CA
SBS-C	STS-5	11-11-82	To STS-8	Head and eye motion (ascent/entry); vestibular function monitoring.	W.E. Thornton	Astronaut office,NASA
SBS-C	STS-5	11-11-82	To STS-8	On-orbit head and eye tracking tasks; motion sickness related studies/counter measures.	W.E. Thornton	Astronaut office, NASA
SBS-C	STS-5	11-11-82	& STS 7,8	Acceleration detection sensitivity; effect of zero-g on gain of otolith organs.	W.E. Thornton	Astronaut office, NASA
SBS-C	STS-5	11-11-82	& STS 7,8	Kinaesthetic ability; changes in neuromuscular system.	W.E. Thornton	Astronaut office, NASA
TDRS-A	STST-6	4-4-83	& STS 7,8	Body fluid shift; multiplanar photogram of lower to upper body fluid shift.	W.E. Thornton	Astronaut office, NASA
SBS-C	STS-5	11-11-82	& STS 6,7,8	Near vision acuity/contrast sensitivity.	J.M. Van der Ploeg	Space Biomed. Res. Inst. (NASA)
TDSR-A	STS-6	4-4-83	& STS 7,8	Audiometry; tympanic membrane compliance.	W.E. Thornton	Astronaut office (NASA, JSC)

Table 13 (continued).

Mission	Flight	Launch	Experiment location	Experiment topic	Investigator	Affiliation
SPAS-01	STS-7	18-6-83	& STS-8	Treadmill operation; cine film.	W.E. Thornton	Astronaut office (NASA, JSC)
SPAS-01	STS-7	18-6-83	& STS-8	Opthalmoscopic measurement of intracranial pressure.	E.L. Shuman	NASA (JSC)
SPAS-01	STS-7	16-8-83	& STS-8	Tissue pressure tonometry; fluid shifts, atrophy, muscle tone study.	W.E. Thornton	Astronaut office (NASA)
SPAS-01	STS-7	16-8-83	& STS-8	In flight countermeasures for SAS; tests of loading devices and of drug efficacy.	W.E. Thornton	Astronaut office (NASA)
PDRS	STS-8	30-8-83		Eye-hand coordination; sensory signals and motion sickness development.	W.E. Thornton	Astronaut office (NASA)
PDRS	STS-8	30-8-83		Anatomical observation; changes observed by palpation, aculation, percussion.	W.E. Thornton	Astronaut office (NASA)
PDRS	STS-8	30-8-83		Study of in flight fluid changes.	W.E. Thornton	Astronaut office (NASA)
PDRS	STS-8	30-8-83		Evoked potentials; auditory and visual measurements on 4 subjects.	W.E. Thornton	Astronaut office (NASA)
PDRS	STS-8	30-8-83		Intraocular pressure; changes measurements with applanation tonometer.	S.M. Pool	Medical Sci. Div. NASA
PDRS	STS-8	30-8-83		Soft contact lens application test; improving location of lens, for ocular motion tests.	W.E. Thornton	Astronaut office (NASA)
PDRS	STS-8	30-8-83	& 41 B	Pre/postflight parallel swing tests for self motion perception.	D.E. Parker	Miami University
PDRS	STS-8	30-8-83		Efficacy of biofeedback in sleep improvement.	W.A. Angelo	Arlington High School

SL-1	STS-9	28-11-83	Spacelab, Pallet	Radiation environment monitoring.	E.V. Benton	University of San Francisco
SL-1	STS-9	28-11-83	Spacelab	Miniature personal electrophysiological tape recorder for data on heart, brain, eyes.	H. Green	Clinical Research Centre, Harrow
SL-1	STS-9	28-11-83	Spacelab	Measurement of central venous pressure & hormones in blood	K.A. Kirsch	Free University of Berlin
SL-1	STS-9	28-11-83	Spacelab	Effects of space flight on erythrokinetics; study of cause of temporary anaemia.	C.S. Leach	Medical Sciences, Div. (NASA)
SL-1	STS-9	28-11-83	Spacelab	Vestibulospinal reflex mechanisms; study of alterations to otolith input and physiological adaptations.	M.F. Reschke	Space Biomed. Res. Inst. (NASA)
SL-1	STS-9	28-11-83	Spacelab	Mass discrimination during weightlessness.	H.E. Ross	University of Stirling
SL-1	STS-9	28-11-83	Spacelab	3-D ballistocardiography in weightlessness; comparison in cardiovascular performances.	A. Scano	University of Rome
SL-1	STS-9	28-11-83	Spacelab	Vestibular function and visual-vestibular interactions.	R. von Baumgarten	University of Mainz
SL-1	STS-9	28-11-83	Spacelab	Zero-g as a stress factor on immune response.	E.W. Voss jr.	University of Illinois
SL-1	STS-9	28-11-83	Spacelab	Sensory-motor adaptation, study of motion sickness	L.R. Young	Mass. Inst. of Technology
OSTA-3	STS-41-G	5-10-84	Mid-deck	Proprioceptive illusions during weightlessness.	D. Watt	McGill University, Montreal
OSTA-3	STS-41-G	5-10-84	Mid-deck	Awareness of limb position.	D. Watt	McGill University, Montreal
OSTA-3	STS-41-G	5-10-84	Mid-deck	Gastro-intestinal gas elimination.	D. Watt	McGill University, Montreal
OSTA-3	STS-41-G	5-10-84	Mid-deck	Measurements of vestibulo-ocular reflex; nervous system adaptation.	D. Watt	McGill University, Montreal

Table 13 (continued).

Mission	Flight	Launch	Experiment location	Experiment topic	Investigator	Affiliation
OSTA-3	STS-41-G	5-10-84	Mid-deck	Time course of space motion sickness.	D. Watt	McGill University, Montreal
OSTA-3	STS-41-G	5-10-84	Mid-deck	Alterations in tactile or proprioceptive functions.	D. Watt	McGill University, Montreal
OSTA-3	STS-41-G	5-10-84	Mid-deck	Changes in taste sensitivity.	D. Watt	McGill University, Montreal
	STS-41G, 41D, 51C, 51B	5-10-84	51B(SL/3)	Action of metoclopramide; relationship of bowel sounds to motion sickness	W.E. Thornton	Astronaut office (NASA)
	STS-41G, 41D, 51C	5-10-84		General visual performance changes in zero-g.	L.V. Genco	Wright-Patterson AFB
	STS-41G, 41D, 51B, C, D	5-10-84		Blood pressure monitoring on re-entry	W.E. Thornton	Astronaut office (NASA)
	STS-51C	24-1-85	Mid-deck	Aggregation of human blood cells.	L. Dintenfass	Kanematsu Inst. Univ. of Sydney
Telesat-1	STS-51D	12-4-85	Mid-deck	American flight echo cardiograph; cardiovascular deconditioning.	M.W. Bungo	Med. Sci. Division, NASA (JSC)
SL-3	STS-51B	29-4-85	Spacelab	Dynamic environment measuring system.	P. Callahan	Life. Sci. Fl. Expts. Office, NASA
SL-3	STS-51B	29-4-85	Spacelab	Autogenic feedback training (AFT) as a countermeasure to space adaptation syndrome.	P.S. Cowings	Neuroscience Branch, NASA (AMES)
SL-3	STS-51B	29-4-85	Mid-deck	Biotelemetry system; tested on rats (body temperature, heart rate, ecg).	C. Schatte	Life. Sci. Fl. Expts. Office, NASA
SL-3	STS-51B & STS-51J	29-4-85		LEG plethysmography; time course of leg volume changes/ fluid shifts.	T.P. Moore	Med. Sci. Division, NASA (JSC)
SL-3	STS-51B & 51J, 61C	29-4-85		Eye-hand coordination during space motion sickness.	W.E. Thornton	Astronaut office (NASA)

SL-3 SL-2	STS-51B & 51F	29-4-85	Spacelab	Combined blood investigation; fluid balance, SMS, and countermeasures	C.S. Leach	Med. Sci. Division, NASA (JSC)
Spin-1	STS-51G	17-6-85	Mid-deck	French postural experiment; sensorimotor adaptation (muscle tone, movement).	A. Berthoz	CNRS (Paris)
Spin-1	STS-51G	17-6-85	Mid-deck	French echocardiograph experiment; cardiovascular data on two crew.	L. Pourcelot	University of Tours
Spin-1	STS-51G & 51D, I, J, 61B	17-6-85	Mid-deck	Clinical characterization of SMS.	W.E. Thornton	Astronaut office (NASA)
SL-2	STS-51F	29-7-85	Mid-deck	Vitamin D metabolites and bone demineralization.	H.K. Schnoes	University of Wisconsin
SL-D1	STS-61A	30-10-85	Spacelab	Body impedance measurement; cardiovascular adaptation.	F. Baisch	DLR, Cologne
SL-D1	STS-61A	30-10-85	Spacelab	Tonometer; eye pressure changes.	J. Draeger	University of Hamburg
SL-D1	STS-61A	30-10-85	Spacelab	Gesture and speech in microgravity; communication aspects of cognitive behaviour.	A.D. Friederici	MPI, Nijmegen
SL-D1	STS-61A	30-10-85	Spacelab	Spatial description in space; effect of perception cues on communication on spatial arrangements.	A.D. Friederici	MPI, Nijmegen
SL-D1	STS-61A	30-10-85	Spacelab	Human reaction time (student experiment).	M. Hoschek	Muehtal
SL-D1	STS-61A	30-10-85	Spacelab	Central venous pressure; blood displacement in upper body.	K. Kirsch	Free University, Berlin
SL-D1	STS-61A	30-10-85	Spacelab	Mass discrimination.	H.E. Ross	University of Stirling
SL-D1	STS-61A	30-10-85	Spacelab	Vestibular research; response to optokinetic stimulation patterns.	R. von Baumgarten	University of Mainz
SL-D1	STS-61A	30-10-85	Spacelab	Vestibular research; perception of linear acceleration.	L.R. Young	MIT, Cambridge
EASE	STS-61-B & 61C	26-11-85	Mid-deck	Salivary acetaminophen. pharmacokinetics.	N.M. Cintron-Trevimo	Med. Sci. Division, NASA (JSC)

Table 13 (continued).

Mission	Flight	Launch	Experiment location	Experiment topic	Investigator	Affiliation
EASE	STS-61B	26-11-85	Mid-deck	Electropuncture in space.	F.Ramirez-Escalano	Secretaria de Communic.
MSL-2	STS-61C	12-1-86		Otolith tilt-translation reinterpretation.	M.F. Reschke	Med. Sci. Division, NASA (JSC)
MSL-2	STS-61C	12-1-86		Changes in total body water in first days.	C.S. Leach	Med. Sci. Division, NASA (JSC)
MSL-2	STS-61C	12-1-86		Non-invasive estimation of central venous pressure.	J.B. Charles	Med. Sci. Division, NASA (JSC)
MSL-2	STS-61C	12-1-86		In flight treadmill stress; documenting frequency of cardiac rhythm abnormalities.	M.W. Bungo	Med. Sci. Division, NASA (JSC)
MSL-2	STS-61C	12-1-86		In flight assessment of renal stone risk factor.	N.M. Cintron-Trevi mo	Med. Sci. Division, NASA (JSC)
MSL-2	STS-61C	12-1-86		Echocardiographic evaluation of cardiovascular deconditioning.	M.W. Bungo	Med. Sci. Division, NASA (JSC)
LDEP-R	STS-32	9-1-90		American flight echocardiograph; size/functioning of heart (also flew on STS-51D).	M.W. Bungo	Med. Sci. Division, NASA (JSC)
SLS-1	STS-40	5-6-91	Spacelab	Weightlessness and autonomic cardiovascular controls.	D.L. Eckberg	Medical College of Virginia
SLS-1	STS-40	5-6-91	Spacelab	In flight study of cardiovascular deconditioning; expiration-rebreathing and gas analysis.	L.E. Farhi	New York State University
SLS-1	STS-40	5-6-91	Spacelab	Motion sickness and inner ear vestibular changes.	L.R. Young	MIT, Cambridge, Mass.
SLS-1	STS-40	5-6-91	Spacelab	Protein metabolism during spaceflight; protein synthesis rates, muscle breakdown rates	T.P. Stein	Univ. of Med. & Dent. New Jersey
SLS-1	STS-40	5-6-91	Spacelab	Fluid-electrolyte regulation in spaceflight; immediate/long term changes to kidney function.	C.Leach-Huntoon	NASA (JSC)

SLS-1	STS-40	5-6-91	Spacelab	Pulmonary function during weightlessness.	J.B. West	Univ. California, San Diego
SLS-1	STS-40	5-6-91	Spacelab	Influence of spaceflight on erythrokinetics; mechanism leading to decrease in erythrocytes (and see SLS-2 = STS-58)	C. Alfrey	Baylor College of Medicine
SLS-1	STS-40	5-6-91	Spacelab	Cardiovascular adaptation to microgravity; catheter observations of fluid redistribution, echocardiograph, leg blood flow/ compliance.	C.G. Blomquist	Southwest Med. Center of Texas
SLS-1	STS-40	5-6-91	Spacelab	Pathophysiology of mineral loss in space flight; levels of calcium metabolism, hormones, and of calcium.	C.D. Arnaud	Univ. California, San Francisco
IML-1	STS-42	22-1-92	Spacelab	Microgravity vestibular investigations; eye-object tracking, rotation perception, visual cues and vestibular response.	M.F. Reschke	NASA (JSC)
IML-1	STS-42	22-1-92	Spacelab	Mental workload and performance experiment.		NASA
IML-1	STS-42	22-1-92	Spacelab	Sled experiment, otolith stimulation/response; rotation experiment, coordination of head/eye movements; visual stimulation experiment, relationship between vision & balance organs; position/movement sensing of body.	D.G.D. Watt	McGill University, Montreal
IML-1	STS-42	22-1-92	Spacelab	Energy expenditure in space flight.	H.G. Parsons	University of Calgary
IML-1	STS-42	22-1-92	Spacelab	Positional and spontaneous nystagmus.	J.A. McClure	London Ear Clinic, Ontario
IML-1	STS-42	22-1-92	Spacelab	Evaluation of an anti-gravity suit; reduction of gravity induced effects on return	R.B. Thirsk	Canadian Space Agency

Table 13 (continued).

Mission	Flight	Launch	Experiment location	Experiment topic	Investigator	Affiliation
IML-1	STS-42	22-1-92	Spacelab	Assessment of back pain in astronaut; alleviate symptoms which result from spine elongation.	P.C. Wing	Univ. British Columbia
DSP	STS-44	24-11-91		Extended duration orbiter medical project; countermeasures and study of readaptive processes.	-	Med. Sci. Division, NASA (JSC)
ATLAS-1	STS-45	24-3-92		Changes in vision parameters: contrast ratio threshold.		Airforce Space Syst. Div. LA
Spacelab-J	STS-47	12-9-92	Spacelab	Autogenic feedback training experiment; biofeedback techniques as countermeasure for SMS.	P. Cowings	NASA (Ames)
Spacelab-J	STS-47	12-9-92	Spacelab	Lower body negative pressure; countermeasure for reducing post-flight orthostatic intolerance.	J.Charles	NASA (JSC)
Spacelab-J	STS-47	12-9-92	Spacelab	Magnetic resonance imaging after exposure to microgravity; muscle/bone changes	A. Leblanc	Methodist Hospital
Spacelab-J	STS-47	12-9-92	Spacelab	Endocrine and metabolic changes	Hismo Seo	Nagoya University, Japan
Spacelab-J	STS-47	12-9-92	Spacelab	Head/eye movements during visual tracking.	Kazuo Koga	Nagoya University, Japan
Spacelab-J	STS-47	12-9-92	Spacelab	Perceptual motor functions in zero-g.	Akira Tada	National Aerospace Laboratory, Japan
Spacelab-J	STS-47	12-9-92	Spacelab	Technology experiment: fluid therapy system; administration of intravenous fluids.		NASA
USML-1	STS-50	25-6-92	Spacelab	Extended duration orbiter medical project.	J.T. Brown	NASA (JSC)
LAGEOS-II	STS-52	22-10-92		Space vision system experiment (SVS).	H.F.L. Ankney	Nat. Res. Council, Canada

LAGEOS-II	STS-52	22=10-92		Space adaptation tests and observations (SATO); vestibular/ocular reflex, body water changes in microgravity, back pains in astronauts, illusion during movement.	D. Watt, H. Parsons, P.C. Wing	McGill Univ., Montreal, Univ. Calgary, Univ. British Columbia
DOD	STS-53			Visual function tester (model II); sensitivity of the eye to image contrast at threshold.	L. Task	Human Eng. Div. US Air Force, Ohio
D-2	STS-55	26-4-93	Spacelab	Peripheral & central haemodynamic adaptation to zero-g.	F. Bonde-Peterson	DAMEC
D-2	STS-55	26-4-93	Spacelab	Regulation of homeostatic volume in microgravity.	R. Gerzer	Univ. Münich, DLR & Univ. Heidelberg
D-2	STS-55	26-4-93	Spacelab	Cardiovascular regulation in microgravity; intravenous saline loading on induced fluid shifts.	G. Blomquist	University of Texas
D-2	STS-55	26-4-93	Spacelab	The central venous pressure in microgravity.	N.Foldager	DAMEC
D-2	STS-55	26-4-93	Spacelab	Leg fluid distribution at rest and under LBNP; changes following dehydration, disuse, and fluid shift.	F. Baisch	DLR (with TNO, Holland)
D-2	STS-55	26-4-93	Spacelab	Determination of segmental fluid content and perfusion.	F. Baisch	DLR (with TNO, Holland & Karolinska Inst. Sweden
D-2	STS-55	26-4-93	Spacelab	Left ventricular function, at rest and under stimulation	L. Beck	DLR (with Univ. Amsterdam & Univ. Tours)
D-2	STS-55	26-4-93	Spacelab	Tonometry -intraocular pressure in microgravity.	J. Draeger	Univ. Hamburg
D-2	STS-55	26-4-93	Spacelab	Tissue thickness & compliance along body axis in microgravity; new method to quantify fluid shifts within superficial tissue.	K.A. Kirsch	Fed. Univ. Berlin

Table 13 (continued).

Mission	Flight	Launch	Experiment location	Experiment topic	Investigator	Affiliation
D-2	STS-55	26-4-93	Spacelab	Influence of microgravity on carotid baroreceptor-cardiac reflex response.	D.L. Eckberg	
D-2	STS-55	26-4-93	Spacelab	Cardiovascular response to LBNP and fluid loading; mechanisms responsible for adverse reactions on landing.	L.Pourcelot	University of Tours, France
D-2	STS-55	26-4-93	Spacelab	Mechanism responsible for the negative nitrogen balance; changes in whole body nitrogen turnover rate, protein synthesis and breakdown.	E.B. Fern	NESTEC Ltd., Research Centre
D-2	STS-55	26-4-93	Spacelab	Effects of microgravity on glucose tolerance.	H.P. Maass	DLR,
D-2	STS-55	26-4-93	Spacelab	Influence of microgravity on endocrine & renal elements of volume homeostasis.	P. Norsk	DAMEC
D-2	STS-55	26-4-93	Spacelab	Effects of space flight on pituitary-gonadatrophin-adrenalin function; blood/urine/saliva analysis for signs of disturbance.	G. Riondino	INCRA, Italy
D-2	STS-55	26-4-93	Spacelab	Regulation of salt balance and blood pressure; role of volume regulatory hormones & plasma proteins.	L. Rocker	Fed. Univ. Berlin
D-2	STS-55	26-4-93	Spacelab	Pulmonary stratification & compartment analysis.	S. Groth	DAMEC
D-2	STS-55	26-4-93	Spacelab	Pulmonary perfusion & ventilation in microgravity.	D. Linnarson	Karolinska Inst.
D-2	STS-55	26-4-93	Spacelab	Ventilation distribution in microgravity.	M.Paiva	Univ. Brussels, Westmead Hospital, Australia & Univ. California
D-2	STS-55	26-4-93	Spacelab	Dynamics of gas exchange, ventilation, & heart rate in submaximal dynamic exercise; assessment of CO_2 kinetics as a monitor of endurance performance.	J. Stegemann	Deutsche Sport-Hochschule

EURECA Spacehab-1	STS-57	21-6-93	Spacehab	Neutral body posture; towards the design of future space facilities.	F.E. Mount	Flight crew support, NASA (JSC)
EURECA Spacehab-1	STS-57	21-6-93	Post-flight	Relation between mission duration and orthostatic functions; orthostatic function during entry, landing, egress.		NASA (JSC)
EURECA Spacehab-1	STS-57	21-6-93	Spacehab	Visual vestibulary integration as a function of adaptation.		NASA
EURECA Spacehab-1	STS-57	21-6-93	Spacehab	Effects of intense exercise during s pace flight; action on aerobic capacity and orthostatic function.		NASA
EURECA Spacehab-1	STS-57	21-6-93		Pre and post flight measurement of cardiorespiratory response.		NASA
SLS-2	STS-58	Oct 93		Intermars tissue equiv. proport. counter.		Med. Sci. Division, NASA (JSC)
SLS-2	STS-58	Oct 93	Spacelab	Air monitoring instrument; evaluation and characterization of atmospheric microbial contaminants.		Med. Sci. Division, NASA (JSC)
SLS-2	STS-58	Oct 93	Spacelab	Energy utilization; caloric requirements for space flight.		Med. Sci. Division, NASA (JSC)
SLS-2	STS-58	Oct 93	Spacelab	Lower body negative pressure; conditioning cardiovascular system prior to re-entry.		Med. Sci. Division, NASA (JSC)
SLS-2	STS-58	Oct 93	Post flight	Orthostatic function during entry, landing, egress (continuing investigation – see STS-57).		Med. Sci. Division, NASA (JSC)
SLS-2	STS-58	Oct 93	Post flight	Evaluation of functional skeletal performance following space flight.		Med. Sci. Division, NASA (JSC)
SLS-2	STS-58	Oct 93	Post flight	Ability of crew to stand following landing		Med. Sci. Division, NASA (JSC)
SLS-2	STS-58	Oct 93	Post flight	Cardio-respiratory response to submaximal exercise.		Med. Sci. Division, NASA (JSC)

Table 13 (continued).

Mission	Flight	Launch	Experiment location	Experiment topic	Investigator	Affiliation
SLS-2	STS-58	Oct93	Spacelab	In flight study of cardiovascular deconditioning.	L.E. Fahri	New York Univ. at Buffalo
SLS-2	STS-58	Oct93	Spacelab	Cardiovascular adaptation to zero gravity.	C.G. Blomquist	S.W. Med. Center, Univ. Texas
SLS-2	STS-58	Oct 93	Spacelab	Pulmonary function during weightlessness.	J.B. West	Univ. California, San Diego
SLS-2	STS-58	Oct 93	Spacelab	Fluid-electrolyte regulation during space flight: changes in kidney action, salt/mineral balance, hormones	C. Leach	NASA (JSC)
SLS-2	STS-58	Oct 93	Spacelab	Influence of spaceflight on erythrokinetics; mechanism leading to decrease in erythrocytes (follow-on from SLS-1).	C. Alfrey	Baylor College of Med., Houston
SLS-2	STS-58	Oct 93	Spacelab	Vestibular experiments in Spacelab; six tests to assess sensory motor adaptation.	L.R. Young	MIT, Cambridge, Mass.
SLS-2	STS-58	Oct 93	Spacelab	Protein metabolism during space flight; study of synthesis/catabolism, & fibrinogen levels	T.P. Stein	N.J. Univ. Med., Dentistry
SLS-2	STS-58	Oct 93	Spacelab	Pathophysiology of mineral loss during space flight; role of vitamin d metabolites and calciotropic hormones	C.D. Arnaud	Univ. California, San Francisco

imperative, however, that national goals guide a research endeavour of this magnitude. Purely academic curiosity is an insufficient rationale for investing tax dollars on this scale.' He went on to comment that '... the primary justification for space life sciences research is a commitment to long term human exploration.'

The NASA long term strategy for life sciences research appears to reflect this viewpoint, up to a certain level. There is indeed a strong emphasis on the human life sciences research and on those research areas in biology which would appear to support the overall development of the human space life sciences programmes. There is also, however, a commitment to basic research, a willingness to allow for 'academic curiosity' and the study of effects down to the most basic level, where these appear to have serious merit as judged by the scientific review process.

Reflecting the official position and emphasis, the NRC Space Studies Board, in their *Space Exploration Initiative* document, presented the following 'order of priority' list for the general research fields in the human life sciences, as far as their importance for future extended space travel is concerned: bone, muscle, and mineral metabolism research; the study of cardiovascular and homeostatic functions; sensorimotor integration processes; psychosocial perturbations and radiation exposure effects.

This, of course, seems to do no more than indicate the likely relative importance of the different processes in reducing the probability of success of an extended space mission, although some might query the relative importance accorded to the radiation effects. Certainly, it is not reflected in the actual distribution of U.S. experiments on the Shuttle for the different research fields. Fig. 3 illustrates that distribution, the data being taken from the information in Table 14.

Evidently, the principal concerns of NASA up to the present time, where most flights – excluding the Skylab mission – have been of short duration, has been with the space sickness and disorientation effects which disturb efficiency during the first few days. With the progress towards longer term missions, on Mir and the International Space Station, that emphasis will presumably change.

The NASA Space Biology programme, operated principally by the NASA Ames establishment, was established across the three main areas of *Gravity Perception*, in animals and plants, *Developmental Biology*, and *Biological Adaptation* – where gravity is used as a variable to examine its effects on organisms and cells. The programme has relied not only upon the Shuttle based experiments,but also upon the opportunities offered by the collaboration with the USSR/Russia in the Biocosmos recoverable satellite series. Fig. 4 shows the relative distribution of U.S. Shuttle based experiments (and European experiments) across the different research disciplines in space biology. The apparently small number of U.S. experiments performed in dosimetry is not a true reflection of the NASA interest in this field since measurements are carried out under other NASA programmes.

Fig. 5 illustrates the distribution, by research discipline, of U.S. experiments carried out on the Russian Biocosmos recoverable spacecraft. Initially, these

Table 14. STS-based biology missions, 1981-1993.

Row No.	Scientific name	Category	Key hardware	Investigator	Affiliation
1	*Helianthus annuus*	Plant	Sealed vials	A.H. Brown	UPA Philadelphia, PA
2	n.a.	macromolecules	Electrophoresls system	R. Snyder	NASA-GSFC, Greenbelt MY
3	*Helianthus annuus*	Plant	Sealed vials	A.H. Brown	UPA Philadelphia PA
4	*Apis mellifera*	Insect	Insect chamber	T.E. Nelson	Southland HS, Adams MN
5	*Musca domestica*	Insect	Insect chamber	T.E. Nelson	Southland HS, Adams MN
6		Insect	Insect chamber	T.E. Nelson	Southland HS, Adams MN
7	*Vigna radiata*	Plant	Plant growth unit (PGU)	J.R. Cowles	U Houston, Houston TX
8	*Avena sativa*	Plant	Plant Growth Unit (PGU)	J.R. Cowles	U Houston, Houston TX
9	*Pinus elliotti*	Plant	Plant growth unit (PGU)	J.R. Cowles	U Houston, Houston TX
10	*Chlorella vulgaris*	Algae, free living	Algae growth chamber	S.M. Walker	Utah State U, Logan UT
11	*Lemna minor*	Plant	Plexiglas growth chamber	K.D. Hunt	Utah State U, Logan UT
12	*Artemia salina*	Invertebrate	Sealed aquatic chamber	B. Moore	Utah State U, Logan UT
13	*Drosophila melanogaster*	Insect	Plexiglas growth chamber	W.M. Moore	Utah State U, Logan UT
14	n.a.	macromolecules	FFE system	D. Clifford	McDonnell Douglas
15	*Microciona porifera*	Invertebrate	Aquarium (2 sections)	A.K. Gillette	Winter Haven HS, Winter Haven FL
16	n.a.	macromolecules	FFE system	D. Clifford	McDonnell Douglas
17	unspecified	Bacteria, free living	GAS canister	K.R. Shriner	USAF Academy, Colorado Springs CO
18	various	Plant	Dacron/Plastic Bags	G.B. Park	Park Seed Co, Greenwood SC
19	n.a.	macromolecules	FFE system	D. Clifford	McDonnell Douglas
20	unspecified	Plant	Seed containers	M. Buchwald	German Youth Fair Program (Jufo)
21		Plant		H. Katzenmeire	German Youth Fair Program (Jufo)
22	*Helianthus* sp.	Plant	Rotating drum	Students	Purdue U, West Lafayette IN
23	*Myrmica* sp.	Insect	Wooden ant farm	Students	Camdery W. Wilson HS, Camden NJ
24	*Raphanus sativus*	Plant	Multiplatter centrifuge	Students	California Institute of Technology, CA
25	n.a.	macromolecules	FFE system	D. Clifford	McDonnell Douglas
26	*Artemia salina*	Invertebrate	HZE tracking sandwich	H. Buecker	DLR Cologne (DFVLR)
27	*Bacillus subtilis*	Bacterial spores	HZE tracking sandwich	H. Buecker	DLR Cologne (DFVLR)
28	*Nicotiana tabacum*	Plant	HZE tracking sandwich	H. Buecker	DLR Cologne (DFVLR)
29	*Arabidopsis thaliana*	Plant	HZE tracking sandwich	H. Buecker	DLR Cologne (DFVLR)
30	*Sordaria fimicola*	Fungus spores	HZE tracking sandwich	H. Buecker	DLR Cologne (DFVLR)
31	*Bacillus subtilis*	Bacterial spores	Open exposure on pallet	G. Horneck	DLR Cologne (DFVLR)

Table 14. STS-based biology missions, 1981-1993.

Row No.	Mission	Flight	Launched	Carrier	Expt.	Topic	Title	Species or donor
1	OSTA-1	STS-2	12 Nov 81	Columbia	HBT	Processing equipment	HEFLEX bioengineering test	Dwarf sunflower
2	OSS-1	STS-3	22 Mar 82	Columbia	EEVT	Analysis equipment	Electrophoresis equipment verification Test	n.a.
3	OSS-1	STS-3	22 Mar 82	Columbia	HBT	Processing equipment	HEFLEX bioengineering test II	Dwarf sunflower
4	OSS-1	STS-3	22 Mar 82	Columbia	SE81-08	Graviperception (orientation)	Insect motion in space flight	Honey bee
5	OSS-1	STS-3	22 Mar 82	Columbia	SE81-08	Graviperception (orientation)	Insect motion in space flight	House fly
6	OSS-1	STS-3	22 Mar 82	Columbia	SE81-08	Graviperception (orientation)	Insect motion in space flight	Bean caterpillar Moth
7	OSS-1	STS-3	22 Mar 82	Columbia	PGU	Growth and metabolism	Lignification in developing plant seedlings	Mung Bean
8	OSS-1	STS-3	22 Mar 82	Columbia	PGU	Growth and metabolism	Lignification in developing plant seedlings	Oat
9	OSS-1	STS-3	22 Mar 82	Columbia	PGU	Growth and metabolism	Lignification in developing plant seedlings	Pine
10	DOD	STS-4	27 Jun 82	Columbia	G-0001	Growth and metabolism	Algal microgravity bioassay experiment	
11	DOD	STS-4	27 Jun 82	Columbia	G-0001	Graviperception (Morphology)	Root growth of duckweed	Duckweed
12	DOD	STS-4	27 Jun 82	Columbia	G-0001	Effect on reproductive units	Microgravity effects on brine shrimp genetics	Brine shrimps
13	DOD	STS-4	27 Jun 82	Columbia	G-0001	Effect on genetic apparatus	Fruit fly experiment	Fruit fly
14	DOD	STS-4	27 Jun 82	Columbia	CFES	Analysis equipment	Continuous flow electrophoresis System	n.a.
15	SBS-c	STS-5	11 Nov 82	Columbia	SE81-02	Graviperception (morphology)	The growth of porifera in microgravity	Sponge
16	TDRS-A	STS-6	04 Apr 83	Challenger	CFES	Analysis equipment	Continuous flow electrophoresis system	n.a.
17	TDRS-A	STS-6	04 Apr 83	Challenger	G-0049	Growth and metabolism	Microorganism development	unspecified
18	TDRS-A	STS-6	04 Apr 83	Challenger	G-0049	Effect on reproductive units	Seeds in space	Fruit and vegetables
19	SPAS-01	STS-7	18 Jun 83	Challenger	CFES	Analysis equipment	Continuous flow electrophoresis system	n.a
20	SPAS-01	STS-7	18 Jun 83	Challenger	G-0002	Radiobiology (mapping)	Biostack	unspecified
21	SPAS-01	STS-7	18 Jun 83	Challenger	G-0002	Metabolism (Cd transport)	Plant contamination by heavy metals	Watercress
22	SPAS-01	STS-7	18 Jun 83	Challenger	G-0002	Graviresponse curve	Geotropism of sunflower sprouts	Sunflower
23	SPAS-01	STS-7	18 Jun 83	Challenger	G-0002	Insect communities	Orbit '81 (ant farm)	Carpenter ant
24	SPAS-01	STS-7	18 Jun 83	Challenger	G-0002	Graviresponse curve	Plant graviperception	Radish
25	PDRS	STS-8	30 Aug 83	Challenger	CFES	Analysis equipment	Continuous flow electrophoresis system	n.a.
26	SL-1	STS-9	28 Nov 83	Columbia	ES-027	Radiobiology (mapping)	Advanced Biostack experiment	Brine Shrimp
27	SL-1	STS-9	28 Nov 83	Columbia	ES-027	Radiobiology (mapping)	Advanced Biostack experiment	
28	SL-1	STS-9	28 Nov 83	Columbia	ES-027	Radiobiology (mapping)	Advanced Biostack experiment	Tobacco
29	SL-1	STS-9	28 Nov 83	Columbia	ES-027	Radiobiology (mapping)	Advanced Biostack experiment	Mouse-ear cress
30	SL-1	STS-9	28 Nov 83	Columbia	ES-027	Radiobiology (mapping)	Advanced Biostack experiment	
31	SL-1	STS-9	28 Nov 83	Columbia	ES-029	Exposure experiment	Microorganisms in hard space environment	

Table 14 (continued).

Row No.	Scientific name	Category	Key hardware	Investigator	Affiliation
32	*Homo sapiens sapiens*	Lymphocytes		A. Cogoli	ETH Zürich
33	*Neurospora crassa*	Fungus	Culture tubes (dark)	F.M. Sulzman	State U of New York, Binghamton NY
34	*Helianthus annuus*	Plant	Videotape recording under IR	A.H. Brown	UPA, Philadelphia PA
35	n.a.	macromolecules	Electrophoresis system		Utah State U, Logan UT
36	*Raphanus sativus*	Plant	Plexiglas growth chamber	Students	Brighton HS, Salt Lake City UT
37	n.a.	macromolecules	Electrophoresis system	M. Bler	University of Arizona Tucson AZ
38	*Rattus rattus*	Mammal	Animal enclosure module (AEM)	D. Weber	Hunter College HS, New York NY
39	*Apis mellifera*	Insect	Bee enclosed module (BEM)	D.M. Poskevich	Waverty Central HS, Waverly TN
40	*Artemia salina*	Invertebrate	HZE tracking sandwich	H. Buecker	DLR Cologne (DFVLR)
41	*Bacillus subtilis*	Bacterial spores	HZE tracking sandwich	H. Buecker	DLR Cologne (DFVLR)
42	*Nicotiana tabacum*	Plant	HZE tracking sandwich	H. Buecker	DLR Cologne (DFVLR)
43	*Arabidopsis thaliana*	Plant	HZE tracking sandwich	H. Buecker	DLR Cologne (DFVLR)
44	*Zea mays*	Plant	HZE tracking sandwich	H. Buecker	DLR Cologne (DFVLR)
45	*Sordaria fimicola*	Fungus spores	HZE tracking sandwich	H. Buecker	DLR Cologne (DFVLR)
46		Plant	Seed containers	D.K. Grigsby	Park Seed Co, Greenwood SC
47	various	Plant	Seed containers	G.B. Park	Park Seed Co, Greenwood SC
48	n.a.	macromolecules	FFE system	D. Clifford	McDonnell Douglas
49	*Raphanus sativus*	Plant		G.A. Smith	UAH, Huntsville AL
50	n.a.	macromolecules	FFE System	D. Clifford	McDonnell Douglas
51	n.a.	macromolecules		D.E. Brookes	University H & S Oregon
52	*Zea mays*	Plant		S.M. Amberg	Seward HS, Seward NE
53	*Musca domestica*	Insect	Fly enclosure module (FEM)	A.I. Fras	Binghamton HS, Binghamton NY
54	*Saimiri sciureus*	Mammal	Research animal holding facility	C. Schatte	NASA-ARC, Moffet Field CA
55	*Rattus rattus*	Mammal	Research animal holding facility	C. Schatte	NASA-ARC, Moffet Field CA
56	*Rattus rattus*	Mammal	Research animal holding facility	C. Schatte	NASA-ARC, Moffet Field CA
57	unspecified	Bacteria, free living	Agar cultures	K. Herman	EI Paso/Ysleta Schools, EI Paso TX
58	*Artemia salina*	Invertebrate	Hatching chamber	D.R. Cake	EI Paso/Ysleta Schools, EI Paso TX
59	*Mucor rouxii*	Fungus, free living	Plexiglas chambers	R. Lopez	EI Paso/Ysleta Schools, EI Paso TX
60	*Lactuca sativa*	Plant		D. Bowden	EI Paso/Ysleta Schools, EI Paso TX
61		Plant	Wet paper towels	P. Campos	EI Paso/Ysleta Schools, EI Paso TX
62	*Hordeum vulgare*	Plant		G. Bryant	EI Paso/Ysleta Schools, EI Paso TX
63	*Chlorella* sp.	Algae, free living		R. Santini	EI Paso/Ysleta Schools, EI Paso TX
64	*Kefir* sp.	Fungus, free living		R. Santini	EI Paso/Ysleta Schools, EI Paso TX

Table 14 (continued).

Row No.	Mission	Flight	Launched	Carrier	Expt.	Topic	Title	Species or donor
32	SL-1	STS-9	28 Nov 83	Columbia	ES-031	Growth and metabolism	Lymphocyte proliferation in weightlessness	Human
33	SL-1	STS-9	28 Nov 83	Columbia	NS-007	Endogenous rhythms	Characteristics of persisting circadian rhythms	
34	SL-1	STS-9	28 Nov 83	Columbia	NS-101	Graviperception (morphology)	Sunflower nutation in microgravity	Dwarf sunflower
35	SPAS-01 A	STS-41 B	03 Feb 84	Challenger	G-0008	Analysis equipment	Electrophoresis flow	n.a.
36	SPAS-01 A	STS-41 B	03 Feb 84	Challenger	G-0008	Gravity and phototropism	Phototropic effect on seedlings in microgravity	Radish
37	SPAS-01 A	STS-41 B	03 Feb 84	Challenger	IEF	Analysis equipment	Isoelectric focusing experiment	n.a.
38	SPAS-01 A	STS-41 B	03 Feb 84	Challenger	SE81-10	Bone and skeleton changes	Arthritis	Rat
39	LDEF-1	STS-41 C	08 Apr 84	Challenger	SE82-17	Insect communities	Comparison of honeycomb structures	Italian honeybee
40	LDEF-1	STS-41 C	08 Apr 84	LDEF		Radiobiology (mapping)	Free-flyer Biostack experiment	Brine shrimp
41	LDEF-1	STS-41 C	08 Apr 84	LDEF		Radiobiology (mapping)	Free-flyer Biostack experiment	
42	LDEF-1	STS-41 C	08 Apr 84	LDEF		Radiobiology (mapping)	Free-flyer Biostack experiment	Tobacco
43	LDEF-1	STS-41 C	08 Apr 84	LDEF		Radiobiology (mapping)	Free-flyer Biostack experiment	Mouse-ear cress
44	LDEF-1	STS-41 C	06 APR 84	LDEF		Radiobiology (mapping)	Free-flyer Biostack experiment	Maize
45	LDEF-1	STS-41 C	06 APR 84	LDEF		Radiobiology (mapping)	Free-flyer Biostack experiment	
46	LDEF-1	STS-41 C	06 APR 84	LDEF	SEEDS	Effect on reproductive units	Space-exposed experiment developed for students	Tomato
47	LDEF-1	STS-41 C	06 APR 84	LDEF		Effect on reproductive units	Seeds in space experiment	various
48	OAST-1	STS-41 D	30 Aug 84	Discovery	CFES	Analysis equipment	Continuous flow electrophoresis System	n.a.
49	OAST-3	STS-41 G	05 Oct 84	Challenger	G-0007	Graviperception (morphology)	Study of the radish root system	Radish
50	TELESAT -1	STS-51 D	12 Apr 85	Discovery	CFES	Analysis equipment	Continuous flow electrophoresis System	n.a.
51	TELESAT -1	STS-51 D	12 Apr 85	Discovery	PPE	Analysis equipment	Phase partitioning experiment	n.a.
52	TELESAT -1	STS-51 D	12 Apr 85	Discovery	SE82-03	Graviperception (morphology)	Statoliths in corn root caps	Maize
53	TELESAT -1	STS-51 D	12 Apr 85	Discovery	SE83-03	Neurovestibular system (brain)	Aging of house fly brains	House fly
54	SL-3	STS-51 B	29 Apr 85	Challenger	RAHF	Holding facility	Research animal holding facility	Squirrel monkey
55	SL-3	STS-51 B	29 Apr 85	Challenger	RAHF	Holding facility	Research animal holding facility	Rat
56	SL-3	STS-51 B	29 Apr 85	Challenger	BTS	Sensors and monitoring	Biotelemetry system	Rat
57	SPTN-1	STS-51 G	17 Jun 85	Discovery	G-0034	Effectiveness of antibiotics	Effectiveness of antibiotics on bacteria	unspecified
58	SPTN-1	STS-51 G	17 Jun 85	Discovery	G-0034	Effect on development	Growth of brine shrimp	Brine shrimp
59	SPTN-1	STS-51 G	17 Jun 85	Discovery	G-0034	Growth of metabolism	Growth of soil mould	Soil mould
60	SPTN-1	STS-51 G	17 Jun 85	Discovery	G-0034	Effect on germination	Growth of lettuce seeds	Lettuce
61	SPTN-1	STS-51 G	17 Jun 85	Discovery	G-0034	Effect on germination	Germination of turnip seeds	Turnip
62	SPTN-1	STS-51 G	17 Jun 85	Discovery	G-0034	Effect on genetic apparatus	Seed germination	Barley
63	SPTN-1	STS-51 G	17 Jun 85	Discovery	G-0034	Growth and metabolism	Symbiotic growth of *Chlorella* and kefir	
64	SPTN-1	STS-51 G	17 Jun 85	Discovery	G-0034	Growth and metabolism	Symbiotic growth of *Chlorella* and kefir	Yeast

Table 14 (continued).

Row No.	Scientific name	Category	Key hardware	Investigator	Affiliation
65	*Dugesia tigrina*	Invertebrate	Growth chamber	M. Chavez	EI Paso/Ysleta Schools, EI Paso TX
66	*Vigna radiata*	Plant	Plant growth unit (PGU)	J.R. Cowles	U Houston, Houston TX
67	*Avena sativa*	Plant	Plant growth unit (PGU)	J.R. Cowles	U Houston, Houston TX
68	*Pinus elliotti*	Plant	Plant growth unit (PGU)	J.R. Cowles	U Houston, Houston TX
69	*Escherichia coli*	Bacteria, free living	Biorack	O. Cifferi	University Pavia
70	Phage K1	Bacteriophage, free living	Biorack	O. Cifferi	University Pavia
71	*Drosophila melanogaster*	Insect	Biorack	R. Marco	Universidad Autonoma Madrid
72	*Physarum polycephalum*	Slime mould, free living	Biorack	V. Sobick	DLR Cologne (DFVLR)
73	*Carausius morosus*	Insect	Biorack	H. Buecker	DLR Cologne (DFVLR)
74	n.a.	Dosimeters	Biorack	H. Buecker	DLR Cologne (DFVLR)
75	*Paramecium caudatum*	Protozoa, free living	Biorack	H. Planel	National Institute of Health, Toulouse
76	*Paramecium caudatum*	Protozoa, free living	Biorack	H. Planel	National Institute of Health, Toulouse
77	*Chlamydomonas reinhardii*	Algae, free living	Biorack	D. Mergenhagen	University Hamburg
78	*Bacillus subtilis*	Bacteria, free living	Biorack	H.D. Mennigmann	University Frankfurt
79	*Homo sapiens sapiens*	Lymphocytes	Biorack	A. Cogoli	ETH Zürich
80	*Homo sapiens sapiens*	Lymphocytes	Biorack	A. Cogoli	ETH Zürich
81	*Lens culinaris*	Plant	Biorack	G. Perbal	University Curie Paris
82		Lymphocytes	Biorack	M. Bouteille	University Paris
83	*Xenopus laevis*	Amphibia	Biorack	G.A. Ubbels	Hubrecht Laboratory Utrecht
84	*Escherichia coli*	Bacteria, free living	Biorack	R. Tixador	National Institute of Health, Toulouse
85	*Lepidium sativum*	Plant	NEXPA facility	D. Volkmann	University Bonn
86	*Zea mays*	Plant	NEXPA facility	J. Gross	University Tübingen
87	*Pimpinella anisum*	Plant Cells	NEXPA facility	R.R. Thelmer	University Münich
88	*Xenopus laevis*	Amphibian	NEXPA facility	J. Neubert	DLR Cologne (DFVLR)
89	n.a.	macromolecules	FFE system	D. Clifford	McDonnell Douglas
90	*Amaranthus hypocondriacae*	Plant		A.J.V. Peluyera	Instituto del Consumidor, Mexico City
91	*Lens esculenta*	Plant		A.J.V. Peluyera	Instituto del Consumidor, Mexico City
92	*Triticum aestivum*	Plant		A.J.V. Peluyera	Instituto del Consumidor, Mexico City
93	*Escherichia coli*	Bacteria, free living		S.E. Flores	University Center, Mexico City
94	Phage lambda CII-68	Bacteriophage, free living		S.E. Flores	University Center, Mexico City
95	*Phaseolus vulgaris*	Plant		I. Ortega	Laboratoria de Cuernava-ca
96	*Zea mays*	Plant		C.L. Wang	Richfield HS, Waco TX
97	*Raphanus sativus*	Plant		G.A. Smith	UAH, Huntsville AL

Table 14 (continued).

Row No.	Mission	Flight	Launched	Carrier	Expt.	Topic	Title	Species or donor
65	SPTN-1	STS-51 G	17 Jun 85	Discovery	G-0034	Regeneration (worm)	Planarian regeneration	Brown planarian
66	SL-2	STS-51 F	29 Jul 85	Challenger	SL2-02	Growth and metabolism	Gravity-influenced lignification in higher plants	Mung bean
67	SL-2	STS-51 F	29 Jul 85	Challenger	SL2-02	Growth and metabolism	Gravity-influenced lignification in higher plants	Oat
68	SL-2	STS-51 F	29 Jul 85	Challenger	SL2-02	Growth and metabolism	Gravity-influenced lignification in higher plants	Pine
69	SL-D1	STS-61 A	30 Oct 85	Challenger	071 DNA	Cell interaction	Microgravity and genetic recombination	
70	SL-D1	STS-61 A	30 Oct 85	Challenger	071 DNA	Cell interaction	Microgravity and genetic recombination	
71	SL-D1	STS-61 A	30 Oct 85	Challenger	15 E Fly	Insect communities	Fruit fly embryogenesis and life span	Fruit fly
72	SL-D1	STS-61 A	30 Oct 85	Challenger	16 D Slime	Endogenous rhythms	Contraction and protoplasmic streaming	
73	SL-D1	STS-61 A	30 Oct 85	Challenger	18 D Biostack	Radiobiology (effects)	Embryogenesis and organogenesis	Stick insect
74	SL-D1	STS-61 A	30 Oct 85	Challenger	19 D Dosimeter	Radiobiology (mapping)	Dosimetric mapping inside Biorack	
75	SL-D1	STS-61 A	30 Oct 85	Challenger	21 F Paramec	Growth and metabolism	Paramecium growth and mineral properties	
76	SL-D1	STS-61 A	30 Oct 85	Challenger	21 F Paramec	Growth and metabolism	Paramecium growth and mineral properties	
77	SL-D1	STS-61 A	30 Oct 85	Challenger	27 D Circadi	Endogenous rhythms	The biological clock of *C. reinhardii* in space	
78	SL-D1	STS-61 A	30 Oct 85	Challenger	28 D Spores	Growth and metabolism	Bacterial growth and differentiation	
79	SL-D1	STS-61 A	30 Oct 85	Challenger	32 CH Blood	Growth and metabolism	Human lymphocyte activation in weightlessness	Human
80	SL-D1	STS-61 A	30 Oct 85	Challenger	33 CH Lympho	Growth and metabolism	Effect of microgravity on lymphocyte action	Human
81	SL-D1	STS-61 A	30 Oct 85	Challenger	39 F Roots	Graviperception (growth)	Graviperception of lentil seedling roots	Lentil
82	SL-D1	STS-61 A	30 Oct 85	Challenger	48 F Plasma	Graviperception (differentiation)	Ultrastructure of mammalian cell polarisation	Mouse
83	SL-D1	STS-61 A	30 Oct 85	Challenger	52 NL Eggs	Fertilisation and embryogenesis	Dorso/ventral axis establishment in embryos	Clawed toad
84	SL-D1	STS-61 A	30 Oct 85	Challenger	58 F Antibio	Effectiveness of antibiotics	Effectiveness of antibiotics under microgravity	
85	SL-D1	STS-61 A	30 Oct 85	Challenger	BOT-01	Graviperception (growth)	Graviperception	Garden cress
86	SL-D1	STS-61 A	30 Oct 85	Challenger	BOT-02	Graviperception (growth)	Geotropism	Maize
87	SL-D1	STS-61 A	30 Oct 85	Challenger	BOT-03	Graviperception (differentiation)	Differentiation of plant cells	Anise
88	SL-D1	STS-61 A	30 Oct 85	Challenger	STA-00	Graviperception (orientation)	Frog statolith experiment (STATEX)	Clawed toad
89	EASE	STS61B	26 Nov 85	Atlantis	CFES	Analysis equipment	Continuous flow electrophoresis system	n.a.
90	EASE	STS61B	26 Nov 85	Atlantis	Seeds	Graviperception (germination)	Effect of weightlessness and light on germination	Amaranth
91	EASE	STS61B	26 Nov 85	Atlantis	Seeds	Graviperception (germination)	Effect of weightlessness and light on germination	Lentil
92	EASE	STS61B	26 Nov 85	Atlantis	Seeds	Graviperception (germination)	Effect of weightlessness and light on germination	Wheat
93	EASE	STS61B	26 Nov 85	Atlantis	Repgrow	Growth and metabolism	Reproduction and growth of bacteria	
94	EASE	STS61B	26 Nov 85	Atlantis	Repgrow	Growth and metabolism	Reproduction and growth of bacteria	
95	EASE	STS61B	26 Nov 85	Atlantis	Transport	Growth and metabolism	Transportation of nutrients in weightlessness	Bean
96	MSL-2	STS-61 C	12 Jan 86	Columbia	SE82-19	Hormone system response	Auxin levels and starch grains in roots	Maize
97	MSL-2	STS-61 C	12 Jan 86	Columbia	G-0007	Graviperception (morphology)	Study of the radish root system	

Table 14 (continued).

Row No.	Scientific name	Category	Key hardware	Investigator	Affiliation
98	*Artemia salina*	Invertebrate	Growth chamber	A. White	B.T. Washington HS Houston TX
99	*Lymantria dispar*	Insect	Nylon mesh on cotton screen	D.K. Hayes	USDA, Beltsville MD
100	*Dermacentor variabilis*	Insect	Nylon mesh on cotton screen	D.K. Hayes	USDA, Beltsville MD
101	n.a.	macromolecules	Liquid chromatography system		Alitech Associates Inc.
102	n.a.	macromolecules	Electrophoresis system	M. Bier	University of Arizona, Tucson AZ
103	n.a.	macromolecules		D.E. Brooks	U of British Columbia, Vancouver
104	*Haplopappus gracilis*	Plant	Plant growth unit (PGU)	A.D. Krikorian	State U of New York, Stony Brook NY
105		Plant	Plant growth unit (PGU)	A.D. Krikorian	State U of New York, Stony Brook NY
106	*Rattus rattus*	Mammal		A.I. Fras	Binghamton HS, Binghamton NY
107	*Gallus domesticus*	Bird	Incubator box	J.C. Vellinger	Jefferson HS, Lafayette IN
108	*Zea mays*	Plant	Plant canisters + N2 freezer	R. Bandursid	University of Michigan
109	*Neurospora* sp.	Fungus		J.S. Ferraro	Southern Illinois U, Carbondale IL
110		Plant	Plant growth unit (PGU)	A.D. Krikorian	State U of New York, Stony Brook NY
111	*Rattus rattus*	Mammal	Animal enclosure modules (AEM)	M. Cornin	CCDS-PSU (CCR) / Genentech, CA
112	various	various	Materials dispersion app. (MDA)	M. Luttgee	CCDS-UCO, Boulder CO (Bioserce)
113	*Rattus rattus*	Mammal	RAHF or AEM	K.M. Baldwin	U of California, Irvine CA
114	*Rattus rattus*	Mammal	RAHF or AEM	E.R. Morey-Holton	NASA-ARC, Moffett Field CA
115	*Rattus rattus*	Mammal	RAHF or AEM	J. Foon Yoong Hoh	NASA-ARC/University of Sydney
116	*Rattus rattus*	Mammal	RAHF or AEM	D.A. Riley	Medical College of Wisc., Milwaukee WI
117	*Rattus rattus*	Mammal	RAHF or AEM	C. Altrey	Baylor College of Medicine, Houston TX
118	*Rattus rattus*	Mammal	RAHF or AEM	R.D. Lange	U of Tennessee, Knoxville TN
119	*Rattus rattus*	Mammal	RAHF or AEM	M. Ross	NASA-ARC, Moffett Field CA
120	*Aurelia aurita*	Invertebrate		D.B. Spangenberg	E. Virginia Medical School, Norfolk VA
121	*Homo sapiens sapiens*	Lymphocytes		A. Cogoli	ETH Zürich
122	various	Plant			Nissho Iwai American Corp., NY
123	various	Plant			Sakana Seeds Corp., Yokohama
124	various	various	Materials dispersion app. (MDA)	M. Luttges	CCDS-UCO, Boulder CO (Bioserve)
125	*Rattus rattus*	Mammal	Animal enclosure modules (AEM)	M. Tischler	University of Arizona, Tucson AZ
126	none	n.a.	STLV bioreactor	G. Spaulding	NASA-JSC, Houston TX

Table 14 (continued).

Row No.	Mission	Flight	Launched	Carrier	Expt.	Topic	Title	Species or donor
98	MSL-2	STS-61 C	12 Jan 86	Columbia	G-0332	Effect on reproductive units	Brine shrimp experiment	Brine shrimp
99	MSL-2	STS-61 C	12 Jan 86	Columbia	G-0470	Effect on reproductive units	Gypsy moth and American Dog ticks	Gypsy moth
100	MSL-2	STS-61 C	12 Jan 86	Columbia	G-0470	Effect on reproductive units	Gypsy moth and American Dog ticks	Dog tick
101	MSL-2	STS-61 C	12 Jan 86	Columbia	G-0446 Joehus	Analysis equipment	High performance liquid chromatography	n.a.
102	TDRS-C	STS-26	29 Sept 88	Discovery	IEF	Analysis equipment	Isoelectric focusing electrophoresis	n.a.
103	TDRS-C	STS-26	29 Sept 88	Discovery	PPE	Analysis equipment	Phase partitioning experiment	n.a.
104	TDRS-D	STS-29	13 Mar 89	Discovery	Chromex	Regeneration (plant roots)	Chromosomes and plant cell division	
105	TDRS-D	STS-29	13 Mar 89	Discovery	Chromex	Regeneration (plant roots)	Chromosomes and plant cell division	Daylily
106	TDRS-D	STS-29	13 Mar 89	Discovery	SE82-08	Bone and skeleton changes	Effects of weightlessness on the healing bone	Rat
107	TDRS-D	STS-29	13 Mar 89	Discovery	SE83-09	Fertilisation and embryogenesis	Chicken embryo development in space	Chicken
108	Galileo	STS-34	18 Oct 89	Atlantis	GHCD	Hormone system response	Growth hormone concentration & distribution	Maize
109	LDEF-R	STS-32	09 Jan 90	Columbia	CNCR	Endogenous rhythms	Characterisation of *Neurospora* circadian rhythms	Pink Bread Mould
110	Ulysses	STS-41	08 Oct 90	Discovery	Chromex-2	Graviperception (differentiation)	Chromosomes and plant cell division	
111	Ulysses	STS-41	08 Oct 90	Discovery	PSE	Effectiveness of medication	Physiological systems experiment	Rat
112	GRO	STS-37	05 Apr 91	Atlantis	BIMDA	Processing equipment	Bioserve I materials dispersion apparatus	various
113	SLS-1	STS-40	05 Jun 91	Columbia		Bone and skeleton changes	Metabolic properties of skeletal muscle	Rat
114	SLS-1	STS-40	05 Jun 91	Columbia		Bone and skeleton changes	Bone, Calcium and space flight	Rat
115	SLS-1	STS-40	05 Jun 91	Columbia		Muscle changes	Skeletal myosin isoenzymes in rats	Rat
116	SLS-1	STS-40	05 Jun 91	Columbia		Muscle changes	Histochemistry and protease activities	Rat
117	SLS-1	STS-40	05 Jun 91	Columbia		Cardiovascular system	Regulation of blood volume during space flight	Rat
118	SLS-1	STS-40	05 Jun 91	Columbia		Cardiovascular system	Regulation of erythropoiesis during space flight	Rat
119	SLS-1	STS-40	05 Jun 91	Columbia		Graviperception (receptor)	Space travel and mammalian gravity receptors	Rat
120	SLS-1	STS-40	05 Jun 91	Columbia		Graviperception (receptor)	*Aurelia* ephyra differentiation and statolith synthesis	Jellyfish
121	SLS-1	STS-40	05 Jun 91	Columbia		Growth and metabolism	Lymphocyte proliferation in weightlessness	Human
122	SLS-1	STS-40	05 Jun 91	Columbia	G-0451	Effect on reproductive units	Flower and vegetable seeds space exposure	various
123	SLS-1	STS-40	05 Jun 91	Columbia	G-0451	Effect on reproductive units	Flower and vegetable seeds space exposure	various
124	TDRS-E	STS-43	05 Jun 91	Atlantis	BIMDA	Processing equipment	Bioserve I materials dispersion apparatus (BIMDA)	various
125	UARS	STS-48	12 Sept 91	Discovery	PARE-01	Muscle changes	Physiological & anatomical rodent experiment	Rat
126	DSP	STS-44	24 Nov 91	Atlantis	BFPT	Processing equipment	Bioreactor flow and particle trajectory	none

Table 14 (continued).

Row No.	Scientific name	Category	Key hardware	Investigator	Affiliation
127		Mammal bone cells	Biorack	J.P. Veldhujzen	Free University Amsterdam
128	*Xenopus laevis*	Amphibia	Biorack	G.A. Ubbells	Hubrecht Laboratory Utrecht
129	*Drosophila melanogaster*	Insect	Biorack	R. Marco	Universidad Autonoma Madrid
130	*Daucus carota*	Plant Cells	Biorack	O. Rasmussen	University of Aarhus
131	*Brassica napus*	Plant cells	Biorack	O. Rasmussen	University of Aarhus
132	*Carausius morosus*	Insect	Biorack	H. Bueker	DLR Cologne
133	none	n.a.	Biorack	G. Reltz	DLR Cologne
134	*Bacillus subtilis*	Bacteria, free living	Biorack	H.D. Mennigmann	University Frankfurt
135	*Bacillus subtilis*	Bacteria spores	HZE tracking sandwich	H. Buecker	DLR Cologne
136	*Saccharomyces cerevisiae*	Fungus spores	HZE tracking sandwich	H. Buecker	DLR Cologne
137	*Sordaria fimicola*	Fungus spores	HZE tracking sandwich	H. Buecker	DLR Cologne
138	*Artemia salina*	Invertebrate	HZE tracking sandwich	H. Buecker	DLR Cologne
139	*Arabidopsis thaliana*	Plant	HZE tracking sandwich	H. Buecker	DLR Cologne
140	*Homo sapiens sapiens*	Erythrocytes	Biorack	A. Cogoli	ETH Zürich
141	*Homo sapiens sapiens*	Lymphocytes	Biorack	A. Cogoli	ETH Zürich
142		Kidney cells	Biorack	A. Cogoli	ETH Zürich
143	*Arabidopsis thalliana*	Plant	Biorack	L.G. Briarty	University of Nottingham, UK
144	*Bacillus subtilis*	Bacteria, spores	HZE tracking sandwich	S. Nagasaki	NASDA, Tokyo
145	*Artemia salina*	Invertebrate	HZE tracking sandwich	S. Nagasaki	NASDA, Tokyo
146	*Zea mays*	Plant	HZE tracking sandwich	S. Nagasaki	NASDA, Tokyo
147	*Lens culinaris*	Plant	Biorack	G. Perbal	Curie University Paris
148	*Physarum polycephalum*	Slime mould, free living	Biorack	I. Block	DLR Cologne
149	*Escherichia coll*	Bacteria, free living	Biorack	R. Tixador	National Institute of Health, Toulouse
150	*Triticum aestivum*	Plant	Gravit. Plant Physical Facility (GPPF)	D.G. Heathcote	UPA Philadelphia PA
151	*Avena sativa*	Plant	Gravit. Plant Physical Facility (GPPF)	A.H. Brown	UPA Philadelphia PA
152	*Caenorhabditis elegans*	Invertebrate	Biorack	G.A. Nelson	NASA-JPL, Pasadena CA
153	*Saccharomyces cerevisiae*	Fungus, free living	Biorack	C.V. Bruschl	L. Berkeley Laboratory, Berkeley CA
154		Mammal limb cells	Biorack	P.J. Duke	University of Texas, Houston TX
155	n.a.	macromolecules		D.E Brools	U of British Columbia, Vancouver
156	*Artemia salina*	Invertebrate			B.T. Washington HS, Houston TX
157	*Rattus rattus*	Mammal Cells	Automated tray assembly	E.R. Morey-Holton	NASA-ARC, Moffett Field CA
158	none	n.a.	Plant growth system (LT)	T.W. Tibbitts	CCDS-UWI Madison WI (WCSAR)

Table 14 (continued).

Row No.	Mission	Flight	Launched	Carrier	Expt.	Topic	Title	Species or donor
127	IML-1	STS-42	22 Jan 92	Discovery	02 NL Bones	Bone and skeleton changes	Mineralization and resorption of fetal bones	Mouse
128	IML-1	STS-42	22 Jan 92	Discovery	05 NL Eggs	Fertilisation and embryogenesis	Dorsal/ventral axis determination in embryos	Clawed toad
129	IML-1	STS-42	22 Jan 92	Discovery	07 E Fly	Insect communities	Development of *Drosophila* in space	Fruit fly
130	IML-1	STS-42	22 Jan 92	Discovery	08 DK Proto	Regeneration (full plant)	Plant differentiation from protoplast	Carrot
131	IML-1	STS-42	22 Jan 92	Discovery	08 DK Proto	Regeneration (full plant)	Plant differentiation from protoplasts	Rape
132	IML-1	STS-42	22 Jan 92	Discovery	10 D Morosus	Radiobiology (effects)	Embryogenesis and organogenesis	Stick insect
133	IML-1	STS-42	22 Jan 92	Discovery	12 D Dosimir	Radiobiology (mapping)	Dosimetric mapping inside Biorack	none
134	IML-1	STS-42	22 Jan 92	Discovery	13 D Spores	Growth and metabolism	Growth and sporulation of *Bacillus subtilis*	
135	IML-1	STS-42	22 Jan 92	Discovery	14 IML-1 Biostack	Radiobiology (mapping)	Dosimetric mapping with the Biostack	
136	IML-1	STS-42	22 Jan 92	Discovery	14 IML-1 Biostack	Radiobiology (mapping)	Dosimetric mapping with the Biostack	Yeast
137	IML-1	STS-42	22 Jan 92	Discovery	14 IML-1 Biostack	Radiobiology (mapping)	Dosimetric mapping with the Biostack	
138	IML-1	STS-42	22 Jan 92	Discovery	14 IML-1 Biostack	Radiobiology (mapping)	Dosimetric mapping with the Biostack	Brine shrimp
139	IML-1	STS-42	22 Jan 92	Discovery	14 IML-1 Biostack	Radiobiology (mapping)	Dosimetric mapping with the Biostack	Mouse-ear cress
140	IML-1	STS-42	22 Jan 92	Discovery	14.1 CH Friend	Growth and metabolism	Friend Leukaemia virus transformed cells	Human
141	IML-1	STS-42	22 Jan 92	Discovery	14.1 CH Friend	Growth and metabolism	Hybridoma cell proliferation and performance	Human
142	IML-1	STS-42	22 Jan 92	Discovery	14.3 CH Culture	Processing equipment	Dynamic cell culture system	Syrian hamster
143	IML-1	STS-42	22 Jan 92	Discovery	15 UK Shoots	Graviperception (growth)	Graviresponse of shoots and cotyledons in microgravity	Mouse-ear cress
144	IML-1	STS-42	22 Jan 92	Discovery	16 IML-1 RMCD	Radiobiology (mapping)	Radiation monitoring container device	
145	IML-1	STS-42	22 Jan 92	Discovery	16 IML-1 RMCD	Radiobiology (mapping)	Radiation monitoring container device	Brine shrimp
146	IML-1	STS-42	22 Jan 92	Discovery	16 IML-1 RMCD	Radiobiology (mapping)	Radiation monitoring container device	Maize
147	IML-1	STS-42	22 Jan 92	Discovery	20 F Roots	Graviperception (growth)	Transmission of the gravity stimulus in roots	Lentil
148	IML-1	STS-42	22 Jan 92	Discovery	22 D Slime	Endogenous rhythms	Gravity related behaviour of *Physarum*	
149	IML-1	STS-42	22 Jan 92	Discovery	23 F Antibio	Effectiveness of antibiotics	Antibiotic penetration of bacterial cells in space	
150	IML-1	STS-42	22 Jan 92	Discovery	6 IML-1 FOTRAN	Graviperception (growth)	Seedling Response to Light Stimulation	Wheat
151	IML-1	STS-42	22 Jan 92	Discovery	6 IML-1 GTHRES	Graviperception (growth)	Gravity sensor threshold of seedlings	Oat
152	IML-1	STS-42	22 Jan 92	Discovery	US-1 Radiat	Radiobiology (mapping)	Genetic and molecular dosimetry of HZE	Nematode
153	IML-1	STS-42	22 Jan 92	Discovery	US-2 Yeast	Effect on genetic material	Microgravity effects on chromosome behaviour	Yeast
154	IML-1	STS-42	22 Jan 92	Discovery	US-3 Cells	Bone and skeleton changes	Chondrogenesis in mouse limb mesenchyme	Mouse
155	IML-1	STS-42	22 Jan 92	Discovery	PPE	Analysis equipment	Phase partitioning experiment	n.a.
156	IML-1	STS-42	22 Jan 92	Discovery	G-0066	Effect on reproductive units	Brine shrimp in microgravity	Brine shrimp
157	ATLAS-1 STS-45	STS-45	24 Mar 92	Atlantis	STL	Effectiveness of medication	Space Tissue Loss	Rat
158	USML-1	SLS-50	25 Jun 92	Columbia	Astro-cul ture	Processing equipment	Astroculture (TM)	none

Table 14 (continued).

Row No.	Scientific name	Category	Key hardware	Investigator	Affiliation
159	various	various	Fluid processing apparatus (FPA)	M.C. Robinson	CCDS-UCO, Boulder CO (Bioserve)
160	*Rattus rattus*	Mammal brain gland cells	Middeck incubator	M.C. Hymer	PA State U, University Park PA
161	*Aspergillus niger*	Fungus, spores	Open exposure on pallet	K. Dose	University Mainz
162	*Aspergillus ochraceus*	Fungus, spores	Open exposure on pallet	K. Dose	University Mainz
163	*Saccharomyces cerevisiae*	Fungus, spores	Open exposure on pallet	J. Kiefer	University Giessen
164	*Bacillus subtilis*	Bacteria, spores	Open exposure on pallet	G. Horneck	DLR Cologne
165	*Deinococcus radiodurans*	Bacteria, spores	Open exposure on pallet	G. Horneck	DLR Cologne
166	*Ditylenchus dipsacii*	Invertebrate	Open exposure on pallet	G. Horneck	DLR Cologne
167	*Neurospora crassa*	Fungus		Y. Miyoshi	Tokyo University
168		Fish	Aquarium (VFEU)	S. Mori	Nagoya University
169	n.a.	Macromolecules	Free flow electrophoresis	M. Kuroda	Osaka University
170		Mammal kidney cells		A. Sato	Tokyo Medical and Dental University
171	*Homo sapiens sapiens*	Lymphocytes		H. Ooka	Tokyo university
172		Plant Cells	Culture chambers (semi-solid agar)	Y. Yamada	Tokyo University
173	*Gallus domesticus*	Bird	Egg holding rack	T. Suda	Showa University
174	*Salmonella typhimurium*	Bacteria, free living	Free flow electrophoresis	T. Yamaguchi	Tokyo Medical and Dental University
175	*Drosophila melanogaster*	Insect	Fly containers	M. Ikenaga	Tokyo University
176	*Zea mays*	Plant	HZE tracking sandwich	S. Nagaoka	NASDA, Tokyo
177	*Bacillus subtilis*	Bacteria, spores	HZE tracking sandwich	S. Nagaoka	NASDA, Tokyo
178	*Artemia salina*	Invertebrate	HZE tracking sandwich	S. Nagaoka	NASDA, Tokyo
179	*Xenopus laevis*	Amphibia	Frog environment unit	K.A. Souza	NASA-ARC Moffett Field CA
180	*Xenopus laevis*	Amphibia	Frog environment unit	K.A. Souza	NASA-ARC Moffett Field CA
181	*Daucus carota*	Plant cells	Culture chambers (semi-solid agar)	A.D. Krikorian	State U of New York, Stony Brook NY
182		Plant cells	Culture chambers (semi-solid agar)	A.D. Krikorian	State U of New York, Stony Brook NY
183		Mammal osteoblast cells		N. Partridge	St. Louis U Medical School, St Louis MO
184	*Vespa orientalis*	Insect	Normal containers	J.S. Ishey	Tel Aviv University
185				Students	Kansas U, Lawrence KS
186		Plant			Kansas U, Lawrence KS
187	various	Various	Material dispersion apparatus (MDA)	M. Lewis	CCDS-UAH, Huntsville AL (CMDS)
188	n.a.	macromolecules		D.E. Brooks	U of British Columbia, Vancouver
189	*Rattus rattus*	Mammal	Animal enclosure modules (AEM)	G. Rodan	CCDS-UCO (Bioserve)/ Merck Inc
190	none	n.a.	ASPEC bioreactor	G. Spaulding	Medical Sciences Div, NASA-JSC
191	various	various	Fluid processing apparatus (FPA)	M. Luttges	CCDS-UCO, Boulder CO (Bioserve)
192	*Arabidopsis thaliana*	Plant	Plant growth unit (PGU)	M. Musgrave	Louisiana State University

Table 14 (continued).

Row No.	Mission	Flight	Launched	Carrier	Expt.	Topic	Title	Species or donor
159	USML-1	STS-50	25 Jun 92	Columbia	GBA	Processing equipment	Generic bioprocessing apparatus	various
160	Eureca	STS-46	01 Aug 92	Atlantis	PHCF	Hormone system responses	Pituitary growth hormone cell function	Rat
161	Eureca	STS-46	01 Aug 92	Eureca	ERA	Exposure experiment	Exobiology and Radiation Assembly (ERA)	
162	Eureca	STS-46	01 Aug 92	Eureca	ERA	Exposure experiment	ERA	
163	Eureca	STS-46	01 Aug 92	Eureca	ERA	Exposure experiment	ERA	Yeast
164	Eureca	STS-46	01 Aug 92	Eureca	ERA	Exposure experiment	ERA	
165	Eureca	STS-46	01 Aug 92	Eureca	ERA	Exposure experiment	ERA	
166	Eureca	STS-46	01 Aug 92	Eureca	ERA	Exposure experiment	ERA	
167	SL-J	STS-47	12 Sept 92	Endeavour	L-12	Endogenous rhythms	Circadian rhythm of conidiation	Pink bread mould
168	SL-J	STS-47	12 Sept 92	Endeavour	L-2	Graviperception (orientation)	Posture and movement of fish in microgravity	Japanese carp
169	SL-J	STS-47	12 Sept 92	Endeavour	L-3	Analysis equipment	Separation of biogenic materials	n.a.
170	SL-J	STS-47	12 Sept 92	Endeavour	L-6A	Growth and metabolism	Ultrastructure and function of mammal cells	Monkey
171	SL-J	STS-47	12 Sept 92	Endeavour	L-6B	Growth and metabolism	Performance of antibody producing cells	Human
172	SL-J	STS-47	12 Sept 92	Endeavour	L-6C	Graviperception (differentiation)	Organ differentiation from plant cells	
173	SL-J	STS-47	12 Sept 92	Endeavour	L-7	Bone and skeleton changes	Calcium metabolism and bone formation	Chicken
174	SL-J	STS-47	12 Sept 92	Endeavour	L-8	Analysis equipment	Separation of cells and cell organelles	
175	SL-J	STS-47	12 Sept 92	Endeavour	L-9	Radiobiology (effects)	Genetic effects of HZE and cosmic radiation	Fruit fly
176	SL-J	STS-47	12 Sept 92	Endeavour	L-11	Radiobiology (mapping)	Biological effects of cosmic radiation	Maize
177	SL-J	STS-47	12 Sept 92	Endeavour	L-11	Radiobiology (mapping)	Biological effects of cosmic radiation	
178	SL-J	STS-47	12 Sept 92	Endeavour	L-11	Radiobiology (mapping)	Biological effects of cosmic radiation	Brine shrimp
179	SL-J	STS-47	12 Sept 92	Endeavour	FEE	Fertilisation and Embryogenesis	Embryology experiment	Clawed toad
180	SL-J	STS-47	12 Sept 92	Endeavour	FEE	Fertilisation and Embryogenesis	Embryology experiment	Clawed toad
181	SL-J	STS-47	12 Sept 92	Endeavour		Effect on genetic material	Plant culture research	Carrot
182	SL-J	STS-47	12 Sept 92	Endeavour		Effect on genetic material	Plant culture research	Daylily
183	SL-J	STS-47	12 Sept 92	Endeavour		Bone and skeleton changes	Bone cell growth and mineralisation	
184	SL-J	STS-47	12 Sept 92	Endeavour	ISAIAH	Insect communities	Investigation about hornets experiment	Oriental hornets
185	SL-J	STS-47	12 Sept 92	Endeavour	G-0255	Graviperception (differentiation)	Studies on cell formation in microgravity	
186	SL-J	STS-47	12 Sept 92	Endeavour	G-0255	Effect on reproductive units	Plant seed germination rates	
187	LAGEOS-11	STS-52	22 Oct 92	Columbia	CMIX	Commercial experiments	Commercial MDA I experiments	various
188	LAGEOS-11	STS-52	22 Oct 92	Columbia	PARLIO	Analysis equipment	Phase partitioning in liquids	n.a.
189	LAGEOS-11	STS-52	22 Oct 92	Columbia	PSE-02	Effectiveness of medication	Physiological systems experiment	Rat
190	TDRS-F	STS-54	13 Jan 93	Endeavour	ASPEC	Processing equipment	Preprogrammed experiment culture system	none
191	TDRS-F	STS-54	13 Jan 93	Endeavour	CGBA	Commercial experiments	Commercial generic bioprocessing apparatus	various
192	TDRS-F	STS-54	13 Jan 93	Endeavour	Chromex-3	Seed production	Chromosomes and plant cell division	Mouse-ear cress

Table 14 (continued).

Row No.	Scientific name	Category	Key hardware	Investigator	Affiliation
193	*Rattus rattus*	Mammal	Animal enclosure modules (AEM)	K.M. Baldwin	U of California, Irvine CA
194	*Xenopus laevis*	Amphibia	Biolabor	E. Horn	University Ulm
195	*Oreochromis mossambicus*	Fish	Biolabor	E. Horn	University Ulm
196	*Xenopus laevis*	Amphibia	Biolabor	J. Neubert	DLR Cologne
197	*Oreochromis mossambicus*	Fish	Biolabor	J. Neubert	DLR Cologne
198	*Xenopus laevis*	Amphibia	Biolabor	H. Rahmann	University of Stuttgart
199	*Oreochromis mossambicus*	Fish	Biolabor	H. Rahmann	University of Stuttgart
200	*Xenopus laevis*	Amphibia	Biolabor	M. Schachner	University Heidelberg
201	*Oreochromis mossambicus*	Fish	Biolabor	M. Schachner	University Heidelberg
202	*Lepidium sativum*	Plant	Biolabor	D. Volkmann	University Karlsruhe
203	*Lepidium sativum*	Plant	Biolabor	M.H. Welsenaeel	University Karlsruhe
204	*Flammulina velutipes*	Fungus	Biolabor	B. Hock	Technical University Mü-nich
205	*Panax ginseng*	Plant cells	Biolabor	H. Jung-Hellger	Plant Laboratory Cells
206	*Saccharomyces uvarum*	Fungus, free living	Biolabor	S. Donhauser	Technical University Mü-nich
207	*Escherichia coli*	Bacteria, free living	Biolabor	H.D. Mennigmann	University Frankfurt
208	*Bacillus subtilis*	Bacteria, free living	Biolabor	H.D. Mennigmann	University Frankfurt
209	*Escherichia coli*	Bacteria, free living	Biolabor	H.D. Mennigmann	University Frankfurt
210	*Bacillus subtilis*	Bacteria, free living	Biolabor	H.D. Mennigmann	University Frankfurt
211	*Homo sapiens sapiens*	Skin fibroblasts	Biolabor	P.K. Mueller	University Lubeck
212	*Homo sapiens sapiens*	T-lymphocytes	Biolabor	K. Reske	University Mainz
213	*Homo sapiens sapiens*	T-lymphocytes	Biolabor	K. Reske	University Mainz
214	*Homo sapiens sapiens*	B-lymphocytes	Biolabor electrofusion	U. Zimmermann	University Wurzburg
215	*Nicotiana tabacum*	Plant cells	Biolabor electrofusion	R. Hampp	University Tübingen
216	*Digitalis* sp.	Plant cells	Biolabor electrofusion	R. Hampp	University Tübingen
217	*Helianthus* sp.	Plant Cells	Biolabor electrofusion	R. Hampp	University Tübingen
218	*Artemia salina*	Invertebrate	Biostack HZE tracking sandwich	H. Buecker	DLR Cologne
219	*Arabidopsis thaliana*	Plant	Biostack HZE tracking sandwich	H. Buecker	DLR Cologne
220	*Sordaria fimicola*	Fungus, spores	Biostack HZE tracking sandwich	H. Buecker	DLR Cologne
221	*Bacillus subtilis*	Bacteria, spores	Biostack HZE tracking sandwich	H. Buecker	DLR Cologne
222	*Saccharomyces cerevisiae*	Fungus, spores	Biostack HZE tracking sandwich	H. Buecker	DLR Cologne
223	*Bacillus subtilis*	Bacteria, spores	Open exposure on pallet	G. Horneck	DLR Cologne
224	*Haemophilus influenzae*	Bacterial isolated DNA	Open exposure on pallet	G. Horneck	DLR Cologne
225	various	various	Bioprocessing modules (BPM)	M. Lewis	CCDS-UAH, Huntsville AL (CMDS)
226	various	various	Materials Dispersion App. (MDA)	M. Lewis	CCDS-UAH, Huntsville AL (CMDS)

Table 14 (continued).

Row No.	Mission	Flight	Launched	Carrier	Expt.	Topic	Title	Species or donor
193	TDRS-F	STS-54	13 Jan 93	Endeavour	PARE-02	Muscle changes	Physiological and anatomical rodent experiment	Rat
194	SL-D2	STS-55	26 Apr 93	Columbia	BB-IZL-Statex 11/1	Graviperception (orientation)	Vestibular reflexes in microgravity	Clawed toad
195	SL-D2	STS-55	26 Apr 93	Columbia	BB-IZL-Statex 11/1	Graviperception (orientation)	Vestibular reflexes in microgravity	Cichlid fish
196	SL-D2	STS-55	26 Apr 93	Columbia	BB-IZL-Statex 11/2	Graviperception (differentiation)	Development of the gravity perceiving organ	Clawed toad
197	SL-D2	STS-55	26 Apr 93	Columbia	BB-IZL-Statex 11/2	Graviperception (differentiation)	Development of the gravity perceiving organ	Cichlid fish
198	SL-D2	STS-55	26 Apr 93	Columbia	BB-IZL-Statex 11/3	Graviperception (differentiation)	CNS neuronal plasticity during ontogenesis	Clawed toad
199	SL-D2	STS-55	26 Apr 93	Columbia	BB-IZL-Statex 11/3	Graviperception (differentiation)	CNS neuronal plasticity during ontogenesis	Cichlid fish
200	SL-D2	STS-55	26 Apr 93	Columbia	BB-IZL-Statex 11/4	Graviperception (differentiation)	Cerebellar development in microgravity	Clawed toad
201	SL-D2	STS-55	26 Apr 93	Columbia	BB-IZL-Statex 11/4	Graviperception (differentiation)	Cerebellar development in microgravity	Cichlid fish
202	SL-D2	STS-55	26 Apr 93	Columbia	BB-IBT-Gravi	Graviperception (growth)	Gravisensitivity of cress roots	Garden cress
203	SL-D2	STS-55	26 Apr 93	Columbia	BB-IBT-Gravi	Graviperception (differentiation)	Cell polarity and gravity	Garden cress
204	SL-D2	STS-55	26 Apr 93	Columbia	BB-IBT-Gravi	Graviperception (differentiation)	Fruiting body development of fungi	
205	SL-D2	STS-55	26 Apr 93	Columbia	BB-IBT-Isa	Growth and metabolism	Secondary metabolites in cell suspensions	
206	SL-D2	STS-55	26 Apr 93	Columbia	BB-IBT-Yeast	Growth and metabolism	Investigation on yeast metabolism	Yeast
207	SL-D2	STS-55	26 Apr 93	Columbia	BB-ICC-Bact	Growth and metabolism	Productivity of bacteria	
208	SL-D2	STS-55	26 Apr 93	Columbia	BB-ICC-Bact	Growth and metabolism	Productivity of bacteria	
209	SL-D2	STS-55	26 Apr 93	Columbia	BB-ICL-Fluc	Effect on genetic material	Fluctuation test on bacterial cultures	
210	SL-D2	STS-55	26 Apr 93	Columbia	BB-ICL-Fluc	Effect on genetic material	Fluctuation test on bacterial cultures	
211	SL-D2	STS-55	26 Apr 93	Columbia	BB-ICC-Meta	Bone and skeleton changes	Collagen synthesis and cell proliferation	Human
212	SL-D2	STS-55	26 Apr 93	Columbia	BB-ICC-Cult	Growth and metabolism	Activation of regulatory T-lymphocytes	Human
213	SL-D2	STS-55	26 Apr 93	Columbia	BB-ICC-Grow	Growth and metabolism	Growth of lymphocytes in microgravity	Human
214	SL-D2	STS-55	26 Apr 93	Columbia	BB-EF-Hybrid	Cell electofusion	Cell fusion and hybridoma production	Human
215	SL-D2	STS-55	26 Apr 93	Columbia	BB-EF-Proto	Cell electrofusion	Electrofusion of plant cell protoplasts	Tobacco
216	SL-D2	STS-55	26 Apr 93	Columbia	BB-EF-Proto	Cell electrofusion	Electrofusion of plant cell protoplasts	Foxglove
217	SL-D2	STS-55	26 Apr 93	Columbia	BB-EF-Proto	Cell electrofusion	Electrofusion of plant cell protoplasts	Sunflower
218	SL-D2	STS-55	26 Apr 93	Columbia	RD-Bios	Radiobiology (mapping)	HZE-particle dosimetry with Biostack	Brine shrimp
219	SL-D2	STS-55	26 Apr 93	Columbia	RD-Bios	Radiobiology (mapping)	HZE-particle dosimetry with Biostack	Mouse-ear cress
220	SL-D2	STS-55	26 Apr 93	Columbia	RD-Bios	Radiobiology (mapping)	HZE-particle dosimetry with Biostack	
221	SL-D2	STS-55	26 Apr 93	Columbia	RD-Bios	Radiobiology (mapping)	HZE-particle dosimetry with Biostack	
222	SL-D2	STS-55	26 Apr 93	Columbia	RD-Bios	Radiobiology (mapping)	HZE-particle dosimetry with Biostack	Yeast
223	SL-D2	STS-55	26 Apr 93	Columbia	RD-Uvrad	Exposure experiment	Responses to solar UV and space vacuum	
224	SL-D2	STS-55	26 Apr 93	Columbia	RD-Uvrad	Exposure experiment	Responses to solar UV and space vacuum	
225	ATLAS-2	STS-56	06 APR 93	Discovery	CMIX-02	Commercial experiments	Commercial MDA I experiments	various
226	ATLAS-2	STS-56	06 Apr 93	Discovery	CMIX-02	Commercial experiments	Commercial MDA I experiments	various

Table 14 (continued).

Row No.	Scientific name	Category	Key hardware	Investigator	Affiliation
227	*Rattus rattus*	Mammal	Animal enclosure modules (AEM)	E.R. Morey-Holton	NASA-ARC, Mountain View CA
228	*Rattus rattus*	Mammal muscle and bone cells	Automated tray assembly	E.R. Morey-Holton	NASA-ARC, Mountain View CA
229	none	n.a.	Plant growth system (LT)	R.J. Bula	CCDS-UWI, Madison WI (WCSAR)
230	*Rhizobium trifolii*	Symbiotic bacteria	Bioprocessing modules (BPM)	M. Luttges	CCDS-UCO, Boulder CO (Bioserve)
231	*Escherichia coli*	Bacteria, free living	Bioprocessing modules (BPM)	M. Luttges	CCDS-UCO, Boulder CO (Bioserve)
232		Amphibian kidney cells	Bioprocessing modules (BPM)	M. Luttges	CCDS-UCO, Boulder CO (Bioserve)
233	various	various	Fluid processing apparatus (FPA)	M. Luttges	CCDS-UCO, Boulder CO (Bioserve)
234	n.a.	macromolecules	Organic separation device	J.M. Van Aistine	CCDS-UAH, Huntsville AL (CMDS)
235	*Rattus rattus*	Mammal	Animal enclosure modules (AEM)	G. Rodan	CCDS-PSU (CCR)/Space Dermat Fnd
236		Mammal colon cells	ASPEC bioreactor	G. Spaulding	Medical Sciences Div. , NASA-JSC
237		Plant		Students	California Central Coast Schools
238		Bacteria		Students	California Central Coast Schools
239		Bacteria		Students	California Central Coast Schools
240		Plant		Students	California Central Coast Schools
241	*Arabidopsis thaliana*	Plant	Plant growth unit (PGU)	M. Musgrave	Louisiana State University
242	*Triticum aestivum*	Plant	Plant growth unit (PGU)	A.D. Krikorian	State U of New York, Stony Brook NY
243	*Triticum aestivum*	Plant	Plant growth unit (PGU)	N. Lewis	Washington State University

Table 14 (continued).

Row No.	Mission	Flight	Launched	Carrier	Expt.	Topic	Title	Species or donor
227	ATLAS-2	STS-56	06 Apr 93	Discovery	PARE-03	Bone and muscle changes	Physiological and anatomical rodent experiment	Rat
228	ATLAS-2	STS-56	06 Apr 93	Discovery	STL-03	Effectiveness	Space tissue loss	Rat
229	Eureca- 1R	STS-57	21 Jun 93	Endeavour	Astro-cul ture	Processing equipment	Astroculture (TM)	none
230	Eureca- 1R	STS-57	21 Jun 93	Endeavour	BPL	Commercial experiments	Bioserve pilot laboratory – experiment 1	
231	Eureca- 1R	STS-57	21 Jun 93	Endeavour	BPL	Commercial experiments	Bioserve pilot laboratory – experiment 2	
232	Eureca- 1R	STS-57	21 Jun 93	Endeavour	BPL	Commercial experiments	Bioserve pilot laboratory – experiment 3	Frog
233	Eureca- 1R	STS-57	21 Jun 93	Endeavour	CGBA	Commercial experiments	Commercial generic bioprocessing apparatus	various
234	Eureca- 1R	STS-57	21 Jun 93	Endeavour	ORSEP	Processing equipment	Organic separation payload	n.a.
235	Eureca- 1R	STS-57	21 Jun 93	Endeavour	PSE-03	Effectiveness of medication	Physiological systems experiment	Rat
236	Eureca- 1R	STS-57	21 Jun 93	Endeavour	ASPEC	Processing equipment	Preprogrammed experiment culture system	
237	Eureca- 1R	STS-57	21 Jun 93	Endeavour	G-450	Graviperception (germination)	Microgravity effects on sprouting seeds	
238	Eureca- 1R	STS-57	21 Jun 93	Endeavour	G-450	Radiobiology (effects)	Effects of radiation on bacteria	
239	Eureca- 1R	STS-57	21 Jun 93	Endeavour	G-450	Radiobiology (effects)	Bacteria survival in radiation	
240	Eureca- 1R	STS-57	21 Jun 93	Endeavour	G-450	Radiobiology (effects)	Effects of radiation on seeds	
241	ACTS-TO S	STS-51		Discovery	Chromex-4	Graviperception (reproduction)	Seed production and forming in microgravity	Mouse-ear cress
242	ACTS-TO S	STS-51		Discovery	Chromex-4	Graviperception (differentiation)	Root and shoot development	Wheat
243	ACTS-TO S	STS-51		Discovery	Chromex-4	Graviperception (differentiation)	Cell wall forming and gene expression	Wheat

Table 15. Recent Russian Mir based life sciences experiments involving ESA member states.

Topic(s)	Principal Investigators	Affiliation
French/Russian *Aragatz* mission, 26 November to 21 December, 1988		
Study of the radiation environment of the Mir space station (CIRCE experiment).	V.M. Petrov *et al.*, V.D. Nguyen, J. Kerlach, M. Siegrist, J.F. Zwilung, A. Akatov *et al.*	Comm. Energ. Atomiqu;, CNES & IBMP, Moscow
Explore constraints on visual image processing and operational performances (VIMINAL experiment); mental rotation of 3D shapes in microgravity.	Y. Matsakis, A. Berthoz, M. Lipschitz, V. Garfinkel	Lab. Physio. Neurosens. & Inst. Prob. Transp. Inform.
Study of postural and locomotor systems and the state of sensorial organs (PHYSALIE experiment).	V.S. Gurfinkel, *et al.* *F. Lestienne, L. Lefort, C. Andre-Deshays et al.*, J.P. Roll *et al.*, K. Popov, M. Lipschitz	Lab. Physio. Neurosens. & Inst. Prob. Transp. Inform., Univ. de Provence
Study of the cardio-vascular system (ECHOGRAPHIE 4 experiment).	P. Arbeille *et al.*, J.L. Cretien, O. Atkov, V. Bystrov *et al.*, G. Fonina *et al.*	INSERM 316, Tours; CNES, Paris, Centre of Cardiology & IBMP, Moscow
Study of water and mineral metabolism and hormonal regulation (MINILAB). Volume regulating hormones, fluid and electrolyte modifications.	G. Gauquelin *et al.*, M.Patrilot, J.L. Cretien, A. Guell, R. KvetnanskY, L. Macho, A. Grigoriev *et al.*	Lab. Environ. Physio, Lyon, Centre Hosp. Lyon Sud; CNES, Toulouse; Inst. Exp. Endocrin, Bratislava & IBMP, Moscow
Influence of space flight on human T-lymphocyte and monocyte functions.	L. Schaffar *et al.* I. Konstantinova *et al.*	INSERM Uzio, Nice & IBMP, Moscow
Study of bone tissue mineralization. Bone densitometric results with the SCOOP miniscanner.	C. Alexandre *et al.* P. Ruegsegger	LBTO, France, Inst. Techn Biomed.
Austrian/Russian *Austromir* mission, 2 to 10 October, 1991		
Investigation of genetic changes in lymphocytes, their possible repair and immunological effects (MIRGEN).	H. Tuschl *et al.*, M.S. Chajdakow, Y.J. Voronkov	öst. Forsch Zent. Siebersdorf & IBMP, Moscow
Disturbances in movement coordination following stimulation with visual, acoustic and proprioceptive stimuli. Eye-head-arm coordination and spinal reflexes in weightlessness (MONOMIR)	M. Burger, F. Gersten Brand *et al.*, E. Hochmair, G. Steinwender, A. Sokolov *et al.*, I.B. Kozlovskaya	Neurology Dept. and Physics Dept. Univ. Innsbruck & IBMP, Moscow
Force-angle velocity relationship of musculature in/against pre-determined translatory movement & correlation with integrated surface EMG.	N. Bachl, R. Baron, K.H. Tschan, M.Mossaheb, W. Bumba, F. Hildebrand, I. Kharitonov	Sports & Exercise Dept., Univ. of Vienna & IBMP, Moscow

Measurement of attention, psychomotor speed, mental feasibility, time estimation, visuospatial perception, memory, function changes. A study of cognitive functions in microgravity (COGIMIR).	T. Benk, F. Gerstenbrand, B. Koserenko, N. Watson	Neurology Dept. Innsbruck Univ.; IBMP, Moscow & Univ. Brit. Columbia
Accuracy of directional hearing and its role in human orientation in microgravity.	A. Persterer, C. Koppensteiner et al., M. Berger, M. Nefjodova	AKG, Vienna, Univ. of Innsbruck & IBMP, Moscow
Examine the vertical vection illusion elicited by an optokinetic stimulator system. Orientational effects from optokinetic stimulation (OPTOVERT).	C. Muller, G. Wiest, L. Deecke, L. Kornilova	Neurology clinic, Univ. Vienna, IBMP, Moscow
Monitoring of cardiovascular parameters during space flight.	M. Moser, E. Gallasch, *et al*, R. Bayevskij, I. Funtowa *et al*	Physiolog. Inst., Univ. of Graz & IBMP, Moscow
Influence of tremor parameters of the changes to muscle loading/ proprioreceptor threshold levels	E. Gallasch, M. Moser *et al.*, I. B. Kozlovskaya, M. Borlov	Physiolog. Inst., Univ. of Graz & IBMP, Moscow
Changes to dynamics of transcapillary fluid shifts in microgravity.	H.G. Hinghofer-Szalkay *et al.*, J. Schmeid, H. Heimel, V.B. Noskov, I. Pestov, A.I. Grogorien	VRSM, Univ. of Graz; Lab. Messtechnik, Graz & IBMP, Moscow
Radiation measurements inside the soviet space station mir (DOSIMIR). Thermoluminescent dosimeters and track etch foils to measure average LET and absorbed dose.	N. Vana, W.Schoewer, M. Fugger, J. Akatov	Atominst., Univ. of Vienna & IBMP, Moscow
German/Russian *Mir 92* mission, 17 to 27 March, 1992		
Monitoring of kidney response from beginning of mission, in comprehensive manner, including stress, drug intake, circadian rhythms.	R. Gerzer, C. Drummer, M. Hear, R.A. Dressendörfer, C.J. Strasburger	Klinik Innestadt, Münich & DLR, Flugmedizin, Cologne
Fluid shifts into/out of superficial tissues under microgravity and terrestrial conditions.	K.A. Kirsch, F.J. Baartz, H.C. Gunga, L. Rocker	Inst. for Physiology, Free Univ., Berlin
Determination extent to which fluid deficit is due to interstitial fluid losses and LBNP effect.	F.J. Baisch	DLR, Inst. Flugmedizin, Cologne
Changes in interocular pressure due to fluid shifts.	J. Draeger, R. Schwartz,	Univ. Hospital, Eppendorf
Influence of microgravity induced fluid loss on heart rate/blood pressure after leg exercise compared to $1 \times g$ dehydration effects.	D. Essfeld, K. Baum, U. Hoffmann, J. Stegemann	Physiol. Institute Deutsche, Sporthochschule, Cologne
Dosimetry in space. Plastic track detectors, emulsions, thermoluminescence detectors combined with chromosomal analysis.	G.Reitz, R. Beaujean, N. Heckley, G. Obe	DLR, Inst. Luft- und Raumfahrtmedizin Cologne, Univ. Kiel; Univ. G.H., Essen

Table 15 (continued).

Topic(s)	Principal Investigators	Affiliation
Sleep and circadian rhythm during a short space mission.	A. Gundel, E.Reucher, M. Vejvodva, J. Zulley, V. Nalishiti	DLR, Inst. Luft- und Raumfahrtmedizin MPI Psychiatrie, Münich & ZPK, Star City
Tests of logical reasoning, decision making, memory retrieval, fine manual control, compared in flight with pre/postflight	D. Manzey, B. Lorenz, A. Schiewe, G. Finell, G. Thiele	DLR, Inst. Luft- und Raumfahrtmedizin Cologne & DLR, Astronautenburo
Illusions of verticality in weightlessness	H. Mittelstaedt, S. Glasauer	MPI Verhaltenphysiologie, Seewiesen; Neurolog. Klinik, Grosshadern, Münich
Separation of canalicular and otolithic contributions to ocular torsion. Vestibulo-oculomotor testing during space flight	A.H. Clarke, W. Teiwes, H. Scherer	HNO Klinik, Klinikum Stieglitz; Free Univ., Berlin
Clarification of interaction between visual, proprioceptive, and vestibular inputs to the equilibrium system.	K. Hoffstetter-Degen J.Wetzig, R. Von Baumgarten	Physiolog. Institut J.G. Univ., Mainz
Test of otolith asymmetry, and relationship to motion sickness	K. Hoffstetter-Degen J.Wetzig, R. Von Baumgarten	Physiolog. Institut J.G. Univ., Mainz
French/Russian *Antares* mission, July to August, 1992		
Determination of ANP and aldosterone modifications, assess role of cortisol as stress indicator, urine output monitoring, HRV and baroflex sensitivity before and after flight and interrelationships.	A. Maillet, C. Gharib, *et al.*, D. Vorobiev *et al.*, A. Soloviev *et al.*, R. Cartier, M. Patricot, M. Tognini	Lab. Phys. Environ., Univ. Lyon; IBMP, CPK, Moscow; Hosp. Cardiologique, Lyon & CNES, Toulouse
To characterize the functional activity of subset of natural killer cells during a long space mission. Mechanisms of antiviral resistance (IMMUNOLOGIE).	I.V. Konstatinum, *et al.*, D. Schmitt, C. Peres	IBMP, Moscow & CNES, Toulouse
To determine whether mental transformation processes, proposed for visual image processing, are influenced by microgravity conditions.	A. Berthoz, G. Leone, M. Lipshits, V. Gurfinkel, Y. Matsakis	Lab. Phys. Neurosens. CNRS; Inst. Info-trans. Prob, Moscow & NEDES, Toulouse
Study of adaptive processes in human proprioceptive functions at cognitive and sensorimotor levels in weightlessness.	J.P. Roll, R. Roll, C. Quoniam, J.C. Gilhodes, O. Charade	Lab. Human Neurobiol., Klinik Provence, Marseille lab. Neurosens, CNRS
Assessment of main cardiac and vascular haemodynamic changes, using an ultrasound space device.	N. Arbielle, J.H. Pottier, A. Roncin, G. Fomina, A. Kotovskaya	Fac. Medicine, Univ. Toulouse & IBMP, Moscow

Study of DNA base lesions upon exposure of cellular DNA to high LET components (BIODOSE)	J. Cadet *et al*, L. Sabatier, L. Nevzgodina	SESAM/DRFMC, DPTE/LCG & IBMP, Moscow
Real dose rate and LET spectrum on Mir using a cylindrical gaseous low pressure and tissue equivalent proportional counter.	V.D. Nguyen *et al.* *C. Andre-deshays, et al.* 3S.B. Koslova, V.M. Petrov	CEA ISPN/DPMD/SDOS, Fontenay; CNES, Toulouse & IBMP, Moscow
French/Russian Altair mission, July to August, 1993 (mostly extensions of studies carried out on the *Antares* mission of 1992)		
Study of human spatial orientation (ILLUSIONS)	J.P. Roll, J.C. Gilhode *et al.*	LNH, Marseille
Study of role of visual motor memory in operational activities (VIMINAL)	A. Berthoz *et al.*	LPN, Paris
Study of cosmic radiation dose distribution laws (NAUSICAA)	M. van Dat	CEA/IPSM
Study of changes in the immune defence system (IMMUNOLOGIE)	D. Schmitt *et al.*	INSERM
Biological dosimeters and the effect of space radiation (BIODOSE)	J. Cadet, L. Sabatier *et al.*	CEA
Study of cardiovascular deconditioning and of hormonal systems involved in blood volume regulation (ORTHOSTATISME)	P. Arbeille, J.M. Pottier, C. Gharib	LBM, Tours; LPE, Lyon
Study of limb/body movements in microgravity (SYNERGIES)	J. Masson, M. Pozzo	LNM, Marseille; LERPS, Dijon

Table 16. Recent Russian (CIS)/ESA-member states collaborations on the Bion spacecraft

Mission	Date	Topic	Authors	Affiliation
Bion 8 (Cosmos 1887)	29-9-87	Study of combined effects of microgravity and cosmic radiation on early stages of development in eggs of the stick insect, *Carausius morosus.*	G. Reitz *et al.*	DLR, Germany & IBMP, Moscow
Bion 8 (Cosmos 1887)	29-9-87	Radiation dosimetry in/outside the BIOKOSMOS (Bion) satellite.	G. Reitz *et al.*	DLR, Germany & IBMP, Moscow
Bion 8 (Cosmos 1887)	29-9-87	Biological damage induced by ionizing cosmic rays in dry *Arabidopsis* seeds.	A.R. Kranz *et al.*	Univ. Frankfurt, IBMP, Moscow & Inst. Developmental Biol., Moscow
Bion 9 (Cosmos 2044)	15-9-89	Study of combined effects of microgravity and cosmic radiation on early stages of development in eggs of *Carausius morosus.*	I.A. Ushakov, A.M. Alpatov & G. Reitz	IBMP, Moscow, DLR, Germany
Bion 9 (Cosmos 2044)	15-9-89	Radiation dosimetry in/outside Bion 9.	G. Reitz *et al.*	DLR, Germany & IBMP, Moscow
Bion 9 (Cosmos 2044)	15-9-89	Biological effects of ionizing cosmic heavy particles on embryonic plant tissues in seeds.	A.R. Kranz *et al.*, K.E. Gartenbach *et al.*	Univ. Frankfurt, IBMP & Inst. Developmental Biol., Moscow
Bion 9 (Cosmos 2044)	15-9-89	Effects of microgravity on regeneration of plants from protoplasts	T.H. Iversen *et al*, O. Rasmussen	Inst. Mol. Biol. Aarhus, Denmark; ETH, Zürich & IBMP, Moscow
Bion 9 (Cosmos 2044)	15-9-89	Microgravity and space radiation effects on *Drosophila melanogaster*: development, ageing, mitotic recombination rate and adaptation.	R. Marco et al.	Univ. Alicante & Univ. Madrid, Spain & IBMP, Moscow
Bion 10 (Cosmos 2229)	29-12-92	Test of the metabolic hypothesis of accelerated ageing in *Drosophila* in space	J. Miguel, R. Marco	Univ. Alicante & Univ. Madrid, Spain & IBMP, Moscow

Bion 10 (Cosmos 2229)	29-12-92	Impact of pre-flight gravity stress on in-flight fitness in *Drosophila melanogaster*	R. Marco, J. Gonzalez-Jurado, M. Maroto, J. Miguel, I.A. Ushakov, A.M. Alpatov	Univ. Alicante & Univ. Madrid, Spain & IBMP, Moscow
Bion 10 (Cosmos 2229)	29-12-92	Ultrastructural changes in dividing *Chlamydomonas monoica* cells due to microgravity.	H. van den Ende, E. van Spronsen, O.V. Gavrilova	Univ. Amsterdam & Univ. St. Petersburg
Bion 10 (Cosmos 2229)	29-12-92	Analysis of chromosomal damage of marker lines of *Arabidopsis* seeds by heavy cosmic particles.	A.R. Kranz, J.U. Schott *et al.*, V.V. Shevchenko, A.M. Marenny *et al.*	Univ. Frankfurt, RCSRS, IBMP Moscow
Bion 10 (Cosmos 2229)	29-12-92	Identification and quantification of incident radiation particles during flight and biological effects on *Wolffia arrhiza.*	G. Reitz, R. Facius *et al.*, L.V. Nevsgodinna, E.N. Maximova, A.M. Marenny *et al.*	DLR, Germany; IBMP, Moscow & RCSRS, Moscow

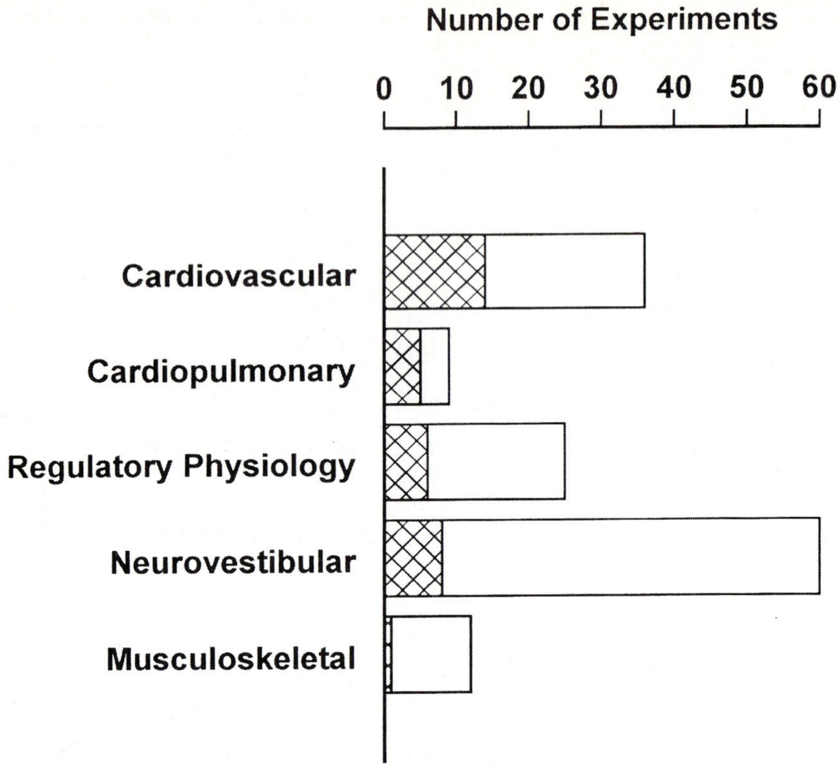

Fig. 3. Human life science experiments performed in each research category on STS missions 1981–1993. Experiments with European Principal Investigators are shown with cross-hatched shading.

experiments were in the form of 'carry-on' packages, requiring no spacecraft services. The more recent experiments, considered here (Bion 6 to 10), were U.S. battery operated instruments, which were integrated into the Russian spacecraft data systems. U.S. experiments were also performed on the rhesus monkeys and rats housed in the Russian animal habitat. The distribution of U.S. research experiments across the disciplines contrasts markedly with the ESA Bion experiments, which were predominantly concerned with dosimetry and cell biology. Concentration on regulatory physiology and musculo-skeletal studies in these missions appears to reflect NASA's primary interests.

Future space research in the life sciences at NASA will be directed increasingly towards exploiting the extended Mir space station following the addition of SPEKTR and PRIRODA modules in 1995. These have been designated by NASA as predominantly for life sciences research, as far as U.S. utilization is concerned. Planning for their outfitting by NASA is presently being carried out. Shuttle, Spacelab and Spacehab flights have been changed accordingly.

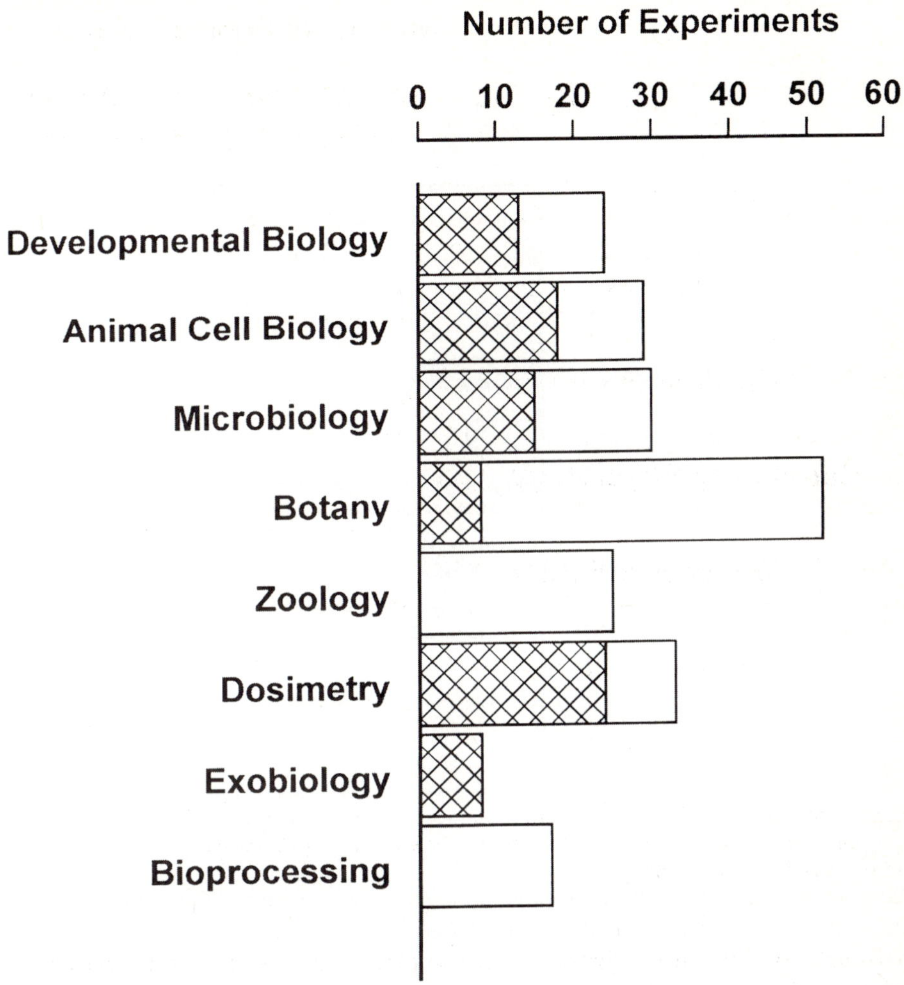

Fig. 4. Biological research experiments performed in each research category on STS missions 1981–1993. Experiments with European Principal Investigators are shown with cross-hatched shading.

Of the 4 independent Spacelab missions remaining in the NASA manifest, only the Neurolab (SLS-4) scheduled for 1997/8 offers a major life sciences mission. The emphasis will be on neurophysiology and it is expected that NASA will be able to make a significant contribution to the experiment hardware. The SL-M mission of May 1995 was the first NASA Shuttle/Spacelab mission to the Mir complex, where it was docked for 4 days. The principal emphasis of this mission was for logistics supply to Mir, but some science research was expected to be carried out with the Spacelab/Mir facilities. Subsequent missions to Mir carrying the Spacelab Long Module may have up to 30% of the resources allocated for scientific research. As yet this is no indica-

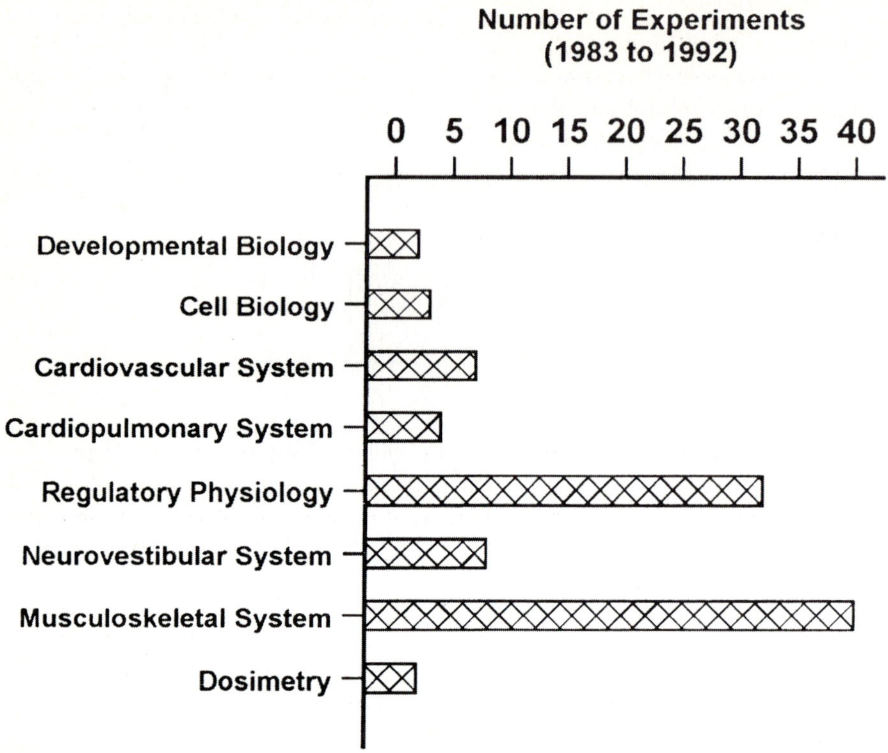

Fig. 5. U.S. life science experiments carried out on the Russian Biocosmos space craft (Bion 6 to 10).

tion as to the likely type of payloads. Due to the very tight schedule for these Shuttle missions to Mir in the 1995–1997 period, coupled with the effects due to the high inclination orbit complicating the rendezvous and limiting severely the launch window each day, there is presently some consideration being given to reducing the number of missions. A limit of 6 may finally result.

Planning for the extension and use of the Mir Space Station is now developing within NASA. The science programme is intended to be conducted in collaboration with Russian scientists. To assist that process, common E.S.-Russian databases and procedures were developed. The Russian crew will be trained to use the U.S. hardware and experiments, so that experimentation can be continued in the absence of U.S. crew. Where possible, NASA will utilize existing Russian Mir science hardware, e.g. plant growth chambers, but it progressively improve the monitoring capabilities, resources and resource utilization.

Over sixty items of experimental hardware have been identified by NASA to be provided to extend the Mir research capabilities. These include: hygiene, sanitation, and radiation monitoring equipment (water and air sampling de-

vices, tissue equivalent proportional counter); Metabolic experiments equipment (blood, urine, saliva collection kits, centrifuge); cardiovascular/ cardiopulmonary experiment equipment (barocuff, continuous blood pressure measurement device, metabolic gas analyzer); sensorimotor and neuromuscular function and exercise equipment; behaviour and performance measuring equipment; fundamental biology research equipment (fixative kits).
The general research programmes to be undertaken on Mir as its capabilities are upgraded are outlined in the following Sections.

8.15.1.1 Outline of U.S. research plans in the life sciences for the Mir extension

Biomedical studies of the consequences of long duration space flight. Studies of the basic mechanisms by which respiration and blood are controlled in healthy individuals on Earth and in microgravity. Studies of fundamental control mechanisms of balance and coordination. Studies of mechanism of maintaining muscle strength and contractility and mechanism of maintaining bone mass and strength. Capacity to perform skilled tasks in microgravity. Drug absorption and action in microgravity. Medical measures to prevent the undesirable effects of microgravity. Evaluate the absorption and utilization of food energy during prolonged exposure to microgravity.

Fundamental biology investigations. Study of plant life cycles, seed to seed to seed. Studies in developmental biology to evaluate how gravity affects organism from fertilization through maturity to aging. Cell structure to evaluate the specific effects of gravity on cells important to biomedical studies and biotechnology.

Metabolism studies. Standardization of protocols for joint experiments. Minimize the total and daily blood volume to be collected from the crew. Define the time and volume of blood to be collected during each flight phase based on the known physiologic changes that occur during the various phases of space flight. Reduce the crew time needed for conducting experiments on Mir (experiment on regulatory mechanisms, metabolic and energy maintenance, pharmacokinetic changes and immunology).

Cardiovascular and pulmonary system experiments. Development of a standard for comparability of hardware and software for data processing, analysis and presentation. US and Russian agreement on scientific content and procedures outlined in the experiment document. Identification of essential hardware for investigations. Specialist have agreed upon a list of in-flight hardware and ground equipment to support joint investigations (experiments: orthostatic tolerance, study of tolerance to physical, physiological responses during descent in the Shuttle).

Sensori-motor and neuromuscular function and exercise experiments. Science required to investigate sensorimotor function must include vestibular, visual-vestibular, and postural protocols to avoid misrepresentation of data and enhance understanding of human spatial orientation. Performance of neuromuscular protocols will add in a synergistic fashion to the sensorimotor protocols, reducing errors of interpretation. No in flight sensorimotor or neuromuscular investigations are currently possible without US-provided hardware (experiments : neuromuscular, sensorimotor - active and passive).

Hygiene, sanitation, and radiation monitoring. Focus on dynamic human-environment interactions during long term space flight, on characterization of the radiation environment, and on biological effects of radiation exposure. Studies will include joint analysis of samples of air and regenerated water, crew microbial biota, microbiota of spacecraft materials and hardware. Studies will be conducted on cytogenetic effects of radiation exposures and joint measurement of the spacecraft's internal radiation environment. Ground-based comparison studies will be conducted with US and Russian microbial and trace chemical air samplers and microbial samplers to facilitate interpretation of flight experiments. US and Russian scientists will work together to develop in flight procedures and equipment for microbial water monitoring (experiments: microbiology, radiological health, traces chemical contamination).

Behaviour and performance studies. Pre-flight psychological preparation and training crew members. In flight monitoring of crew member's psychological condition. In flight provision of psychological support to crew members (experiments: psychological dynamics and efficiency to countermeasures during extended space flight, effectiveness of manual control during simulation of flight tasks).

Fundamental biology research. Study of ontogenesis, including embryogenesis, and reproductive capability of Japanese quail during space flight. Study of complete ontogenic cycle of higher plants during space flight (experiment 'Seed to seed'). Study of photosynthesis of higher plants during space flight. U.S. and Russian representatives agree on the need for more post-flight studies of biological materials received on Earth following the investigations during Mir-Shuttle mission (experiments: incubator, greenhouse).

Biotechnology experiments. Electrophoresis experiments, using Russian hardware initially. Cell culture, using the U.S. developed Bioreactor. Gaseous nitrogen freezer equipment provided, for sample storage and return.
As mentioned earlier, NASA has reached an agreement with the RKA to take a major share of the resources offered by the two new modules SPEKTR and PRIRODA, and to dedicate these to the life sciences. The SPEKTR will accommodate about 1000 kg and PRIRODA 840 kg of useful payload for this purpose, primarily for human physiology experiments. Much of this equip-

ment will probably derive from that used earlier on the SLS-1 and SLS-2 Spacelab missions.

8.15.1.2 The U.S.–Russian co-operation agreement

According to this agreement, there will be three phases to the future developments. Phase 1 will start during 1995 when the two Russian experiment modules PRIRODA and SPEKTR will be launched and docked with the Mir-1 Space Station. NASA expects to use both of these modules, in cooperation with Russia, primarily for life sciences research. They will contain extensive crew exercise equipment and also accommodate EXPRESS racks, designed originally for the Space Station Freedom. These are intended for largely self contained experiments, located in standard drawer units. Phase 1 also covers the Shuttle missions to Mir, in the time frame from 1995 to 1997. During this phase U.S. crews will be on Mir during mission numbers Mir 18A (March – June 1995), Mir 20B (March – May 1996), Mir 21 (May – October 1996), Mir 22 (November 1996 – April 1997), with a further flight, Mir 23 (April – August 1997), under discussion. In addition, there will be an international ESA crew on flight Mir 19 (September 1995 – February 1996, Euromir 95) and French astronauts will be part of the Mir 21 crew (Cassiopea mission in 1996).

Phase 1 will involve extending the orbital lifetime of Mir up to at least the end of 1997. One of the first objectives of Phase 1 is therefore to replace or improve some of the essential sub-systems of Mir, such as the life support system, the solar power generator and the attitude control system. NASA will provide up to 10 Shuttle missions to Mir-1 during this period (to the end of 1997). Five of these missions are scheduled to involve the Spacelab, with the first mission in May 1995.

Phase 2 of the U.S.-Russian co-operation agreement, will involve the assembly of a new Core Station, using a new upgraded Mir module, with improved power and control systems, and data handling facilities. Russia will first provide an Energy 'Block', serving for initial station attitude control and reboost, together with a Service Module, similar to the Mir-1 pressurized module and containing basic housekeeping functions, life support, power and thermal control, and performing limited payload operations. Further flights will provide docking nodes (as seen in the Mir-1 configuration) as well as a solar array, including the truss structures. Finally, Phase 2 developments will be completed with the launch of the US Laboratory and its outfitting in the period 1997/1998. Phase 2 will be operated on a 'man tended' basis.

Phase 3, covering 1998 – 2002, will see the completion of the Space Station, followed by the progressive addition of elements derived from the international partners, Canada, Japan and ESA and the U.S. SpaceHab module, to provide a permanently inhabited Laboratory system, around 2002.

At the present time, the US/Russian planning for the utilization of the Station through Phase 2 and up to the addition of contributions from the other Partners, foresees only a US/Russian activity. This situation may well change however, as far as ESA is concerned. There are offers from NASA for ESA to contribute to the NASA outfitting programme, and to derive experimentation opportunities in return, during this period. The NASA approach to experiment hardware for the Station is seen to be undergoing a significant change. The original concept of facility structure, occupying double racks, is being replaced with a more flexible approach which is based upon the apparent success of the Mid-deck Locker type of independent experiment. The EXPRESS racks to be fitted into SPEKTR and PRIRODA are evidence of this change.

8.15.1.3 NASA life science research facilities for the International Space Station

Three major life science facilities are currently foreseen for the outfitting of the U.S. laboratory on the Station. These are the Centrifuge Facility, the Human Research Facility – for which the Johnson Space Center is now responsible, and the Gravitational Biology Facility, which is supported by the Ames Research Center.

The Gravitational Biology Facility will occupy two Station racks and contains the modules and items listed in Table 17. The purpose of this facility is to support the growth and development of a variety of specimens, including cell and tissue cultures, higher plants, insects, fish, avian embryos, and small mammals. It is also intended to allow basic specimen manipulation processes and analyses. The first of the two racks, containing the Bioculture system, plant research unit, and insect habitat, together with the temperature controlled chamber and slow freeze accessories, is intended for a 1998 launch to the Station. The second rack, containing the rest of the equipment would be on Station in 2000.

The Human Research Facility (HRF) has three basic objectives: to develop countermeasure procedures and protocols to maintain crew health and performance in space, to provide a basic research capability which is focused on earth based health problems, and to develop the capability to integrate human experiment data sets, from past, current, and future programmes.

The facility is likely to contain the following basic items: (i) automatic blood pressure system; (ii) core body temperature measurement system; (iii) experiment control computer system; (iv) echocardiographic/doppler blood flow; (v) carotid sinus baroreceptor stimulator.

The three phase programme for the development and deployment of the facility equipment (see Section 8.15.1.2) is correlated with the US/Russian flight schedule. During Phase 1, hardware will be located in the SPEKTR and PRIRODA modules, and also involve the STS-82 flight in 1996. Much of the scientific effort during this period will be directed towards the establishment of basic data sets relating to long term (90 day) space missions and the correlation with existing Russian data. Countermeasure procedures evaluation and the testing of prototype hardware will also take place.

Table 17. Components of the Gravitational Biology Facility planned for the International Space Station.

Integral to GBF development		Provided by other programmes	
Habitats	Multi-disciplinary items	Habitats	Multi-disciplinary items
Bio-culture system Plant research unit Insect habitat Aquatic habitat Avian egg incubator Rodent birthing & rearing habitat	Temperature controlled chamber Dissection equipment Slow freeze accessories Temperature controlled laboratory centrifuge Compound microscope	Large plant habitat (CFP)* Rodent breeding habitat (CFP) Rodent habitat (CFP)	Passive dosimeter (SS)* Quick/snap freezer (SS) –196° C freezer (SS) Life sciences glovebox (CFP) Dissecting microscope (SS) Fluid handling tools (SS) –70° C freezer (SS) Refrigerator (4° C) (SS) –20° C freezer (SS) Freeze dryer (SS) Mass measuring device (SS)

*CFP = Centrifuge Facility Project, SS = Space Station Programme.

During Phase 2, the objectives will be to continue Phase 1 efforts to develop science risk assessment protocols and countermeasure hardware and prescriptions for extended duration missions. Gather and analyze science and hardware data for use in future HRF phases and life sciences programmes. Assess Shuttle/Mir science and hardware data to enhance subsequent HRF phases. Establish acceptable levels of crew health and performance. Develop operational hardware that will occupy 0.8 rack in the US Lab.

The equipment rack to support Phase 2 will be launched in 1998, and will be accommodated in the U.S. Lab. Phase 3 of the HRF programme will be concerned with the development of countermeasures for missions of 180 days, or greater, and with a programme of basic human life sciences research. The hardware will occupy 2 Station racks, and be located in the U.S. Laboratory during 2000.

The Centrifuge Facility provides a variable gravity environment (0.01 to 2 × g) for research with plants and animals in the Station. The limit to animal size would be a small primate.

The facility will include: centrifuge system; habitat holding system (1–2 racks); rodent and plant habitats; life sciences glovebox; specimen service system; rodent transporter.

The requirements placed upon the design of the centrifuge are to support a variety of experiments and specimen types for weeks to months (including multiple specimen types simultaneously with flexible accommodation growth; near

Table 18. Current launch manifest for NASA life sciences facilities.

Science discipline	Payload	1998	1999	2000	2001	2002	2003	2004
Life and biomedical sciences and applications	Gravitational biology facility	1*	1	2*	2	2	2	2
	Human research facility	0.8*	0.8	2*	2	2	2	2
	Habitat holding system						1*	1
	Centrifuge rotor							Launch after 2004
	Life sciences glovebox					1*	1	1
	Service system					0.5*	0.5	0.5
	Stowage			0.5	0.5	0.2*	0.2	0.2
Cumulative total of racks on station		1.8	1.8	4.5	4.5	6.7	6.7	6.7

The figures show the number of Space Station racks, those with an asterisk indicate when the equipment items are to be launched for deployment in the Station.

identical conditions from microgravity to 2 × g, with multiple gravitational levels simultaneously and precisely controllable environments; access to specimens while maintaining bioisolation and biocontainment, able to accommodate hazardous materials and isolate disturbances from the Space Station.

The accommodation requirements for experimental organisms which are to be met by the facilities are as follows.

For *mice and rats:* group or single housing; temperature range 18 to 27°C (to within 1°); relative humidity range 40 to 70%; controllable light cycle; light and dark cycle video; food and water available continuously.

For *plants*: one large (1300 cm^2) or three small growing area(s) per habitat; growing height to 35 cm; temperature range 15 to 30°C (to within 1°); relative humidity range 40 to 80% (to within 10%); CO_2 controlled within the range 300 to 2000 ppm with an accuracy of 10 to 50 ppm; nutrient solution to have controllable pH and conductivity; controllable light cycle levels with light and dark cycle video.

The Holding Unit and the Habitats are expected to be launched in 2002, with the Centrifuge itself following in 2004. The current launch schedule for the Life Science Facilities is shown in Tables 18 and 19.

Table 19. NASA Spacelab missions, with life sciences and materials sciences payloads.

Missions	1983	1984	1985	1986	1987	1988	1989	1990	1991	1992	1993	1994	1995	1996	1997	
SL-1 (9 days)	▲	Neurovestibular, Blood Studies, Space Biology, Materials and Fluid Sciences														
SL-3 (7 days) 2			▲	Metabolic, Neurovestibular, Hardware Verification, Materials and Fluid Sciences												
SL-2 (8 days)			▲	Musculoskeletal, Plant Biology												
SL-D1 (7 days)			▲	Cardiovascular, Neurovestibular, Space Biology, Plant Biology, Material and Fluid Sciences												
SLS-1* (9 days)									▲	Cardiovascular/Cardiopulmonary, Metabolic, Musculoskeletal, Neurovestibular, Medical Technolo gy, Combustion						
IML-1 (7 days)										▲	Neurovestibular, Human Performance, Radiation, Cell Biology, Plant Biology, Materials and Fluid Sciences, Critical Point Studies, Protein Crystal Growth					
USML-1** (13 days)										▲	Materials and Fluid Sciences, Protein Crystal Growth, Combustion					
SL-J (7 days)										▲	Cardiovascular Countermeasures , Gravitational Biology, Materials and Fluid Sciences, Protein, Crystal Growth					
SL-D2 (10 days)										▲	Cardiovascular/ Fluids, Cell Fusion, Materials and Fluid Sciences					
SLS-2* (13 days)											▲	Cardiovascular, Metabolic, Musculoskeletal , Neurovestibular				
IML-2 (13 days)												▲	Performance, Space Biology, Protein Crystal Growth, Electrophoresis, Vibration Isolation Studies, Materials and Fluid Sciences			
USML-2** (16 days)													▲	Materials and Fluid Sciences Protein Crystal Growth		
MSL-1															▲	Experiment Selection
SLS-4*															▲	(Anthrorack, Biorack)? (Rhesus Facility) Neurophysiology

*Life Sciences Dedicated Missions; ** Material Sciences Dedicated Missions

8.15.2 Canada

The research programme in the space life sciences in Canada effectively started in 1984, with the emphasis initially on neuro-vestibular physiology. Canadian life sciences experimenters participated in a series of Spacelab missions (SL-1, D-1, D-2, SLS-1, SLS-2, IML-1), and a Canadian life sciences payload specialist, Dr R. Bondar, flew on the IML-1 mission.

Six Canadian experiments, in physiology, were performed on IML-1, covering the space adaptation syndrome (Dr D. Watt, McGill Univ.), positional and spontaneous nystagmus (Mr. J. McClure), energy expenditure in space flight (Dr H. Parsons, Univ. of Calgary), measurement of venous compliance (Dr B. Thirsk, CSA), back pain in astronauts (Dr P. Wing, UBC) and phase positioning experiment (Dr D. Brooks, UBC). Dr Watt, Dr Wing, and Dr Parsons, were similarly involved in experiments on STS-52 in October 1992. Canadian participation in other Shuttle and Spacelab flights are given within Tables 13 and 14.

Collaboration with the USSR began in 1987 and has continued, with collaborations on Russian satellite missions and an experiment set carried out on Mir. The 14 day Bion-10 mission included two Canadian experiments. One, by Dr R. Wassersug, was to investigate the development of the lungs and vestibular organs of amphibians in microgravity. Dr H. Ing carried out experiments on neutron radiation effects.

The Canadian programme in life sciences research currently supports work in the principal areas: *human physiology*, covering neurophysiology, cardiovascular physiology, metabolism and endocrinology, and musculoskeletal physiology; *space biology*, including gravitational biology, developmental biology, and physiology; *radiation*, which covers radiation (dosage) measurements, the effects on humans, and radiation biology.

Some 20 groups are currently supported for the various flight experiments, involving about 75 scientists, in total. Interest in continuing experimentation is high, with some 65 proposed experiments awaiting flight opportunities.

Future research activities will involve missions in the Spacelab, in the Mir space station, and the international space station. IML-2 had a Canadian experiment on board entitled *Spinal Changes in Microgravity* (Dr J. Ledsome, of UBC), to study changes in the nervous system response and the spinal changes during spaceflight. A *Torso Rotation Experiment*, intended to study vestibular-eye-neck coordination, (Dr D. Watt, of McGill) is intended to fly on the Shuttle as a Mid-deck Secondary Payload, during 1994. An *Aquatic Research Facility* is being constructed in Montreal, as a joint CSA/NASA project. It will allow the study of samples within a 100 ml water volume, under microgravity, and within a $1 \times g$ control volume. The first flight of the ARF is scheduled for 1995, with annual flights thereafter. The Principal Investigators for the first flight are: Dr B. Crawford (UBC), studying *The Effect of Microgravity on Early Starfish Morphogenesis* and Dr R. O'Dor (Dalhousie) *The Role of Gravity in Feeding and Development of Bivalve Larvae.*

Neurolab, intended for a 1997 launch, will probably have one or more Canadian experiments included. 12 Proposals were submitted for review. Russian missions and cooperations will continue. Radiation studies on Mir are planned to continue (Dr H. Ing and Dr I. Thompson), and studies on metabolism (Dr H. Parsons) with Russian experimenters are intended for Mir missions. Dr B. Glickman (Univ. of Victoria) will continue studies on the mutations that occur in cosmonauts after spaceflight.

8.15.3 Russia

The political and financial problems of the Russian Commonwealth of Independent States in recent times appears to have caused a slowing down of the space life sciences research programmes. Despite this, space experiments in this field continue to be undertaken, many now involving collaborative ventures with other European countries, and also with America.

In addition to the RKA, the Russian Space Agency, there are several space enterprises which are acting as commercial companies and arrange for spaceflights involving the Mir space station and also specialized satellites.

An agreement between the RKA and ESA was made in October 1992, which included microgravity research activities and which committed ESA to two missions to Mir by European astronauts. These were planned to occupy 1 month in 1994, and 4 to 5 months in 1995, and will permit the use of Mir facilities. Arrangements for flights of this type have also been made by several ESA Member States (e.g. France, Germany and Austria), and have included substantial programmes of life sciences research. A summary of these flights, and associated life sciences programmes, is given in Table 15.

In addition to these programmes involving the Mir space station, which have a emphasis on human physiology, ESA and individual European countries have participated in the Bion series of recoverable satellites, in order to carry out space biological research. The Bion Programme is based upon the Foton spacecraft system, the first of which was launched in 1985, and was used on that occasion for experiments in materials science. The spacecraft has three distinct units: the power supply module, the instrument module,and the descent vehicle containing the payload. The latter is separated just prior to re-entry, and is equipped for an eventual soft landing.

For the Bion programme, the descent capsule has a pressurized unit, and is designed to allow for late access. In addition to this pressurized unit, special containers are attached to the outside of the descent vehicle to provide the opportunity for experimentation under outer space conditions, to support radiation and exobiology studies.

Basic characteristics of the Bion system are: an orbit of 220 to 300 km with an inclination of 62.8°; payload mass of 700 kg with a volume of 4.7 m^3; temperature range 15 to 28° C; pressure, 730 to 770 mm Hg; pO_2, 140 to 180 mm Hg; pCO_2, less than 2 mm Hg; humidity 60%. The vehicle provides experiments with an average power supply of 400 W, the maximum being 700 W for 1.5 h d^{-1}. Microgravity duration is 14 to 16 days, the quality being better than $10^{-4} \times g$. Late access is available up to 8 to 10 h before launch, and with early retrieval the experimental material is accessible 12 h after landing.

The NIKA-T is a new re-entry satellite , with 1995 as the original planned commissioning date. The mission duration is intended to be 3 to 4 months. It will provide a capsule specifically designed for medical and space biology experiments, with considerable improvements on the Bion design. Payload mass is expected to be some 1,200 kg, and the power supply capability is of the order of 4 to 5 kW.

The Bion Scientific Programme has been operated by the Institute for Bio-Medical Problems (IBMP), Moscow. The Bion 8 mission, of September 1987, contained three autonomous experiments supported by ESA, all of which were concerned with the effects of cosmic radiation and microgravity. The Bion 9 mission, of September, 9 1989, contained 5 ESA supported experiments, and the Bion 10 mission of Dec. 1992, had eight ESA experiments and involved some 50 investigators. As in the previous missions, the Bion 10 experiments involved a considerable amount of collaboration with Russian experimenters, many from the IBMP. Details of these experiments are provided in Table 16. This series of Bion missions also contained monkeys as subjects for research into the physiological changes and adaptations to weightlessness during the early phases of space flight.

The Mir Space Station complex consists of the basic module, which has two axial and four side docking units, a scientific module, and the manned and unmanned transport spacecraft. The actual configuration is modified as required to support the emphasis of ongoing programmes by the introduction of programme dedicated flight modules. A large module was originally intended for biological and medical research programmes. Called the Medilab, this had been under design for several years, and was originally intended for a 1995 launch. It now looks highly unlikely that it will be developed.

Within the basic module of Mir are provisions for the working and the relaxation of crew, together with the control of Station elements, power, and data transmission systems. The RKA also has within Mir various facilities for microgravity research which are made available to users on a commercial basis, for example, the units Biokryst and Ainur, used for crystallization of proteins, the Ruchei electrophoresis apparatus, the Rekomb reactor for cell hybridization, and the microcultivator Vita.

Well over 80 studies in space biology have been performed in the Mir station over the past 3 years. The Svetoblock-M, Magnetogravistat, Vazon, and other equipment have been used for studies on the growth and development of higher plants, involving some 20 species. The Bioterm-10 and Rost-4M apparatus were used to study changes in the development of plant tissue cultures, exposed to open space conditions. The equipment designated Inkubator 2, Ovanzhereya, and Svet, were used for the study of modifications to the embryonic and post-embryonic development of birds. Human physiology and medical experiment have also been a major feature of the Mir research programmes. Planning for the progressive extension of Mir capabilities and the development of an International Space Station is proceeding now (see Section 8.15.1.2).

The Russian life sciences programme for the future International Space Station has only been defined in rough outline. It is clear however, that it is heavily biased towards the practical aspects of manned space flight, together with more basic research programmes which are intended to contribute towards Earth based research, often of an applied nature. In *space medicine*, the emphasis is upon establishing and maintaining optimal crew health, following the adaptation process, determining the most effective residence period in space, and exploring the results of extended space exposure, comparable with the expected duration of a manned Mars mission. Equipment will include an ultrasound cardiograph, 'biopotentials' indicator, centrifuge, cryogenic (O_2) refrigerator, cardiac recorder, heat and mass measuring devices.

The *biology* research programme will extend from that performed with the Salyut and Mir Space Stations. No details have been given, merely a statement that it will cover the subcellular through to organism range. There is, however, a strong emphasis on research connected to life support systems, biotechnology, and applied biology. The equipment foreseen is primarily a greenhouse, with a growth/lighting section, and control unit, and a centrifuge, together with microscope, gas analyzer, and sample preparation equipment. *Biotechnology* covers the biological dynamics of cells in cultivation, the mechanisms of cell fusion, separation and purification of genetically engineered virus proteins and their extraction by electrophoresis and other methods. Experimental and pilot production of biological materials will be undertaken, including pharmaceutical, diagnostic, and prophylactic preparations, hybrid cells and micro-organisms, together with microcrystals, biopolymers, and viruses. The equipment will include the Ruchei-2M electrophoresis units, a Recomb-2 unit, the Luch-2M biocrystallization unit, the Biocryst-2M crystallization unit, Bioproduct analysis system. The total mass of equipment for biotechnology, accommodated in habitation modules, is about 2,500 kg, and the consumables throughput is estimated at 4,500 kg y^{-1}.

8.15.4 Japan

A substantial space programme is being progressively established by Japan, through its agency NASDA. Research in the life sciences is considered an important element, with some emphasis on the human physiology aspects being evident, due to the interest in manned space activities. The other areas of microgravity life science research are also being supported however, as indicated by the complement of Japanese experiments on the joint NASA-NASDA Spacelab-J (STS-47) mission, which took place in September 1992. These are listed below.

Human physiology. Health monitoring of the Japanese payload specialist, monitoring of heart rate, blood pressure, and respiration rate; and blood (principal investigator Dr Sekiguchi, NASDA) and urine sampling to study endocrine and metabolic changes in the payload specialist (principal investigator Dr N. Matsui, Nagoya University).

Neurophysiology. Comparative measurement of visual stability on Earth and in microgravity, eye, head and neck movements and their relationship to spatial coordination (principal investigator Dr K. Koga, Nagoya University); research and perceptual motor functions in microgravity, tracking and handling operations studied to explore potential future improvements by automation (principal investigator Dr A. Tada, National Aerospace Laboratory); neurophysiological study of visuo–vestibular control of posture and movement in fish during adaptation to microgravity, response of fish with the gravity sensing organs removed are compared to the behaviour of normal fish, to contribute to the study of the conflict theory of space motion sickness (principal investigator Dr S. Mori, Nagoya University).

Cell biology. Effects of microgravity on the ultrastructure and functions of cultured mammalian cells, investigation of the changes in internal organisation and enzyme production of cultured cells in microgravity (principal investigator Dr A. Sato, Tokyo Medical and Dental Univ.); study of the effects of microgravity on cell growth of human antibody producing cells and their secretions, changes in growth rate, antibody producing capability, and morphology of immune hybridoma cells (principal investigator Dr H. Ooka, Tokyo Medical and Dental Hospital); organ differentiation from cultured plant cells under microgravity, effects on cell division, cell differentiation, and the growth of plant cells (principal investigator Dr Y. Yamada, Kyoto University); circadian rhythm of conidiation in *Neurospora crassa*, effect of microgravity on the rhythm of this mould in order to determine if this is endogenous or due to geophysical factors (principal investigator Dr Y. Miyoshi, Tokyo University).

Developmental biology. Effects of microgravity on calcium metabolism and bone formation, investigation of bone formation in chicken embryos (principal investigator Dr T. Suda, Showa University).

Radiation biology. Genetic effects of HZE and cosmic radiation, study of effects on *Drosophila melanogaster* (fruit flies) (principal investigator Dr M. Ikenaga, Kyoto University); study of the biological effect of cosmic radiation and the development of radiation protection technology. Radiation track detectors and biological specimen, to study the effects of HZE particles (principal investigator Dr S. Nagaoka, NASDA).

Biotechnology. Separation of biogenic materials by electrophoresis, free flow electrophoresis to separate and purify a sample of mixed proteins (principal investigator Dr M. Kuroda, Osaka University); separation of animal cells and cellular organelle by free flow electrophoresis, study of the effectiveness of electrophoresis to obtain a pure strain of bacteria (principal investigator Dr T. Yamaguchi, Tokyo Medical and Dental Inst.); crystal growth of enzymes, protein crystal growth for structural analysis (principal investigator Dr Y. Morita, Kyoto University).

Japanese life scientists were also involved with experiments on the Spacelab SLS-2 mission, in October 1993. Three post flight experiments were carried out, covering the study of bone metabolism in microgravity, histochemical and biochemical studies of heart muscle changes in rats, and the effects on - adrenoreceptors in rat hind limb muscles.

Collaborative programmes with NASA have continued to be a feature of Japanese planning. The IML-2 Spacelab mission contained Japanese experiments. These were principally materials science experiments, but included studies of calcium metabolism, motion sickness, radiation experiments on biological samples, chromosome separation experiments, and protein crystallization studies.

Planning for participation in the SLS-4 (Neurolab) mission, in 1998, is now under way. Some 21 life science experiments proposals have been submitted from Japan. Of these 3 are concerned with behaviour and cognition, 5 with cellular and neurobiology, 3 with developmental neurobiology, 4 with homeostasis of the nervous system, and 6 with the sensory and motor system.

Small experiment payloads, destined for the Mid-Deck locker location in the Shuttle, are also being planned. They cover the general fields of space biology, physiology, and countermeasures. Small short term experiments are also performed using a DAS MU-300 aircraft for parabolic flights.

The JEM pressurized module will be a major contribution by Japan to the International Space Station. Due to be in place by the year 2002, it will open up a major opportunity to expand the Japanese space life sciences programme.

In the meantime, Japan is evidently moving to establish a considerable degree of autonomy for its overall space programme in the future. This is based upon a range of Japanese launchers, the largest of which, H-II, will be ready for service in 1994. This vehicle will have a payload capability similar to that of the Ariane IV (9 ton into low earth orbit, 2 ton into geosynchronous orbit). The intention has been to develop an unmanned space plane, the HOPE, to be launched into orbit by an uprated H-II, in order to provide logistics support to a manned space station element, such as the JEM. It is not clear at present if that plan will continue.

The first flight of the so-called Space Flyer Unit is due to occur in 1995. This is a re-usable, free flying platform which is designed to operate autonomously in space over a period of months. The launch will take place using the H-II rocket; with recovery by the Shuttle. Like the ESA Eureca platform, the SFU can be used for longer term biology experimentation. The first flight will include automated experiments on chromosome separation and protein crystallization.

A specific programme of 3 missions dedicated to space biology is planned for the period 1996 through 1998. This programme will be based upon the

Japanese J–1 three stage rocket, which has a payload capacity of some 800 kg, into low earth orbit. The intention is to develop a recoverable biology payload unit, the J–Lifesat, somewhat along the lines of the Russian Bion space modules.

8.15.4.1 LIFESAT missions

Each of the three LIFESAT missions will be focused on one scientific goal. Reference experiments have been outlined within that framework and these may be summarized as follows.

Mission 1. Study of the biological effects of space radiation relevant to JEM crew. The experimental objectives are broadly to collect cosmic radiation environment data in low earth orbit, both inside and outside the vehicle, conducting long term exposure of a mammal (mouse) to obtain data on the chromosomal change and tumour formation. The orbit altitude will be 300 to 400 km, the mission will last up to 60 days and the experiment will feature real time radiation monitoring. Carcinogen will be administered before flight to raise the cancer rate and in flight tumor induction at the cellular level, chromosomal changes and damage and induction of somatic mutation will all be monitored. The experimental animals will be 20 male or female mice. Ground controls will feature identical animals given controlled exposure to gamma–rays and X–rays and there will be simulation of launch vibration for the tumour induction test group. Postflight analysis will include studies of the cellular functions, chromosomal mutations and aberrations, pathogenic anatomical and molecular biological analysis of tumours, haemocytometer analysis, and mineral bone density measurements.

Mission 2. Study of bone, and muscle activity in microgravity. Experiment objectives in this case are study of calcium desorption or demineralization in mouse bone, post–flight mineral recovery, bone formation, muscle atrophy, and otolith deformation. Again 20 male or female mice will be used and some will be subjected to pre–flight drug administration to study pharmacological countermeasures for osteoporosis prevention. The mission period will be up to 60 days with full animal life support and physiological data (heart rate, temperature, electromyogram) acquisition by telemetry. Ground controls will include immobilization of hind limb muscles. Postflight analysis of the bones will include bone stroma proteins, calcium balance, dynamic tests, morphology, biochemical, bone mineral density, amino acid and mineral contents. Muscle analysis will cover the stretching and soleus muscle characterization together with muscle protein characterization. The studies will also include blood mineral analysis, and urea/faeces analysis, and a detailed study of the morphology and biochemistry of the vestibular apparatus.

Mission 3. Pathological physiology in microgravity. Experiment objectives will be to obtain data on the immunological and haematopoietic functions, drug metabolism, infection and care for health maintenance in space. The flight details will be the same as the previous two missions; duration will be similar. 20 male/female mice will be used, with heart rate and temperature monitoring. Photographic records will be made of colour (still) images at the rate of 1 frame h^{-1}. Postflight analysis will examine immunological functions in depth, including the pathological test of immunological organs, functional tests of lymphocytes and the study of bacterial infection, including the intestinal bacteria. Haemostatic studies, coagulation and fibrinolytic factor determinations, and the analysis of haematopoietic functions will be carried out.

Apart from the Lifesat programme, the major effort in the Japanese long term plan for experimentation in the life sciences lies in collaborations on future Spacelab flights with NASA (especially IML-2), leading to the introduction of the JEM onto the Space Station.

8.16 ESA space life sciences programme

ESA experiments are defined as experiments solicited in the member states through ESA Announcements of Opportunity and, when recommended for flight by ESA science peer groups, are funded by their respective national authorities. ESA provides then for the mission opportunity and larger research facilities which can be used by more than one investigator (multiuser facilities), or for different research topics (multipurpose facilities).

Initially, the ESA Life Sciences experiments were largely exploratory in nature, as is to be expected with a new research activity. Progressively the move has been towards the establishment of reliable data sets on the different phenomena that have been observed. More recently this emphasis has been changing again, as interest focuses on in depth studies of the different phenomena. It is the recognition of that change which has prompted ESA to initiate this examination of the world-wide trends in the microgravity life sciences research. The intention is to try to define more clearly the future subjects for study, in order to concentrate more effectively the limited resources which ESA has at its disposal. The development of the ESA life sciences programme can be seen in Tables 20 and 21 which cover the unmanned, principally biology, research missions and then the manned missions, which include also the human physiology research activities.

Details of the individual experiments which have been flown can be found in Tables 16 and 22 which, respectively, list the joint Russian/European Bion, Biocosmos and Foton series of flights, and the European experiments flown on the Shuttle.

Table 20. Development of ESA's unmanned life sciences experiments.

Mission	Date	Facility	Research topics
Biocosmos 8 (USSR-ESA)	1987 (5 days)	Biorack type 1 container	Radiation biology Dosimetry
Biocosmos 9 (USSR-ESA)	1989 (10 days)	Biorack type 1 container	Cell developmental biology Radiation biology
Foton (Russia-ESA)	1992 (11 days)	Biopan	Test flight: minimum shielding
EURECA (ESA)	1992-1993 (9 months)	ERA	Radiation biology, Dosimetry Solar UV radiation Survival of primitive life forms
Biocosmos 10 (Russia-ESA)	1992-1993 (12 days)	Biobox	Cell molecular biology Calcium metabolism Radiation biology Dosimetry
Foton (ESA)	1993 (14 days)	Biopan	8Exobiology Cell-biology Radiation biology
Foton (ESA)	1994 (14 days)	Biobox	Cell molecular biology Developmental biology Radiation effects on metabolism
Foton (ESA)	1996-1997 (28 days)	Biobox Biopan	To be decided To be decided

The formal ESA Microgravity Research Programme (EMIR) is now defined within a framework of political, financial, and general research objectives. The EMIR Programme requires that, in the planning equal opportunity shall be given to the life sciences and to the materials/fluid physics sciences; with the general guideline that scientific quality shall be the final determinant when setting priorities. As far as the Life Sciences are concerned, the programme contains the general provisions outlined in the following Sections.

8.16.1 Recent and ESA life sciences experiments

IML–2. This 8 day Spacelab mission in 1994 had 18 ESA life science experiments on board (Table 23), Biorack being flown to support these, and other, experiments.

Euromir Missions. The ESA experiments for the Euromir Missions are predominantly in the life sciences – in large part due to the return capability for samples being restricted to 10 kg. Table 24 lists the experiments selected for the Euromir 94 and 95 missions.

Foton 10: the ESA Biopan and Biobox. The Biopan is a small multi–user experiment facility designed to support research into the combined effects of cosmic

Table 21. Development of ESA's life sciences programme on manned laboratories.

Mission (partners)	Date (duration)	Facility	Research topics
Spacelab SL-1 (NASA/ESA)	1983 (9 days)	Sled/BRS	Neuro-vestibular physiology Cardiovascular physiology Cell biology, Dosimetry
Spacelab D-1 (Germany/ESA)	1985 (10 days)	Biorack Sled	Cell/molecular/developmental biology, Radiation biology Neurovestibular physiology
Spacelab SLS-1 (NASA/ESA)	1991 (10 days)	Incubator	Cell biology/Immunology
Spacelab IML-1 (NASA-ESA-CNES-CSA-NASDA-DARA)	1992 (9 days)	Biorack	Cell, molecular & developmental biology
MIR-92 (Germany-ESA-Russia)	1992 (7 days)	Vesta	Oculo-vestibular interaction
Spacelab D-2 (DARA-ESA-NASA)	1993 (9 days)	Anthrorack	Central/peripheral haemodynamics; cardio-pulmonary physiology; fluid volume regulation; metabolic processes
Spacelab IML-2 (NASA-ESA-NASDA-CNES-DARA-CSA)	1994 (14 days)	Biorack	Cell molecular biology Radiation biology Developmental processes
EUROMIR 94 (ESA)	1994 (30 days)	Freezer, Centrifuge	Physiology, biology, radiation
EUROMIR 95	1995 (135 days)	Bone densitometer, VOG, RMS II	Pulmonary, cardiovascular, musculo-skeletal, neuro-vestibular
Spacelab LMS-1 (NASA/ESA)	1996 (16 days)	Torque Velocity Dynamometer	Musculo-skeletal
Spacelab/Mir	1996 (3 missions)	Biorack	Biology, radiation
Spacelab Neurolab (NASA-ESA-CNES-DARA-CSA-NASDA)	1998 (16 days)	BRS-Rotator Anthrorack-derived hardware EDEN	Neurological/behavioural

radiation, solar U.V., space vacuum, and microgravity on biological samples. It therefore serves the interests of exobiology, radiation biology, radiation physics, and dosimetry. Attached to the outside of the Foton spacecraft, the lid of the pan-shaped container is opened in space. Closure is effected prior to re-entry. Biopan first flew in October 1992, on a technical evaluation. It was originally scheduled for the first scientific mission in December, 1993, but was delayed, the Biopan-2 flight being rescheduled to September 1994 (but eventually launched on 16 February 1995). Several experiments from the first flight will be reflown on Biopan-2. Experiments recommended for Biopan-2 were as follows: *Survival of organic molecules under space conditions. Catalysis of reactions by solar UV* (A. Brack); *Radiation (HZE) induced base modifications in the cell*

Table 22. Experiments in biology and human life sciences (physiology and medicine) on the space shuttle (1981-1993) with European Principal Investigators.

Mission	Flight	Launched	Title	Species used	Investigator	Affiliation	Country
SPAS-01	STS-7	18 Jun 83	Radiobiology mapping in Biostack.	Unspecified plant	M. Buchvald	German Youth Fair Programme (Jufo)	D
SPAS-01	STS-7	18 Jun 83	Plant contamination by heavy metals (Cd transport).	Watercress	H. Katzenmeier	German Youth Fair Programme (Jufo)	D
SL-1	STS-9	28 Nov 83	Radiobiology mapping: advanced biostack experiment.	Brine shrimp, *Artemia salina*	H. Buecker	DLR Cologne (DFVLR)	D
SL-1	STS-9	28 Nov 83	Radiobiology mapping: advanced biostack experiment.	*Bacillus subtilis*	H. Buecker	DLR Cologne (DFVLR)	D
SL-1	STS-9	28 Nov 83	Radiobiology mapping: advanced biostack experiment.	Tobacco, *Nicotiana tabacum*	H. Buecker	DLR Cologne (DFVLR)	D
SL-1	STS-9	28 Nov 83	Radiobiology mapping: advanced biostack experiment.	Mouse-ear cress, *Arabidopsis thaliana*	H. Buecker	DLR Cologne (DFVLR)	D
SL-1	STS-9	28 Nov 83	Radiobiology mapping: advanced biostack experiment.	Fungus spores (*Sordaria fimicola*)	H. Buecker	DLR Cologne (DFVLR)	D
SL-1	STS-9	28 Nov 83	Microorganisms in hard space environment	*Bacillus subtilis*, spores	G. Horneck	DLR Cologne (DFVLR)	D
SL-1	STS-9	28 Nov 83	Lymphocyte proliferation in weightlessness	Human	A. Cogoli	ETH Zürich	CH
LDEF-1	STS-41C	06 Apr 84	Radiobiology in the free-flyer Biostack experiment	Brine shrimp	H. Buecker	DLR Cologne (DFVLR)	D
LDEF-1	STS-41C	06 Apr 84	Radiobiology in the free-flyer Biostack experiment	*Bacillus subtilis*, spores	H. Buecker	DLR Cologne (DFVLR)	D
LDEF-1	STS-41C	06 Apr 84	Radiobiology in the free-flyer Biostack experiment	Tobacco	H. Buecker	DLR Cologne (DFVLR)	D

LDEF-1	STS-41C	06 Apr 84	Radiobiology in the free-flyer Biostack experiment	*Arabidopsis thaliana*	H. Buecker	DLR Cologne (DFVLR)	D
LDEF-1	STS-41C	06 Apr 84	Radiobiology in the free-flyer Biostack experiment	Maize (*Zea mays*)	H. Buecker	DLR Cologne (DFVLR)	D
LDEF-1	STS-41C	06 Apr 84	Radiobiology mapping: advanced biostack experiment.	Fungus spores (*Sordaria fimicola*)	H. Buecker	DLR Cologne (DFVLR)	D
SL-D1	STS-61A	30 Oct 85	Microgravity and genetic recombination	*Escherichia coli*	O. Cifferi	University Pavia	I
SL-D1	STS-61A	30 Oct 85	Microgravity and genetic recombination.	Bacteriophage K1	O. Cifferi	University Pavia	I
SL-D1	STS-61A	30 Oct 85	Fruit fly embryogenesis and life span.	Fruit fly, *Drosophila melanogaster*	R. Marco	Universidad Autonoma Madrid	E
SL-D1	STS-61A	30 Oct 85	Endogenous rhythms in contraction and protoplasmic streaming.	Slime mould, *Physarum polycephalum*	V. Sobick	DLR Cologne (DFVLR)	D
SL-D1	STS-61A	30 Oct 85	Radiobiology of embryogenesis and organogenesis	Stick insect, *Carausius morosus*	H. Buecker	DLR Cologne (DFVLR)	D
SL-D1	STS-61A	30 Oct 85	Dosimetric mapping inside Biorack	Stick insect, *Carausius morosus*	H. Buecker	DLR Cologne (DFVLR)	D
SL-D1	STS-61A	30 Oct 85	*Paramecium* growth and mineral properties	*Paramecium caudatum*	H. Planel	National Institute of Health, Toulouse	F
SL-D1	STS-61A	30 Oct 85	Bacterial growth and mineral properties	*Aerobacter aerogenes*	H. Planel	National Institute of Health, Toulouse	F
SL-D1	STS-61A	30 Oct 85	The biological clock of *C. reinhardii* in space.	*Chlamydomonas reinhardii*	D. Mergenhagen	University Hamburg	D
SL-D1	STS-61A	30 Oct 85	Bacterial growth and differentiation.	*Bacillus subtilis*	H.D. Mennigmann	University Frankfurt	D
SL-D1	STS-61A	30 Oct 85	Human lymphocyte activation in weightlessness.	Human	A. Cogoli	ETH Zürich	CH

Table 22 (continued).

Mission	Flight	Launched	Title	Species used	Investigator	Affiliation	Country
SL-D1	STS-61A	30 Oct 85	Effect of microgravity on lymphocyte action	Human	A. Cogoli	ETH Zürich	CH
SL-D1	STS-61A	30 Oct 85	Graviperception of lentil seedling roots	*Lens culinaris*	G. Perbal	University Curie Paris	F
SL-D1	STS-61A	30 Oct 85	Ultrastructure of mammalian cell polarisation	Mouse lymphocytes	M. Boutille	University Paris	F
SL-D1	STS-61A	30 Oct 85	Fertilisation and embryogenesis; dorso/ventral axis establishment in embryos.	Clawed toad, *Xenopus laevis*	G.A. Ubbels	Hubrecht Laboratory, Utrecht	NL
SL-D1	STS-61A	30 Oct 85	Effectiveness of antibiotics in microgravity.	*Escherichia coli*	R. Tixador	National Institute of Health, Toulouse	F
SL-D1	STS-61A	30 Oct 85	Graviperception.	Garden cress, *Lepidium sativum*	D. Volkmann	University Bonn	D
SL-D1	STS-61A	30 Oct 85	Graviperception and geotropism.	Maize, *Zea mays*	J. Gross	University Tübingen	D
SL-D1	STS-61A	30 Oct 85	Graviperception and differentiation of plant cells	Anise, *Pimpinella anisum*	R.R. Thelmer	University Münich	D
SL-D1	STS-61A	30 Oct 85	Graviperception (orientation); amphibian statolith experiment	Clawed toad, *Xenopus laevis*	J. Neubert	DLR Cologne (DFVLR)	D
SLS-1	STS-40	05 Jun 91	Lymphocyte proliferation in weightlessness	Human	A. Cogoli	ETH Zürich	CH
IML-1	STS-42	22 Jan 92	Bone and skeleton changes; mineralization and resorption of fetal bones	Mouse bone cells	J.P. Veldhuijzen	Free University Amsterdam	NL
IML-1	STS-42	22 Jan 92	Fertilisation and Embryogenesis Dorsal/ventral axis determination in embryos	Clawed toad, *Xenopus laevis*	G.A. Ubbels	Hubrecht Laboratory, Utrecht	NL

IML-1	STS-42	22 Jan 92	Development of *Drosophila* in space	Fruit fly, *Drosophila melanogaster*	R. Marco	Universidad Autonoma Madrid	E
IML-1	STS-42	22 Jan 92	Plant differentiation and regeneration from protoplasts	Carrot, *Daucus carota*	O. Rasmussen	University of Aarhus	DK
IML-1	STS-42	22 Jan 92	Plant differentiation and regeneration from protoplasts	Rape, *Brassica napus*	O. Rasmussen	University of Aarhus	DK
IML-1	STS-42	22 Jan 92	Radiobiology of embryogenesis and organogenesis of *C. morosus*	Stick insect, *Carausius morosus*	H. Buecker	DLR Cologne	D
IML-1	STS-42	22 Jan 92	Dosimetric mapping inside Biorack	none	G. Reitz	DLR Cologne	D
IML-1	STS-42	22 Jan 92	Growth and sporulation of *Bacillus subtilis*	Bacillus subtilis	H.D. Mennigmann	University Frankfurt	D
IML-1	STS-42	22 Jan 92	Dosimetric mapping with the Biostack	*Bacillus subtilis*	H. Buecker	DLR Cologne	D
IML-1	STS-42	22 Jan 92	Dosimetric mapping with the Biostack	Yeast spores, *Saccharomyces cerevisiae*	H. Buecker	DLR Cologne	D
IML-1	STS-42	22 Jan 92	Dosimetric mapping with the Biostack	*Sordaria fimicola*, spores	H. Buecker	DLR Cologne	D
IML-1	STS-42	22 Jan 92	Dosimetric mapping with the Biostack	Brine shrimp, *Artemia salina*	H. Buecker	DLR Cologne	D
IML-1	STS-42	22 Jan 92	Dosimetric mapping with the Biostack	Mouse-ear cress, *Arabidopsis thaliana*	H. Buecker	DLR Cologne	D
IML-1	STS-42	22 Jan 92	Friend Leukaemia virus transformed cells	Human erythrocytes	A. Cogoli	ETH Zürich	CH
IML-1	STS-42	22 Jan 92	Hybridoma cell proliferation and performance	Human lymphocytes	A. Cogoli	ETH Zürich	CH
IML-1	STS-42	22 Jan 92	Dynamic cell culture system	Syrian hamster kidney cells	A. Cogoli	ETH Zürich	CH
IML-1	STS-42	22 Jan 92	Graviresponse of shoots and cotyledons in microgravity.	Mouse-ear cress, *Arabidopsis thaliana*	L.G. Briarty	University of Nottingham	UK

Table 22 (continued).

Mission	Flight	Launched	Title	Species used	Investigator	Affiliation	Country
IML-1	STS-42	22 Jan 92	Transmission of the gravity stimulus in roots	Lentil, *Lens culinaris*	G. Perbal	Curie University Paris	F
IML-1	STS-42	22 Jan 92	Gravity related behaviour of *Physarum*	Physarum polycephalum	I. Block	DLR, Cologne	D
IML-1	STS-42	22 Jan 92	Antibiotic penetration of bacterial cells in space	*Escherichia coli*	R. Tixador	National Institute of Health, Toulouse	F
EURECA	STS-46	01 Aug 92	Exposure experiment on the exobiology and radiation assembly (ERA)	*Aspergillus niger* spores	K. Dose	University of Mainz	D
EURECA	STS-46	01 Aug 92	Exposure experiment on ERA	*Aspergillus ochraceus* spores	K. Dose	University of Mainz	D
EURECA	STS-46	01 Aug 92	Exposure experiment on ERA	Yeast spores, *Saccharomyces cerevisiae*	J. Kiefer	University of Giessen	D
EURECA	STS-46	01 Aug 92	Exposure experiment on ERA	*Bacillus subtilis* spores	G. Horneck	DLR, Cologne	D
EURECA	STS-46	01 Aug 92	Exposure experiment on ERA	*Deinococcus radiodurans* spores	G. Horneck	DLR, Cologne	D
EURECA	STS-46	01 Aug 92	Exposure experiment on ERA	*Ditlenchus dipsacii*	G. Horneck	DLR, Cologne	D
SL-D2	STS-55	26 Apr 93	Dosimetric mapping	Dosimeters	G. Reitz	DLR, Cologne	D
SL-D2	STS-55	26 Apr 93	Dosimetric mapping with the Biostack HZE tracking sandwich	Brine shrimp, *Arabidopsis*, fungal and bacterial spores	H. Buecker	DLR, Cologne	D
SL-D2	STS-55	26 Apr 93	Vestibular reflexes in microgravity	Clawed toad, *Xenopus laevis*	E. Horn	University Ulm	D
SL-D2	STS-55	26 Apr 93	Vestibular reflexes in microgravity	Cichlid fish, *Oreochromis mossambicus*	E. Horn	University Ulm	D

SL-D2	STS-55	26 Apr 93	Development of the gravity perceiving organ	Clawed toad, *Xenopus laevis*	J. Neubert	DLR, Cologne	D
SL-D2	STS-55	26 Apr 93	Development of the gravity perceiving organ.	Cichlid fish, *Oreochromis mossambicus*	J. Neubert	DLR, Cologne	D
SL-D2	STS-55	26 Apr 93	CNS neuronal plasticity during ontogenesis.	Clawed toad, *Xenopus laevis*	H. Rahmann	University of Stuttgart	D
SL-D2	STS-55	26 Apr 93	CNS neuronal plasticity during ontogenesis.	Cichlid fish, *Oreochromis mossambicus*	H. Rahmann	University of Stuttgart	D
SL-D2	STS-55	26 Apr 93	Cerebellar development in microgravity	Clawed toad, *Xenopus laevis*	M. Schachner	University Heidelberg	D
SL-D2	STS-55	26 Apr 93	Cerebellar development in microgravity	Cichlid fish, *Oreochromis mossambicus*	M. Schachner	University Heidelberg	D
SL-D2	STS-55	26 Apr 93	Gravisensitivity of cress roots	Garden cress, *Lepidium sativum*	D. Volkmann	University Bonn	D
SL-D2	STS-55	26 Apr 93	Cell polarity and gravity	Garden cress, *Lepidium sativum*	M.H. Weisen-seel	University Karlsruhe	D
SL-D2	STS-55	26 Apr 93	Fruiting body development of fungi	*Flammulina velutipes*	B. Hock	Technical University Münich	D
SL-D2	STS-55	26 Apr 93	Secondary metabolites in plant cell suspensions	*Panax ginseng*	H. Jung-Heiliger	Plant Laboratory	D
SL-D2	STS-55	26 Apr 93	Investigations on yeast metabolism	*Saccharomyces uvarum*	S. Donhauser	Technical University Münich	D
SL-D2	STS-55	26 Apr 93	Productivity of bacteria	*Escherichia coli*	H.D. Mennig-mann	University Frankfurt	D
SL-D2	STS-55	26 Apr 93	Productivity of bacteria	*Bacillus subtilis*	H.D. Mennig-mann	University Frankfurt	D
SL-D2	STS-55	26 Apr 93	Effect on genetic material; fluctuation test on bacterial cultures	*Escherichia coli*	H.D. Mennig-mann	University Frankfurt	D

Table 22 (continued).

Mission	Flight	Launched	Title	Species used	Investigator	Affiliation	Country
SL-D2	STS-55	26 Apr 93	Effect on genetic material; fluctuation test on bacterial cultures	*Bacillus subtilis*	H.D. Mennigmann	University Frankfurt	D
SL-D2	STS-55	26 Apr 93	Bone and skeleton changes; collagen synthesis and cell proliferation	Human, skin fibroblasts	P.K. Mueller	University Lübeck	D
SL-D2	STS-55	26 Apr 93	Activation of regulatory T-lymphocytes	Human T-lymphocytes	K. Reske	University Mainz	D
SL-D2	STS-55	26 Apr 93	Growth of lymphocytes under microgravity	Human T-lymphocytes	K. Reske	University Mainz	D
SL-D2	STS-55	26 Apr 93	Cell fusion and hybridoma production	Human B-Lymphocytes	U. Zimmermann	University Würzburg	D
SL-D2	STS-55	26 Apr 93	Electrofusion of plant cell protoplasts	Tobacco, *Nicotiana tabacum*	R. Hampp	University Tübingen	D
SL-D2	STS-55	26 Apr 93	Electrofusion of plant cell protoplasts	Foxglove, *Digitalis* sp.	R. Hampp	University Tübingen	D
SL-D2	STS-55	26 Apr 93	Electrofusion of plant cell protoplasts	Sunflower, *Helianthus* sp.	R. Hampp	University Tübingen	D
SL-D2	STS-55	26 Apr 93	Radiobiology (mapping); HZE-particle dosimetry with Biostack	Mouse-ear cress, *Arabidopsis thaliana*	H. Buecker	DLR Cologne	D
SL-D2	STS-55	26 Apr 93	Radiobiology (mapping); HZE-particle dosimetry with Biostack	*Artemia salina*	H. Buecker	DLR Cologne	D
SL-D2	STS-55	26 Apr 93	Radiobiology (mapping); HZE-particle dosimetry with Biostack	*Sordaria fimicola*, spores	H. Buecker	DLR Cologne	D
SL-D2	STS-55	26 Apr 93	Radiobiology (mapping); HZE-particle dosimetry with Biostack	Yeast spores, *Saccharomyces cerevisiae*	H. Buecker	DLR Cologne	D
SL-D2	STS-55	26 Apr 93	Exposure experiment: responses to solar UV and space vacuum	*Bacillus subtilis*, spores	H. Buecker	DLR Cologne	D

SL-D2	STS-55	26 Apr 93	Exposure experiment: responses to solar UV and space vacuum	*Bacillus subtilis*, spores	G. Horneck	DLR Cologne	D
SL-D2	STS-55	26 Apr 93	Exposure experiment: responses to solar UV and space vacuum	Isolated DNA of the bacterium *Haemophilus influenzae.*	G. Horneck	DLR Cologne	D
Human physiology and medical experiments							
SL-1	STS-9	28-11-83	Personal miniature electrophysiological tape recorder; data on heart, brain, eyes	Human	H. Green	Clinical Research Centre, Harrow	UK
SL-1	STS-9	28-11-83	Measurements of central venous pressure and hormones in blood; data on body fluid shifts	Human	K.A. Kirsch	Free University of Berlin	D
SL-1	STS-9	28-11-83	Mass discrimination during weightlessness	Human	H.E. Ross	University of Stirling	UK
SL-1	STS-9	28-11-83	Comparison in cardio-vascular performances; 3D ballistocardiography in weightlessness.	Human	A. Scano	University of Rome	I
SL-1	STS-9	28-11-83	Vestibular function and visual-vestibular interactions.	Human	R. von Baumgarten	University of Mainz	D
SPIN-1	STS-51G	17-6-85	Sensorimotor adaptation (muscle tone, movement)	Human	A. Berthoz	Paris CNRS	F
SPIN-1	STS-51G	17-6-85	Cardiovascular data on two crew; French echocardio-graph experiment	Human	L. Pourcelot	University of Tours	F
SL-D1	STS-61A	30-10-85	Cardiovascular adaptation; body impedance measure-ment	Human	F. Baisch	DLR Cologne	D

Table 22 (continued).

Mission	Flight	Launched	Title	Species used	Investigator	Affiliation	Country
SL-D1	STS-61A	30-10-85	Eye pressure changes measured with a tonometer.	Human	J. Draeger	University of Hamburg	D
SL-D1	STS-61A	30-10-85	Communication aspects of cognitive behaviour; gesture and speech in microgravity.	Human	A.D. Friederici	MPI, Nijmegen	NL
SL-D1	STS-61A	30-10-85	Spatial description in space: effect of perception cues on communication of spatial arrangements.	Human	A.D. Friederici	MPI, Nijmegen	NL
SL-D1	STS-61A	30-10-85	Human reactions in micro-gravity: reaction time (student experiment)	Human	M. Hoschek	Mühltal	D
SL-D1	STS-61A	30-10-85	Blood displacement in upper body: central venous pressure	Human	K. Kirsch	Free University of Berlin	D
SL-D1	STS-61A	30-10-85	Effects on cognitive behaviour: mass discrimination (also on SL-1)	Human	H.E. Ross	University of Stirling	UK
SL-D1	STS-61A	30-10-85	Vestibular research: response to optokinetic stimulation patterns (also on SL-1).	Human	R. von Baum-garten	University of Mainz	D
D-2	STS-55	26-4-93	Peripheral & central haemo-dynamics adaptation to microgravity: cardiovascular reflexes.	Human	F. Bonde-Petersen	DAMEC Copenhagen	DK
D-2	STS-55	26-4-93	Regulation of homeostatic volume in microgravity.	Human	R. Gerzer	Univ. of Mü-nich (+ DLR & Univ. of Heidelberg)	D
D-2	STS-55	26-4-93	The central venous pressure in microgravity: study at launch & in microgravity during adaptation	Human	N. Foldager	DAMEC Copenhagen	DK

D-2	STS-55	26-4-93	Leg fluid distribution at rest and under LBNP: changes following dehydration, disuse, and fluid shift.	Human	F. Baisch	DLR (with TNO, Holland)	D NL
D-2	STS-55	26-4-93	Determination of segmental fluid content and perfusion	Human	F. Baisch	DLR (with TNO, Holland & Karolinska Inst. Sweden)	D NL S
D-2	STS-55	26-4-93	Cardiovascular adaptation: left ventricular function, at rest and under stimulation	Human	L. Beck	DLR(with U. of Amsterdam & Univ. of Tours)	D NL F
D-2	STS-55	26-4-93	Tonometric measurement of intraocular pressure in microgravity: increase in intraocular pressure, peak values, adaptation.	Human	J. Draeger	University of Hamburg	D
D-2	STS-55	26-4-93	Tissue thickness & compliance along body axis in microgravity; new method to quantify fluid shifts within superficial tissue.	Human	K. Kirsch	Free University of Berlin	D
D-2	STS-55	26-4-93	Cardiovascular response to LBNP and fluid loading.	Human	L. Pourcelot	University of Tours	F
D-2	STS-55	26-4-93	Changes in whole body nitrogen turnover rate, protein synthesis & breakdown.	Human	E.B. Fern	NESTEC Ltd, Research Centre	CH
D-2	STS-55	26-4-93	Effects of microgravity on glucose tolerance.	Human	H.P. Maass	DLR Cologne	D
D-2	STS-55	26-4-93	Influence of microgravity on endocrine & renal elements of volume homeostasis.	Human	P. Norsk	DAMEC Copenhagen	DK

Table 22 (continued).

Mission	Flight	Launched	Title	Species used	Investigator	Affiliation	Country
D-2	STS-55	26-4-93	Effects of space flight on pituitary-gonad-adrenalin function.	Human	I.G. Riondino	NCRA, Rome	I
D-2	STS-55	26-4-93	Regulation of salt balance and blood pressure: role of volume regulating hormones & plasma proteins.	Human	L. Röcker	Free Univ. Berlin	D
D-2	STS-55	26-4-93	Pulmonary stratification and compartment analysis	Human	S. Groth	DAMEC Copenhagen	DK
D-2	STS-55	26-4-93	Pulmonary perfusion & ventilation in microgravity	Human	D. Linnarsson	Karolinska Inst.	S
D-2	STS-55	26-4-93	Effect of gravity on ventilation pattern in the lung.	Human	M. Paiva	Univ. of Brussels (+Westmead Hosp., Australia, + Univ. of California)	B (AUSTRALIA, USA)
D-2	STS-55	26-4-93	Dynamics of gas exchange, ventilation, & heart rate in submaximal dynamic exercise: to determine if CO_2 kinetics qualifies as a monitor of endurance performance.	Human	J. Stegemann	Deutsche Sport-Hochschule Köln	D
D-2	STS-55	26-4-93	Biological dosimetry	Human leucocytes	G. Obe	University of Essen	D
D-2	STS-55	26-4-93	Personal dosimetry	Dosimeters	G. Reitz	DLR Cologne	D
D-2	STS-55	26-4-93	Dosimetric mapping	Dosimeters	G. Reitz	DLR Cologne	D

Table 23. European Experiments on board IML-2 (STS-65, 1994).

Experiment code name	Description	Principal Investigator(s) and affiliation
ADHESION	Lymphocyte activation, differentiation, and adhesion-dependence of activation.	A. Cogoli, University of Sassari
ANTIGEN	Antigen presentation and T-cell proliferation in microgravity.	A. Cogoli & D. Morrison, University of Sassari & NASA-JSC
MOTION	Lymphocyte movements and interactions (Observation in NIZEMI).	A. Cogoli & M. Cogoli, University of Sassari
PHORBOL and CYTOKINES	Effect of microgravity on cellular action: the role of cytokines.	D. Schmitt, University of Toulouse
BIOREACTOR	Effect of stirring and mixing in a bioreactor experiment in microgravity. resume	A. Cogoli & B. Bechler, ETH Weltraumbiologie, Zürich
KINETICS	Radiation repair kinetics in eukaryotes.	G. Horneck, DLR Cologne
REPAIR	Efficiency of radiation repair in prokaryotes.	G. Horneck, DLR Cologne
DROSOPHILA	Investigation of the mechanism involved in the effects of space microgravity on *Drosophila* development, behaviour and aging.	R. Marco, Independent University of Madrid
DOSIMETRY	Dosimetric mapping inside Biorack.	G. Reitz, DLR Cologne
SIGNAL	Cell microenvironment and membrane signal transduction in microgravity.	P. Bouloc, University of Paris 7
AGGREGATE	Molecular biological investigations of animal multi-cell aggregates reconstituted under microgravity.	U. Heinlein, University of Düsseldorf
MOUSE	Regulation of cell growth and differentiation by microgravity: retinoic acid-induced cell differentiation.	S. de Laat (P.Rijken), Hubrecht Laboratorium, Utrecht
RANDOM	Plant growth and random walk.	A. Johnsson, University of Trondheim
LENTIL	Effect of microgravity on lentil morphogenesis.	G. Perbal (D. Driss-Ecole), University of Paris 6
TRANSFORM	Root orientation, growth regulation, adaptation, and agravitropic behaviour of genetically transformed roots.	T.H. Iversen, University of Trondheim
EGGS	The role of gravity in the establishment of embryonic axes in the amphibian embryo.	G. Ubbels, Hubrecht Laboratorium, Utrecht
URCHIN	The sea urchin larva, a potential model for studying biomineralization and demineralization processes in space.	H.J. Marthy, Laboratoire Arago, Banyuls sur Mer
BONES	The effects of microgravity on varying $1 \times g$ exposure periods on bone resorption: an *in vitro* experiment.	P. Veldhuijzen, Free University of Amsterdam

Table 24. The ESA experiments on the Euromir 94 and 95 missions.

Experiment description	Principal Investigator	Affiliation
Euromir 94		
Circadian rhythms and sleep during a 30 day mission using Oxford Medilog System	A. Gundel	DLR Flugmedizin, Cologne
Spatial orientation and space sickness.	H. Mittelstadt	MPI für Verhaltens-physiologie Seewiesen, Germany
Adaptation of basic vestibulo-oculomotor mechanisms to altered gravity conditions (using VOG).	H. Scherer	Universitätsklinik, F.U. Berlin
Posture and movement, (using accelerometer camera from ALTAIR).	A. Berthoz	LPN CNRS, Paris
STAMP (Cognitive Experiment), using VIMINAL.	A. Berthoz	LPN CNRS, Paris
Otolith adaptation to different levels of gravity (ground experiment).	W. Bles	TNO Inst. for Perception, Soesterberg, Holland
Correlation of eye torsion changes with the time course of the space adaptation mechanism.	C.H. Markham	UCLA School of Medicine, Los Angeles
Non-invasive stress monitoring in space flight by hormone measurement in saliva, requiring ergometer.	C.J. Strasburger	Medizinische Klinik, Univ. Münich
Gastro-enteropancreatic peptides during zero-g and their involvement in space motion sickness.	R.L. Riepl	Medizinische Klinik, Univ. Münich
Fluid and electrolyte balance during weightlessness and possibilities of their regulation.	C. Drummer	DLR, Inst. Luft- und Raumfahrtmedizin
Fluid shifts into and out of superficial tissue stability along the body axis under microgravity in man (using ultrasonic pulse echo equipment, USD).	K.A. Kirsch	Physiology Department, Free University, Berlin
Influence of microgravity on osmo- and volume regulation in man: dynamic responses to isotonic and hypertonic loads.	P. Bie	Dept. of Medical Physiology, Univ. of Copenhagen, Denmark
Space flight related orthostatic intolerance: the role of autonomic nervous system, water balance, volume regulation hormones and energy expenditure, (requiring Russian LBNP device, ECG, haematocrit).	C. Gharib	Lab. de Physiologie de l'Environ, Lyon, France
Influence of space flight on energy metabolism and its circadian variation.	H. Demaria Pesce	CNRS/INSERM, Inst. Biologie, College de France, Paris
Effects of changes in central venous pressure on the erythropoietic system, under 1 × g and microgravity conditions.	H.C. Gunga	Physiology Dept. Free Univ. Berlin
Central venous pressure during weightlessness.	N. Foldager	DAMEC, Univ. Hospital, Copenhagen, Denmark

Table 24 (continued).

Experiment description	Principal Investigator	Affiliation
Radiation health during prolonged space flight.	G. Reitz	DLR, Inst. Luft- und Raumfahrtmedizin
Magnetic resonance spectroscopy, imaging of human muscles before and after space flight (ground experiment).	J. Zange	DLR, Inst. Luft- und Raumfahrtmedizin
Changes in mechanical properties of human muscle during space flight (ground experiment).	F. Goubel	Univ. de Techn., Compiegne, France
Bone mass and structure changes and bone remodelling in space.	C. Alexandre	Lab. Biolog. du Tissu Osseux, St. Etienne, France
Euromir 95		
Influence of microgravity on renal fluid excretion in humans.	P. Norsk	DAMEC Research Righshospitalet, Copenhagen
Regulation of cardiovascular response to exercise in humans.	G. Ferretti	Fac. Medicine/Physiologie, Geneva, Switzerland
Non-invasive monitoring of drug metabolism and drug effect during prolonged weightlessness.	R. Gerzer	DLR, Inst. Luft- und Raumfahrtmedizin Köln
Differential effects of otolith input on ocular laterpulsion, cyclorotation, perceived visual vertical and ahead, and tonic neck reflexes in man.	H. Dietrich	Dept. of Neurology, Univ. of Münich
Correlation of eye torsion changes with time course of space adaptation syndrome. Binocular disconjugacy.	C.H. Markham	UCLA, School of Medicine, Los Angeles, U.S.A.
Bone mass and structure measurement during long term space flight using an ultrasound bone densitometer.	C. Alexandre	Lab. de Biologie du Tissu Osseux, Faculté de Médicine, St. Etienne, France
The application of mechanical stimulation to prevent loss of bone mass in long term space flight using simulated heel strike transients.	A. Goodship	Department of Anatomy, Univ. of Bristol, U.K.
Radiation health during prolonged space flight: environmental and personal dosimetry.	G. Reitz	DLR, Inst. Luft- und Raumfahrtmedizin
Chromosomal aberrations in peripheral lymphocytes of astronauts (ground experiment).	G. Obe	Dept. of Genetics, Univ. of Essen, Germany.
Influence of gravity on the preparation and execution of voluntary movements (ground experiment).	D.G. Ruegg	Inst. of Physiology, Univ. of Fribourg, Switzerland
Effects of short term/long term microgravity on the pulmonary gas exchange, respiratory and cardiovascular control during rest and exercise.	D. Linnarsson	Karolinska Inst. Stockholm, Sweden

Table 24 (continued).

Experiment description	Principal Investigator	Affiliation
Pulmonary function in microgravity.	M. Paiva	Inst. de Recherche Interdisciplinaire, Univ. Libre, Bruxelles, Belgium
Interstitial fluid balance under microgravity with special reference to pulmonary mechanics.	D. Negrini	Institute of Physiology, Univ. of Milan
Effects of microgravity on the biomechanical and bioenergetic characteristics of human skeletal muscle (ground experiment).	P. Di Prampero	Institute of Biology, University of Udine, Italy
Changes in the mechanical properties of humans during space flight (ground experiment).	F. Goubel	Univ. de Technologie, Compiegne, France
Magnetic resonance spectroscopy and imaging of human muscles and bones before and after spaceflight (ground experiment).	J. Zange	DLR, Inst. Luft- und Raumfahrtmedizin Köln
Effect of Vitamin K supplement on bone mass during microgravity conditions.	C. Vermeer	Biochemistry Department, Univ. of Limburg, Holland
The effect of venous pressure on bone mineral density in weightlessness conditions	I.D. McCarthy	Dept. of Surgery, Hammersmith Hospital, London

genome (J. Cadet); *Effects of cosmic radiation and microgravity on dormant shrimp embryos* (A. Hernandorena); *Effects of cosmic radiation on isolated organic macromolecules. Effectiveness of radioprotective drugs* (J.P. Moatti); *Relative contributions of individual components of the cosmic radiation spectrum and effectiveness of radioprotective drugs* (G. Reitz); *Exposure of bacteria and isolated DNA to solar UV, to study limits to survival and concepts for organisms on meteorites* (G. Horneck).

The Biobox is an incubator able to maintain selected temperatures between 4°C and 37°C, using an integral battery power supply. Limited telecommunication and telemetry facilities are available. It is designed to be accommodated within the Foton spacecraft, and first flew, with problems, on the Bion–10 mission. The experiments from that mission were reflown on Foton–10. These are as follows: *To test whether in vitro cultures of osteoblast–like cells are gravity dependant in proliferation and activity* (C. Alexandre); *The gravity dependence of cell differentiation in vitro of bone marrow cultures* (G. Schoeters); *Growth factor production and sensitivity and bone tissue development under gravitational stress* (J.P. Veldhuizen).

8.16.2 Spacelab facilities and flights

LMS Spacelab Mission. ESA's Advanced Protein Crystallization Facility (APCF) and the Torque Velocity Dynamometer (TVD) will fly on the Life and Microgravity Spacelab (LMS) mission scheduled for launch on 27 June 1996 with a mission duration of 16 days. All facilities will be flown on a co-operative basis with NASA. The TVD is a new instrument intended for testing skeletal muscle performance and muscle disuse atrophy. It was originally developed for flight on Spacelab SLS-3, which was planned for flight in 1996 but subsequently cancelled.

Shuttle Missions to Mir (S/MM). NASA is planning at least seven Shuttle missions to the Russian Space Station Mir (see Section 8.15.1.2). Some of these missions will carry the Spacehab Module and ESA's Biorack will be accommodated in this Spacehab Module and flown on the following three missions: S/MM-03 (March 1996), S/MM-05 (December 1996) and S/MM-06 (May 1997). It is now also planned to fly the APCF on S/MM-06 (May 1997).

8.16.2.1 Biorack

Biorack is an existing multi-purpose facility, previously flown on D-1, IML-1 and IML-2, whose main components are incubators, a cooler/freezer unit, and a glovebox. It also provides special canisters for sample transport and maintenance under defined thermal conditions. Reflight refurbishment activities for the flight of Biorack on Shuttle missions to Mir are in progress. In particular, the Biorack incubators are being equipped with improved centrifuges providing variable gravitational acceleration.

8.16.2.2 Advanced Protein Crystallization Facility (APCF)

APCF is to be reflown on the USML-2 Mission (16 days duration, launch on 21 September 1995), the LMS Mission (27 June 1996) and on the Shuttle Mission to Mir S/MM-06 (May 1997). Cooperation with NASA is also anticipated for use of both APCF units on the LMS mission scheduled for flight on 27 June 1996.

8.16.2.3 Glove Box

Will be reflown on the USML-2 Spacelab Mission (21 September 1995).

8.16.2.4 Anthrorack

This human physiology research facility was originally earmarked for reflight on the E-1 mission, having previously flown on the D-2 mission (26 April to 6 May 1993). Since the E-1 mission was cancelled, the majority of hardware items of Anthrorack were placed in storage. Some items are expected to be re-used on the Neurolab Mission (26 February 1998), in modified form, to support neurophysiology experiments - being then designated EDEN (ESA Developed Elements for Neurolab). This would contain a support computer/data handling system, automated blood pressure monitor, signal processing capabilities for EMG, EOG, ECG, EEG, and dynamometer outputs, and a vestibulo-ocular interaction recording system. Initial development was completed in late November 1994. An area of major uncertainty throughout development was the scientific requirement for the Anthrorack Lower Body Negative Pressure Device (LBNPD), intended to perform micro-neurography during the application of lower body negative pressure. This is a delicate measurement, imposing design requirements which the Anthrorack LBNPD had not been built to meet. Because of the high cost of redesign it was not possible to implement the changes to the LBNPD to achieve compatibility with the micro-neurography objectives.

8.16.3 Euromir missions

8.16.3.1 Euromir 94 flight

The post mission Baseline Data Collection (BDC) programme was finalised on 16 November 1994 and transport of the BDC equipment from Star City back to Western Europe was initiated on 16 November 1994 and completed in early February 1995. The life science samples (blood samples, etc.) from pre- and post-mission BDC and flight were stored under controlled conditions at Star City and transported (hand-carried) to DLR in Germany around 20 November 1994 where they were put into further storage. Distribution of the samples to investigators started early 1995.

8.16.3.2 Euromir 95

Selection of experiments for Euromir 95 was based, in part, on the mass allocation for orbital transfer (105 kg up and 5 kg down). Excess return mass may be returned via Shuttle missions to Mir instead of the Russian Soyuz vehicle. Another consideration in experiment choice and design is that experiments depend on availability of astronaut time. For the whole Euromir 95 payload, 405 h from the ESA astronaut are available, 210 h being set aside for life science experiments. Crew time involvement in experiments needs to be subjected to detailed assessment.

ESA Microgravity multi-user facilities which are planned to be used on the Euromir 95 mission in the framework of the EMIR-1 Programme include the Respiratory Gas Analyzer, also referred to as RMS-II; the Bone Densitometer (BDM) and the Freezer. The Respiratory Gas Analyzer is a new instrument and is based on an opto-acoustic, excitation/detection system, which has comparable sensitivity for many gases to the mass spectrometer, but a substantially smaller mass/volume and power requirement, and it does not require a vacuum. Although problems have been encountered with the operating system software, the schedule for crew training, science verification and flight qualification has been maintained. Integration of all models of the Bone Densitometer was completed on schedule. The freezer for Euromir 95, which is an inflatable unit, has been newly developed for this mission.

8.16.4 Mini-missions

There will continue to be a programme of short duration low-gravity experiments based upon sounding rockets, aircraft parabolic flights, and drop towers, since they have proven to be of value for exploratory work and equipment testing.

Sounding rockets, in particular, have played a significant role in microgravity experimentation. Table 25 shows the range of Life Sciences rocket borne experiments which have been carried out through the years by ESA and individual Member States, using the Texus and Maser rockets. The longer microgravity duration (14 min.) and larger payload (440 kg) available with the new MAXUS rocket may further increase the usefulness of these systems.

Parabolic aircraft flights have been used recently for precursor experiments, for equipment checkout, and for experiment optimization in the life sciences. They have also been used for the study of the 'instantaneous' physiological reactions of humans to transient changes in gravitational acceleration, as it cycles from about $10^{-2} \times g$ through to some $2 \times g$. Consequently a continuing programme of activity, amounting to two campaigns, each of 90 parabolas of 20 seconds duration, per year is planned for such research.

8.17 Conclusions

Particularly for the European nations, the flight opportunities in the past have been very restricted; indeed, they still are. For the scientists, this lack of continuity of experimentation has, previously, often been combined with limited facilities for controls or *in situ* analysis. Evidently, the transition from exploratory research to a programme of rigorous, in depth research on a frequent routine basis, has yet to be made for many subject areas. Perhaps

Table 25. European sounding rocket biology experiments (to 1993).

Row No.	Flight	Launch	Discipline	Topic	Title	Species
1	Texus 11	27 Apr 85	Cell biology	Cell electrofusion	Electrofusion of yeast cells	Yeast
2	Texus 13	30 Apr 86	Cell biology	Cell electrofusion	Electrofusion of yeast cells	Yeast
3	Texus 13	30 Apr 86	Cell biology	Cell electrofusion	Electrofusion of plant cells	Tobacco
4	Texus 13	30 Apr 86	Cell biology	Cell electrofusion	Electrofusion of plant cells	Oat
5	Texus 14B	03 May 87	Cell biology	Cell electrofusion	Fusion of yeast cells	Yeast
6	Texus 14B	03 May 87	Cell biology	Transfection (gene transfer)	Electric field induced gene transfer	Human
7	Texus 17	02 May 88	Cell biology	Cell electrofusion	Electrofusion of plant cells	Tobacco
8	Texus 17	02 May 88	Developmental biology	Fertilisation and embryo polarity	Amphibian embryo	Clawed toad
9	Texus 18	06 May 88	Botany	Graviperception (cell structure)	Cress germination	Garden cress
10	Texus 18	06 May 88	Technology	Electrophoresis (FFE)	Electrophoretic separation	Rabbit
11	Texus 18	06 May 88	Technology	Electrophoresis (FFE)	Electrophoretic separation	Rat
12	Texus 18	06 May 88	Technology	Electrophoresis (FFE)	Electrophoretic separation	Guinea Pig
13	Texus 18	06 May 88	Cell biology	Cell electrofusion	Electrofusion of yeast cells	Yeast
14	Texus 18	06 May 88	Cell biology	Transfection (gene transfer)	Electric field induced gene Transfer	Human
15	Texus 19	28 Nov 88	Botany	Graviperception (cell structure)	Cress germination	Garden cress
16	Texus 19	28 Nov 88	Zoology	Motility	Motility of bull spermatozoa	Bull
17	Texus 20	02 Dec 88	Technology	Electrophoresis (FFE)	Electrophoretic separation	Rabbit
18	Texus 20	02 Dec 88	Technology	Electrophoresis (FFE)	Electrophoretic separation	Rat
19	Texus 20	02 Dec 88	Technology	Electrophoresis (FFE)	Electrophoretic separation	Guinea pig
20	Texus 20	02 Dec 88	Technology	Electrophoresis (FFE)	Electrophoretic separation	Chicken
21	Texus 20	02 Dec 88	Technology	Electrophoresis (FFE)	Electrophoretic separation	Human
22	Texus 20	02 Dec 88	Technology	Electrophoresis (FFE)	Electrophoretic separation	Cattle
23	Texus 20	02 Dec 88	Technology	Electrophoresis (FFE)	Electrophoretic Separation	Mouse
24	Maser 3	10 Apr 89	Development biology	Fertilisation and embryo polarity	Dorsoventral axis development	Clawed toad
25	Maser 3	10 Apr 89	Botany	Membrane functions	During sexual reproduction	
26	Maser 3	10 Apr 89	Cell biology	Membrane and metabolism	Cell growth and differentiation	Human
27	Maser 3	10 Apr 89	Cell biology	Membrane functions	Membrane binding of Concanavalin A	Human
28	Texus 21	30 Apr 89	Cell biology	Cell electrofusion	Electrofusion of plant protoplasts	Tobacco
29	Texus 21	30 Ape 89	Botany	Graviperception (cell structure)	Protoplasm streaming in *Chara* rhizoids	
30	Texus 22	03 May 89	Cell biology	Cell electrofusion	Electrofusion of yeast cells	Yeast
31	Texus 22	03 May 89	Cell biology	Transfection (gene transfer)	Electric field induced gene transfer	Human
32	Texus 23	25 Nov 89	Botany	Graviperception (cell structure)	Cress germination	Garden cress
33	Texus 24	25 Nov 89	Botany	Graviperception (orientation)	Mobility of *Euglena*	
34	Texus 24	06 Dec 89	Cell biology	Electrophoresis (FFE)	Electrophoretic separation	
35	Maser 4	29 Mar 90	Cell biology	Membrane and metabolism	Lymphocytes in microgravity	Human
36	Maser 4	29 Mar 90	Developmental biology	Fertilisation and embryo polarity	Fertilisation of sea urchin eggs	Sea urchin
37	Maser 4	29 Mar 90	Cell biology	Membrane and metabolism	Regulation of cell growth and differentiation	Human
38	Maser 4	29 Mar 90	Cell biology	Membrane and metabolism	Regulation of cell growth and differentiation	Human
39	Maser 4	29 Mar 90	Technology	Electrophoresis (CFE)	Electrophoretic orientation of DNA molecules	
40	Texus 25	13 May 90	Botany	Graviperception (cell structure)	Statolith function in *Chara* rhizoids	
41	Texus 25	13 May 90	Cell biology	Cell electrofusion	Electrofusion of plant protoplasts	Tobacco
42	Texus 26	15 May 90	Zoology	Motility	Motility of spermatozoa	Bull
43	Texus 26	15 May 90	Cell biology	Cell electrofusion	Electrofusion of animal cells	Human
44	Texus 27	15 Nov 90	Technology	Electrophoresis (CFE)	Electrophoresis visualisation EXP2 (haemoglobin)	Bull
45	Texus 27	15 Nov 90	Microbiology	Graviperception (Orientation)	Role of gravity for the spatial orientation of cells	

Table 25. European sounding rocket biology experiments (to 1993) .

Row No.	Specific name	Category	Cell type	Key hardware	Investigator	Affiliation	Country
1	*Saccharomyces* sp.	Fungus	Free living	TEM 06-5	U. Zimmermann	University Wurzburg	D
2	*Saccharomyces* sp.	Fungus	Free living	TEM 06-5	U. Zimmermann	University Wurzburg	D
3	*Nicotiana tabacum*	Plant cells	Leaf cells	TEM 06-5	R. Hampp	University Tübingen	D
4	*Avena sativa*	Plant cells	Leaf cells	TEM 06-5	R. Hampp	University Tübingen	D
5	*Saccharomyces cerevisiae*	Fungus	Free living	TEM 06-11	U. Zimmermann	University Wurzburg	D
6	*Homo sapiens sapiens*	Mammal cells	Lymphocytes	TEM 06-11	U. Zimmermann	University Wurzburg	D
7	*Nicotiana tabacum*	Plant cells	Isolated	TEM 06-5	R. Hampp	University Tübingen	D
8	*Xenopus laevis*	Amphibian		TEM 06-15	G. Ubbeis	Hubrecht Laboratory Utrecht	NL
9	*Lepidium sativum*	Plants		TEM KT	D. Volkmann	University Bonn	D
10		Mammal cells	Erythrocytes	TEM 06-13	K. Hanning	MPI Martinsried	D
11	*Rattus rattus*	Mammal cells	Erythrocytes	TEM 06-13	K. Hanning	MPI Martinsried	D
12		Mammal cells	Erythrocytes	TEM 06-13	K. Hanning	MPI Martinsried	D
13	*Saccharomyces cerevisiae*	Fungus	Free living	TEM 06-11	U. Zimmermann	University Wurzburg	D
14	*Homo sapiens sapiens*	Mammal cells	Lymphocytes	TEM 06-11	U. Zimmermann	University Wurzburg	D
15	*Lepidium sativum*	Plants		TEM KT	D. Volkmann	University Bonn	D
16	*Bovinus*	Mammal cells	Spermatozoa	TEM 06-5	U. Engelmann	University Münich	D
17		Mammal cells	Erythrocytes	TEM 06-13	K. Hanning	MPI Martinsried	D
18	*Rattus rattus*	Mammal cells	Erythrocytes	TEM 06-13	K. Hanning	MPI Martinsried	D
19		Mammal cells	Erythrocytes	TEM 06-13	K. Hanning	MPI Martinsried	D
20	*Gallus domesticus*	Avian cells	Erythrocytes	TEM 06-13	K. Hanning	MPI Martinsried	D
21	*Homo sapiens sapiens*	Mammal cells	Erythrocytes	TEM 06-13	K. Hanning	MPI Martinsried	D
22		Mammal cells	Erythrocytes	TEM 06-13	K. Hanning	MPI Martinsried	D
23		Mammal cells	Erythrocytes	TEM 06-13	K. Hanning	MPI Martinsried	D
24	*Xenopus laevis*	Amphibian (E)		TEM 06-15	G. Ubbeis	Hubrecht Laboratory Utrecht	NL
25	*Chlamydomonas sp.*	Algae	Free living	CIS-1	P. Van den Ende	University Amsterdam	NL
26	*Homo sapiens sapiens*	Mammal cells	A431 Carcinoma	CIS -1	S.W. De Laat	Hubrecht Laboratory Utrecht	NL
27	*Homo sapiens sapiens*	Mammal cells	Lymphocytes	CIS-1	A. Cogoli	ETH Zürich	CH
28	*Nicotiana tabacum*	Plant cells	Isolated	TEM 06-5	R. Hammp	University Tübingen	D
29	*Chara* sp.	Algae		TEM 06-16	A. Sievers	University Bonn	D
30	*Saccharomyces cerevisiae*	Fungus	Free living	TEM 06-11	U. Zimmermann	University Wurzburg	D
31	*Homo sapiens sapiens*	Mammal cells	Lymphocytes	TEM 06-11	U. Zimmermann	University Wurzburg	D
32	*Lepidium sativum*	Plant		TEM BIO	D. Volkmann	University Bonn	D
33	*Euglena* sp.	Algae	Free living	TEM 06-19	D.P. Haeder	University Erlangen	D
34	*Homo sapiens sapiens*	Mammal cells	Erythrocytes	TEM 04-1 (06-13)	K. Hanning	MPI Martinsried	D
35	*Homo sapiens sapiens*	Mammal cells	Lymphocytes	CIS-2	A. Cogoli	ETH Zürich	CH
36	*Paracentrotus lividus*	Echinoderm		CIS-2	H.J. Marthy	Lab Arago/University Paris Curie	F
37	*Homo sapiens sapiens*	Mammal cells	A431 Carcinoma	CIS-2	S.W. De Laat	Hubrecht Laboratory Utrecht	NL
38	*Homo sapiens sapiens*	Mammal cells	A431 Carcinoma	CIS-2	A. Verkleij	State University Utrecht	NL
39	n.a.	macro-molecules		CIS-2	B. Norden	Chalmers University Gothenburg	S
40	*Chara* sp.	Algae		TEM 06-16	A. Sievers	University Bonn	D
41	*Nicotiana tabacum*	Plant cells	Isolated	TEM 06-5	R. Hampp	University Tübingen	D
42		Mammal cells	Spermatozoa	TEM 06-19	P. Schill	University Giessen	D
43	*Homo sapiens sapiens*	Mammal cells	Lymphocytes	TEM 06-11	U. Zimmermann	University Wurzburg	D
44		macro-molecules		TEM04-2	H. Balman	University Toulouse	F
45	*Paramecium* sp.	Protozoa (Ciliates)	Free living	TEM 06-19	R. Hemmers-bach-Krause	DLR Cologne	D

Table 25 (continued).

Row No.	Flight	Launch	Discipline	Topic	Title	Species
46	Texus 28	23 Nov 91	Technology	Electrophoresis (CFE)	Electrophoresis visualisation EXP2 (haemoglobin)	Bull
47	Texus 28	23 Nov 91	Botany	Graviperception (cell structure)	*Chara* cytoskeleton	
48	Texus 28	23 Nov 91	Botany	Graviperception (orientation)	Gravitaxis and phototaxis in flagellates	
49	Texus 28	23 Nov 91	Microbiology	Graviperception (orientation)	Role of gravity for the spatial orientation of cells	
50	Maser 5	09 Apr 92	Cell biology	Cell electrofusion	Cell fusion	Human
51	Maser 5	09 Apr 92	Cell biology	Membrane functions	Responses to protein kinase C signal transduction	Human
52	Maser 5	09 Apr 92	Developmental biology	Fertilisation and embryo polarity	Fertilisation of sea urchin eggs	Sea urchin
53	Texus 29	05 Sep 92	Botany	Graviperception (orientation)	Gravitaxis and phototaxis in flagellates	
54	Texus 29	05 Sep 92	Botany	Graviperception (cell structure)	Cytoplasm streaming in *Chara* rhizoids	
55	Maxus 1B	08 Nov 92	Cell biology	Membrane and metabolism	Lymphocytes in microgravity	Human
56	Maxus 1B	08 Nov 92	Technology	Electrophoresis (CFE)	Electrophoretic orientation experiment (DNA)	
57	Texus 30	01 May 93	Cell biology	Cell metabolism	Protein pattern in mesophyll protoplasts	
58	Texus 30	01 May 93	Cell biology	Cell metabolism	Gravitation and energy metabolism of plant cells	Tobacco
59	Texus 30	01 May 93	Botany	Graviperception (orientation)	Gravitaxis and phototaxis in flagellates	
60	Texus 30	01 May 93	Botany	Graviperception (cell structure)	Cytoplasm streaming in *Chara* rhizoids	

that is the principal change yet to be made, and something upon which effort should now be focused.

Three other general conclusions can also be made from this survey. First, it is clear that ESA member states have established a solid and competitive position in all of the main life science research fields. It is especially true in the space biology area. That position is likely to be maintained, provided a facility of the quality of the Spacelab can be flown on a regular basis, so as to afford frequent follow-up experiments, adequate controls, and the necessary sample handling and preservation facilities.

Second, it seems to be that the area of musculo-skeletal studies has received relatively limited attention. Interest in the earlier flights has naturally centred on the immediate and short term effects of microgravity, on the cardiovascular and the neuro-vestibular research areas. That situation is obviously changing, with the reality of long duration missions, but evidently there is much still to be done in the musculo-skeletal research field. The problem for ESA will be to provide for the necessary controlled experimentation on long duration missions on a routine basis during the next few years.

Third, it is seen that the field of biotechnology research has been strongly emphasized in the outline of the future Russian programme, with considerable interest in such research also being expressed by NASA. By contrast, ESA has not emphasized this research area, as far as space experiments are concerned. Neither do the ESA Member States appear to have developed

Table 25 (continued)

Row No.	Specific name	Category	Cell type	Key hardware	Investigator	Affiliation	Country
46		macro-molecules		TEM04-2	H. Balman	University Toulouse	F
47	*Chara* sp.	Algae		TEM 06-16	A. Sievers	University Bonn	D
48	*Euglena gracilis*	Algae	Free Living	TEM 06-5	D.P. Haeder	University Erlangen	D
49	*Paramecium* sp.	Protozoa (Ciliates)	Free living	TEM 06-5	R. Hemmers-bach-Krause	DLR Cologne	D
50	*Homo sapiens sapiens*	Mammal cells	Fibroblasts	CIS-3	A. Cogoli	ETH Zürich	CH
51	*Homo sapiens sapiens*	Mammal cells	A431 Carcinoma	CIS-3	S.W. De Laat	Hubrecht Laboratory Utrecht	NL
52	*Paracentrotus lividus*	Echinoderm		CIS-3	H.J. Marthy	Lab Arago/University Paris Curie	F
53	*Euglena gracilis*	Algae	Free living	TEM 06-5	D.P. Haeder	University Erlangen	D
54	*Chara* sp.	Algae		TEM 06-5	A. Sievers	University Bonn	D
55	*Homo sapiens sapiens*	mammal cells	Lymphocytes	TEM 06-5M	A. Cogoli	ETH Zürich	CH
56	n.a.	macro-molecules		MEO	B. Norden	Chalmers University Gothenburg	S
57	*Vicia faba*	Plant cells	Mesophyll	TEM 06-21	H. Schnabel	University Bonn	D
58	*Nicotiana tabacum*	Plant cells	Isolated	TEM 06-21	R. Hampp	University Tübingen	D
59	*Euglena gracilis*	Algae	Free living	TEM 06-5	D. P. Haeder	University Erlangen	D
60	*Chara* sp.	Algae		TEM 06-5	A. Sievers	University Bonn	D

much interest in this topic, with the exception of Germany. In so far as the closed bioregenerative life support system developments fall within this heading, it would be surprising if that research were not to be strongly pursued by the principals in the Space Station programme in the future. NASA official policy in space life science research follows from a commitment to the long term goal of extended manned space missions. With that objective in mind, much of the NASA research programme concentrates on human physiology, with supporting studies down to the cellular level in biology. That relative emphasis looks set to continue.

It is interesting to note in passing that, as part of its longer term strategy, NASA has signed a cooperative agreement with the U.S. National Institutes of Health. That agreement is significant, for it recognizes and formalizes the essential link between the space and ground based life science research. It is intended to bring the NASA life science space research work more readily within the critical purview of the NIH researchers, whilst at the same time encouraging the exchange of information and personnel between the two domains.

To conclude then, it is clear that ESA and the Member States have established a very good research record in space life sciences research, and must be commended for having achieved so much, despite so many practical constraints. The future for the ESA Life Sciences programme must surely lie in exploiting vigorously these current strengths.

Appendix

Some comments on biological aspects of life support systems

HELMUT G. HINGHOFER-SZALKAY & DAVID MOORE

A.1 Introduction

In this Section we discuss, in outline, some of the aspects of the space life sciences that relate to the problems of extended manned space missions. This is not intended to be a fully detailed review, but rather will indicate some of the relevant literature by describing selected research results and introducing the main problems that still exist and need to be addressed.

A.2 Bioregenerative life support systems

Bioregenerative life support systems (BLS) are intended to recycle CO_2, waste, including faeces, and to produce oxygen and edible biomass. Principally, the systems are composed of several interconnected compartments which form a more or less closed loop. The recycling component mainly consists of microorganisms, whereas other compartments differ in the complexity of their constituents (from enzymes to plants and animals). For the International Space Station, a facility to grow fresh vegetables on a continuous basis (so-called 'salad machine') is foreseen and can be considered as a step towards the development of more complex BLS which serve aesthetic and psychological, as well as nutritional, purposes.

Both recycling and production must be maintained reliably for long periods of time. The interdependency of BLS compartments renders them subject to system instability if even a single compartment fails, so any perturbation in one element may affect the whole system. Dynamic changes within the system may, on the other hand, lead to self-regulation if internal or environmental factors vary. Technological control, essentially based on selective handling of flows between compartments, must prevent damage becoming irreversible and potentially endangering the whole system.

Reliable, solid, and robust technology; high productivity and low waste production; low maintenance, high automatization; and compatibility with microgravity conditions, are the most important criteria for BLS. Stability and redundancy are additional valuable features.

The substances participating in BLS reactions may be grouped as substrates, products, catalysts, and inert matter. Products of one compartment may serve as substrates in another so that in complex interconnections there is a network of bioprocesses in which individual pathways may be temporarily reversed for the sake of integrated system functioning.

Within the compartment generating them, substrates often display significant inhibitory action on catalysts (feedback inhibition) and hence must be eliminated or kept at an acceptable level. Undesirable compounds (like toxic gases, proteins inducing foam formation, polymers increasing viscosity) need to be closely monitored.

Life support systems are generally classified as physicochemical or bioregenerative; however, the former quickly become biological due to microbial growth which in certain instances turn out to be beneficial. As an example, water purification systems using granulated activated carbon rapidly become colonized with bacteria. This was actually found to increase efficiency of removal of particular organic contaminants and also to extend the effective life of the purification system.

Recycling is the main criterion of BLS, whereas *provisioned* systems need *reclamation,* which means that consumables have to be regularly restored to acceptable quality standards prior to reuse. Partially closed BLS need relatively little reclamation; most systems vary over time with respect to their degree of closure.

Source and sink size are critical determinants in the design and specifications of BLS. Evidently, an outside energy source is always needed to keep the system alive. A sink is any element (or organism) which functionally removes any substance from the larger system. This can be achieved through biochemical transformation to compounds more acceptable to the recipients, and/or through trapping and storing for later removal.

It has been demonstrated that a system with larger sinks is more flexible and allows design specifications to be less precise. The largest test bed so far is Biosphere-2 in Arizona, with more than 3000 species interacting as a broad network of interdependent sinks. The finding of an unexpectedly active oxygen sink in the soil, which then required that oxygen be injected into the system after little more than 1 year's operation, demonstrated how much is still to be learned about complex BLS. Self-regulation is a common feature of natural ecological systems, but stability cannot be taken for granted. Even our planet, the BLS with the largest sinks known, has bioregenerative limits. Therefore, the sheer magnitude and complexity of a BLS does not necessarily guarantee long-term stability.

Colonization of future space settlements (Moon, Mars) will inevitably call for stable, reliable, and complex BLS; and to arrive at this goal, much more research will be necessary to get sufficient knowledge on the critical points within a complete BLS. Biosphere-2, in particular, is an unique test bed having sufficient magnitude for this type of investigation, but no matter how impressive such achievements appear, some enormous problems in basic biology remain. There is a small but active group of people carrying out research on various aspects of BLS and CELLS (closed ecological life support systems). Atmosphere recycling being a prime design criterion, a variety of plants are being studied for their potential as atmospheric 'scrubbers'. These range from algae, through water weeds (e.g. *Lemna* spp. – the duckweeds), to established crop plants like wheat and potatoes. This is not the place for a detailed review. The topic was dealt with by MacElroy *et al.* (1985) and in a relatively recent COSPAR Symposium (published in *Advances in Space Research*, volume 9 (part 8), 1989).

What needs to be emphasised here is that whilst constructions like Biosphere-2 and the Soviet BIOS-3 demonstrate that completely sealed closed ecological systems can be made to work on Earth, we are still too ignorant of the effects of long term exposure to space flight conditions to be confident that similar systems would continue to work in space. On Earth, microgravity can be simulated (parabolic flight, drop towers) for such short periods of time that they are simply not applicable to testing BLS. Neither can the radiation pattern encountered in space be simulated on Earth. We quote Krikorian & Levine (1991): "...CELSS-related activities directed towards finding practical solutions for life support in the space environment are dependent on basic science. Any CELSS effort aimed at solving practical problems and providing practical solutions to difficulties encountered in growing plants in space will of necessity rely on the answers generated by basic research performed by gravitational plant biologists..."

To this statement we would add the fact which is stressed in Chapter 1 (Section 1.6): that only one plant and one insect have been taken through a complete generation cycle in orbit. Both exhibited a high frequency of developmental abnormalities in the progeny produced in microgravity that were of a sort which could adversely affect the ability of those progeny to form a further generation. Since no multicellular animal or plant has yet been taken through more than one generation in orbit, the stability of even single populations, let alone mixed, competing populations in an ecological microcosm, is seriously in doubt.

A.3 Food, nutrition, metabolism

A.3.1 Background

Only limited data are available on the adaptive role of metabolic readjustment in man during spaceflight. With increasing mission duration and distance from earth, there is a corresponding increase in the scientific and logistical problems of ensuring adequate nutritional supply. The range of available food and drink must guarantee that all essential nutrients are supplied in the correct balance to maintain the physiological range of life functions and well-being under the conditions of working and living in a spaceship or extraterrestrial habitat. The database available today from experiences collected so far do not allow for precise projections of nutritional problems which are likely to arise during missions of extended range from Earth, like flights to Mars. Adaptive changes which improve human performance and well-being in space might have detrimental consequences upon return to Earth (Seddon *et al.*, 1994).

Inadequate nutrient supply to the tissues results in nutritional deficiency. Human diet is composed of macronutrients – water, proteins, carbohydrates, fat, nucleic acids – and micronutrients – trace elements and vitamins – which regulate metabolic processes by acting as coenzymes or as essential elemental constituents of enzyme complexes which regulate the utilization of proteins, carbohydrates and fats. It is conceivable that future long-term flights will require fine balancing of the nutritional adequacy of the diet.

From the 20 amino acids which are supplied by dietary protein, 8 to 10 are essential in the diet of humans because they cannot be synthesized. The requirements of essential amino acids range from 3 to 15 mg kg^{-1} day^{-1} in adults. Animal protein generally provides good proportions of essential amino acids (Linkswiler 1982). The recommended daily allowance (RDA) for protein, assuming a mixture of animal and vegetable protein, is about 60 g day^{-1} for men, and about 50 g day^{-1} for women.

Dietary protein intake influences albumin metabolism (**James & Hay 1968**), blood volume, fluid balance, cardiovascular functioning, and body composition. A major microgravity-related effect is deconditioning of the musculoskeletal system, which is made evident as atrophy of 'antigravity' muscles, with loss of muscle mass (and nitrogen). Bodily fat stores are also usually diminished (Forbes 1987), and decreased mechanical load is thought to result in bone catabolism as well, with consequential loss of protein and calcium (Stein & Gaprindashrili 1994).

Nutritional fat is a vehicle for flavours, fat-soluble vitamins, gives texture, delays gastric emptying, and cushions body organs (Coniglio 1984; **Flatt 1987**). The caloric value is the highest of all nutrients. Fat consumption furnishes essential fatty acids; the needs can be met by a daily intake of 15–25 g of proper dietary fat. Fat is of primary importance for buffering energetic needs: day-to-day fluctuations in energy balance is met primarily by body fat rather than by carbohydrate or protein (**Abbott *et al.* 1988**). A dietary source of polyunsaturated fatty acids is required by humans; these are linoleic acid, α-linoleic acid, and arachidonic acid. The amount of linoleic acid required by humans is about 1–2% of dietary energy; linoleic acid is a prominent component of the dietary fats. Essential fatty acids and their derivatives are precursors of prostaglandins, thromboxanes and prostacyclins. They are involved in membrane structures and play a beneficial role in cardiovascular and blood pressure regulation (Iacono *et al.* 1982).

Carbohydrates are not essential in the sense that essential amino acids or fatty acids are, but digestible carbohydrates are the most important source of food energy, and indigestible polysaccharides contribute to dietary fibre. The brain and the complement of red blood cells, which depend almost entirely on carbohydrates for energy metabolism, together consume 150–200 g glucose d^{-1}. About half of dietary energy is provided by carbohydrates. This corresponds to 5 to 6 MJ metabolizable energy from 300 to 350 g digestible carbohydrate; a minimum of about 2 g kg^{-1} (normally about 150 g) is required to prevent ketosis.

The main source of carbohydrate, as well as the only source of dietary fibre, is vegetable food. Foods which are rich in starch also contain dietary fibre and a variety of essential nutrients. Consumption of fibre is desirable from a preventive medical point of view because it increases stool mass, decreases turnover time of bowel contents (Eastwood 1984), and binds potential carcinogenic compounds. The effect on stool weight may be directly due to water-holding capacity (fibres resist mucosal absorption of water), or indirect as a result of bacterial growth or increase in metabolites (e.g., volatile fatty acids). Fibre is subject to bacterial decomposition in the large intestine and gives rise to flatulence; dietary fibre in space food requires careful consideration.

With regard to minerals, the estimated safe and adequate daily dietary intake of sodium is 1.1 to 3.3 g in adults. Sodium intake is best assessed by urinary sodium output. Usual daily potassium intake ranges between 3 and 11 g. A daily minimum need of 1 g can be reasonably assumed; estimates range from 0.8 to 3 g d^{-1}. With usual diet, there is no danger of insufficient supply of potassium. Calcium constitutes 1.5 to 2% of the total body mass, with >99% present in bones, where the ratio to phosphorus is somewhat greater than 2:1. About 85% of body phosphorus (0.7 g) is in bone as calcium phosphate and hydroxyapatite, 15% in cells and extracellular fluids. In the usual diet, the phosphorus:calcium ratio is high, e.g. 15 to 20 in meat, about 2 in eggs, grains, and nuts. In contrast, most green leafy vegetables contain more calcium than phosphorus. The recommended Ca:P ratio of 1:1 is impossible to achieve in diets high in protein.

A multitude of trace elements are essential for life, serving as cofactors in enzyme reactions, binding of oxygen, and structural components of non-enzymatic macromolecules (Linder 1985). The number of indices that are presently available to assess accurately the status of the majority of the trace elements is limited. Circulating levels may not reflect the status of the element available for nutritional needs.

Tissue stores of a trace element may not be available to meet the needs during deficient supply because they may be bound to enzyme proteins from which cellular stores cannot be mobilized. The action of trace elements depends upon nutrition, metabolism, age, and gender. Trace elements play an essential role for immunity; immobilization, isolation, weightlessness, and/or altered gravitational acceleration may decrease defence capabilities *via* altered trace element status. A particularly important role may be played by nutritional fluoride (**Boivin *et al.* 1993**) in long-term flight.

Iron status is best assessed by measuring plasma ferritin which normally ranges from 20 to 200 μg l^{-1}; each μg l^{-1} corresponds to 140 μg kg^{-1} BW of iron stores (**Finch & Hubers 1982**). Zinc status may be assessed by leukocyte zinc levels (**Bro *et al.* 1988**), or erythrocyte Cu, Zn superoxide dismutase, and serum alkaline phosphatase activity (Hambidge *et al.* 1989). Copper status seems to be indicated by the specific activity of ceruloplasmin, as estimated by the ratio of enzymatic to immunoreactive ceruloplasmin, erythrocyte Cu, Zn superoxide dismutase, or saliva copper concentration (**Milne *et al.* 1990; Bales *et al.* 1990**).

Selenium is measured in whole blood and plasma, erythrocytes, platelets and leucocytes; plasma and platelet glutathione peroxidase is sensitive to selenium intake and can be used to assess selenium needs (**Sandström *et al.* 1990**). Urinary selenium excretion is used to estimate selenium intake. Serum and urinary *manganese* concentration is not sensitive to dietary alterations (**Greger & Davis 1990**), but probably reflects manganese status.

Vitamin A (retinol) is essential for vision, growth, epithelial cell differentiation, reproduction, and the integrity of the immune system. Retinol also influences calcium metabolism and calcification. Various carotenoids are biologically active as vitamin A. Vitamin D (D–hormone) and its metabolites regulate calcium, phosphorus, and bone metabolism. Parathyroid hormone causes formation of the active hormone in the kidney. D–hormone stimulates active transport of calcium across the small intestine. Vitamin E (tocopherol) serves as an antioxidant, preventing (per)oxidation of polyunsaturated fatty acids in cell membranes, may play a role as anticarcinogenic as well as erythropoietic factor, and is thought to slow down aging processes (Linder 1985). Vitamin C (ascorbic acid) is an antioxidant as well, affecting the body's redox potential, and is involved in hydroxylation reactions, procollagen synthesis, and transport and absorption of iron. It enhances the response of neutrophils to chemotactic stimuli and also may exert other positive effects on the immune system.

A.3.2 Earth–gravity and simulated microgravity

A major factor in space flight nutrition is connected to deconditioning, catabolic state, and calcium loss. There is controversy how much calcium is needed for maintaining body stores, i.e. bone mass. On the basis of calcium balance studies conducted with groups of individuals accustomed to adequate intakes of foods high in calcium, the daily allowance is 800 mg for individuals 18 years of age and older (Lutwak 1982). Absorption of calcium from food sources is not influenced by solubility, it is determined mainly by other food components (**Heaney *et al.* 1990**), and is, besides hormonal influences, decreased by stress and immobilization. Vitamin D increases the gastrointestinal calcium absorption (**Orwoll *et al.* 1988**), underlining the significance of sufficient vitamin D intake in astronauts. The 0.8 g d^{-1} given to Shuttle astronauts may be insufficient and could be increased to 1.2 to 1.5 g d^{-1} in supplements or calcium–rich foods; this is a level beyond which additional calcium is not usually absorbed. However, this may raise concerns about hypercalciuria and possible renal stone formation.

Experimental studies on the requirement for calcium, phosphorus, and vitamin D in relation to hormonal reactions (e.g. calcitonin) and bone metabolism could be useful; phosphate intake would have to be standardized in this connection, because of possible depletion effects on sodium and zinc. If calcium supplements were provided on a continuous basis, e.g. starting 2 weeks prior to flight, urinary/faecal excretion studies would show which percentage of dietary calcium is retained by the body under circumstances of space flight.

Both in simulation studies and during prolonged space flight, the renal loss of calcium and potassium reaches a maximum by the second month, most probably caused by a reduced capacity of tissues to retain electrolytes. This phenomenon of diminished mineral pool can be attenuated but not completely avoided by the use of countermeasures, like exercise, LBNP, and salt-water supplement (**Gazenko *et al.* 1982; Grigoriev 1983**). It is an open question whether reduced gravity as on Mars (0.4 × g) or Moon (0.16 × g) is sufficient to prevent bone loss and if supplements in calcium, phosphorus, vitamin D, fluoride, or hormones will be needed in addition to exercise countermeasures during long-term stay on extraterrestrial bases.

About 60% of body magnesium is located in the skeleton. More than 300 enzymatic systems are magnesium-dependent, such as ATP-dependent enzyme reactions. Magnesium plays a key role in neuromuscular transmission and activity; it is essential for the normal metabolism of calcium and potassium in adult humans, and for the mobilization of calcium from bone (**Shils 1985**). It is the mineral ion of chlorophyll, and green vegetables are an important source. The intake of magnesium varies widely, the average range is probably 180 to 480 mg d^{-1}, the recommended daily allowance being 300 to 350 mg for adults. Two-thirds of ingested magnesium is excreted in the stool. A diminished extracellular Mg concentration may lead to dysfunction of excitable tissues within the neuromuscular and cardiovascular systems giving rise to tetany and neurological disorders. Serum magnesium indicates extracellular Mg status but not total body magnesium. Peripheral mononuclear blood cell Mg seems to be a good indicator for total body stores (Hook *et al.* 1987).

Trace elements are absorbed by the gut to a very different fraction (1 to 100%) of what is present in the diet. Dietary input of any nutrient may considerably exceed the fraction actually available to the body. The efficiency of absorption mechanisms correspond to the bodys needs and metabolic status. Bioavailability also depends on the presence of other substances in the food. The mechanisms of digestion, absorption, and metabolism may be disrupted by physiological stress factors, such as motion sickness, nervousness and anxiety. Eating while agitated, fatigued, or worried may give rise to gastrointestinal disturbances. Nutrient utilization depends upon a great number of psychological and neurohormonal control mechanisms, the quantitative interaction of which needs to be studied in more detail, and with use of novel laboratory as well as mathematical simulation techniques.

A.3.3 In flight experimental evidence

Space diets have been designed according to the recommended daily allowance (RDA) values as published by the NRCs Food and Nutrition Board. The

RDAs were developed for terrestrial use, and the actual needs of astronauts may differ, at least in terms of catabolic shifts due to bone/muscle degradation in microgravity. Energy and protein supply have been gradually increased in space diets, the highest figure being reached with the Voskhod cosmonaut's diet containing 3600 kcal d^{-1}.

Most space food today is dehydrated to reduce mass and transportation costs. Dehydration of solid food results in a mass/volume reduction by a factor of 3 (meat) to 10 (components extremely rich in water) due to the removal of solvent or bound water. Eggs may be compacted to about 30% of initial volume, potatoes to about 25% initial volume and fruit to 20–25% of initial volume.

During space flight, loss of body protein occurs through muscle atrophy (Ushakov 1980); physical activity is not only the major variable affecting caloric expenditure but also determines the nutritional need for amino acids. In the Apollo and Shuttle missions, protein consumption was about 80 g d^{-1}, based on ingredients used and on amino acid analyses. Great individual differences were observed (up to 160 g d^{-1}). The Russians have prescribed a higher proportion of protein, in an effort to counteract muscle degradation. For U.S. astronauts, protein provided 19% of nutrient energy (Leonard 1982; Popov 1975; Sauer & Calley 1975), and 23% in Soviet cosmonauts (Popov 1975). After Salyut missions, a decrease in amino acids was found in blood plasma, particularly in essential amino acids, and it has been concluded that astronaut diets should be supplemented with certain amino acids before, during and after flight (Popov & Latskevich 1984).

Fat intake of U.S. astronauts (Gemini, Apollo, Shuttle) decreased in flight, compared to the consumption on Earth (**Altman & Talbot 1987**). Data from Apollo, Skylab, and STS suggest that the energy costs of vigorous exercise such as EVAs are consistent with those associated with comparable physical activity on the ground. An assessment of fat mass by different methods indicated that in three Skylab flights, fat loss contributed 45% to total body weight loss, and lean mass 55% (**Leonard *et al.* 1983**).

Carbohydrate consumption was higher in astronauts in flight (averaging 400 g d^{-1}) than under control conditions (350 g d^{-1}) whereas crude fibre intake was the same in flight and on the ground (5 to 10 g d^{-1}). During Gemini, Apollo, and Skylab missions, a deterioration of glucose tolerance in flight was observed (Leach & Rambaut 1977), and with the first Shuttle missions, post-flight blood glucose increased more than 20%, with great differences between individuals. Data should therefore be considered on an individual basis rather than as mean values in future investigations.

As an average figure from Skylab, a net loss of approximately 100 mmol of sodium (corresponding to 0.4% of the overall body stores) occurred from the extracellular space, stabilizing at a lower level after initial disturbance in flight (Leach *et al.* 1975; Leonard 1985). Evidence from Gemini 7, Apollo 15–17, and Skylab (Leach & Rambaut 1977) showed that body *potassium* stores steadily decrease in flights exceeding one month in duration; e.g. by 240 mmol in 30 days, representing a 0.7% diminution of the body stores despite adequate potassium ingestion throughout these missions (**Leonard *et al.* 1983**).

In weightlessness or with reduced gravitational load, there is decreased mechanical stress in the trabeculae of the lower body bones. This is not balanced by sustained or even increased mechanical demand in the upper parts of the body. Despite some net shift from the legs to the bones of the skull, neck, and arms, the calcium content in the astronaut's body decreases continually even after several months in flight. The negative calcium balance (high urinary output which is not met by equivalent calcium resorption from the gut) cannot be compensated by increased dietary calcium supply. Blood plasma calcium concentration is significantly elevated in flight, from 9.5 to 10.2 mg dl^{-1} (**Leonard *et al.* 1983**). Urinary calcium excretion is increased about twofold after several weeks mission duration. The renal excretion of phosphate was found to be increased in astronauts by some 20%, due to continuing reduction in bone mass. Some decrease in serum magnesium has been reported post flight, evidence of significant in flight magnesium change seems to be lacking.

In U.S. space missions, vitamin A intake exceeded the RDA by up to 200% (**Altman & Talbot 1987**). Vitamin D was provided to 100% RDA in U.S. space foods by multivitamin supplements. This was considered advisable considering the in flight absence of UV light. In the Shuttle OFT programme, 15 IU d^{-1} vitamin E (about 15 mg tocopherol equivalent) was provided by the standard menu.

Evidence indicates that high dietary protein induces calcium loss (**Hegsted 1981; Kim & Linkswiler 1979; Linkswiler *et al.* 1981;** Schuette & Linkswiler 1984; Zemel 1988) as it may be desirable for space menus. Commonly used complex dietary proteins do not have these effects in strictly controlled long term human studies (**Spencer *et al.* 1983**) in which large amounts of dietary protein with a high phosphorus content were used. An average balance on protein, calcium, and phosphorus intake (**Spencer *et al.* 1988**) may therefore be desirable for space travellers.

A.3.4 Conclusions

Future long term exposure of humans to reduced gravitational acceleration will show nutritional and metabolic adaptation to microgravity conditions. The kinetics of protein and amino acids in the body need investigation in flight to understand properly the underlying adaptive mechanisms. It is reasonable to assume that muscle adapts immediately to the reduced work load in microgravity by losing protein. The hormonal background is still far from being understood. Insulin resistance occurs, as it does in ground-based studies of muscular wasting due to bed rest (Shangraw *et al.* 1988), resulting in decreased glucose uptake. Growth hormone activity also seems to be depressed during space flight (Sawchenko *et al.* 1992). The etiology of protein loss is most probably multifactorial and includes stress responses, under-loading and under-nutrition.

The balance of catabolic and anabolic pathways in bone and muscle depends upon the amount and time-profile of mechanical load. This load can be seen to be composed of basically two factors: Acceleration forces acting on the body as a whole (gravitational load); and resistant forces to be counteracted (movement, isometric exercise). The combination of these two factors determines the metabolic balance and nutrient demand of the musculoskeletal system. A reduction of the average forces to which this system is subjected leads to atrophy which cannot be counteracted by nutritional measures alone. Mechanical devices, such as the 'penguin' suit, seem to be effective in preventing protein catabolism and muscle atrophy. A loss of 1000 g muscle tissue would constitute a 200 g loss of protein, since muscle is 20% protein. There is a need to relate the demands of physical performance in space with preservation of muscle mass; dietary strategies can assist in preventing amino acid and protein loss from the astronaut's body.

Purified proteins and amino acid mixtures as they are used in experimental diets lead to calcium loss in humans and could theoretically exaggerate the problem of negative calcium balance with long lasting periods of microgravity.

It is still not known how disuse osteoporosis during space missions might be prevented, nor what the steady-state level of bone calcium might be after long duration flights. Dietary countermeasures, i.e. oral calcium supplements, are limited for several reasons. Absorption is regulated according to actual needs which are determined by local mechanic stress and strain in the bone; calcium overloading may result in formation of calcium-containing kidney stones, and calcium supplementation interferes with absorption of other minerals and trace elements.

The mechanisms of nutrient transfer from the intestines to the blood are functionally modified by the bodys metabolic status, so the degree to which a nutrient is available for absorption and assimilation depends on the physiological condition of the body. Matrix effects influence the degree to which nutrients (mostly minerals and trace elements) are taken up by the intestinal mucosa. The resulting problem of reduced bioavailability may become important in long-term space travel if diets become imbalanced. Fluoride is required for proper functioning of bone and may be a candidate for possible decalcification countermeasures. Careful dosage is essential because chronic fluoride load exceeding demand by a factor of about 10 results in toxicity symptoms. The net result showed increased skeletal mass and decreased negative calcium balance. There are conflicting results on dosage and duration of the supplementation with or without combination with calcium administration. Investigations are needed to define optimal fluoridation in astronauts because no countermeasure yet applied were sufficient to prevent weightlessness induced bone loss.

Manganese is linked to central nervous system function, so investigations into manganese balance in astronauts will be mandatory in extended missions; no information on metabolism of this important trace element in flight is yet available. Another interesting question is whether ascorbic acid, a vitamin which is nontoxic even with very high dosage (up to 100 times RDA), would provide increased immunological defence capacity as a countermeasure to possibly decreased immunological capability in astronauts during flight. Vitamin C (and B) was present in U.S. space menus at almost 10 times RDA, following findings of low blood concentrations of vitamin B and C after various stresses.

Another important problem is which lighting system will be employed in future space habitats in order to meet biological needs. Unless there is full-spectrum light or quartz windows to permit penetration of solar UV rays, it is likely that there will be insufficient endogenous vitamin D production. Vitamin D has special significance in the space environment because it is a primary regulator of calcium absorption. The amount of vitamin D required to alleviate insufficiency needs to be determined. At the same time, toxic levels should be considered; a safe range of exogenous vitamin intake on long term missions has not yet been established.

Thiamin status requires close monitoring in any long lasting mission; it can be assessed by measuring renal vitamin excretion and the transketolase activity in erythrocytes. Similarly, cobalamin (vitamin B-12) will probably have to be supplied by self-contained biological life support systems in order to avoid symptoms of deficiency. Macrocytic anaemia is one of the early consequences of low folate supply. Inter-relationships of the B-complex vitamins are essential in the performance of metabolic and catabolic reactions in the body, thus B-vitamins were generally supplied on a high level in U.S. space menus. The 10–15% decrease in the astronaut's red cell mass prompted the recommendation of a folic acid supply of 200% RDA in flight. The OFT Shuttle standard menu supplied 400 μg folate which is identical with RDA folate intake. It is yet to be decided whether higher folate in astronaut diets will boost red cell synthesis, and if so, if increased erythropoiesis is beneficial in terms of cardiovascular adaptation to weightlessness. It might be hypothesized that on the level of blood viscosity and microcirculatory exchange, the adapted state of 'space anaemia' provides more advantage than a 'normal' red cell mass.

Space diets may be designed according to commonly recommended nutrient intake values which were developed for terrestrial use. Actual needs of astronauts may, however, deviate significantly from these general guidelines. The major differences between weightless and $1 \times g$ conditions as they influence metabolic needs are in the cardiovascular/fluid electrolyte and musculo-skeletal systems. For astronauts, it is particularly important to maintain an optimal fluid and electrolyte balance. The need for potable water depends on the net fluid balance of the body which differs from ground control during space flight. The physiology of fluid volumes and electrolyte concentrations during space flight is closely connected to the regulation of energy metabolism, biomechanics, bone remodelling, endocrinology, kidney physiology, and cardiovascular functioning. There is still a lack of understanding how body subsystems interact quantitatively along the time-line of microgravity adaptation and $1 \times g$ readaptation processes.

For long term space missions, it will be necessary to completely regenerate life support materials; as space habitats get farther from earth, resupply of food and nutrients will become increasingly difficult. The design of any such system has to be based on the very wide knowledge of food science and technology. In order to minimize the production load and waste treatments, a maximum 'index of nutritional quality' needs to be achieved. Nutritional quality and consistency may be optimized with respect to any stress conditions or crew requirements which have to be specified on an individual basis. To date, no database exists on features of such systems on sufficiently reliable quantitative grounds.

Food acceptance is of major importance in any situation where humans are in a remote and isolated, challenging situation since mood and motivation heavily depend on the hedonistic, rewarding character of eating. There are enormous research needs to be met before questions of nutritional support under space flight and extraterrestrial conditions are defined and can be answered clearly. Food of plant origin will probably contribute to a major part to the menus. Cereals, vegetables, fruits, peas, beans, lentils, nuts,and seeds will likely be the cornerstones of future space diet. Unless food of animal origin is added for longer missions (i.e. months or years), nutrient deficiencies may occur. Data on amino acid content of different edible proteins can be used for planning of balanced protein supply with vegetarian diets. Deficiencies may occur with a prolonged vegetarian diet, e.g. with calcium, iron, zinc, and other trace elements. Careful planning of balanced diets can avoid long term nutrition problems even with minimal sources of animal food added to the diet. Any combination of food, resulting in certain diets, should be monitored continuously for all essential nutrients. This can be done by using computer software which not only analyzes meal composition but also allows for automatic checking if individual needs are met by certain intake patterns. Such type of supply monitoring needs to be complemented by regular nutritional assessments.

A.4 Countermeasures

A.4.1 Background

The body adapts to weightlessness by musculoskeletal and cardiovascular deconditioning; unlike other physiologic adaptations, calcium loss occurring in the weight bearing bones during microgravity does not reach a plateau during the first 6 or so mission months. Cosmonauts who have remained in space for extended time periods required several weeks of recovery time on Earth to return to their preflight physical health (Gazenko *et al.* 1980). It is not known whether longer stays in microgravity will result in any type of permanent damage either to the cardiovascular or musculoskeletal system.

Artificial gravity has been proposed as the most effective way to prevent deconditioning for long duration missions (more than one year), but such rotational forces will also produce Coriolis acceleration. Currently, three approaches are being used or considered as continuing countermeasures to the deleterious effects of microgravity on the crew: physical exercise, lower body negative pressure (LBNP), and drug therapy. There is still little knowledge on the relative merits of these conditioning efforts and none, of course address the similar problems which would afflict other organisms which might be components of a biological life support system.

A.4.2 Earth gravity and microgravity simulation

Bed rest studies have often been employed as a substitute to study the effects of simulated weightlessness (e.g. **Chobanian *et al.* 1974; Greenleaf & Kozlowski 1982**; Greenleaf 1983; Dallman *et al.* 1984; **Pequignot *et al.* 1985; Annat *et al.* 1986; Martin *et al.* 1986; Beckett *et al.* 1986; Blamick *et al.* 1988; Jansson *et al.* 1988; Shangraw *et al.* 1988; Leblanc *et al.* 1990; Stuart *et al.* 1990; Fortney *et al.* 1991; van der Wiel *et al.* 1991; Pannier *et al.* 1991; Arnaud *et al.* 1992; Greenleaf *et al.* 1992; Palle *et al.* 1992; Frey *et al.* 1993; Kawai *et al.* 1993; Lacolley *et al.* 1993; Hartikainen *et al.* 1993; Melchior & Fortney 1993; Vernikos *et al.* 1993**).

Early bedrest and LBNP studies showed that oral rehydration provided a large increase in plasma volume and a reduction in stressed heart rate response (stress by LBNP or gravity), but only for a period of a few hours. These results developed into the simple procedure of oral rehydration shortly before re-entry. Some of the more recent results from bed rest studies have been mentioned in the previous chapter; others were, for example:

(i) An early (several days) increment in resorption parameters, like fasting urinary calcium/creatinine and hydroxyproline/creatinine ratios, or serum calcium and phosphate levels, and a decrease in serum 1,25-dihydroxyvitamin D (**van der Wiel *et al.* 1991**).

(ii) decreased whole-body protein synthesis with low dietary protein (**Stuart *et al.* 1990**); indication of the involvement of other factors than leg compliance, as altered cardiac or autonomic nervous function, or blood volume, to orthostatic intolerance after bed rest (**Melchior, Fortney 1993**).

(iii) an increased LBNP-induced sympathetic activity with regard to deficient vascular resistance after 1 day of head-down bed rest (**Lacolley *et al.* 1993**).

(iv) benefits of isotonic exercise for the maintenance of plasma volume (**Greenleaf *et al.* 1992**).

(v) no change in (**Frey *et al.* 1993**), or an (early) increase (**Kawai *et al.* 1993**) in cerebral blood flow with head-down tilt bed rest.

Ingestion of isotonic salt solution combined with LBNP proved successful in returning plasma volume to baseline during extended bed rest (**Hyatt & West 1977**). Reduced chronotropic response and increased plasma volume with orthostatic challenge has also been observed with several hours of hydration plus LBNP (**Fortney 1991**). A recent investigation has shown that saline ingestion during pseudo-orthostatic challenge (1 l isotonic NaCl, 30 mm Hg LBNP) reduces plasma volume loss (from 14 to 6%) and less than 70% of the ingested 'extra' volume went into the extravascular space of the lower body (**Aratow *et al.* 1993**).

Artificial gravity may solve most of the problems caused by weightlessness, but within a rotating environment, stimulations of the vestibular, visual, and proprioceptor systems may produce symptoms of vertigo, disorientation, lassitude, postural aberrations, or nausea (motion sickness). Test subjects of the Pensacola Slow Rotation Room responded with a deterioration in well-being, frequently associated with a feeling of increasing lethargy. Adaptation and subsequent tolerance has been shown to occur for rotation speeds up to 10 rpm; no symptoms developed at 1 rpm or less (Green *et al.* 1971).

The effects of rotation on walking abilities on circular and flat walls was studied at the Langley Rotating Space Station Simulator, a circular platform of 12 m diameter with a 1.8 m vertical wall at the periphery. Subjects were suspended horizontally by a servomechanized boom and allowed to walk on the walls. Rotation speeds varied from 3 to 10.5 rpm, corresponding to gravitational accelerations from 0.05 × g to 0.75 × g at the level of the subjects feet. At 0.05 × g, while it was possible to walk in the direction of rotation, the test persons were unable to walk against the direction of rotation. Walking in the direction of rotation at a level between 0.17 and 0.3 × g was most comfortable. At higher gravitational accelerations, the subjects reported sensations of leg and body heaviness which were disturbing above 0.5 × g (Letko & Spady 1968).

For a rotating station of more than 20 m radius, the gravity gradient is not critical; motion sickness and the deviation of limbs during movement pose the major problems. Even subjects highly susceptible to vestibular side effects are not handicapped on sudden exposure to 1 rpm; some individuals adapt and function in environments rotating at rates as high as 10 rpm, but the ability to adapt varies from person to person, taking as long as 2 weeks to occur (Graybiel 1975). The only adaptation schedule which was able to provide a symptom-free adaptation to 10 rpm was a 9-step increase over a period of 25 days (Graybiel 1968). These findings suggest that the time required to effect adaptation can be greatly shortened by setting up an adaptation schedule for controlled head movements and stepwise increases in angular velocity.

A.4.3 In flight experimental evidence

LBNP (up to −50 mm Hg) was used extensively during Skylab missions to stress the cardiovascular system, to determine the extent and time course of deconditioning, and to determine whether in flight data on LBNP would be useful in predicting post flight orthostatic intolerance. In the Soviet programme, typical usage of the LBNP (up to −65 mm Hg in the 'Chibis' suit) was one 20 minute session every 4 days, and for 50 min d^{-1} during the final two days of the mission (Gazenko *et al.* 1980).

The treadmill is a device which places high loads on leg muscles by using bungee cords to simulate gravity pulling the individuals towards the walking surface. All Soviet space stations provided treadmills. Rapid (about 10 min) fatigue precludes use of the treadmill as an aerobic exercise device. However, Skylab treadmill proved to be a suited countermeasure to muscular deconditioning (Nicogossian *et al.* 1989).

Ergometers for use in flight are bicycle-like devices which can be pedalled with the hands or feet. Heart rate and metabolic level can be studied as a function of generated power (W h^{-1}). The ergometer used on Skylab operates within the range of 25 to 300 W, and at 20 to 80 rpm. The bicycle ergometer does not generate force levels approaching those used during ambulation. It is not thought to be an effective musculoskeletal or cardiovascular countermeasure on its own (Rambaut 1988).

Antigravity suits are standard equipment for fighter pilots. They are also used aboard the Space Shuttle during return to Earth to overcome symptoms of cardiovascular deconditioning (orthostatic intolerance). Standard antigravity suits work by inflating bladders in the leg and abdominal regions; the suit provides an increase in peripheral resistance, prevents fluid loss into the tissue spaces, increases venous return, and supports the diaphragm and thus the heart. The antigravity suit improves venous return and postflight orthostatic stability (Gazenko *et al.* 1981).

The penguin suit which is worn about 16 h during the normal day's activities, including exercise, provides passive stress on antigravity muscles (Rambaut 1988). The design of the suit is such that the cosmonaut can adjust the degree of tension in the elastic elements in order to stress specific muscle groups.

Studies conducted during the early Shuttle flights on a total of 26 trials (17 subjects utilizing the saline countermeasure and nine subject not using any countermeasure) demonstrated protection from orthostatic intolerance during post flight Stand Tests 1 to 2 h after landing (**Bungo *et al.* 1985**). A tendency toward reduced efficiency of salt water loading was noted in the Shuttle programme (Nicogossian 1989; Nicogossian *et al.* 1994).

A.4.4 Conclusions

Careful selection of countermeasures alone or in combination, and definition of timing, is needed to optimize the readaptation capacity of astronauts upon return to 1 × g (or landing on other celestial bodies after long term flights). The potential of possible side effects is not yet clear. Sufficient scientific information can only be derived from well-designed investigations in flight with a suitably large number of single observations. Because there are many potential variables in each mission, erroneous conclusions can be drawn by combining data from dissimilar test situations, and standardisation of conditions needs to be more firmly applied to future investigations.

Two of the host of major questions surrounding the concept of artificial gravity are: how much artificial gravity is needed? and what are the physiological limits to radius and angular velocity? Advantages of large scale rotating habitats, in addition to the medical ones, include ease of fluid and waste management, simpler designs of ecological life support systems, and potential benefits to crew psychological well being and social interactions. The decision to build large rotating habitats of significant complexity and cost might be reasonable if it is shown that the pathological effects of long term space habitation are not preventable through other less expensive means. Economically, there will be a trade-off between the development & construction costs of the habitat versus the rehabilitation costs associated with the recovery of crews following long duration flights. Rationalizing the development of a rotating space habitat is based significantly on the assumption that artificial gravity can stress and recondition the various systems of the body. Rotating habitats may be necessary even in facilities on planetary surfaces such as Mars or the Moon, where reduced gravity levels may lead to physiological deconditioning similar to that experienced in microgravity. The maximum angular velocity to which an average group of individuals might be able to adapt, the time-scale of adaptation, and the degree of discomfort involved are all completely unknown. We also have yet to establish the minimum gravitational acceleration which is required to maintain normal physiological functioning of each of the major systems in the human body, and the exposure time which must be spent at that acceleration to ensure normal functioning. Proposed facilities to investigate physiological effects of artificial gravity include the following.

(i) The artificial gravity sleeper (AGS) concept, which is a 2 m radius centrifuge to rotate a subject about an axis near the head, providing a centripetal acceleration gradient which is maximal in the feet and near zero at the head. It has been hypothesized that rotation during sleep will help reverse fluid shifts, maintain vascular tone, and provide acceleration for musculoskeletal system, lymphocytes, etc.

(ii) A fractional gravity facility, with a $1 \times g$ jogging track, in which the crew would live, work, and sleep in $0.5 \times g$, but for 0.5 to 1 h during the day they would jog around a track in the direction of rotation of the habitat, in effect increasing their angular velocity and thereby providing additional gravitational stress.

A.5 References

Abbott WGH, Howard BU, Christin L, Freymond D, Lillioja S, Boyce VL, Anderson TE, Bogardus C & Ravussin E. Short-term energy balance: relationship with protein, carbohydrate, and fat balances. Amer. J. Physiol. 255: E332–7, 1988.

Altman PL & Talbot JM. Nutrition and metabolism in spaceflight. J. Nutrit. 117: 421–7, 1987.

Annat G, Güell A, Gauquelin G, Vincent M, Mayet MH, Bizollon CA, Legros JJ, Pottier M & Gharib C. Plasma vasopressin, neurophysin, renin and aldosterone dur-

ing a 4-day head-down bed rest with and without exercise. Eur. J. Appl. Physiol. 55: 59–63, 1986.

Aratow M, Fortney SM, Watenpaugh DE, Crenshaw AG & Hargens AR. Transcapillary fluid responses to lower body negative pressure. J. Appl. Physiol. 74: 2763–70, 1993.

Arnaud SB, Sherrard DJ, Maloney N, Whalen RT & Fung P. Effects of 1-week head-down tilt bed rest on bone formation and the calcium endocrine system. Aviat. Space Environ. Med. 63: 14–20, 1992.

Bales CW, Freeland-Graves JH, Askey S, Behmardi F, Podocik, RS, Fickel JJ & Greenlee P. Magnesium, copper and protein concentrations in human saliva; age- and sex-related differences. Amer. J. Clin. Nutrit. 51: 462–9, 1990.

Beckett WS, Vroman NB, Nigro D, Thompson-Gorman S, Wilkerson JE & Fortney SM. Effect of prolonged bed rest on lung volume in normal individuals. J. Appl. Physiol. 61: 919–25, 1986.

Blamick CA, Goldwater DJ & Convertino VA. Leg vascular responsiveness during acute orthostasis following simulated weightlessness. Aviat. Space Environ. Med. 59: 40–3, 1988.

Boivin G, Dupuis J & Meunier PJ. Fluoride and osteoporosis. World Rev. Nutrit. Diet. 73: 80–103, 1993.

Bro S, Pedersen NT, Jørgensen PJ & Hørder M. A method for determination of zinc in mononuclear leucocytes. J. Trace Elem. Elect. Health Dis. 2: 143–8, 1988.

Bungo MW, Charles JB & Johnson PC. Cardiovascular deconditioning during space flight and the use of saline as a countermeasure to orthostatic intolerance. Aviat. Space Environ. Med. 56: 985–90, 1985.

Chobanian AV, Lille RD, Tercyak A & Blevins P. The metabolic and hemodynamic effects of prolonged bed rest in normal subjects. Circulation 49: 551–9, 1974.

Coniglio JG. Fat. In: *Present Knowledge in Nutrition,* 5th edn. Ed. by Olson RE, Broquist HP, Chichester CO, Darby WJ, Kolbey AC & Stalvey RM. The Nutrition Foundation, Washington D.C., pp. 79–89, 1984.

Dallman MF, Vernikos J, Keil LC, OHara D & Convertino V. Hormonal, fluid and electrolyte responses to 6° anti-orthostatic bed rest in healthy male subjects. In: *Stress: The Role of Catecholamines and Neurotransmitters.* Ed. by Usdin E, Kvetnansky R & Axelrod J. Gordon & Breach, pp. 1057–77, 1984.

Eastwood M. Dietary fibre. In: *Present Knowledge in Nutrition,* 5th edn. Ed. by Olson RE, Broquist HP, Chichester CO, Darby WJ, Kolbey AC & Stalvey RM. The Nutrition Foundation, Washington D.C., pp. 156–75, 1984.

Finch CA & Hubers H. Perspectives in iron metabolism. New England J. Med. 306: 1520–8, 1982.

Flatt JP, Dietary fat, carbohydrate balance, and weight maintenance: effects of exercise. Amer. J. Clin. Nutrit. 45: 296–306, 1987.

Forbes GB. *Human Body Composition: Growth, Ageing, Nutrition and Activity.* Springer-Verlag, New York, 1987.

Fortney SM. Development of lower body negative pressure as a counter-measure for orthostatic intolerance. J. Clin. Pharmacol. 31: 888–92, 1991.

Fortney SM, Hyatt KH, Davis JE & Vogel JM. Changes in body fluid compartments during a 28-day bed rest. Aviat. Space Environ. Med. 62: 97–104, 1991.

Frey MAB, Mader TH, Bagian JP, Charles JB & Meehan RT. Cerebral blood velocity and other cardiovascular responses to 2 days of head-down tilt. J. Appl. Physiol. 74: 319–25, 1993.

Gazenko OG, Il'in YA, Genin AM, Kotovskaya AR, Korol'kov VI, Tjigranyan RA & Portugalov VV. Principal results of physiological experiments with mammals aboard the Cosmos 936 biosatellite. Kosmich. Biol. Aviakosmich. Med. 14: 33–37, 1980.

Gazenko OG, Genin AM & Yegorov AD, Major medical results of the Salyut-6-Soyuz 185-day space flight. *International Aviation Federation XXXII Congress,* 1981.

Gazenko OG, Shumakov VI, Kakurin LI, Katkov VE, Chestukhin VV, Nikolayenko EM, Guodzev SV, Rumyanzev VV & Vasilyev VK. Effects of various countermea-

sures against the adverse effects of weightlessness on central circulation in the healthy man. Aviat. Space Environ. Med. 53: 523–30, 1982.
Graybiel A. Angular velocities, angular accelerations, and Coriolis accelerations. In: *Foundations of Space Biology and Medicine, Vol. II, Book 1*. Ed. by Calvin M & Gazenko OG, pp. 247–304, 1975.
Graybiel A. *The problem of artificial gravity in space flight*. NASA PR–20, 1968.
Green JA, Peacock JL & Holm AP. *A study of human performance in a rotating environment*. Research Report No. SD 7Q456. NASA CR 111866, 1971.
Greenleaf JE. Bed–rest studies: fluid and electrolyte responses. In: *Space Physiology*. Ed. by Garcia J, Guerin M & Laverlochere C. Cepadus, Toulouse, pp. 335–48, 1983.
Greenleaf JE & Kozlowski S. Reduction of peak oxygen uptake after prolonged bed rest. Med. Sci. Sports Exercise 14: 477–80, 1982.
Greenleaf JE, Vernikos J, Wade CE & Barnes PR. Effect of leg exercise training on vascular volumes during 30 days of 6° head–down bed rest. J. Appl. Physiol. 72: 1887–94, 1992.
Greger JL & Davis CD. Intake, serum concentrations and urinary excretion of manganese in adult males. Amer. J. Clin. Nutrit. 51: 457–61, 1990.
Grigoriev AI. Correction of changes in fluid–electrolyte metabolism in manned space flights. Aviat. Space Environ. Med. 54: 318–23, 1983.
Hambridge KM, Casey CE & Krebs NF. Zinc. In: *Trace Elements in Human and Animal Nutrition*, 3rd edn. Vol 2. Ed. by Mertz W. Academic Press, Inc., New York & London, pp. 1–137, 1986.
Hartikainen J, Tarkiainen I, Tahvanainen K, Mäntysaari M, Länsimies E & Pyörälä K. Circadian variation of cardiac autonomic regulation during 24–h bed rest. Clin. Physiol. 13: 185–96, 1993.
Heaney RP, Recken RR & Weaver CM. Absorbability of calcium sources. The limited role of solubility. Calcif. Tissue 46: 300–4, 1990.
Hegsted M. Urinary calcium and calcium balance in young men as affected by the level of protein and phosphate intake. J. Nutrit. 111: 553–62, 1981.
Hook GR, Hosseini JM & Elin RJ. Quantitative Mg analysis of individual mononuclear cells by electron probe X–Ray micro analysis. In: *Magnesium in Cellular Processes & Medicine*, 4th International Symposium on Magnesium. Ed. by Altura A, Durlach D & Seelip S. Karger, Basel, pp. 142–50, 1987.
Hyatt KH & West DA. Reversal of bedrest–induced orthostatic intolerance by lower body negative pressure and saline. Aviat. Space Environ. Med. 48: 120–4, 1977.
Iacono JM, Judd JT, Marshall MW, Dougherty RM, Canary JJ, Mackin JF, Binder RA & Weinland RT. Role of dietary polyunsaturated fatty acids and prostaglandins in reducing blood pressure and improving thrombogenic indices. In: *Animal Products in Human Nutrition*. Academic Press, New York & London, pp. 479–500, 1982.
James WPT & Hay AM. Albumin metabolism: effect of the nutritional state and the dietary protein intake. J. Clin. Invest. 47: 1958–72, 1968.
Jansson E, Sylvén C, Arvidsson I & Eriksson E. Increase in myoglobin content and decrease in oxidative enzyme activities by leg muscle immobilization in man. Acta Physiol. Scand. 132: 515–7, 1988.
Kawai Y, Murthy G, Watenpaugh DE, Breit GA, Deroshia CW & Hargens AR. Cerebral blood flow velocity in humans exposed to 24 h of head–down tilt. J. Appl. Physiol. 74: 3046–510, 1993.
Kim Y & Linkswiler HM. Effect of level of protein intake on calcium metabolism and on parathyroid on renal function in the adult human male. J. Nutrit. 109: 1199–404, 1979.
Krikorian, AD &Levine, HG. Development and growth in space. In: *Plant Physiology, a Treatise*. Vol. X. Growth and Development. Ed. by Bidwell RGS. Academic Press, Inc., San Diego, California, U.S.A., pp. 491–555. 1991.
Lacolley PJ, Pannier BM, Cuche JL, Hermida JS, Laurent S, Maisonblanche P, Duchier JL, Levy BL & Safar ME. Microgravity and orthostatic intolerance: carotid hemodynamics and peripheral responses. Amer. J. Physiol. 264: H588–94, 1993.

Leach, CS, Alexander WC & Johnson PC. Endocrine, electrolyte, and fluid volume changes associated with Apollo missions. In: *Biomedical Results of Apollo*. Ed. by Johnston RS, Dietlein LF & Berry CA. NASA SP–368, pp. 163–84, 1975.

Leach, CS & Rambaut PC. Biochemical responses of the Skylab crewmen: an overview. In: *Biomedical Results from Skylab*. Ed. by Johnston, RS & Dietlein. NASA SP–377, pp. 204–16, 1977.

Leblanc AD, Schneider VS, Evans HJ, Engelbretson DA & Krebs JM. Bone mineral loss and recovery after 17 weeks of bed rest. J. Bone Miner. Res. 5: 843–50, 1990.

Leonard JI. *Energy balance and the composition of weight loss during prolonged space flight*. Report prepared for NASA. Management and Technical Services Company, Houston, 1982.

Leonard JI. Understanding metabolic alterations in space flight using quantitative models: fluid and energy balance. *36th Congress of the International Aviation Federation*, Pergamon Press, 1985.

Leonard JI, Leach CS & Rambaut PC. Quantitation of tissue loss during prolonged space flight. Amer. J. Clin. Nutrit. 38: 667–79, 1983.

Letko W & Spady, jr. AA. Walking in simulated lunar gravity. In: *Proceedings, 4th Symposium on the Role of the Vestibular Organs in Space Exploration*. Naval Aerospace Medical Research Laboratory, Naval Aerospace Medical Institute, Naval Aerospace Medical Center, Pensacola, FL. NASA SP–187, pp. 347–351, 1968.

Linder MC. *Nutritional Biochemistry and Metabolism*.. Elsevier, New York, pp. 69–197, 1985.

Linkswiler HM, Zemel MB, Hegsted M & Schuette S. Protein induced hypercalciuria. Fed. Proc. 40: 2429–33, 1981.

Linkswiler HM. Importance of animal protein in human nutrition. In: *Animal Products in Human Nutrition*. Ed. by ????. Academic Press, New York & London, pp. 263–73, 1982.

Lutwak L, Dietary calcium: sources, interaction with other nutrients, and relationship to dental, bone, and kidney diseases. In: *Animal Products in Human Nutrition*. Academic Press, New York & London, pp. 275–84, 1982.

MacElroy RD, Smernoff DT & Klein HP. (eds.) *Controlled Ecological Life Support System: Life Support Systems in Space Travel*. NASA CP 2378, Moffett Field, CA, 1985.

Martin DG, Convertino VA, Goldwater D & Ferguson EW. Plasma viscosity elevations with simulated weightlessness. Aviat. Space Environ. Med. 57: 426–31, 1986.

Melchior FM & Fortney SM. Orthostatic intolerance during a 13-day bed rest does not result from increased leg compliance. J. Appl. Physiol. 74: 285–92, 1993.

Milne DB, Johnson P, Klevay L & Sandstead H. Effect of copper intake on balance absorption and status indices of copper in men. Nutrit. Res. 10: 975–86, 1990.

Mitchell CA. Bioregenerative life support systems. Amer. J. Clin. Nutrit. 60: 820S–45, 1994.

Nicogossian AE, Countermeasures to space deconditioning. In: *Space Physiology and Medicine*, 2nd edn. Ed. by Nicogossian AE, Huntoon CS & Pool SL, Lea & Febiger, Philadelphia, pp. 294–311, 1989.

Nicogossian AE, Huntoon CS & Pool SL. *Space Physiology and Medicine*, 2nd edn, Lea & Febiger, Philadelphia, 1989.

Nicogossian AE, Sawin CF & Grigoriev AI. Countermeasures to space deconditioning. In: *Space Physiology and Medicine*, 3rd ed. Ed. by Nicogossian E, Leach Huntoon C & Pool SL. Lea & Febiger, pp. 447–67, 1994.

Nicogossian AE, Sulzman F, Radtke M & Bungo M. Assessment of the efficacy of medical countermeasures in space flight. Acta Astronautica 17: 195–98, 1988.

Orwoll ES, Weigel RM, Oviatt, SK, McClung, MR & Deftos LJ. Calcium and cholecalciferol: effects of small supplements in normal men. Amer J. Clin. Nutrit. 48: 127–30, 1988.

Palle S, Vico L, Bourrin S & Alexandre C. Bone tissue response to four-month antiorthostatic bedrest: a bone histomorphometric study. Calcif. Tissue Int. 51: 189–94, 1992.

Pannier BM, Lacolley PJ, Gharib C, London GM, Cuche JL, Duchier JL, Levy BI & Safar ME. Twenty-four hours of bed rest with head-down tilt: venous and arteriolar changes of limbs. Amer. J. Physiol. 260: H1043-50, 1991.

Pequignot JM, Guell A, Gauquelin G, Jarsaillon E, Annat G, Bes A, Peyrin L & Gharib C. Epinephrine, norepinephrine, and dopamine during a 4-day head-down bed rest. J. Appl. Physiol. 58: 157-63, 1985.

Popov GI. Food and water supply. In: *Foundations of Space Biology and Medicine*, Vol. III. Ed. by Calvin M & Gazenko OG. Washington, DC.: NASA., pp. 22-55, 1975.

Popov IG & Latskevich AA. Blood amino acids of cosmonauts before and after 211-day spaceflight. Space Biol. Aerospace Med. 18: 10-17, 1984.

Rambaut P. The prevention of adverse physiological change in Space Station crew members. Acta Astronautica 17: 199-202, 1988.

Sandström B, Davidsson L, Eriksson R & Alpsten M. Effect of long-term element supplementation on blood trace element levels and absorption of (^{95}Se), (^{54}Mn) and (^{165}Zn). J. Trace Elem. Elect. Health Dis. 4: 65-72, 1990.

Sawchenko DE, Arias C & Krasnov I. Effects of space flight on hypothalamic peptide systems controlling pituitary growth hormone dynamics. J. Appl. Pysiol. 73: 1585-655, 1992.

Sauer RL & Calley DJ. Portable water supply. In: *Biomedical Results of Apollo.* Ed. by Johnston, RS, Dietlein LF & Berry CA. NASA SP-368, pp. 495-515, 1975.

Schuette SA & Linkswiler HM. Calcium. In: *Present Knowledge in Nutrition,* 5th edn. Ed. by Olson RE, Broquist HP, Chichester CO, Darby WJ, Kolbey AC & Stalvey RM. The Nutrition Foundation, Washington D.C., pp. 400-12, 1984.

Seddon MR, Fettman, MJ & Phillips RW. Practical and clinical nutritional concerns during space flight. Amer. J. Clin. Nutrit. 60: 8255-305, 1994.

Shangraw RE, Stuart CA, Prince MJ, Peters EJ & Wolfe RR. Insulin responsiveness of protein metabolism in vivo following bedrest in humans. Amer. J. Physiol. 225: E548-58, 1988.

Shils ME. Magnesium, calcium, and parathyroid hormone interactions. Ann. N. Y. Acad. Sci. 355: 165-80, 1985.

Spencer H, Kramer L & Osis T. Further studies of the effect of a high protein diet as meat on calcium metabolism. Amer. J. Clin. Nutrit. 37: 924-9, 1983.

Spencer H, Kramer L & Osis T. Do protein and phosphorus cause calcium loss? J. Nutrit. 118: 657-60, 1988.

Stein TP & Gaprindashvili T. Space flight and protein metabolism, with special reference to humans. Amer. J. Clin. Nutrit. 60: 8065-195, 1994.

Stuart CA, Shangraw RE, Peters EJ & Wolfe RR. Effect of dietary protein on bed-rest-related changes in whole-body-protein synthesis. Amer. J. Clin. Nutrit. 52: 509-14, 1990.

Ushakov AS. *Nutrition during long flight.* NASA TM-76436, 1980.

van der Wiel HE, Lips P, Nauta J, Netelenbos JC & Hazenberg GJ. Biochemical parameters of bone turnover during ten days of bed rest and subsequent mobilization. Bone and Mineral. 13: 123-29, 1991.

Vernikos J, Dallman MF, Keil LC, OHara D & Convertino V. Gender differences in endocrine responses to posture and 7 days of 6° head-down bed rest. Amer. J. Physiol. 265: E153-61, 1993.

Zemel MB. Calcium utilization: effect of varying level and source of dietary protein. Amer. J. Clin. Nutrit. 48: 880-3, 1988.

Subject Index

T

U

V

W

X

Z